Mathematical Programming and Games

EDWARD L. KAPLAN

Department of Mathematics
Oregon State University

John Wiley & Sons
New York Chichester Brisbane Toronto

Library of Congress Cataloging in Publication Data:

Kaplan, Edward L. (Edward Lynn), 1920–
Mathematical programming and games.

Bibliography: p.
Includes index.
1. Programming (Mathematics) 2. Game theory.
I. Title.
QA402.5.K327 519.7 81-2990
ISBN 0-471-03632-3 AACR2

Printed in the United States of America

10 9 8 7 6 5 4 3 2 1

To
Vern and Eunice,
Jeanette, Tom, Ken,
Karen, Sheila and Andy

Contents by Chapter

A. Basic Ideas and Applications

1. Introduction 1
2. Solution Sets, Duality and Pivoting 8
3. Feasible Linear Programs 55
4. Matrix Games 105
5. Bimatrix and *n*-Person Games 149

B. General Methods in Linear Programming

6. The Two-Phase Simplex and Bounded Variables 229
7. The Revised Simplex; Matrices and Determinants 261
8. Parameters and Multiple Objectives 315

C. Networks and Special Structures

9. Optimal Paths and Networks 385
10. Dynamic Programming, Decisions and Games 428
11. Simplex Transportation Algorithms 468

Answers to Exercises 515

Bibliography 560

Index 575

Notes for Instructors and Other Readers

This book gives a relatively complete treatment of linear programming together with those portions of linear algebra, game theory and dynamic programming that are related to it. By suitably selecting and ordering the material, the book can be used for a variety of college or university courses of various lengths and levels. Any effort that this selection may involve can be balanced against the freedom and flexibility it makes possible, the labor-saving features discussed hereafter in the preface, and the advantage of having supplementary material readily at hand to answer many of the questions that inevitably arise. The checklist immediately preceding Chapter 1 illustrates possible selections of topics for courses of one, two or three terms, and the indicated pages suggest a point after which the continued study of the topic becomes optional.

The prerequisites are elementary courses in linear algebra and calculus, the latter only for the sake of maturity. The relevant topics in linear algebra are reviewed as they are needed in Chapter 2 and in Section 7.1.

The preface also contains an interdependence diagram for the chapters, which has the following implications:

1. Sections 9.1 to 10.4, on shortest paths and dynamic programming, are relatively independent of the others and can be studied at any time. Since the easiest algorithms in the book occur here, they can be used as an introduction, or to utilize small blocks of time, or to make a one-year course more feasible by providing a change of pace somewhere during the course.

2. Chapter 11, on the transportation problem, also can be studied at any time, when preceded by 9.1. If necessary the occasional references to earlier chapters can be ignored. By beginning with 9.1 and Chapter 11, the reader can get some of the flavor of the simplex method while postponing consideration of its linear-algebraic aspects (Chapter 2). Another traditional reason for considering this approach has been the arithmetic simplicity of the transportation algorithm, but this simplicity is nearly matched by the all-integer pivot procedure of 2.5.

3. Chapters 4 and 5, in that order, can be studied at any time after Chapter 3; that is, the game theory can be postponed or omitted altogether. As indicated in the checklist, the author's practice is to give priority to 6.1–6.2, as if they came at the end of Chapter 3. At any rate game theory is an application

of the simplex method of Chapter 3; it has much interest in its own right, as the study of conflicting attempts at optimization; and it can bring diversity and increased motivation into a course in linear programming. The most interesting applications of game theory occur in 4.7 and in Chapter 5. Beyond these, articles such as Bennett and Dando (1979) and Kristol (1980) illustrate the intense practical importance of game-theoretic concepts with little or no calculation required.

No doubt some readers are like the author in being put off temporarily by a new treatment of material already familiar in a somewhat different form. It is obvious, however, that students beginning a new subject do not have this problem. *Every* treatment is new to them in content as well as form, and it is only a matter of choosing the best presentation for which they have the prerequisites.

E. L. K.

Preface

The mathematics in this book is "finite" in the sense of finite dimensionality; finite convergence, and rationality, that is, it has a theoretical ability to obtain in a finite number of steps a precise optimum in terms of rational numbers when the input data are rational. The topics considered (together with applications) include various forms of the simplex method of linear programming, related topics in linear algebra games (in normal, characteristic-function and extensive forms), shortest-path algorithms, dynamic programming, and the transportation problem. A set of exercises with answers is provided for each of 74 sections.

Certain branches of mathematics are sometimes thought of as being intrinsically pure or applied; however, in many cases the distinction resides in the manner in which the subject is presented and the motives that attract people to it. Some like mathematics to be relevant, or capable of helping them to understand and to cope with the physical and the cultural worlds in which they live; these are the motives of application. The "pure" motives are concerned with subjective matters such as the intellectual challenge of mathematics, its order and elegance, its articulation of the specific and the general, its confirmations and astonishments, and its internal consistency in a world of opinion, conflict and change.

The oldest and most widely studied areas of mathematics naturally are among those able to appeal to both the pure and the applied motives. These are the arithmetic, geometry, algebra and calculus of our schools and colleges. In more recent times the list has been extended to include other areas such as probability theory and mathematical programming, which are the major mathematical components of operations research.

In the spirit of mathematical programming itself, one aim of this book is to try to minimize the effort required of the reader, subject to nontrivial lower bounds on the amount of information provided. Another aim is to consolidate in one work many of the most interesting and important finite methods of mathematical programming. Accordingly some variation in the level of difficulty is to be expected even within individual chapters. While the majority of the sections are intended for a senior or first-year graduate course, others are accessible to less advanced students. Calculus is a prerequisite for the sake of maturity, but is actually used in very few instances.

Among other things students may come to appreciate that, like the calculus, mathematical programming has elegant theories and techniques as well as immediate practical applications, and perhaps even a few philosophical connotations. The game theory in Chapter 5 is a natural context for the consideration of many economic and social problems. In the Marriage Game (Exercise 12 of 5.1) and the Prisoner's Trilemma ((16) of 5.2) the pure strategies are identified with psychological orientations.

The following are some of the special features that contribute to the usefulness of the book.

1. *The all-integer pivot procedure* in a condensed tableau is represented by formula (1) of Section 2.5:

$$\begin{array}{c|cc} \delta & x_l & x_j \\ \hline x_k & \pi & \beta \\ x_i & \gamma & \alpha \end{array} \quad \rightarrow \quad \begin{array}{c|cc} \pi & x_k & x_j \\ \hline x_l & \delta & \beta \\ x_i & -\gamma & (\pi\alpha-\beta\gamma)/\delta \text{ (integer)} \end{array}$$

Here the arrow represents the effect of the pivot step with π as the pivot, δ (later π) is the detached common denominator, the other numerical entries are the numerators, and i and j represent the various indices other than the pivotal k and l. It is found (and proved in 7.8) that if a linear program is stated in terms of integers, or polynomial functions of a parameter as in 8.5, with unit coefficients for the slack or artificial variables, then this procedure insures that all the entries including the quotients $(\pi\alpha-\beta\gamma)/\delta$ will be integers or polynomials after every pivot step.

Thus in the integer case there is no difficulty in optimizing small linear programs with pencil and paper without getting involved in fractional arithmetic or rounding errors. On the other hand, if computer programming or usage is part of the course, the more than fifty algorithms in the book (with theory, description and examples) provide a wealth of potential computer projects from which to choose at a variety of levels.

2. *Duality theory* is developed at the outset and used consistently throughout; the condensed tableau makes this easy to do. Rather than being presented as a *fait accompli*, it is here "discovered" by three different routes, via orthogonal-complementary vector spaces in 2.3, the computational rules for pivot steps in 2.4 or 2.5, and linear programs themselves in 3.1 and 3.6. Duality is seen to be as simple and as fundamental as the negative transpose of a matrix. It is the *negative* transpose for essentially the same reason that the slopes of perpendicular lines are negative reciprocals—to produce a scalar product of zero. Six reasons why the dual problem is important are listed in 3.3 (e_1)–(e_6).

3. *Notation* is part of the communication medium of mathematics and thus virtually a part of the mathematical structure itself. Optimization has been attempted in this area subject to the constraints of well-established usage. For both linear programs and games this book uses

Vector of Labels for	Rows	Columns	Objective
Row equations	$\mathbf{x}$ or $\mathbf{x}_M$	$\mathbf{y}$ or $\mathbf{x}_N$	$\min z=\mathbf{c}\mathbf{x}_N+d$
Column equations	$\mathbf{u}$ or $\mathbf{u}_M$	$\mathbf{v}$ or $\mathbf{u}_N$	$\max \bar{w}=\mathbf{b}\mathbf{u}_M+d$

thus precisely reflecting the duality theory in the notation. Complex-function theory has long enjoyed the nice standard notation $u+iv=w=f(x+iy)=f(z)$; one would like linear programming to be as well equipped.

Section 3.8 presents the evidence for and against one of the less orthodox conventions in the book, concerning the sign of the constant terms b_i. Among other things this convention is necessary to achieve the consistency with game theory noted in (8) of 4.1, and to validate the following assertion in 3.7: when the constraints are inequalities in nonnegative variables, negative parameters a_{ij}, b_i or c_j are favorable to the primal objective and positive parameters are unfavorable. (In the dual the reverse is true.)

The following assignments generally hold except as regards the alphabetic position of Greek letters, which tend to be used on an *ad hoc* basis:

Scalars, variable nodes: lightface italic lowercase
Vectors: boldface roman lowercase
Matrices: boldface roman capitals
Sets, determinants, fixed nodes, players (I, J) of a game: lightface italic capitals
Indices and integers: middle of the alphabet
Other constants: early in the alphabet
Other variables: late in the alphabet

However, the omission of the lightface-boldface distinction in hand or typewritten material will rarely cause ambiguity.

A bar over a symbol is either equivalent to a minus sign (with or without an additive constant) or else denotes the complement of a set. The symbol $\overset{\triangle}{=}$, meaning "equals by definition," is used when the introduction of a new letter symbol is not otherwise announced. For example,

$$a_0 \overset{\triangle}{=} \max_i \min_j a_{ij} \overset{\triangle}{=} \min_j a_{kj} \overset{\triangle}{=} a_{kl}$$

defines a_0 explicitly and k and l implicitly, the a_{ij} being given.

As an aid to communication, some redundancy without undue repetition is provided via tabular summaries and the parallel use of English and mathematical symbolism.

In the interest of progress toward a consensus in terminology and notation, no copyright protection is claimed for the vocabulary of individual words and symbols used herein. Of course the majority of such selections are themselves drawn from the existing literature.

4. *Linear algebra* as a major area of separate study has not been assumed as a prerequisite, although a previous acquaintance with vector and matrix notations and concepts is essential for most students. Such notations have been used to clarify the exposition but are not permitted to monopolize it or to obscure the duality principle. Much of the notational and logical burden that they otherwise would bear is carried by the more tangible formats of the condensed tableau and the Tucker schema and their solution sets. The rather limited requirements in linear algebra are reviewed as the need arises in Chapters 2 and 7; the most essential sections are 2.3–2.5 and 7.1. Linear algebra (LA) and linear programming (LP) make a happy combination; some

LA is needed for a full understanding of LP, and LP is needed to motivate LA, which otherwise seems excessively abstract to most students.

5. *Logical motivation* as well as motivation via *applications* is provided, and relations among the various topics are emphasized. Examples are the heuristic treatments of duality and of determinants (the latter in 7.7), and the use of pivot steps and Dantzig's simplex method to obtain transparent finite derivations of the elementary facts of linear algebra in 2.7, the duality theorems of linear programming in 3.3, and the minimax theorem for games in 4.4–4.5. Practical applications are given special attention in Chapters 1, 5, 9 and 10 and in Sections 3.6, 3.7, 4.7, 8.1–8.3, 8.9 and 11.1. These include gasoline blending in 3.7, electric power production in 8.1–8.3, scheduling and warehousing in 9.4–9.5, a simple poker game in 10.6, and decision problems involving the choice of a career or an investment in 4.7(d)–(e), 8.9 and 10.4.

6. *Examples and exercises with answers* are included in substantial numbers, section by section. Each example is carefully chosen and organized in a form suitable for the student's own use in doing exercises. A majority of the exercises are routine applications of the material in the text; the others permit the student to exercise his or her general mathematical background. A very few exercises, which are starred, are included for the purpose of supplementing the theory in the text.

7. *New material and recent developments* not available in other textbooks include the all-integer pivot procedure already mentioned, new methods for solving small systems of linear equations in 2.6(f) (several of them suitable for freshmen), the evaluation of the eight possible sign conventions in 3.8, the classification of bimatrix games in 5.2, the analysis and generalizations of the Shapley and Banzhaf values for games in 5.7 and 5.8, and the treatment of multiple objective linear programming in 8.8 and 8.9, which brings together several streams of research that thus far have developed rather independently. One of these is concerned with the sheer enumeration of extreme points; another involves the conditions for efficiency, culminating in Theorem 8 of 8.8.

Depending on its length, a senior-level course may emphasize some of the following sections without necessarily being limited to them: 1.2, 2.3–2.6, 3.3, 3.4, 3.6, 4.1, 4.2, 4.4, 4.7, 5.1, 5.3, 5.6, 6.1, 6.2, 7.1, 7.3, 8.3, 9.1–9.4, 10.1–10.4, 11.1–11.4. A more detailed selection of topics follows the table of contents. The dependence relations among the chapters are approximately as indicated in the figure. Here 7.3, for example, stands for 7.1–7.3, which is understood to be a prerequisite for the remainder of the chapter (except 7.7–7.8 in this case). The dashed arrows indicate that the earlier material is related but not indispensable to the later material. Sections 5.1 to 5.8 are relatively independent but 5.1 is a prerequisite for 5.2 to 5.4 and possibly 5.7.

Most of the sources used by the author are cited in the text and listed in the bibliography but inevitably these represent only a small sample of the vast literature that exists. Like any student of the subject, the author is especially indebted to the creative work of George B. Dantzig, both mathematical and organizational; his book published in 1963 also contains much bibliographic and historical information not readily available elsewhere. The ideas of Albert W. Tucker have greatly influenced

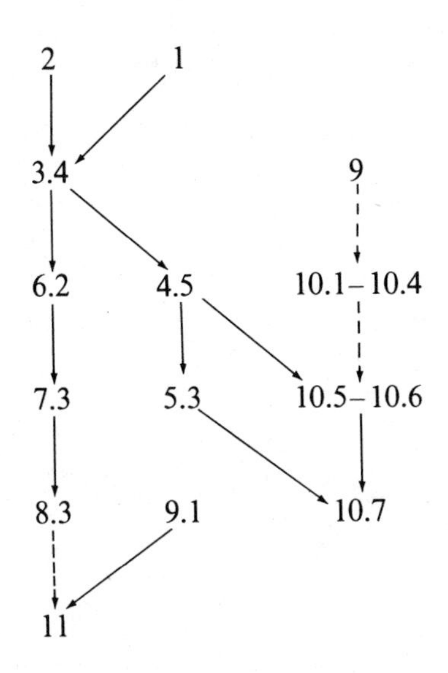

the earlier chapters as well as 7.8. Arvid T. Lonseth and the succeeding chairmen of the Mathematics Department at Oregon State University provided vital encouragement. Two sabbatical years granted by the university were devoted to the project. The author's students have made helpful suggestions since 1962. The manuscript was skillfully typed by Clover Redfern, Deanna L. Cramer, Jolan Eross, Donna Templeton, Lora Wixom and Sadie Airth.

Comments and corrections from readers will be carefully noted and at least the more substantial contributions gratefully acknowledged.

The cover design is a larger permuted form of (22) of 2.3, or of the tableau that results when an ordinary transportation problem is stated in transshipment form. It is also related to the assignment problem (11.4) and to two topics in integer programming, namely congruences (mod 2) and the traveling salesman problem. The formula below it is (1) of 2.5, cited in item 1 above.

Edward L. Kaplan

Contents

Notes for Instructors and Other Readers ix
Preface xi
Checklist of Objectives for a One-Year Course xxi

1. Introduction 1

1.1 The Nature of the Problem 1
1.2 An Example of Resource Allocation 2
1.3 Applications to Transportation 3
1.4 Games and Dual Problems 4
1.5 Relation to the Calculus 5

2. Solution Sets, Duality and Pivoting 8

2.1 Linear Equations and Vector Spaces 8
2.2 Linear Inequalities and Convex Sets 15
2.3 Orthogonality and Duality 20
2.4 Tableaus and Pivot Steps 27
2.5 Hand Calculations in Terms of Integers 33
2.6 Numerical Applications of Pivoting 36
2.7 Proving Theorems by Pivoting 48

3. Feasible Linear Programs 55

3.1 Dual Programs in Nonnegative Variables 55
3.2 The Simplex Algorithm: Introduction 60
3.3 The Simplex Algorithm: Theory and Use 64
3.4 Examples for the Simplex Algorithm 75
3.5 Checking and Revising Optimal Solutions 84
3.6 Some Simple Models and Their Duals 88
3.7 Other Applications 95
3.8 Sign Conventions in the Schema and Tableau 100

4. Matrix Games 105

4.1 Pure Strategies 105
4.2 Mixed Strategies and Graphical Solutions 111

4.3 Domination; Fictitious Play 119
4.4 Solving Matrix Games as Linear Programs 123
4.5 The Minimax Theorem; Kernels 127
4.6 Other Relations Between Games and Programs 134
4.7 Games Against Nature 140

5. Bimatrix and *n*-Person Games 149

5.1 Introduction to Bimatrix Games 149
5.2 Analysis of Bimatrix Games 158
5.3 Cooperative Solutions (Nash) 170
5.4 The Calculation of Equilibrium Points (Lemke) 180
5.5 A Simple Production Model 188
5.6 A Bimatrix Model of Economic Growth 190
5.7 The Shapley Value for *n*-Person Games 198
5.8 A Priori Calculation of Voting Power 213

6. The Two-Phase Simplex and Bounded Variables 229

6.1 Equality Constraints Without Slack Variables 229
6.2 Unrestricted Variables 236
6.3 Column Selection by Ratio Criteria 239
6.4 Freer Row Selection in Phases 1a and 1b 244
6.5 Bounded Variables 249
6.6 Degeneracy and Lexicography 257

7. The Revised Simplex; Matrices and Determinants 261

7.1 Matrix Algebra and Pivoting 261
7.2 The Explicit-Inverse Method 268
7.3 The Product Form of Inverse 272
7.4 Gaussian Elimination: $\mathbf{B}=\mathbf{LU}$ 282
7.5 Speed and Precision 289
7.6 Pivoting for Sparseness or Size 295
7.7 Introduction to Determinants 300
7.8 Determinants and Pivoting 306

8. Parameters and Multiple Objectives 315

8.1 Add, Drop or Change a Column or Row 315
8.2 Ranges for Individual Scalars 322
8.3 One Parameter in $\mathbf{b}$ and $\mathbf{c}$ ($T_0^* \neq \varnothing$) 336
8.4 One Parameter in $\mathbf{b}$ and $\mathbf{c}$ ($T_0^* = \varnothing$) 346
8.5 One Parameter in a Column or Row of $\mathbf{A}$ 349
8.6 Several Parameters in $\mathbf{b}$ and $\mathbf{c}$ 354
8.7 Nonlinear Parameters in $\mathbf{b}$ and $\mathbf{c}$ 357
8.8 Multiple-Objective Linear Programming 364
8.9 Multiple Objectives: Examples and Variations 376

9. Optimal Paths and Networks 385

9.1 Graphs, Trees and Networks 385
9.2 Shortest Spanning Trees ($c_{ij} = c_{ji}$) 390
9.3 Shortest Paths from one Node ($c_{ij} \geq 0$) 400
9.4 Critical Paths in Acyclic Scheduling Networks 407
9.5 The Warehouse, Replacement, Assortment 414
9.6 Shortest Paths from One Node (General) 418
9.7 All the Shortest Paths 422
9.8 Maximum-Capacity Paths 424

10. Dynamic Programming, Decisions and Games 428

10.1 Examples of Dynamic Programming 428
10.2 Dynamic Programming Models 432
10.3 Remarks on Dynamic Programming 436
10.4 Decision Networks 440
10.5 Zero-Sum Extensive Games 447
10.6 Poker and Behavior Strategies 454
10.7 Variable-Sum Extensive Games 462

11. Simplex Transportation Algorithms 468
11.1 The Transportation Problem and Its Dual 468
11.2 Basic Solutions, Triangularity and Trees 473
11.3 Pivot Steps and Pricing Methods 480
11.4 Uncapacitated Transportation Algorithms 484
11.5 Additional Constraints and Variables 492
11.6 The Capacitated Transportation Algorithm 497
11.7 Sensitivity Analysis 504
11.8 One Parameter in **a**, **b** and **C** 510

Answers to Exercises 515

Bibliography 560

Index 575

Checklist of Objectives for a One-Year Course

The second term has the first as a prerequisite; otherwise any selection or permutation of the three terms is possible.

FIRST TERM

Section		**Pages**
1.2	Graphical solution of linear programs (Figure 1.2)	2
2.2	Convex solution sets (Figure 2.2D)	19
2.3	Orthogonal complements (14)–(21)	23–25
2.4	Primal and dual tableaus (Table 2.4)	28
	Computer rules for pivoting 1. to 5.	30
	Invariance of primal and dual solution sets	31
2.5	Rules for pivoting with integers	34
2.6	Matrix inversion (Table 2.6A) (a)	38
	Solving linear equations (condensed tableau) (b)	37
	Solving linear equations (extended tableau) (f)	41–45
3.1	LP tableaus and notation (Table 3.1)	60
3.2	Convex sets and basic solutions in LP	60–64
3.3	Simplex algorithm (summary)	65–68
	Primal-dual orthogonality (Figure 3.3A)	69–70
	Optimality test (summary, Theorem 7*)	71
	Why the dual problem is important	71
3.4	Simplex algorithm (Examples 1 to 4)	75–79
3.5	Optimality test (a) (example)	84–86
3.6	Models 1 and 2 (Table 3.6)	88–92
3.7	A production problem	95–96
	Formulation of linear programs	97
6.1	Two-phase simplex (Table 6.1, Figure 6.1)	231–232
	Example of two-phase simplex	229–235
6.2	Complementary pairs of variables (Table 6.2)	236–238
4.1	Game concepts, saddle points, maximin	105–110
4.2	Mixed strategies	112–118
	Graphical solutions	112–116
4.3	Domination (ii)	120
4.4	LP solution of matrix games	123–127
4.5	Minimax Theorem (1), (2a)	127–129
	Kernel of a game (9)	129–131

SECOND TERM

4.7	Games against nature (Table 4.7)	140–145
5.1	Feasibility, domination, efficiency, equilibria	149–154
	Prisoner's Dilemma	155, 167
5.2	Axis, utilities	168
5.3	Nash cooperative solution and three others	170–176
5.5	Leontief models	188–189
5.6	Economic growth (Examples 2 and 3)	190–195
5.7	Shapley value of *n*-person game	198–203
7.1	Matrix algebra (9), (11), (16) to (18)	261–268
7.3	Forward and backward transformations, pricing out (2) to (8)	272–274
	Revised simplex method (Tables 7.3, 7.3A, B)	275–278
7.4	Gaussian elimination $\mathbf{B}=\mathbf{LU}$ (Tables 7.4A, B)	282–287
7.5	Numbers of operations (2)	290
	Control of round-off errors (b_1) to (b_4)	293–294
8.1	Changing a column or row (a) to (c), Table 8.1	315–322
8.2	Ranging (Tables 8.2A, B)	322–329
8.3	One parameter in **b** and **c** (Examples 1 to 4)	336–344
8.5	One parameter in a column of **A** (Tables 8.5A, B)	349–353
8.8–8.9	Multiple objectives	364–366; 369–373; 376–383

THIRD TERM (Independent)

9.1	Graphs and trees	385–389
9.2	Shortest spanning tree (Tables 9.2B, C)	391–395
9.3	Shortest paths from one node ($c_{ij} \geq 0$)	400–404
9.4	Critical paths in scheduling (Table 9.4)	407–412
9.5	Warehouse problem (a) (Table 9.5)	414–416
	Equipment replacement (3), (4); assortment (c)	417
9.7	All the shortest paths (Table 9.7)	422–424
9.8	Maximum-capacity paths (2)	425
10.1	Dynamic programming (examples (a) to (d))	428–431
10.2	DP examples (e) to (g)	432–434
	DP examples summarized (Table 10.2)	434
10.3	DP remarks	436–440
10.4	Decision networks (Figure 10.4A, B, D)	440–446
10.5	Extensive games (a) (more if 4.1–4.2 studied)	447–449
11.1	Transportation problem	468–471
11.2	Basic solutions, triangularity, and trees	473–479
11.3	Δx loop (Fig. 11.3), Δc partition (4)	480–483
	Methods for pricing out (a) to (d)	483
11.4	Simplex transportation algorithm (Tables 11.4A, B)	484–490
	Assignment problem (Table 11.4C)	490–491
11.5	Restricted transportation problem (Table 11.5)	493–494
	Bottleneck transportation problem (Exercises 14 to 20)	497, 557
11.6	Capacitated transportation (Tables 11.6 A, B)	497–503

CHAPTER ONE

Introduction

1.1 THE NATURE OF THE PROBLEM

Mathematical programming is concerned with *optimization*, which involves making plans or decisions such that the "best" outcome is obtained. Evidently this requires a prior decision as to the criterion by which the best outcome is to be defined or identified, as well as a recognition of what actions are within the realm of possibility.

Mathematical analysis is most useful in situations that can be described at least approximately by a quantitative model in which the possible actions correspond to values assumed by a set of real variables x_j or y_j; the criterion of goodness or lack thereof is a real-valued function of these variables, called the objective function, that is to be maximized or minimized; and the realm of possibility (called the feasible region) is defined by a system of constraints (equations and/or inequalities) involving these variables. While some decision-making situations are probably too complex and too subjective ever to be usefully described in such a way, many practical problems of production, transportation, consumption, marketing, economic policy and military tactics can be approximated by such a model. The necessary theory and techniques have been developed only in the last few decades and are attractive objects of study.

To permit a more thorough treatment, the scope of this book is limited to problems that may be large or small but are solvable in finite terms by methods that are usually either linear or enumerative or both. The related mathematical topics also are treated. *Linear programming* problems form an important subclass in which both the objective function and the constraints involve only linear functions of the variables. Here "programming" means planning and decision making but it is also true that in practice electronic computers with suitable programs are needed to perform the calculations. For instructional purposes, however, small problems and examples are naturally very useful.

The remainder of this chapter gives a small sample of illustrative applications. More will appear in Chapters 5, 9, 10 and elsewhere as noted in the preface, after the treatment of linear programs in Chapter 3 has made possible a more profitable discussion. Chapter 2 paves the way with some underlying ideas and the fundamental arithmetical procedure of pivoting.

1.2 AN EXAMPLE OF RESOURCE ALLOCATION

An agricultural chemical firm has 170 tons of nitrogen fertilizer N, 100 tons of phosphate P and 60 tons of potash K. It is able to sell a 5:4:1 mixture of these substances at a profit of \$10 per ton and it can sell a 4:2:4 mixture at a profit of \$20 per ton. How many tons y_1, y_2 of the two mixtures should be prepared to obtain the maximum profit? y_1 and y_2 cannot be negative since the firm is contemplating only sales, not purchases. Since the amounts of nitrogen, phosphate and potash used are not to exceed the amounts available, the constraints are $y_1 \geq 0$, $y_2 \geq 0$,

$$\begin{aligned}
&\text{Nitrogen used} = .5y_1 + .4y_2 \leq 170 \text{ available.}\\
&\text{Phosphate used} = .4y_1 + .2y_2 \leq 100 \text{ available.}\\
&\text{Potash used} = .1y_1 + .4y_2 \leq 60 \text{ available.}
\end{aligned}$$

We are to maximize the profit $10y_1 + 20y_2$, which will be denoted by $\bar{z}$.

Since there are only two independent variables y_1 and y_2, a graphical solution is possible as in Figure 1.2. The set of possible points (y_1, y_2) (called the *feasible region*) is the quadrilateral $OABC$. It is the intersection of five half-planes lying on the proper sides of the lines $y_1 = 0$, $y_2 = 0$, $.5y_1 + .4y_2 = 170$, $.4y_1 + .2y_2 = 100$, and $.1y_1 + .4y_2 = 60$. The small triangles or carets point into the desired half-planes, which are determined by testing some convenient point such as $(0, 0)$.

To see which feasible point will maximize $\bar{z} = 10y_1 + 20y_2$, note that for a given constant value of $\bar{z}$ (say 3500), the graph of this equation is a straight line of slope $-\frac{1}{2}$. As $\bar{z}$ increases, the line will move parallel to itself in a roughly "northeasterly" direction. (Three such dashed lines are shown in Figure 1.2.) As it does so, we

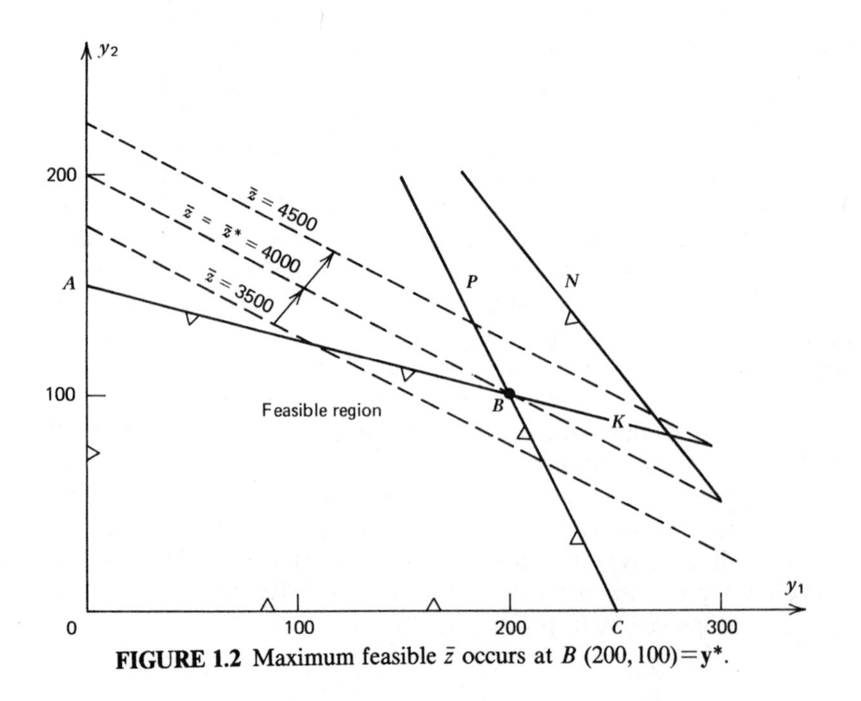

FIGURE 1.2 Maximum feasible $\bar{z}$ occurs at B $(200, 100) = \mathbf{y}^*$.

observe that the last point at which it touches the feasible region is $B=(200,100)$, where $y_1=200$ tons, $y_2=100$ tons, $\bar{z}=\$4000$. Evidently this is the maximum.

The graph only needs to be accurate enough to show that the optimum lies at the intersection of the lines $.4y_1+.2y_2=100$ and $.1y_1+.4y_2=60$, which then can be solved algebraically. The nitrogen line turns out to be superfluous; there are 30 tons left over. The fact that the optimum is obtained at a corner of the feasible region is an important property of these problems.

1.3 APPLICATIONS TO TRANSPORTATION

A simple example of a transportation problem is the following. A steel scrap dealer has scrapyards at Atlanta, Baltimore, and Chicago containing 8000, 9000, and 13000 tons of scrap respectively. The dealer has orders for 12000 tons in Cincinnati, 10000 tons in Detroit, and 7000 in Pittsburgh. The shipping distances in miles between cities are the following:

	Destinations		
Sources	1. Cin	2. Det	3. Pit
1. Atl	450	710	700
2. Bal	500	510	230
3. Chi	320	290	460

It is desired to fill the orders at the lowest possible cost.

The problem involves nine unknowns x_{ij}, the amount shipped from source i to destination j, and six constraints, one for each source and destination, in addition to $x_{ij}\geq 0$. If we naively assume that the shipping cost per ton is proportional to the distance, then the objective function to be minimized can be taken as the total amount of shipping in ton-miles.

Because it is a somewhat atypical special case, the transportation problem is postponed until Chapter 11. Nevertheless it is an important and approachable topic, and the reader who wishes to study a simplex algorithm as soon as possible may go to 11.1 to 11.4 at any time after reading Chapter 1 and 9.1. In that event the occasional references to earlier chapters may be ignored. Chapter 9 and 10.1 to 10.4 are also immediately accessible.

Instead of shipment from one place to another, consider shipment from one time point to another. Shipment into the future is called storage or saving; this requires that the commodity not be too perishable. Shipment from the future into the present is called borrowing. Evidently temporal shipment differs from spatial in having quite different costs in the two directions; indeed it may be possible in only one direction. Origins and destinations may specify both places and times if desired.

In another interpretation x_{ij} may represent the number of machines of type i assigned to perform jobs of type j; the cost or advantage of each such assignment and the numbers of machines and jobs of each type are given. A related but more general type of problem arises when such an assignment is accompanied by the transformation of the original commodity or service into something else, at a rate that varies from one assignment to another. An example is the use of a given fleet of aircraft to satisfy estimated needs for passenger or cargo service on various routes. The variable rate just mentioned stems from the various sizes and speeds of the aircraft and the various lengths of the routes.

When shipments need not or cannot be made directly from source to destination but are sometimes routed via intermediate points, the transportation problem leads into transshipment and network theory. Networks are also used extensively for the scheduling of the many tasks that must be performed in the proper sequence in order to complete a very complex project. Every transportation facility, whether it is an urban expressway, a fleet of trucks, a telephone channel, a microwave link, a pipeline or an electric transmission line, has a limited carrying capacity, which is an important parameter in practice.

1.4 GAMES AND DUAL PROBLEMS

Perhaps the simplest example of a game is that of "matching pennies," described as follows:

		Second player's choice	
		Heads	Tails
First player's choice	Heads	+1	−1
	Tails	−1	+1

+1 means that the first player I wins; −1 means that I loses (something of equal value). As everyone knows, the safe way to play the game is to choose heads or tails at random, with equal probabilities. However, a game (a *matrix* game, to be more specific) can be much more complicated; in general it has the following structure:

		J's choice				
		1	$\dots$	j	$\dots$	n
I's choice	1	a_{11}	$\dots$	a_{1j}	$\dots$	a_{1n}
	$\vdots$	$\vdots$		$\vdots$		$\vdots$
	i	a_{i1}	$\dots$	a_{ij}	$\dots$	a_{in}
	$\vdots$	$\vdots$		$\vdots$		$\vdots$
	m	a_{m1}	$\dots$	a_{mj}	$\dots$	a_{mn}

Here the number a_{ij} represents the amount of something desirable that I receives from J in the event that I and J independently (perhaps simultaneously) choose row i and column j respectively. I and J each have the problem of deciding how to make their selections. The problem need not arise from a parlor game; it may represent a competitive business situation or a military conflict as in 5.1, a problem in statistical decision theory as in 4.7, or a mathematical equivalent of any linear program as in 4.6. In 4.4 it is shown that a matrix game can be reduced to a linear programming problem; other methods for solving small games are considered in 4.2 to 4.6.

In 1947, while working for the U.S. Air Force, George B. Dantzig began the development of an elegant numerical procedure (the simplex method) for solving problems like those above. This method has the most interesting property that in solving the original problem (called the primal), it also automatically solves a quite different yet intimately related problem called the dual. The dual of the problem in 1.2 is

$$
\begin{aligned}
u_1, u_2, u_3 &\geq 0 \\
.5u_1 + .4u_2 + .1u_3 &\geq 10 \\
.4u_1 + .2u_2 + .4u_3 &\geq 20 \\
170u_1 + 100u_2 + 60u_3 &\text{ to be minimized}
\end{aligned}
$$

This type of problem will be discussed in 3.6 where it will be shown that u_1, u_2, u_3 can be interpreted as the values to the firm per ton of nitrate, phosphate and potash, respectively, calculated as if these materials had no possible use except to fill the two orders being considered in the problem. The optimal dual solution is needed to calculate the effect of changes in the primal problem and to confirm the optimality of the primal solution, and vice versa (the primal is the dual of the dual). In game theory the dual and the primal solutions determine the strategies for the first and the second player respectively. The transportation problem also has a dual, which is important in its solution.

1.5 RELATION TO THE CALCULUS

The differential calculus is not a prerequisite for this book from a logical point of view but is desirable for the sake of mathematical maturity. Thus most readers will probably already have had some experience with the use of calculus in finding maxima and minima, and some may wonder if mathematical programming is just an alternative name or procedure for doing the same thing. The answer is that the two subjects have much in common as to purpose, and their concepts come together in the area of nonlinear programming (the Kuhn-Tucker conditions). However, the calculus has little to contribute to the problems studied in *linear* programming.

Let us recall how the calculus is used. If the function f of one or more variables is defined on a nonempty region R of finite extent that is closed (includes its boundary) and if f is continuous everywhere in R, then it is true that f has a maximum and a minimum in R and the points where the maximum and the minimum occur must be of the following types:

(a) A point where (all) the first-order derivative(s) of f are zero.
(b) A point where (some of) the first-order derivative(s) of f do(es) not exist.
(c) A point on the boundary of the region R.

Now linear programming considers only those functions f that are linear, say $c_1x_1 + \cdots + c_nx_n$. These functions everywhere have partial derivatives, which are simply the constant coefficients $c_1, \ldots, c_n$. Thus there are no points of type (b). Also, these coefficients are never all zero unless f itself is constant, in which case every point of R, including the boundary points of type (c), yields both the maximum and the minimum. Consequently points of type (a) need not be considered either.

Thus we conclude that if a linear function has a maximum or a minimum, it must have it on the boundary of the feasible region R. Since the boundary may consist of as many pieces as there are subsets of constraints, a systematic method of searching is essential, and that is just what the simplex method of linear programming provides. The derivatives still determine the location of the maximum and the minimum even on the boundary but since these derivatives are just the constant coefficients $c_1, \ldots, c_n$ as noted before, it is not surprising that linear programming can get along without the concepts and methods of the calculus.

In linear programming problems a maximum or a minimum may fail to exist for either of two reasons: (i) The constraints may be inconsistent so that R is empty and the maximum and minimum are meaningless. (ii) R may be of infinite extent, in which case the maximum and the minimum may be either finite or infinite. If infinite, it is likely that the constraints are incorrect or incomplete; in any case it is customary to apply the names maximum and minimum to finite values only.

A third possibility, that the maximum or minimum be approachable as a limit but not attainable because R is not closed, is excluded by avoiding the use of strict inequalities in defining R.

EXERCISES 1.5

Use the graphical method of 1.2 to optimize the following linear programming problems or to show that no optimum exists.

1. Min $z = 2x_1 + 3x_2$ subject to $x_1 \geq 0$, $x_2 \geq 0$, $3x_1 + 2x_2 \geq 2$, $2x_1 + x_2 \geq 4$, $x_1 + 4x_2 \geq 3$.

2. Max $\bar{w} = 4u_1 - 6u_2$ subject to $u_1 \geq 0$, $u_2 \geq 0$, $u_1 - u_2 - 3 \leq 0$, $u_1 \leq 3u_2$, $-u_1 + 2u_2 + 1 \geq 0$.

3. Min $z = 2x_1 + x_2$ subject to $x_1 \geq 0$, $x_2 \geq 0$, $x_1 + 2x_2 \geq 3$, $3x_1 + 2x_2 \geq 2$, $3x_1 + x_2 \leq 1$.

4. Max $\bar{w} = u_1 - u_2$ subject to $u_1 \geq 0$, $u_2 \geq 0$, $2u_1 + u_2 \geq 4$, $u_1 - 2u_2 + 2 \geq 0$, $u_2 \leq 2$.

5. Max $\bar{z} = 4x_1 + 3x_2$ subject to $x_1^2 + x_2^2 \leq 100$.

6. Min $z = x_1^2 + x_2^2$ subject to $4x_1 + 3x_2 \geq 50$.

7. Max $\bar{z} = x_1^2 + x_2^2$ subject to $4x_1 + 3x_2 \leq 50$.

GENERAL REFERENCES†

Mathematical Programming

Abadie (1970)
Antosiewicz (1955)
Aronofsky (1969)
Arrow-Hurwicz-Uzawa (1958)
Balinski (1974)
Balinski-Hellerman (1975)
Balinski-Lemarechal (1978)
Bazaraa-Jarvis (1977)
Beale (1968, 1969)
Charnes-Cooper (1961)
Cooper-Steinberg (1974)
Dantzig (1963)
Dantzig-Veinott (1968)
Gale (1960)
Garvin (1960)
Gass (1958, 1969)
Geoffrion (1972)
Gill-Murray-Wright (1981)
Graves-Wolfe (1963)
Hadley (1962)
Hillier-Lieberman (1967)
Karlin (1959)
Knuth (1968–1973)
Koopmans (1951)
Kuhn (1970)
Kuhn-Tucker (1956)
Luenberger (1969)
Murty (1976)
Orchard-Hays (1968)
Sasieni-Yaspan-Friedman (1959)
Simonnard (1966)
Singleton-Tyndall (1974)
Spivey-Thrall (1970)
Vajda (1958–1972)
Van de Panne (1975)
Whittle (1971)
Zoutendijk (1976)

Games

Blackwell-Girshick (1954)
Dresher-Shapley-Tucker (1964)
Dresher-Tucker-Wolfe (1957)
Gale (1960)
Harsanyi (1977)
Karlin (1959)
Kuhn-Tucker (1950, 1953)
Luce-Raiffa (1957)
McKinsey (1952)
Owen (1968)
Parthasarathy-Raghavan (1971)
Rapoport (1966, 1970)
Shubik (1959–1980)
Tucker-Luce (1959)
Von Neumann-Morgenstern (1944, 1953)

†Usually not repeated in the lists for individual chapters. Dynamic programming references follow Chapter 10.

CHAPTER TWO

Solutions Sets, Duality and Pivoting

2.1 LINEAR EQUATIONS AND VECTOR SPACES

Consider the system of linear equations

$$2x_1 + x_2 = 7$$

$$x_1 - x_2 = 2 \tag{1}$$

Eliminating x_2 by adding the corresponding members of the two equations gives $3x_1 = 9$ or $x_1 = 3$. Subtracting twice the second equation from the first gives $3x_2 = 3$ or $x_2 = 1$. This procedure has accomplished two things: it has proved that (1) cannot have any solution other than $x_1 = 3$, $x_2 = 1$, and it has provided the numbers 3 and 1, which we may prove to be a solution by substituting them in (1). Later we will see that this verification by substitution is not logically necessary when dealing with linear equations, although it is a desirable protection against numerical errors in hand calculation.

Although the solution $x_1 = 3$, $x_2 = 1$ involves two numbers, it is only one solution and it is represented by a single point in the Cartesian plane, at the intersection of the graphs of the equations in Figure 2.1A. The solution point, denoted by $(3, 1)$, is an example of a vector, which is said to have the scalars 3 and 1 as its components. The order in which the components are written down is important, since $(1, 3)$ is not the same point as $(3, 1)$, and $x_1 = 1$, $x_2 = 3$ is not a solution of (1). One also may associate the vector $(3, 1)$ with the directed line segment (the dotted arrow in Figure 2.1A) from the origin $(0, 0)$ to the point $(3, 1)$.

The solution set of a system of equations is the set of all points that satisfy all the equations. Thus the solution set of (1) consists of exactly one point $(3, 1)$. If the two straight lines in Figure 2.1 had been parallel and distinct, the solution set would have contained no points at all; the solution set would be called *empty* and denoted by $\varnothing$ and the equations would be called inconsistent. A single linear equation in two variables, such as $2x_1 + x_2 = 7$, has an infinite number of points in its solution set; this set is depicted by the *graph* of the equation, which is a straight line extending infinitely far in both directions. We will find that the solution set for a system of linear equations always contains either no points, or exactly one point, or an infinite number of points.

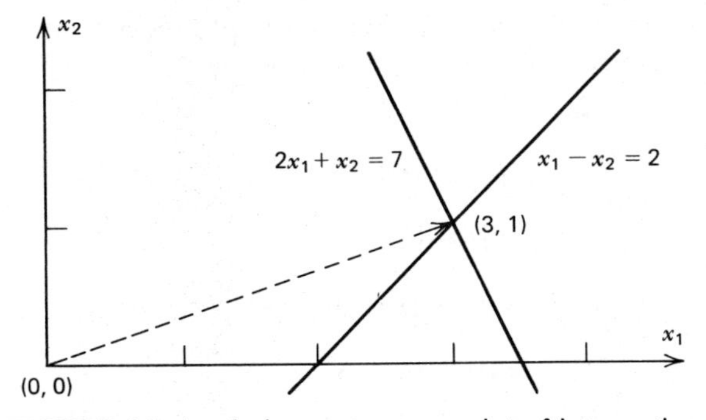

FIGURE 2.1 A solution vector as a point of intersection.

At most three variables can be represented geometrically but in theory and practice we must be able to handle more than that. For the purposes of this book it suffices to say that a *vector is an ordered n-tuple of real numbers*, written

$$\mathbf{x} = (x_1, x_2, \ldots, x_n) \tag{2}$$

which is subject to the operations described in the following paragraph. The real numbers $x_1, \ldots, x_n$ (also called *scalars*) are the *components* of the vector. In other areas of mathematics this definition may be broadened in one or more ways: the scalars may be complex numbers as well as real, n may be infinite, or the components may have no unique identities, being dependent on an arbitrary selection of a basis or coordinate system.

Two vectors $\mathbf{x} = (x_1, \ldots, x_n)$ and $\mathbf{y} = (y_1, \ldots, y_n)$ are said to be equal if and only if $x_j = y_j$ for $j = 1, \ldots, n$. One also defines $\mathbf{x} + \mathbf{y} = (x_1 + y_1, \ldots, x_n + y_n)$ and $t\mathbf{x} = (tx_1, \ldots, tx_n)$, where t is any scalar. $\mathbf{0}$ represents a vector all of whose components are zero. Thus it is true that $\mathbf{x} + \mathbf{0} = \mathbf{x}$ and $\mathbf{x} + \mathbf{x} = 2\mathbf{x}$, for example. These definitions will be useful in describing the solution sets of systems of linear equations in any number n of variables, in terms of a finite number of particular solutions.

Homogeneous Equations

For a more general study of linear equations and their solution sets it is convenient to begin with homogeneous equations such as

$$x_1 + 2x_2 - 3x_3 + 5x_4 = 0 \tag{3}$$

which are characterized by having the constant term equal to zero. Any system of such equations always admits the trivial solution $\mathbf{x} = \mathbf{0}$. Particular nontrivial solutions, if they exist, may be found by solving for a suitable fixed subset of the variables, called the basic variables, after assuming that some one nonbasic variable is equal to 1 and the others are equal to 0.

For example, let us solve (3) for x_1 and choose x_2, x_3, x_4 in turn as the variable set equal to 1. This yields three particular solutions of (3) which we denote by

$$\xi^1 = (-2, 1, 0, 0), \qquad \xi^2 = (3, 0, 1, 0), \qquad \xi^3 = (-5, 0, 0, 1) \tag{4}$$

(The superscripts are not exponents but subscripts moved upstairs to avoid any suggestion of components of vectors.) Now it is easy to verify by substitution that for any numbers t_1, t_2, t_3,

$$\begin{aligned} x_1 &= -2t_1 + 3t_2 - 5t_3 && (5') \\ x_2 &= 1t_1 + 0t_2 + 0t_3 = t_1 \\ x_3 &= 0t_1 + 1t_2 + 0t_3 = t_2 && (5'') \\ x_4 &= 0t_1 + 0t_2 + 1t_3 = t_3 \end{aligned} \qquad (5)$$

is also a solution of (3). This is written more compactly as

$$\mathbf{x} = t_1\xi^1 + t_2\xi^2 + t_3\xi^3 \qquad (6)$$

Notice that each of the solution vectors in (4) appears as a column of coefficients in (5).

Evidently any number of particular solutions could have been combined as in (5) or (6), and other solutions could have been chosen; the result would still satisfy (3). The actual form of (4) and (6) was chosen to yield two further properties:

(a) Every solution of (3) is obtainable from (6).
(b) Three is the smallest number of terms in (6) that will give property (a).

The proof of (a) for the single equation (3) is very simple. If (x_1, x_2, x_3, x_4) is a solution of (3), then

$$x_1 = -2x_2 + 3x_3 - 5x_4$$

By (5″) we may and must choose t_1, t_2, t_3 equal to x_2, x_3, x_4 respectively. This makes (5′) identical with (3), and so (5) has produced the solution. The proof of (b), and a similar analysis for a set of equations, requires simple calculations that will be explained in the following sections.

What has been asserted above is usually described by saying that the solution set of (3) is a (linear) vector space of rank or dimension three for which the set of vectors ξ^1, ξ^2, ξ^3 forms a basis (only one of many possible bases, any one of which will work in (6)). This vector space is contained in the four-dimensional vector space of all 4-tuples (x_1, x_2, x_3, x_4), for which the simplest basis consists of the unit vectors

$$(1,0,0,0), \qquad (0,1,0,0), \qquad (0,0,1,0), \qquad (0,0,0,1) \qquad (7)$$

More formally, we have the following definitions and theorems.

A (linear) vector space is a set X of vectors (n-tuples) including the origin $\mathbf{0}$ such that whenever $\mathbf{x}$ and $\mathbf{y}$ belong to the set, so does $s\mathbf{x}+t\mathbf{y}$, where s and t are any scalars.

In other words, the vector space must be closed under multiplication by scalars (which do not themselves belong to the space) and closed under the addition of vectors. A third version is that a vector space is a linear variety or flat (defined below) that contains $\mathbf{0}$.

A set of vectors $\xi^1,\ldots,\xi^p$ is said to generate or *span* a vector space if every vector $\mathbf{x}$ in the space can be obtained in the form

$$\mathbf{x}=t_1\xi^1+\cdots+t_p\xi^p \tag{8}$$

for at least one set of scalars $t_1,\ldots,t_p$.

A *basis* of a vector space is a spanning set containing a minimum number of vectors, or equivalently by Theorem 1 of 2.7, a maximum set of linearly independent vectors, meaning that none is in the space spanned by the others. These sets will always be finite in this book. The theorem cited shows that every spanning set contains a basis.

The *rank* or *dimension* of a vector space is the number of vectors in any of its bases. It will be shown in Theorem 6 of 2.7 that the dimension of the solution set of a system of homogeneous linear equations is at least equal to the number of variables minus the number of equations.

The vector spaces that can be visualized in ordinary three-dimensional space, numbered according to their rank, are the following: (0) The point at the origin. (1) Any straight line through the origin. (2) Any plane through the origin. (3) The entire three-dimensional space.

We leave to the reader the easy verification that a vector space of n-tuples as defined above satisfies the following axioms for an abstract vector space X:

A1. If $\mathbf{x}$ and $\mathbf{y}$ are in X, so is $\mathbf{x}+\mathbf{y}$ (closure).
A2. $(\mathbf{x}+\mathbf{y})+\mathbf{z}=\mathbf{x}+(\mathbf{y}+\mathbf{z})$ (associative law).
A3. X contains a unique $\mathbf{0}$ (the origin) such that $\mathbf{x}+\mathbf{0}=\mathbf{0}+\mathbf{x}=\mathbf{x}$ for all $\mathbf{x}$ in X.
A4. For every $\mathbf{x}$ in X there is a unique $-\mathbf{x}$ in X such that $\mathbf{x}+(-\mathbf{x})=-\mathbf{x}+\mathbf{x}=\mathbf{0}$.
A5. $\mathbf{x}+\mathbf{y}=\mathbf{y}+\mathbf{x}$ (commutative law).
B1. If $\mathbf{x}$ is in X and t is a scalar, $t\mathbf{x}$ is in X (closure). ($\mathbf{x}t$ would be just another notation for the same thing. $(-1)\mathbf{x}=-\mathbf{x}$.)
B2. $s(t\mathbf{x})=(st)\mathbf{x}$ (associative law for multiplication).
B3. $1\mathbf{x}=\mathbf{x}$ (1 is the multiplicative unit).
B4. $t(\mathbf{x}+\mathbf{y})=t\mathbf{x}+t\mathbf{y}$
B5. $(s+t)\mathbf{x}=s\mathbf{x}+t\mathbf{x}$

For a given field of scalars, these axioms together with the rank completely determine the abstract structure of the vector space; the rank could be infinite. Axioms A1 to A5 are equivalent to the statement that a vector space is a commutative group with respect to the operation of addition. *A subset of a vector space is itself a vector space if it satisfies A1, A3, and B1*, since then the other axioms are satisfied automatically, as in the following proof.

THEOREM 1 The set of all solutions of the system of homogeneous linear equations

$$\sum_{j=1}^{n} a_{ij}x_j=0 \qquad (i=1,\ldots,m) \tag{9}$$

forms a vector space.

Proof Let $\mathbf{x}$ represent one solution and $\mathbf{y}$ another, so that we also have

$$\sum_{j=1}^{n} a_{ij}y_j=0 \qquad (i=1,\ldots,m) \tag{10}$$

Multiplying (9) by s and (10) by t and adding pairs of equations having the same value of i gives

$$\sum_{j=1}^{n} a_{ij}(sx_j+ty_j)=0 \qquad (i=1,\ldots,m) \tag{11}$$

which shows that $s\mathbf{x}+t\mathbf{y}$ is also a solution. Since $\mathbf{0}$ is obviously a solution, this proves that the solutions form a vector space.

Nonhomogeneous Equations

If we attempt to extend the proof of Theorem 1 to the system of nonhomogeneous linear equations

$$\sum_{j=1}^{n} a_{ij}x_j=b_i \qquad (i=1,\ldots,m) \tag{12}$$

we obtain in place of (11) the result

$$\sum_{j=1}^{n} a_{ij}(sx_j+ty_j)=(s+t)b_i \qquad (i=1,\ldots,m) \tag{13}$$

so that if some $b_i \neq 0$, then $s\mathbf{x}+t\mathbf{y}$ is a solution if and only if $s+t=1$, or $s=1-t$.

Another way to characterize the solution set of (12) is to translate the origin to some point $\boldsymbol{\xi}$ that satisfies (12), if such a point exists. Then we have

$$\sum_{j=1}^{n} a_{ij}\xi_j=b_i \qquad (i=1,\ldots,m)$$

and subtracting each of these equations from the equation in (12) having the same value of i gives

$$\sum_{j=1}^{n} a_{ij}(x_j - \xi_j) = 0 \qquad (i = 1, \ldots, m) \tag{14}$$

By the preceding Theorem 1, the set of values of $\mathbf{x} - \boldsymbol{\xi}$ satisfying (14) forms a vector space, or $\mathbf{x} = \boldsymbol{\xi} +$ an element of a vector space.

The discussion of (13) motivates the following definition; other terms that appear in the literature are *flat* and *affine subspace*.

A linear variety is a set X' of vectors (possibly empty) such that if $\mathbf{x}$ *and* $\mathbf{y}$ *belong to the set, so does* $(1-t)\mathbf{x} + t\mathbf{y}$ *for any scalar* t.

As t varies, $(1-t)\mathbf{x} + t\mathbf{y}$ describes the infinite straight line through $\mathbf{x}$ and $\mathbf{y}$. Thus the discussion of (13) proved the following.

THEOREM 2 The solution set of a system of linear equations (not necessarily homogeneous) is a linear variety.

If the system consists of only one equation, the solution set is called a plane. Unlike a vector space, a linear variety may be empty; if it is not empty, then in ordinary three-dimensional space it consists of a point, a line or a plane in any position, or else the whole space.

THEOREM 3 If the vectors $\mathbf{x}^1, \ldots, \mathbf{x}^p$ belong to a linear variety X', so does $t_1\mathbf{x}^1 + \cdots + t_p\mathbf{x}^p$ for any scalars t_i such that $t_1 + \cdots + t_p = 1$.

Proof The theorem is trivial for $p = 1$; the proof really begins with $p = 2$, for which the theorem is true by definition. According to the principle of mathematical induction, the proof may be completed by showing that the truth of the theorem as stated for any particular integer $p = 2, 3, \ldots,$ implies its truth for the next larger integer $p+1$. Thus we have to show that if $t_1 + \cdots + t_{p+1} = 1$, then $\mathbf{x} = t_1\mathbf{x}^1 + \cdots + t_{p+1}\mathbf{x}^{p+1}$ belongs to X'. Since $p \geq 2$, not all of the $t_i = 1$; suppose the indices are chosen so that $t_{p+1} \neq 1$. Then the combination may be written as $\mathbf{x} = (1 - t_{p+1})\mathbf{y} + t_{p+1}\mathbf{x}^{p+1}$, where

$$\mathbf{y} = \left(t_1\mathbf{x}^1 + \cdots + t_p\mathbf{x}^p\right) / \left(t_1 + \cdots + t_p\right) = \left(t_1\mathbf{x}^1 + \cdots + t_p x^p\right) / \left(1 - t_{p+1}\right).$$

Now $\mathbf{y}$ is in X' because it is an appropriate linear combination of only p vectors in X', for which the theorem is supposed already proved. Then $\mathbf{x}$ also is in X' by the definition of linear variety.

THEOREM 4 A set X' of vectors is a nonempty linear variety if and only if the set can be obtained by translating (adding the same constant vector $\boldsymbol{\xi}$ to) the vectors in a vector space X.

Proof Since a vector space always contains $\mathbf{0}$, the set of translates will contain $\xi+\mathbf{0}=\xi$ and so be nonempty. It is a linear variety because if $\mathbf{x}'=\mathbf{x}+\xi$ and $\mathbf{y}'=\mathbf{y}+\xi$ are two of its points, coming from $\mathbf{x}$ and $\mathbf{y}$ in X, then

$$(1-t)\mathbf{x}'+t\mathbf{y}'=(1-t)\mathbf{x}+t\mathbf{y}+\xi \tag{15}$$

is a translate of $(1-t)\mathbf{x}+t\mathbf{y}$ in X and so (15) is in X' as required.

Conversely, let ξ be any point in a nonempty linear variety X' and let $\mathbf{x}$ and $\mathbf{y}$ be any two points of X', possibly equal to one another or to ξ. Define X as the set of vectors of the form $\mathbf{x}'-\xi$, where $\mathbf{x}'$ is in X'. Then $\mathbf{x}-\xi$ and $\mathbf{y}-\xi$ represent any two points of X, which will be a vector space if $s(\mathbf{x}-\xi)+t(\mathbf{y}-\xi)$ is in X for all scalars s and t. This is equivalent to having $s\mathbf{x}+t\mathbf{y}+(-s-t+1)\xi$ in X', which is true by Theorem 3.

It can be shown that the converses of Theorems 1 and 2 are true in the sense that if one is given a vector space or a linear variety contained in (but not identical with) the vector space of all n-tuples of scalars, then the given space is the solution set of some system of linear equations, homogeneous or nonhomogeneous, respectively.

EXERCISES 2.1

1. Describe the differences between vector spaces and other linear varieties in terms of (a) their relation to the origin and (b) the systems of equations of which they can be the solutions sets.

2. If $\mathbf{x}$ and $\mathbf{y}$ are distinct points of (a) a vector space X or (b) a linear variety X', what else must X or X' contain? (Describe in symbols, words and pictures.)

3. Simplify $2(3,1,2)-3(-2,1,3)$.

4. What values of s and t make $s(1,4)+t(2,-3)=(0,11)$?

In 5 to 9 describe (a) the smallest vector space and (b) the smallest linear variety containing the indicated points.

5. $(1,2)$.

6. $(1,2)$, $(2,4)$.

7. $(1,2)$, $(2,1)$.

8. $(1,0,0)$, $(0,1,0)$.

9. $(1,0,0)$, $(0,1,0)$, $(0,0,1)$.

In 10 and 11 find a basis for the solution set or for the vector space X of which it is a translate (as in (4) and Theorem 4).

10. $3x_1-2x_2+4x_3=0$.

11. $2x_1+5x_2-x_3+3x_4=10$.

12. If $(1,1,2)$ and $(1,-1,0)$ belong to a two-dimensional vector subspace, does $(-2,0,-2)$ belong to it? Does $(1,0,0)$?

13. If $(1,1,2)$ and $(1,-1,0)$ belong to a linear variety, does $(1,3,4)$ belong to it? Does $(-2,0,-2)$?

14. The intersection of a vector space and a linear variety is always what? Prove your assertion.

2.2 LINEAR INEQUALITIES AND CONVEX SETS

The introduction of inequality constraints in general leads to a more complicated solution set which will now be referred to as the feasible region, in order to exclude specifically those (infeasible) solutions that satisfy the equality constraints but not the inequalities. An example of such a feasible region is the polygon $OABC$ in Figure 1.2 of Chapter 1. It has only four sides because one of the five inequality constraints (that for potash) is redundant, being implied by the other four. When the constraints are linear we will show that the feasible region always has the following property.

A set C of vectors (points) is said to be *convex* if whenever the set contains two points $\mathbf{x}$ and $\mathbf{y}$, it also contains every point of the form $(1-t)\mathbf{x}+t\mathbf{y}$ for $0 \leq t \leq 1$, which describes the straight line segment joining $\mathbf{x}$ and $\mathbf{y}$ (limited to $\mathbf{x}$ and $\mathbf{y}$ and points in between). Figure 2.2A shows two sets that are convex and two that are not.

The *intersection* of a collection of sets $C_1, \ldots, C_m$ is a set that includes precisely those objects that belong to every set in the collection. It is denoted by $C_1 \cap C_2 \cap \cdots \cap C_m$ or $\cap_1^m C_i$. The objects that belong to a set are called its elements or members; they will be vectors in our applications. If $\mathbf{x}$ belongs to C, this can be symbolized by $\mathbf{x} \in C$. Intersections are important for us because the feasible region for a collection of conditions (equality and/or inequality constraints) is obviously the intersection of the feasible regions for the conditions taken one at a time. We will occasionally have use for the closely related notion of the *union* of a collection of sets, which includes precisely those objects that belong to at least one set in the collection. The union of the sets M and N is denoted by $M \cup N$. The *difference* set $M-K$ consists of those elements that are in (belong to) the set M but not the set K.

THEOREM 1 The intersection of any number of convex sets is again a convex set.

Proof If $\mathbf{x}$ and $\mathbf{y}$ belong to the intersection, they belong to each of the convex sets individually. Thus by the convexity $(1-t)\mathbf{x}+t\mathbf{y}$ (for $0 \leq t \leq 1$) also belongs to each of the convex sets and hence to their intersection. This proves the intersection is convex. A similar sort of theorem and proof are valid for vector spaces and for linear varieties. Note that the convex intersection may happen to be empty. (Some writers do not regard the empty set as convex.)

THEOREM 2 The feasible region determined by a system of linear equations and inequalities is convex.

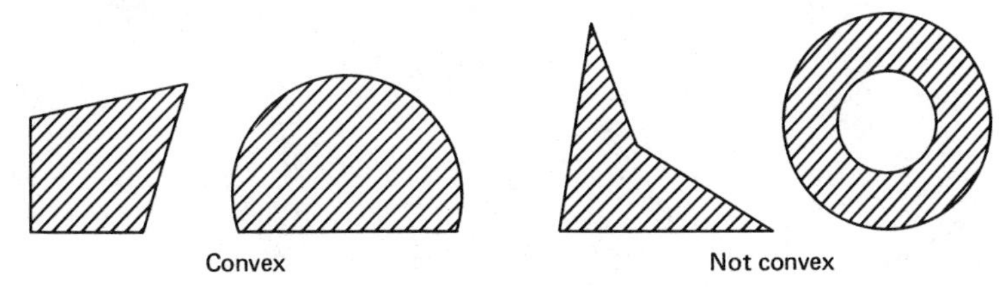

FIGURE 2.2A

Proof Since a linear variety is clearly convex by definition, Theorem 2 of 2.1 shows that each linear equation has a convex solution set. Any linear inequality can be written as $\Sigma_1^n a_j x_j + b \leq 0$; suppose that for a second point $\mathbf{y}$ we also have $\Sigma_1^n a_j y_j + b \leq 0$. Since $1-t$ and t are both nonnegative in the definition of convexity, these numbers may be multiplied into the respective inequalities without reversing them. Then adding the results gives

$$\sum_{i=1}^{n} a_i[(1-t)x_i + ty_i] + (1-t+t)b \leq 0$$

which shows that $(1-t)\mathbf{x}+t\mathbf{y}$ belongs to the solution set, and it is therefore convex. The feasible region is the intersection of the convex solution sets, one for each individual constraint; it is convex by Theorem 1.

With inequalities as with equalities, the homogeneous case (all constant terms b_i equal to zero) is especially simple. In the proof of Theorem 2 there would then be no use for the fact that $1-t$ and t sum to 1. Thus we have the following definition and theorem.

A *convex cone* is a set Λ of vectors including $\mathbf{0}$ such that whenever $\mathbf{x}$ and $\mathbf{y}$ belong to the set, so does $s\mathbf{x}+t\mathbf{y}$ for all $s \geq 0$ and $t \geq 0$. In other words, the convex cone must be closed under multiplication by nonnegative scalars and closed under the addition of vectors. In geometrical terms, a convex cone contains the origin, every ray (infinite half line) from the origin through any other point of the cone, and every line segment joining any two points of the cone. Every vector space is a convex cone; other convex cones may have a circular disc, a triangle or a convex polygon as their cross section. Figure 2.2B shows some convex cones.

THEOREM 3 The feasible region determined by a system of homogeneous linear equations and inequalities is a convex cone. (Proof suggested above.)

The converses of Theorems 2 and 3 are false. Convex sets may have curved boundaries that cannot be defined by a finite number of linear constraints. Convex sets that can be so defined are called *polyhedral*.

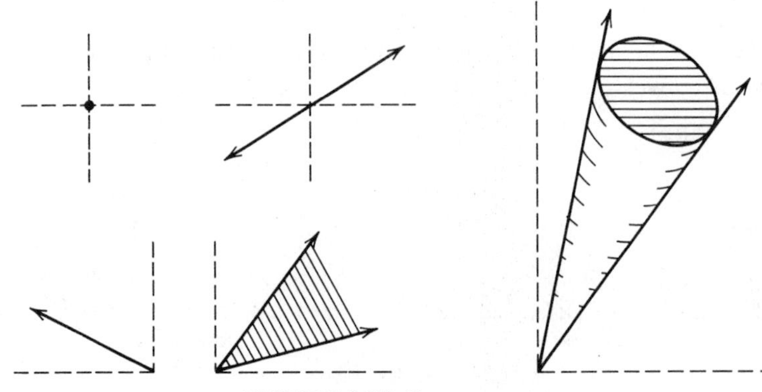

FIGURE 2.2B Convex cones.

A *convex combination* of the points $\mathbf{x}^1,\ldots,\mathbf{x}^p$ is any point of the form

$$\sum_{i=1}^{p} t_i x^i \qquad \text{with all } t_i \geq 0 \qquad \text{and} \qquad \sum_{i=1}^{p} t_i = 1 \tag{1}$$

The *convex hull* of a set S of points is the set of all convex combinations of all selections of (finite numbers of) points $\mathbf{x}^i$ from S.

The following theorem gives an equivalent definition.

THEOREM 4 The convex hull $\langle S \rangle$ of any set S is a convex set and is the intersection (the "smallest") of all the convex sets containing S.

Proof The proof of Theorem 3 of 2.1 requires no change if one requires all $t_i \geq 0$ as now desired and shows that a set is convex if and only if it contains all convex combinations of its points. Thus every convex set containing S, and hence the intersection of all such sets, contains the convex hull $\langle S \rangle$. The proof is completed by showing that $\langle S \rangle$ is convex, is thus itself one of the intersecting sets, and therefore contains the intersection. To show that $\langle S \rangle$ is convex, let $\Sigma_{i=1}^{p} s_i \mathbf{x}^i$ and $\Sigma_{i=1}^{p} t_i \mathbf{x}^i$ be two convex combinations of points of S; there is no loss of generality in assuming that the same points are involved in both combinations since any point that occurs in one can be introduced with a coefficient of zero in the other. If $0 \leq r_1 = 1 - r_2 \leq 1$, we must show that

$$r_1 \sum_{i=1}^{p} s_i \mathbf{x}^i + r_2 \sum_{i=1}^{p} t_i \mathbf{x}^i = \sum_{i=1}^{p} (r_1 s_i + r_2 t_i) \mathbf{x}^i$$

is a convex combination, which it is since $r_1 s_i + r_2 t_i \geq 0$ and

$$\sum_{i=1}^{p} (r_1 s_i + r_2 t_i) = r_1 \sum_{i=1}^{p} s_i + r_2 \sum_{i=1}^{p} t_i = r_1 + r_2 = 1$$

In Figure 2.2C the boundary and interior of the quadrilateral is the convex hull of the six heavy dots. The convex hull of a circle or sphere is its boundary and interior.

An *extreme point* of a convex set C is a point $\mathbf{x}^0$ which is not a convex combination of any two (or more) points of C that are different from $\mathbf{x}^0$. Equivalent definitions are that $\mathbf{x}^0$ is not midway between any two distinct points of C or that C remains convex after $\mathbf{x}^0$ has been removed from C. Thus an extreme point is either a "corner" point or it lies on a curved portion of the boundary. For example, the extreme points of a triangle are its three corners and the extreme points of a cube are its eight corners, while every point on the surface of a sphere is an extreme point of the sphere (surface and interior). In Theorem 9 of 3.3 we will see that extreme points suffice to yield the optimal solutions of linear programs.

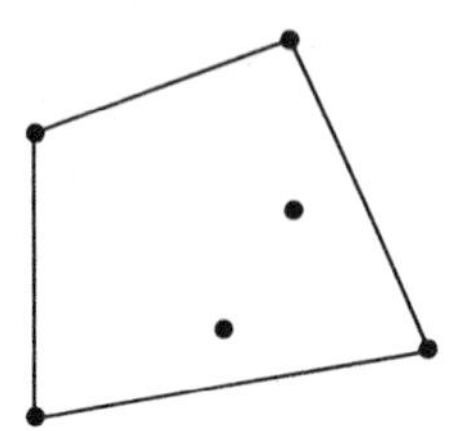

FIGURE 2.2C The convex hull of six points.

The properties of the five types of sets considered in this section and the preceding are summarized in Figure 2.2D. From the nature of the defining constraints it is clear that one has

$$\begin{array}{ccccccc} & \nearrow & \Lambda & \searrow & & & \\ X & & & & A & \longrightarrow & C \\ & \searrow & X' & \nearrow & & & \end{array} \qquad (2)$$

where $X \to X'$ means "every vector space is a linear variety" and so on. Also the intersection of two or more sets of the same type is again a set of that same type.

EXERCISES 2.2

In 1 to 10 indicate whether the given set is convex. If not, describe its convex hull. If it is convex, then (a) classify it into one of the categories in Figure 2.2D (the most restrictive to which it belongs), (b) specify what extreme points it has. Graphing may help.

1. $\{(x_1, x_2)\colon x_1 \geq 0,\ x_2 \geq 0\}$ = the set of points (x_1, x_2) such that $x_1 \geq 0$ and $x_2 \geq 0$.
2. $\{(x_1, x_2)\colon x_1 \geq 0,\ x_2 \geq 0,\ x_1 + x_2 \geq 1\}$.
3. $\{(x_1, 0)\colon 0 \leq x_1 \leq 1\}$.
4. $\{(x_1, 0)\colon |x_1| \geq 1\}$.
5. $\{(x_1, 0)\colon 0 < x_1 < 1\}$.
6. $\{(0,0)\}$.
7. $\{(0,0), (1,1)\}$.
8. $\{(x_1, 1)\colon -\infty < x_1 < \infty\}$.
9. $\{(x_1, x_2)\colon x_1 = x_2\}$.
10. $\{(x_1, x_2)\colon x_1^2 + x_2^2 \leq 1\}$.
11. If $(1,1,2)$ and $(1,-1,0)$ are extreme points of a convex set, does $(1,3,4)$ belong to the set? Does $(1,0,1)$?
12. What convex combination of $(0,0)$, $(0,3)$, (3.0) yields $(1,1)$?
13. What is the convex hull of the set $\{(5,0), (2,1), (0,0), (0,4), (1,2)\}$? (Draw a graph.)
14. Prove that $\{(x_1, x_2)\colon x_1^2 + x_2^2 \leq 1\}$ is the convex hull of its circular boundary.
15. Prove that $\{(x_1, x_2)\colon |x_1| \leq 1,\ |x_2| \leq 1\}$ is the convex hull of $\{(1,1), (1,-1), (-1,1), (-1,-1)\}$.
16. The intersection of a linear variety with a convex set is always what? Prove your assertion.

Symbol	Example	Constraints	Solution Set	If Set Contains **x** and **y** It Also Contains
X	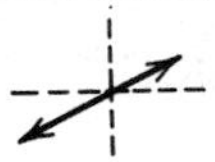	Homogeneous linear equations	Vector space	**0** and infinite plane (or line or point) $s\mathbf{x}+t\mathbf{y}$ through **0**, **x** and **y**
X'		Nonhomogeneous linear equations	Linear variety or flat	Infinite line $(1-t)\mathbf{x}+t\mathbf{y}$ through **x** and **y**
Λ		Homogeneous linear equations and/or inequalities	Convex polyhedral cone	**0** and portion of plane through **0**, **x** and **y** given by $s\mathbf{x}+t\mathbf{y}$ $(s\geq 0,\ t\geq 0)$
A		Nonhomogeneous linear equations and/or inequalities	Convex polyhedral set	Line segment $(1-t)\mathbf{x}+t\mathbf{y}$ $(0\leq t\leq 1)$ joining **x** and **y**
C	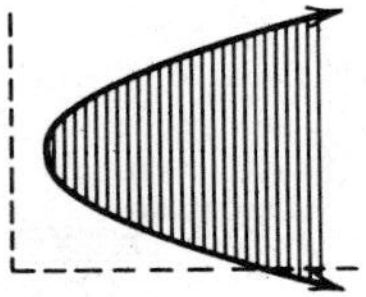	Linear equations and/or linear or nonlinear inequalities of the type: convex function $\leq b$	Convex set	Line segment $(1-t)\mathbf{x}+t\mathbf{y}$ $(0\leq t\leq 1)$ joining **x** and **y**

FIGURE 2.2D Properties of convex solution sets.

2.3 ORTHOGONALITY AND DUALITY

In this section and the next, as it is customary in linear programming, the number of components in the vectors will be denoted by $q=m+n$ rather than n.

Consider again equation (3) of 2.1:

$$x_1+2x_2-3x_3+5x_4=0 \tag{1}$$

The homogeneous linear combination of x_1, x_2, x_3, x_4 appearing in the left member is called a *linear functional* of (or a linear form in) the vector $\mathbf{x}=(x_1, x_2, x_3, x_4)$, and the values of $\mathbf{x}$ that make it zero are called its kernel; this is the same as the solution set of (1), which by Theorem 1 of 2.1 is a vector space.

Heretofore the only vectors we have used are those representing the points $\mathbf{x}$. However, in looking at (1), another quadruplet of numbers is clearly before our eyes —namely $(1,2,-3,5)$—which we denote by $\mathbf{u}$, formed from the coefficients appearing in the linear functional. We preferred this to saying "in the equation," since if both sides of (1) are multiplied by 2, to give

$$2x_1+4x_2-6x_3+10x_4=0 \tag{2}$$

we might question whether the equation has changed significantly (certainly its solution set has not), but there is no doubt that the linear functional has changed.

The formation of (2) illustrates the fact that the multiplication of a vector by a scalar makes as much sense when the vector represents a linear functional as when it is a "point." The same is true of the addition of vectors; if they are linear functionals, their addition corresponds to the addition of the values taken on by the functionals, thus:

$$\begin{aligned} f(x)&=u_1x_1+u_2x_2+u_3x_3\\ g(x)&=v_1x_1+v_2x_2+v_3x_3\\ \hline f(x)+g(x)&=(u_1+v_1)x_1+(u_2+v_2)x_2+(u_3+v_3)x_3 \end{aligned} \tag{3}$$

Both of these operations are used in the solution of linear equations by the successive elimination of unknowns.

Thus we see that the linear functionals over a vector space of q-tuples are themselves a vector space of q-tuples; of course there is a zero functional all of whose coefficients are zero. Since the two vector spaces can be represented in the same way, this raises the interesting question whether they can or should be regarded as identical.[†] The fact is that they can, with consequences to be noted briefly, and for simplicity this is often done. The visual imagination is aided and no harm should be done if it is realized that some of the resulting structure may be irrelevant. If the point $\mathbf{x}$ and the linear functional $\mathbf{u}$ belonged to the same space, we should expect to

[†] In analytic geometry the X and Y axes can be represented in the same way (by real numbers) but they are not regarded as identical, because they are distinct subsets of the plane.

be able to form their sum $\mathbf{x}+\mathbf{u}$. Certainly we can add the q-tuples but the operation seems to be meaningless in linear programming and will not actually occur. This may be an accidental characteristic of the methods employed or it may reflect the fact that in applications x_i and u_i generally have different physical dimensions.

Whether or not one can add $\mathbf{x}$ and $\mathbf{u}$, one can certainly combine them to obtain the scalar value that the functional $\mathbf{u}$ associates with the point $\mathbf{x}$, which is denoted simply by $\mathbf{ux}$ or $\mathbf{xu}$ and called a *(linear) form* or *scalar product*

$$\mathbf{ux}=(u_1,\ldots,u_q)(x_1,\ldots,x_q)=u_1x_1+\cdots+u_qx_q. \tag{4}$$

The easy verification of the following formal properties of this product is left to the reader.

$$\mathbf{ux}=\mathbf{xu} \qquad \text{(commutative law).} \tag{5}$$

$$\mathbf{u}(\mathbf{x}+\mathbf{y})=\mathbf{ux}+\mathbf{uy}, \qquad (\mathbf{u}+\mathbf{v})\mathbf{x}=\mathbf{ux}+\mathbf{vx} \qquad \text{(distributive laws or bilinearity)} \tag{6}$$

$$s(\mathbf{ux})=(s\mathbf{u})\mathbf{x}=\mathbf{u}(s\mathbf{x})=(\mathbf{u}s)\mathbf{x}=(\mathbf{ux})s=\mathbf{u}(\mathbf{x}s) \qquad (s=\text{any scalar}) \tag{7}$$

$$(\mathbf{xu})\mathbf{y}\neq\mathbf{x}(\mathbf{uy}) \qquad (\mathbf{ux})\mathbf{v}\neq\mathbf{u}(\mathbf{xv}) \qquad \text{in general.} \tag{8}$$

Equation (5) would be modified if complex numbers were being admitted as scalars or if row and column vectors were being distinguished. In the latter case the scalar product would be written as

$$\mathbf{u}^T\mathbf{x}=[u_1,\ldots,u_q]\begin{bmatrix} x_1 \\ \vdots \\ x_q \end{bmatrix}=u_1x_1+\cdots+u_qx_q \tag{9}$$

whereas the reverse order would give the matrix product

$$\mathbf{xu}^T=\begin{bmatrix} x_1 \\ \vdots \\ x_m \end{bmatrix}[u_1,\ldots,u_n]=\begin{bmatrix} x_1u_1 & \ldots & x_1u_n \\ \vdots & & \vdots \\ x_mu_1 & \ldots & x_mu_n \end{bmatrix} \tag{10}$$

a rectangular matrix of mn elements x_iy_j. Here $\mathbf{x}$ and $\mathbf{u}$ may have differing numbers of components. The notation $\mathbf{u}^T$ (T for "transpose") indicates that the column vector $\mathbf{u}$ is to be written as a row. The general definition of matrix product follows (17) hereafter.

On the few occasions when a product like (10) occurs in this book it will be denoted by $\mathbf{x}\otimes\mathbf{u}$, which is the transpose of $\mathbf{u}\otimes\mathbf{x}$. Otherwise in expressions such as $\mathbf{ux}$, $\mathbf{uA}$, $\mathbf{Ax}$, or $\mathbf{uAy}$ we may assume (if we care) that $\mathbf{u}$ is a row vector because it is on the left and $\mathbf{x}$ or $\mathbf{y}$ is a column vector because it is on the right. Thus we do not explicitly distinguish between row and column vectors, although we must distinguish between any nonsymmetric matrix $\mathbf{A}$ and its transpose $\mathbf{A}^T$, in which rows and columns are interchanged.

As for (8), it is the inevitable consequence of the two different kinds of multiplication involved. To be sure, the more conventional notation has the property not only of distinguishing (rather subtly) between (9) and (10) but also of making multiple products associative *insofar as they are defined*. Thus it distinguishes (**xu**)**y** from **x**(**uy**) not by grouping the factors but by writing them as $\mathbf{x}^T\mathbf{u}\mathbf{y}^T$ and $\mathbf{x}\mathbf{u}^T\mathbf{y}$, respectively. This unlimited associativity is indispensable in handling the more intricate vector and matrix formulas but it is not needed in this book.

The vector space of linear functionals **u** is called the dual space of the vector space of points **x** on which they are defined. However, the properties (4) to (7) are symmetric in **u** and **x**; it can be shown that the original space of points can be regarded as the dual of the space of functionals. If the scalar product (4) is zero, **u** and **x** are said to be orthogonal, without implying that they belong to the same space.

If the two dual spaces of vectors **x** and **u** are identified, the resulting space is called self-dual. If in addition the scalars are real numbers, the space is called Euclidean and the linear form (4) is called the *inner product*, the dot product, or the scalar product, although we have already used the latter term in the more general sense. It is now possible to take the inner product of a vector by itself:

$$\mathbf{x}\mathbf{x} = \|\mathbf{x}\|^2 = x_1^2 + \cdots + x_q^2 \tag{11}$$

By the Pythagorean theorem, this is the square of the distance from the origin to the point **x** if $q=2$ or 3. In general, the distance $d(\mathbf{x},\mathbf{y})$ between **x** and **y** is defined as $\|\mathbf{x}-\mathbf{y}\|$ and the angle θ between two nonzero vectors **x** and **y** is defined by

$$\cos\theta = \mathbf{x}\mathbf{y}/\|\mathbf{x}\|\,\|\mathbf{y}\| \tag{12}$$

Thus $\theta = 90°$ for orthogonal vectors.

The concepts of Euclidean distance and angle have no place in Dantzig's original simplex method but the former is used in the work cited at (9) of 6.3 and the end of 7.6.

Orthogonal Complements

Consider a system of m linear homogeneous equations in q variables; (1) is an example with $m=1$, $q=4$. Several vector spaces are associated with it as follows:

1. The solution set of the system, say X_{q-m}, is a vector space by Theorem 1 of 2.1. It is contained in the vector space X_q of all the q-tuples.
2. The vector space U_m generated by the vectors of coefficients [such as $(1,2,-3,5)$ in (1)] in the equations or linear functionals. It is contained in the vector space U_q of all such functionals.

Suppose the m equations in the system are linearly independent; that is, no one of them is equal to a linear combination (a sum of positive or negative multiples) of some of the others. Then m is the rank of U_m and by Theorem 6 of 2.7, $q-m$ is the rank of X_{q-m}. U_m and X_{q-m} are said to be *orthogonal vector spaces* because the linear equations insure that every vector in one of the spaces is orthogonal to every vector in the other space. In addition to this, *their ranks m and q−m add up to q*, the rank of the spaces containing them; hence they are called *orthogonal complements*. Each is uniquely determined by the other as the solution set associated with it, thus:

$$U_m^{\perp} \triangleq X_{q-m} = \{\mathbf{x} \text{ in } X_q \text{ with } \mathbf{ux}=0 \text{ for all } \mathbf{u} \text{ in } U_m\} \tag{13}$$

An example of orthogonal complements in the plane $X_2 = U_2$ is a pair of perpendicular lines through the origin. In three dimensions, two such lines are still orthogonal but not complementary because their dimensions do not add up to the total: $1+1 \neq 3$; instead, a plane and a perpendicular line intersecting it at the origin are orthogonal complements. Also the origin alone and the whole space are always orthogonal complements of a trivial sort.

Like any sets, two vector subspaces can be intersected and the result is obviously a vector space; e.g., two distinct planes through the origin intersect in a line through the origin. Another operation is the sum $X+Y$ of two vector subspaces X and Y; it is the vector subspace consisting of all vectors $\mathbf{x}+\mathbf{y}$ where $\mathbf{x} \in X$ and $\mathbf{y} \in Y$. It is also the convex hull of the union $X \cup Y$. Thus the sum of two distinct lines through the origin in three dimensions is the plane containing the lines. Orthogonal complementarity relates the two operations as follows:

$$(X+Y)^{\perp} = X^{\perp} \cap Y^{\perp} \tag{13'}$$

because each set consists of all the vectors orthogonal to every vector in X and orthogonal to every vector in Y.

If U_q and X_q are identified and U_m and X_n are orthogonal, then $\text{rank}(U_m + X_n) = \text{rank}\, U_m + \text{rank}\, X_n$. For a proof, choose bases $\mathbf{u}^1, \ldots, \mathbf{u}^m$ for U_m and $\mathbf{x}^1, \ldots, \mathbf{x}^n$ for X_n and suppose that

$$b_1\mathbf{u}^1 + \cdots + b_m\mathbf{u}^m = c_1\mathbf{x}^1 + \cdots + c_n\mathbf{x}^n \triangleq \mathbf{y}$$

Then $|\mathbf{y}|^2 = \Sigma_i b_i \mathbf{u}^i \cdot \Sigma_j c_j \mathbf{x}^j = 0$ because all $\mathbf{u}^i\mathbf{x}^j = 0$. This implies $\mathbf{y} = \mathbf{0}$, all $b_i = 0$ (because the $\mathbf{u}^i$ are linearly independent), and all $c_j = 0$ (because the $\mathbf{x}^j$ are independent). Hence the $\mathbf{u}^i$ and $\mathbf{x}^j$ are a basis for $U_m + X_n$.

Now consider two homogeneous equations in five variables $x_1, \ldots, x_5$. To obtain some solutions we suppose they have been solved for two of the variables, say x_1 and x_2, in terms of the others, so that they take the form

$$1 \cdot x_1 \qquad + a_{13}x_3 + a_{14}x_4 + a_{15}x_5 = 0$$

$$1 \cdot x_2 + a_{23}x_3 + a_{24}x_4 + a_{25}x_5 = 0 \tag{14}$$

By substituting 1 for each of x_3, x_4, x_5 in turn and 0 for the other two, and calculating x_1 and x_2 in each case, we obtain a spanning set (which can be shown to be a basis)

$$\begin{array}{l}(-a_{13}, -a_{23}, 1, 0, 0)\\(-a_{14}, -a_{24}, 0, 1, 0)\\(-a_{15}, -a_{25}, 0, 0, 1)\end{array} \tag{15}$$

for the solution set X_3 of (14). However, by interpreting (14) as orthogonality conditions, we see that X_3 is the orthogonal complement of the vector space U_2 generated by the vectors of coefficients in (14); that is, U_2 has a basis

$$\begin{array}{l}(1, 0, a_{13}, a_{14}, a_{15})\\(0, 1, a_{23}, a_{24}, a_{25})\end{array} \tag{16}$$

each of whose vectors is orthogonal to each of the vectors in (15).

The bases (15) and (16) have been arranged to display special forms in that each array contains a square of the largest possible size (called a *unit matrix* **I**) having ones on the diagonal and zeros elsewhere. Such a basis, or any obtainable from it by rearranging rows and columns, is called a *canonical basis*, and the pair (15) and (16) of canonical bases are called *dual bases*. The derivation of a canonical system such as (14) from an arbitrary system of linear equations will be illustrated in (9) of 2.6 and (6), (8) and (10) of 2.7.

The elements a_{ij} outside the unit matrices naturally have to change in sign in going from (15) to (16) in order to make the orthogonality condition work; in addition to that, rows are changed to columns and vice versa. These two facts are expressed by the statement that each of the arrays

$$\begin{bmatrix} -a_{13} & -a_{23} \\ -a_{14} & -a_{24} \\ -a_{15} & -a_{25} \end{bmatrix} \quad \text{and} \quad \begin{bmatrix} a_{13} & a_{14} & a_{15} \\ a_{23} & a_{24} & a_{25} \end{bmatrix} = \mathbf{A} \tag{17}$$

is the *negative transpose* of the other.

Let us recall that two matrices are multiplied together by forming the scalar products of the rows of the first with the columns of the second, so that $\mathbf{C}=\mathbf{AB}$ means that $c_{ik}=\Sigma_j a_{ij} b_{jk}$. (There are more details in 7.1.) Then the orthogonality of (16) and (15) may be concisely verified by writing

$$[\mathbf{I}_2, \mathbf{A}]\begin{bmatrix} -\mathbf{A} \\ \mathbf{I}_3 \end{bmatrix} = -\mathbf{I}_2\mathbf{A} + \mathbf{A}\mathbf{I}_3 = -\mathbf{A} + \mathbf{A} = \mathbf{0} \tag{18}$$

where

$$\mathbf{I}_2 = \begin{bmatrix} 1 & 0 \\ 0 & 1 \end{bmatrix}, \quad \mathbf{I}_3 = \begin{bmatrix} 1 & 0 & 0 \\ 0 & 1 & 0 \\ 0 & 0 & 1 \end{bmatrix}, \quad \mathbf{0} = \begin{bmatrix} 0 & 0 & 0 \\ 0 & 0 & 0 \end{bmatrix}$$

To complete the symmetry of the situation, we write down equations whose coefficients are the vectors in (15), just as equations (14) have coefficients in (16):

$$
\begin{aligned}
-a_{13}u_1 - a_{23}u_2 + u_3 \qquad\qquad &= 0 \\
-a_{14}u_1 - a_{24}u_2 \qquad + u_4 \qquad &= 0 \\
-a_{15}u_1 - a_{25}u_2 \qquad\qquad + u_5 &= 0
\end{aligned} \tag{19}
$$

(14) and (19) are called *dual systems* of equations.

If we define $M=\{1,2\}$, $N=\{3,4,5\}$, $\mathbf{x}_M=(x_1,x_2)$, $\mathbf{x}_N=(x_3,x_4,x_5)$, $\mathbf{u}_M=(u_1,u_2)$, $\mathbf{u}_N=(u_3,u_4,u_5)$, (14) and (19) may be written in matrix notation as

$$\mathbf{x}_M+\mathbf{A}x_N=\mathbf{0}, \qquad -\mathbf{u}_M\mathbf{A}+\mathbf{u}_N=\mathbf{0} \tag{20}$$

where it may be inferred that $\mathbf{x}_M$ and $\mathbf{x}_N$ are columns and $\mathbf{u}_M$ and $\mathbf{u}_N$ are rows. Then if $\mathbf{x}=(\mathbf{x}_M,\mathbf{x}_N)$ and $\mathbf{u}=(\mathbf{u}_M,\mathbf{u}_N)$ satisfy (20) (i.e., (14) and (19) respectively), we may obtain another proof of orthogonality thus:

$$\mathbf{ux}=\mathbf{u}_M\mathbf{x}_M+\mathbf{u}_N\mathbf{x}_N=\mathbf{u}_M(-\mathbf{Ax}_N)+(\mathbf{u}_M\mathbf{A})\mathbf{x}_N=0 \tag{21}$$

since the triple products are equal except for sign. This may be compared with the corresponding calculation for perpendicular lines (in the plane) having slopes μ and $-1/\mu$:

$$(1,\mu)(1,-1/\mu)=0 \qquad \text{or} \qquad (1,\mu)(-\mu,1)=0$$

In each case the change of sign is essential in making the scalar product equal to zero.

The independent variables (the components of $\mathbf{x}_N$ and $\mathbf{u}_M$) are commonly called *nonbasic variables* while the dependent variables (the components of $\mathbf{x}_M$ and $\mathbf{u}_N$, which are solved for) are called *basic variables*. M is the primal basis and N is the dual basis.

We have now used the term *dual* in two different but related senses in this section, first in connection with the dual spaces of points $\mathbf{x}$ and linear functionals $\mathbf{u}$, both being of full rank q, and now in connection with their orthogonal-complementary subspaces, in the example spanned by (15) and (16) respectively, whose two ranks add to $q=5$. It is the latter more specialized form of duality that permeates linear programming, as we shall see again in the next section and in 3.1.

The remainder of this section is optional. Bases of the forms illustrated by (15) and (16), which are dual in the second sense, can easily be expanded to bases which are dual in a specially strong form of the first sense. This is accomplished by the following square matrix

$$\mathbf{Q}=\begin{bmatrix} 1 & 0 & a_{13} & a_{14} & a_{15} \\ 0 & 1 & a_{23} & a_{24} & a_{25} \\ 0 & 0 & -1 & 0 & 0 \\ 0 & 0 & 0 & -1 & 0 \\ 0 & 0 & 0 & 0 & -1 \end{bmatrix} \tag{22}$$

whose last three columns are the negatives of the basis vectors (15) for the solution space, and whose first two rows are the basis (16) for the dual space spanned by the linear functionals in (14). This form is used by Harris (1973) in describing the Devex linear programming code.

Bases $\{\xi^1,\ldots,\xi^q\}$ and $\{\omega^1,\ldots,\omega^q\}$ for a pair of dual spaces are called *dual bases* in general if

$$\omega^i\xi^j=\delta_{ij} \tag{23}$$

where $\delta_{ij}=1$ if $i=j$ and is zero otherwise. It is simply an equivalent statement of this definition to say that if a matrix $\boldsymbol{\Omega}$ is formed whose ith row contains the components of ω^i $(i=1,\ldots,q)$, and a matrix $\boldsymbol{\Xi}$ formed whose jth column is ξ^j then $\boldsymbol{\Omega}$ and $\boldsymbol{\Xi}$ are inverse matrices that in general are different. It is interesting that (22) is its own inverse, being both $\boldsymbol{\Omega}$ and $\boldsymbol{\Xi}$. Thus $\mathbf{Q}^2=\mathbf{I}$; any such square matrix is said to be *involutory*.

With a suitable partitioning and rearrangement of rows and columns, an involutory matrix can be written in either of the forms

$$\mathbf{T}\begin{bmatrix}\mathbf{I}_m & \mathbf{0}\\ \mathbf{0} & -\mathbf{I}_n\end{bmatrix}\mathbf{T}^{-1} \tag{24}$$

$$\begin{aligned}\begin{bmatrix}\mathbf{I}_m+\mathbf{AC} & \mathbf{A}\\ -2\mathbf{C}-\mathbf{CAC} & -\mathbf{I}_n-\mathbf{CA}\end{bmatrix}&=\begin{bmatrix}\mathbf{I}_m & \mathbf{0}\\ -\mathbf{C} & \mathbf{I}_n\end{bmatrix}\begin{bmatrix}\mathbf{I}_m & \mathbf{A}\\ \mathbf{0} & -\mathbf{I}_n\end{bmatrix}\begin{bmatrix}\mathbf{I}_m & \mathbf{0}\\ \mathbf{C} & \mathbf{I}_n\end{bmatrix}\\ &=\begin{bmatrix}\mathbf{I}_m & \mathbf{0}\\ -\mathbf{C} & -\mathbf{I}_n\end{bmatrix}\begin{bmatrix}\mathbf{I}_m & -\mathbf{A}\\ \mathbf{0} & -\mathbf{I}_n\end{bmatrix}\begin{bmatrix}\mathbf{I}_m & \mathbf{0}\\ -\mathbf{C} & -\mathbf{I}_n\end{bmatrix}\end{aligned} \tag{25}$$

To obtain a matrix like (22) one takes $\mathbf{C}=\mathbf{0}$ and

$$\mathbf{T}=\mathbf{T}^{-1}=\begin{bmatrix}\mathbf{I}_m & \mathbf{A}/2\\ \mathbf{0} & -\mathbf{I}_n\end{bmatrix}$$

EXERCISES 2.3

1. Evaluate $(3,-1,2,1)(-2,1,-2,1)$. Which of these vectors may appropriately be regarded as a column vector?
2. What is the equation of the plane passing through $(1,1,1)$ and perpendicular to the line joining the origin to the point $(1,3,-2)$?
3. Express $-8x_1+9x_2$ as a linear combination of the functionals $2x_1-x_2$ and $-x_1+3x_2$.
4. Rewrite $\mathbf{xu}+\mathbf{yv}+\mathbf{uy}+\mathbf{vx}$ so as to indicate how it could most easily be evaluated arithmetically.
5. What is the (Euclidean) distance between the points $(3,-1,5)$ and $(0,11,1)$?

In 6 to 9 find the angle between the given pair of vectors.

6. $(2,-3,1)$, $(2,3,5)$. **7.** $(1,0,-2)$, $(2,0,-4)$.
8. $(-1,0,1)$, $(0,1,-1)$. **9.** $(1,-1,1)$, $(-2,2,-2)$.

In 10 to 12 give a basis for the solution set of the given equations and a dual basis for its orthogonal complement.

10. $3x_1+4x_2+x_3=0,\ -2x_1+5x_2+x_4=0.$
11. $2x_1+x_2-3x_3+x_4+2x_5=0,\ x_1+x_2+2x_3-3x_4-x_5=0.$
(Solve for x_1 and x_2.)
12. $x_1-2x_2=0,\ x_1+4x_3=0,\ x_1-3x_4=0.$
13. Prove the sum X_1+X_2 of vector spaces is a vector space.
14. When **A** has two rows and two columns, verify that $\mathbf{u}(\mathbf{A}\mathbf{x})=(\mathbf{u}\mathbf{A})\mathbf{x}$.
15. Verify that (24) and (25) are involutory.

2.4 TABLEAUS AND PIVOT STEPS

It is desirable to have a compact representation for the primal equations (14) and the dual equations (19) of 2.3. These are copied as (1) and (3) in Table 2.4 together with several kinds of tableaus and the Tucker schema. The forms (2), (4) and (5) each have the advantage of permitting a single array to represent the primal equations (1) as row equations and the dual equations (3) as column equations, because (2) and (4) essentially differ only in the choice of the letter x or u. *Otherwise the difference between (1) and (3) comes in the manner of reading the tableau.* The broken line is associated with the labels for the column equations. We shall use (2) the most but not exclusively.

Equation (3′) will be useful for theoretical purposes, to show in (9) that the procedures applicable to (1) are simultaneously relevant to (3) as well. The matrix in (3′) is $-\mathbf{A}^T$, the negative transpose of the matrix **A** in (2), (4) and (5). The labels in (3′) also are transposed but they do not change sign.

To give a little preview of how tableaus will be used in linear programming, let us reconsider the fertilizer problem of 1.2. By defining the slack (basic) variable $x_1=170-.5y-.4y_2$, the original inequality $.5y_1+.4y_2\leq 100$ can be replaced by $x_1\geq 0$ and the equation $x_1+.5y_1+.4y_2-100=0$. Slack variables x_2 and x_3 are introduced similarly, and one of the nonbasic variables is replaced by 1 to accommodate the constant terms. The resulting tableau is (6).

1	y_1	y_2	1		
x_1	.5	.4	-170	all $x_i\geq 0$	(6)
x_2	.4	.2	-100	$y_j\geq 0$	
x_3	.1	.4	-60		
z	-10	-20	0		

TABLE 2.4

The primal (row) equations

$$\begin{aligned} x_1 \quad + a_{13}x_3 + a_{14}x_4 + a_{15}x_5 &= 0 \\ x_2 + a_{23}x_3 + a_{24}x_4 + a_{25}x_5 &= 0 \end{aligned} \tag{1}$$

are represented by the primal tableau

$$\begin{array}{c|ccc} 1 & x_3 & x_4 & x_5 \\ \hline x_1 & a_{13} & a_{14} & a_{15} \\ x_2 & a_{23} & a_{24} & a_{25} \end{array} \tag{2}$$

The dual (column) equations

$$\begin{aligned} u_3 &= a_{13}u_1 + a_{23}u_2 \\ u_4 &= a_{14}u_1 + a_{24}u_2 \\ u_5 &= a_{15}u_1 + a_{25}u_2 \end{aligned} \tag{3}$$

are represented by (3′) or more conveniently by (4):

(row equations) (column equations)

$$\begin{array}{ccc} 1 & u_1 & u_2 \\ \hline u_3 & -a_{13} & -a_{23} \\ u_4 & -a_{14} & -a_{24} \\ u_5 & -a_{15} & -a_{25} \end{array} \quad (3') \qquad \begin{array}{cccc} 1 & u_3 & u_4 & u_5 \\ \hdashline u_1 & a_{13} & a_{14} & a_{15} \\ u_2 & a_{23} & a_{24} & a_{25} \end{array} \tag{4}$$

(dual tableau)

The Tucker schema combines (2) and (4):

$$\begin{array}{c|ccc|c} & x_3 & x_4 & x_5 & \\ \hdashline u_1 & a_{13} & a_{14} & a_{15} & = -x_1 \\ u_2 & a_{23} & a_{24} & a_{25} & = -x_2 \\ \hline & \| & \| & \| & \\ & u_3 & u_4 & u_5 & \end{array} \tag{5}$$

Only one of (2), (4) or (5) needs to be written out.

The equations are related to the scalar product of the first row or column (of labels) with the other rows or columns. The horizontal line in (2) or (4) is a reminder to obtain the row equations (1) by adding all the way across and setting the result equal to zero, and to obtain the column equations (3) by inserting equality signs where the horizontal line is crossed. Variables x_1, x_2, u_3, u_4, u_5 are basic, the other variables nonbasic.

By putting the nonbasic y_1 and $y_2=0$ one obtains what is called the basic solution associated with the tableau, that is, $x_1=170$, $x_2=100$, $x_3=60$, $\bar{z}=0$, which happens to be feasible but not optimal. Nevertheless it will be shown that an optimal solution can always be obtained as the basic solution for *some* tableau, which is nearly always different from the initial tableau. Hence it is very important to be able to change the basis. This is normally done one variable at a time and is facilitated by the tableau format.

Pivot Steps

Consider the row equations (1), which have the basis $\{x_1, x_2\}$. To go to the basis $\{x_3, x_2\}$ (dropping x_1 and adding x_3), one has to take the equation involving x_1, solve it for x_3, and use the result to eliminate x_3 from the remaining equations. This may be started by dividing the first equation by a_{13}, which must be nonzero (otherwise $\{x_3, x_2\}$ could not be a basis). This gives

$$\frac{1}{a_{13}}x_1 + x_3 + \frac{a_{14}}{a_{13}}x_4 + \frac{a_{15}}{a_{13}}x_5 = 0$$

Now multiply this result by $-a_{23}$ and add to the second equation to get

$$-\frac{a_{23}}{a_{13}}x_1 + x_2 + \left(a_{24} - \frac{a_{23}a_{14}}{a_{13}}\right)x_4 + \left(a_{25} - \frac{a_{23}a_{15}}{a_{13}}\right)x_5 = 0$$

Remembering that the basic (dependent) variables go on the left in the tableau for the row equations, we see that the two new row equations just obtained have the tableau

1	x_1	x_4	x_5
x_3	$\dfrac{1}{a_{13}}$	$\dfrac{a_{14}}{a_{13}}$	$\dfrac{a_{15}}{a_{13}}$
x_2	$-\dfrac{a_{23}}{a_{13}}$	$a_{24}-\dfrac{a_{23}a_{14}}{a_{13}}$	$a_{25}-\dfrac{a_{23}a_{15}}{a_{13}}$

(7)

Since x_3 has replaced x_1 in the basis, and x_1 has replaced x_3 among the nonbasic variables, (7) has been arranged to have the same relationship to (2). The transition from (2) to (7) is called *pivoting*, a pivot transformation, a pivot step or an exchange of variables. The nonzero element a_{13} is called the *pivot*; it is distinguished by an asterisk and denoted by a_{kl} in general. By comparing (2) and (7), one deduces the following.

General Rules for Pivoting (Exchanging Variables)

1. Exchange the label (variable with subscript k, corresponding to a constraint) in the pivotal row with that in the pivotal column (subscript l), to reflect the change of basis. Copy the rest of the labels, and the 1 in the corner, without change.

2. Replace the pivot a_{kl} by its reciprocal.

$$a'_{lk} = 1/a_{kl} \qquad (a_{kl} \neq 0)$$

3. Divide the other numbers a_{kj} in the pivotal row by the pivot.

$$a'_{lj} = a_{kj}/a_{kl} \qquad (j \neq l)$$

4. Divide the other numbers a_{il} in the pivotal column by the negative of the pivot.

$$a'_{ik} = -a_{il}/a_{kl} \qquad (i \neq k)$$

5. $a'_{ij} = a_{ij} - a_{kj}a_{il}/a_{kl} = a_{ij} - a_{il}a'_{lj} \qquad (i \neq k, j \neq l)$. The latter may be described thus for each $i \neq k$. Store $-a_{il}$ in memory (a_{il} is in the old pivotal column l and row i). Use it to multiply each a'_{lj} in the new pivotal row, and (for each nonpivotal j) add the result to a_{ij} in the old row i to obtain a'_{ij} in the corresponding new row.

If the matrix $\mathbf{A}$ has m rows and n columns, then there are $n-1$ possible values for j (two values, 4 or 5, in the example) in rules 3 and 5, and $m-1$ possible values for i ($=2$) in rules 4 and 5. Neither of these sets of values includes either of the indices $k=1$ or $l=3$.

These rules are in the form suitable for problems large enough to require electronic circuitry and the acceptance of roundoff errors. Rules adapted to small exercises suitable for hand calculation will be derived in the following section. The reader can easily convince himself that the rules apply to a tableau of any size; the essential thing is that the example have at least two basic and two nonbasic variables, so that the tableau contains at least one example of each type of element mentioned in the rules. Thus the most compact presentation of the rules is

1	x_l	x_j		1	x_k	x_j	
			$\rightarrow$				
x_k	p^*	b		x_l	$1/p$	b/p	(8)
x_i	c	a		x_i	$-c/p$	$a-(bc/p)$	

Here the arrow represents the operation of pivoting on $p \triangleq a_{kl}$ ($\triangleq$ means "equals by definition"); x_i represents each of the basic variables other than x_k in the first tableau or x_l in the second; x_j represents each of the nonbasic variables other than x_l or x_k; $b = a_{kj}$; $c = a_{il}$; $a = a_{ij}$, any of the possibly many elements outside the pivotal row and column. The suppression of most of the subscripts may make the essentials of the operation more prominent. Also the resulting reliance on the physical position (on the paper or in the computer) of the new entries $b/p = a'_{lj}$ and $-c/p = a'_{ik}$ (which replace $b = a_{kj}$ and $c = a_{il}$ respectively) makes unnecessary the recollection (in interpreting rules 3 and 4) that in the new tableau rule 1 has interchanged the labels x_k and x_l in the pivotal row and column.

The nature of the above rules provides the second clue to the principle of duality: the rules treat the rows and the columns of the tableau in exactly the same way, with the sole exception of the change of sign required in the pivotal column but not in the pivotal row. Thus one is not surprised that the following theorem is true.

Primal-Dual Invariance Theorem for Pivot Steps When the foregoing five rules are used to transform a tableau, the solution set for the row equations is unchanged and the (orthogonal) solution set for the column equations is also unchanged.

Proof The derivation of the rules showed that every vector $\mathbf{x}$ that satisfies the original row equations (such as (1)) also satisfies the row equations (such as (7)) obtained by pivoting. However, by applying the same rules to the new row equations and using $1/a_{13}$ in (7) or $1/p$ in (8) as the pivot, it is easily verified that the original row equations are recovered again. Thus the solution sets must be the same.

It is left as an exercise for the reader to give a similar derivation of the same rules 1 to 5 for the column equations (3) by solving for u_1, u_4, u_5 in terms of u_3 and u_2. Alternatively the proof may be completed by writing the equations before and after the pivot step both as columns and as rows (as in (4) and (3′)), and using the rules 1 to 5 to transform the latter. Thus if $\equiv$ connects column and row versions of the same equations and $\rightarrow$ denotes the pivot step, one obtains (9).

$$
\begin{array}{ccc}
1 & u_l & u_j \\
\hline
u_k & p^* & b \\
u_i & c & a
\end{array}
\quad \equiv \quad
\begin{array}{ccc}
1 & u_k & u_i \\
\hline
u_l & -p^* & -c \\
u_j & -b & -a
\end{array}
\quad \rightarrow
$$

$$
\begin{array}{ccc}
1 & u_l & u_i \\
\hline
u_k & -1/p & c/p \\
u_j & -b/p & -a+(bc/p)
\end{array}
\quad \equiv \quad
\begin{array}{ccc}
1 & u_k & u_j \\
\hline
u_l & 1/p & b/p \\
u_i & -c/p & a-(bc/p)
\end{array}
\tag{9}
$$

In going from the first to the last of these tableaus the numerical entries p, a, b, c are transformed just as they were in (8).

A given tableau together with all those obtainable from it by one or more pivot steps forms what Tucker has called a *combinatorial equivalence class* (or combivalence class) of tableaus. By the above theorem the tableaus are all equivalent in the sense that they all have identical pairs of solution sets. Also any two tableaus in the class are equivalent in the general sense that the relation between them is reflexive (a tableau is equivalent to itself), symmetric (if a sequence of pivot steps takes tableau T_1 into tableau T_2 then the reverse sequence of reverse pivot steps will take T_2 and T_1), and transitive (if one can get from T_1 to T_2 and from T_2 to T_3 by pivoting, then one can get from T_1 to T_3).

There are no more tableaus in the equivalence class than there are ways of selecting the basic x_i or the nonbasic u_i (say m) out of the total number of x_h or u_h (say $m+n$), which is the binomial coefficient $(m+n)!/m!n!$. There may be fewer than this as a result of the vanishing of some prospective pivots. Of course there may be as many as $(m+n)! = 1 \cdot 2 \cdot 3 \cdots (m+n-1)(m+n)$ equivalent tableaus if mere permutation of the rows (and the columns) of a tableau were regarded as producing new tableaus, and this number could be doubled by including negative transposes. An example of an equivalence class of tableaus is given near the end of the next section.

Condensed Versus Extended Tableaus

The condensed tableau (2) offers a unique combination of compactness and symmetry (the row and the column equations are read in virtually the same way), but it does have one limitation that is illustrated by the pivot step

$$\begin{array}{c|cc} 1 & x_3 & x_4 \\ \hline x_1 & 1^* & 1 \\ x_2 & 1 & 1 \end{array} \quad \rightarrow \quad \begin{array}{c|cc} 1 & x_1 & x_4 \\ \hline x_3 & 1 & 1 \\ x_2 & -1 & 0 \end{array} \tag{10}$$

The 2 by 2 matrices are of ranks 1 and 2 respectively and therefore the second could not possibly be obtained from the first by multiplication by some other matrix (Theorem 1 of 7.1).

In order to be able to pivot by matrix multiplication as in Chapter 7, one may write out the primal tableau in *extended* form by providing a separate column for the coefficients of each basic variable. Thus (2) and (7) are replaced by (11) and (12).

$$\begin{array}{c|ccccc} & x_1 & x_2 & x_3 & x_4 & x_5 \\ \hline (x_1) & 1 & 0 & a_{13}^* & a_{14} & a_{15} \\ (x_2) & 0 & 1 & a_{23} & a_{24} & a_{25} \end{array} \tag{11}$$

$$\begin{array}{c|ccccc} & x_1 & x_2 & x_3 & x_4 & x_5 \\ \hline (x_3) & \dfrac{1^*}{a_{13}} & 0 & 1 & \dfrac{a_{14}}{a_{13}} & \dfrac{a_{15}}{a_{13}} \\ (x_2) & -\dfrac{a_{23}}{a_{13}} & 1 & 0 & a_{24}-\dfrac{a_{14}a_{23}}{a_{13}} & a_{25}-\dfrac{a_{15}a_{23}}{a_{13}} \end{array} \tag{12}$$

Note that the columns have been kept in standard order 1 to 5 but the rows in (12) have the same order (x_3, x_2) as in (7). Although the row labels remain convenient, they are no longer used in reading off the equations, and they could be determined from the fact that a row and column of the same label always intersect at one of the 1's of the unit matrix.

Both types of tableau have the property that the matrix of numbers in the tableau after a pivot step can be obtained by subtracting a matrix of rank one from the original matrix. The matrix to be subtracted is the matrix product ((10) of 2.3) of the pivotal column and the pivotal row (slightly modified), divided by the pivot a_{13}:

$$(4)-(7) = a_{13}^{-1}\begin{bmatrix} a_{13}-1 \\ a_{23} \end{bmatrix}\begin{bmatrix} a_{13}+1 & a_{14} & a_{15} \end{bmatrix} \tag{13}$$

$$(11)-(12) = a_{13}^{-1}\begin{bmatrix} a_{13}-1 \\ a_{23} \end{bmatrix}\begin{bmatrix} 1 & 0 & a_{13} & a_{14} & a_{15} \end{bmatrix} \tag{14}$$

Table 3.4 hereafter illustrates both the extended tableau (11) and a compact way to stack the condensed tableaus (2). An alternative interpretation of the extended tableau, as specifying the relations among *constant vectors* (the columns in any tableau, but usually the initial tableau) rather than *variable scalars* as above, is indicated in Theorem 4 of 7.1.

EXERCISES 2.4

1. Write the primal tableau and the schema for the row equations $3x_1+4x_2+x_3=0$, $-2x_1+5x_2+x_4=0$. Use them to write out the corresponding column equations.
2. Write the dual tableau and the schema for the column equations $3u_1+u_2-2u_3=0$, $-2u_1+5u_3+u_4=0$, $u_1-4u_3-u_5=0$. Use them to write out the corresponding row equations.
3. Perform two pivot steps to exchange first x_1 and x_3 and then x_2 and x_4 in the tableau. Obtain a solution for the initial tableau by setting $x_3=x_4=x_5=1$; check the final tableau by showing that this solution also satisfies it.

1	x_3	x_4	x_5
x_1	1	3	-2
x_2	1	2	1

4. Solve the following equations for x_1 and x_2:

$$2x_1+x_2-3x_3+\ x_4+2x_5=0$$

$$x_1+x_2+2x_3-3x_4-\ x_5=0$$

Write out the tableau and perform two pivot steps to obtain the basis $\{x_3, x_5\}$. Write out the row and column equations in the final tableau. Obtain an initial dual solution by setting $u_1=u_2=1$ and show that it satisfies the final column equations.
5. Write the initial tableaus in 1 and 3 in extended form.
6. Write the Tucker schema for 3. Write a tableau in which the given row equations are column equations.
7. Prove the validity of the pivot rules for the column equations in (5) by solving the first equation for u_1 and eliminating u_1 from the other equations.

2.5 HAND CALCULATIONS IN TERMS OF INTEGERS

For the hand calculation of the small examples to be worked in this book it is desirable to recast the rules for pivoting so as to avoid the explicit occurrence of fractions. This is accomplished by using a detached common denominator, denoted by δ, which is recorded in the upper left-hand corner of the tableau in place of the 1 previously used. In forming both row and column equations in the way already described, δ will multiply all the basic (dependent) variables and no others; this is exactly what a common denominator should do after the equations are cleared of fractions by multiplying by the common denominator.

The new rules are easily deduced by writing the tableaus before and after the pivot step both with and without a detached denominator, and using (8) or rules 1 to 5 in 2.4 to transform the latter. Thus if $\equiv$ connects tableaus differing only in the treatment of the denominator and $\rightarrow$ denotes the pivot step, one has

$$\begin{array}{c|cc} \delta & x_l & x_j \\ \hline x_k & \pi^* & \beta \\ x_i & \gamma & \alpha \end{array} \equiv \begin{array}{c|cc} 1 & x_l & x_j \\ \hline x_k & (\pi/\delta)^* & \beta/\delta \\ x_i & \gamma/\delta & \alpha/\delta \end{array} \rightarrow \tag{1}$$

$$\begin{array}{c|cc} 1 & x_k & x_j \\ \hline x_l & \delta/\pi & \beta/\pi \\ x_i & -\gamma/\pi & (\pi\alpha-\beta\gamma)/\delta\pi \end{array} \equiv \begin{array}{c|cc} \pi & x_k & x_j \\ \hline x_l & \delta & \beta \\ x_i & -\gamma & (\pi\alpha-\beta\gamma)/\delta \end{array}$$

It may seem that a fraction is still present in the last tableau in (1). Before rebutting this one should remark that since the basis is changed by only one variable at a time, we will generally be interested in sequences of equivalent tableaus, not just two of them. Now it is a pleasant fact that if the sequence of equivalent tableaus contains a tableau (usually the initial tableau, but any equivalent tableau will do) whose entries are all integers and whose common detached denominator is 1, then the quotients $(\pi\alpha-\beta\gamma)/\delta$ in (1) will always be integers for every possible pivot step in every equivalent tableau. This fact is proved in Theorem 16 of 7.7 and discussed further below. By comparing the first and last tableaus in (1) we obtain the following.

Rules for Pivoting with Integers

0′. Mark the pivot with an asterisk.

1′. Interchange the label (variable) in the pivotal row k with that in the pivotal column l. Copy the symbols for the other variables without change.

2′. Interchange the numerator α_{kl} of the pivot ($\neq 0$) with the old denominator δ.

3′. Copy the rest of the pivotal row as is.

4′. Copy the rest of the numerators in the pivotal column with reversed signs.

5′. Replace each of the other numerators α_{ij} in the tableau by the quotient $(\alpha_{ij}\alpha_{kl}-\alpha_{il}\alpha_{kj})/\delta$, which will always be an exact integer if any tableau in the equivalence class has denominator 1 and integer numerators. Many readers will recognize $\alpha_{ij}\alpha_{kl}-\alpha_{il}\alpha_{kj}$ as $(\pm)$ the value of a 2 by 2 determinant, namely, the product of the old numerator α_{ij} and the pivotal numerator α_{kl}, minus the product of the old numerators α_{il} and α_{kj} at the ends of the other diagonal of the rectangle formed by the four α's.

The exact division principle could be invalidated if δ were reduced (but not to 1) by cancelling out a common factor, as one could do in the second of the equivalent tableaus

1	x_4	x_5	x_6
x_1	-1	1	1
x_2	1	-1	1
x_3	1	1	-1

and

4	x_1	x_2	x_3
x_4	0	2	2
x_5	2	0	2
x_6	2	2	0

(2)

On the other hand, if one pivot step (on a pivot whose numerator is not divisible by any factor of the common denominator) goes through without introducing fractions into the tableau, the same will be true of all possible pivot steps, as in the following complete equivalence class:

Check	6	x_3	x_4
15	x_1	5*	4
33	x_2	15	12

Check	5	x_1	x_4
15	x_3	6	4*
-10	x_2	-15	0

(3)

4	x_1	x_3
x_4	6	5
x_2	-12*	0

-12	x_2	x_3
x_4	-6	-15*
x_1	4	0

-15	x_2	x_4
x_3	-6*	-12
x_1	5	0

The occurrence of the zero shows that $\{x_3, x_4\}$ cannot be the basic variables. Each of the above tableaus yields 15 others when the rows or columns are permuted, or the numerators and denominators all changed in sign, or the negative transpose taken, but these are trivial changes. Thus the first tableau in (3) is not significantly different from any of

6	x_4	x_3
x_2	12	15
x_1	4	5

-6	x_4	x_3
x_2	-12	-15
x_1	-4	-5

-6	x_2	x_1
x_4	12	4
x_3	15	5

(4)

The last pivot -6 in (3) leads back to the first tableau again and thus provides some assurance that all are correct. However, the usual method for *checking the arithmetic* is to provide an additional column as in the first two tableaus of (3). These numbers give the values of the left members of the row equations when every variable is set equal to 1; thus

$$15=6+5+4, \qquad 33=6+15+12 \tag{5}$$

In effect the new column provides coefficients for a new variable that always remains nonbasic and has the value ± 1 (the sign depends on whether one chooses to put it in the left or the right member). Thus rules 1′ to 5′ can be applied to the check values; e.g., in the second tableau in (3), $-10=(5\cdot 33-15\cdot 15)/6$. The check consists in reverting to the original method of calculation in the new tableau ($15=5+6+4$, $-10=5-15+0$), which the theorem of 2.4 justifies. A common error is to forget to divide by δ in rule 5′, since we usually begin with $\delta=1$ and have no divisions at the first pivot step.

To save space, check values are sometimes omitted in this book but students are advised to use them consistently.

For illustrative purposes there is little if anything to gain by choosing an example or exercise having a denominator $\neq 1$ in the initial tableau, but in fact there is no difficulty (apart from the size of the integers that may occur) in handling any tableau of rational numbers in terms of integers alone. If the initial common denominator is $\delta \neq 1$, one can replace it by 1, solve the problem so modified, and then in the final tableau multiply δ times every element whose row and column labels have both been exchanged, and divide δ (not without fractions in general) into every element neither of whose row and column labels have been exchanged. The other elements are unchanged. (See Exercise 4.)

EXERCISES 2.5

Use the rules of 2.5 with check values to exchange all the variables in the following tableaus. (In 2.6 this is shown to yield the inverse of the matrix.)

1.

	1	y_1	y_2	y_3
3	x_1	0	3	−1
9	x_2	2	2	4
−5	x_3	−3	2	−5

2.

	1	y_1	y_2	y_3
5	x_1	3	−1	2
2	x_2	0	2	−1
2	x_3	2	−3	2

3. Use the pivoting to solve for x_1, x_2, x_3: $y_1=3x_2-x_3, y_2=2x_1+x_3, y_3=5x_1+2x_2$.

4. By using δx_i rather than x_i as the basic variables, prove the validity of the procedure described in the last paragraph of the section.

5. Modify the proof given in (9) of 2.4 to show that the rules of 2.5 leave unchanged the solution set of the column equations.

2.6 NUMERICAL APPLICATIONS OF PIVOTING

The most important application for our purposes is the optimization of linear programs beginning in 3.2 and 3.3. The present section considers the simplest applications of pivoting, such as the inversion of matrices and the solution of systems of linear equations, which may be more or less familiar to many readers. These operations are important in themselves and provide practice in using the rules of 2.4 or 2.5. The applications considered in this book are listed in Table 2.6C at the end of this section.

(a) *Matrix inversion*. Consider a tableau whose row equations are $\mathbf{x}+\mathbf{A}\mathbf{y}=\mathbf{0}$ or $\mathbf{x}=-\mathbf{A}\mathbf{y}$, where $\mathbf{A}$ is an m by m square matrix. If it is possible to pivot exactly once in each row and in each column of the tableau (m pivots in all), then all the x_i will be exchanged with all the y_j and a solution $\mathbf{y}=-\mathbf{A}^{-1}\mathbf{x}$ will be obtained, where $\mathbf{A}^{-1}$ is another m by m matrix called the *inverse* of $\mathbf{A}$. (More details are given in 7.1 and 7.5.) Table 2.6A is an example.

TABLE 2.6A Matrix Inversion

Check	1	y_1	y_2	y_3	y_4		Check	1	y_1	x_3	y_3	y_4
4	x_1	0	4	1	−2	→	0	x_1	12	−4	1	−10
1	x_2	2	0	−3	1		1	x_2	2*	0	−3	1
1	x_3	−3	1*	0	2		1	y_2	−3	1	0	2
6	x_4	−1	2	4	0		4	x_4	5	−2	4	−4

Check	2	x_2	x_3	y_3	y_4		Check	−13	x_2	x_3	y_3	x_4
−12	x_1	−12	−8	38	−32	→	126	x_1	−2	−12	121*	32
1	y_1	1	0	−3	1		−8	y_1	−4	2	8	−1
5	y_2	3	2	−9	7		−43	y_2	−2	1	−22	−7
3	x_4	−5	−4	23	−13*		3	y_4	−5	−4	23	2

Check	121	x_2	x_3	x_1	x_4			121	x_1	x_2	x_3	x_4
126	y_3	−2	−12	−13	32	≡		y_1	−8	36	−26	29
152	y_1	36	−26	−8	29			y_2	22	22	11	11
187	y_2	22	11	22	11			y_3	−13	−2	−12	32
195	y_4	43	16	−23	38			y_4	−23	43	16	38

The pivots may be chosen in various ways; a computing machine should prefer large pivots to reduce roundoff error while a person using pencil and paper and the rules of 2.5 will prefer pivots that are small in absolute value (but not zero). If the pivots are not confined to the diagonal positions, it will be necessary to rearrange the rows and columns (after the pivoting is completed) so that the labels x_i and y_i are in their original sequences. For example, the first and the last matrices in Table 2.6A are inverses (the last over the denominator 121).

Apart from the sign, the value of the determinant of the original matrix is the final denominator 121 or the product of all the pivots according as the rules of 2.5 or 2.4 are used. This number remains available if the pivotal rows and columns are omitted as in (c) or (d).

(b) *Solving linear equations* (Jordan elimination). Let us solve

$$x_1 - 2x_2 - x_3 + 3x_4 + 1 = 0$$

$$-x_1 - x_2 + 2x_3 + x_4 - 2 = 0$$

$$2x_1 + x_2 + x_3 + x_4 - 3 = 0$$

for x_1, x_2, x_3 in terms of x_4.

Although thus far the labels in the tableaus have always been letters representing variables, the pivoting process remains valid if some of these variables are replaced by constants. In writing the above equations as a tableau, it is advisable to let the *zeros* take the place of the basic variables, and to accommodate the constant terms by putting a 1 in the place of an additional nonbasic variable. One then has the following three pivot steps:

Check	1	x_1	x_2	x_3	x_4	1
2	0	1*	−2	−1	3	1
−1	0	−1	−1	2	1	−2
2	0	2	1	1	1	−3

Check	1	x_2	x_3	x_4	1
2	x_1	−2	−1	3	1
1	0	−3	1*	4	−1
−2	0	5	3	−5	−5

Check	1	x_2	x_4	1
3	x_1	−5	7	0
1	x_3	−3	4	−1
−5	0	14*	−17	−2

Check	14	x_4	1
17	x_1	13	−10
−1	x_3	5	−20
−5	x_2	−17	−2

(1)

The solution is

$$x_1 = \tfrac{10}{14} - \tfrac{13}{14}x_4, \qquad x_2 = \tfrac{2}{14} + \tfrac{17}{14}x_4, \qquad x_3 = \tfrac{20}{14} - \tfrac{5}{14}x_4 \tag{2}$$

or $\mathbf{x} = (10, 2, 20, 0)/14 + (-13, 17, -5, 14)x_4/14$, obtained from the row equations of the last tableau. The pivots are chosen so as to make x_1, x_2, x_3 basic and the zeros nonbasic; as fast as the latter occurs, the corresponding columns of coefficients (which then have nothing but 0 to multiply) are dropped from the tableau. The check values are used as before except that the denominator is *not* added to the row sum in those rows where a zero occupies the position of a basic variable.

The following special situations may occur:

(b_1) The variables (if any) that remain nonbasic (e.g., x_4 in the example) must be free to assume any assigned values. If the specified basis contradicts this principle the procedure will fail because one or more required pivots are zero.

(b_2) An extreme form of (b_1) occurs if the equations are *inconsistent*. Then a row is obtained in which every entry is zero except the constant term; thus no set of variables can be solved for.

(b_3) If one or more equations are *redundant*, then a corresponding number of rows will be obtained in which every entry is zero. Such rows may be dropped.

(c) *Gaussian elimination* in linear equations differs from the Jordan elimination in (b) in that the transforms of pivotal rows as well as pivotal columns are dropped, although the pivotal rows themselves must be stored for later use. Then the labels or basic "variables" in the remaining rows are all zero and therefore omitted. If each tableau is placed below the preceding, the top row of column labels in the first tableau can serve for all. The phrase "over δ" to the right of each tableau specifies the common denominator δ. Thus (1) is replaced by

Check	x_1	x_2	x_3	x_4	1		
2	1*	−2	−1	3	1		(3)
−1	−1	−1	2	1	−2	over 1	
2	2	1	1	1	−3		
1		−3	1*	4	−1	over 1	
−2		5	3	−5	−5		
−5		14		−17	−2	over 1	

The last line shows that $14x_2 - 17x_4 - 2 = 0$ or

$$x_2 = \tfrac{2}{14} + \tfrac{17}{14}x_4$$

Substituting this in one of the equations of the preceding tableau, preferably the pivotal row, and solving for x_3 gives

$$x_3 = \tfrac{20}{14} - \tfrac{5}{14}x_4$$

Substituting both of these in one of the original equations and solving for x_1 completes the solution by *back substitution*. The solution can be checked by substitution in the remaining original equations. This procedure is more complicated than Jordan elimination but for large problems it is faster and it can be systematized as in 7.4.

(d) *The rank of a matrix and its row and column bases* are discussed in the following section. They can be obtained from doubly contracting tableaus as in (c), but one may not need to label the rows and columns. Instead, as illustrated in (4), each pivotal position is marked twice, once in the original matrix and once in the current tableau. The rank is simply the maximum number of pivot steps, given that the pivotal row and column are dropped (not transformed or replaced) at each step. The rows and columns of the original matrix that contain the pivotal positions form row and column bases, while the pivotal rows and columns provide equivalent bases in echelon form. The calculation in (4) is an example.

$$
\begin{array}{r|rrrrr}
\text{Check} & & & & & \\
13 & 3^{A} & -1 & 4 & 2 & 5 \\
-12 & -6 & 2 & -5^{B} & 0 & -3 \\
5 & 1 & -2^{C} & 4 & 3 & -1 \\
6 & -2 & -1 & 3 & 5 & 1 \\
7 & 2 & 3 & -1 & 1^{D} & 2
\end{array}
\rightarrow
\begin{array}{r|rrrr}
\text{Check} & & & & \\
42 & 0 & 9^{B} & 12 & 21 \\
2 & -5 & 8 & 7 & -8 \\
44 & -5 & 17 & 19 & 13 \\
-5 & 11 & -11 & -1 & -4 \\
 & & \text{over } 3 & &
\end{array}
\downarrow
\tag{4}
$$

$$
\begin{array}{r|rr}
\text{Check} & & \\
0 & 0 & 0 \\
157 & -28^{D} & 185 \\
 & \text{over } -15 &
\end{array}
\leftarrow
\begin{array}{r|rrr}
\text{Check} & & & \\
-106 & -15^{C} & -11 & -80 \\
-106 & -15 & -11 & -80 \\
139 & 33 & 41 & 65 \\
 & & \text{over } 9 &
\end{array}
$$

The letters A, B, C, D mark the pivotal positions in the original matrix and the pivots in the successive tableaus. The rank if 4; rows $1, 2, 3, 5$ in the original matrix form a basis as do columns $1, 2, 3, 4$. The following (pivotal) rows and columns form echelon bases for the row and the column spaces:

$$
\begin{array}{l}
(3^*, \quad -1, \; 4, \quad 2, \quad 5) \\
(0, \quad 0, \; 9^*, \quad 12, \quad 21) \\
(0, \; -15^*, \; 0, \; -11, \; -80) \\
(0, \quad 0, \; 0, \; -28^*, \; 185)
\end{array}
\tag{5}
$$

$$
\begin{bmatrix} 3 \\ -6 \\ 1 \\ -2 \\ 2 \end{bmatrix}
\begin{bmatrix} 0 \\ 9 \\ 8 \\ 17 \\ -11 \end{bmatrix}
\begin{bmatrix} 0 \\ 0 \\ -15 \\ -15 \\ 33 \end{bmatrix}
\begin{bmatrix} 0 \\ 0 \\ 0 \\ 0 \\ -28 \end{bmatrix}
\tag{6}
$$

Evidently results of this sort can be obtained as by-products of the solution of linear equations as in (b) and (c). For example, the row and the column space of the 3 by 4 matrix of coefficients in (1) or (3) have the echelon bases

$$
\begin{array}{l}
(1, \quad -2, \quad -1, \quad 3) \\
(0, \quad -3, \quad 1, \quad 4) \\
(0, \quad 14, \quad 0, \quad -17)
\end{array}
\tag{7}
$$

$$
\begin{bmatrix} 1 \\ -1 \\ 2 \end{bmatrix}
\begin{bmatrix} 0 \\ 1 \\ 3 \end{bmatrix}
\begin{bmatrix} 0 \\ 0 \\ 14 \end{bmatrix}
\tag{8}
$$

(1) also yields canonical bases for the rows

$$(1,0,0,13/14), \qquad (0,1,0,-17/14), \qquad (0,0,1,5/14) \tag{9}$$

and for the kernel

$$(-13,17,-5,14)/14 \tag{10}$$

(e) *Statistical regression analysis* attempts to estimate or predict the values of certain (dependent) variables by using linear functions of certain other (independent) variables. Each such linear function is determined so as to minimize the sums of squares of the residuals, which are the differences between the calculated and the observed values; this gives rise to a set of linear equations to be solved for the unknown coefficients of the linear function. The statistician who is uncertain as to which independent variables are worth considering may wish to introduce them one at a time by means of pivot steps.

More specifically, if one pivots on a diagonal element a_{kk} in the matrix **A** in which a_{ij} = the sum of the observed values of $x_i x_j$, and then disregards the kth row and column, the result is a matrix like **A** but referring to the residuals obtained when x_k is used to estimate the remaining variables; among other things it yields values for the partial correlation coefficients. Stiefel (1963) may be consulted for details. A related application is the diagonalization of a quadratic form.

(f) *Elementary methods for linear equations*. The methods of (b) and (c) above were chosen as introductions to the study of linear programming. Efficient methods for electronic computers will be discussed in 7.4. Some readers may be interested in methods suitable for freshman algebra, or wherever the students get their second exposure to systems of linear equations. For this purpose the usual *Jordan elimination in unlabeled extended tableaus* can be recommended. The recognition that several possibilities exist for the choice of the divisors of the various rows gives rise to methods (f_1) to (f_4), but a common procedure for pivot selection will be described.

If linear programming is not involved, the constant terms can be isolated in the right-hand members. The check column is the sum of the other columns as before, but it is put on the right beside the column of constant terms. Thus if the check column is included and there are m consistent and independent equations in n unknowns ($n \geq m$), there will be $m+1$ (unlabeled) tableaus each having m rows and $n+2$ columns; these tableaus are denoted by T_0 (the coefficients given initially), $T_1, \ldots, T_{m-1}$, and T_m (from which the solution is read off). The columns are kept in order, to identify the variables, but it is convenient to be able to permute the rows. The rth pivotal row appears first in parentheses in T_{r-1} and then (usually after division) in T_r with a starred pivot.

The selection of a pivot implies that the other entries in the pivotal column will be replaced by zeros, and they must remain zero in all the following pivot steps. Therefore in the tableaus they could be denoted by ϕ, and 0 reserved for "accidental" zeros that need not remain zero. Then the pivot that is selected in T_{r-1} and starred in T_r must be nonzero and lie in the first n columns; it must have $r-1$ (rather than

$r-2$) ϕ's in its row and no ϕ in its column. (After pivoting it has $m-1$ new ϕ's in its column). This ensures that one never has two pivots in the same row or column.

The preceding paragraph permits any possible sequence of pivots to be used, but many students will appreciate a more specific procedure that obviates the use of the ϕ's and is still capable in principle of handling any problem. The following rules are used in the examples hereafter.

Rules for Solving Equations in Extended Tableaus

1. Take as the pivot the *first nonzero* entry in the *last row* of the current tableau T_{r-1} $(r-1=0,1,\dots)$, but never in the column of constants or check values. This old pivotal row is parenthesized. If no such pivot exists, end the calculation by reading off the solution or concluding that no solution exists.

2. The new pivotal row (immediately below the old, except perhaps in (f_1) hereafter) is the *first row of the new tableau* T_r and is obtained by dividing the old pivotal row by its greatest common divisor in (f_1) or (f_3), by the pivot in (f_2), or by the δ_r of (11) in (f_4). The new pivot is starred.

3. Excepting the zeros in the pivotal column l, the elements in the remaining $m-1$ rows of the new tableau are calculated as follows, where a_{1l}^* is the starred pivot, $j\neq l$, i is not the pivotal row, and elements of the new tableau are distinguished by an accent or a star.

(f_1) $\quad a'_{ij}=a_{1l}^*a_{ij}-a'_{1j}a_{il}$

(f_2) $\quad a'_{ij}=a_{ij}-a'_{1j}a_{il}$ $\quad$ (rule 5 of 2.4, but a'_{lj} is denoted by a'_{1j})

$(f_3),(f_4)$ $\quad a'_{ij}=(a_{1l}^*a_{ij}-a'_{1j}a_{il})/\pi_{r-1}$ $\quad$ (rule 5′ of 2.5)

where π_{r-1} is the starred pivot (in T_{r-1}) that immediately preceded a_{1l}^*. Fractions are routine in (f_2) and occasional in (f_3).

4. In the new tableau verify that each check value is the sum of the other elements in its row. Go back to rule 1.

The first row is stipulated in rule 2 because it is the first to be written in the new tableau; it is convenient to have it close to the old tableau, which it must transform; and it guarantees that the last row of the old tableau, specified in rule 1, is always eligible to be pivotal. By shifting the rth pivotal row from the last position in T_{r-1} to the first position in T_r, these rules permute the rows cyclically and finally restore them to their original order. If the occurrence of zeros does not prevent pivoting the columns in the natural order $1,2,\dots,m$, the components of $\mathbf{x}$ will be obtained in this same order by reading up from the bottom in the last tableau.

Rule 1 specifies the first nonzero entry to minimize the risk of confusion with the columns of constants and check values on the right. If one remembers to exclude these columns, any nonzero entry in the last row will do. In particular one could choose the last nonzero entry each time, thus pivoting up the principal diagonal. This would be advantageous if these methods (f), instead of (a), were used for matrix inversion.

There is an inability to complete m pivot steps if and only if there is redundancy (one or more rows reduce entirely to zeros and hence may be discarded) or inconsistency (at least one row reduces to $0=$nonzero) among the equations.

(f_1) is the simplest procedure for *small exercises* that do not demand the use of a calculator and the acceptance of roundoff errors; there are no fractions in the tableaus. It differs from 2.5 in that any common integer factor is divided out of the pivotal row, but the use of the divisors (called δ in 2.5, but π_{r-1} here) in rule 5′ is omitted. Instead there is an option to rewrite some additional (nonpivotal) rows after dividing each by its greatest common divisor; when using rules 1 and 2 one should take care that the new set of nonpivotal rows (without parentheses) is in the same order as it was before. As usual the check column (written last) is the sum of the other columns.

The tableaus T_0, T_1 (on the left), T_2, T_3 for solving $-6x_1+4x_2+x_3=1$, $4x_1+10x_2+x_3=9$, $8x_1+2x_2-x_3=3$ then might be as follows:

−6	4	1	1	0		0	6*	1	5	12
4	10	1	9	24		(48	0	−8	8	48)
(8	2	−1	3	12)		(0	0	−32	−64	−96)
8*	2	−1	3	12		6	0	−1	1	6
0	44	2	26	72		0	0	1*	2	3
(0	72	12	60	144)		0	6	0	3	9
0	6*	1	5	12		6	0	0	3	9

In T_2 (upper right) the pivotal row (0 0 −32 −64 −96) is divided by −32 to obtain the first row 0 0 1* 2 3 in T_3, and the row above it is divided by 8 to give 6 0 −1 1 6, which remains the last nonpivotal row in T_2. Both rows (as they were before division) are parenthesized since neither will be transformed by the pivot step on 1*. Since there are no divisions in the tableau except by (greatest) common divisors, no fractions can occur until the value of $\mathbf{x}=(3/6, 3/6, 2/1)=(1/2, 1/2, 2)$ is read off from the last tableau. For example, the entry −64 in the third row on the right is obtained simply as 6*(26)−(44)5, the numerator $\pi\alpha-\beta\gamma$ in rule 5′ of 2.5. Evidently 6 and −44 are multipliers for the fifth and the seventh rows on the left such that x_2 is eliminated from the sum; zero is obtained in the second column and preserved in the first column.

(f_2) uses rules 3 and 5 of 2.4. This popular method is the best of (f_1) to (f_4) if a *calculator or computer is used* and roundoff errors are accepted, but if not, distracting hand calculations may be required to process the fractional entries. Division in the pivotal row reduces the pivot to 1; −1 might be more convenient here, but +1 is preferred for consistency with 2.4 and linear programming. The method is illustrated in the first four columns of Table 2.6B. (The check column is omitted.)

Rule 5 of 2.4 with $a'_{ij}=a_{ij}-a'_{1j}a_{il}$ may be paraphrased as follows, for each nonpivotal i. If a calculator is used, store $-a_{il}$ (from the old pivotal column l) in the memory. For each nonpivotal j, use $-a_{il}$ to multiply a'_{1j} in the new pivotal row, and add the product to a_{ij} in the row i to obtain a'_{ij} in the corresponding new row. For example, −4/3 in the fourth column of Table 2.6B is obtained as $(-11/2)(5/6)+(13/4)$.

TABLE 2.6B Three Methods for Solving $-6x_1+4x_2+x_3=1,\quad 4x_1+10x_2+x_3=9,\quad 8x_1+2x_2-x_3=3$

Divide rth pivotal row by	(f₂) Pivot			(f₃) GCD of Row				(f₄)	$\delta_2=3=(1\cdot 12/(12,8),12)$ $\delta_3=8=(4\cdot 16/(64,24),16)$				
-6	4	1	1	-6	4	1	1	T_0	-6	4	1	1	0
4	10	1	9	4	10	1	9		4	10	1	9	24
(8	2	-1	3)	(8	2	-1	3)		(8	2	-1	3	12)
1*	$\frac{2}{8}$	$-\frac{1}{8}$	$\frac{3}{8}$	8*	2	-1	3	T_1	8*	2	-1	3	12
0	$\frac{11}{2}$	$\frac{1}{4}$	$\frac{13}{4}$	0	44	2	26		0	44	2	26	72
(0	9	$\frac{3}{2}$	$\frac{15}{2}$)	(0	72	12	60)		(0	72	12	60	144)
0	1*	$\frac{1}{6}$	$\frac{5}{6}$	0	6*	1	5	T_2	0	24*	4	20	48
1	0	$-\frac{1}{6}$	$\frac{1}{6}$	6	0	-1	1		24	0	-4	4	24
(0	0	$-\frac{2}{3}$	$-\frac{4}{3}$)	(0	0	-4	-8)		(0	0	-16	-32	-48)
0	0	1*	2	0	0	1*	2	T_3	0	0	2*	4	6
0	1	0	$\frac{1}{2}$	0	1	0	$\frac{1}{2}$		0	2	0	1	3
1	0	0	$\frac{1}{2}$	1	0	0	$\frac{1}{2}$		2	0	0	1	3

$\mathbf{x}=(1/2,\ 1/2,\ 2)$. See the text for method (f_1).
(f_4) uses (11); the last column illustrates the checking procedure.

(f_3) follows 2.5 but allows the following option. Before each pivot step the parenthesized (pivotal) row in T_{r-1} can be divided by its greatest common divisor d_r to obtain the starred pivotal row in T_r. The latter, acting through rule 5′ of 2.5, converts the other $m-1$ rows of T_{r-1} into the other rows of T_r. (They have zeros in the pivotal column.) The price of this simplicity is that rule 5′ may occasionally produce fractions, as in the last tableau in the middle section of Table 2.6B. If they are to be eliminated by multiplication, it may be desirable to multiply every row in the tableau by the same constant. In the constants column in T_2, the entry -8 is obtained as $(6\cdot 26-44\cdot 5)/8$, where 8 is the previous pivot.

(f_4) To be certain of avoiding fractions in (f_3), the divisor of the (parenthesized) pivotal row in T_{r-1} $(1\le r\le m)$ may be taken to be

$$\delta_r \triangleq \left(\frac{c_r d_r}{(c_r d_r, \pi_{r-1})}, d_r\right) \tag{11}$$

where d_r = the greatest common divisor of this row,
c_r = the GCD of all the other entries in T_{r-1} (which include π_{r-1}),
π_{r-1} = the preceding pivot (the starred entry in T_{r-1}),
(a, b) = the greatest common divisor of a and b. If $r=1$ one has $\pi_0=1$, $\delta_1=d_1$. If we choose to disregard c_r, then $c_r=1$ and (11) becomes

$$\delta_r = d_r/(d_r, \pi_{r-1}) \tag{12}$$

which is less than or equal to (11).

Equation (11) begins with $c_r d_r$, which is a divisor of every numerator $\pi_r\alpha-\beta\gamma$ that is divided by π_{r-1} in 2.5. This $c_r d_r$ is divided by $(c_r d_r, \pi_{r-1})$, the factors that may be lost in the division by π_{r-1}. Finally additional factors are discarded if necessary to make δ_r a divisor of d_r and so a divisor of the pivotal row. Thus in Table 2.6B one has $\delta_2=3=(1\cdot 12/(12,8),12)$; $\delta_3=8=(4\cdot 16/(64,24),16)$. The -32 in the constants column of T_2 is obtained as $(24\cdot 26-44\cdot 20)/8$, where 8 is the previous pivot.

The next two paragraphs, which prove that fractions are not encountered, are optional reading.

The integer property IP (the avoidance of fractions in all the tableaus) is required here only in a sense weaker than in 2.5, because pivotal columns are discarded as in 2.6(b) (or rather, equivalently, each is reduced to zero in all but one position), no pivot step is reversed, and the common denominator is no longer a denominator but occurs once in each pivoted column. It continues to be used as an indication of divisibility at the following pivot step by rule 5′ of 2.5. To show that (11) or (12) preserves IP, we use induction on the number r of divisors $\delta_1,\dots,\delta_r$. At the first step $\pi_0=1$ and the division of the first row by $\delta_1=d_1$ could be incorporated in the formulation of the problem; Theorem 16 of 7.8 with unit initial denominator establishes IP for this initial case $r=1$.

To determine the effect of dividing by δ_r (thus going from $r-1$ to r divisors) we regard T_{r-1} as the initial tableau and use (16) of 7.8 with the common denominator and each numerator multiplied by $\delta^{\lambda-1}=\pi_{r-1}^{\lambda-1}$. This shows subsequent tableaus can be written in the form

Λ	x_k exchanged	y_j not exchanged
y_l exchanged	$\pm\pi_{r-1}\Lambda_{kl}$	$\pm\Lambda^{\cdot j}_{\cdot l}$
x_i not exchanged	$\pm\Lambda^{i\cdot}_{k\cdot}$	Λ^{ij}/π_{r-1}

(13)

Here the x_k columns are omitted and only the determinants $\Lambda^{\cdot j}_{\cdot l}$ and Λ^{ij}/π_{r-1} are significant; they are all integers by the induction hypothesis for $r-1$ divisors. Inspection of (13) or of rule 5′ of 2.5 shows that the sole effect of dividing the pivotal row by δ_r is to divide every element $\Lambda^{\cdot j}_{\cdot l}$ and Λ^{ij}/π_{r-1} and every denominator Λ of every subsequent tableau by δ_r. This never produces fractions because each of these determinants contains the initial (i.e., rth) pivotal row k, which is divisible by d_r and hence by δ_r (see (11)), and every Λ^{ij} also contains a row $i \neq k$, which is divisible by c_r. Thus Λ^{ij} is divisible by both π_{r-1} and $c_r d_r$, therefore by $c_r d_r \pi_{r-1}/(c_r d_r, \pi_{r-1})$, and therefore by $\delta_r \pi_{r-1}$, which (11) shows is a divisor of the preceding. Thus the IP is proved for r divisors.

The following table summarizes some estimates of the relative advantages of the methods (f_1) to (f_4):

	(f_1)	(f_2)	(f_3)	(f_4)
Best with calculator?	No	Yes	No	No
Max. size of problem	Small	Large	Medium	Medium
Avoids fractions?	Yes	No	Mostly	Yes
Simple procedure?	Yes	Yes	Yes	Less
Derived from section	—	2.4	2.5	2.5

It should be added that (f_1) is simplest for small systems of linear equations, (f_2) frequently appears in the general literature, and (f_3) is closest to the methods of 2.5 that are employed elsewhere in this book.

Table 2.6C lists and classifies the numerical applications of pivoting in this book, and a few others.

EXERCISES 2.6

Use the rules of 2.5 to invert 1 to 4 (or conclude that no inverse exists) and find their determinants.

1.
$$\begin{matrix} 0 & -3 & -4 \\ 2 & 0 & 2 \\ -3 & 2 & 0 \end{matrix}$$

2.
$$\begin{matrix} 2 & -4 & 2 \\ -3 & 4 & -3 \\ 5 & 2 & 4 \end{matrix}$$

3.
$$\begin{matrix} 1 & -2 & 0 & 1 \\ 2 & 1 & -1 & -1 \\ -1 & 0 & 2 & 1 \\ 2 & -1 & 1 & 1 \end{matrix}$$

4.
$$\begin{matrix} 1 & 3 & 0 & 0 \\ 0 & -4 & 0 & 0 \\ 0 & 2 & 1 & 0 \\ 0 & 5 & 0 & 1 \end{matrix}$$

TABLE 2.6C Numerical Applications of Pivoting

Notes: *A canonical basis* contains (a permutation of) a unit or diagonal submatrix of maximum possible size as in 2.3; it is in the final tableau. A canonical basis is a special case of an *echelon basis*. The latter is triangular or trapezoidal; its vectors can be arranged so that the jth vector has zeros in $j-1$ positions, including the $j-2$ position specified for the preceding vector. The linear independence of the vectors in either of these types of bases is obvious (Theorem 2 of 2.7). The row space and the kernel are orthogonal complements.

1. Tableaus of constant size.

Matrix inversion 2.6(a), Chapter 7.

Canonical bases for the column space (image), the row space and the kernel. (5)–(11) of 2.7.

†Linear programming 3.2, 3.3 and thereafter.

†Block pivoting 7.8

2. Singly contracting tableaus (the transform of the pivotal column is omitted after each pivot step).

Jordan solution of linear equations 2.6(b), 7.5.

Canonical bases for the row space and kernel (9) of 2.6; (6), (8), (10) of 2.7.

3. Doubly contracting tableaus (pivotal rows and columns are dropped).

Evaluation of determinants 2.6(a), 7.5(a_5), 7.7, 7.8.

Gaussian solution of linear equations 2.6(c), 7.4, 7.5.

Rank of a matrix 2.6(d), 2.7.

Echelon or original row and column bases formed from the pivotal rows and columns 2.6(d).

†Partial correlation and regression after elimination of variables already regressed on 2.6(e).

4. Other applications.

†Revised simplex methods 7.1 to 7.3.

†Integer, linearized and quadratic programming.

†Excepting these, all the applications listed make the maximum possible number of exchanges as in (6) of 2.7.

If possible, solve 5 to 8 by Jordan elimination. If the solution is not unique, solve for the variables having the smallest subscripts. If not possible, explain why.

5.
$$\begin{aligned} -y_1+2y_2+4y_3&=1\\ 3y_1-y_2+2y_3&=-2\\ -4y_1+3y_2+y_3&=4 \end{aligned}$$

6.
$$\begin{aligned} x_1-4x_2-x_3+2x_4&=3\\ -2x_1+x_2+3x_3-x_4&=2\\ -3x_1-2x_2+5x_3+x_4&=8 \end{aligned}$$

7.
$$\begin{aligned} 3y_1-y_2+2y_3&=4\\ y_1+2y_2+y_3&=5\\ 7y_1-7y_2+4y_3&=2 \end{aligned}$$

8.
$$\begin{aligned} 2x_1+x_2-3x_3&=2\\ -x_1+2x_2+x_3&=3\\ x_1+8x_2-3x_3&=12 \end{aligned}$$

Determine the ranks and row and column bases for the following matrices.

9.
$$\begin{matrix} 2 & 5 & -1 & 6 & 3\\ 1 & -2 & 4 & -3 & -1\\ 4 & 1 & 7 & 0 & 1\\ -5 & -8 & -2 & -9 & -5 \end{matrix}$$

10.
$$\begin{matrix} 4 & -1 & 2 & 7 & 5\\ 0 & 2 & 3 & -6 & 4\\ 0 & 0 & 0 & 5 & -1\\ 0 & 0 & 0 & 0 & 0 \end{matrix}$$

2.7 PROVING THEOREMS BY PIVOTING

The theorems of this section are concerned with the concepts of rank and basis for a vector space or a matrix. The methods of linear programming take care of these matters more or less automatically and the attention is centered on the computational procedure and its rationale. Hence this section may be omitted by readers who wish to get on to linear programming as soon as possible. Nevertheless these fundamental ideas from linear algebra are very important for mathematics and computation in general. The usefulness of pivoting in this connection has been emphasized by Tucker (1960 et seq.) and Stiefel (1963). The latter attributes it to Steinitz and cites it as evidence of the unity of pure and applied mathematics.

Let us recall that vectors $\boldsymbol{\xi}^1,\ldots,\boldsymbol{\xi}^p$ are said to be *linearly independent* if the only homogeneous linear relation

$$u_1\boldsymbol{\xi}^1+\cdots+u_p\boldsymbol{\xi}^p=\mathbf{0} \tag{1}$$

among them is the trivial one in which all $u_i=0$. On the other hand, if $u_1\neq 0$, for example, we may solve for $\boldsymbol{\xi}^1$ in terms of the others; then $\boldsymbol{\xi}^1$ is in the space spanned by $\boldsymbol{\xi}^2,\ldots,\boldsymbol{\xi}^p$ and is said to be linearly dependent on them.

One way to obtain a basis for a vector space is to start with a spanning set and eliminate (as often as required) one or more vectors that are linearly dependent on those that remain, thus obtaining a spanning set that is minimal at least relative to this particular procedure. Another way is to start with a (possibly empty) set of linearly independent vectors and expand it until a maximal linearly independent set is obtained. In both cases the resulting basis is equivalent by definition to a linearly independent spanning set. Now if two different people independently seek a basis for the same space, we do not expect that they will necessarily choose the same basis, but we would like the number of vectors in each basis to be the same, since that number is defined to be the rank or dimension of the space.

To show that the numbers are the same, let us compare (one of) the largest with (one of) the smallest. The largest, say $\{\boldsymbol{\eta}^1,\ldots,\boldsymbol{\eta}^q\}$, must still be linearly independent, and the smallest, say $\{\boldsymbol{\xi}^1,\ldots,\boldsymbol{\xi}^p\}$, must still be a spanning set. The latter fact implies that the $\boldsymbol{\eta}^j$ must be linearly dependent on the $\boldsymbol{\xi}^i$ as indicated by the column equations of (2). This is the only occasion we will have to

$$\begin{array}{c|ccc} 1 & \boldsymbol{\eta}^1 & \cdots & \boldsymbol{\eta}^q \\ \hline \boldsymbol{\xi}^1 & a_{11} & \cdots & a_{1q} \\ \vdots & \vdots & & \vdots \\ \boldsymbol{\xi}^p & a_{p1} & \cdots & a_{pq} \end{array} \qquad \text{with } q\geq p \tag{2}$$

use vectors as labels for *individual* rows and columns, but one can easily verify that all of the algebra of pivoting goes through unchanged; of course the a_{ij} are still scalars. Elsewhere a vector label will represent an ordered set of scalar labels; this will be confirmed by the occurrence of a matrix at the intersection of a row and column having vector labels.

Now the procedure is to pivot as many times as possible but never twice in the same row or column. When it is impossible to go further, the tableau will have one of the following three forms, aside from probable permutations of the rows and of the columns:

$$\begin{array}{c|cc} 1 & \xi\text{'s} & \eta\text{'s} \\ \hline \eta\text{'s} & \mathbf{A}_1 & \mathbf{A}_2 \\ \xi\text{'s} & \mathbf{A}_3 & \mathbf{0} \end{array} \qquad \begin{array}{c|cc} 1 & \xi\text{'s} & \eta\text{'s} \\ \hline \eta\text{'s} & \mathbf{A}_1 & \mathbf{A}_2 \end{array} \qquad \begin{array}{c|c} 1 & \xi\text{'s} \\ \hline \eta\text{'s} & \mathbf{A}^{-1} \end{array} \tag{3}$$

All entries must be zero in the block labeled **0** in the first form to prevent further pivoting. The first two forms cannot actually occur because the columns on the right assert that some of the $\boldsymbol{\eta}^j$ are dependent on the others, contrary to hypothesis. Only the third form is possible, in which $p=q$ and the $\boldsymbol{\xi}^i$ are now expressed in terms of the $\boldsymbol{\eta}^j$. If the larger basis were infinite, a suitable finite subset thereof could be used as the $\boldsymbol{\eta}^j$. Thus we have proved the following.

THEOREM 1 All bases (linearly independent spanning sets) of a vector space have the same rank (number of vectors).

We will prove that a set of vectors are linearly independent if they are the nonzero rows of a (*row*)-*echelon* matrix. This can mean that each row begins with a longer string of zeros than its predecessor, except that several rows consisting entirely of zeros may occur at the bottom, as in the following example:

$$\begin{array}{rrrrrr} -5 & 0 & 2 & 0 & 0 & 0 \\ 0 & 0 & 4 & 0 & 3 & 1 \\ 0 & 0 & 0 & 2 & 6 & 0 \\ 0 & 0 & 0 & 0 & 0 & 0 \\ 0 & 0 & 0 & 0 & 0 & 0 \end{array} \tag{4}$$

However, we will allow the rows and columns to be permuted; the proof of the theorem will define and identify this situation.

THEOREM 2 The nonzero rows of an echelon matrix are linearly independent and hence are a basis for the space that they span. (The canonical bases (15) and (16) of 2.3 are special cases.)

Proof Let the nonzero rows $\xi^1, \xi^2, \ldots, \xi^m$ be placed first. We have to show that the linear relation (1) is possible only if all $u_i=0$. If $m \geq 1$, the echelon property requires that at least one of the columns, say l_1, contain exactly one nonzero element $\xi(k_1, l_1)$. Then (1) implies that $u_i=0$ for $i=k_1$. After row k_1 and column l_1 have been deleted, the echelon property requires that at least one column l_2 have exactly one nonzero element $\xi(k_2, l_2)$ in the remaining matrix. Then (1) implies $u_i=0$ for $i=k_2$, row k_2 and column l_2 are deleted, etc. The process continues until one has $u_i=0$ for every nonzero row $i=1, \ldots, m$. After the rows and columns are arranged as in (4), $\xi(k_h, l_h)$ will be the first nonzero element in row h.

At this point the reader should recall how the tableau was defined in 2.4 so that its row and column equations had the dual canonical bases (15) and (16) of 2.3 as coefficients. The unit matrices in the latter appear implicitly in the tableau as the coefficients of the basic variables, but in effect these unit matrices may be eliminated in the initial tableau by replacing the basic variables by zeros as in 2.6(b). The pivot steps are still reversible in principle, regardless of whether or not in practice we retain all of the tableau entries that would tell us how to make an actual reversal. Thus we can assert the following, which is really only a restatement and slight generalization of the primal-dual invariance theorem of 2.4.

THEOREM 3 The vector space spanned by the coefficients of the row equations of a tableau, including the basic variables, is unchanged by pivot steps even after any selected components of the vectors have been consistently deleted (perhaps because the corresponding "variables" are required to be zero, but this is immaterial). A similar statement holds for the coefficients of the column equations.

Completion of the Proof Since the pivoting process involves nothing more than replacing the original (possibly curtailed) vectors of coefficients by linear combinations thereof, it is clear that the old coefficients span everything that the new coefficients do. However, since the pivot transformations are reversible, the words "new" and "old" may be interchanged, and the two spaces must be the same.

Fully Exchanged Tableaus

Consider the linear mapping $\mathbf{x}=-\mathbf{A}\mathbf{y}$ and the corresponding tableau

$$\begin{array}{cc} 1 & \mathbf{y} \\ \hline \mathbf{x} & \mathbf{A} \end{array} \qquad \text{or} \qquad \begin{array}{l} \mathbf{x}+\mathbf{A}\mathbf{y}=\mathbf{0}, \\ \mathbf{v}=\mathbf{u}\mathbf{A} \end{array} \tag{5}$$

After the maximum possible number ρ of exchanges have been made and the rows and columns have been rearranged if necessary, the final tableau has the form

$$\begin{array}{cc|cc} & & \rho & n-\rho \\ & 1 & \mathbf{x}_K & \mathbf{y}_{\bar{L}} \\ \hline \rho & \mathbf{y}_L & \mathbf{\Lambda}^{-1} & \mathbf{A}_1 \\ m-\rho & \mathbf{x}_{\bar{K}} & \mathbf{A}_2 & \mathbf{0} \end{array} \qquad \text{or} \qquad \begin{array}{ccc} & \rho & n-\rho \\ 1 & \mathbf{u}_K & \mathbf{v}_{\bar{L}} \\ \hdashline \mathbf{v}_L & \mathbf{\Lambda}^{-1} & \mathbf{A}_1 \\ \mathbf{u}_{\bar{K}} & \mathbf{A}_2 & \mathbf{0} \end{array} \tag{6}$$

Here there are ρ and $m-\rho$ rows and ρ and $n-\rho$ columns. K and L are the sets of the indices of those x_i and y_j respectively (ρ pairs) that have been exchanged; unless $\mathbf{A}=\mathbf{0}$ they are nonempty and usually not unique. The argument of 2.6(a) shows that the submatrix $\mathbf{\Lambda}^{-1}$ in (6) is the inverse of the matrix $\mathbf{\Lambda}$ in the rows K and columns L of (5). The vector $\mathbf{x}_K$, for example, has as its components precisely those x_i for which i is in K. The sets $\bar{K}$ and $\bar{L}$ (complementary to K and L) contain the indices of the variables that have not been exchanged. If neither is empty, they must intersect in a block of zeros in (6), as otherwise one or more additional exchanges would be possible.

If $\bar{L}$ is empty, (6) gives $\mathbf{y}=\mathbf{y}_L=-\boldsymbol{\Lambda}^{-1}\mathbf{x}_K$. Thus the mapping is invertible in the restricted sense that the original vectors $\mathbf{y}$ can be recovered but only for those $\mathbf{x}$ satisfying $\mathbf{x}_{\bar{K}}+\mathbf{A}_2\mathbf{x}_K=\mathbf{0}$. Such a mapping is called *one-to-one* or *injective*. If $\bar{K}$ is empty then every $\mathbf{x}=\mathbf{x}_K$ is the image of some $\mathbf{y}$, such as $\mathbf{y}_{\bar{L}}=\mathbf{0}$, $\mathbf{y}_L=-\boldsymbol{\Lambda}^{-1}\mathbf{x}$, and the mapping is called *onto* or *surjective*. The original matrix $\mathbf{A}$ is invertible if and only if $\bar{K}$ and $\bar{L}$ are both empty; then the mapping is called *invertible* or *bijective*.

By ignoring $\mathbf{x}$ one obtains in (5) the rows of $\mathbf{A}$ and in (6) the rows of

$$\begin{bmatrix} \mathbf{I}_\rho & \mathbf{A}_1 \\ \mathbf{0} & \mathbf{0} \end{bmatrix} \tag{7}$$

where $\mathbf{I}_\rho$ is a ρ by ρ matrix with ones on the diagonal and zeros elsewhere. Since the rows of zeros contribute nothing, (7) and Theorem 3 show that the rows of $[\mathbf{I}_\rho,\mathbf{A}_1]$ form a canonical basis for the space spanned by the rows of $\mathbf{A}$, which is called the *row space* of $\mathbf{A}$. Similarly, ignoring $\mathbf{v}$ in the column equations gives the canonical basis (9) for the *column space* of $\mathbf{A}$. The column space is also called the *image* of either of the linear transformations $\mathbf{Ay}=\mathbf{x}$ or $-\mathbf{Ay}=\mathbf{x}$, because they express $\mathbf{x}$ as a linear combination of the columns of $\mathbf{A}$.

These bases have the usual orthogonal complements (10) and (11). The space spanned by (10) is called the *kernel* of the linear transformation; left multiplication by $\mathbf{A}$ maps the columns of (10) into $\mathbf{0}$ because the rows of (7) or (8) do the same.

Row space spanned by rows of

$$[\mathbf{I}_\rho,\mathbf{A}_1] \tag{8}$$

Column space (image) spanned by columns of

$$\begin{bmatrix} -\mathbf{I}_\rho \\ \mathbf{A}_2 \end{bmatrix} \tag{9}$$

Kernel spanned by columns of

$$\begin{bmatrix} \mathbf{A}_1 \\ -\mathbf{I}_{n-\rho} \end{bmatrix} \tag{10}$$

Dual kernel spanned by rows of

$$[\mathbf{A}_2,\mathbf{I}_{m-\rho}] \tag{11}$$

Although $\boldsymbol{\Lambda}^{-1}$ does not appear explicitly in any of these, it is implicit in both $\mathbf{A}_1$ and $\mathbf{A}_2$ as (4) of 7.8 shows.

A much more extensive treatment of these matters and of (12) is contained in the book by Gewirtz, Sitomer and Tucker (1974).

THEOREM 4 The row and column spaces of $\mathbf{A}$ have the same rank ρ, called the *rank of the matrix* $\mathbf{A}$, which is equal to the maximum number of pairs of variables that can be exchanged in a tableau of matrix $\mathbf{A}$. $\rho\le\min(m,n)$.

Proof Theorems 2 and 3 show that (8) and (9) are bases for the row and column spaces; each has ρ vectors. Theorem 1 shows that any other basis for either space will also contain ρ vectors.

THEOREM 5 The rows K of $\mathbf{A}$ form a basis for its row space and the columns L of $\mathbf{A}$ form a basis for its column space. (The ρ pivot steps can be performed in these rows or in these columns alone, after the others have been discarded, and still the basis (8) or (9) is obtained.)

THEOREM 6 Let ρ be the rank of $\mathbf{A}$ (m by n) and ρ' the rank of $[\mathbf{A},\mathbf{b}]$. Then the solution set of the system of linear equations $\mathbf{Ay}=\mathbf{b}$ is empty if and only if $\rho'=\rho+1$, and it is a nonempty linear variety of dimension $n-\rho$ if and only if $\rho'=\rho$.

Proof If $\mathbf{Ay}=\mathbf{b}$ has a solution $\mathbf{y}$, the equation asserts that $\mathbf{b}$ is linearly dependent on the columns of $\mathbf{A}$ so that $\rho'=\rho$. Otherwise $\mathbf{b}$ is not linearly dependent and can be adjoined to any column basis of $\mathbf{A}$ so that $\rho'=\rho+1$. If $\mathbf{Ay}=\mathbf{b}$ has a nonempty solution set, its dimension is defined to be that of the vector space of $\mathbf{y}$ satisfying $\mathbf{Ay}=\mathbf{0}$ (cf. Theorem 4 of 2.1); comparison of (8) and (10) shows that this is $n-\rho$.

In applying Theorems 4 to 6, one may drop the pivotal row and column at each step, thus obtaining smaller and smaller tableaus. According to 2.6(d) the same is true if echelon bases for the row and column spaces are desired. However, to find $\mathbf{A}_1$ in (8) or (10) one must keep the pivotal rows and to find $\mathbf{A}_2$ in (9) or (11) one must keep the pivotal columns.

Unsymmetrical duality is a property of the schema

$$\begin{array}{c|c|c} & \mathbf{y} & \\ \hline \boldsymbol{\pi} & \mathbf{A} & =\mathbf{0} \\ \hline & \| & \\ & \mathbf{v} & \end{array} \tag{12}$$

discussed by Gewirtz et al. The row equations $\mathbf{Ay}=\mathbf{0}$ are said to be in *standard* (i.e., general) *form* and the column equations $\mathbf{v}=\boldsymbol{\pi}\mathbf{A}$ in *parametric form*. Obviously $\mathbf{vy}=0$ so that the solution sets for $\mathbf{y}$ and $\mathbf{v}$ are orthogonal as usual; the latter is spanned by the rows of $\mathbf{A}$.

One can reduce (12) to the symmetrical canonical form by eliminating the π_i and the zeros. As in 2.6(b) this is done by pivoting in different rows and columns. The transform of each pivotal column is omitted because in the row equations it is multiplied by a zero that has become nonbasic and in the column equations it merely serves to define a basic π_i, which may be considered extraneous. However, if $\mathbf{A}$ includes unit vectors among its columns, some or all of the π_i will be included among the v_j. Evidently a canonical tableau can be considered to be in standard form or parametric form or both, and pivotal reduction to a canonical form is the method employed to change from a standard to a parametric form or vice versa.

THEOREM 7 Every $\mathbf{u}$ satisfying $\mathbf{uA}=\mathbf{0}$ also satisfies $\mathbf{ub}=0$ (if and) only if the latter equation is a linear combination of the former (i.e., there is a $\mathbf{y}$ such that $\mathbf{Ay}=\mathbf{b}$).

THEOREM 7′ $\mathbf{Ay}=\mathbf{b}$ has no solution $\mathbf{y}$ (if and) only if there is a linear combination of the equations that reduces to the impossible equation $0=1$ (i.e., there is a $\mathbf{u}$ such that $\mathbf{uA}=\mathbf{0}$ but $\mathbf{ub}\neq 0$).

Proof Theorems 7 and 7′ are equivalent because they are contrapositives. In each case the "if" statement is obvious. The problem is formulated in the schema

$$\begin{array}{c|cc|c} & \mathbf{y} & 1 & \\ \hline \mathbf{u} & \mathbf{A} & -\mathbf{b} & =\mathbf{0} \\ \hline & \| & \| & \\ & \mathbf{0} & \bar{w} & \end{array} \tag{13}$$

We have to show that a solution $\mathbf{y}$ exists if and only if $\bar{w}$ is always zero.

As in (6) we exchange as many variables and zeros as possible to get

$$\begin{array}{c|ccc|c} & \mathbf{0} & \mathbf{y}_{\bar{L}} & 1 & \\ \hline \mathbf{0} & \mathbf{A}^{-1} & \mathbf{A}_1 & \mathbf{b}'_L & =-\mathbf{y}_L \\ \mathbf{u}_{\bar{K}} & \mathbf{A}_2 & \mathbf{0} & \mathbf{b}'_{\bar{K}} & =\mathbf{0} \\ \hline & \| & \| & \| & \\ & \mathbf{u}_K & \mathbf{0} & \bar{w} & \end{array} \tag{14}$$

If $\bar{K}$ is empty or if $\mathbf{b}'_{\bar{K}}=\mathbf{0}$, both statements are true in Theorem 7 and false in Theorem 7′. If $\bar{K}$ is not empty and $\mathbf{b}'_{\bar{K}}\neq\mathbf{0}$, both statements are false in Theorem 7 and true in Theorem 7′ because (13) and (14) have the same solution set. This completes the proof.

THEOREM 8 If the values of the variables in a set M' are uniquely determined by the values of the remaining variables in a system of linear equations, then 2.6(b) can be used to reduce the system to a tableau with a full set of basic variables including all the members of M'.

Proof We begin by choosing only members of M' to enter the basis. If we were to find an $x_l \in M'$ that could not be made basic because it had zero coefficients in all the equations (if any) that do not yet have basic variables, this would contradict the hypothesis because x_l could vary while the nonmembers of M' did not. Thus all the members of M' can be made basic. Then if some equations still lack basic variables, this means that the nonmembers of M' cannot all vary independently and some of them have to become basic also. Since the system is consistent by hypothesis, a full set of basic variables can be obtained.

EXERCISES 2.7

1. Use a pivoting procedure to express $(2,1,10,-3)$ in terms of $(2,-1,3,1)$, $(1,4,2,-3)$, $(3,1,-2,2)$ if possible.

2,3. Use 2.6(d) (doubly contracting tableaus) to find echelon matrices having the same row space as

$$\begin{bmatrix} 3 & -1 & 4 & 2 & 1 \\ 1 & 5 & -2 & 3 & -1 \\ -1 & 2 & 1 & 0 & -4 \\ 4 & -5 & 11 & 1 & -1 \end{bmatrix}, \quad \begin{bmatrix} 2 & 1 & -3 & 5 & -1 \\ 1 & -3 & 2 & 1 & 4 \\ 5 & -1 & -4 & 11 & 0 \\ 4 & -5 & 1 & 7 & 7 \end{bmatrix}$$

4. How would you determine whether the vectors in a given set are linearly dependent? Prove that m vectors each having n components are always dependent when $m>n$.

5. Let $\rho_1 = \operatorname{rank} \mathbf{A}_1$, $\rho_2 = \operatorname{rank} \mathbf{A}_2$, $\rho =$ rank of $\mathbf{A}$, consisting of precisely the rows of $\mathbf{A}_1$ and the rows of $\mathbf{A}_2$. In terms of row spaces, what is the significance of (a) $\rho = \rho_1$ (b) $\rho = \rho_1 = \rho_2$? (c) How could (a) be tested by a single sequence of pivot steps?

6. Explain how Theorem 6 applies to (14) and (15) of 2.3.

7. The planes $3x_1 - x_2 + x_3 = 0$, $-x_1 + 2x_2 + 3x_3 = 0$, $x_1 + 3x_2 + 7x_3 = 0$ intersect in a straight line. Prove this and find a vector lying in the line. What does Theorem 6 say about these three equations?

8. Find canonical bases for the image and the kernel of

$$\begin{matrix} -3 & 1 & 1 & -2 \\ 1 & 2 & -1 & 3 \\ -1 & 5 & -1 & 4 \end{matrix}$$

See (13′) of 2.3 for 9 and 10, which are otherwise independent.

9.* Find a basis for the intersection of the vector spaces having $\{(2,-1,-2), (-1,2,7)\}$ and $\{(3,1,9), (1,2,8)\}$ as bases.

10.* Show that $\operatorname{rank}(X+Y) + \operatorname{rank}(X \cap Y) = \operatorname{rank} X + \operatorname{rank} Y$.

REFERENCES (CHAPTER 2)

Beale (1970)
Gewirtz-Sitomer-Tucker (1974)
Nering (1963)
Stiefel (1963)
Tucker (1963a)
Zelinsky (1968, 1973)

CHAPTER THREE

Feasible Linear Programs

3.1 DUAL PROGRAMS IN NONNEGATIVE VARIABLES

In 2.1 we found that the solution set of a system of linear equations is a linear variety X', which may be empty, or a point, a line that is infinite in both directions, a plane, etc. To maximize or minimize a linear function $\mathbf{cx}$ over such a set is trivial or pointless; either the linear function is constant over the set or it is unbounded both above and below. To show this, let $\mathbf{x}^1$ and $\mathbf{x}^2$ be two points of X' such that $\mathbf{cx}^1 - \mathbf{cx}^2 = d \neq 0$. Then for every scalar t, the point $\mathbf{x} = t\mathbf{x}^1 + (1-t)\mathbf{x}^2$ belongs to X', and $\mathbf{cx} = \mathbf{cx}^2 + dt$, which is unbounded since $d \neq 0$ and t is unrestricted. Figure 3.1 illustrates this for the case max $(x_1 + x_2)$ subject to $x_1 + 2x_2 = 2$ and no inequality constraints.

Thus linear programming problems invariably include inequality constraints. In fact, it is convenient to introduce the basic ideas of linear programming in connection with problems in which all the given constraints can be represented as inequalities and all the individual variables themselves (except the objective function z or $\bar{w}$ that is to be optimized) are required to be nonnegative. By multiplying by -1 where necessary to reverse the sense of an inequality or to convert maximization into minimization, such a problem or *linear program* may be formulated as follows:

Subject to the constraints $y_j \geq 0$ and

$$\begin{array}{l} a_{11}y_1 + \cdots + a_{1n}y_n + b_1 \leq 0 \\ \vdots \\ a_{m1}y_1 + \cdots + a_{mn}y_n + b_m \leq 0 \end{array} \tag{1}$$

determine $y_1, \ldots, y_n$ so as to minimize

$$c_1y_1 + \cdots + c_ny_n + d = z$$

By introducing slack (dependent) variables $x_i \geq 0$ that are defined so that $-x_1, \ldots, -x_m$ are equal to the left sides of the inequalities in (1), one gets all the inequalities in the form of nonnegativity conditions, and the other constraints

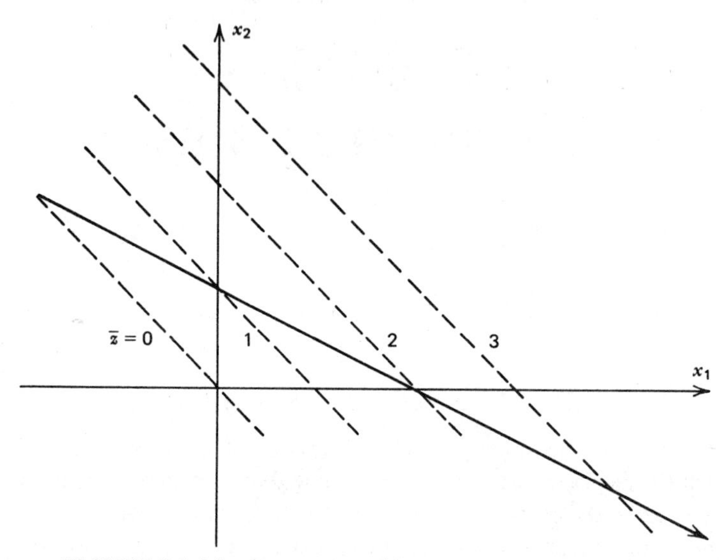

FIGURE 3.1 Max $\bar{z}=x_1+x_2$ subject to $x_1+2x_2=2$ is $+\infty$.

become equalities. Then (1) takes the form

Subject to all $x_i \geq 0$ and $y_j \geq 0$ and

$$\begin{aligned} x_1 + \quad + a_{11}y_1 + \cdots + a_{1n}y_n + b_1 &= 0 \\ \vdots \qquad\qquad\quad \vdots \quad \vdots & \\ x_m + a_{m1}y_1 + \cdots + a_{mn}y_n + b_m &= 0 \end{aligned} \tag{2}$$

minimize $z = c_1 y_1 + \cdots + c_n y_n + d$

Here the x_i are basic variables. Any $\mathbf{x},\mathbf{y}$ that satisfies (2) and minimizes z is called an *optimal solution*.

If it happens that all $b_i \leq 0$ and all $c_j \geq 0$ it is easy to see that the solution $x_i = -b_i$ and $y_j = 0$ for all i and j not only satisfies the constraints but also makes min $z = d$, because $c_j \geq 0$ and $y_j \geq 0$ implies that $z \geq d$. If this happy state of affairs does not exist, we must attempt to bring it about by solving the equations (2) for some set of basic variables other than the present set $\{x_1, x_2, \ldots, x_m, z\}$.

The determination of the optimal set of basic variables by means of a sequence of pivot steps will be considered in the next two sections. For the present we note that the y_j in the optimal basis or in any other basis must be eliminated from z, and if this is indeed possible it can always be accomplished by adding to z the appropriate multiples $u_1, \ldots, u_m$ of the left members of (2). Since these left members are required to equal zero, such an operation will change the appearance of the problem but not the optimality of any $\mathbf{x}, \mathbf{y}$. The result is that z takes the form

$$\begin{aligned} z = & u_1 x_1 + \cdots + u_m x_m \\ & + (u_1 a_{11} + \cdots + u_m a_{m1} + c_1) y_1 \\ & + \cdots \\ & + (u_1 a_{1n} + \cdots + u_m a_{mn} + c_n) y_n \\ & + u_1 b_1 + \cdots + u_m b_m + d \\ = & \mathbf{ux} + (\mathbf{uA} + \mathbf{c})\mathbf{y} + \mathbf{ub} + d \end{aligned} \tag{3}$$

in vector notation. (Note that $\mathbf{uAy} = \Sigma_{i=1}^{m} \Sigma_{j=1}^{n} u_i a_{ij} y_j$.)

The immediate object is to produce nonnegative coefficients for the x_i and y_j in (3). If this is accomplished, then (3) implies that $z \geq \mathbf{ub} + d$ because all $x_i \geq 0$ and $y_j \geq 0$. It does not yet imply that $\min z = \mathbf{ub} + d$, because the x_i and y_j are not all independent but must still satisfy (2). For this reason it is not useful to minimize $\mathbf{ub} + d$. Quite to the contrary, the conclusion $z \geq \mathbf{ub} + d$ is stronger, the larger $\mathbf{ub} + d$ is, and in fact we will find that *maximizing* $\mathbf{ub} + d$ serves to establish the equality $\min z = \max(\mathbf{ub} + d)$ whenever either of these quantities is finite.

Hence we are led to consider the linear program (4), in which the slack variables v_j represent the coefficients of y_j in the new expression (3) for z.

$$\begin{aligned} & \text{Subject to all } u_i \geq 0 \quad \text{and} \quad v_j \geq 0 \quad \text{and} \\ & v_1 = u_1 a_{11} + \cdots + u_m a_{m1} + c_1 \geq 0 \\ & \vdots \\ & v_n = u_1 a_{1n} + \cdots + u_m a_{mn} + c_n \geq 0 \\ \text{maximize} \quad & \bar{w} = u_1 b_1 + \cdots + u_m b_m + d \end{aligned} \tag{4}$$

This linear program is called the *dual* program corresponding to the *primal* program (1) or (2). We will see that both are solved simultaneously and that $\min z = \max \bar{w}$, if either of these exist.

Note that if some of the x_i were absent in (2), then there would be no need to require that the corresponding $u_i \geq 0$; in other words, inequalities have nonnegative multipliers u_i but the multipliers for equations are not sign restricted.

The programs (2) and (4) are said to be in *canonical form*. In the tableau notation introduced in 2.4, the primal (2) is represented by the row equations of

1	y_1	$\cdots$	y_n	1	
x_1	a_{11}	$\cdots$	a_{1n}	b_1	
$\vdots$	$\vdots$		$\vdots$	$\vdots$	
x_m	a_{m1}	$\cdots$	a_{mn}	b_m	All $x_i, y_j \geq 0$
$-z$	c_1	$\cdots$	c_n	d	Min z

(5)

Note the devices employed here: the constant terms b_i are accounted for by inserting the constant 1 where otherwise another independent variable y_{n+1} might have appeared; the defining relation for z is written as

$$-z+c_1y_1+\cdots+c_ny_n+d=0$$

so that $-z$ can be regarded as one of the basic (dependent) variables; and each row equation is obtained by summing $n+2$ products all the way across and setting the result equal to zero.

Let us write (4) as a tableau also, but in terms of the column equations, since it is the dual. The result is

$$\begin{array}{ccccc l}
1 & v_1 & \cdots & v_n & \bar{w} & \\
\hdashline
u_1 & a_1 & \cdots & a_{1n} & b_1 & \\
\vdots & \vdots & & \vdots & \vdots & \\
u_m & a_{m1} & \cdots & a_{mn} & b_m & \text{All } u_i, v_j \geq 0 \\
1 & c_1 & \cdots & c_n & d & \text{Max } \bar{w}
\end{array} \tag{6}$$

This is identical with (5) except that the x's and y's have been replaced by u's and v's, respectively, $-z$ has been replaced by 1, and the 1 in the upper right corner has been replaced by $\bar{w}$; z was to be minimized, $\bar{w}$ is to be maximized. The position of the objective function label $\bar{w}$ suffices (except in 4.4 and 5.4) to show that the column equations are being used, but we continue to use the dashed line also.

Both (5) and (6) above are contained in the Tucker schema (7).

$$\begin{array}{c|cccc:l l}
 & y_1 & \cdots & y_n & 1 & \text{Primal} & \text{All} \\
\hdashline
u_1 & a_{11} & \cdots & a_{1n} & b_1 & =-x_1\leq 0 & u_i\geq 0 \\
\vdots & \vdots & & \vdots & \vdots & \quad\vdots & v_j\geq 0 \\
u_m & a_{m1} & \cdots & a_{mn} & b_m & =-x_m\leq 0 & x_i\geq 0 \\
1 & c_1 & \cdots & c_n & d & =z\min & y_j\geq 0 \\
\hline
 & \| & & \| & \| & & \\
\text{Dual} & v_1 & \cdots & v_n & \bar{w}\max & & \\
 & \text{VI} & & \text{VI} & & & \\
 & 0 & & 0 & & &
\end{array} \tag{7}$$

Except for the manner in which the letters $u,\ldots,z$ are assigned to the variables, (7) uses the notation of Balinski and Tucker (1969).

In practice it suffices to write out only one of (5), (6) or (7). The last of these may be simplified by omitting "$=-x_i$" and "$=v_j$" but keeping "≤ 0" and "≥ 0." Thus (7) asserts that the linear functions such as $a_{i1}y_1+\cdots+a_{in}y_n+b_i$ in the *rows are negative* (or zero) or minimized, and the linear functions such as $u_1a_{1j}+\cdots+u_ma_{mj}+c_j$ in the *columns are positive* (or zero) or maximized. This emphasizes the relationship of linear programming to game theory as in (10) of 4.6.

After pivot steps have been performed in (5), say, it acquires a somewhat jumbled form such as

1	y_1	x_1	1
y_2	$a'(y_2, y_1)$	$a'(y_2, x_1)$	$b'(y_2)$
x_2	$a'(x_2, y_1)$	$a'(x_2, x_1)$	$b'(x_2)$
$-z$	$c'(y_1)$	$c'(x_1)$	d'

(8)

This may be acceptable or even preferable in that (as in (a) of 2.6) it makes clear which variables have been exchanged (x_1 and y_2 in (8)); also in applications the x_i and y_j may have very different interpretations (e.g., in (6) of 2.4, 3.6, and 4.4).

On the other hand, the mixture of x_i and y_j as in (8) would be very inconvenient when one wants to give a general description of the pivoting procedure as in 2.4, 3.2 and 3.3. Therefore (9) and (10) of Table 3.1 show an alternate notation in which the combined row vector $(\mathbf{u},\mathbf{v})$ and column vector $(\mathbf{x},\mathbf{y})$ are replaced by $\mathbf{u}=(\mathbf{u}_M,\mathbf{u}_N)$ and $\mathbf{x}=(\mathbf{x}_M,\mathbf{x}_N)$, respectively. Then M and N are redefined after each pivot step so that they always correspond to rows and columns respectively. The index h represents an arbitrary member of the union $M \cup N=\{1,2,\ldots,m+n\}$.

In Table 3.1 equations (1) to (7) are reproduced in compact form for easy reference; (9) is the equivalent of (7) in the alternate notation; (10) summarizes these and a few other notational conventions of this book.

In Table 3.1 it is assumed, as elsewhere in this book if not otherwise stated, that the primal equations correspond to the rows of the tableau and the dual equations to the columns. This is customary and usually does not conflict with other connotations that the words often have, such as (a) the primal is the first of the problems to be considered; (b) the c_j are the costs and the dual variables are shadow prices as in 3.6; (c) the primal has more nonbasic variables than the dual has, so that $n>m$, as in the revised simplex method of 7.1 to 7.3; (d) the primal as given initially may involve equations that lack basic variables while the dual variable may be permitted to be negative as well as positive, as in 6.1 and 6.2.

EXERCISES 3.1

Write out a tableau, a schema, and the dual (for 1 and 2) or the primal (for 3) linear program for the following. DO NOT SOLVE.

1. Minimize $-3y_1 - y_2 + 2y_3$ subject to $\mathbf{y} \geq \mathbf{0}$, $2y_1 + 5y_2 - y_3 \leq 8$, and $-y_1 + 3y_2 + 4y_3 \leq 10$.
2. Maximize $x_1 - x_2 - x_3$ subject to $\mathbf{x} \geq \mathbf{0}$, $4x_1 - 3x_2 + 2x_3 - x_4 = 8$, $-2x_1 + x_2 - x_3 \geq 4$.
3. Maximize $u_1 + 2u_2$ subject to $\mathbf{u} \geq \mathbf{0}$, $3u_1 + u_2 \leq 5$, $u_1 + u_2 \leq 3$, $u_1 + 4u_2 \leq 6$. (Regard this as a dual linear program.)
4. Write down the tableau whose *row* equations are given by (4).
5. Imitate the argument that derived (4) from (2) and thus derive (2) by using $y_1, \ldots, y_n$ as multipliers for the equations in (4).

TABLE 3.1. Summary of Equations and Notations

Primal (row equations) $\mathbf{x}, \mathbf{y} \geq \mathbf{0}$		Dual (column equations) $\mathbf{u}, \mathbf{v} \geq \mathbf{0}$	
$\mathbf{Ay} + \mathbf{b} \leq \mathbf{0}$		$(\mathbf{u}, 1)$ times columns of (2) gives	
$\text{Min } z = \mathbf{cy} + d$	(1)	$\min z = \mathbf{ux} + (\mathbf{uA} + \mathbf{c})\mathbf{y} + \mathbf{ub} + d$	(3)
$\mathbf{x} + \mathbf{Ay} + \mathbf{b} = \mathbf{0}$		$\mathbf{v} = \mathbf{uA} + \mathbf{c}$	
$\min z = \mathbf{cy} + d$	(2)	$\max \bar{w} = \mathbf{ub} + \mathbf{c}$	(4)

$$\begin{array}{c|cc} 1 & y_j & 1 \\ \hline x_i & a_{ij} & b_i \\ -z & c_j & d \end{array} \qquad (5) \qquad\qquad \begin{array}{c|cc} 1 & v_j & \bar{w}(\max) \\ \hline u_i & a_{ij} & b_i \\ 1 & c_j & d \end{array} \qquad (6)$$

$$\begin{array}{c|cc|l} & y_j & 1 & \\ \hline u_i & a_{ij} & b_i & = -x_i \leq 0 \\ 1 & c_j & d & = z \min \\ \hline & \| & \| & \\ & 0 \leq v_j & \bar{w}\max & \end{array} \qquad (7) \qquad\qquad \begin{array}{c|cc|l} & \mathbf{x}_N & 1 & \\ \hline \mathbf{u}_M & \mathbf{A} & \mathbf{b} & = -\mathbf{x}_M \\ 1 & \mathbf{c} & \mathbf{d} & = z \min \\ \hline & \| & \| & \\ & \mathbf{u}_N & \bar{w}\max & \end{array} \qquad (9)$$

Labels for	Rows	Columns	Objective Function	
Primal variables	$\mathbf{x}$ or $\mathbf{x}_M$	$\mathbf{y}$ or $\mathbf{x}_N$	$z = \mathbf{cx}_N + d$ (min)	
Dual variables	$\mathbf{u}$ or $\mathbf{u}_M$	$\mathbf{v}$ or $\mathbf{u}_N$	$\bar{w} = \mathbf{bu}_M + d$ (max)	
Nonbasic variables	$\mathbf{u}$ or $\mathbf{u}_M$	$\mathbf{y}$ or $\mathbf{x}_N$		
Constant terms	b	c		
Arbitrary subscripts	i	j		(10)
Specific subscripts	k	l		
Sets of pivotal subscripts	K	L		
Full sets of subscripts	M	N		
Numbers of subscripts	m	n		

3.2 THE SIMPLEX ALGORITHM: INTRODUCTION

The determination of the x_h and z in the primal of (9) of Table 3.1 will be discussed; then by using the dual (column) interpretation of the final tableau the u_h and $\bar{w}$ also can be read off. In general the complete solution involves two phases:

Phase 1. A feasible solution for the x_h (satisfying all the constraints) is found if possible.

Phase 2. This solution is altered as many times as necessary until it is impossible to make z any smaller or else it is shown that z is unbounded below. The process can be guaranteed to terminate after a finite number of steps.

Only Phase 2 is discussed in this chapter. It is treated first because in a number of interesting problems a feasible solution to one of the problems (primal or dual) is available at a glance, and in other cases the Phase 1 procedure is derived from that of Phase 2. The combined procedure will be discussed in Chapter 6 and thereafter.

The Phase 2 procedure is called the *simplex algorithm* and the two-phase procedure and other variants are called the *simplex method*. These were developed by George B. Dantzig from 1947 onward. Section 7-3 of Dantzig (1963) explains how the name was derived from geometry; a simplex is the simplest of the polyhedra, being the convex hull of $n+1$ points in n dimensions, not contained in any linear variety of less than n dimensions. Nowadays the name also suggests the growth of a great body of theory and practice from the simple idea of constructing a sequence of basic solutions or extreme points such that the value of the objective function changes in one direction only.

For each of the two problems defined by the schema (9) of 3.1, three sets of points $(x_1,\ldots,x_{m+n})$ or $(u_1,\ldots,u_{m+n})$, respectively, may be distinguished:

(a) The *solution set* (a linear variety X' or U') obtained by disregarding the conditions $x_h \geq 0$ or $u_h \geq 0$ and disregarding the optimization of z or $\bar{w}$.

(b) The *feasible region* (a convex polyhedral set X^+ contained in X' or U^+ in U') obtained by disregarding only the optimization.

(c) The *set of optimal solutions* (a convex polyhedral set X^* contained in X^+ or U^* in U^+) obtained by considering all requirements. X^+ and/or X^* may be empty but X' is not. If the optimal solution is unique, X^* consists of a single point $\mathbf{x}^*$.

The simplest way to obtain a point in X' is just to set all the nonbasic variables equal to zero. In the primal of (9) of 3.1 this gives the point x^B with

$$x_i^B = -b_i \quad \text{for } i \text{ in } M, \qquad x_j^B = 0 \quad \text{for } j \text{ in } N, \qquad z = d \tag{1}$$

which is called the primal *basic solution* associated with the schema or the tableau or the basis M. Similarly the dual basic solution u^B has

$$u_i^B = 0 \quad \text{for } i \text{ in } M, \qquad u_j^B = c_j \quad \text{for } j \text{ in } N, \qquad \bar{w} = d \tag{2}$$

If $M = \{1,\ldots,m\}$, (1) and (2) become

$$\left(x_1^B,\ldots,x_m^B,x_{m+1}^B,\ldots,x_{m+n}^B\right) = (-b_1,\ldots,-b_m,0,\ldots,0)$$

$$\left(u_1^B,\ldots,u_m^B,u_{m+1}^B,\ldots,u_{m+n}^B\right) = (0,\ldots,0,c_{m+1},\ldots,c_{m+n})$$

Notice that these primal and dual basic solutions, derived from the same schema, satisfy the following condition, called *complementary slackness*:

$$u_h^B x_h^B = 0 \left(x_h^B = 0 \quad \text{or} \quad u_h^B = 0 \quad \text{or both}\right)(h = 1,\ldots,m+n) \tag{3}$$

The values of the symbol B are in one-to-one correspondence with those of the set M or N in (9) of 3.1 but B is used for the sake of symmetry and to suggest the word "basic."

If a basic solution $\mathbf{x}^B$ also satisfies the nonnegativity requirements $x_h^B \geq 0$, so that it belongs to the feasible region X^+, it is called a *feasible basic solution*. In the present chapter it is assumed that one of the basic solutions (1) and (2) is feasible; that is, that

$$\begin{aligned} &\text{All} \quad b_i \leq 0 \quad \text{(primal feasibility) or} \\ &\text{All} \quad c_j \geq 0 \quad \text{(dual feasibility)}, \end{aligned} \tag{4}$$

so that no Phase 1 is required.

Note the difference in the signs in (4), which is typical of the passage from the primal to the dual and vice versa. The case $b_i \leq 0$ will be discussed in detail and the other will then follow immediately.

In Phase 2 the problem is to move from one feasible solution to another chosen in order to reduce the value of z. All the feasible solutions, and usually many infeasible solutions as well, are obtained by considering all possible nonnegative values for the nonbasic variables x_j with j in N. Since the latter are all zero in the current basic solution (1), the only possible change is to increase one or more nonbasic variables from zero to some positive values.

For simplicity, such increases are made in only one variable at a time, say x_l. Since the schema or tableau expresses z in terms of the (independent) nonbasic variables x_j alone, with coefficients c_j, it is clear that $c_l < 0$ is a necessary and sufficient condition that an increase in x_l will effect a decrease in z. To obtain very simply a more specific selection among the various negative c_j that may be available, it is customary to select (one of) the "most negative" of the c_j:

$$c_l = \min_j \{c_j < 0\} \tag{5}$$

If all $c_j \geq 0$, then it is clear that no feasible changes in the nonbasic variables can decrease z, the current feasible basic solution is an optimal solution, and the calculation is ended. If all $c_j > 0$, it is clear that any feasible change will in fact increase z, and so the existing optimal solution is unique.

If a $c_l < 0$ has been selected, the next question is how large x_l should become. The larger the better, insofar as z is concerned, but the requirement that the solution remain feasible must be considered. Since all the nonbasic x_j (other than x_l) are to remain at zero, the solution is a function of x_l alone and can be read off the row equations of (9) of 3.1:

$$x_i = -b_i - a_{il} x_l \geq 0 \qquad (i \text{ in } M) \tag{6}$$

Here i runs through the set M of subscripts of the basic variables and the feasibility condition $x_i \geq 0$ has been included in (6).

The m conditions (6) divide into two classes distinguished by the sign of a_{il}:

(θ) If $a_{il} \leq 0$, (6) is automatically satisfied and may be ignored, because we are assuming $-b_i \geq 0$ and $x_l \geq 0$.

($+$) If $a_{il} > 0$, (6) is equivalent to $-b_i/a_{il} \geq x_l$, which is consistent with $x_l \geq 0$ since $-b_i/a_{il} \geq 0$. Thus there are two possible conclusions, depending on whether the class ($+$) is empty or not:

(i) If ($+$) is empty, (6) imposes no restriction on x_l, which can be made as large as one pleases, while z becomes as small as one pleases. The calculation is ended. The pure mathematician may regret not being able to exhibit any point $\mathbf{x}$ that yields a minimum value of z. The practical user of linear programming will probably be forced to conclude that there is an error or an inadequacy in his model or in the calculation. Notice that this case is characterized in the tableau by a column l that is exclusively nonpositive, while c_l is in fact strictly negative; this may be described by the scheme

$$\begin{array}{c} \theta \\ \vdots \\ \theta \\ - \end{array}$$

in which the symbol θ means "zero or negative." Notice also that such a column corresponds to a dual equation that cannot have any feasible solution. Similarly, a nonnegative row k having $a_{kj} \geq 0$ for all j and $b_k > 0$, symbolized by

$$\oplus \cdots \oplus +$$

corresponds to a primal equation that admits no feasible solution, while in the dual it implies that $\bar{w}$ can be increased indefinitely by increasing u_k.

(ii) If the class ($+$) is not empty, (6) is equivalent to

$$x_l \leq -b_k/a_{kl} \triangleq \min_i \{-b_i/a_{il} \text{ with } i \text{ in } M \text{ and } a_{il} > 0\} \tag{7}$$

where it is assumed that all $b_i \leq 0$ and k is the index of (one of) the row(s) that yields the minimum ratio. Since x_l should be as large as possible to minimize z, the equality should hold in (7). This means that x_k will be reduced to (or remain at) zero (see (6)) while x_l goes from zero to $|b_k|/a_{kl}$. The values of the other basic variables in the solution will usually change also but will remain nonnegative. The other nonbasic variables remain at zero.

In general, the type of calculation described under (ii) has to be repeated a number of times in order to give nonbasic variables other than x_l a chance to decrease z if they can. The calculation was begun with a tableau and a basic feasible solution uniquely related by the fact that the nonbasic variables in the tableau were zero in the solution. To reestablish such a relationship for a new solution, if it differs from the old, requires a new tableau. Since x_l hopefully has become positive (though it may remain at zero), it should become basic, while x_k (now being zero) can become nonbasic. This is accomplished by pivoting on a_{kl} as in 2.4 or 2.5. Thus we have proved the first two theorems of the following section.

This solution procedure differs from that applied to systems of linear equations in 2.6(b) in that here we are not told (but have to discover by observing the objective function) what set of variables are to be solved for (made basic). Also no columns are dropped here; the tableau retains its original size.

EXERCISES 3.2

In 1 and 2 give a graphical or geometrical determination of the sets X', X^+, and X^*:

1. Min $x_1 - x_2$ subject to $x_1 \geq 0$, $x_2 \geq 0$, $x_1 + x_2 = 2$.
2. Max $x_1 + x_2$ subject to all $x_h \geq 0$, $x_1 + x_2 + x_3 = 3$.

In 3 to 5 determine whether the given solution is basic; if it is, obtain the associated tableau (without any objective function).

3. (3,2,0) in
$$\begin{aligned} -x_1 + 4x_2 + 2x_3 &= 5 \\ 2x_1 - x_2 + 3x_3 &= 4 \end{aligned}$$
4. (0,2,1,1) in
$$\begin{aligned} x_1 + 3x_2 - x_3 + 2x_4 &= 7 \\ 2x_1 + x_2 + x_3 - x_4 &= 2 \\ x_1 - x_2 + 3x_3 - 4x_4 &= -3 \end{aligned}$$
5. (0,0,11,9) in
$$\begin{aligned} x_1 + 3x_2 - x_3 + 2x_4 &= 7 \\ 2x_1 + x_2 + x_3 - x_4 &= 2 \\ 3x_1 - x_2 + 3x_3 - 4x_4 &= -3 \end{aligned}$$
6. Give a discussion similar to that from (5) to (7) but dealing with the maximization of $\overline{w}$ in the dual problem.
7. Explain why $a_{14} = 3$ is selected as the pivot in this tableau, and perform this pivot step not by the five rules of 2.4 or 2.5 but by algebraic operations on the row equations written out.

1	x_3	x_4	1
x_1	2	3*	−1
x_2	−1	1	−2
$-z$	1	−2	0

3.3 THE SIMPLEX ALGORITHM: THEORY AND USE

First the primal and the dual tableau will be written down in compact form for easy reference:

1	$\mathbf{x}_N$	1
$\mathbf{x}_M$	$\mathbf{A}$	$\mathbf{b}$
$-z$	$\mathbf{c}$	d

(row equations)

1	$\mathbf{u}_N$	$\overline{w}$
$\mathbf{u}_M$	$\mathbf{A}$	$\mathbf{b}$
1	$\mathbf{c}$	d

(column equations)

(1)

with $\mathbf{x} \geq 0$ and $\mathbf{u} \geq 0$. In terms of a few individual elements, including the pivot a_{kl}, these become

$$\begin{array}{c|ccc} 1 & x_l & x_j & 1 \\ \hline x_k & a_{kl}^* & a_{kj} & b_k \\ x_i & a_{il} & a_{ij} & b_i \\ -z & c_l & c_j & d \end{array} \qquad \begin{array}{cccc} 1 & u_l & u_j & \bar{w} \\ \hdashline u_k & a_{kl}^* & a_{kj} & b_k \\ u_i & a_{il} & a_{ij} & b_i \\ 1 & c_l & c_j & d \end{array} \tag{1'}$$

Here $z = \mathbf{c}\mathbf{x}_N + d$ is to be minimized and $\bar{w} = \mathbf{b}\mathbf{u}_M + d$ is to be maximized. The basic variables are $-z$, $\bar{w}$, the components x_i of $\mathbf{x}_M$, and the components u_j of $\mathbf{u}_N$.

Of the following theorems, 1 and 2′ have been proved in the preceding section, and 2″ is proved analogously or by applying 2′ to the negative transpose of the tableau. Theorems 1, 2′, 2″, 8, and 9 depend on the constraints $\mathbf{x} \geq \mathbf{0}$, $\mathbf{u} \geq \mathbf{0}$ assumed in this chapter; the other theorems do not, although the proof of this fact may require later ideas from 6.1 and 6.6.

THEOREM 1 There is *no optimal solution* for either primal or dual if one encounters a tableau containing either:

(a_1) A nonpositive column l (other than the last column $\mathbf{b}$, d) having $c_l < 0$ and $a_{il} \leq 0$ for all i in M; or

(a_2) A nonnegative row k (other than the last row $\mathbf{c}, d$) having $b_k > 0$ and $a_{kj} \geq 0$ for all j in N; or

(a_3) Both of these.

These conditions are depicted by

$$\begin{array}{c} \theta \\ \vdots \\ \vdots \\ \theta \\ - \end{array} \quad \text{or} \quad \oplus \cdots \oplus + \quad \text{or} \quad \begin{array}{c} \theta \\ \vdots \\ \oplus \cdots 0 \cdots \oplus + \\ \vdots \\ \theta \\ - \end{array} \tag{2}$$

respectively. (a_1) or (a_2) always occurs whenever (3) or (5) following is unable to supply a pivot, and they may occur (and should be heeded) on other occasions as well. More general criteria not requiring $\mathbf{x} \geq 0$ and $\mathbf{u} \geq 0$ will be given in Chapter 6.

Nonnegative columns and nonpositive rows (the opposite of (2)) represent redundant constraints that could be omitted; however, this is not recommended except in the initial tableau.

THEOREM 2′. (Primal) Simplex Algorithm (Dantzig) If all $b_i \leq 0$, z is decreased from d to $d - b_k c_l / a_{kl}$ (or left unchanged, if $b_k = 0$) while primal feasibility is preserved by pivoting† on $a_{kl} > 0$. Here the pivotal column l with $c_l < 0$ is selected first according to $\min c_j$ or otherwise, and then the pivotal row k is selected

†See the note on page 66.

according to the criterion

$$-b_k/a_{kl} \triangleq \min_i \{-b_i/a_{il} \text{ with } a_{il}>0\} \tag{3}$$

The pattern of signs is: or more briefly

$$\begin{array}{lccc} & & \theta & \\ \text{Pivot: } + & & \theta & b_i \\ & & \theta & \\ c_j\text{: } \pm & - & \pm & \end{array} \qquad \begin{array}{cc} + & \theta \\ - & \end{array} \tag{4}$$

The calculation stops if $a_{il} \leq 0$ for all i in M or if $c_j \geq 0$ all for j in N.

THEOREM 2″. Dual Simplex Algorithm (Lemke, 1954) If all $c_j \geq 0$, $\bar{w}$ is increased from d to $d - b_k c_l/a_{kl}$ (or left unchanged, if $c_l = 0$) while the dual feasibility $c_j \geq 0$ is preserved by pivoting† on $a_{kl} < 0$. Here the pivotal row k with $b_k > 0$ is selected first according to $\max b_i$ or otherwise, and then the pivotal column l is selected according to the criterion

$$c_l/(-a_{kl}) \triangleq \min_j \{c_j/(-a_{kj}) \text{ with } a_{kj}<0\} \tag{5}$$

The pattern of signs is or more briefly

$$\begin{array}{lc} & \pm \\ \text{Pivot: } - & + \, b_k \\ & \pm \\ c_j\text{: } \oplus\oplus\oplus & \end{array} \qquad \begin{array}{cc} - & + \\ \oplus & \end{array} \tag{6}$$

The calculation stops if $a_{kj} \geq 0$ for all j in N or if $b_i \leq 0$ for all i in M.

Remark. Since $z = \bar{w} = d$, Theorem 2″ will increase z also. This is unavoidable since the primal solution is not yet feasible in this case, while every reasonable change increases z, or leaves it unchanged, because all $c_j \geq 0$.

There is no voluntary choice between Theorems 2′ and 2″; the first must be used throughout a series of pivot steps that begins with $\mathbf{b} \leq 0$, and the second must be used throughout when $\mathbf{c} \geq 0$. In discussing the rationale of the method one should use the primal (row equation) labels x, z in Theorem 2′ and the dual labels $u, \bar{w}$ in Theorem 2″. However, in solving exercises there is no need to continue this restriction; either kind of labels will be used with either theorems.

In 2.5 a method was described for checking the arithmetic of pivoting. Obviously this method says nothing about the appropriateness of the pivot that is selected. In applying the primal simplex algorithm, if primal feasibility is lost as a result of some b_i becoming positive, this indicates that (3) has not been used correctly in determining k; however, it is commonplace for some previously feasible c_j to become negative. Similarly in the dual simplex, if dual feasibility is lost because some c_j

†See 2.4 and 2.5. Note that no b_i, c_j, or d is ever considered as a pivot, and that the actual value of d never has any influence on the selection of the pivot. Variables z and $\bar{w}$ should remain basic because they are not sign restricted and they are unlikely to be zero in the optimal solution, as nonbasic variables are.

becomes negative, (5) has not been used correctly in determining l; however, a previously feasible b_i may very well become positive. Equation (3) or (5) does not necessarily determine k or l uniquely; ties may have to be broken by an arbitrary decision or as in 6.6.

It may well happen that Theorems 1 and 2 (2′ or 2″) are both found to be applicable to the same tableau; then of course Theorem 1 takes precedence and the calculation is stopped. To be sure, computer programs usually find it inconvenient to apply Theorem 1; they keep going until Theorem 2 fails to yield any more pivots.

It may be helpful to have a statement that combines Theorems 2′ and 2″, giving a description that applies equally to the primal or dual form of the simplex algorithm.

The Simplex Algorithm

(b_1) It is assumed that the constraints include $\mathbf{x} \geq \mathbf{0}$ and $\mathbf{u} \geq \mathbf{0}$, and that either the primal basic solution $\mathbf{x}_M = -\mathbf{b}$, $\mathbf{x}_N = \mathbf{0}$ or the dual basic solution $\mathbf{u}_M = \mathbf{0}$, $\mathbf{u}_N = \mathbf{c}$ satisfies these constraints. Thus $\mathbf{b} \leq \mathbf{0}$ *or* $\mathbf{c} \geq \mathbf{0}$. More general problems will be considered in Chapter 6.

(b_2) Try to select an infeasible $c_l < 0$ or $b_k > 0$, usually the *most infeasible* (furthest from zero). If no such element exists, the tableau is optimal; *stop*.

(b_3) Try to select a pivot $a_{kl} \neq 0$, of sign *opposite* to the sign of the infeasible c_l or b_k, and in its column l or row k. If no such pivot exists, there is no optimal solution; *stop*. Note that the ratio criterion (3) or (5) has the prospective pivots in the denominator (that's why they can't be zero) and selects the *smallest nonnegative ratio*, the absolute values of the negative b_i or a_{kj} being used. The patterns of signs are

a_{kl}	b_k	$=$	$+$	θ	or	$-$	$+$
c_l			$-$			$\oplus$	

(b_4) Pivot on a_{kl} and return to (b_2). If all has been done correctly, (b_1) will still be true with the same kind of feasibility.

At this point the reader has the option of working some of the exercises in the following Section 3.4 before proceeding with the (rather easy) theory in the present Section 3.3.

THEOREM 3 The simplex algorithm can be made to terminate after a *finite number of pivot steps.*

Proof Since the number of essentially different tableaus is at most $(m+n)!/m!n!$, the theorem will be true if no tableau is obtained more than once (or even a finite number of times). This is almost guaranteed by the fact that the tableau entry d (the common value of z and $\bar{w}$) never increases in Theorem 2′ and never decreases in Theorem 2″, and one or the other is used for the whole procedure. The only exception is that d may indeed remain forever at the same nonoptimal value. Section

6.6 describes a way to exclude this phenomenon, but it is usually considered unnecessary in practice. Section 6.1 will eliminate the dependence on $\mathbf{x}\geq\mathbf{0}$, $\mathbf{u}\geq\mathbf{0}$ and on $\mathbf{b}\leq\mathbf{0}$ or $\mathbf{c}\geq\mathbf{0}$.

THEOREM 4 The following statements are logically equivalent (all true or all false):

(c_1) The primal problem has an optimal solution.
(c_2) The dual problem has an optimal solution.
(c_3) Each problem has a feasible solution of some sort.
(c_4) Both problems have basic optimal solutions $\mathbf{x}^*$ and $\mathbf{u}^*$ which are obtained from the same optimal tableau and make $z^*=\bar{w}^*$, thus:

	Primal	Dual
Row indices i	$x_i^*=-b_i^*\geq 0$	$u_i^*=0$
Column indices j	$x_j^*=0$	$u_j^*=c_j^*\geq 0$
Objective function	$z^*=d^*$	$\bar{w}^*=d^*$

(7)

Proof Each statement is true if the simplex algorithm ends with an optimal tableau (having $\mathbf{b}\leq\mathbf{0}$ *and* $\mathbf{c}\geq\mathbf{0}$); each statement is false if the algorithm ends as in Theorem 1. There is no other possibility since Theorem 3 guarantees termination.

The signs in an optimal tableau may be depicted as

$$\begin{array}{cccc} \pm & \cdots & \pm & \theta \\ \vdots & & \vdots & \vdots \\ \pm & \cdots & \pm & \theta \\ \oplus & \cdots & \oplus & \pm \end{array} \quad \mathbf{b}\leq\mathbf{0} \qquad \text{with } \mathbf{c}\geq\mathbf{0} \tag{8}$$

Note that its negative transpose is again optimal, as it must be.

It is sometimes convenient to say that $z^*=+\infty$ (the worst possible result in a minimization problem) when no feasible primal solution $\mathbf{x}$ exists, and to say that $z^*=-\infty$ when feasible primal solutions do exist and make z unbounded below. Since $\pm\infty$ are not numbers in the ordinary sense (and no $\mathbf{x}^*$ is associated with them), these conventions are regarded as consistent with the assertion that z^* does not exist in such cases. In the dual problem $\bar{w}^*=-\infty$ means that no feasible $\mathbf{u}$ exists, and $\bar{w}^*=+\infty$ means that $\bar{w}$ is unbounded above. This still makes $z^*=\bar{w}^*$ except when neither the primal nor the dual has a feasible solution; then one might say $z^*=\pm\infty$.

The next three theorems indicate ways in which the concept of *orthogonality* applies to linear programs.

THEOREM 5 Let the linear varieties X' and U' be the set of solutions (not necessarily feasible) for the primal and the dual problems respectively, and let $\mathbf{x}^0$ be a point of X' and $\mathbf{u}^0$ a point of U'. Then

$$X = X' - \mathbf{x}^0 = \{\mathbf{x} - \mathbf{x}^0 \text{ with } \mathbf{x} \text{ in } X'\} \text{ and}$$

$$U = U' - \mathbf{u}^0 = \{\mathbf{u} - \mathbf{u}^0 \text{ with } \mathbf{u} \text{ in } U'\} \tag{9}$$

are vector spaces that are the solution sets for the homogeneous constraints obtained by replacing the constant terms b_i and c_j by zero. Then X and U are perpendicular (orthogonal) and the same description is applied to X' and U'. (For example, a line and a plane in three dimensions can be perpendicular without containing the origin.) Equation (21) of 2.3 gives the proof.

THEOREM 6 The basic solutions (1) and (2) of 3.2 from the same tableau, and any pair of primal and dual optimal solutions, whether basic or not, are orthogonal vectors without any need to shift the origin to eliminate the constants as in Theorem 5, nor to include the constants in the matrix as in Theorem 7. In fact a stronger result holds: each term in the scalar product is zero, thus:

$$\text{All } u_h^B x_h^B = u_h^* x_h^* = 0 \ (u_h = 0 \text{ or } x_h = 0 \text{ or both}). \tag{10}$$

(Theorem 1 of 5.6 gives a still stronger result.)

The property (10) is called *complementary slackness* because the set M of indices h for which x_h^B is slack (basic, possibly positive) and the set N for which u_h^B is slack are complementary subsets of $1, \ldots, m+n$. Equation (10) is obvious for $u_h^B x_h^B$; for $u_h^* x_h^*$ it follows from (12) and Theorem 7* below. Figure 3.3A illustrates Theorems 5 and 6 in two and three dimensions.

If the matrix $\mathbf{A}$ is enlarged by including the elements b_i, c_j and d, and the basic variables $-z$ and $\bar{w}$ and the nonbasic constants 1 are included in the vectors, then the orthogonality condition becomes

$$(\mathbf{u}_M, \mathbf{u}_N, 1, \bar{w})(\mathbf{x}_M, \mathbf{x}_N, -z, 1) = \mathbf{ux} - z + \bar{w} \tag{11}$$

$$= (\mathbf{u}_M, \mathbf{u}_M\mathbf{A} + \mathbf{c}, 1, \mathbf{u}_M\mathbf{b} + d)(-\mathbf{A}\mathbf{x}_N - \mathbf{b}, \mathbf{x}_N, -\mathbf{c}\mathbf{x}_N - d, 1)$$

$$= -\mathbf{u}_M\mathbf{A}\mathbf{x}_N - \mathbf{u}_M\mathbf{b} \quad + \mathbf{u}_M\mathbf{A}\mathbf{x}_N + \mathbf{c}\mathbf{x}_N \quad - \mathbf{c}\mathbf{x}_N - d \quad + \mathbf{u}_M\mathbf{b} + d \quad = 0$$

Here the first vector in each product is a row and the last is a column. This proves Theorem 7.

THEOREM 7 If $\mathbf{x}$ and $\mathbf{u}$ are feasible solutions (of any sort) of the primal and the dual problems respectively, then

$$\mathbf{ux} = z(\mathbf{x}) - \bar{w}(\mathbf{u}) = \mathbf{c}\mathbf{x}_N - \mathbf{u}_M\mathbf{b} \geq 0 \tag{12}$$

Here $\mathbf{b}$ and $\mathbf{c}$ are the entries in any one tableau whose row and column indices are M and N, respectively.

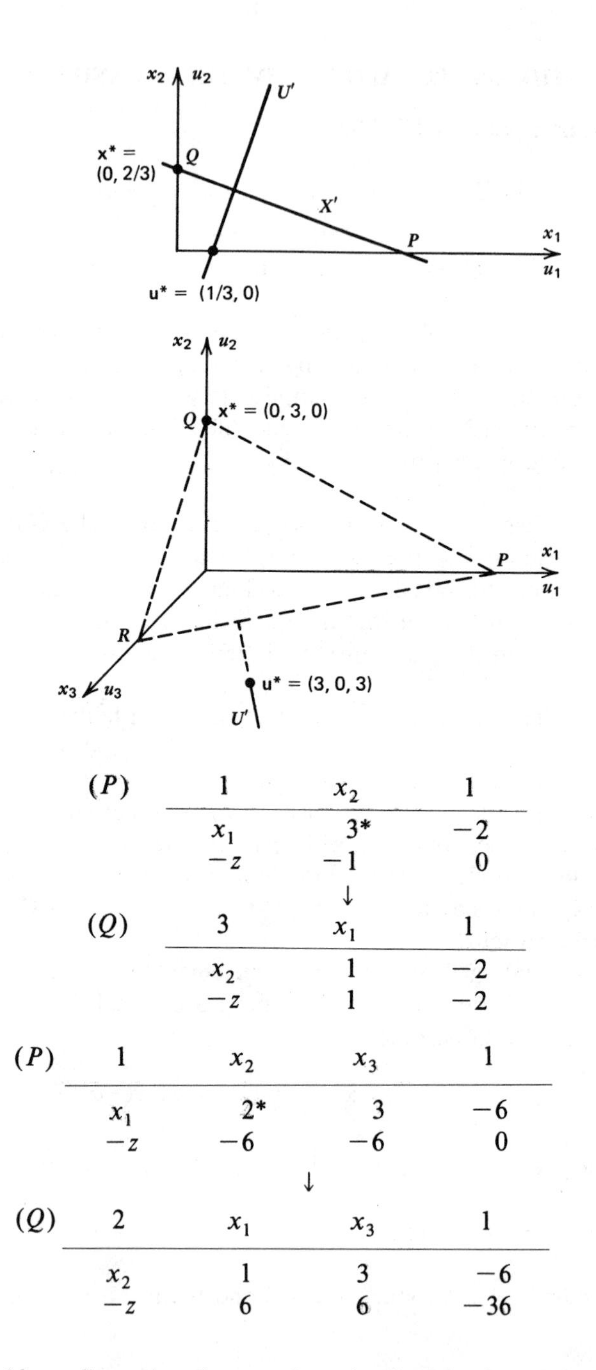

(P)	1	x_2	1
	x_1	3*	−2
	$-z$	−1	0

↓

(Q)	3	x_1	1
	x_2	1	−2
	$-z$	1	−2

(P)	1	x_2	x_3	1
	x_1	2*	3	−6
	$-z$	−6	−6	0

↓

(Q)	2	x_1	x_3	1
	x_2	1	3	−6
	$-z$	6	6	−36

In the second (three-dimensional) example, primal X' is plane PQR; feasible region is triangle PQR. Dual U' is line through $\mathbf{u}^*$ orthogonal to plane PQR, which it appears to intersect below the u_1u_3 plane, in which optimal $\mathbf{u}^*$ lies. However, X' and U' really lie in different spaces, and so their apparent point of intersection $(18, -6, 12)/7$ has no significance. Points P and Q are the basic primal solutions for the tableaus.

FIGURE 3.3A Primal-dual orthogonality (Theorems 5 and 6).

THEOREM 7* Feasible solutions $\mathbf{x}$ for the primal and $\mathbf{u}$ for the dual are optimal if and only if they make $z(\mathbf{x})=\bar{w}(\mathbf{u})$ or equivalently $\mathbf{bu}_M=\mathbf{cx}_N$.

More specifically, $\mathbf{x}$, z and $\mathbf{u}, \bar{w}$ are optimal solutions for the primal and the dual respectively of any tableau (1) in this chapter if and only if the following are all satisfied:

(d_1) $x_h \geq 0$ and $u_h \geq 0$ for $h=1,\ldots,m+n$.

(d_2) $u_h x_h = 0$ for $h=1,\ldots,m+n$.

(d_3) $\mathbf{cx}_N + d = z = \bar{w} = \mathbf{bu}_M + d$.

(d_4) $\mathbf{x}$ satisfies the row equations and $\mathbf{u}$ satisfies the column equations. (The last row and column are covered by (d_3)).

Proof (d_1) and (d_4) are the constraints of the linear program. (d_3) follows from Theorem 4 and the definition of z and $\bar{w}$. (d_2) follows from the other conditions and Theorem 7. Conversely, if $\mathbf{u}$ and $\mathbf{x}$ are feasible by (d_1) and (d_4), then

$$\bar{w}(\mathbf{u}) \leq \max \bar{w} = \min z \leq z(\mathbf{x}) \tag{13}$$

Evidently (d_3) implies that all the terms in (13) are equal and the solutions are optimal.

Theorem 7 shows that (d_2) is equivalent to part of (d_3) (i.e., $\mathbf{cx}_N = \mathbf{bu}_M$) when ($d_1$) and ($d_4$) are satisfied but not in general. In any case it is desirable to check all four conditions. If one of the simpler conditions (d_1) to (d_3) is found to be violated, there will be no need to check the more complex condition (d_4). A compact format for verifying (d_3) and (d_4) will be illustrated in 3.5. It could happen that $\mathbf{x}$ has more than m positive components or $\mathbf{u}$ more than n. This indicates that $\mathbf{x}$ or $\mathbf{u}$ is a nonbasic solution and, if optimal, it is not the only optimal solution.

Why the Dual Problem Is Important

(e_1) $\mathbf{u}^*$ as well as $\mathbf{x}^*$ is needed to apply Theorem 7*, which is the easiest available method for checking the alleged optimality of a solution, basic or nonbasic. (The latter is important in 8.8.)

(e_2) The relation $z^* = \bar{w}^* = \mathbf{bu}_M^* + d$ shows how z^* is affected by changes in $\mathbf{b}$, provided $\mathbf{A}$ and $\mathbf{c}$ are unchanged and the change in $\mathbf{b}$ is small enough to leave the optimal basis M^* unchanged. (Then $\mathbf{u}^*$ is unchanged.)

(e_3) If $\mathbf{u}$ is only known to be feasible and not necessarily optimal, $\bar{w}(\mathbf{u}) = \mathbf{bu}_M + d$ is still a lower bound for z^* according to (13). Similarly a feasible $\mathbf{x}$ gives $\mathbf{cx}_N + d$ as an upper bound.

(e_4) In 4.4 we will find that the dual problem for a matrix game yields the optimal strategy for the player who chooses the rows in the matrix.

(e_5) The dual problem underlies the theory, though not the application, of the dual simplex algorithm (Theorem 2″).

(e_6) While solving the primal problem, the simplex method also solves the dual with no additional work.

(e_1) and (e_2) are discussed further in 3.5 and 3.6.

The next theorem identifies basic feasible solutions and extreme points (defined in 2.2).

THEOREM 8 For each of the following four types, the set of all nonnegative solutions forms a convex set and has any point $\mathbf{x}^B$ as an *extreme point* if and only if $\mathbf{x}^B$ is a *basic feasible solution* of the specified type: primal feasible, dual feasible, primal optimal, dual optimal.

Proof The analogy between the primal and the dual makes it sufficient to consider the primal. If the set in question is empty, the theorem is trivially true. Otherwise $z^* = \min z$ is unique if it exists. Then the set of optimal solutions is the feasible region obtained by adding the constraint $z = \mathbf{c}\mathbf{x}_N + d = z^*$ to the others. Thus it suffices to consider nonempty primal feasible regions, which are convex by Theorem 2 of 2.2, and to show that a nonnegative solution is not basic if and only if it is not extreme.

The proof uses the facts that the nonbasic variables are zero in the basic solution; these nonbasic variables uniquely determine the values of the basic variables; and the nonbasic value 0 is an extreme point for the values of the (nonnegative) variables. Thus the zero components of a basic solution must uniquely determine the others. Now suppose that $\mathbf{x} \geq \mathbf{0}$ is not extreme and hence $\mathbf{x} = (\mathbf{x}' + \mathbf{x}'')/2$ where $\mathbf{x}'$ and $\mathbf{x}''$ are distinct feasible points. Since $\mathbf{x}' \geq \mathbf{0}$ and $\mathbf{x}'' \geq \mathbf{0}$, they must have zero components wherever $\mathbf{x}$ does, but then the zero components of $\mathbf{x}$ do not uniquely determine the remaining components and $\mathbf{x}$ is not basic.

Conversely, suppose $\mathbf{x}$ is not basic. Then Theorem 8 of 2.7 (with M' = the set of nonzero components of $\mathbf{x}$) implies there is a feasible $\mathbf{x}' \neq \mathbf{x}$ such that $x'_h = 0$ whenever $x_h = 0$. Then the ray from $\mathbf{x}'$ through $\mathbf{x}$ will contain feasible points $\mathbf{x}''$ beyond $\mathbf{x}$ (there may or may not be a most distant feasible $\mathbf{x}''$). This is true by Theorem 2 or 3 of 2.1 with $\mathbf{x}'' = \mathbf{x}' + t(\mathbf{x} - \mathbf{x}')$ and $t > 1$. When $t = 1$ one obtains $\mathbf{x}$; the components that are zero in $\mathbf{x}$ remain zero for all t while the components that are positive in $\mathbf{x}$ remain nonnegative for some $t > 1$. Thus $\mathbf{x}$ lies between the feasible points $\mathbf{x}'$ and $\mathbf{x}''$, and $\mathbf{x}$ is not extreme. As illustrated in Figure 3.3B, $\mathbf{x}'$ or $\mathbf{x}''$ may or may not have more zero components than $\mathbf{x}$ has; they cannot have nonzeros where $\mathbf{x}$ has a zero.

THEOREM 9 If a linear program in nonnegative variables has an optimal solution (primal or dual), it has an optimal solution at an extreme point of the feasible region.

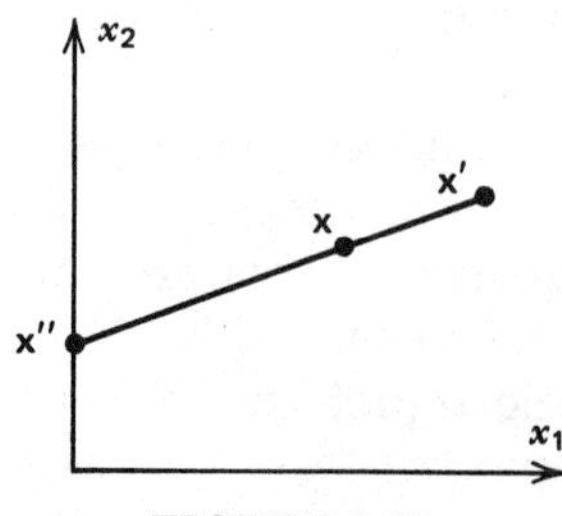

FIGURE 3.3B

This is an immediate corollary of Theorems 4 and 8. It is also true for the minimization of a concave function. The conclusion is that in seeking an optimum one need only look at the extreme points. Since the extreme points of a linear program are finite in number and any one of them is relatively easy to calculate and visualize, the conclusion has a strong appeal, but a *caveat* is in order. Whether or not the fact is explicitly recognized, complex real-life situations may well involve multiple objective functions (considered briefly in 8.8) or the minimization of convex or more general functions, in which cases one cannot restrict attention to extreme points. Of course, the omission of relevant constraints or variables may also have serious consequences.

EXERCISES 3.3

By inspection of each of the following tableaus determine whether it is already optimal or whether Theorem 1, 2′ or 2″ is immediately applicable. If Theorem 2′ or 2″, determine the pivotal indices k and l. All $x_h \geq 0$, $u_h \geq 0$.

1.

1	x_3	x_4	1
x_1	2	−3	2
x_2	−1	1	4
$-z$	−3	−1	0

2.

1	x_3	x_4	1
x_1	−1	−3	3
x_2	−2	1	2
$-z$	4	1	1

3.

1	x_3	x_4	1
x_1	−4	2	−2
x_2	3	0	1
$-z$	−1	−2	3

4.

1	u_3	u_4	$\overline{w}$
u_1	1	6	−3
u_2	−4	−3	−2
1	−2	4	−1

5.

1	u_3	u_4	$\overline{w}$
u_1	2	1	−4
u_2	−1	0	−5
1	3	2	3

6.

1	x_3	x_4	1
x_1	1	−1	−3
x_2	−2	0	−1
$-z$	1	−2	−1

By testing the given values in the given tableau, what does (13) permit you to conclude about z^*?

7. $\mathbf{x}_{3,4,5} = (3, 1, 2)$
$\mathbf{u}_{1,2} = (2, 3)$

1	x_3	x_4	x_5	1
x_1	2	1	−3	0
x_2	−1	−3	−2	8
$-z$	−1	8	13	0

8. $\mathbf{x}_{3,4,5} = (3, 1, 2)$
$\mathbf{u}_{1,2} = (1, 3)$

1	u_3	u_4	u_5	$\overline{w}$
u_1	1	3	−4	1
u_2	−1	1	−1	2
1	3	−4	8	2

Which of the following pairs of solutions are optimal for the given linear program? If not, why not? (Use Theorem 7*, not pivoting.)

9. (a) $\mathbf{x}=(1,0,2,4,0,0)$
$\mathbf{u}=(0,2,0,1,3,0)$
(b) $\mathbf{x}_{4,5,6}=(1,1,1)$
$\mathbf{u}_{1,2,3}=(1,3,1)$
(c) $\mathbf{x}=(0,0,0,1,2,1)$
$\mathbf{u}=(3,2,2,0,0,0)$

1	x_4	x_5	x_6	1
x_1	3	-1	2	-3
x_2	1	2	-3	-2
x_3	-2	-1	2	2
$-z$	-8	2	-5	0

10. (a) $\mathbf{x}=(0,0,0,19,16)/13$
$\mathbf{u}=(47,10,57,0,0)/13$
(b) $\mathbf{x}=(0,0,-1.6,3.8,0)$
$\mathbf{u}=(2.6,0.1,0,0,0)$
(c) $\mathbf{x}=(0,0,1,0,2)$, $\mathbf{u}=(2,1,0,3,0)$.

1	u_3	u_4	u_5	$\overline{w}$
u_1	3	1	-2	1
u_2	2	4	5	-12
1	-8	-3	-1	0

Optimize 11 and 12 by calculating z at each *feasible* basic solution and choosing the smallest. (Theorem 4.)

11. Min $2x_1+x_2$ subject to $x_1\geq 0$, $x_2\geq 0$, $x_1+x_2\geq 3$, $x_1+3x_2\geq 5$.
12. Min $z=-x_1-x_2-x_3$ subject to $\mathbf{x}\geq 0$, $x_1+2x_2+x_4=5$, $x_2+3x_3+x_5=7$.
13. By inspection find simple feasible primal and dual solutions of the given tableau. What can you conclude?

1	x_3	x_4	x_5	1
x_1	3	1	3	-2
x_2	2	-2	1	1
$-z$	-2	-1	-2	0

For each $\mathbf{x},\mathbf{u}$ pair, determine whether it could possibly be optimal, that is, in case it also satisfies the (unspecified) row and column equations. If not, state why. These are nine independent problems despite the shared data:

$\mathbf{b}_{4,5,6}=(2,-3,1)$ $\qquad$ $\mathbf{c}_{1,2,3}=(3,1,2)$ in 14, 16
$\mathbf{b}_{1,2}=(3,2)$ $\qquad$ $\mathbf{c}_{3,4,5}=(1,4,5)$ in 15

	$\mathbf{x}$	$\mathbf{u}$
14.	(a) $(1,0,2,0,0,3)$	$(0,1,0,2,-1,0)$
	(b) $(0,2,1,0,3,0)$	$(3,0,0,1,0,1)$
	(c) $(1,0,1,2,0,0)$	$(0,1,1,0,0,5)$
	(d) $(1,1,1,0,1,0)$	$(0,0,0,2,0,2)$
15.	(a) $(0,0,1,1,1)$	$(2,2,0,0,0)$
	(b) $(-3,-2,0,0,0)$	$(0,0,1,4,5)$
	(c) $(2,0,0,1,0)$	$(0,2,1,0,3)$
	(d) $(0,3,0,4,0)$	$(1,0,3,0,5)$

Exercise 15(d) can't satisfy the row and column equations: why?

16.	$(0,0,0,0,0,1)$	$(0,0,0,3,2,0)$

17. Briefly list the facts that establish the validity of the (primal and dual) simplex algorithms.

3.4 EXAMPLES FOR THE SIMPLEX ALGORITHM

This section gives six numerical examples of the use of the tableau and the simplex algorithm. It is understood that all variables except $z=-\bar{z}$ and $\bar{w}=-w$ are required to be nonnegative. Practical applications will be discussed in 3.6 and 3.7.

Example 1

Maximize $\bar{z}=x_1+x_2-3x_3+2$ subject to $-3x_1-5x_2+2x_3+6\geq 0$, $2x_1\leq 1$.

Slack variables $x_4, x_5\geq 0$ are inserted into the inequalities to convert them into equations. If each is inserted on the small side of the inequality sign, it will have coefficient $+1$; if there are some terms on the other side, they should be shifted over with reversed signs. This will give the equation in the form in which it is copied into the tableau. Of course an equation can always be multiplied throughout by -1 when necessary to conform to the conventions of the tableau, and this is done to the equation for $\bar{z}$ to convert it to a minimizing problem.

The resulting program and tableau are

minimize $z=-x_1-x_2+3x_3-2$
subject to $x_4+3x_1+5x_2-2x_3-6=0$
and $x_5+2x_1 \qquad -1=0$

Check	1	x_1	x_2	x_3	1
1	x_4	3	5	−2	−6
2	x_5	2*	0	0	−1
0	$-z$	−1	−1	3	−2

As in 2.5, the check column is the sum of all the terms obtained by putting every variable (including $-z$) equal to 1.

This tableau has $b_4=-6\leq 0$ and $b_5=-1\leq 0$ and so the primal simplex algorithm (Theorem 2′ of 3.3) applies. The first two columns tie for $\min c_j=-1$, and the first (that of x_1) is chosen arbitrarily. Within this column the pivot 2* is determined by $\min\{6/3, 1/2\}=1/2$. Performing the pivot step as in 2.5 gives

Check	2	x_5	x_2	x_3	1
−4	x_4	−3	10*	−4	−9
2	x_1	1	0	0	−1
2	$-z$	1	−2	6	−5

In this tableau the check column is calculated both by the rules of 2.5 and also by putting all variables equal to 1. Thus in the last row one has

$$2=[2\cdot 0-2\cdot(-1)]/1=2\cdot 1+1\cdot 1+1\cdot(-2)+1\cdot 6+1\cdot(-5)$$

There is now only one $c_j<0$, that is, $c_2=-2/2$, and in this column only one positive pivot 10*. This pivot gives

Check	10	x_5	x_4	x_3	1
−4	x_2	−3	2	−4	−9
10	x_1	5	0	0	−5
6	$-z$	2	2	26	−34

Since now all $c_j \geq 0$ (and all $b_i \leq 0$ as they always must be in the primal simplex algorithm), this tableau is optimal, with $z^*=-34/10$. The remainder of the last column, with the minus signs dropped and divided by the denominator 10, gives the nonzero components of $\mathbf{x}^*=(.5,.9,0,0,0)$. The last row gives the dual solution $\mathbf{u}^*=(0,0,2.6,.2,.2)$. The original problem had $\bar{z}=-z$ so that $\bar{z}^*=3.4$. ■

Example 2

Minimize $z=x_3+3x_5-2$ subject to $x_1+2x_3+x_4-x_5=1$ and $x_3+3x_4+x_5 \geq 4$. After introducing a nonnegative x_2 to take up the slack in the second constraint one has

$$x_1 \quad +2x_3+ \ x_4-x_5-1=0$$

$$x_2- \ x_3-3x_4-x_5+4=0$$

These are now in the form of row constraints with basic variables x_1 and x_2. Thus the initial tableau is

Check	1	x_3	x_4	x_5	1
2	x_1	2	1	−1	−1
0	x_2	−1	−3*	−1	4
3	$-z$	1	0	3	−2

Since all the c_j in the last row are nonnegative, the problem is dual feasible and Theorem 2″ of 3.3 applies. We have to look for the largest (positive) b_i, which is $b_2=4$ in this case. In this row x_2 we must find a pivot of opposite sign, of which three are available. The minimum value of $c_j/|a_{2j}|$ for $a_{2j}<0$ is zero, occurring in column x_4. Thus $a_{24}=-3$ is the pivot. Applying the pivot rules in Section 2.5 gives the next tableau

Check	−3	x_3	x_2	x_5	1
−6	x_1	−5	−1	4	−1
0	x_4	−1	1	−1	4
−9	$-z$	−3	0	−9	6

Since the pivot rules depend very much on the signs of the elements, the negative denominator −3 could be confusing in a hand calculation. Hence we reverse the signs of the denominator and all its numerators, which includes every number (but no labels) in the tableau except the 1 that labels the constants column. In practice this requires neither rewriting the tableau nor erasing, but only unwavering attention while one systematically changes the signs, inserting minus signs before the previously positive elements, and making vertical strokes through the old minus signs to convert them to pluses. The result is

Check	+3	x_3	x_2	x_5	1
+6	x_1	+5	+1	−4*	+1
0	x_4	+1	−1	+1	−4
+9	$-z$	+3	0	+9	−6

Now $b_1 = 1/3$ is positive, and the only negative pivot in this row is $a_{15} = -4/3$. Performing the pivot operation and changing signs again gives

Check	+4	x_3	x_2	x_1	1
−6	x_5	−5	−1	−3	−1
+2	x_4	+3	−1	+1	−5
+30	$-z$	+19	+3	+9	−5

Since now all $b_i \leq 0$ (and all c_j have remained nonnegative, as they must if no mistake is made), the optimal tableau has been obtained. It shows that $z^* = \bar{w}^* = -5/4$, $\mathbf{x}^* = (0,0,0,5,1)/4$, $\mathbf{u}^* = (9,3,19,0,0)/4$. ■

Example 3

Optimize

Check	1	x_3	x_4	x_5	1
−4	x_1	−3	−2	1*	−1
8	x_2	7	3	−2	−1
−1	$-z$	1	1	−1	−3

This tableau fails to be optimal only because of the negative $c_5 = -1$, and Theorem 2′ applies. The only pivot of opposite sign available in this column is the starred element $a_{15} = 1$, which gives

Check	1	x_3	x_4	x_1	1
−4	x_5	−3	−2	1	−1
0	x_2	1	−1	2	−3
−5	$-z$	−2	−1	1	−4

Primal feasibility ($b_i \leq 0$) has been preserved as it should be. The most negative c_j is $c_3 = -2$, which seems to call for $a_{23} = 1$ as a pivot. However, $c_4 = -1$ is also negative and has no pivot of opposite sign in its (nonpositive) column. This terminates the calculation according to Theorem 1 of 3.3; z is unbounded below and the dual has no feasible solution.

The nonnegative column x_1 is also worthy of note. It corresponds to the dual equation

$$u_1 = u_5 + 2u_2 + 1 \geq 0$$

which is automatically satisfied by all nonnegative u_2, u_5. Since $u_1 \geq 1$, x_1 would be zero in any basic optimal solution, if one existed, by complementary slackness (Theorem 6). Another argument is that increasing x_1 would increase z and also make the primal constraints harder to satisfy. So the zero variable x_1 and its nonnegative column could be deleted without changing the optimal values of any of the remaining variables. However, this is probably not worthwhile since it would usually require recalculation of the check values and a separate calculation to find u_1. ■

Example 4

The problem of 1.2 was

$$\begin{aligned} .5y_1 + .4y_2 &\leq 170 \\ .4y_1 + .2y_2 &\leq 100 \\ .1y_1 + .4y_2 &\leq 60 \\ 10y_1 + 20y_2 &\max \end{aligned}$$

or

1	y_1	y_2	1
x_1	.5	.4	−170
x_2	.4	.2	−100
x_3	.1	.4	−60
$-z$	−10	−20	0

where the slack variables x_i are the unused quantities of the unmixed fertilizers.

Of course this can be solved as it stands. However, one may make the entries more comparable in magnitude by multiplying the first three rows by 10, dividing the last row by 10, and extracting a factor of 100 from the last column to obtain

Check	1	y_1	y_2	100
−7	$10x_1$	5	4	−17
−3	$10x_2$	4	2	−10
0	$10x_3$	1	4*	−6
−2	$-.1z$	−1	−2	0

or

1	v_1	v_2	$.01\overline{w}$
$.1u_1$	5	4	−17
$.1u_2$	4	2	−10
$.1u_3$	1	4	−6
10	−1	−2	0

Note that the dual labels are changed in the opposite senses. To obtain small integral check values, all labels (even and especially the "100") are replaced by 1, so that $-7 = 1 + 5 + 4 - 17$, etc. Two pivot steps give

	4	y_1	$10x_3$	100
−28	$10x_1$	16	−4	−44
−12	$10x_2$	14*	−2	−28
0	y_2	1	1	−6
−8	$-.1z$	−2	2	−12

	14	$10x_2$	$10x_3$	100
−50	$10x_1$	−16	−6	−42
−12	y_1	4	−2	−28
3	y_2	−1	4	−14
−34	$-.1z$	2	6	−56

Then one has $14(-.1z^*)+100(-56)=0$ or $-z^*=\bar{z}^*=5600/1.4=4000$, $\mathbf{y}^*=(200,100)$, $.1\mathbf{u}^*=10(0,2,6)/14$ or $\mathbf{u}^*=(0,100,300)/7$.

This rescaling is as useful in large computer problems as it is in small exercises. ■

Slow Solutions and Max Δz

If the optimal basis M^* were somehow known in advance, one could obtain the final tableau in the minimum number of pivot steps, one for each variable in M^* but not in the initial basis. In practice the actual number of steps is often not greatly in excess of this number, but sometimes it is. To illustrate the latter possibility the remaining two examples have been constructed with malice aforethought, although they are solved by a method that tends to be somewhat more effective.

Recall the two steps (b_2) and (b_3) in pivot selection in 3.3 and how they relate to the convergence proof in Theorem 3 of 3.3. The second, involving the minimum ratio, is often uniquely defined and then not subject to any possible improvement. The first merely requires selecting an infeasible b_i or c_j; the selection of the *most* infeasible is plausible but not necessary.

Since convergence depends on the irreversible movement of the value of z toward the final z^*, it is even more plausible that the change in z, which is $\Delta z=-b_k c_l/a_{kl}$ by rule 5 of 2.4, should have its absolute value maximized:

$$\max|\Delta z|=|b_k c_l/a_{kl}| \tag{1}$$

This maximization is over l in the primal simplex (Example 6) and over k in the dual simplex (Example 5). Meanwhile the other index k or l is a function of l or k, respectively, determined by the minimum-ratio rule. Thus a considerable part of the tableau may have to be scanned, and (1) is too laborious to use in large problems.

Graphical interpretations of Examples 5 and 6 are given in Figures 3.4A and B. In both cases it is the column constraints that are graphed in terms of the nonbasic u_1 and u_2.

Example 5

This example illustrates the option to express the given constraints in terms of column equations; hence they are written in terms of u_i, $\bar{w}$ and slack v_j:

$$\begin{aligned} 6u_1-8u_2&\le 1\\ u_1+2u_2&\le 1\\ -2u_1+4u_2&\le 1\\ -4u_1+5u_2&\le 1\\ u_1+\ u_2&=\bar{w}\max \end{aligned}$$

	1	v_1	v_2	v_3	v_4	$\bar{w}$
	u_1	-6	-1	2	4	1
	u_2	8	-2	-4	-5^*	1
	1	1	1	1	1	0

It is the b_i that are infeasible and there is a tie for max b_i. $\Delta z=1/6$ for $kl=11$ and $\Delta z=1/5$ for $kl=24$; thus (1) selects $a_{24}=-5$ as the first pivot. Thereafter the pivots are uniquely determined without further recourse to (1), as follows:

5	v_1	v_2	v_3	u_2	$\bar{w}$
u_1	2	−13	−6*	4	9
v_4	−8	2	4	−1	−1
1	13	3	1	1	1

6	v_1	v_2	u_1	u_2	$\bar{w}$
v_3	−2	13	−5	−4	−9
v_4	−8	−8*	4	2	6
1	16	1	1	2	3

8	v_1	v_4	u_1	u_2	$\bar{w}$
v_3	−20*	13	2	−1	1
v_2	8	−6	−4	−2	−6
1	20	1	2	3	5

20	v_3	v_4	u_1	u_2	$\bar{w}$
v_1	−8	−13	−2	1	−1
v_2	8	−2	−8	−6	−14
1	20	35	10	5	15

In each tableau, all signs have been reversed in order to keep the denominator positive. Here the primal problem is dual-feasible.

The dual variables u_1 and u_2 have changed from nonbasic to basic and have the optimal values $10/20$ and $5/20$. Also $\bar{w}^* = 15/20$. In Figure 3.4A the successive basic solutions are labeled 0, I, II, III, IV. They are so numerous

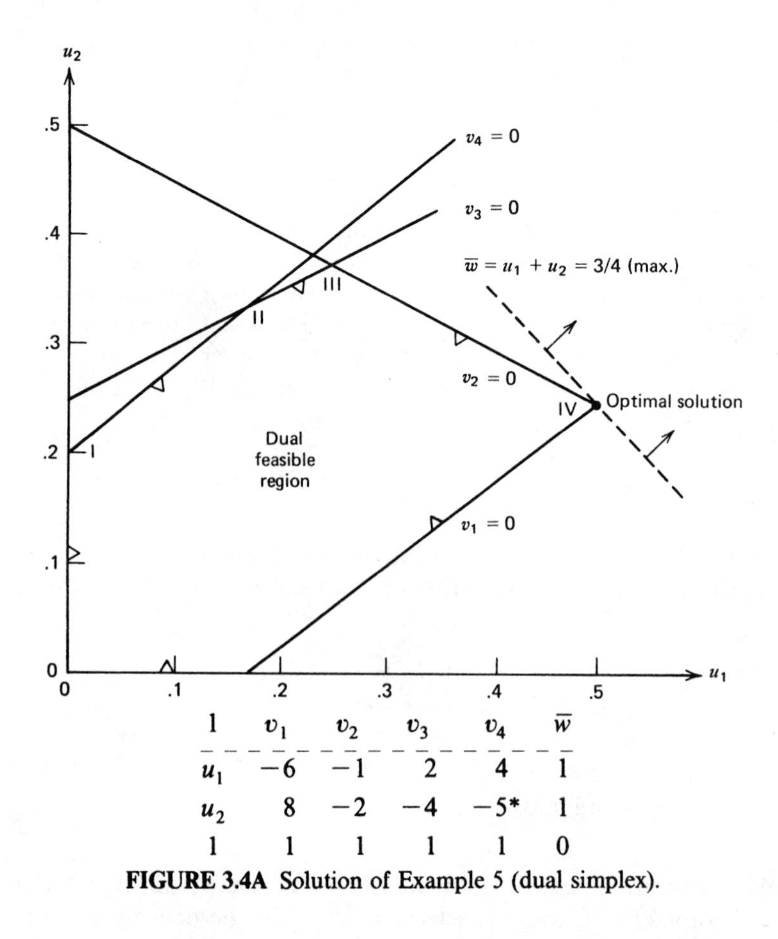

1	v_1	v_2	v_3	v_4	$\bar{w}$
u_1	−6	−1	2	4	1
u_2	8	−2	−4	−5*	1
1	1	1	1	1	0

FIGURE 3.4A Solution of Example 5 (dual simplex).

because of the necessity of circling halfway around the feasible region, and it was not obvious that it would be better to circle in the opposite direction.

This example is solved in two pivot steps by the methods of 6.1 or 6.4, which tend to be better than the dual simplex when $n>m$. By choosing the pivotal column first, they allow more scope for shrewdness in this step, which is the crucial one when the columns are more numerous. ■

Example 6

The solution of this primal-feasible problem according to Theorem 2′ is recorded graphically in Figure 3.4B and numerically in the six tableaus of Table 3.4. In the

TABLE 3.4 (Example 6)

Check	Dual nonbasis, denom.	u_3	u_4	u_5	u_6	u_7	$-w$
46	u_1	21*	5	20	−1	1	−1
39	u_2	18	15	5	2	−1	−1
−106	1	−42	−27	−36	−2	0	0
46	u_3	1	5	20	−1	1	−1
−9	u_2	−18	225*	−255	60	−39	−3
−294	21	42	−357	84	−84	42	−42
495	u_3	15	−5	275*	−25	20	−10
−9	u_4	−18	21	−255	60	−39	−3
−3303	225	144	357	−3435	120	−213	−501
495	u_5	15	−5	225	−25	20	−10
550	u_4	−5	20	255	45*	−25	−15
3520	275	405	360	3435	−235	45	−765
131	u_5	2	1	60	25	1*	−3
550	u_6	−5	20	255	275	−25	−15
1046	45	62	76	780	235	−14	−138
131	u_7	2	1	60	25	45	−3
85	u_6	1	1	39	20	25	−2
64	1	2	2	36	13	14	−4
Check	Nonbasis, denominator	u_1	u_2	u_3	u_4	u_5	$-w$

Second tableau in extended form:

	x_1	x_2	x_3	x_4	x_5	x_6	x_7	1
46	1	0	21	5	20	−1	1	−1
−9	−18	21	0	225*	−255	60	−39	−3
−315	42	0	0	−357	84	−84	42	−42

graph of the dual constraints the solution begins at the origin 0 as usual, where it is separated from the dual feasible region in the upper right by the three redundant constraints $u_3 \geq 0$, $u_4 \geq 0$, $u_5 \geq 0$. Such constraints do not necessarily result in extra iterations, but here they do because each of u_3, u_4, u_5 yield larger $|\Delta z|$ when they are made basic than either of the optimal basic variables u_6, u_7 does by itself. (The criteria $\max|\Delta z|$ and $\min c_j$ yield the same pivot selections in this problem.) ■

The tableau remains primal feasible throughout. In the dual graph, the primal feasibility of the successive solution points (0,0), I, II, III, IV, V is recognizable by the following property: if the constraints corresponding to the two intersecting lines that determine the point were the only constraints in the problem, then the point in question would be not only feasible but optimal as well. To facilitate this test, the small triangles pointing toward the allowable sides have been placed close to the solution points.

Except in the first and last tableaus, the rows of column labels have been omitted in Table 3.4 for the sake of compactness. The label for column j must agree with the row label for the immediately preceding starred pivot in column j; if there is none, then the initial column label is still valid. In the absence of any other available space, the common denominator is recorded in place of $-z$ (which never changes its position in the basis) or the corresponding dual label 1.

For a comparison, the bottom of Table 3.4 shows the second tableau in extended form, with the primal labels normally used in this form. The basic primal variables are identified by the fact that they have only one nonzero coefficient, which is the common denominator. When all entries in the tableau are divided by this denominator, the basic variables acquire their usual unit coefficients. Since $-z$ is always basic, its column is omitted; this changes the check value in row $-z$. To perform a pivot step, all entries are transformed by rule 5′ of 2.5, with the following exceptions: the pivot and the pivotal row are copied without change, though the row acquires the label of the pivotal column, if row labels are recorded; the remainder of the pivotal column is replaced by zeros, with no change of label.

Example 6 will be optimized in two pivot steps in 6.3, essentially by a comparison of intercepts on some line drawn out from the origin $u_1 = 0$, $u_2 = 0$. The negativity of a_{16} and a_{27}, which leaves each of a_{26} and a_{17} the only pivot in its column, is also a possible clue.

A family of linear programs requiring many more than the minimum number of pivot steps is given by Klee (1968) and Klee and Minty (1972).

EXERCISES 3.4

Obtain an optimal tableau and optimal $\mathbf{x}^*, \mathbf{u}^*, z^*$ for the following, or show that none exists. Use of a check column is recommended. Note that the procedure is unaffected by the option to show column equations in $\mathbf{u}$ instead of the row equations in $\mathbf{x}$.

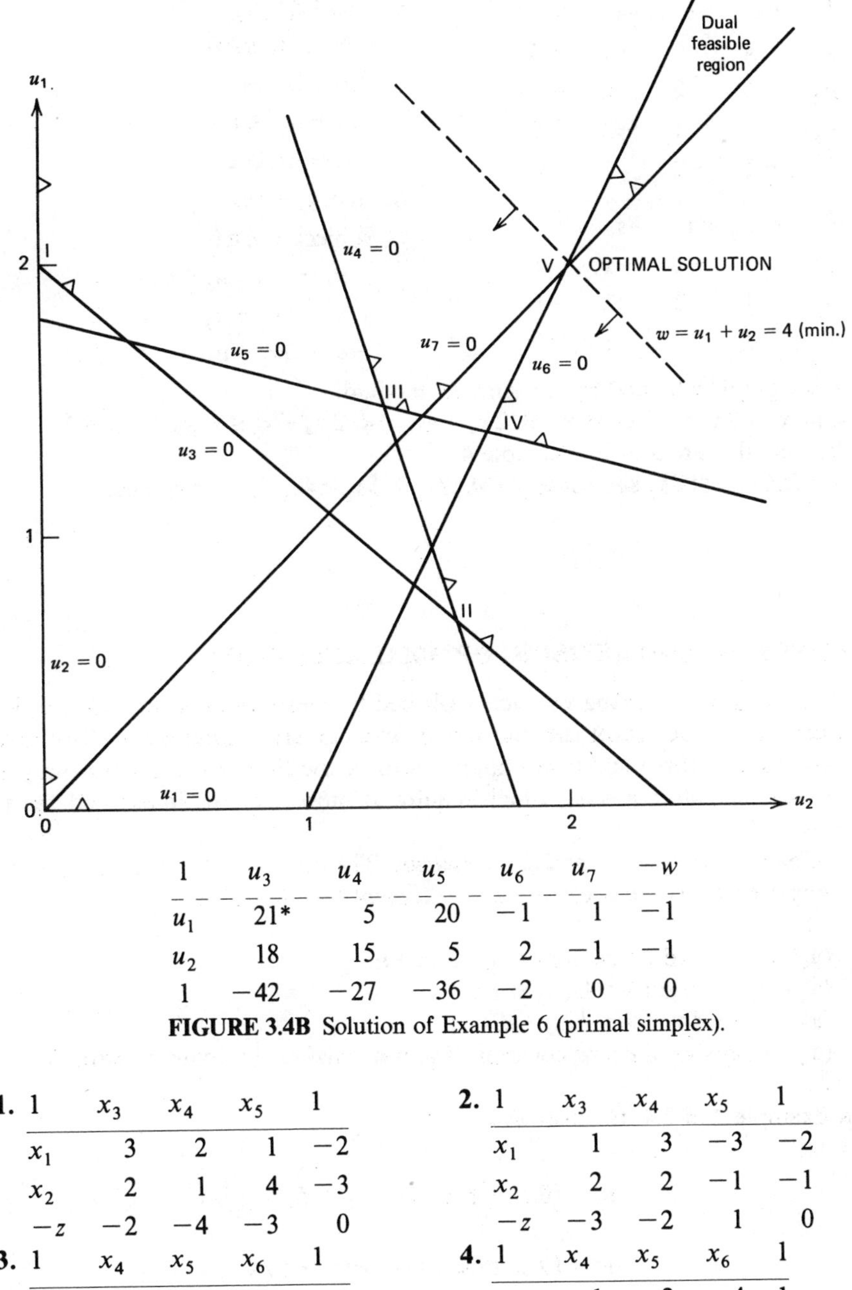

1	u_3	u_4	u_5	u_6	u_7	$-w$
u_1	21*	5	20	−1	1	−1
u_2	18	15	5	2	−1	−1
1	−42	−27	−36	−2	0	0

FIGURE 3.4B Solution of Example 6 (primal simplex).

1.

1	x_3	x_4	x_5	1
x_1	3	2	1	−2
x_2	2	1	4	−3
$-z$	−2	−4	−3	0

2.

1	x_3	x_4	x_5	1
x_1	1	3	−3	−2
x_2	2	2	−1	−1
$-z$	−3	−2	1	0

3.

1	x_4	x_5	x_6	1
x_1	2	−3	1	2
x_2	−3	1	2	−4
x_3	−2	2	−3	1
$-z$	1	1	2	0

4.

1	x_4	x_5	x_6	1
x_1	−1	−2	4	1
x_2	3	−1	−1	2
x_3	−1	3	−2	1
$-z$	3	2	1	0

5.

1	u_4	u_5	u_6	$\overline{w}$
u_1	4	0	−1	−4
u_2	−3	2	1	−1
u_3	1	1	−2	−5
1	−1	−2	1	0

6. $\min 4u_1+u_2$
subject to $u\geq 0$,
$2u_1+3u_2\leq 1$
$u_1+\ u_2\geq 1$
$5u_1+2u_2\geq 2$

7.

1	u_3	u_4	u_5	$\overline{w}$
u_1	−1	−1	−1	4
u_2	1	2	2	−6
1	3	2	1	0

8. $\max u_1-2u_3$
subject to $u\geq 0$
$u_1-2u_2-u_3+u_4=\ 1$
$u_1-4u_2+u_3\qquad\geq -2$
$-u_1-3u_2+u_3\qquad\leq\ 3$

9. Solve graphically and by the simplex method:
$\max x_4+2x_5$ subject to $\mathbf{x}\geq\mathbf{0}$, $2x_4+3x_5\leq 4$, $2x_4+x_5\leq 2$, $x_4+3x_5\leq 3$.

10. Rescale the labels as in Example 4:
$\min 200y_1+500y_2$ subject to $\mathbf{y}\geq\mathbf{0}$, $.3y_1+.5y_2\geq 40$, $.7y_1+.5y_2\geq 60$.

3.5 CHECKING AND REVISING OPTIMAL SOLUTIONS

In 2.5, 2.6 and 3.4 checking was accomplished by means of an extra column; it has to be updated at each pivot step but it will detect an error immediately. The present section considers three additional applications of the theorem of 2.4 (pivoting does not change the solution sets), which require at most only the initial and the final tableaus.

(a) *Checking a pair of optimal solutions.* Theorem 7* of 3.3 asserts that the following conditions are necessary and sufficient for optimality:

(a_1) $x_h\geq 0$ and $u_h\geq 0$ for $h=1,\ldots,m+n$.
(a_2) $u_hx_h=0$ for $h=1,\ldots,m+n$.
(a_3) $cx_N+d=z=\overline{w}=bu_M+d$.
(a_4) $\mathbf{x}$ satisfies the row constraints and $\mathbf{u}$ satisfies the column restraints.

In Example 2 of 3.4, to check that

$$\mathbf{x}^*=(0,0,0,5,1)/4,\ z^*=-5/4$$

$$\mathbf{u}^*=(9,3,19,0,0)/4,\ w^*=-5/4$$

are indeed optimal, one verifies (a_1) and (a_2) by inspection and then substitutes the primal solution in the three row equations and the dual solution in the four column equations. The work is facilitated by writing the primal and the dual values to be substituted next to their labels, thus:

1		0	5	1	4	1		19	0	0	-5	
	1	x_3	x_4	x_5	1		1	u_3	u_4	u_5	$\bar{w}$	
0	x_1	2	1	-1	-1	9	u_1	2	1	-1	-1	(1)
0	x_2	-1	-3	-1	4	3	u_2	-1	-3	-1	4	
5	$-z$	1	0	3	-2	4	1	1	0	3	-2	

Note that fractions are avoided as usual, the numerators appearing by themselves. The denominator 1 of the tableau is written down again in the upper left corner where it will multiply the numerators of the values of the basic variables; the denominator 4 of the solutions must multiply the constants column in the primal, the constants row in the dual. If one wants to avoid any exception to the general routine, the value of the label $-z$ (with numerator 5) is used in the primal tableau. The first row in the body of the first tableau gives

$$0\cdot 1+2\cdot 0+1\cdot 5-1\cdot 1-1\cdot 4=0$$

The first column in the body of the second tableau gives

$$1\cdot 4-1\cdot 3+2\cdot 9=19\cdot 1$$

(It is convenient to read up from the bottom.) There are two more rows and three more columns to verify in this manner.

In practice the initial tableau need not be recopied once, much less twice. The complementary slackness of the two basic solutions offers the possibility of writing both of them in one row and one column, if some device is adopted to distinguish the one set of numbers from the other; then whenever a number from the "other" set is encountered, it is understood that it is to be replaced by zero as in (1). In the following combined form of (1), the dual values are enclosed in parentheses. The common denominator 4 occurs as a primal value in the last column and a dual value in the last row.

1		(19)	5	1	4		
	1	x_3	x_4	x_5	1		
(9)	x_1	2	1	-1	-1		(2)
(3)	x_2	-1	-3	-1	4		
(4)	$-z$	1	0	3	-2	$=-5$	

Since (2) affords no convenient place to enter the value of $z^*=\bar{w}^*$ within the tableau, its numerator -5 is entered by itself at the end of the last row, where z would appear after being shifted to the right member. This -5 is both a primal and a dual value. In the last row the modified sum is not 0 but -5 thus:

$$0\cdot 1+1\cdot 0+0\cdot 5+3\cdot 1-2\cdot 4=-5$$

Similarly the last column gives

$$-2\cdot 4+4\cdot 3-1\cdot 9=-5$$

the columns being read from the bottom up as before.

(b) *Revising an optimal solution.* More general treatments will be given in Chapters 7 and 8. Here we suppose that after a linear program has been solved, it is desired to solve another one having the same **A** but different **b** and/or **c** (d also may change). If the new values are utterly different, the previous solution may not be helpful, but this is not the typical situation. In any case, so long as the final basis is not changed, the rules of 2.4 or 2.5 show that the same class of elements change in the final tableau as in the initial. For this class there are three possibilities: the elements in column **b**, or those in row **c**, or both.

The method can be explained very simply as a reversal of what was done in (a). Here the new initial basic solution(s) could be substituted in the existing final tableau, in which the old d^* and $\mathbf{b}^*$ and/or $\mathbf{c}^*$ are deleted or ignored, and new values calculated so as to cause the equations to be satisfied; each row equation determines one such revised b_i and each column equation determines one c_j. Then there are two equations to determine d, which gives only a partial check.

For the sake of internal consistency and clarity it seems better to work with the new values of the b_i^0 and/or c_j^0 from the initial tableau rather than the corresponding (nonoptimal) basic solution. After all the sign changes are accounted for, it is found to be appropriate to enter $-c_j^0$ on the left of the final tableau and c_j^0 and b_i^0 everywhere else, and then to calculate the final b_i and c_j without any further changes of sign. The initial common denominator of $\mathbf{b}^0, \mathbf{c}^0, d^0$ is not entered anywhere. Usually it is 1, as here; otherwise it would be divided into each new element as it is calculated below.

Suppose that the initial $\mathbf{b}^0, \mathbf{c}^0, d^0$ in Example 2 of 3.4 are changed to

$$\mathbf{b}^0=(-2,3,0,0,0), \qquad \mathbf{c}^0=(0,0,2,1,2), \qquad d^0=6$$

These values are applied to the final tableau, with letters representing the numerators of the b_i and c_j to be determined and $-c_4^0=-1$ and $-c_5^0=-2$ entered on the left as just described, to obtain (3):

4		(2)	3	−2		(6)
	4	x_3	x_2	x_1	1	
(−2)	x_5	−5	−1	−3	β_5	3
(−1)	x_4	3	−1	1	β_4	−5
	$-z$	γ_3	γ_2	γ_1	α_{00}	
6		15	3	5		23

(3)

The column equations use the parenthesized values. Thus

$$\gamma_3=4c_3=4\cdot 2-2(-5)-1\cdot 3=15$$

$$\beta_5=4b_5=4\cdot 0+0\cdot(-5)+3(-1)-2(-3)=3$$

and so on. The tableau (3) is not optimal; pivoting on $a_{51}=-3/4$ produces the optimal tableau (5) below. The basic solution for (3) or (5) can be checked by substitution into the initial tableau as in (2).

The calculation as in (3) is no substitute for (2); (3) is not an adequate check for either the original or the revised optimal solution. To see this, suppose an error is made in calculating some a_{kl} in an intermediate tableau in which neither x_k nor x_l has yet been exchanged, although later they are. Then the optimal solution will generally be erroneous but consistent with the original $\mathbf{b}^0$ and $\mathbf{c}^0$ and an altered $\mathbf{A}^0$; unfortunately $\mathbf{A}^0$ is not used in (3). (When the pivots are reversed, only erroneous a_{ij} in pivotal rows or columns produce errors in $\mathbf{b}^0$ or $\mathbf{c}^0$.)

The calculation in (3) may be summarized thus:

$$\mathbf{b}=\mathbf{b}_M=\mathbf{b}_M^0+\mathbf{A}\mathbf{b}_N^0=[\mathbf{I}_m,\mathbf{A}]\begin{bmatrix}\mathbf{b}_M^0\\ \mathbf{b}_N^0\end{bmatrix}=\mathbf{B}^{-1}\mathbf{b}_{M_0}^0$$

$$\mathbf{c}=\mathbf{c}_N=\mathbf{c}_N^0-\mathbf{c}_M^0\mathbf{A}=-\left[\mathbf{c}_M^0,\mathbf{c}_N^0\right]\begin{bmatrix}\mathbf{A}\\ -\mathbf{I}_n\end{bmatrix}=-\mathbf{c}_{N_0}^0\mathbf{C}^{-1} \tag{4}$$

$$d=d^0+\mathbf{c}_N\mathbf{b}_N^0=d^0-\mathbf{c}_M^0\mathbf{b}_M$$

The superscript zero refers to the initial tableau. M and N are the sets of row and column indices for $\mathbf{A}$ in the current or final tableau. $\mathbf{b}_M^0, \mathbf{b}_N^0, \mathbf{c}_M^0, \mathbf{c}_N^0$ have components defined for i in M or j in N as indicated; these are zero unless i is in M_0 (the row indices in the initial tableau) and j is in N_0. They are arranged in the order in which they occur in the current tableau.

$\mathbf{I}_m$ is an m by m unit matrix.
$\mathbf{B}^{-1}$ consists of the columns of $[\mathbf{I}_m,\mathbf{A}]$ with indices in M_0.
$\mathbf{C}^{-1}$ consists of the rows of $\begin{bmatrix}\mathbf{A}\\ -\mathbf{I}_n\end{bmatrix}$ with indices in N_0.

(c) *Checking the equivalence of two tableaus*. For this either the rows or the columns alone should suffice; the latter avoid the need for any changes of sign. The procedure is to construct a solution for one tableau by setting the nonbasic variables equal to 1 or other nonzero values and verifying that this point is a solution for the other tableau. Negative values are quite acceptable but no variable that is nonbasic in either tableau can be permitted to be zero; otherwise the corresponding row or column of the tableau would be multiplied by zero and hence not be checked.

Putting $x_1=x_2=x_3=1$ in (3) gives the solution $\mathbf{x}^1=(4,4,4,2,6)/4$, $-z=-46/4$. Equation (5) shows the substitution of these values into an equivalent tableau.

3		4	4	6	4 = denominator of $\mathbf{x}^1$
	3	x_3	x_2	x_5	1
4	x_1	5	1	−4	−3
2	x_4	1	−1	1	−3
−46	$-z$	5	1	5	21

(5)

For example the last row gives

$$-46\cdot3+5\cdot4+1\cdot4+5\cdot6+21\cdot4=0$$

EXERCISES 3.5

1, 3, 5, 7, 9. Apply the checking method of (1) or (2) to the like-numbered exercises in 3.4.

2. Without pivoting, prove that $\mathbf{y}_{1,2}=(1,10)/14$, $\mathbf{u}=(9,0,2)/14$, $\bar{z}^*=31/14$ are optimal solutions for max $\bar{z}=y_1+3y_2$ subject to $y_1, y_2\geq0$, $2y_1+4y_2\leq3$, $-y_1+2y_2+1\geq0$, $2y_1-3y_2+2\geq0$.

In the following (incomplete) pairs of initial and final tableaus, determine the missing entries in the latter (as in (3), without pivoting).

4.

1	x_1	x_2	x_3	1
x_4	-2	-1	1	1
x_5	1	3	2	-3
$-z$	1	2	-1	3

5	x_4	x_2	x_5	1
x_1	-2	5	1	
x_3	1	5	2	
$-z$				

6.

1	x_3	x_4	x_5	1
x_1	-2	1	1	-3
x_2	1	1	-1	-4
$-z$	0	-1	3	2

3	x_2	x_1	x_5	1
x_4	2	1	-1	
x_3	1	-1	-2	
$-z$				

8. As in (5), show that the initial and the final tableaus for Example 4 of 3.4 are equivalent.

10, 11. With no more pivoting than necessary, obtain the new optimal solutions when the initial tableaus are changed so as to have $\mathbf{b}_{1,2,3}=(-3,1,-4)$ in Exercise 5 and $\mathbf{c}_{4,5,6}=(2,-1,1)$ in Exercise 3 of 3.4.

12. All the methods of this section rely on the easily remembered theorem of 2.4. However, what details is it helpful to keep in mind?

3.6 SOME SIMPLE MODELS AND THEIR DUALS

Whenever applications of linear programming are discussed, variants of one or both of two simple types of problems are invariably given as examples. The first of these, called *Model 1*, is illustrated by the problem of 1.2, which the reader may review at this time. It was solved by the simplex method in Example 4 of 3.4.

The illustration of *Model 2* involves the same two mixtures of fertilizers N, P, K in the proportions 5:4:1 and 4:2:4. However, we are now concerned not with a supplier but with a consumer (a farmer) who buys y_1 tons of 5:4:1 at \$100 per ton and y_2 tons of 4:2:4 at \$60 per ton in order to satisfy a need of at least 175 tons of N, 110 tons of P, and 60 tons of K. Instead of maximizing a profit the farmer wants to minimize the cost.

The two problems evidently reflect different market situations as well as different lines of business. In Model 1 at least one mixture must be selling at a premium over the value of its pure constituents; otherwise no profit could be made. Since the pure raw materials are mined or synthesized separately, this would be a reasonable expectation if a demand for the mixtures exists. In the Model 2 problem the mixtures are assumed to be cheaper or more readily available than the pure materials, although this assumption is subject to possible rebuttal.

Thus the conditions are $y_1 \geq 0$, $y_2 \geq 0$,

$$\text{N supplied} = .5y_1 + .4y_2 \geq 175 \text{ required.} \quad (1)$$

$$\text{P supplied} = .4y_1 + .2y_2 \geq 110 \text{ required.} \quad (2)$$

$$\text{K supplied} = .1y_1 + .4y_2 \geq 60 \text{ required.} \quad (3)$$

$$\text{Cost} = 100y_1 + 60y_2 = z \text{ to be min.} \quad (4)$$

In Figure 3.6 the problem is graphed in the style of Figure 1.2. The feasible region lies above and to the right of the y_1 axis, the line $ABCD$ and the y_2 axis. Since we are

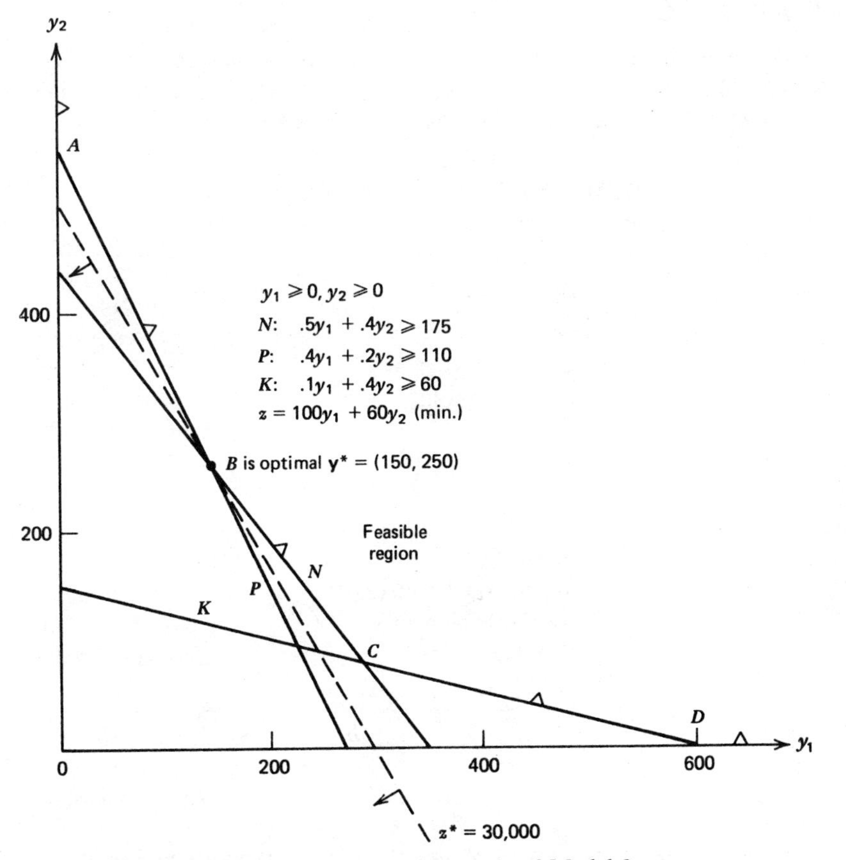

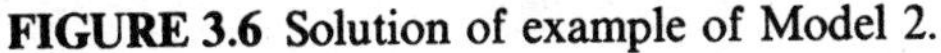

FIGURE 3.6 Solution of example of Model 2.

minimizing $100y_1 + 60y_2$, the dashed line moves toward the origin until it last contacts the feasible region at the optimal solution B.

The initial and the final tableaus are

1	y_1	y_2	1
x_1	$-.5$	$-.4$	175
x_2	$-.4$	$-.2$	110
x_3	$-.1$	$-.4$	60
$-z$	100	60	0

6	x_2	x_1	1
y_2	50	-40	-1500
y_1	-40	20	-900
x_3	16	-14	-330
$-z$	1000	400	180,000

(5)

Thus $\mathbf{y}^* = (150, 250)$ tons, $\mathbf{u}^* = (200/3, 500/3, 0)\$/\text{ton}$, $z^* = \$30{,}000$.

The characteristics of the two models are summarized in Table 3.6, but the descriptions should not be limited to a literal or materialistic interpretation. As in the exercises below, the "elements" may be time periods during which various pieces of equipment are used, space available in several different locations, quantities of energy of various kinds, or the three velocity components of a spacecraft.

The Significance of the Dual

The two parts of this discussion correspond to items (e_1) and (e_2) in 3.3. The Model 2 example will be used because its data are close at hand. The reader is asked to give a similar treatment of the Model 1 example in Exercise 8.

TABLE 3.6 Comparison of Models 1 and 2

	Model 1 (1.2)	Model 2 (3.6)
Objective	$\bar{z} = \bar{\mathbf{c}}\mathbf{y}$ (max)	$z = \mathbf{c}\mathbf{y}$ (min)
Inputs	$\bar{b}_i - x_i$ (available elements consumed)	y_j (mixtures bought)
Outputs	y_j (mixtures sold)	$b_i + x_i$ (needed elements supplied)
Proportion of element i in mixture j	a_{ij}	$\bar{a}_{ij}$
Type of situation	Max profit for supplier, who mixes available elements. Synthesis.	Min cost for consumer, who needs elements but buys mixtures. Analysis.
Initial tableau ($\mathbf{x} \geq \mathbf{0}$, $\mathbf{y} \geq \mathbf{0}$)	$\begin{array}{c\|cc} 1 & y_j & 1 \\ \hline x_i & a_{ij} & -\bar{b}_i \\ \bar{z} & -\bar{c}_j & 0 \end{array}$ (primal-feasible)	$\begin{array}{c\|cc} 1 & y_j & 1 \\ \hline x_i & -\bar{a}_{ij} & b_i \\ -z & c_j & 0 \end{array}$ (dual-feasible)

$\bar{a}_{ij}, \bar{b}_i, \bar{c}_j$ denote absolute values when a_{ij}, b_i, or $c_j < 0$. Model 2 is not the dual of Model 1 despite resemblance.

An important observation is that in both models one has

$$z^* = \mathbf{c}\mathbf{y}^* + d = \mathbf{b}\mathbf{u}^* + d \tag{6}$$

with z^* measured in dollars and b_i measured in tons of pure fertilizer $i = \text{N}, \text{P}, \text{K}$. Consequently u_i^* has the nature of a price (called *shadow price*) in dollars per ton for pure fertilizer i. Since it is calculated within a very restricted context, it cannot be expected to agree with a more broadly based market price. Indeed in (5) the shadow price of potash K is zero because it is being supplied in excess of its requirement by $x_3 = 330/6 = 55$ tons. On the other hand the consumer would be better off to buy *some* of any pure fertilizer that he or she could get for less than the shadow price.

(i) The dual of the Model 2 example resembles the primal of Model 1 in being a *maximizing* problem of interest to a *supplier* of chemical fertilizers. However, the variables and the input data have quite different interpretations. In the Model 1 primal the quantities of the two mixtures have to be determined, whereas the Model 2 dual is concerned with the shadow prices u_i of the three pure fertilizers in the *consumer's* problem. However, the supplier and the consumer each have a reason to be interested in the other's problem because of the possibility of doing business with one another.

To discuss the dual of Model 2 we suppose that the consumer knows a little algebra but not necessarily any linear programming. The supplier is assumed to be able to supply pure fertilizers at competitive prices, but what should these prices be? They could be the consumer's shadow prices.

To close the deal, the supplier leads the consumer through the argument of 3.1. Equations (1), (2), and (3) are multiplied by the consumer's shadow prices $200/3, 500/3$, 0, respectively, and added to (4). The result is

$$\begin{aligned}(5\cdot 20/3 + 4\cdot 50/3)y_1 + (4\cdot 20/3 + 2\cdot 50/3)y_2 + z \\ \geq 175\cdot 200/3 + 110\cdot 500/3 + 100y_1 + 60y_2\end{aligned} \tag{7}$$

The terms in y_1 and y_2 cancel out and there remains $z \geq 30{,}000$. Thus the consumer, without necessarily knowing his $\mathbf{y}^*$, is persuaded that he cannot do better than to let the supplier take care of his requirements at a total cost of $30,000. Since the dual problem is maximizing, this is the largest price that the supplier could justify by such a procedure.

(ii) Equation (6) provides two ways of calculating $z^* = \$30{,}000$ and two ways of breaking this cost down into parts. One can attribute $c_1 y_1^* = c_2 y_2^* = \$15{,}000$ to the purchase of each mixture, or $b_1 u_1^* = \$35{,}000/3$ to the satisfaction of the N requirement and $b_2 u_2^* = \$55{,}000/3$ to P; K is regarded as free of charge since it is in excess.

Section 3.5 showed how to revise an optimal solution when $\mathbf{b}$ and/or $\mathbf{c}$ changed, but this required the knowledge of the entire optimal tableau or something equivalent to it. What can one get from (6)? From 3.5 or the fact that one never pivots in $\mathbf{b}$ or $\mathbf{c}$, it is apparent that $\mathbf{y}^*$ *does not depend on* $\mathbf{c}$ *and* $\mathbf{u}^*$ *does not depend on* $\mathbf{b}$, *except insofar as* $\mathbf{b}$ *and* $\mathbf{c}$ *help to determine the optimal basis* M^*.

Thus if $\mathbf{c}'$ is close enough to $\mathbf{c}$ to leave M^* and $\mathbf{y}^*$ unchanged, the new z^* can be calculated by $z^{*\prime} = \mathbf{c}'\mathbf{y}^* + d$; and if $\mathbf{b}'$ is close enough to $\mathbf{b}$ to leave the dual solution

$\mathbf{u}^*$ unchanged, the new $z^{*\prime}=\mathbf{b}'\mathbf{u}^*+d$. Subject to the preservation of M^*, one can even calculate z^* *approximately* when both $\mathbf{b}$ and $\mathbf{c}$ change by the formula

$$z^{*\prime} \doteq \mathbf{c}'\mathbf{y}^* + \mathbf{b}'\mathbf{u}^* + 2d - z^* \tag{8}$$

Unfortunately, if one has only $\mathbf{x}^*$, $\mathbf{y}^*$, $\mathbf{u}^*$, $\mathbf{v}^*$, z^* and the initial tableau available, it is not easy to determine how small such changes must be, but such questions can be investigated as in 8.2 before the optimal tableau is discarded.

To illustrate these ideas, suppose that in (5) one wishes to change $\mathbf{b}$ to (200, 130, 70) and/or change $\mathbf{c}$ to (110, 70). Figure 3.6 or the final tableau in (5) may be used to verify that $M^*=\{1,2\}=\{\text{N},\text{P}\}$ still but will not be used otherwise. If only $\mathbf{b}$ is changed one has $z^{*\prime}=(200,130,70)(200,500,0)/3=35{,}000$. If only $\mathbf{c}$ is changed one has $z^{*\prime}=(110,70)(150,250)=34{,}000$. If both are changed, $z^{*\prime}=35{,}000+34{,}000-30{,}000=39{,}000$. In the latter case the exact values are $z^{*\prime}=39{,}500$, $\mathbf{y}^*=(200,250)$, $\mathbf{u}^*=(100,150,0)$, $x_3^*=50$.

Model 2a is like Model 2 except that initially it has no slack variables in the primal. It is typified by a blending problem such as the following. A manufacturer can purchase any amount of four alloys of lead, tin, and zinc, thus:

Proportion in Blend	% lead	% tin	% zinc	Price ($/lb)
y_1	30	50	20	1.8
y_2	60	40	0	1.6
y_3	50	0	50	.2
y_4	0	60	40	2.0

By blending some of the four alloys, the manufacturer wishes to produce, as cheaply as possible, an alloy containing 40 percent lead, 40 percent tin, and 20 percent zinc. Thus we have, taking $z=$ the cost in dollars for 10 pounds of the blend:

$$\begin{aligned}
y_1 + y_2 + y_3 + y_4 &= 1 \\
.3y_1 + .6y_2 + .5y_3 &= .4 \\
.5y_1 + .4y_2 + .6y_4 &= .4 \\
.2y_1 + .5y_3 + .4y_4 &= .2 \\
18y_1 + 16y_2 + 2y_3 + 20y_4 &= z \text{ to be min}
\end{aligned} \tag{9}$$

One of the constraints may be omitted, since the first is the sum of the other three. The extreme points of the feasible line are (0, 7, 2, 4)/13 and (8, 3, 2, 0)/13, and both make $z^*=196/13$ as it happens. Unlike Models 1 and 2, this model does not have to have a solution. (For example, a pure metal cannot be obtained by mixing alloys.)

Now suppose that pure lead, tin, and zinc are also available at prices of \$.20, \$4.00 and \$.40 per pound and used in the blend in proportions x_1, x_2, x_3. The problem becomes

x_1	x_2	x_3	y_1	y_2	y_3	y_4	
1			.3	.6	.5		$=.4$
	1		.5	.4		.6	$=.4$
		1	.2		.5	.4	$=.2$
2	40	4	18	16	2	20	$=z(\min)$

(10)

The x_i can serve as slack variables provided they are eliminated from z by multiplying the constraints in (10) by -2, -40, and -4 respectively and adding to the last line. The result is the primal-feasible tableau

1	y_1	y_2	y_3	y_4	1
x_1	.3	.6	.5	0	$-.4$
x_2	.5	.4	0	.6	$-.4$
x_3	.2	0	.5	.4	$-.2$
$-z$	-3.4	-1.2	-1.0	-5.6	17.6

(11)

Here 17.6 is the cost of making 10 pounds of blend out of pure metals, and $3.4y_1 + 1.2y_2 + 1.0y_3 + 5.6y_4$ is the saving achieved by substituting certain of the available alloys in place of some of the pure metals. The optimal solutions use 25 percent pure lead even though all the alloys are selling for less than the value of their pure constituents.

Equation (11) has the form of Model 1 but with an interpretation somewhat different from Table 3.6. Here the x_i (elements) and y_j (mixture) are all quantities of inputs that are purchased, while the output is a specified quantity of one mixture (obtained by combining the inputs) whose composition is specified by the $\bar{b}_i$. The operator seeks to maximize the saving in cost resulting from the inclusion of the y_j mixtures as inputs.

Every feasible solution of (9) is also a feasible solution of (10) or (11) having all $x_i = 0$. Hence if such a solution is optimal for (11) it must also be optimal for (9). Conversely, if (9) has an optimal solution, it can be made optimal for a problem like (11) by choosing the prices of the pure metals so high that one cannot afford to use them. Thus one can infer a method for solving linear programs like (9) whose constraints are equations without slack variables.

EXERCISES 3.6

Define the nonbasic variables and set up the initial tableau. State whether it is Model 1, 2 or 2a or none of these. Solutions are not required.

1. A student in a work-study program needs \$1200 and 38 credits in order to graduate. There are three jobs available, which for one hour's work yield \$2.00, \$2.50 and \$2.80 and .09, .08 and .06 credits, respectively. What program will require the least expenditure of time?

2. A printer has 15 hours to spare on a printing press and 12 hours on a bookbinder. There are three jobs available with the following characteristics (per 1000 items):

	Printing Time (hr)	Binding Time	Profit ($)
Blank books	0	1.5	95
Paperbacks	7	2	360
Handbills	0.2	0	8

3. A farmer wants each animal to receive at least 3 kilograms of protein and 15,000 calories and at most 2 kilograms of roughage per day. Three kinds of feed are available with the following characteristics per unit:

Cost	Protein (kg)	1000 Calories	Roughage (kg)
$8	18	60	6
7	10	85	5
3	6	50	7

4. A firm can make three kinds of cloth on which the profit is $1, $1.60 and $2 per yard. One yard requires
.17, .24, and .29 lb of red yarn (4300 lb available),
.27, .25, and .19 lb of yellow yarn (3900 lb available), and
.30, .26, and .25 lb of blue yarn (4700 lb available).

5. A construction company has 1400 bags of cement at the home office (0,0) (rectangular coordinates) and 3500 bags at its warehouse at (130,240). It needs 1200 bags at (−100,170), 1500 at (50, −120), and 1800 at (270,60). Assume that the shipping cost per bag from (x_i, y_i) to (x_j, y_j) is proportional to $|x_i - x_j| + |y_i - y_j|$. (The roads go north-south and east-west only.)

6. A wholesaler has to package as many 220-pound lots of a certain material as convenient on short notice. There is a large supply of 50-pound cartons, and in fact at least 400 of them must be shipped out to make room for incoming inventory. Beside this there are only 500 of the 40-pound size and 300 of the 60-pound size. Every carton used must be completely filled. (Hint: list the four ways in which 220 pounds can be packed.)

7. A certain individual likes liquid refreshment consisting of 17 percent alcohol, 2 percent sugar, 1 percent carbon dioxide and 80 percent water. How should he mix it from the following ingredients? (Water may be added.)

	% Alcohol	% Sugar	% CO_2	Cost/lb ($)
Soda water	0	0	3	0.30
Sparkling wine	12	0	2	1.00
Sweet wine	20	4	0	1.65
Vodka	50	0	0	4.00
Sugar	0	100	0	0.25

8. Discuss the Model 1 problem of 1.2 as in (i) and (ii).

3.7 OTHER APPLICATIONS

This section considers two more examples of possible applications of linear programming and provides a few hints for the formulation of such problems. Here and in the exercises it is not assumed that the problem is necessarily solvable by the methods of 3.3.

A Production Problem

Suppose that a sand and gravel firm has contracts for daily quantities b_1 and b_2 of sand and gravel, respectively. It has two pits, operated at rates y_1 and y_2 to be determined, which result in a daily cost of $c_1y_1+c_2y_2$, a daily production of quantities $\bar{a}_{11}y_1+\bar{a}_{12}y_2$ of sand and $\bar{a}_{21}y_1+\bar{a}_{22}y_2$ of gravel, and a need for a trucking capacity of $a_{31}y_1+a_{32}y_2$, of which $\bar{b}_3$ is supplied from the firm's own personnel and equipment. Two other activities are the renting of additional trucking capacity y_3 from others at a cost of c_3y_3, and the renting of idle trucking capacity y_4 to others for a return of $\bar{c}_4y_4$; usually $c_3>\bar{c}_4$ and both are not done at once ($y_3y_4=0$). The value of unused sand or gravel is neglected for simplicity. This gives the tableau

1	y_1	y_2	y_3	y_4	1
x_1	$-\bar{a}_{11}$	$-\bar{a}_{12}$	0	0	b_1
x_2	$-\bar{a}_{21}$	$-\bar{a}_{22}$	0	0	b_2
0	a_{31}	a_{32}	-1	1	$-\bar{b}_3$
$-z$	c_1	c_2	c_3	$-\bar{c}_4$	0

(1)

in which all the a's, b's, and c's are nonnegative before the indicated signs are prefixed, as are the x_i and y_j.

The sign conventions associated with a schema lend themselves to a consistent interpretation. Consider the schema with primal labels only:

y_1	$\cdots$	y_n	1	
a_{11}	$\cdots$	a_{1n}	b_1	$=-x_1\le 0$ (all $y_j>0$)
$\vdots$		$\vdots$	$\vdots$	
a_{m1}	$\cdots$	a_{mn}	b_m	$=-x_m\le 0$
c_1	$\cdots$	c_n	0	$=z$ min

(2)

The significance of the signs of the entries is as follows:

$b_i>0$ for outputs required.
$b_i<0$ for inputs (resources) available.
$c_j>0$ for costly activities, needed to achieve feasibility.
$c_j<0$ for profitable activities, limited only by the availability of inputs.
$a_{ij}>0$ if activity j tightens constraint i, by consuming the input or output i.
$a_{ij}<0$ if activity j relaxes constraint i, by producing the input or output i.

Because of the fact that all the rows $a_{i1}y_1 + \cdots + a_{in}y_n + b_i$ are to be nonpositive and the last row **cy** is to be minimized, all of the above statements can be combined into a pair of equivalent statements, thus:

Homogeneous sign convention for primal linear programs having inequality constraints and nonnegative variables:
$b_i, c_j, a_{ij} < 0$ (or barred parameters $\bar{b}_i, \bar{c}_j, \bar{a}_{ij} > 0$) are favorable to primal feasibility and objective (minimization); that is, they correspond to available resources, profitable or productive activities. Conversely,
$b_i, c_j, a_{ij} > 0$ are unfavorable to primal feasibility and objective function minimization; that is, they correspond to demands that have to be satisfied, costly or consumptive activities.

This convention is the opposite of what would seem to be natural, but it is the inevitable consequence of the customary preference for the matrix $[\mathbf{I}, \mathbf{A}]$ rather than $[-\mathbf{I}, \mathbf{A}]$ in 7.1 to 7.3 (see also 3.8). In any case the objection is muted by the facts that the "natural" convention comes into play in the dual problem, in which the words "favorable" and "unfavorable" have to be interchanged; and the demands $b_i > 0$ are favorable to the firm in the larger sense of giving it a job to do, the return for which is regarded as constant and therefore not included in the formulation of the problem.

It should be recognized that if an equation contains no slack variable, or if the equation is to be obtained by means of a pivot step or other algebraic manipulation, one cannot necessarily anticipate which of its coefficients will represent "favorable" effects. (For example, see Exercise 10 of 11.7.)

Certainly the easiest way to formulate a given problem as a linear program is to *recognize that a familiar model will fit it*, when that is possible. Otherwise one must proceed more slowly and methodically, perhaps as in Table 3.7A.

Gasoline Blending

This problem was discussed by Charnes, Cooper and Mellon (1952) and Garvin (1960). A small hypothetical example will be considered here. Suppose that an oil refinery produces three "streams" of raw gasolines that are characterized by performance numbers (octane ratings) of 75, 92, and 106, vapor pressures (a measure of volatility) of 5. 8 and 10 pounds per square inch and availabilities of 3500, 2800 and 3100 barrels per day; these can be sold for $10, $12 and $15 per barrel, respectively.

In addition the firm has the options of mixing these raw gasolines to make either regular or premium motor fuel. These must have octane ratings ≥ 78 and 95 and vapor pressures ≤ 9 and 7, and they sell for $11 and $14 per barrel.

Let y_{ij} denote the number of barrels of raw gasoline $i = 1,2,3$ used in producing motor fuel $j = 1,2$, while y_{i0} is the quantity of i sold unmixed. These variables are nonnegative. If there is no change of volume and no substantial expense involved in mixing, the daily gross revenue to be maximized is (in dollars)

$$\begin{aligned} \bar{z} = \; & 10y_{10} + 11y_{11} + 14y_{12} \\ & + 12y_{20} + 11y_{21} + 14y_{22} \\ & + 15y_{30} + 11y_{31} + 14y_{32} \end{aligned} \tag{3}$$

TABLE 3.7A Formulation of a Linear Program

---

0. Can you recognize that a familiar model is suitable? If not:
1. Extract or infer the phrase describing the objective function z (to be minimized) or $\bar{z}=-z$ (to be maximized).
2. Assign symbols $y_1, y_2, \ldots$ to the nonbasic (independent, nonslack) variables on which z or $\bar{z}$ depends, meanwhile defining each variable precisely in words and giving its unit of measurement. In some problems more than one letter or more than one subscript may be desirable.
3. Determine the coefficients c_j or $\bar{c}_j$ with the correct signs. Write down the resulting expression for z or $\bar{z}$, respectively.
4. Extract the constants b_i or $|b_i|$.
5. Write down the constraint corresponding to each b_i, with correct signs. (This may require additional variables, or a new approach with new variables may be indicated.) Consider very carefully whether it uses $=$, $\leq$, or $\geq$.
6. Ask if there are other constraints for which no b_i is given explicitly, either because it has some obvious value such as 0, 1, or -1, or because it has to be replaced by (a linear function of) one or more variables. Are all the $y_j \geq 0$? Are any data or conditions still unaccounted for? Are some variables superfluous?
7. Assign indices i to the constraints and insert a slack $x_i \geq 0$ on the low side of each inequality. Set up the tableau (perhaps with zeros in place of basic variables in some rows) and check the signs where possible by using the homogeneous sign convention above (for the primal).

The availabilities of the raw gasolines yield the equations

$$
\begin{aligned}
y_{10}+y_{11}+y_{12}&=3500\\
y_{20}+y_{21}+y_{22}&=2800\\
y_{30}+y_{31}+y_{32}&=3100
\end{aligned}
\tag{4}
$$

It is assumed that the octane rating and vapor pressure of a mixture are the volume-weighted averages of the constituents so that we have

$$\frac{75y_{11}+92y_{21}+106y_{31}}{y_{11}+y_{21}+y_{31}} \geq 78$$

and so on. Clearing of fractions gives

$$
\begin{aligned}
75y_{11}+92y_{21}+106y_{31} &\geq 78(y_{11}+y_{21}+y_{31})\\
5y_{11}+\ 8y_{21}+\ 10y_{31} &\leq\ 9(y_{11}+y_{21}+y_{31})\\
75y_{12}+92y_{22}+106y_{32} &\geq 95(y_{12}+y_{22}+y_{32})\\
5y_{12}+\ 8y_{22}+\ 10y_{32} &\leq\ 7(y_{12}+y_{22}+y_{32})
\end{aligned}
$$

TABLE 3.7B Gasoline Blending ($x_{ij}, y_{ij} \geq 0$)

1	y_{11}	y_{21}	y_{31}	y_{12}	y_{22}	y_{32}	1
y_{10}	1			1			−3,500
y_{20}		1			1		−2,800
y_{30}			1			1	−3,100
x_{11}	3	−14	−28				0
x_{21}	−4	−1	1				0
x_{12}				20	3	−11	0
x_{22}				−2	1	3	0
$-z$	−1	1	4	−4	−2	1	−115,100

Collecting terms and inserting slack variables x_{hj} gives

$$
\begin{aligned}
x_{11} + 3y_{11} - 14y_{21} - 28y_{31} &= 0 \\
x_{21} - 4y_{11} - y_{21} + y_{31} &= 0 \\
x_{12} + 20y_{12} + 3y_{22} - 11y_{32} &= 0 \\
x_{22} - 2y_{12} + y_{22} + 3y_{32} &= 0
\end{aligned}
\tag{5}
$$

Finally we may use (4) to eliminate the y_{i0} from $\bar{z}$. This gives

$$
\begin{aligned}
\bar{z} = {} & 35{,}000 + y_{11} + 4y_{12} \\
& + 33{,}600 - y_{21} + 2y_{22} \\
& + 46{,}500 - 4y_{31} - y_{32}
\end{aligned}
\tag{6}
$$

The resulting initial tableau is shown in Table 3.7B in which the omitted coefficients are zero.

The reader may wish to tackle this problem with a calculator or a computer, or to try to ascertain by inspection which variables are positive and so reduce it to a smaller problem.

EXERCISES 3.7

Define the nonbasic variables and set up the initial tableau. No solution required.

1. When process P_1 is operated for one hour it costs \$60, uses amounts 14 and 9 of materials M_1 and M_2 and produces amounts 15 and 25 of M_3 and M_4. For process P_2 these numbers are \$80; 10, 18; 26, 13. Only 70 units of each of M_1 and M_2 are available. At least 100 units of each of M_3 and M_4 should be produced.
2. A dairy wishes to produce milk containing exactly 4 percent of butterfat, at least 3 percent of protein and at most 6 percent of lactose. It can buy milk from three

breeds of cows:

Breed	% Fat	% Protein	% Lactose	Cost/100 lb
1	3	6	7	$6.80
2	4	5	5	7.50
3	5	4	6	8.30

Assume that water may be added.

3. A firm has two machines P_1 and P_2. P_1 can produce 3000 wood screws (WS) or 2500 machine screws (MS) per hour, while P_2 can produce 5000 WS or 4000 MS per hour. A change from WS to MS or vice versa can be quickly made. P_1 and P_2 cost $20 and $25 per hour to operate. WS sell for $.02 each and MS for $.03. Demand exists for at most 3500 WS and 3500 MS per hour and excess production is not tolerated.
4. Some years before the energy crisis there was an oil refiner who could sell no more than 58,000 gallons per day of gasoline G_1 at 23¢ per gallon and 73,000 gallons of G_2 at 19¢ per gallon. (An excess could be disposed of at 15¢ per gallon.) The refiner could buy and refine three types of crude oil at $3.70, $2.60, and $4.10 per barrel. One barrel yielded 7, 6, or 12 gallons of G_1 *and* 12, 7, or 8 gallons of G_2.
5. A desert settlement must obtain daily at least 40,000 gallons of purified water containing at most 0.02 percent salt from a well that yields only 50,000 gallons of salty water. Three purification processes have efficiencies of 100 percent, 50 percent and 40 percent (=purified water produced/salty water consumed) and costs of 5¢, 4¢ and 1¢ per gallon of purified water, which contains 0 percent, 0.01 percent and 0.03 percent salt, respectively.
6. A dairy farm produces daily 50,000 pounds of milk M_1, 40,000 pounds of M_2 and 70,000 pounds of M_3. The percentages of butterfat are 6, 5, 3 and of other solids 12, 9, 7. The dairy's only profitable use for this production is to make grade A milk containing at least 4 percent butter and 8 percent other solids and grade AA containing at least 5 percent butter and 10 percent other solids. No water is added. The profit is 3.3¢ per pound for A and 4.2¢ per pound for AA.
7. One has to ship 30 tons of material M_1 and 40 tons of material M_2, having volumes of 1200 cubic feet and 600 cubic feet. Two modes of transportation T_1 and T_2 are available, having weight limits of 35 tons and 50 tons, volume limits of 1000 cubic feet each and costs of $15 and $12 per ton, respectively.
8. A spacecraft has several jets of different orientations and thrusts, which are assumed not to vary with time, but they can be programmed to fire independently for any desired periods of time, while fuel remains available. It is required to change the three velocity components of the center of gravity of the spacecraft by specified amounts with a minimum consumption of fuel. (See Cooper and Steinberg, 1974, pp. 36–37.)

3.8 SIGN CONVENTIONS IN THE SCHEMA AND TABLEAU

If each of the s_h and σ_h represents $+1$ or -1, the possible sign conventions for the schema are included in

$$\begin{array}{c|cc|l} & \mathbf{y} & s_1 & \\ \hline \mathbf{u} & \mathbf{A} & \mathbf{b} & =s_2\mathbf{x} \\ \sigma_1 & \mathbf{c} & d & =s_3 z \\ \hline & \| & \| & \\ & \sigma_2\mathbf{v} & \sigma_3\bar{w} & \end{array} \tag{1}$$

As usual we want $\min z$ and $\max \bar{w}$, and $\mathbf{u}, \mathbf{v}, \mathbf{x}, \mathbf{y} \geq \mathbf{0}$. Since the dual problem is uniquely determined by the tableau and the sign restrictions that define the primal, the s_h must determine the σ_h; in fact one has

$$\sigma_1 = -s_2 s_3 \qquad \sigma_2 = -s_2 \qquad \sigma_3 = -s_1 s_2 \tag{2}$$

Thus (1) represents eight possible schemas, designated as follows:

	s_1	s_2	s_3	σ_1	σ_2	σ_3	Correct Signs			
H	$+$	$-$	$+$	$+$	$+$	$+$	$z,$	$\bar{w},$		$\mathbf{v}$
H'	$+$	$+$	$-$	$+$	$-$	$-$	$\bar{z},$	$w,$	$\mathbf{x}$	
$-H$	$-$	$-$	$-$	$-$	$+$	$-$			$\mathbf{x}$	
$-H'$	$-$	$+$	$+$	$-$	$-$	$+$				$\mathbf{v}$
P	$-$	$-$	$+$	$+$	$+$	$-$		$w,$	$\mathbf{x},$	$\mathbf{v}$
P'	$+$	$+$	$+$	$-$	$-$	$-$	$z,$		$\mathbf{x},$	$\mathbf{v}$
$-P$	$+$	$-$	$-$	$-$	$+$	$+$	$\bar{z}$			
$-P'$	$-$	$+$	$-$	$+$	$-$	$+$		$\bar{w}$		

(3)

The symbols for these schemas have the following interpretations:

(i) The letter H (for homogeneous) refers to the fact that in schemas H and H' one has $s_1 = \sigma_1 = +1$, which are thus nonnegative like the other nonbasic quantities.

(ii) The letter P (for prevalent positive) refers to the fact that in P and P' the basic solutions are $\mathbf{x} = \mathbf{b}$ and $\mathbf{v} = \mathbf{c}$ without any sign changes, and at optimality $\mathbf{b}^* \geq \mathbf{0}$ and $\mathbf{c}^* \geq \mathbf{0}$. Thus P and P' have

$$s_1 s_2 = -\sigma_3 = \sigma_1 \sigma_2 = s_3 = +1$$

(iii) The accented symbols are those for which $s_2 = +1$, so that the matrix of the extended primal tableau of 2.4 and 7.1 is $[-\mathbf{I}, \mathbf{A}]$ or $[\mathbf{I}, -\mathbf{A}]$ instead of the customary $[\mathbf{I}, \mathbf{A}]$. These schemas are obtained from the unaccented by transposing the matrix containing $\mathbf{A}, \mathbf{b}, \mathbf{c}, d$, or equivalently, by exchanging s_1, s_2, s_3 with $\sigma_1, \sigma_2, -\sigma_3$ respectively, since z is comparable with $w = -\bar{w}$. The result is to change the signs of $\mathbf{A}, \mathbf{b}, \mathbf{c}, d$ in H' and $-H'$ and to change the signs of $\mathbf{A}$ and d alone in P' and $-P'$.

(iv) A minus sign preceding a symbol indicates that $\mathbf{b}$ and $\mathbf{c}$ (or equivalently, $s_1, s_3, \sigma_1, \sigma_3$) have been changed in sign. The resulting schemas are neither homogeneous nor positive.

TABLE 3.8 Comparison of Sign Conventions

Total of Scores Below (maximum = $8\frac{1}{2}$)		6	$4\frac{1}{2}$	4	2	$4\frac{1}{2}$	$2\frac{1}{2}$	
Advantages	Scores for	H	H'	$-H$	$-H'$	P	P'	Conditions
†Found without sign changes: (positive signs)	x_i	.	1	1	.	1	1	$s_1 s_2 = +1$
	y_l	.	.	1	.	1	.	$s_1 = s_2 = -1$
	v_j	$\frac{1}{2}$	.	.	$\frac{1}{2}$	$\frac{1}{2}$	$\frac{1}{2}$	$\sigma_1 \sigma_2 = +1$
	u_k	.	.	.	$\frac{1}{2}$	.	$\frac{1}{2}$	$\sigma_1 = \sigma_2 = -1$
Shown without sign changes:	z or $\bar{z}$	$\frac{1}{2}$	$\frac{1}{2}$	.	.	.	$\frac{1}{2}$	$s_1 = +1$
	$\bar{w}$ or w	$\frac{1}{2}$	$\frac{1}{2}$	.	.	$\frac{1}{2}$	.	$\sigma_1 = +1$
All + or all − row or column is inconsistent or redundant; a_{ij}, b_i, c_j have common orientations (3.7); all functions convex or all concave in convex nonlinear programming		$1\frac{1}{2}$	$1\frac{1}{2}$	.	.	$\frac{1}{2}$	.	$s_1 = \sigma_1 = +1$
Avoids use of $-\mathbf{I}$ or $-\mathbf{A}$ (7.1)		1	.	1	.	1	.	$s_2 = -1$
Negative transpose dualizes		1	1	1	1	.	.	$s_1 \sigma_1 = +1$
Program is $m+1$ by $n+1$ game (4.6)		$\frac{1}{2}$	.	.	.	.	.	H only
Ease of identification (if $-H'$ is disregarded, only H has feasible $\mathbf{b} \leq \mathbf{0}$)		$\frac{1}{2}$	.	.	.	.	.	H only

†x_i and v_j are basic; y_l and u_k are nonbasic, entering the basis.

Most of the schemas or tableaus in (3) have appeared in the literature, but the last two may be safely disregarded. The comparative advantages of the others are listed in Table 3.8.

The highest-scoring convention H has been used by Mills (1956), Tucker (1963a), Balinski and Gomory (1963), and Balinski and Tucker (1969). It is used throughout this book, although in quadratic programming there is good reason for the customary use of P'. Before adopting H, the author (like most others) used P in numerical calculations and $-H$ in discussing duality theory. However, with or without P as an alternate, further discussion seems to confirm the superiority of H over $-H$.

While it may be argued that the assignment of relative weights to the various advantages is a very subjective matter, this question becomes significant only as regards the first group of four items versus the others, since *H has perfect scores for all the latter*. Although the first group certainly deserves attention, it is lacking in structural significance, and no schema has a perfect score in this group; P does best here but it fails to show z correctly without sign changes (the constant d must be changed in sign). In fact the only schema that displays the complete basic solution $\mathbf{x}$, $\mathbf{v}$, z or $\bar{z}$ without any change of sign is P', the transpose of P. This P' serves well in Beale's quadratic algorithm but is little used in linear programming, as it has only one other small advantage. The only schemas that display the primal basic variables $\mathbf{x}$ and z or $\bar{z}$ without change of sign are H' and P'.

Although Table 3.8 looks ahead to Chapter 7 and even to nonlinear programming, it omits several considerations for each of which the effects seem to cancel one another out.

(a) If equality constraints requiring artificial variables are present as in 6.1 and 6.2, one must drop some of the sign restrictions on the c_j^* and u_j but not on b_i^* and x_i. The result is that once the initial tableau has been properly set up in 6.1, this tableau and all its successors can immediately be identified as of type H (rather than P, P', $-H$ or H') by virtue of having $\mathbf{b}\leq 0$, whereas the other four types can easily be confused, since each has $\mathbf{b}\geq 0$ and probably arbitrary signs in $\mathbf{c}$ and d. Thus in 6.1 the significance of the last line of Table 3.8 tends to be increased and that of the preceding two lines decreased.

(b) $-H$ has the attractive inequalities

$$\bar{z}=\mathbf{cy}-d\leq\mathbf{uAy}-d\leq\mathbf{bu}-d=w \tag{4}$$

slightly marred by the $-d$, and the unattractive dual constraints

$$\mathbf{v}+\mathbf{c}=\mathbf{uA} \tag{5}$$

which would be much improved by having $\mathbf{c}$ on the other side.

(c) As illustrated in 3.6 and 3.7, no convention need prevent one from defining the parameters of any particular model so that they are nonnegative. With the convention H it is obvious that every nontrivial solvable schema must contain both positive and negative entries, but any a priori assumption about the arrangement of these signs will necessarily have only limited applicability. In Table 3.6 convention $-H$ makes all the tableau entries positive in Model 1, but then it makes them all negative in Model 2, which is not the dual of Model 1.

The convention H or H' is widely used in nonlinear programming, namely, whenever the Lagrangian function is defined as $\phi(\mathbf{x})=f_0(\mathbf{x})+\boldsymbol{\lambda}\mathbf{f}(\mathbf{x})$ and the problem is either to find max $f_0(\mathbf{x})$ subject to $\mathbf{f}(\mathbf{x})\geq\mathbf{0}$, or to find min $f_0(\mathbf{x})$ subject to $\mathbf{f}(\mathbf{x})\leq\mathbf{0}$. Then one has $\boldsymbol{\lambda}\geq\mathbf{0}$ and in the more tractable cases the functions f_0, f_i and ϕ are either all concave (for max $f_0(\mathbf{x})$) or all convex (for min $f_0(\mathbf{x})$). Conventions corresponding to $-H$ or $-H'$ involve the nuisance of specifying one property (convex or concave) for f_0 and ϕ and the other for the f_i. P is also consistent with this idea in going from primal to dual but not the reverse.

In short, H has many significant advantages, while its disadvantages can be largely summarized and disposed of in one brief sentence.

When ascertaining a nonnegative value for any primal $\mathrm{x_i}$ *or* $\mathrm{y_j}$, *basic or nonbasic, discard the minus signs preceding the numerical values of the* $\mathrm{b_i}$.

The structure of linear programs demands that the b_i and the c_j have opposite signs in an optimal tableau, and various factors have led to the choice $\mathbf{b}^*\leq\mathbf{0}$, $\mathbf{c}^*\geq\mathbf{0}$. This differs from the popular P only in the signs of $\mathbf{b}$ and d (the constants column). Most of the disadvantages of $-H$ stem from its unnatural selection of a basis such as $(1,0,0)$, $(0,1,0)$, $(0,0,-1)$ for a partially ordered primal or dual vector space. After a pivot is selected, the *arithmetic* procedure of pivoting (by any method whatever) is entirely unaffected by the choice of the sign convention, if $\varepsilon_4=\varepsilon_1\varepsilon_2\varepsilon_3$ below.

Equation (1) presented the schemas as they are written in practice, with $\mathbf{A}$, $\mathbf{b}$, $\mathbf{c}$ and d representing the actual numerical entries appearing therein. To identify the notation used in any publication it may be more convenient to write its equations (primal or dual or both) in the purely symbolic or alphabetic form

$$\begin{array}{c|cc|l} & \mathbf{y} & 1 & \\ \hline \mathbf{u} & \varepsilon_1\mathbf{A} & \varepsilon_2\mathbf{b} & =-\mathbf{x} \\ 1 & \varepsilon_3\mathbf{c} & \varepsilon_4 d & =z(\min) \\ \hline & \| & \| & \\ & \mathbf{v} & \bar{w}(\max) & \end{array} \tag{6}$$

Here each ε_h is $+1$ or -1; if all are $+1$, (6) is in the homogeneous form H. Comparison with (1) shows that

$$\begin{aligned} &s_1=\varepsilon_1\varepsilon_2 \qquad \sigma_1=\varepsilon_1\varepsilon_3 \\ &\sigma_2=-s_2=\varepsilon_1 \qquad \sigma_3=\varepsilon_2 \qquad s_3=\varepsilon_3 \\ &\varepsilon_4=s_1s_3=\sigma_1\sigma_3=\varepsilon_1\varepsilon_2\varepsilon_3 \end{aligned} \tag{7}$$

Since $\varepsilon_4=\varepsilon_1\varepsilon_2\varepsilon_3$, the eight kinds of schemas are determined by $(\varepsilon_1,\varepsilon_2,\varepsilon_3)=\boldsymbol{\varepsilon}$ thus:

$$\begin{array}{lcccccccc} \boldsymbol{\varepsilon}= & +++ & --- & +-- & -++ & +-+ & --+ & ++- & -+- \\ & H & H' & -H & -H' & P & P' & -P & -P' \end{array} \tag{8}$$

CHAPTER FOUR

Matrix Games

Game theory emerged as a recognized branch of mathematics in 1944 (a few years before linear programming did the same) with the publication of the book by Von Neumann and Morgenstern, although Von Neumann's basic theorem appeared as early as 1928. Thereafter Dantzig, Gale, Kuhn, Tucker and others studied the close relationship between matrix games and linear programs, which contributes to the understanding of both areas. Extensive discussions in relatively nontechnical terms can be found in Luce and Raiffa (1957).

Sections 4.2 and 4.4 constitute the core of this chapter. An elementary course may consider 4.1 to 4.5, followed or preceded by 6.1 and 6.2, and possibly including 4.7 (games against nature), 5.1 (bimatrix games) and 10.5 (extensive games).

4.1 PURE STRATEGIES

The word "game" brings to mind activities such as basketball, chess and bridge, in which the essential element for our purposes is competition or conflict rather than amusement. Hence the more serious contests arising in social, economic, political and military conflicts can also be regarded as games. We rightly expect that these contests will differ in their susceptibility to quantitative analysis, which requires that the following elements be defined:

(a) The *players* are the competitors in the game. A player may be an individual, a temporary coalition, a team, a corporation, a political party, a nation, etc. Except in 5.7 and 5.8, this book considers only *two-person games*; they have two players denoted by I and J.

(b) A *move* in a game may be thought of as a time point (possibly coinciding with some other such points) at which one of the players is required to make a decision. Games in which a player has more than one move are considered in 10.5. In this chapter each of the two players is assumed to have only one move. The order in which the players move will be considered in (e) following.

(c) A player's *pure strategies* are the alternatives (finite in number in this book) among which he is supposed to choose when it is his move. In the two-person one-move case the pure strategies for I and J are denoted by $i=1,\dots,m$ and $j=1,\dots,n$, respectively. Instead of choosing a pure strategy outright, a player may assign probabilities to his pure strategies; this specifies a *mixed strategy*.

(d) The *payoffs* are numbers measuring the desirability of the possible outcomes of the game; e.g., they may be amounts of money. In the two-person one-move case, if I chooses i and J chooses j, the payoffs to I and J are denoted by a_{ij} and b_{ij}, respectively.

In this chapter the game is assumed to be *zero-sum*, meaning $a_{ij}+b_{ij}=0$ or $b_{ij}=-a_{ij}$; what one player gains, the other loses. In this case the m by n matrix $\mathbf{A}$ of elements a_{ij} defines a *matrix game*. In such a game it is important to remember that the object is to *choose the row* i *to try to maximize* a_{ij} *and to choose the column* j *to try to minimize* a_{ij}.

The class of two-person games like chess or matching pennies that seemingly have no numerical payoffs, being simply won, lost or drawn, can be considered as zero-sum games in which $a_{ij}=1$ if I wins, $a_{ij}=-1$ if J wins, and $a_{ij}=0$ if there is a draw.

(e) A *solution* of a game specifies the optimal strategies that rational players will use and the payoff(s) that will result. Much of this chapter will be devoted to developing precise definitions of the solution. The optimal strategies are not necessarily unique. The optimal payoff of a matrix game (called its *value*) is unique after one has specified the order in which the players move, or in any case if mixed strategies are used. As indicated in Table 4.1, the six possibilities yield at most only three different values, of which a^* is much the most important.

The first three versions (symbolized by lower-case i and j) use pure strategies; the last three (with capital I and J) use mixed strategies. The symbol $i\to j$ or $I\to J$ indicates that player I moves first; player J observes I's move and then makes his own choice. Since braces are the standard notation for an (unordered) set, the symbol $\{i, j\}$ or $\{I, J\}$ is used when the order of the moves is unspecified in the sense that each player makes his choice without knowing the other's choice, beyond what he can deduce by logical analysis. If and only if there is coincidence among the three different ways to play a game, denoted by $i\to j$, $j\to i$ and $\{I, J\}$, they are an example of $\{i, j\}$ (a saddle point), which otherwise is not solvable.

Applications of game theory that hopefully have real practical or philosophical interest will be given later in 4.7, 5.1, 5.2, 5.6, 5.7 and 10.5. For the present we will be content with a few simple examples to illustrate the ideas. Many more are available in Williams (1954) and elsewhere. This section considers the pure strategies.

TABLE 4.1 Versions of Matrix Games

Symbol	Value	Name	Comment
$i\to j$	a_0	Pure maximin	$a_0\le a^*\le a^0$
$j\to i$	a^0	Pure minimax	
$\{i, j\}$	a^*	Saddle point	Exists iff $a_0=a^0$
$I\to J$	a^*	Mixed maximin	Dual LP
$J\to I$	a^*	Mixed minimax	Primal LP
$\{I, J\}$	a^*	Matrix game (mixed strategies)	Primal and dual LP; standard version

Example 1

Consider the 2 by 3 game **A** in the rectangle in (1).

$$\begin{array}{r|ccc|cl} a_{ij}= & 3 & 1^* & 8 & 1 & \left.\right\} \max= \\ & 4^* & 10 & 0 & 0 & \quad a_0=1 \\ \hline \max_i a_{ij}= & 4 & 10 & 8 & \min_j a_{ij} & \\ & \multicolumn{3}{c}{\underbrace{\qquad\qquad}_{\min=a^0=4}} & & \end{array} \tag{1}$$

The right-hand column (1,0) shows the values of $\min_j a_{ij}$ = the smallest of $a_{i1}, a_{i2}, \ldots, a_{in}$ in row i, which is 1 for $i=1$ and 0 for $i=2$. (It does not depend on j, which is a "dummy" variable.) The larger of these numbers, $\max\{1,0\}=1$, is defined as the value of the game $i \to j$, which is denoted by

$$\text{val}\, i \to j \stackrel{\Delta}{=} \max_i \min_j a_{ij} \stackrel{\Delta}{=} a_0 \tag{2}$$

The Δ above the equality signs indicates we are defining the new symbols $\text{val}\, i \to j$ and a_0.

I moves first in (2). He knows that if he chooses the first row, J then will choose the smallest payoff in that row, which is 1, and if I chooses the second row, J will choose the smallest payoff 0 in it. Since I is maximizing, he chooses the first row, which gives him 1 instead of 0. Similarly one calculates below the rectangle in (1) that

$$\text{val}\, j \to i \stackrel{\Delta}{=} \min_j \max_i a_{ij} \stackrel{\Delta}{=} a^0 \tag{3}$$

equals $\min\{4,10,8\}=4$, where 4, 10 and 8 are the maximum payoffs $\max_i a_{ij}$ in the columns $j=1,2,3$. ■

Thus the versions $i \to j$ and $j \to i$ have different solutions, which are marked by stars in (1). This means that the version $\{i, j\}$ has no solution; when the players of this game are restricted to pure strategies, the order in which they move is essential. This is the principal reason why mixed strategies will be introduced presently.

In evaluating an expression such as $\log \sin x$ it is understood that one proceeds from right to left in the order $x \to \sin x \to \log \sin x$. Similarly in (2) and (3) *the operation on the right* (*min or max*) *is performed first*, although it is performed by the player who *moves second*! (His letter j or i likewise appears *on the right* in the symbol $i \to j$ or $j \to i$ for the version of the game.) The explanation is that the first optimization requires that the other player's pure strategy be known. Steps (a) to (c) in 4.2 give more details.

Example 2. A Strictly Determined Game

I and J are partners in a trading or investment firm. J is more creative in his thinking but he has a tendency to go overboard. Hence they have agreed that J

will decide which commodity is suitable for purchase or sale, while I will decide whether 1, 2 or 3 lots of the commodity will be traded. Currently I and J have opposite opinions; I thinks the firm should increase its cash position while J thinks less cash would be desirable. The firms' research staff has suggested three possibilities: to sell commodity $j=1$ at $10,000 per lot, to buy $j=2$ at $15,000 per lot, and to sell $j=3$ at $20,000 per lot.

Thus in terms of units of $1000, the payoffs (the changes in the cash assets) are $a_{ij}=\alpha_i\beta_j$ where $\boldsymbol{\alpha}=(1,2,3)$, $\boldsymbol{\beta}=(10,-15,20)$.

$$\begin{array}{c} \quad j=1 \quad j=2 \quad j=3 \\ \mathbf{A}=\begin{bmatrix} 10 & -15 & 20 \\ 20 & -30 & 40 \\ 30 & -45 & 60 \end{bmatrix} \end{array} \quad \begin{array}{l} (1 \text{ lot})\ (i=1) \\ (2 \text{ lots})\ (i=2) \\ (3 \text{ lots})\ (i=3) \end{array} \tag{4}$$

The element 40, for example, means that 2 lots of commodity 3 have been sold at $20,000 each, thus adding $40,000 to the cash on hand. The solutions for the versions $i\to j$ and $j\to i$ are calculated in (5).

$$\begin{array}{c|ccc|c|l} a_{ij} & 10 & -15^* & 20 & -15 & \text{max} \\ & 20 & -30 & 40 & -30 & =a_0 \\ & 30 & -45 & 60 & -45 & =-15 \\ \hline \max_i a_{ij} & 30 & -15 & 60 & \min_j a_{ij} & \\ & \multicolumn{3}{c}{\min=a^0=-15} & & \end{array} \tag{5}$$

These two solutions are the same; each chooses $i=1$, $j=2$ for a payoff of -15. Hence this common solution, which is called a saddle point, is accepted as the solution of the game $\{i, j\}$ with simultaneous or independent moves. It is also the solution of $\{I, J\}$, because the pure strategies are included among the mixed strategies. ■

A game like (5) (having $a^0=a_0$) is said to be *strictly determined* or to have a *saddle point* or a *solution in pure strategies*, the unordered version $\{i, j\}$ being understood. Two equivalent conditions for this are the equality of (2) and (3), that is,

$$\max_i \min_j a_{ij} = \min_j \max_i a_{ij} \left(\triangleq a_{kl}, \text{ say} \right) \tag{6}$$

and the existence of some a_{kl} that is the minimum over j in its row $i=k$ and the maximum over i in its column $j=l$, thus:

$$\max_i \min_j a_{ij} \geq \min_j a_{kj} = a_{kl} = \max_i a_{il} \geq \min_j \max_i a_{ij} \tag{7}$$

It is obvious that (6) implies (7); conversely (7) implies (6) with the help of (11) below. Equation (7) explains the term "saddle point"; there is a point near the center of a saddle that is lowest in the fore-and-aft direction and highest in the left-and-right direction. In the game matrix the horse is headed east.

In Chapter 3 we had to remember that $\mathbf{b}\leq\mathbf{0}$ and $\mathbf{c}\geq\mathbf{0}$ for feasibility or optimality; in this section we must remember to maximize over i and minimize over j. Equation (7) ties these conventions together as follows. A linear programming tableau in nonnegative variables

$$\begin{matrix} \mathbf{A} & \mathbf{b} \\ \mathbf{c} & 0 \end{matrix} \tag{8}$$

(in which the corner element $d=0$) is *optimal* if and only if it is the matrix of a zero-sum game in which the element $d=0$ is a *saddle point*. Beginners in game theory should check this out a few times until the conventions are thoroughly learned. Further details given in 3.8 and 4.6 are optional.

The following table is another aid in remembering the conventions.

Problem	Dual	Primal
Equations	Columns	Rows
Optimum	Maximum	Minimum
Variables	$u, v, \bar{w}$	x, y, z
Strategies	i or $\mathbf{u}$	j or $\mathbf{y}$
Player	I	J

(8a)

In each row of (8a), the word or symbol in the second column alphabetically precedes that in the third column. Of course this property would be lost if i and j were exchanged with their equivalents "row" and "column."

In many games, such as chess and war, we are accustomed to think it advantageous to have the first move. In such cases the payoff matrix $\mathbf{A}$ depends on who moves first. For example, let I receive \$2 if he is the first to choose heads and let J receive \$1 if he is the first to choose heads; there is no payment if both players choose tails ($i=j=2$). Then the payoffs are

$$\mathbf{A}=\begin{matrix} 2 & 2 \\ -1 & 0 \end{matrix} \quad \text{or} \quad \begin{matrix} -1 & 2 \\ -1 & 0 \end{matrix} \tag{9}$$

according as I or J moves first.

With a *fixed* matrix $\mathbf{A}$, if it makes any difference at all, it is advantageous to be the *second* to move in a matrix game with pure strategies only; the second player to move is able to match his strategy to the first but not vice versa. For a formal demonstration, let us write

$$a_0 \triangleq a_{kl} \triangleq \max_i \min_j a_{ij} = \min_j a_{kj} \leq a_{kq} \tag{10}$$

$$a^0 \triangleq a_{pq} \triangleq \min_j \max_i a_{ij} = \max_i a_{iq} \geq a_{kq}$$

corresponding to I and J, respectively, as the first to move. In each line the two equalities on the left are definitions and the rest follows directly from these and the meaning of max and min. Equation (10) implies

$$a_0 = \max_i \min_j a_{ij} \leq \min_j \max_i a_{ij} = a^0 \tag{11}$$

In short, the *maximin*$\leq$*minimax*; *the operation* (*min or max*) *on the right is the more powerful.* (As noted above, this is the first optimization but the second move.) In words the proof is that $a_0 = a_{kl} \leq a_{kq} \leq a_{pq} = a^0$ because a_{kq} is in the row of a_0 and in the column of a^0, and a_0 is the smallest element in its row k and a^0 is the largest element in its column q.

A notable characteristic of a solution of $\{i, j\}$ or $\{I, J\}$ is that it does not matter if one of the players is obliged to reveal his strategy in advance to the other player, provided that both know game theory. Many games, even the simplest, lack this property when only pure strategies are considered. In matching pennies, if one player is required to announce his choice of head or tail in advance, he is put at the mercy of his opponent. (The payoff matrix is

$$\begin{array}{ccc} & H & T \\ H & 1 & -1 \\ T & -1 & 1 \end{array} \tag{12}$$

if a match makes I the winner.)

Everyone knows the solution of this game. The optimal (mixed) strategy for each player consists in choosing head or tail at random with a probability of $1/2$ for each; the average payoff is $1 \cdot 1/2 - 1 \cdot 1/2 = 0$. Either of these optimal strategies could indeed be announced in advance without incurring any disadvantage. Not only does a random (mixed) strategy serve to hinder the opponent, but it is the classic method for dividing what would otherwise be indivisible and making continuous what was originally discrete. In 4.2 we study this in detail and in 4.4 we show that every matrix game has a solution in terms of mixed strategies.

EXERCISES 4.1

In terms of pure strategies, determine the saddle points (if any) or else the minimax and the maximin for 1 to 3.

1. $\begin{matrix} 2 & -4 & -1 \\ 2 & 5 & 2 \\ -6 & 8 & 1 \end{matrix}$ **2.** $\begin{matrix} 3 & -2 & 0 \\ -1 & 4 & 1 \\ 7 & -5 & 2 \end{matrix}$ **3.** $\begin{matrix} -2 & 3 & -1 \\ 5 & 0 & 4 \\ 3 & 1 & 2 \end{matrix}$

4. Use (11) to show that if there is more than one saddle point, all give the same payoff.

Solve 5 to 7 in the form (a) $i \to j$ (b) $j \to i$ (c) $\{I, J\}$.

5. $\begin{matrix} 7 & 1 \\ 1 & 7 \end{matrix}$ (see (12)) **6.** $\begin{matrix} 4 & 2 \\ 0 & 1 \end{matrix}$ **7.** $\begin{matrix} -1 & 4 \\ 3 & 3 \end{matrix}$

Set up the game matrix for 8 and 9.

8. Each player has the pure strategies *paper*=1, *rock*=2, and *scissors*=3. Identical choices have $a_{ii}=0$; otherwise the winner is paid 1. The winner is indicated by "paper covers rock, rock breaks scissors, scissors cut paper."
9. A watchman I inside a building can guard only one of its three entrances, but his position cannot be seen from outside by the robber J, whose capture is worth \$500. If J breaks into the guarded entrance, he is captured with probability .3, .6 or .7, depending on the entrance, or else he escapes without any loot. If he breaks into an unguarded entrance, he will surely escape with loot worth \$200, 300 or 400, respectively.

4.2 MIXED STRATEGIES AND GRAPHICAL SOLUTIONS

Consider the matrix game

$$\begin{array}{r|cc|ll} & 1^* & 90 & 1 & \rbrace \; \max \\ & 11^* & 0 & 0 & \; =1 \\ \hline \max\limits_i a_{ij} = & 11 & 90 & \min\limits_j a_{ij} & \\ & \multicolumn{2}{c}{\underbrace{\hphantom{11 \quad 90}}_{\min=11}} & & \end{array} \tag{1}$$

If the moves are unordered, I might nevertheless begin conservatively by solving $i \to j$, which gives him at least $a_0=1$; and J might begin by solving $j \to i$, which insures that he will pay at most $a^0=11$. When the resulting strategies confront one another, any payoff in the closed interval $[a_0, a^0]$ might result in general, but here it is $a_0=1$.

If this game is played repeatedly, it is clear that I will soon venture to choose $i=2$; if J continues to choose $j=1$ (as he is most likely to do because of his fear of having to pay 90), this will increase I's return from 1 to 11. However, if I always chooses $i=2$, J will presently acquire the courage to choose $j=2$ and so reduce the payoff to 0 more and more frequently.

Therefore let u_2 denote the probability with which I chooses $i=2$, independently of whatever J may do, so that in a large number r of plays of the game, I chooses $i=1$ approximately $r(1-u_2)$ times and $i=2$ approximately ru_2 times. If J always chooses $j=1$ the total payoff will be approximately

$$1 \cdot r(1-u_2)+11ru_2=r(1+10u_2) \tag{2}$$

If J always chooses $j=2$ it is $r \cdot 90(1-u_2)$. When $u_2=.89$ these totals are the same, that is, $9.9r$ or an average of 9.9 per play regardless of what J may do. The approximations involved can be made as good as desired, in a probabilistic sense, by choosing r large enough.

A similar calculation for J, with y_2 denoting the probability of choosing $j=2$, gives

$$1 \cdot (1-y_2)+90 \cdot y_2=11(1-y_2)+0 \cdot y_2 \tag{3}$$

Then $y_2 = .1$ and the average payoff is again 9.9 per play, regardless of what I may do. This type of analysis will be studied more carefully hereafter. However, it is plausible that the equality of the payoffs in (2) and (3) is advantageous because it deprives the other player of any meaningful choice.

In general the mixed strategies chosen by I and J are *probability vectors* $\mathbf{u}$ and $\mathbf{y}$ respectively. This means

$$\mathbf{u} \geq \mathbf{0} \qquad (\text{all } u_i \geq 0), \qquad u_1 + u_2 + \cdots + u_m = 1$$

$$\mathbf{y} \geq \mathbf{0} \qquad (\text{all } y_j \geq 0), \qquad y_1 + y_2 + \cdots + y_n = 1 \tag{4}$$

The two players' probabilities are assumed to be *independent*, so that the product $u_i y_j$ is the joint probability that I chooses i and J chooses j in the same play of the game. When these $u_i y_j$ are used as weights in averaging the corresponding payoffs a_{ij}, the *expected payoff*

$$\sum_{i=1}^{m} \sum_{j=1}^{n} u_i a_{ij} y_j \triangleq \mathbf{uAy} \tag{5}$$

is obtained. As usual it represents a transfer from J to I. Of course the sum of the probabilities $u_i y_j$ themselves is 1: $\Sigma_i \Sigma_j u_i y_j = (\Sigma_i u_i)(\Sigma_j y_j) = 1$. This notation is consistent with our usage in linear programming; in the literature of game theory the letter $\mathbf{x}$ is generally used instead of $\mathbf{u}$. To be sure, in our primal tableaus in 4.4, the labels x_i will replace the dual u_i as they usually do.

The definition (5) is in accord with the interpretation that if the game is played a large number of times, the proportion of the plays in which a_{ij} occurs will approach $u_i y_j$ and so the average payoff per play will approach (5). One does not have to employ this interpretation, however. Mathematically it is natural to think of a mixed strategy as a new kind of pure strategy, which like any pure strategy can be chosen and used (via (5)) in a single play of the game, although as a matter of terminology it is convenient to maintain the distinction between pure and mixed strategies. Note that the set of all mixed strategies $\mathbf{u}$ forms a simplex of dimension $m-1$ as defined early in 3.2 and the set of $\mathbf{y}$ forms a simplex of dimension $n-1$. In dimensions 1 and 2 they appear as shown in Figure 4.2A, in isolation in the upper drawings, and imbedded in a two- or three-dimensional coordinate system in the lower drawings. In the lower right-hand drawing the simplex is the equilateral triangle with corners $(1,0,0)$, $(0,1,0)$ and $(0,0,1)$.

Consider the version $I \rightarrow J$ in which I is the first to choose a mixed strategy. Since he has to be concerned about J's response, I might imagine a trial-and-error investigation as follows:

(a) I chooses a tentative mixed strategy $\mathbf{u}^0$.

(b) J chooses a $\mathbf{y} = \mathbf{y}^0$ to minimize $\mathbf{u}^0\mathbf{Ay}$ while $\mathbf{u}^0$ is constant.

(c) After J has thus reacted to every possible $\mathbf{u}^0$, I chooses the $\mathbf{u}^0$ that maximizes $\mathbf{u}^0\mathbf{Ay}^0$, wherein $\mathbf{y}^0$ is a function of $\mathbf{u}^0$ as described in (b).

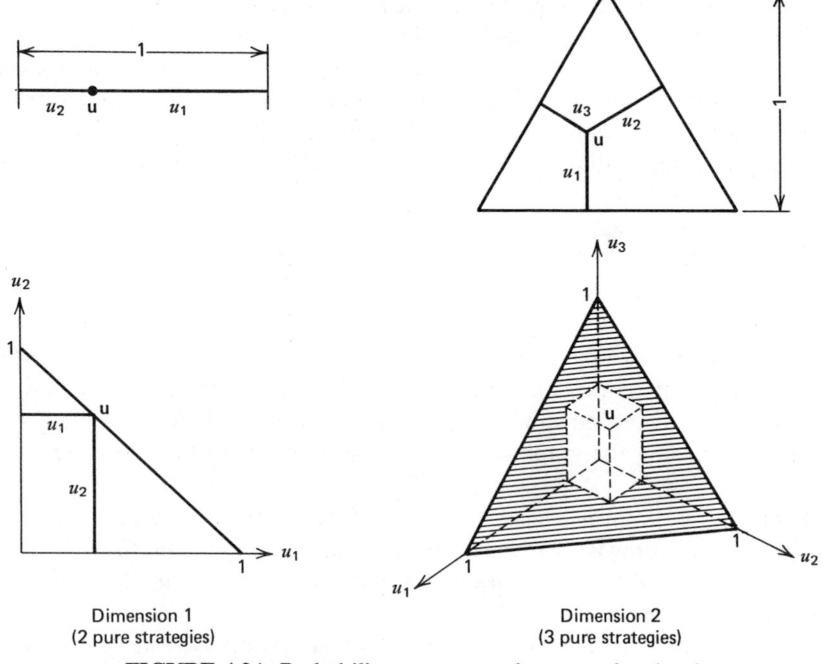

FIGURE 4.2A Probability vectors as elements of a simplex.

Obviously step (c) is more complicated than step (b), though to be sure it has to be performed only once. The result is that while it will usually be essential to consider all the mixed strategies $\mathbf{u}^0$ in step (c), it is easy to see that for the one-sided purposes of I's analysis, $\mathbf{y}^0$ could be restricted to pure strategies j without loss of generality; in step (b) one is simply minimizing (with respect to nonnegative weights y_j) the average of the n components $\Sigma_i u_i^0 a_{ij}$ $(j=1,\ldots,n)$ of the constant vector $\mathbf{u}^0\mathbf{A}$, and such an average can be as small as, but no smaller than, the smallest of the individual components.

Thus I's calculation is summarized by the expression

$$\bar{w}^* = \max_{\mathbf{u}} \left[\min_{j|\mathbf{u}} \sum_{i=1}^{m} u_i a_{ij} \right] \tag{6}$$

The notation $\min_{j|\mathbf{u}}$ means the minimum obtained by varying j while $\mathbf{u}$ is held fixed. Note that the letters $\mathbf{u}\, j|\mathbf{u}$ below the "max min," when read from right to left, correspond to the steps (a), (b) and (c), respectively. Of course there is a similar analysis when J chooses first in the version $J \to I$, that is,

$$z^* = \min_{\mathbf{y}} \left[\max_{i|\mathbf{y}} \sum_{j=1}^{n} a_{ij} y_j \right] \tag{7}$$

In 4.4 we will show that $\bar{w}^* = z^*$ and it doesn't matter who moves first; (6) and (7) yield a dual pair of linear programs. (If the players are limited to pure strategies, (6) and (7) reduce to the usually unequal values (2) and (3) of 4.1.)

As an example to illustrate the procedure, suppose the players I and J are tennis clubs that are simultaneously choosing individual representatives for an exhibition match. I has two candidates $i=1,2$ available and J has four, $j=1,2,3,4$. The estimated probabilities that i will defeat j are as follows, in percent:

i	$j=1$	2	3	4
1	30	10	60	25
2	40	60	30	55

(8)

When I has only two pure strategies as in (8), (6) can be evaluated graphically as in Figure 4.2B. This is possible because only the n pure strategies of J have to be considered, while I's mixed strategies depend on a single real number, say $u_2=1-u_1$. The steps in drawing the graph correspond more or less to the foregoing steps (a), (b) and (c).

(a′) Plot $(1-u_2)a_{1j}+u_2a_{2j}$ as a function of u_2 for $0\le u_2\le 1$ and $j=1,2,3,4$. For each j this involves drawing a (dashed) straight line from the point $(0, a_{1j})$ on the left to $(1, a_{2j})$ on the right. These lines give the values of $\Sigma_i u_i a_{ij}$ in (6); each must be labeled with its value of j.

(b′) Over portions of the dashed lines, draw the solid polygonal line showing for each u_2 the smallest amount that J can pay. It gives the values of $\min_j \Sigma_i u_i a_{ij}$ in (6). To get it, J uses $j=2$, 1 or 3, depending on the value of u_2.

(c′) Select the highest point on this solid line, which occurs at $u_2=.75$, $\bar{w}=37.5$. Thus the mixed strategy $\mathbf{u}^*=(.25,.75)$ guarantees that I will gain at least 37.5 (the percentage chance that I will win the match). This 37.5 is the numerical value of (6).

The precision of the solution thus obtained is not limited to the precision of the graph, as long as the latter clearly indicates which two of J's pure strategies (in this case $j=1$ and 3) he will use. The value $u_2=.75$ was obtained by solving

$$30(1-u_2)+40u_2=60(1-u_2)+30u_2$$

which results after eliminating u_1 and $\bar{w}$ from the equations

$$\begin{aligned} \bar{w}&=u_1a_{11}+u_2a_{21}=30u_1+40u_2 \\ \bar{w}&=u_1a_{13}+u_2a_{23}=60u_1+30u_2 \\ 1&=u_1+u_2 \end{aligned} \tag{9}$$

As long as I uses his optimal strategy (.25,.75), J may use any strategy involving $j=1$ or 3. However, J should protect himself by preventing I from changing to a less conservative strategy. If J always chose $j=3$, I could increase his payoff to 60 by taking $u_2=0$; if J always chose $j=1$, I could get 40 by taking $u_2=1$. To prevent this, J must assign probabilities y_1 and y_3 to $j=1$ and 3 so that the resulting payoff is independent of u_2; i.e., it is the same for both $i=1$ and $i=2$.

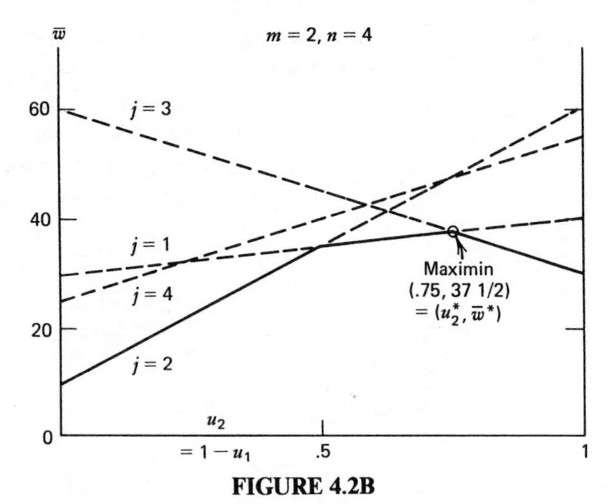

FIGURE 4.2B

This can be done much as in (9), by solving

$$30(1-y_3)+60y_3=40(1-y_3)+30y_3$$

which results after eliminating y_1 and z from

$$\begin{aligned} z&=a_{11}y_1+a_{13}y_3=30y_1+60y_3\\ z&=a_{21}y_1+a_{23}y_3=40y_1+30y_3\\ 1&=y_1+y_3 \end{aligned} \tag{10}$$

This gives $\mathbf{y}^*=(.75,0,.25,0)$ as J's optimal strategy. The results of 4.4 show that each player has an optimal strategy involving at most the smaller of the numbers m, n of their respective pure strategies.

This completes the solution of the game, but to give an example of a graphical solution from J's point of view, this is done in Figure 4.2C for the probabilities y_3

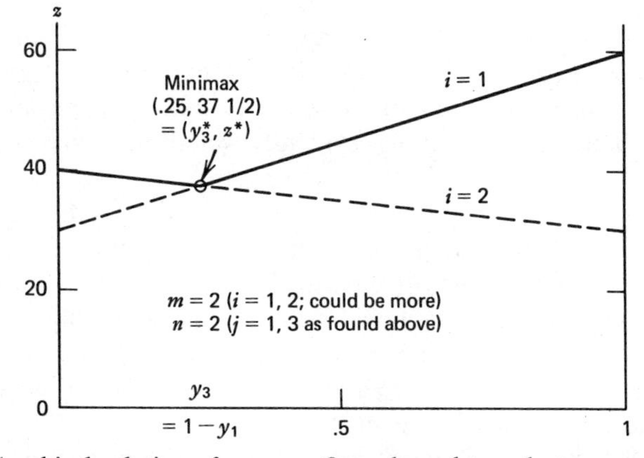

FIGURE 4.2C Graphical solution of a game. One player has only two pure strategies (one probability u_2 or y_3 is to be found).

and $y_1 = 1 - y_3$ that have already been found to be the only positive ones for J. Then I chooses the higher of the two lines in every case, and J chooses $y_3 = .25$ as before to minimize the resulting payoff graphed by the solid polygonal line.

To check the optimality of a pair of probability vectors $\mathbf{u}^*$ and $\mathbf{y}^*$, each of which has nonnegative components whose sum is one, the expected payoff for each pure strategy of each player against the optimal mixed strategy of his opponent should be calculated as follows:

$$\begin{array}{c|cccc|cc}
u_i^* & y_j^* = .75 & 0 & .25 & 0 & \Sigma_j a_{ij} y_j^* & \\
\hline
.25 & 30 & 10 & 60 & 25 & 37.5 & \text{max} = \\
.75 & 40 & 60 & 30 & 55 & 37.5 & 37.5 \\
\hline
\Sigma_i u_i^* a_{ij} = & 37.5 & 47.5 & 37.5 & 47.5 & & \| \\
 & & \text{min} = 37.5 & & & = & a^*
\end{array} \tag{11}$$

These expected payoffs are the scalar products of $\mathbf{u}^*$ with the columns of $\mathbf{A}$ and the products of $\mathbf{y}^*$ with the rows of $\mathbf{A}$. If the minimum of the former (in the new row at the bottom of (11)) equals the maximum of the latter (in the new column at the right), the common value a^* is called the *value of the game*.

This test is the game-theory analogue of Theorem 7* of 3.3. I's maximization in (11) shows him that $a^* \leq 37.5$; he cannot get more unless J changes to a less effective strategy, which he will not do. Similarly J's minimization shows that $a^* \geq 37.5$; he cannot pay less unless I is foolish. Hence $a^* = 37.5$ and each player has done the best that he can.

To summarize, $\mathbf{u}^*$ and $\mathbf{y}^*$ are optimal strategies for I and J respectively if and only if they satisfy (4) and

$$\min_j \sum_{i=1}^{m} u_i^* a_{ij} = \max_i \sum_{j=1}^{n} a_{ij} y_j^* \tag{12}$$

(Compare with (6) and (7)).

The value a^* always has the same interpretation as the a_{ij} in the matrix. In a sense the array $\mathbf{A}$ of payoffs a_{ij} is replaced by a single payoff a^*, on the assumption that both players know and use their optimal strategies. In the example a value of 50 (a 50 percent chance of I's winning) evidently would mean that the two clubs were evenly matched.

If a_{ij} is a transfer of money or some other commodity from J to I, then a^* is the average amount transferred per game in a long series of plays. Thus a payment of a^* from J to I should compensate for the cancellation of a game previously agreed on, and a payment of a^* from I to J should make the players willing to schedule a game. In this way any game can be made fair, although it is customary to say that $a^* = 0$ characterizes a *fair game*. Of course a game may represent an inescapable situation and as such have to be played, fair or not.

Figure 4.2D shows the relationship of the value of the game a^* and the optimal strategies $\mathbf{u}^*$ and $\mathbf{y}^*$ to the contours of constant values of the average payoff $\mathbf{uAy}$

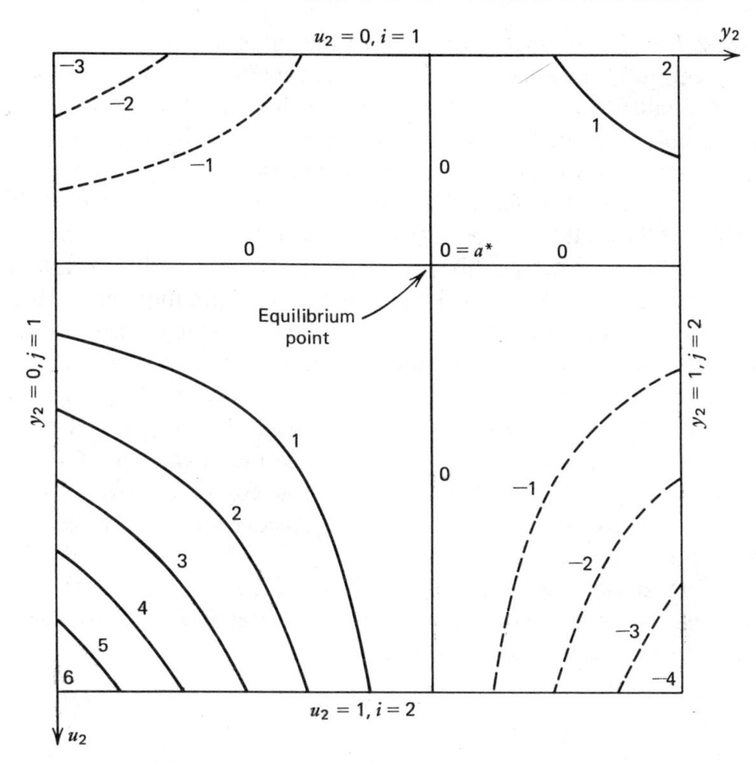

FIGURE 4.2D Equilibrium point and contours of values of **uAy** for the game having

$$\mathbf{A}=\begin{bmatrix} -3 & 2 \\ 6 & -4 \end{bmatrix}, \qquad a^*=0, \qquad \begin{matrix} \mathbf{u}^*=(2,1)/3, \\ \mathbf{y}^*=(2,3)/5. \end{matrix}$$

considered as a function of $y_2 = 1 - y_1$ and $u_2 = 1 - u_1$ for a particular 2 by 2 game. It displays the optimal solution as an equilibrium point (a saddle point in terms of mixed strategies).

The properties of optimal mixed strategies seem to justify the random selection of pure strategies with the appropriate probabilities not only in repeated plays of a game but even in a single play that does not involve intolerable risks. In support of this viewpoint one may cite the type of experience represented by Lloyd's of London, who reputedly will make a bet on almost anything by putting it in the form of an insurance policy; many different games, each played once, may still heed the law of averages. However, when little or no repetition is involved and the risk is substantial, one or both of two difficulties may arise that may be called problems of scaling and of mutual confidence, respectively. The position of player J in (1) above, with his potential loss of 90 units, will serve to illustrate both of these problems.

The *problem of scaling* becomes especially troublesome when small probabilities of very large gains or losses are involved. If J's loss of 90 represents years subtracted from his lifetime or the loss of his entire fortune of \$90,000,000, he may well feel that his distaste for this outcome is not 90 but virtually infinite, and he will refuse to assign any positive probability to his second pure strategy. Conversely, in some circumstances I might feel that 90 overstated the value of this payoff to him. These objections seem quite legitimate. While there are mathematical theorems about the possibility of transforming the payoffs to so-called utility scales that hopefully eliminate the objection, such a transformation may be difficult to find in practice and in any case would probably destroy the zero-sum property of the game.

The *problem of mutual confidence* arises for J when he is the agent of others, for whom the loss of 90 would be substantial but not so large as to raise the problem of scaling. Then J may feel that his professional integrity calls for him to use a randomized strategy, but what will happen to his reputation and career if the unlucky outcome is the loss of 90 and there are no replays to enable him to recoup? Even if those who pass judgment on him have an appreciation of game theory, there may remain the nuisance of being prepared to prove, if challenged after the unfortunate fact, that he was using the optimal probability value $y_2 = .1$ rather than a stupid or traitorous one such as $y_2 = .5$ or 1. Since the advantage of $y_2 = .1$ is very modest anyway, yielding an expected loss of 9.9 as compared with 11 (or conceivably only 1) resulting from $y_2 = 0$, J might well consider the randomized strategy not worth the trouble.

It seems fair to say, however, that in many if not most cases the theoretical appropriateness of a mixed strategy will imply its practical appropriateness, if the game in fact describes the practical situation.

EXERCISES 4.2

Solve 1 to 4 and check your solutions as in (11) and (12).

1. $\begin{matrix} 0 & 3 & 5 & 1 \\ 4 & 0 & -1 & 2 \end{matrix}$

2. $\begin{matrix} 1 & 4 & -2 & 3 \\ 2 & -1 & 3 & 1 \end{matrix}$

3. $\begin{matrix} 0 & 4 \\ 3 & 0 \\ 5 & -1 \\ 1 & 2 \end{matrix}$

4. $\begin{matrix} 1 & 2 \\ 4 & -1 \\ -2 & 3 \\ 3 & 1 \end{matrix}$

5. A player tosses a coin before playing. If he gets a head, he uses the mixed strategy (.3,.4,.3); if he gets a tail, he uses (.1,.2,.7). Is this a new type of strategy? Explain.

Use (11) to (12) to find which are optimal in 6 to 7.

6. (a) $\mathbf{u}=(3,4,0)/7$, $\mathbf{y}=(0,4,3)/7$
(b) $\mathbf{u}=(2,1,0)/3$, $\mathbf{y}=(1,0,0)$
(c) $\mathbf{u}=(1,1,0)/2$, $\mathbf{y}=(1,0,0)$

$$\mathbf{A}= \begin{matrix} 0 & -1 & 3 \\ 0 & 2 & -1 \\ -2 & 1 & 1 \end{matrix}$$

7. (a) $\mathbf{u}=(2,5,0)/7$, $\mathbf{y}=(6,0,1)/7$
(b) $\mathbf{u}=(1,2,0)/3$, $\mathbf{y}=(0,1,1)/2$
(c) $\mathbf{u}=(1,0,3)/4$, $\mathbf{y}=(2,2,0)/4$

$$\mathbf{A}= \begin{matrix} 4 & 7 & -1 \\ 3 & 1 & 5 \\ 6 & 5 & 0 \end{matrix}$$

8. How do $\mathbf{u}=(48,34,23)/115$
$\mathbf{y}=(23,51,31)/115$
relate to the game?

$$\mathbf{A}= \begin{matrix} 0 & 5 & 1 \\ 3 & 0 & 7 \\ 8 & 2 & 0 \end{matrix}$$

9. Let $\begin{bmatrix} a & b \\ c & d \end{bmatrix}$ be a 2 by 2 game matrix. Show that it has a saddle point if $(a-b)(c-d)\geq 0$ or $(a-c)(b-d)\geq 0$, and otherwise the solution involves the mixed strategies

$$\mathbf{u}^* = \left(\frac{d-c}{a+d-b-c}, \frac{a-b}{a+d-b-c} \right), \qquad \mathbf{y}^* = \left(\frac{d-b}{a+d-b-c}, \frac{a-c}{a+d-b-c} \right)$$

$$a^* = \frac{ad-bc}{a+d-b-c} = a + (b-a) y_2^*$$

In the latter case show that the diagonals of the matrix are *separated*; i.e., either a and d are both larger than b and c, or vice versa.

10. Let $\alpha \le \beta \le \gamma \le \delta$ be the entries in a 2 by 2 game matrix, in any arrangement. Show that the value of the game always lies in the closed interval $[\beta, \gamma]$.

11. Show that the symmetric game

$$\begin{matrix} 0 & c_3 & -c_2 \\ -c_3 & 0 & c_1 \\ c_2 & -c_1 & 0 \end{matrix}$$

has $\mathbf{u}^* = \mathbf{y}^* = (c_1, c_2, c_3)/(c_1 + c_2 + c_3)$ if c_1, c_2, c_3 all have the same sign (i.e., the nonzero payoffs in each row and column have opposite signs). Otherwise it has a saddle point.

4.3 DOMINATION; FICTITIOUS PLAY

This section briefly considers a few miscellaneous topics. Beside their possible usefulness they permit the reader to acquire more familiarity with games before beginning the general theory of the following section.

(i) *Linear transformation of the payoffs*. By using the foregoing test (12), one can easily see that if $\mathbf{u}^*$ and $\mathbf{y}^*$ are optimal for payoffs a_{ij}, they are also optimal for payoffs $sa_{ij} + a$, where $s > 0$ but otherwise s and a are arbitrary constants; the value is changed from a^* to $sa^* + a$. This is not surprising since a is a fixed payment not affected by the choice of strategies and s in effect merely changes the units in which the payments are made. In 4.6 and 5.4 the constant a will be used to insure that the value of the resulting game has a specified sign.

If $s < 0$, a new game is obtained that may have no relation to the old, because the effect is to reverse the action of the max and min. Since the same reversal results from interchanging rows and columns, however, one sees that when $s > 0$, the optimal strategies for I and J in the game (a_{ij}) are the same as those for J and I, respectively, in the game $(-sa_{ji} + a)$, and the latter has the value $-sa^* + a$. When $s = 1$ and $a = 0$, one obtains the negative transpose $-\mathbf{A}^T$ of $\mathbf{A}$; in effect this operation simply exchanges the labels I and J that are applied to the two players so that the values in general are related by $\operatorname{val} \mathbf{A} = -\operatorname{val}(-\mathbf{A}^T)$. If $\mathbf{A} = -\mathbf{A}^T$ this reduces to $a^* = -a^* = 0$ and the game is called *symmetric*.

More generally a matrix $\mathbf{A}$ may be regarded as defining a symmetric game if a permutation of its rows and/or a permutation of its columns suffices to convert $\mathbf{A}$ into $-\mathbf{A}^T$ (not $-\mathbf{A}$ or $\mathbf{A}^T$). This agrees with our expectation that matching pennies is a symmetric game, because interchanging the two rows of $\begin{bmatrix} 1 & -1 \\ -1 & 1 \end{bmatrix}$ gives $\begin{bmatrix} -1 & 1 \\ 1 & -1 \end{bmatrix}$, which is the negative transpose. This game cannot be arranged to have $a_{ij} = -a_{ji}$. Shapley (1964) gives several other examples.

(ii) The *domination* of one row or column of **A** by another may permit one to reduce the size of a game matrix by discarding the dominat*ed* row or the dominat*ing* column, as in the game

i	$j=1$	2	3	4	5
1	−6	−1	2	4	−1
2	8	−2	−4	−5	−2
3	−6	−2	1	2	0

(1)

One may first observe that $\mathbf{a}_{.2} \leq \mathbf{a}_{.5}$, by which we mean that every element in the second column vector is less than or equal to the corresponding element in the fifth column vector. Thus if J ever used $j=5$, he could use $j=2$ instead and be sure of doing at least as well. Therefore the fifth column $\mathbf{a}_{.5}$ can be discarded without changing the value of the game. Conceivably there may be some so-called optimal strategies having $y_5 > 0$, which will be lost. By construction they are as good as any other optimal strategies if I is rational, but if they exist at all they will be inferior in the event of certain irrational actions on the part of I, such as choosing $u_3 > 0$. Similarly one may discard the third row $\mathbf{a}_{3.}$ because $\mathbf{a}_{3.} \leq \mathbf{a}_{1.}$ after $\mathbf{a}_{.5}$ has been discarded. The result is the smaller equivalent game matrix

i	$j=1$	2	3	4
1	−6	−1	2	4
2	8	−2	−4	−5

(2)

which would admit a graphical solution.

Pure strategies can be discarded equally well by comparison with mixed strategies. Thus if (2) contained a column $(-2,3) = \mathbf{a}_{.6}$, say, it could be discarded because $\mathbf{a}_{.6} \geq (\mathbf{a}_{.1} + \mathbf{a}_{.3})/2 = (-2,2)$. Also it is easy to see that with the sign conventions used for the b_i and c_j in this book, rows and columns in a linear program in basic $x_i \geq 0$ and nonbasic $x_j \geq 0$ can be discarded in just the same way, with the added possibility (not present in games) of multiplying the rows and columns by arbitrary positive numbers before comparing or adding them.

In a large matrix, a search for dominance relations is unlikely to be worth the trouble unless it is assisted by theoretical or practical insights that suggest where dominance is likely to occur.

(iii) *Fictitious play.* If two players play a game repeatedly without knowing their optimal strategies, it would be natural for each to keep a record of the past actions of the other and to be influenced by that record in making a choice in the current play. For computational purposes, although probably not in actual play, one may suppose that the current choice of a pure strategy is precisely that which would have yielded the best total payoff, had it been used against all the past choices of the opponent. For simplicity the fictional players choose alternately rather than simultaneously. Brown (1951) suggested the resulting "fictitious play" as a means for getting an approximate solution of a game.

Robinson (1951) proved that the procedure yields a sequence converging to the value of the game, but convergence may be very slow, no doubt reflecting the fact that the optimal strategy for one player is not determined by the optimal strategy for the other but by potential (in this case, actual) departures from the latter. Von Neumann (1954) undertook to improve Brown's procedure by using mixed strategies as responses in each play and optimizing the weighting of the data, but still the error in the value decreases only as the inverse square root of the number of iterations. If the optimum strategies are not unique, both procedures may be indecisive as to which optimum strategy is being approximated. Shapley (1964) showed that the method fails for some general classes of variable-sum games.

Whatever its computational merits, the original procedure of Brown is an easy way to get practice in applying the elementary ideas of game theory. Consider the game (1) or (3).

(a) Begin by having J, say, choose a minimax pure strategy $j=2$. This insures that a saddle point will be discovered if any exists.

(b) In the resulting column vector of payoffs $(-1,-2,-2)$, I chooses $i=1$ to obtain the largest, -1, which is an upper limit on the value a^* of the game.

(c) In the resulting row $(-6,-1,2,4,-1)$, J chooses $j=1$ to obtain the smallest, -6, which is a lower limit on a^*.

(d) The resulting column $(-6,8,-6)$ is added to the preceding to obtain the total $(-7,6,-8)$. I chooses $i=2$ to obtain the largest, 6. When this is divided by the number of plays, 2, one obtains $a^* \leq 3$, a weaker result than that already obtained in (b).

(e) The resulting row $(8,-2,-4,-5,-2)$ is added to the preceding to obtain the total $(2,-3,-2,-1,-3)$. J chooses either $j=2$ or $j=5$ to obtain the smallest, -3. Dividing by the number of plays, 2, gives $a^* \geq -3/2$. Strategy $j=2$ is chosen because it is dominated by $j=5$ (the most important reason), or because it was more often chosen previously, or because $2<5$.

(f) The resulting column $(-1,-2,-2)$ is added to the previous total to obtain $(-8,4,-10)$ and $a^* \leq 4/3$, etc.

The foregoing calculations may be arranged in the following compact form. The asterisks record the successive selections of pure strategies; the last column has none because it has not yet generated a row.

						(b)	(d)	(f)	
	−6	−1	2	4	−1	−1*	−7	−8	
	8	−2	−4	−5	−2	−2	6*	4	(3)
	−6	−2	1	2	0	−2	−8	−10	
(a)		*							
(c)	−6*	−1	2	4	−1				
(e)	2	−3*	−2	−1	−3				

Already we have found that $-3/2 \leq a^* \leq -1$, and have a suggestion that perhaps each player will use his first two pure strategies. In fact, for any $\mathbf{u}$ and $\mathbf{y}$ one has

$$\min_j \sum_i u_i a_{ij} \leq \mathbf{uAy} \leq \max_i \sum_j a_{ij} y_j \tag{4}$$

$$\min_j \sum_i u_i a_{ij} \leq \max_{\mathbf{u}'} \min_j \sum_i u_i' a_{ij} = a^* \tag{5}$$

$$a^* = \min_{\mathbf{y}'} \max_i \sum_j a_{ij} y_j' \leq \max_i \sum_j a_{ij} y_j$$

the latter being established in 4.4.

The game

$$\begin{matrix} -5 & 4 & -1 \\ 6 & -5 & 1 \end{matrix} \tag{6}$$

illustrates the limitations of fictitious play. The optimal strategies $\mathbf{u}^* = (6,5)/11$ and $\mathbf{y}^* = (0,2,9)/11$ use $j=2$ and 3, whereas fictitious play has a long-persisting initial preference for $j=1$ and 2 because these columns contain $\min_j a_{ij} = -5$. In a figure like 4.2B, the linear graphs for the three columns very nearly have a point in common.

EXERCISES 4.3

Use the idea of domination to reduce the size of the matrix in 1 and 2. Then solve graphically.

1. $\begin{matrix} 2 & 4 & 1 \\ 6 & -3 & 6 \\ 0 & 3 & -1 \end{matrix}$

2. $\begin{matrix} 4 & -3 & 3 & 1 \\ 1 & 0 & -1 & 0 \\ -2 & 5 & -4 & 6 \end{matrix}$

Tabulate fictitious plays (six for each player) for the games in 3 to 5.

3. (8) of 4.2

4. $\begin{matrix} 6 & -6 & 4 & -3 \\ -5 & 6 & -7 & 4 \\ -3 & 4 & 6 & -7 \end{matrix}$

5. (6) of 4.3

Are 6 and 7 fair games? Solve by domination and 4.2.

6. $\begin{matrix} 0 & 1 & -1 \\ -1 & 3 & 1 \\ 1 & -1 & -3 \end{matrix}$

7. $\begin{matrix} 0 & 1 & 2 \\ 1 & 0 & -1 \\ -2 & -1 & 0 \end{matrix}$

Are 8 and 9 symmetrical games? Demonstrate.

8. $\begin{matrix} 0 & a & -a \\ a & b & -b \\ -a & -b & b \end{matrix}$

9. $\begin{matrix} a & -a & b & -c \\ -a & a & -c & b \\ c & -b & d & -d \\ -b & c & -d & d \end{matrix}$

10. Show that $\Sigma_i \Sigma_j a_{ij} = 0$ for a symmetric game, even in the generalized sense.

11. Compare the sense of the inequalities in (3) or (5) of this section with that in (1) of 4.1. Explain.

For $-\infty < t < \infty$, express the value of the game as a function of t. (First find what values of t, if any, make each element a saddle point. Then use Exercise 9 of 4.2.)

12. $\begin{matrix} t & 2 \\ -1 & 1 \end{matrix}$ **13.** $\begin{matrix} 7t+8 & 1 \\ t+2 & 4t+5 \end{matrix}$ **14.** $\begin{matrix} t & 2 \\ 1 & -t \end{matrix}$

4.4 SOLVING MATRIX GAMES AS LINEAR PROGRAMS

In 4.2 the analysis (a) to (c) of the game from I's point of view led to the definition (6) for $\bar{w}^*$, that is,

$$\bar{w}^* = \max_{\mathbf{u}} \min_{\mathbf{y}} \mathbf{uAy} = \max_{\mathbf{u}} \min_{j} \sum_{i=1}^{m} u_i a_{ij} \tag{1}$$

$\bar{w}^*$ is called the maximin payoff for mixed strategies; it is the same as the value a^* of the game $\mathbf{A}$. Any probability vector $\mathbf{u}=\mathbf{u}^*$ that yields the maximum in (1) is called an optimal strategy for I. The optimal strategy $\mathbf{y}^*$ for J is determined not by (1) but by (3) to (5) below.

The equality of the two minima in (1) is easily given a formal proof. Since $\mathbf{uAy}=\sum_{i=1}^{m} u_i a_{ij}$ when $y_j=1$ and all other components of $\mathbf{y}$ are zero, one has

$$\min_{\mathbf{y}} \mathbf{uAy} \leq \min_{j} \sum_{i=1}^{m} u_i a_{ij} \tag{1'}$$

The reverse inequality follows by summing

$$y_l \sum_{i=1}^{m} u_i a_{il} \geq y_l \min_{j} \sum_{i=1}^{m} u_i a_{ij}$$

over l, and then minimizing the resulting left member $\mathbf{uAy}$ over $\mathbf{y}$, which no longer appears on the right.

In order to reduce (1) to a linear program we note that the minimum over j may also be described as the *greatest lower bound*, so that (1) may be written as

$$\bar{w}^* = \max_{\mathbf{u}} \bar{w} \qquad \text{such that } \bar{w} \leq \sum_{i=1}^{m} u_i a_{ij} \qquad \text{and } \sum_{i=1}^{m} u_i = 1 \tag{2}$$

for all $j=1,\ldots,n$ and all $u_i \geq 0$. The maximization of $\bar{w}$ in (2) not only achieves player I's objective in selecting $\mathbf{u}$ but is also needed to complete the description of J's objective (minimizing over j), by converting the lower bound $\bar{w}$ into the greatest lower bound or $\min_j \sum_i u_i a_{ij}$. It is obvious that (2) has an optimal solution since any probability vector $\mathbf{u}$ is feasible for it, and $\mathbf{u} \geq 0$ and $0 \leq u_i \leq 1$ implies that $\bar{w}$ is bounded above by $\max_{i,j} a_{ij}$.

A similar analysis from the point of view of player J choosing $\mathbf{y}$ first leads to the definition

$$z^* = \min_{\mathbf{y}} \max_{\mathbf{u}} \mathbf{uAy} = \min_{\mathbf{y}} \max_{i} \sum_{j} a_{ij} y_j \tag{3}$$

and the linear program

$$z^* = \min_{\mathbf{y}} z \qquad \text{such that } z \geq \sum_{j=1}^{n} a_{ij} y_j \qquad \text{and} \qquad \sum_{j=1}^{n} y_j = 1 \tag{4}$$

for all $i=1,\ldots,m$ and all $y_j \geq 0$. This minimization of z describes J's objective and also converts the upper bound z into the least upper bound or $\max_i \Sigma_j a_{ij} y_j$.

After slack variables v_j and x_i are inserted, (2) and (4) become the dual and the primal problem, respectively, of the schema

$$\begin{array}{c|cccc|l}
 & y_1 & \cdots & y_n & -z & \\
\hline
u_1 & a_{11} & \cdots & a_{1n} & 1 & = -x_1 \\
\vdots & & & \vdots & & \quad \vdots \\
u_m & a_{m1} & \cdots & a_{mn} & 1 & = -x_m \\
\overline{w} & -1 & \cdots & -1 & 0 & = -1 \\
\hline
 & \| & & \| & \| & \\
 & v_1 & \cdots & v_n & 1 &
\end{array} \tag{5}$$

which has the labels $-z$, $\overline{w}$, 1 and -1 in unusual locations. Here all variables are required to be nonnegative except $\overline{w}$, which is to be maximized as usual, and z, which is to be minimized. Hence the basic solutions $x_i = -1$ and $v_j = -1$ in (5) are infeasible. Note that the problem of the player who chooses *rows* is given by the *column* equations and vice versa. Tucker (1960b) applied his schema in essentially this way; see also Balinski and Tucker (1969).

The scheme (5) is peculiar in having $\overline{w}$ and z nonbasic and the constant 1 basic whereas the reverse is normally true (e.g., in (7) of 3.1). An exchange of labels will remedy the abnormality, but this cannot be done in a single pivot step because the pivotal position (in the lower right corner) has 0 as its entry. Instead we must determine (perhaps as suggested hereafter) some x_k and y_l to be exchanged also. For setting down the algebra and for programming a computer it is a little better to exchange y_l with 1 and then $-z$ with x_k since this makes both pivots equal to ± 1.

For solving games with pencil and paper the preceding method has the slight disadvantage of locating the new constants $\mathbf{b}$ and $\mathbf{c}$ in the interior of the tableau, thus making it desirable to move them out. Therefore we begin by pivoting on a_{kl} itself to exchange x_k and y_l and then pivot on 1 (over the denominator a_{kl}) in the lower right corner to exchange $-z$ with 1. We indicate later how to choose $a_{kl} \neq 0$.

If i and j denote variable indices as usual, but with $i \neq k$ and $j \neq l$, the initial primal tableau and its two successors may be written as

$$\begin{array}{cccc}
1 & y_j & y_l & -z \\
\hline
x_i & a_{ij} & a_{il} & 1 \\
x_k & a_{kj} & a_{kl}^* & 1 \\
1 & -1 & -1 & 0
\end{array} \tag{6}$$

a_{kl}	y_j	x_k	$-z$
x_i	$a_{ij}a_{kl}-a_{il}a_{kj}$	$-a_{il}$	$a_{kl}-a_{il}$
y_l	a_{kj}	1	1
1	$a_{kj}-a_{kl}$	1	1*

1	y_j	x_k	1
x_i	$a_{ij}+a_{kl}-a_{il}-a_{kj}$	-1	$a_{il}-a_{kl}$
y_l	1	0	-1
$-z$	$a_{kj}-a_{kl}$	1	a_{kl}

(7)

(7) is a tableau of standard type, though of course it shows the effects of the pivot steps that produce it. Thus no more pivots will be chosen in the last row or column. (7) could be used as a model for going directly from (6) to (7), but students should not do this unless they fully understand the procedure using two pivot steps. The latter are checked in the usual way with an additional column, but in (7) there is also the check that for all $i \neq k$ and $j \neq l$,

$$a_{ij}=a_{kl}+(a_{kj}-a_{kl})+(a_{il}-a_{kl})+(a_{ij}+a_{kl}-a_{il}-a_{kj}) \tag{8}$$

the sum of four entries at the corners of a rectangle in (7).

In order to solve (5) by the methods of 3.3, (7) must be either primal or dual feasible. Inspection of (7) shows that for this to be true the pivot a_{kl} must be either the smallest a_{ij} in its row or the largest in its column; that is, it must satisfy one or the other of the two conditions that determine a saddle point in (7) of 4.1. This together with $a_{kl} \neq 0$ suffices to make a_{kl} acceptable as a pivot. However, it is possible to be more specific.

The tableau (7) is primal feasible if $a_{kl}=\max_i a_{il}$. Since it makes $z=a_{kl}$ and z is to be minimized, it is reasonable to choose

$a_{kl}=\min_j \max_i a_{ij} \triangleq a^0$. The other alternative is to choose

$a_{kl}=\max_i \min_j a_{ij} \triangleq a_0$ and have (7) dual feasible. Either of these choices will discover a saddle point if one exists.

Evidently the selection of the pivotal columns will be more critical when they are the more numerous; i.e., when $n>m$. In this case no one row alone (unless there is a saddle point) should be allowed to dominate the selection of the pivotal column(s), as happens if $a_{kl}=a_0$ or the tableau is dual-feasible. Therefore the following is suggested:

If $m \leq n$, let $a_{kl}=a^0=\min_j \max_i a_{ij}$.
If $m>n$, let $a_{kl}=a_0=\max_i \min_j a_{ij}$. (9)

In other words, one prefers to *optimize the a_{ij} in the short direction first*. By choosing the less effective direction for the more effective optimization (the first of the two), one may hope to obtain that one of a^0 and a_0 that is closer to a^*.

There are two ways available to insure that $a_{kl} \neq 0$. One is to observe that (7) is unchanged if, before pivoting in (6), one adds $1-a_{kl}$ to every a_{ij} (thus making the pivot$=1$), and then after (7) is obtained, one replaces the lower right-hand element (which will be the initial pivot$=1$) by the original value of a_{kl}. The alternative is simply to change the selection made in (9) when necessary to insure $a_{kl} \neq 0$. In the latter case the following procedure for reducing a game to a linear program is obtained.

Reduction of a Game to a Linear Program

(a) Calculate a^0 or a_0 as in (9). If a^0 is the smallest in its row or a_0 the largest in its column, it is a saddle point; *stop*. If a^0 or $a_0 = 0$, calculate the other for use as the first pivot a_{kl}.

(b) If there is no saddle point, form a tableau by adjoining the following to the game matrix **A**: a first row of labels 1, **y**, $-z$ and a first column 1, **x**, 1 (these follow the usual convention except that $-z$ and the nonbasic 1 have changed places); a value $d=0$ in the lower right corner; all $b_i = 1$ in the remainder of the last column and all $c_j = -1$ in the remainder of the last row. (In a conventional tableau all these b_i and c_j would be described as infeasible; obviously they go with the given labels.)

(c) Pivot first on the selected a_{kl} and then on the entry (1 over the denominator a_{kl}) in the lower right corner. The result is a conventional tableau that is primal feasible if $a_{kl} = a^0$ and dual feasible if $a_{kl} = a_0$.

(d) One or more additional pivot steps are required to optimize this conventional tableau, unless a saddle point was overlooked in step (a). J's probability vector is $\mathbf{y}^*$ and I's probability vector is $\mathbf{u}^*$, the dual counterpart of $\mathbf{x}$.

Let us apply this procedure to the game

$$\mathbf{A} = \begin{bmatrix} -6 & -1 & 2 \\ 8 & -2 & -4 \\ 10 & -3 & 3 \end{bmatrix} \tag{10}$$

$$\max_i a_{ij} = \underbrace{10 \quad -1 \quad 3}_{\min = a^0 = -1}$$

Since the matrix is square, there is no "shorter" direction, but (9) calls for optimizing vertically first by finding the largest a_{ij} in each column. The following tableaus result from the initial pivot $a_{kl} = a^0 = a_{12} = -1$; this is not a saddle point because it is not the smallest in its row. The use of a check column is recommended.

1	y_1	y_2	y_3	$-z$
x_1	-6	-1^*	2	1
x_2	8	-2	-4	1
x_3	10	-3	3	1
1	-1	-1	-1	0

-1	y_1	x_1	y_3	$-z$
y_2	-6	1	2	1
x_2	-20	2	8	1
x_3	-28	3	3	2
1	-5	1	3	1^*

1	y_1	x_1	y_3	1
y_2	1	0	1	-1
x_2	15^*	-1	-5	-1
x_3	18	-1	3	-2
$-z$	-5	1	3	-1

15	x_2	x_1	y_3	1
y_2	-1	1	20	-14
y_1	1	-1	-5	-1
x_3	-18	3	135	-12
$-z$	5	10	20	-20

The primal simplex rules of 3.3 give the last pivot 15.

The last tableau gives $z^* = a^* = -20/15$. The column equations with u_j in place of x_j give $\mathbf{u}^* = (10,5,0)/15$, and the row equations give $\mathbf{y}^* = (1,14,0)/15$. The slack variables x_i and v_j are of little interest. The probabilities being nonnegative and adding up to 1, they are checked as in (11) and (12) of 4.2:

$$
\begin{array}{c|ccc|cl}
u_i^* & y_j^* = \frac{1}{15} & \frac{14}{15} & 0 & \Sigma_j a_{ij} y_j^* & \\
\hline
\frac{2}{3} & -6 & -1 & 2 & -\frac{4}{3} & \\
\frac{1}{3} & 8 & -2 & -4 & -\frac{4}{3} & \text{max} \\
0 & 10 & -3 & 3 & -\frac{32}{15} & = -4/3 \\
\hline
\Sigma_i u_i^* a_{ij} = & -\frac{4}{3} & -\frac{4}{3} & 0 & & \| \\
 & \multicolumn{3}{c}{\text{min} = -4/3} & = & a^*
\end{array}
\tag{11}
$$

Other methods of solution will be considered in 4.5 and 4.6.

EXERCISES 4.4

Solve the games in 1 through 6 by linear programming.

1. $\begin{matrix} 3 & 1 & -5 \\ -1 & -2 & 2 \\ 1 & 3 & -1 \end{matrix}$ **2.** $\begin{matrix} 4 & 1 & -1 & -4 \\ 0 & -6 & 2 & -3 \\ -2 & -1 & 0 & 2 \end{matrix}$ **3.** $\begin{matrix} 0 & 3 & 4 \\ 7 & 0 & 2 \\ 1 & 2 & -1 \end{matrix}$

4. $\begin{matrix} 4 & 3 & 2 \\ 2 & 3 & 3 \\ 1 & 4 & 5 \end{matrix}$ **5.** $\begin{matrix} 6 & -1 & 1 & -2 \\ 4 & 5 & 2 & 3 \\ -2 & 3 & 0 & -1 \end{matrix}$ **6.** $\begin{matrix} 1 & 0 & 2 \\ 0 & 2 & 0 \\ 2 & 0 & -1 \end{matrix}$

7. If you have good reason to believe that $u_p^* > 0$ and $y_q^* > 0$, how might the algorithm be modified?

8. Go from (6) to (7) by first exchanging y_l with 1.

9. Use (1)=(3) to show that $a_0 \le a^* \le a^0$ (see (11) of 4.1).

4.5 THE MINIMAX THEOREM; KERNELS

The linear-programming proof that $a^* = \overline{w}^* = z^*$ in 4.4 establishes the following famous result, first proved by Von Neumann (1928) by much less elementary methods.

THEOREM 1 If $\mathbf{u}$ and $\mathbf{y}$ are arbitrary probability vectors with m and n components and $\mathbf{A}$ is a real m by n matrix, one has

$$
\begin{aligned}
\max_{\mathbf{u}} \min_{\mathbf{y}} \mathbf{uAy} &= \min_{\mathbf{y}} \max_{\mathbf{u}} \mathbf{uAy} \\
&= \max_{\mathbf{u}} \min_{j} \sum_{i=1}^{m} u_i a_{ij} = \min_{\mathbf{y}} \max_{i} \sum_{j=1}^{n} a_{ij} y_j
\end{aligned}
\tag{1}
$$

(the maximin $\bar{w}^*$ = the minimax z^*). The common value in (1) uniquely defines the *value* a^* of the game **A**. The *optimal strategies* $\mathbf{u}^*$ and $\mathbf{y}^*$, which need not be unique, are determined by the second optimization (on the left in each member). If it is omitted, then one has

$$\min_j \sum_{i=1}^{m} u_i a_{ij} \leq a^* \leq \max_i \sum_{j=1}^{n} a_{ij} y_j \tag{1'}$$

for any probability vectors $\mathbf{u}$ and $\mathbf{y}$.

THEOREM 2 Probability vectors $\mathbf{u}^*$ and $\mathbf{y}^*$ are optimal if and only if any one of the equivalent properties (2a) to (2d) holds. Equations (1), (2a) and (2b) have the common value $a^* = \mathbf{u}^*\mathbf{A}\mathbf{y}^*$.

$$\min_j \sum_{i=1}^{m} u_i^* a_{ij} = \max_i \sum_{j=1}^{n} a_{ij} y_j^* \tag{2a}$$

$$\min_{\mathbf{y}} \mathbf{u}^*\mathbf{A}\mathbf{y} = \mathbf{u}^*\mathbf{A}\mathbf{y}^* = \max_{\mathbf{u}} \mathbf{u}\mathbf{A}\mathbf{y}^* \tag{2b}$$

$$\mathbf{u}^*\mathbf{A}\mathbf{y} \geq \mathbf{u}^*\mathbf{A}\mathbf{y}^* \geq \mathbf{u}\mathbf{A}\mathbf{y}^* \tag{2c}$$

for all probability vectors $\mathbf{u}$ and $\mathbf{y}$ ($\mathbf{u}^*, \mathbf{y}^*$ is an *equilibrium point*).

$$\sum_{i=1}^{m} u_i^* a_{ij} \geq \mathbf{u}^*\mathbf{A}\mathbf{y}^* \geq \sum_{j=1}^{n} a_{ij} y_j^* \text{ for all } i, j \tag{2d}$$

Proof If $\mathbf{u}^*$ and $\mathbf{y}^*$ are optimal, by definition they may replace $\mathbf{u}$ and $\mathbf{y}$ in the second line of (1). The $\max_u$ and $\min_y$ being thereby accomplished, they may be omitted and so (2a) is obtained with value a^*. Conversely, if (2a) is true, write it as $\bar{w} = z$. Then

$$\max_{\mathbf{u}} \min_j \sum_{i=1}^{m} u_i a_{ij} \geq \bar{w} = z \geq \min_{\mathbf{y}} \max_i \sum_{j=1}^{n} a_{ij} y_j \tag{3}$$

By (1) or the analogue of (11) of 4.1 for mixed strategies, equality must hold in (3) and so $\mathbf{u}^*$ and $\mathbf{y}^*$ are optimal.

It remains to prove the equivalence of (2a) to (2d) by establishing the circle of implications

$$(2c) \Rightarrow (2b) \Rightarrow (2a) \Rightarrow (2d) \Rightarrow (2c) \tag{4}$$

The first is obvious; the second follows from the equality in (1′) of 4.4. For the rest, let a denote the common value in (2a), which implies

$$\sum_{i=1}^{m} u_i^* a_{ij} \geq a \geq \sum_{j=1}^{n} a_{ij} y_j^* \text{ for all } i, j \tag{5}$$

This will reduce to (2d) when we have shown that $a=\mathbf{u}^*\mathbf{A}\mathbf{y}^*$. Now multiply the left inequality in (5) by y_j and the right inequality by u_i and sum over j and i respectively. The result is $\mathbf{u}^*\mathbf{A}\mathbf{y}\geq a\geq\mathbf{u}\mathbf{A}\mathbf{y}^*$ or (2c), since setting $\mathbf{u}=\mathbf{u}^*$ and $\mathbf{y}=\mathbf{y}^*$ gives $a=\mathbf{u}^*\mathbf{A}\mathbf{y}^*$. This completes the proof.

THEOREM 3 Equality holds in (2d) for every j for which $y_j^*>0$ and every i for which $u_i^*>0$; i.e., all $u_i^*x_i^*=v_j^*y_j^*=0$.

Proof In (2d) multiply the left inequality by y_j^* and the right inequality by u_i^* and sum over j and i respectively. The result is

$$\mathbf{u}^*\mathbf{A}\mathbf{y}^*\geq\mathbf{u}^*\mathbf{A}\mathbf{y}^*\geq\mathbf{u}^*\mathbf{A}\mathbf{y}^*$$

but if the theorem failed to be true in even one instance, the impossible strict inequality $\mathbf{u}^*\mathbf{A}\mathbf{y}^*>\mathbf{u}^*\mathbf{A}\mathbf{y}^*$ would be obtained. This is the game-theory version of complementary slackness (Theorem 6 of 3.3).

THEOREM 4 The set of all optimal pairs $(\mathbf{u}^*,\mathbf{y}^*)$ is convex polyhedral and is the Cartesian product of the set of $\mathbf{u}^*$ and the set of $\mathbf{y}^*$.

Proof The Cartesian product means that if $(\mathbf{u}^*,\mathbf{y}^*)$ and $(\mathbf{u}^0,\mathbf{y}^0)$ are optimal, so are $(\mathbf{u}^*,\mathbf{y}^0)$ and $(\mathbf{u}^0,\mathbf{y}^*)$; any optimal $\mathbf{u}$ can be paired with any optimal $\mathbf{y}$. To show this, replace $\mathbf{u}$ and $\mathbf{y}$ in (2c) by $\mathbf{u}^0$ and $\mathbf{y}^0$, exchange the roles of * and 0, and combine the results to show

$$\mathbf{u}^*\mathbf{A}\mathbf{y}^0=\mathbf{u}^*\mathbf{A}\mathbf{y}^*=\mathbf{u}^0\mathbf{A}\mathbf{y}^*=\mathbf{u}^0\mathbf{A}\mathbf{y}^0 \tag{6}$$

Then note that in applying the optimality test (2d), each vector $\mathbf{u}$ or $\mathbf{y}$ can now be checked independently of the other.

The individual sets and their product are convex polyhedral because they are defined by linear equations and inequalities. Other properties of these sets are given by Gale and Sherman (1950) and Bohnenblust, Karlin and Shapley (1950).

Kernels

The material in this subsection has diverse motivations. It leads to a possibly simpler method for solving small games, say with $m+n\leq 6$. It also provides a method for attacking rather difficult problems such as determining all optimal solutions or solving a game depending on a parameter as in Example 3 of 5.6, (10) and (11) of 5.3, and Exercises 13 and 14 of this section.

An (optimal) *kernel* $\mathbf{\Lambda}$ of a game may be defined as the square matrix containing those payoffs a_{ij} whose labels x_i and y_j have been exchanged in some particular optimal tableau. Thus $\mathbf{\Lambda}$ includes all the a_{ij} such that $u_i^*>0$ and $y_j^*>0$ in the corresponding basic optimal solution, while all the slacks $x_i=0$ and $v_j=0$ in the equations involving $\mathbf{\Lambda}$. Briefly, all the positive slacks are outside the kernel and all the positive probabilities are inside the kernel and uniquely determined thereby.

Then the row and column equations in (5) of 4.4 reduce to

$$\begin{aligned}
&u_1^* a_{1j} + \cdots + u_\lambda^* a_{\lambda j} = a^* && \text{for } j=1,\ldots,\lambda \\
&u_1^* + u_2^* + \cdots + u_\lambda^* = 1, && u_i^* = 0 \text{ for } i>\lambda \\
&a_{i1} y_1^* + \cdots + a_{i\lambda} y_\lambda^* = a^* && \text{for } i=1,\ldots,\lambda \\
&y_1^* + y_2^* + \cdots + y_\lambda^* = 1, && y_j^* = 0 \text{ for } j>\lambda
\end{aligned} \tag{7}$$

where $\bar{w}=z=a^*$ and the notation has been simplified by supposing that the kernel includes precisely the values $1,2,\ldots,\lambda$ for i and j. If solved correctly for both $\mathbf{u}^*$ and $\mathbf{y}^*$, (7) insures that $x_1=x_2=\cdots=x_\lambda=0$ and $v_1=v_2=\cdots=v_\lambda=0$ but *all the other nonnegativity conditions on u_i, v_j, x_i, y_j must be confirmed* after (7) is solved, in order to assure that the prospective kernel is in fact optimal. As usual, if $\mathbf{u}^*$ and $\mathbf{y}^*$ are probability vectors satisfying (2a), their optimality is guaranteed. It is notable that in (7) and (2a) the a_{ij} such that neither i nor j is in the alleged kernel are irrelevant; they could be changed arbitrarily without affecting the result of the test, although the existence of optimal strategies derived from other kernels could be affected.

Linear programs also have kernels; the corresponding equations are

$$\begin{aligned}
&u_1^* a_{1j} + \cdots + u_\lambda^* a_{\lambda j} + c_j = 0 && \text{for } j=1,\ldots,\lambda \\
&a_{i1} y_1^* + \cdots + a_{i\lambda} y_\lambda^* + b_i = 0 && \text{for } i=1,\ldots,\lambda
\end{aligned} \tag{8}$$

As before, $u_i^*=0$ for $i>\lambda$, $y_j^*=0$ for $j>\lambda$, both sets of equations must be solved, and the solutions must be checked by Theorem 7* of 3.3.

This approach depends on identifying an optimal kernel. Equation (11) following shows that the idea of domination (4.3 (ii)) will not always suffice for this purpose. If all the optimal solutions are desired via Theorem 7 following, all the square submatrices have to be considered as prospective kernels. Plausible suggestions for identifying a kernel in a small game of unknown character will be mentioned later. Another similar game already solved may provide a clue. If $\mathbf{u}^*$ is known and $\mathbf{y}^*$ is not, the column equations of (5) of 4.4 give $\mathbf{v}^*$. If there are equal numbers of $u_i^*>0$ and $v_j^*=0$, these uniquely determine the corresponding kernel; otherwise some guessing may still be required in bringing in enough additional rows or columns (not both) to yield a square submatrix. Finally the following theorems may determine the kernel.

THEOREM 5 **A** itself is its own unique optimal kernel if **A** is m by m and there exists a constant a such that **A** or $-\mathbf{A}$ satisfies $a_{ij}>a$ if $i=j$, $a_{ij}<a$ if $i\neq j$, and either $\Sigma_i a_{ij} \geq ma$ for all j or $\Sigma_j a_{ij} \geq ma$ for all i.

Remark. This theorem and several related ones are stated and proved by Bohnenblust, Karlin and Shapley (1950). In their convenient terminology it is stated thus: a dominant and separated diagonal in a square matrix implies that the game is completely mixed (i.e., has a unique solution $\mathbf{u}^*>0$, $\mathbf{y}^*>0$). The rows or columns may be permuted and/or the signs of the a_{ij} reversed in order to satisfy the conditions. The following is a special case of another of their theorems.

THEOREM 6 A 3 by 3 game $\mathbf{A}$ is its own unique kernel if the three sets of numbers $\{a_{11}, a_{22}, a_{33}\}$, $\{a_{12}, a_{23}, a_{31}\}$ and $\{a_{13}, a_{21}, a_{32}\}$ lie in disjoint intervals. (In a larger matrix these intervals must be ordered as the diagonals are in the matrix.) After possible permutation, this is equivalent to having the three largest entries uniquely defined and lying in different rows and columns, and the three smallest entries likewise.

The solution of (7) will now be considered. When $\lambda=1$ a saddle point is being sought; the solution for $\lambda=2$ is given in Exercise 9 of 4.2. In general the following procedure is derived from Cramer's rule (Theorem 8 of 7.7, where background information on determinants is reviewed). Let $\mathbf{\Lambda}$ be the kernel; let $\operatorname{tradj}\mathbf{\Lambda}$ be its (transposed) adjoint, obtained from $\mathbf{\Lambda}$ by replacing each a_{ij} by its cofactor in the determinant of $\mathbf{\Lambda}$; and let K and L be the sets of row and column indices in $\mathbf{\Lambda}$. Then

$$
\begin{aligned}
\mathbf{u}_K^* &= \sigma^{-1}\cdot(\text{sum of column vectors of } \operatorname{tradj}\mathbf{\Lambda})\\
\mathbf{y}_L^* &= \sigma^{-1}\cdot(\text{sum of row vectors of } \operatorname{tradj}\mathbf{\Lambda})\\
a^* &= \sigma^{-1}\cdot\text{determinant of } \mathbf{\Lambda} \qquad (9)\\
\sigma &= \text{sum of all elements of } \operatorname{tradj}\mathbf{\Lambda}\\
&= \text{sum of the components of } \sigma\mathbf{u}_K^* \text{ or } \sigma\mathbf{y}_L^*
\end{aligned}
$$

The other components of $\mathbf{u}^*$ and $\mathbf{y}^*$ are zero. It remains to verify that $\mathbf{u}^*$ and $\mathbf{y}^*$ are probability vectors satisfying (2a).

For example, the game matrix

$$
\mathbf{A}=\begin{bmatrix} 1 & 4 & -2\\ -1 & 0 & 3\\ 3 & -3 & 0 \end{bmatrix} \text{ with } \operatorname{tradj}\mathbf{A}=\begin{bmatrix} 9 & 9 & 3\\ 6 & 6 & 15\\ 12 & -1 & 4 \end{bmatrix}
$$

is itself its only kernel by Theorem 6. Here $15=(-1)^{2+3}\begin{vmatrix} 1 & 4\\ 3 & -3\end{vmatrix}=-1\cdot(-3)+3\cdot 4$, etc. The sums of the column and the row vectors of $\operatorname{tradj}\mathbf{A}$ are $(21,27,15)$ and $(27,14,22)$. Then $\sigma=63$, $\mathbf{u}^*=(21,27,15)/63$, $\mathbf{y}^*=(27,14,22)/63$, $a^*=(-2\cdot 3+3\cdot 15+0\cdot 4)/63=13/21$ (using the third column).

Equations (9) solve (7) if and only if $\sigma\neq 0$. Two applications of Theorem 5 of 7.7 show that $\pm\sigma$ is the determinant $|\mathbf{\Lambda}'|$ of the matrix $\mathbf{\Lambda}'$ obtained from $\mathbf{\Lambda}$ by adjoining a column of 1's and a row of -1's with 0 at their intersection, as in (5) of 4.4, which has $\mathbf{\Lambda}'$ as a submatrix. Theorem 7 of 7.7 shows that $|\mathbf{\Lambda}'|\neq 0$ is a necessary and sufficient condition to permit the exchange of all the labels in the rows and columns of $\mathbf{\Lambda}'$ by pivot steps. This will exhibit (9) as dual and primal basic solutions of (5) of 4.4; by Theorem 8 of 3.3, this is possible if and only if the optimal solutions $\mathbf{u}^*$ and $\mathbf{y}^*$ given by (9) are extreme points of the sets of all optimal strategies for I and J respectively. Thus we obtain the following result given by Shapley and Snow (1950); also see McKinsey (1952).

THEOREM 7 Optimal strategies $\mathbf{u}^*$ and $\mathbf{y}^*$ for a game $\mathbf{A}$ are extreme points for the sets of all optimal strategies for I and J respectively if and only if there is a kernel $\boldsymbol{\Lambda}$ (a square submatrix of $\mathbf{A}$) such that (9) yields $\mathbf{u}^*$ and $\mathbf{y}^*$ with $\sigma \neq 0$. (The components not in $\mathbf{u}_K^*$ or $\mathbf{y}_L^*$ are zero.) Note that if $\boldsymbol{\Lambda}$ is 1 by 1, then $\operatorname{tradj} \boldsymbol{\Lambda} = [1]$ and $\sigma = 1$.

The following table presents estimates of the minimum number of multiplications or divisions required to solve (7) without checking the optimality of the kernel. Naturally multiplication or division by 1 is not counted; no checking or operations recognized as redundant are included; no allowance is made for some $a_{ij} = 0$ or 1; the first two pivots are ± 1 in 4.4, which involves more addition and subtraction than the other methods; the first pivot is arbitrary in 4.6.

$\lambda=$	2	3	4	5	6	Formula
4.4	3	13	34	70	125	$(4\lambda^3 - 3\lambda^2 - \lambda)/6$
4.5(9)	4	25	82			
4.6(5)	9	25	54	100	167	$(4\lambda^3 + 3\lambda^2 + 5\lambda)/6$
7.5(9)	11	25	47	79	123	$(2\lambda^3 + 6\lambda^2 + 16\lambda - 6)/6$

(10)

Their simplicity recommends (9) of 4.5 for $\lambda = 3$ and the symmetric form of Exercise 9 of 4.2 for $\lambda = 2$ even though the number of multiplications or divisions is not minimized. For large kernels the Gaussian elimination method of 2.6(c), 7.4 and 7.5 is the most efficient—if the optimal K and L are known.

To use (9) to solve a 3 by 3 game with pencil and paper, the following procedure is suggested.

(a) Check for a saddle point. (Find the largest element in each column. Is it the smallest in its row?)

(b) Are the three largest (or the three smallest) payoffs uniquely defined and located in different rows and columns? If so, see if Theorem 5 or 6 applies.

(c) See if there is a dominated row or a dominating column; if so, it can be *definitely* eliminated.

(d) Look for a row having decisively the smallest $\Sigma_j a_{ij}$, and/or a column having decisively the largest $\Sigma_i a_{ij}$. This may *tentatively* eliminate row k and column l and determine a prospective 2 by 2 kernel $\mathbf{A}_{kl}$. The latter can be eliminated at once if it has a dominant row or column (of two elements).

(e) If the solution remains undiscovered, especially if Theorem 5 or 6 nearly applies, solve the 3 by 3 kernel. If this doesn't work, the optimal kernel being $\mathbf{A}_{kl}$ (2 by 2), the resulting values of u_k and y_l will have opposite signs ($u_k y_l \leq 0$) except in degenerate cases (Exercise 11), so that some information about k and l has been obtained. Such information is equally applicable to $\mathbf{A}$ and to $-\mathbf{A}$ (hence its ambiguity), because the square $\mathbf{A}$ as a kernel makes no distinction between these cases, whereas the smaller kernels do.

(f) The alleged optimal strategies must be probability vectors satisfying (2a). If **A** is its own kernel, (2a) is automatically satisfied if the arithmetic is correct.

For example, consider the game

$$\mathbf{A}=\begin{bmatrix}1 & 3 & 0\\ 2 & 0 & 5\\ 0 & 5 & -4\end{bmatrix} \qquad \operatorname{tradj}\mathbf{A}=\begin{array}{rrr|r} -25 & 8 & 10 & -7\\ 12 & -4 & -5 & 3\\ 15 & -5 & -6 & 4\\ \hline 2 & -1 & -1 & 0\end{array} \tag{11}$$

Steps (a) to (c) give nothing; there is no domination even by convex combinations. In (d) the small row sum 1 and large column sum 8 tentatively select the cofactor of a_{32} as a kernel, but it has a saddle point. In fact the elimination of $j=2$ cannot yield a 2 by 2 kernel because it leaves the remainders of the rows ordered by domination: $(2,5)>(1,0)>(0,-4)$. Persisting with the elimination of $i=3$, we find that $j=3$ is now tentatively eliminated by its maximum sum $0+5$. The cofactor $|\mathbf{A}_{33}|$ of a_{33} proves to be the optimal kernel and Exercise 9 of 4.2 gives $\mathbf{u}^*=(.5,.5,0)$, $\mathbf{y}^*=(.75,.25,0)$, $a^*=1.5$.

For the sake of illustration (9) will be applied to the 3 by 3 matrix. Equation (11) also shows $\operatorname{tradj}\mathbf{A}$ with its row and column sums. Since these sums are of mixed signs, it hardly matters that the overall sum σ happens to be zero so that no **u** and **y** can be found, nonnegative or otherwise. We can still look for opposed signs as required by (e). There are $1+2^2=5$ ways of getting these, among which the case $kl=11$ has a certain distinction. However, $\mathbf{A}_{11}$ is the kernel for $-\mathbf{A}$, not **A**. Its unsuccessful test is shown below, preceded by the successful test of $\mathbf{A}_{33}$.

	.75	.25	0	
.5	1	3	0	1.5
.5	2	0	5	1.5
0	0	5	−4	1.25
	1.5	1.5	2.5	↑

min = 1.5 = max

	0	$\frac{9}{14}$	$\frac{5}{14}$	
0	1	3	0	$\frac{27}{14}$
$\frac{9}{14}$	2	0	5	$\frac{25}{14}$
$\frac{5}{14}$	0	5	−4	$\frac{25}{14}$
	$\frac{18}{14}$	$\frac{25}{14}$	$\frac{25}{14}$	↑

$\min=\frac{18}{14}\neq\frac{27}{14}=\max$

EXERCISES 4.5

Solve 1 to 8 by using (9) and (a) to (f) as needed.

1.
$$\begin{array}{rrr} 2 & 3 & -2\\ 0 & -1 & -1\\ 1 & -2 & 3\end{array}$$
$(u_2^*=y_1^*=0)$

2.
$$\begin{array}{rrrr} -2 & 1 & 7 & 0\\ 1 & -1 & 0 & -7\\ 3 & -2 & 1 & -1\\ -6 & 4 & -1 & 3\end{array}$$
$(\mathbf{y}^*=(.4,.6,0,0))$

3.
$$\begin{array}{rrr} 3 & 8 & 1\\ 5 & 4 & 3\\ 4 & 6 & 5\\ 7 & 0 & 4\end{array}$$
$(\mathbf{y}^*=(2,1,0)/3)$

4. Exercise 4 of 4.3 with $L=\{2,3,4\}$. (Use a calculator.)
5. Let **A** be square with $a_{ij}=0$ for $i\neq j$. Show that $a^*=0$ (a saddle point) if some $a_{ii}a_{jj}\leq 0$; otherwise **A** is the kernel and $a^*=1/\Sigma a_{ii}^{-1}$.

6.

1	4	1
1	1	4
5	1	0

7.

5	−10	9
−2	3	−3
−7	12	−4

8.

1	0	−1
2	1	−3
−2	0	4

9. What can you conclude about a^* (a) by using a^0 and a_0 (pure strategies), (b) by considering only $\mathbf{u}=\mathbf{y}=(1,1,1)/3$?

2	3	1
0	4	−6
5	1	10

10.* In (10) verify the numbers 25 and 82 for 4.5 (9) in (10).

11.* Prove the assertion in step (e) in the text.

12. For $\lambda=3$, prove in detail that (9) is the solution of (7). (Refer to 7.7 if necessary).

Find the value a^* as a function of t for $-\infty<t<\infty$, by checking for saddle points and three 2 by 2 kernels.

13.

$t-1$	-2	3
3	$t+4$	-1

14.

t	$6-t$	$1+t$
2	1	0

15. Show that the maximum number of kernels in an m by n matrix is $\binom{m+n}{m}-1$ $=(m+n)!/m!n!-1$.

16. Show that the game has one kernel contained in another.

1	1	−2
0	1	−1
0	−1	1

17. Devise a 2 by 3 game such that J's optimal strategy is unique and pure but I has no optimal pure strategy.

4.6 OTHER RELATIONS BETWEEN GAMES AND PROGRAMS

An alternative way of reducing the linear programs (2) and (4) of 4.4 to canonical form begins by adding to every payoff a_{ij} a constant a such that the sign of the value a^*+a of the game with payoffs $a_{ij}+a$ is obvious; e.g., one could make all $a_{ij}+a>0$. We may suppose without loss of generality that $\overline{w}+a$ is restricted to have the sign of $\overline{w}^*+a=a^*+a$. After the insertion of the a and division by $|\overline{w}+a|$, the constraints of (2) of 4.4 become

$$\frac{\overline{w}+a}{|\overline{w}+a|}=\pm 1\leq \sum_{i=1}^{m}\frac{u_i}{|\overline{w}+a|}(a_{ij}+a) \qquad \text{for } j=1,\ldots,n$$

$$\frac{u_1}{|\overline{w}+a|}+\cdots+\frac{u_m}{|\overline{w}+a|}=\frac{1}{|\overline{w}+a|}=\mp\overline{w}'$$

If we define

$$u_i/|\bar{w}+a|=u_i'\geq 0 \qquad -(\bar{w}+a)^{-1}=\bar{w}' \tag{1}$$

and let $\mp$ denote the sign opposed to that of a^*+a, the program takes the form

$$\text{maximize} \qquad \bar{w}'=\mp(u_1'+\cdots+u_m') \tag{2}$$

$$\text{subject to} \qquad \sum_{i=1}^{m} u_i'(a_{ij}+a)\mp 1\geq 0 \qquad \text{for } j=1,\ldots,n$$

A similar treatment of (4) of 4.4 gives

$$y_j/|z+a|=y_j'\geq 0, \qquad -(z+a)^{-1}=z' \tag{3}$$

$$\text{minimize} \qquad z'=\mp(y_1'+\cdots+y_n') \tag{4}$$

$$\text{subject to} \qquad \sum_{j=1}^{n} y_j'(a_{ij}+a)\mp 1\leq 0 \qquad \text{for } i=1,\ldots,m$$

After slack variables v_j' and x_i' are inserted, (2) and (4) appear as the dual and the primal problems respectively in the schema

$$\begin{array}{c|cccc|l}
 & y_1' & \cdots & y_n' & 1 & \\
\hline
u_1' & a_{11}+a & \cdots & a_{1n}+a & \mp 1 & =-x_1' \\
\vdots & \vdots & & \vdots & \vdots & = \ \vdots \\
u_m' & a_{m1}+a & \cdots & a_{mn}+a & \mp 1 & =-x_m' \\
1 & \mp 1 & \cdots & \mp 1 & 0 & =z'\min \\
\hline
 & \| & & \| & \| & \\
 & v_1' & \cdots & v_n' & \bar{w}'\max &
\end{array} \tag{5}$$

in which all variables are nonnegative except perhaps $\bar{w}'$ and z'. As before the problem of the player who chooses rows is given by the column equations and vice versa. The schema (5) is either primal or dual feasible according to the choice of the sign $\mp$, which must be opposed to that of a^*+a; otherwise (5) can have no optimal solutions at all, in order to avoid contradicting (1) or (3). The relative signs of the entries in (5) are similar to those in Models 1 and 2 in Table 3.6. In Exercise 11 the reader is asked to compare the setup (5) with (5) of 4.4.

Schemas (5) and (8) following suffice to indicate the general procedure. What follows are some specific suggestions for the choice of the constant a and the first pivot in the simplex method. The position of the pivot is given by (9) of 4.4 (optimize first in the short direction), and its sign always agrees with a^*+a and is opposed to $b_i=c_j=\mp 1$. As for a, one sometimes has a choice between putting $a=0$ as in (b) below or following the general procedure (c) that chooses a so that the first pivot is ± 1. The latter favors mixed signs and small absolute values for the $a_{ij}+a$, which is desirable for both electronic and hand calculation.

(a) Begin by calculating $a^0=\min_j \max_i a_{ij}$ if $m\leq n$ and $a_0=\max_i \min_j a_{ij}$ if $m>n$. If a^0 is smallest in its row or a_0 largest in its column, a saddle point has been found; *stop*.

(b) Put $a=0$ if **A** has a positive row or a negative column. An equivalent condition is that a^0 and a_0 have the same sign; then a^* also has this sign by Exercise 9 of 4.4. In this case one may pivot on a^0 if $m<n$, on a_0 if $m>n$, and on $\min(|a^0|,|a_0|)$ if $m=n$. More generally, pivot on either a^0 or a_0 if the pivot is known to have the sign of a^*.

(c) The general procedure is as follows.

(c_1) If $m\leq n$, take $a=-1-a^0$, $b_i=c_j=1$ and pivot on $a^0+a=-1$ in the position of a^0. The tableau changes from dual feasible to primal feasible.

(c_2) If $m>n$, take $a=1-a_0$, $b_i=c_j=-1$ and pivot on $a_0+a=1$ in the position of a_0. The tableau changes from primal feasible to dual feasible.

Let us apply this procedure to the game in (6) as follows. Step (b) does not apply. By (c_1), adding $a=-1-a^0=-2$ to each payoff gives the game (10) in 4.4. With the present method we get the following tableaus.

$$\mathbf{A}=\begin{bmatrix}-4 & 1 & 4\\ 10 & 0 & -2\\ 12 & -1 & 5\end{bmatrix}$$

$$\max_i a_{ij} = \underbrace{12 \quad 1 \quad 5}_{\min=a^0=1}$$

1	y_1'	y_2'	y_3'	1
x_1'	-6	-1^*	2	1
x_2'	8	-2	-4	1
x_3'	10	-3	3	1
$-z'$	1	1	1	0

(6)

$+1$	y_1'	x_1'	y_3'	1
y_2'	$+6$	-1	-2	-1
x_2'	$+20^*$	-2	-8	-1
x_3'	$+28$	-3	-3	-2
$-z'$	-5	$+1$	$+3$	$+1$

20	x_2'	x_1'	y_3'	1
y_2'	-6	-8	8	-14
y_1'	1	-2	-8	-1
x_3'	-28	-4	164	-12
$-z'$	5	10	20	15

The column equations of the last (the optimal) tableau, with u' and v' replacing x' and y', respectively, give

$$\mathbf{u}'^*=(10,5,0)/20, \qquad \bar{w}'^*=15/20$$

and the row equations give

$$\mathbf{y}'^*=(1,14,0)/20, \qquad z'^*=15/20$$

Division by $|\overline{w}'^*|$ or $|z'^*|$ gives the probability vectors

$$\mathbf{u}^* = (10,5,0)/15 \qquad \mathbf{y}^* = (1,14,0)/15$$

$$a^* = -a - (z'^*)^{-1} = 2 - 20/15 = 10/15 \tag{7}$$

This solution can be checked as in (11) of 4.4.

In general, apart from permutations of rows and columns, the way the desired data occur in the final tableau can be represented as follows, with K, L= the sets of indices exchanged and $\overline{K}$, $\overline{L}$= the sets of indices not exchanged.

δ^*	$\mathbf{x}'_K$	$\mathbf{y}'_{\overline{L}}$	1		
$\mathbf{y}'_L$			$\pm d^*\mathbf{y}^*_L$		
$\mathbf{x}'_{\overline{K}}$			slacks	$(\mathbf{u}^*_{\overline{K}} = \mathbf{0})$	(8)
$-z$	$\mp d^*\mathbf{u}^*_K$	slacks	d^*		
		$(\mathbf{y}^*_{\overline{L}} = \mathbf{0})$		$a^* = -a - (\delta^*/d^*)$	

The game

$$\begin{array}{|rr|rr} \hline 3 & -3 & 1 & 3 \\ -3 & 3 & 1 & 3 \\ \hline -3 & -3 & 2^* & 0 \\ -1 & -1 & 0 & -2^* \end{array} \tag{9}$$

is an example in which (a) to (c) does not work well. The initial pivotal positions a^0 or a_0 that it prescribes are starred, but the desirable positions (the kernel of the game) are those within the square.

Exercise 9 of 8.4 combined with the invariance theorem of 2.4 shows that any sequence of pivot steps in a matrix $\mathbf{A}$ or $\begin{bmatrix} \mathbf{A} & \mathbf{b} \\ \mathbf{c} & d \end{bmatrix}$ produces no change in the sign ($+$, $-$, or 0) of the value of the game having this matrix.

Conversion of Linear Programs into Games

Such conversion completes the demonstration of the equivalence of matrix games and those canonical linear programs that possess optimal solutions.

(i) If the linear program happens to have all $b_i = c_j = 1$ or all $b_i = c_j = -1$, the conversion is at once indicated by (5) above; when we begin with the program, we may always take $a = 0$. The game of matrix $\mathbf{A}$ will have a value whose sign is opposed to that of the b_i and c_j if and only if the program has an optimal solution. More generally, if all b_i and c_j are nonzero and have the same sign, the program can be reduced to the form (5) by rescaling the variables so as to divide the entries in each row i by $|b_i|$ and in each column j by $|c_j|$.

(ii) A conversion to an $m+1$ by $n+1$ game of value zero is indicated by Theorem 6 in Gale, Kuhn and Tucker (1951). One has to proceed as if the optimal value $z^*=\bar{w}^*$ of the objective functions were known in advance, so that it can be included in the corner entry of the schema, thus:

$$\begin{array}{c|cc|l} & y_j^* & 1 & \\ \hline u_i^* & a_{ij} & b_i & \leq 0 \\ 1 & c_j & d-z^* & =0 \\ \hline & \text{VI} & \| & \\ & 0 & 0 & \end{array} \qquad \begin{array}{l} \text{All } u_i^* \geq 0 \\ \quad y_j^* \geq 0 \\ i=1,\ldots,m \\ j=1,\ldots,n \end{array} \qquad (10)$$

In (iii) we describe a device that in effect does make the value of z^* known in advance.

The schema (10) asserts that the game matrix within the rectangle has the value 0 and the optimal strategies

$$\frac{(u_1^*,\ldots,u_m^*,1)}{u_1^*+\cdots+u_m^*+1} \qquad \frac{(y_1^*,\ldots,y_n^*,1)}{y_1^*+\cdots+y_n^*+1} \qquad (11)$$

the denominators being such as to make the probabilities add to one. To confirm this statement, the optimality test may be applied as in (2a) of 4.4, while noting that $z^*=bu^*+d=cy^*+d$.

In (11) note that the pure strategies corresponding to the row of **c** and the column of **b** must be used with positive probabilities; if no such optimal strategies exist, the original program could not have an optimal solution. If the original schema is already optimal, the optimality conditions $b_i \leq 0$ and $c_j \geq 0$ are just what is required to make $d-z^*=0$ a saddle point of the corresponding game.

(iii) Conversion to a symmetric $m+n+1$ by $m+n+1$ game is possible for any canonical program, as Dantzig (1951) showed. To motivate this procedure, note that for any square game matrix with $a_{ij}=-a_{ji}$ for all i and j, the value a^* is known to be 0 by (i) of 4.3; such a game is said to be symmetric. Similarly any square schema with $a_{ij}=-a_{ji}$ and $b_i=-c_i$ for all i and j has $z^*=d$, if it can be optimized. Its primal and dual problems are identical, except perhaps the names of the variables.

Now one can convert any schema (12) into such a self-dual schema (13) by writing all the constraints of (12), both primal and dual, as primal constraints of (13), thus:

$$\begin{array}{c|cc|l} & \mathbf{y} & 1 & \\ \hline \mathbf{u} & \mathbf{A} & \mathbf{b} & =-\mathbf{x} \\ 1 & \mathbf{c} & d & =z\min \\ \hline & \| & \| & \\ & \mathbf{v} & \bar{w}\max & \end{array} \qquad \mathbf{u},\mathbf{v},\mathbf{x},\mathbf{y} \geq 0 \qquad (12)$$

$$\begin{array}{c|ccc|l} & \mathbf{u} & \mathbf{y} & 1 & \\ \hline \mathbf{u} & \mathbf{0} & \mathbf{A} & \mathbf{b} & =-\mathbf{x} \\ \mathbf{y} & -\mathbf{A}^T & \mathbf{0} & -\mathbf{c}^T & =-\mathbf{v} \\ 1 & -\mathbf{b}^T & \mathbf{c} & 0 & =-\bar{t} \\ \hline & \| & \| & \| & \\ & \mathbf{x} & \mathbf{v} & \bar{t}\max & \end{array} \qquad (13)$$

Here T denotes the transpose (omitted from the variable vectors). The last row within the square, defining $-\bar{t}$, might be added by symmetry at first, but is seen to be just right. It requires us to minimize $\mathbf{cy}-\mathbf{bu}=z-\bar{w}$; by Theorems 7 and 7* of 3.3, this minimum is zero and its attainment insures that the feasible primal and dual solutions are optimal.

Since $\bar{t}^*=0$, the game matrix in (13), like that in (10), has the value 0. Both players have the same optimal strategy

$$\frac{(u_1^*,\ldots,u_m^*,y_1^*,\ldots,y_n^*,1)}{u_1^*+\cdots+u_m^*+y_1^*+\cdots+y_n^*+1} \tag{14}$$

Since this strategy uses the row and column $(\mathbf{b},-\mathbf{c}^T)$ with positive probability, it is clear that the nonexistence of such an optimal strategy for the game matrix (13) must imply that (12) has no optimal solution. However, this conclusion does not follow in case (13) merely happens to have *some* optimal strategies that do not use the last row and column.

EXERCISES 4.6

1.–6. Solve the corresponding exercises of 4.4 by using (5) and (a) to (c).

7. Use a shortcut method to solve the game

$$\begin{matrix} 0 & 0 & 2 & -2 & 1 & -3 \\ 0 & 0 & -1 & 3 & 2 & -1 \\ -2 & 1 & 0 & 0 & 0 & -1 \\ 2 & -3 & 0 & 0 & 0 & 1 \\ -1 & -2 & 0 & 0 & 0 & 4 \\ 3 & 1 & 1 & -1 & -4 & 0 \end{matrix}$$

8. Interpret this program as a game:

1	y_1	y_2	y_3	1
x_1	0	-4	1	-3
x_2	4	0	-5	2
x_3	-1	5	0	-1
$-z$	3	-2	1	1

9. Interpret the linear program of Exercise 7 of 3.4 as a 3 by 4 game. Write down its matrix and derive its solution from the given solution of the linear program.

10. Models 1 and 2 of Table 3.6 can be converted into games by a third method not used in 8 or 9. Explain.

11. List the differences between the initial tableaus (5) of 4.4 and (5) of 4.6.

4.7 GAMES AGAINST NATURE

In the games considered so far each player's ignorance about the other's actions is mitigated by the assumption that each is rational with a known motivation (I to maximize, J to minimize); thus each can calculate the other's optimal strategy as easily as his own. This is no longer true if I, say, represents chance, a passive opponent, or a more or less inscrutable physical, social or biological environment. These agencies, referred to as "nature," are presumably unaware of J's existence and are neither hostile nor cooperative toward him, but only indifferent. In some cases the fiction of a hostile opponent may still be useful but more often it requires modification.

Nature's pure strategies are called *states of nature* and the resulting problem is often described as *decision making under uncertainty*. A very practical response from a decision maker J who is ignorant of the state of nature is to make observations or experiments, which in effect become a part of the game. These are designed to yield information (usually only probabilistic) about the state of nature, thus leading into statistical decision theory. The latter is outside the scope of this book, but its concern with the costs of errors and of experimentation, as well as the concept of regret discussed hereafter, has determined the convention that the decision maker is the minimizer J. The available procedures are illustrated here without the complications of experimentation.

(a) *Bayes solution*. Observational data may indicate the relative frequencies of nature's pure strategies in the past, thus approximately determining I's mixed strategy. Lacking this, J may be willing to guess that I's pure strategies are equally likely or whatever. *J chooses the pure strategy that minimizes the expected payoff* (a loss to J): $\min_j \Sigma_i u_i^0 a_{ij}$. Here the u_i^0 are estimated or guessed or postulated but not optimized.

Consider the following simplified version of the example at the end of 10.4. I determines the weather and J is a Rotary Club whose estimated losses from a clambake are as follows:

Have the clambake?	Yes	No	Probability
Fair weather	−100	0	.7
Bad weather	200	0	.3
Average	− 10	0	

(1)

Here $-10 = -100(.7) + 200(.3)$ is the minimum loss to J; by going ahead with the clambake, the club will have \$10 as its average profit. The probability of fair weather (.7) is presumably determined from the weather records for the locality.

The Bayes solution can be recommended when nature's probabilities are well established and the stakes are small enough to make it plausible that they are proportional to the (probably subjective) utilities.

(b) *Minimax loss*. If large stakes are involved or there is some threshold of loss that must not be exceeded for reasons of success, survival or legality, the expected-value calculation of (a) may be rejected in favor of the fiction of a *hostile nature* even if nature's actual strategy is more or less known. Consider the decision on whether to carry automobile liability insurance.

	Insured	Not Insured	Probability	
No accident	100	0	.9995	(2)
Accident	4000*	100,000	.0005	
Average	102	50		

Method (a) would advise against carrying insurance, but many motorists are unwilling to accept even a small risk of a $100,000 loss in order to save $52 on the average. Considered as a *matrix game*, (2) has a saddle point at the asterisk. While it is probably untrue that nature will send an accident, it is probably better for the driver to have insurance.

This approach involves being "prepared for the worst"; its advisability depends on a comparison of the costs with the benefits. Consider the game against nature

$$\begin{matrix} 1 & 100 \\ 102 & 101^* \end{matrix} \tag{3}$$

This has a saddle point at the asterisk; however, if I is not known to favor $i=2$, J would be foolish to choose $j=2$ because $j=1$ is much better in the first row and makes little difference in the second. Essentially the same conclusion is reached via (a) or (c).

(c) *The minimax regret* criterion of Savage (1951) may be motivated in several ways. Comparing (2) with (3) we note that the absolute level of any row or its dominance over other rows is not necessarily relevant; the more important row for J is the one in which he has the greater control over the payoff. If some of the differences among rows were eliminated, the false assumption of a hostile nature should have less effect. Since J has no control over the selection of the row, he should calculate his loss in terms of its excess over the minimum in the corresponding row, thus:

$$r_{ij} \triangleq a_{ij} - \min_{l} a_{il} \triangleq \text{regret} \tag{4}$$

The matrix game defined by the r_{ij} then determines J's optimal strategy. The corresponding strategy for I is useful for checking as in (2a) of 4.5 and determining the kernel.

The regret matrices for (2) and (3) are

$$\begin{matrix} 100 & 0 \\ 0 & 96{,}000 \end{matrix} \quad \text{and} \quad \begin{matrix} 0 & 99 \\ 1 & 0 \end{matrix}$$

Solving these games, we find that J is advised to choose $j=1$ with probabilities .999 and .99, respectively. Since there is no question of deceiving or deterring nature, these may as well be interpreted as prescribing $j=1$ all the time as concluded above. However, regret gives a probability of only $1/3$ for having the clambake in (1) because it ignores the information in the weather records.

It is easy to see that the regret matrix will have a saddle point if and only if one column of $\mathbf{A}$ is dominated by all the other columns (Exercise 8). In that case all the methods of this section choose the dominated column exclusively. However, J's optimal strategy in $\mathbf{R}$ may be pure even though there is no saddle point in $\mathbf{R}$ (see Exercise 7 of 4.7 or 17 of 4.5).

(d) *Investment problems* may plausibly be treated by various methods. Whereas a pure strategy as in (a) may be preferred as the solution for some games, a mixed strategy (meaning a diversification of investments) may be welcomed by the investor J.

Suppose J believes the following scores are proportional to the probable losses in the purchasing power of his savings for three future states of nature i and five different investments j:

$$\begin{matrix} \text{Prosperity} & -6 & -4 & 0 & -1 & -1 \\ \text{Inflation} & 2 & 2 & 0 & 2 & 4 \\ \text{Deflation} & 5 & 4 & 3 & 2 & -1 \end{matrix} \tag{5}$$

For the Bayes solution we assume that $\mathbf{u}^0=(1,1,1)/3$. Since these probabilities are controversial, J's life savings are at stake, and the Bayes solution in its simplest form (a) does not provide diversification, it must be regarded with suspicion. In fact it recommends that all assets be invested in $j=1$; this is the most speculative because its results are the most variable, ranging from a profit of 6 to a loss of 5. This gives the minimum average loss equal to $(-6+2+5)/3=1/3$. (Diversification *within* the category $j=1$ would still be possible and desirable but not comparable with diversification among values of j.)

The minimax loss criterion (b) solves (5) as a matrix game to obtain $\mathbf{y}^*=(0,0,5,0,3)/8$, $\mathbf{u}^*=(0,1,1)/2$ and $a^*=3/2$. Since the other rows dominate $i=1$, the methods of 4.2 could be used in the last two rows. Thus the state of prosperity has no influence on the solution, and J has to renounce much (though not all) of the possible benefit from prosperity in order to secure maximum protection against the undesirable states of nature; in fact his loss is $-3/8$ in the event of prosperity and $3/2$ otherwise. The Bayes average is $7/8$.

The regret matrix for (5) is nonnegative as always, with zeros at $ij=11,23,35$. These positions correctly suggest the kernel, so that (9) of 4.5 can be used to obtain $\mathbf{y}^*=(14,0,2,0,16)/32$, $\mathbf{u}^*=(8,13,11)/32$, and minimax regret$=92/32=2.875$. It is interesting to apply these probabilities to the original losses (5):

$$\begin{array}{c|ccccc|c}
 & 14 & & 2 & & 16 & \\
\hline
8 & -6 & -4 & 0 & -1 & -1 & -100 \\
13 & 2 & 2 & 0 & 2 & 4 & 92 \\
11 & 5 & 4 & 3 & 2 & -1 & 60 \\
\hline
 & 33 & 38 & 33 & 40 & 33 & \text{over } 32
\end{array} \qquad (6)$$

Thus there will be a profit of $100/32$ in the event of prosperity and a loss of $92/32$ or $60/32$ otherwise. The Bayes average is $52/96=.542$.

(e) *Bayes-constrained minimax loss.* In example (d) we observed that the Bayes solution and the minimax loss were utterly different in their underlying philosophies and in the solutions they prescribed. Nevertheless these methods can be combined to yield a continuous spectrum of methods bridging the gulf between them, thus:

$$\min_{\mathbf{y}} \max_{i} \sum_{j} a_{ij} y_j \triangleq z^*(\alpha) \text{ subject to}$$

$$\mathbf{u}^0\mathbf{A}\mathbf{y}=\alpha \qquad \Sigma y_j=1 \qquad \mathbf{y}\geq\mathbf{0} \qquad (7)$$

Equation (7) describes the minimax loss procedure (b) except that it includes the Bayes constraint $\mathbf{u}^0\mathbf{A}\mathbf{y}=\alpha$. Here $\mathbf{u}^0$ is the Bayes strategy attributed to nature, $\mathbf{y}$ is J's strategy, and α is the resulting loss, which is used as a parameter. The variable $\mathbf{y}$ is optimized to minimize the *maximum* loss $z(\alpha)$, not the average loss α, which is regarded as constant.

Naturally α can be varied from one optimization to another, $\mathbf{A}$ remaining the same. The smallest possible value of α is $\min_j \Sigma_j u_i^0 a_{ij} \triangleq \alpha_0$, the minimum Bayes loss, for which (7) reduces to the Bayes solution (a). The largest interesting value of α is $\mathbf{u}^0\mathbf{A}\mathbf{y}^* \triangleq \alpha_1$, the Bayes loss when J uses the strategy $\mathbf{y}^*$ appropriate for the (unconstrained) minimax loss (b), to which (7) then reduces. J would like to have a small α and a small $z^*(\alpha)$; however, $z^*(\alpha)$ decreases as α increases, so that J has to decide their relative importance, perhaps by choosing an exchange rate.

If (7) is applied to (5), the initial tableau analogous to (5) of 4.4 is

Check	1	y_1	y_2	y_3	y_4	y_5	1	$-z$
-10	x_1	-6	-4	0	-1	-1	0	1
12	x_2	2	2	0	2	4	0	1
15	x_3	5	4	3	2	-1^*	0	1
8	0	1	2	3	3	2	-3α	0
4	0	1	1	1	1	1	-1	0

(8)

Row x_3 is pivotal because it yields the largest loss (i.e., 5) for the Bayes strategy $\mathbf{y}=(1,0,0,0,0)$. The smallest a_{ij} in this row is chosen because this will give a dual-feasible tableau after the second pivot step, which exchanges $-z$ with the basic zero in the last row; the last column is then discarded. The third pivot exchanges the remaining basic zero with y_1; this could be a dual-simplex step (-1 is the only negative element in that row), or it could be chosen because $y_1>0$ in the Bayes solution.

After the three pivot steps the resulting tableau is

Check	1	y_2	y_3	y_4	x_3	1	
-25	x_1	-8	-14	-14	-1	$-22+33\alpha$	(9)
-23	x_2	-7	-16^*	-13	-1	$-11+24\alpha$	
4	y_5	1	2	2	0	$1-3\alpha$	
0	y_1	0	-1	-1	0	$-2+3\alpha$	
19	$-z$	5	10	9	1	$11-18\alpha$	

For $1/3\le\alpha\le 11/24$, this gives the optimal solution $\mathbf{y}^*=(2-3\alpha,0,0,0,-1+3\alpha)$, $z^*=11-18\alpha$. As expected, the third row shows there is no feasible solution for $\alpha<1/3$. For $\alpha>11/24$, x_2 becomes infeasible and Theorem 2″ of 3.3 selects a pivot in row x_2 by the criterion $\min(5/7,10/16,9/13,1/1)$. (This is an example of the parametric programming of 8.3.) The next tableau is

Check	16	y_2	x_2	y_4	x_3	1	
-78	x_1	-30	-14	-42	-2	$-198+192\alpha$	(10)
23	y_3	7	-1	13	1	$11-24\alpha$	
18	y_5	2	2	6	-2	-6	
23	y_1	7	-1	-3^*	1	$-21+24\alpha$	
74	$-z$	10	10	14	6	$66-48\alpha$	

For $11/24\le\alpha\le 7/8$, this gives the optimal solution $\mathbf{y}^*=(21-24\alpha,0,-11+24\alpha,0,6)/16$, $z^*=(33-24\alpha)/8$. For $\alpha>7/8$, y_1 becomes infeasible and should be exchanged with y_4. However, this is not carried out because one observes that it makes $z^*=4\alpha-2$, so that z^* is now increasing as α increases and will continue to do so according to the theorem (b) of 8.7 (z^* is a convex function of α). Thus we have $1/3\le\alpha\le 7/8$. Note that the last row of (10) does not give the unconstrained $\mathbf{u}^*=(0,.5,.5)$.

The relation between the Bayes loss $\alpha=\mathbf{u}^0\mathbf{Ay}$ and the corresponding minimax loss $z^*(\alpha)$ in (7) may be summarized as follows:

$$\begin{array}{ll} \alpha: & \tfrac{1}{3}\le \quad \alpha \quad \le\tfrac{11}{24}\le \quad \alpha \quad \le\tfrac{7}{8} \\ z^*(\alpha): & 5\ge 11-18\alpha\ge\tfrac{11}{4}\ge 4.125-3\alpha\ge 1.5 \end{array} \tag{11}$$

The objective z^* decreases 18 times as fast as α increases in the first interval and 3 times as fast in the second. Thus if the exchange ratio between z^* and α is used as the criterion, it may suffice to consider only three values of α: 1/3 if α is more than 18 times as important as z^*, 7/8 if α is less than 3 times as important as z^* and 11/24 otherwise.

We have now obtained four particular solutions of the investment problem:

	24α	$24z^*$	$16y^*$
Pure Bayes	8	120	(16,0,0,0,0)
Minimax Bayes	11	66	(10,0,0,0,6)
Minimax regret	13	69	(7,0,1,0,8)
Minimax loss	21	36	(0,0,10,0,6)

(12)

In this problem the minimax Bayes and regret solutions are rather similar; the latter is easier to define and calculate but gives slightly larger values for both the losses α and z^*. The minimin solution below coincides with the pure Bayes in (12).

(f) *Minimin loss and summary*. If a hostile nature is sometimes a useful fiction, is the same ever true of a cooperative nature? No doubt it is, not only for the gambler who lives on hope but also for others who have a (probably ill-defined) threshold of success to cross in establishing their position or achieving a breakthrough in research or the like. Of course this threshold is different from the one mentioned in (b) above. There J wanted to *stay* below the threshold to avoid large losses; now he hopes to *get* below to get large gains. In the language of 8.7, J's disutility was a convex function of the nominal loss in (b); here it is a concave function.

To be sure, if $i=2$ is not known to be preferred by I, the pure minimin loss $\min_i \min_j a_{ij}$ is just as foolish in the game

$$\begin{matrix} 1 & 100 \\ 1 & 0 \end{matrix} \qquad (13)$$

in which it recommends $j=2$, as the minimax loss was in (3). However, each has a place in the collection of methods represented by Table 4.7.

TABLE 4.7

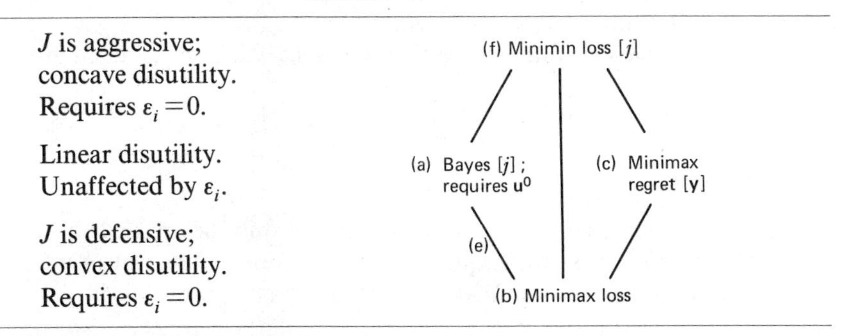

(a) to (f) refer to the appropriate subsection.

[j] means J's optimal strategy is usually pure.

[y] means J's optimal strategy is usually mixed.

$\mathbf{u}^0$ is the strategy attributed to I (nature).

ε_i are (undetermined) errors in the a_{ij}, resulting from a failure to evaluate the intrinsic worth of the states of nature, as in Exercise 4. (See also Exercises 2, 7 and 10.)

The five line segments suggest possibilities for obtaining spectra of intermediate methods.

All these methods have the desirable property of ignoring any column of $\mathbf{A}$ that dominates another column. (b) and (f) ignore dominated and dominating rows, respectively.

A simple way to combine two of the four basic methods in Table 4.7 is to form a weighted average of the two criteria before minimizing over j or $\mathbf{y}$. In addition to (e) this gives for the pure Bayes and the minimax loss

$$\min_j\left[(1-\theta)\Sigma_i u_i^0 a_{ij} + \theta \max_i a_{ij}\right] \qquad (0 \leq \theta \leq 1) \tag{14}$$

Similar classes of methods are obtained by using the standard (root-mean-square) deviation thus:

$$\min_j\left[\Sigma_i u_i^0 a_{ij} + t\left(\Sigma_i u_i^0 a_{ij}^2 - \left(\Sigma_i u_i^0 a_{ij}\right)^2\right)^{1/2}\right] \tag{15}$$

$$\min_y \Sigma_i u_i^0 (\Sigma_j a_{ij} y_j)^2 \qquad \text{or} \qquad \min_y \Sigma_i \Sigma_j u_i^0 a_{ij}^2 y_j \tag{16}$$

$$\text{subject to } \mathbf{u}^0\mathbf{A}\mathbf{y} = \alpha \qquad \Sigma y_j = 1 \qquad \mathbf{y} \geq \mathbf{0}$$

The minimand in (16) is $\sigma^2 + \alpha^2$ where σ is the standard deviation of the realized losses when I uses $\mathbf{u}^0$. Equation (16) requires quadratic programming methods; the first form is appropriate for investment problems.

The Bayes constrained minimin loss is

$$\min_i \min_y \Sigma_j a_{ij} y_j \triangleq z^*(\alpha) \qquad \text{subject to } \mathbf{u}^0\mathbf{A}\mathbf{y} \leq \alpha. \tag{17}$$

For each value of i one has to solve a linear program involving only the two constraints $\mathbf{u}^0\mathbf{A}\mathbf{y} \leq \alpha$ and $\Sigma y_j = 1$ in addition to $\mathbf{y} \geq \mathbf{0}$. The inequality is needed with α because a minimum of convex functions is not necessarily convex. This method and others are illustrated in Exercises 13 and 9.

The related topic of multiple objective functions will be considered in 8.8 and 8.9.

EXERCISES 4.7

For the games 1 to 7, set up the loss matrix $\mathbf{A}$ and find the optimal strategy for J by the method of (a) Bayes (with equally likely states of nature, unless information to the contrary is given), (b) minimax loss, (c) minimax regret.

1. A student can solve a problem in 60 minutes unaided and in 10 minutes if she has a certain reference book that may or may not be in the library. Going to the library and trying to locate the book requires 20 minutes.
2. One thousand lottery tickets will be sold at \$1 each. There are two prizes, one of \$400 and one of \$100. The purchase of one ticket is being considered.
3. A student with a grade-point average of 3 is deciding whether to elect SU grading (only two grades, satisfactory or unsatisfactory) for a certain course for

which he is registering. He scores the possibilities thus:

Grade	A	B	C	D	F
Regular grading	4	3	2	1	−1
SU grading	3	3	3	0	0
Probability u_i^0	.3	.3	.2	.1	.1

4. A consultant has accepted an assignment in a new location for an unknown period. His housing cost will be \$100/month if he buys a house, \$200/month if he rents, and \$400/month is he stays at a hotel. However, he expects to get \$2000 less than he paid if he resells a house, and to forfeit \$300 if he gives up a rented house. To simplify matters he proposes to consider durations of 1, 6 and 30 months with probabilities .6, .3 and .1, but not to change the accommodation that he first decides on.

5. A research project has an estimated success probability of 10 percent. If it is undertaken successfully, the gross benefit is expected to be 20 times the cost.

6. Without surgery a patient has an estimated life expectancy of one year. Surgery is expected to be fatal or futile (zero expectancy) with probability one-half and to yield a life expectancy of three years with probability one-half.

7. A firm has four courses of action whose associated losses are believed to depend on which party controls the government, provided a commitment is made in advance of the election.

Democratic control	2	−4	−3	3
Divided control	−2	2	−1	0
Republican control	3	−5	4	−1

Assume that the firm deletes the last row, believing that Republican control is very unlikely.

8. Prove the saddle point assertion at the end of (c).

9. Give two examples of compromise methods in Table 4.7 not involving the Bayes $\mathbf{u}^0$.

10. List some of the considerations that determine the nature and the applicability of the methods of this section.

Find the Bayes-constrained minimax loss (7).

11.
$$\begin{matrix} 3 & 6 \\ 8 & 6 \end{matrix}$$

12.
$$\begin{matrix} 5 & 2 & 1 \\ 0 & 2 & -3 \\ -4 & 0 & 4 \end{matrix}$$

13.
$$\begin{matrix} -5 & 6 & 1 \\ 4 & -3 & 0 \\ 5 & -1 & -1 \end{matrix}$$

In 13 also get the minimin loss (17).

14. Suppose that a \$4000 car has a probability of .995 of escaping damage and a probability of .005 of being destroyed in an accident. An insurance policy covering all but the first \$100 of damage costs \$50.

REFERENCES (CHAPTER 4)

Balinski-Tucker (1969)
Benayoun-Montgolfier-Tergny-Laritchev (1971)
Bohnenblust-Karlin-Shapley (1950)
Borel (1921–1927)
Brown (1951)
Danskin (1967)
Dantzig (1951a, 1956)
Dorfman (1951)
Dresher (1961)
Gale-Kuhn-Tucker (1950a, b; 1951)
Gale-Sherman (1950)
Hurwicz (1951)
Kakutani (1941)
Keeney-Raiffa (1976)
Motzkin-Raiffa-Thompson-Thrall (1953)
Robinson (1951)
Roy (1971)
Savage (1951)
Shapley (1964)
Shapley-Snow (1950)
Starr-Zeleny (1977)
Thrall-Coombs-Davis (1954)
Tucker (1960b)
Von Neumann (1928, 1954)
Williams (1954)
Zeleny-Cochrane (1973)

CHAPTER FIVE

Bimatrix and *n*-Person Games

The sections of this chapter have much independence, but 5.1 is a prerequisite for 5.2, 5.3, 5.4 and perhaps 5.7.

5.1 INTRODUCTION TO BIMATRIX GAMES

The relevant definitions for these games have already been given in 4.1. If there are two players I and J, their respective payoffs are denoted by a_{ij} and b_{ij}. I chooses i or $\mathbf{u}$ with the object of maximizing a_{ij} or $\mathbf{uAy}$ and J chooses j or $\mathbf{y}$ with the object of maximizing b_{ij} or $\mathbf{uBy}$. J is no longer minimizing, because now he is looking at his actual payoffs rather than their negatives. This is called a *bimatrix game* because it is defined by the two m by n matrices $\mathbf{A}$ and $\mathbf{B}$ whose elements are a_{ij} and b_{ij}.

Sometimes it is convenient to use the payoffs a_{ij} and b_{ij} as values of two variables α and β in a Cartesian plane, so that one obtains mn payoff points (a_{ij}, b_{ij}). In order to define the game by such a *payoff diagram*, one would need to attach the values of i and j as labels to each point, but this is unnecessary for the purposes for which such diagrams will usually be used.

The matrix games considered in 4.1 to 4.6 may be described as those two-person games whose points lie on the line $\alpha+\beta=0$ through the origin. However, in (i) of 4.3 we noted that a linear transformation of the payoffs, which replaces a_{ij} by $sa_{ij}+a$, makes only a trivial change in the matrix game if $s>0$. The same is true of a bimatrix game, with the result that it can be solved as a matrix game (by considering $\mathbf{A}$ or $-\mathbf{B}$ alone) whenever all its points (a_{ij}, b_{ij}) lie on any straight line of negative slope, such as $s_1\alpha+s_2\beta+s_0=0$ $(s_1 s_2>0)$. Hence in this chapter new problems arise only in the case where *these points are not all collinear*, not merely nonzero-sum.

Example (a)

Isabel (I) and John (J) are newlyweds who have to decide where to work. John has offers in cities 1, 2 and 3 within commuting distance of one another and Isabel has offers in cities 1 and 2. Each develops his or her own procedure for estimating the desirability of his or her own alternatives, something like the following: annual salary in units of \$250 minus hours per week for commuting plus estimated points for job satisfaction and prospects for the future. They arrive at the following payoffs:

$$\begin{array}{ccc} (52,50) & (44,44) & (44,41) \\ (42,42) & (46,49) & (39,43) \end{array} \qquad (1)$$

This bimatrix game has the matrices

$$\mathbf{A}=\begin{bmatrix} 52 & 44 & 44 \\ 42 & 46 & 39 \end{bmatrix} \qquad \mathbf{B}=\begin{bmatrix} 50 & 44 & 41 \\ 42 & 49 & 43 \end{bmatrix}$$

of payoffs to I and J, respectively, but it is more convenient to write them together as in (1).

It is clear that I and J should both work in the first city, since this gives I the greatest possible satisfaction $\max_{i,j} a_{ij}=a_{11}=52$ units, and also gives J the maximum satisfaction $\max_{i,j} b_{ij}=b_{11}=50$ units. Once the table (1) has been established and understood, there is no need for the players to persuade one another to cooperate. The same was true of zero-sum games, but for a different reason. There neither player could avoid hurting the other in serving his or her own interests; in (1) neither player can avoid helping the other. (Later examples will be more complex.) In the terminology that will be introduced following, what distinguishes this solution $(a_{11}, b_{11})=(52,50)$ is that it is the only efficient point in the game, or equivalently, it dominates all the other points. ■

Feasible Points

A payoff point (α, β) is said to be *feasible* for a game if it can be obtained by cooperation among the players, using a mixed (joint) strategy if necessary. The latter is a matrix of probabilities π_{ij}, where π_{ij} is the probability of selecting the pair of pure strategies i and j; then the feasible points are those (α, β) obtainable as average values

$$\alpha=\sum_i \sum_j \pi_{ij} a_{ij} \qquad \beta=\sum_i \sum_j \pi_{ij} b_{ij} \qquad (2)$$

where all $\pi_{ij}\geq 0$ and $\Sigma_i \Sigma_j \pi_{ij}=1$. *Thus the feasible region or prospect space* Ω *is precisely the convex hull of the points* (a_{ij}, b_{ij}) as defined in 2.2. However, this is not a linear program because the players have different objective functions α and β.

If only two of the π_{ij} are nonzero, say π_{kl} and $\pi_{pq}=1-\pi_{kl}$, then as π_{kl} varies between 0 and 1, (2) generates the points on the straight line segment joining (a_{kl}, b_{kl}) and (a_{pq}, b_{pq}). If the players are unable or unwilling to cooperate, the most they can do is to use independent individual probability vectors $\mathbf{u}$ and $\mathbf{y}$, so that $\pi_{ij}=u_i y_j$; of course this includes the case where $u_k=y_l=\pi_{kl}=1$ and the other probabilities are zero (the pure strategies k and l are used).

The feasible region could be expanded by allowing side payments of money or transferable utility from one player to another beyond what the utilities a_{ij} and b_{ij} themselves prescribe. However, this actually specializes the model rather than generalizes it; (9) of 5.3 shows how easily side payments can be incorporated into the matrices **A** and **B**. In some situations side payments (then called bribes) are illegal or dishonorable.

Domination and Efficient Points

If two vectors **x** and **x**′ have equal numbers of components, which satisfy $x_i \geq x_i'$ for all i, we say that **x** *dominates* **x**′ and write $\mathbf{x} \geq \mathbf{x}'$. When this idea was used in 4.3(ii), the vectors in question were the rows (or the columns) of **A**. In this section also we are interested in dominating rows of **A** and dominating columns of **B**, but examples such as (d) following warn against the deletion of dominated rows or columns in the present context. When one vector dominates *all* the others, we use the words *dominance* and *dominant*.

A new application of domination (to the vectors (α, β)) arises in this section. If one payoff point (α, β) dominates another, (α_0, β_0), this means that $\alpha \geq \alpha_0$ and $\beta \geq \beta_0$. Thus the dominating payoffs (α, β) are acceptable to *both* players in place of (α_0, β_0). Each of two points can dominate the other if and only if they are identical. Neither of two points will dominate the other if and only if their *differences* $\alpha' - \alpha, \beta' - \beta$ (not α and β themselves) have opposite signs or lie in the second or fourth quadrant ($(\alpha' - \alpha)(\beta' - \beta) < 0$), as illustrated in Figure 5.1A. Stricter forms of domination would exclude $(\alpha, \beta) = (\alpha_0, \beta_0)$ or even $\alpha = \alpha_0$ or $\beta = \beta_0$. In the literature one encounters more complex definitions given by Von Neumann and Morgenstern (1944) and Aumann (1961).

A feasible payoff point (α_0, β_0) is said to be *efficient* or jointly maximal or Pareto optimal if *it is not dominated by any other feasible point*. One payoff α or β can be increased only if the other is decreased. There is no other feasible point in the (closed) "northeast" or positive quadrant with its corner at (α_0, β_0).

The set of all efficient points, denoted by Ω', may be described as the "*northeast*" *boundary* of the feasible region Ω. For a bimatrix game it consists of a connected sequence of straight line segments whose negative slopes become steeper as α increases. Thus each efficient point can be obtained as a convex combination (2) of at most two of the points (a_{ij}, b_{ij}). In Figures 5.1B and C the efficient points lie on two and on one line segment respectively, shown as solid lines.

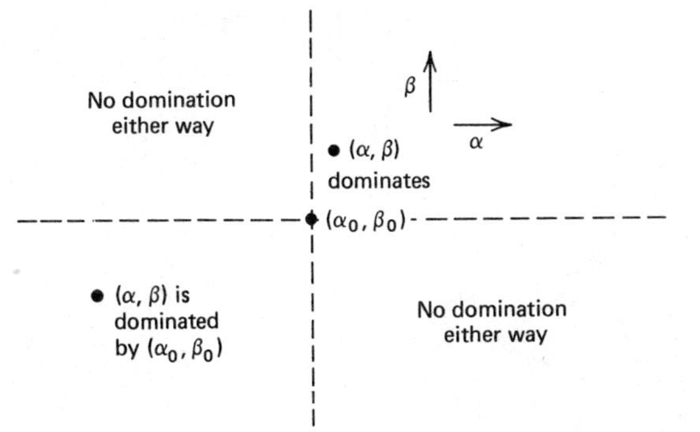

FIGURE 5.1A Domination

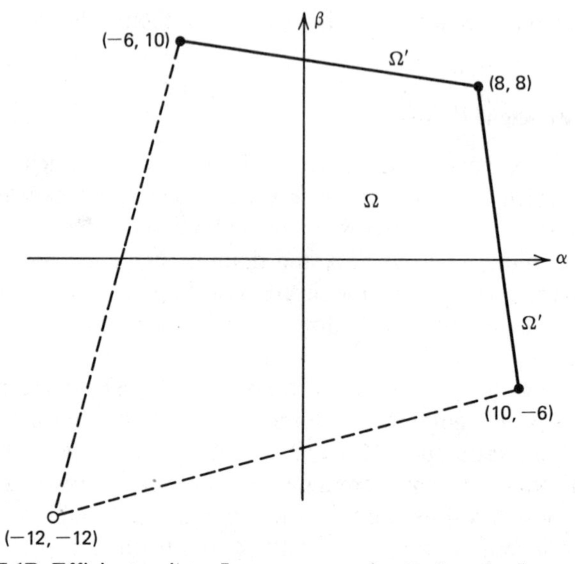

FIGURE 5.1B Efficient points: In pure strategies ●. In mixed strategies Ω'.

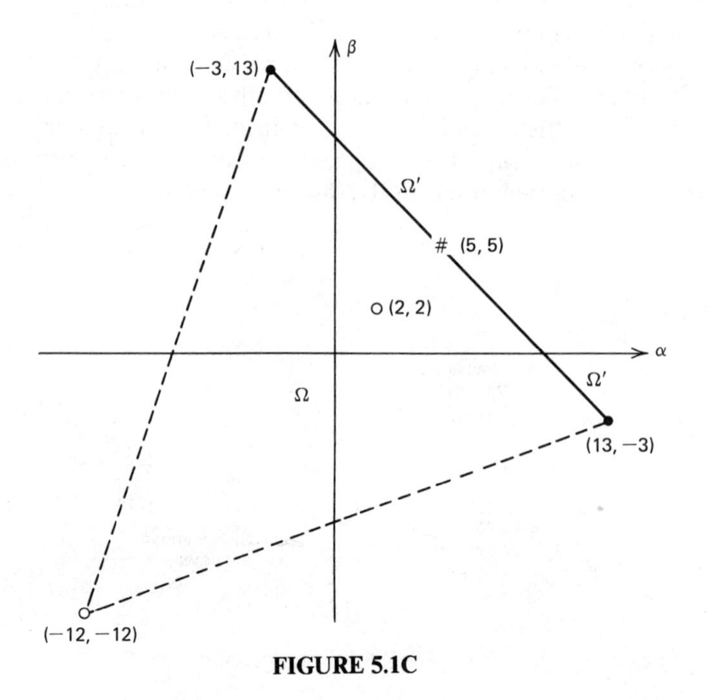

FIGURE 5.1C

The concept of efficiency obviously can be extended to games having more than two players and also to the situation of a single decision maker seeking to reconcile several different criteria of value, such as profit and safety. This is a standard method for the analysis of portfolios of securities. (Markowitz, 1959).

Example (b)

I and J are two reporters who work independently although they are employed by the same newspaper. They have to decide which one will cover an event in which their editor is assumed to be very much interested. They estimate their payoffs as follows in terms of the editor's approval:

	J covers	J doesn't	
I covers	$(8,8)$	$(10,-6)$	(3)
I doesn't	$(-6,10)$	$(-12,-12)$	

The payoff diagram for this game is shown in Figure 5.1B. The efficient region Ω' consists of three of the four original points (all except $(-12,-12)$) and two of the line segments joining them (shown as solid lines). The dashed lines are the remainder of the boundary of the feasible region Ω.

Although there are other efficient points, $(8,8)$ is distinguished by its symmetrical location in Figure 5.1B and the fact that no matter what J does, I will get a larger a_{ij} if he covers the story, and vice versa. This *dominance* of the first row of $\mathbf{A}$ ($a_{1j} \geq a_{ij}$ for all i and j) and the first column of $\mathbf{B}$ insures that $ij=11$ with payoffs $(8,8)$ is an equilibrium point as defined below, and it will be a unique equilibrium point unless several a_{ij} are equal or several b_{ij} are equal (Exercise 19 of 5.2). ■

Example (c)

Let us modify the previous example by assuming that the editor considers it rather wasteful for two reporters to cover the same event, though still better than no coverage at all. Then we might have

	J covers	J doesn't	
I covers	$(2,2)$	$(13,-3)$	(4)
I doesn't	$(-3,13)$	$(-12,-12)$	

The payoff diagram for this game is shown in Figure 5.1C. Although the first row and column are still dominant as before, the resulting point $(2,2)$ is no longer efficient, in agreement with the editor's opinion. If the two reporters will toss a coin to decide who covers the story, their average payoffs will be $(-3,13)/2+(13,-3)/2=(5,5)$, and each is better off than at $(2,2)$. In view of the *symmetry* of the game ($a_{ij}=b_{ji}$), $(5,5)$ is a reasonable cooperative solution. ■

For the present the payoff diagram may be used to determine the efficient points; (6) of 5.3 gives a simple numerical procedure.

Equilibrium Points

For any number of players an *equilibrium point* is defined as a selection of *individual* pure or mixed strategies (one probability vector for each player) such that if any one player changes his strategy, he cannot thereby obtain a larger payoff. For two players the mixed strategies $\mathbf{u}^*$ and $\mathbf{y}^*$ define an equilibrium point provided

$$\alpha^* \triangleq \mathbf{u}^*\mathbf{A}\mathbf{y}^* \geq \mathbf{u}\mathbf{A}\mathbf{y}^* \text{ for all strategies } \mathbf{u}.$$

$$\beta^* \triangleq \mathbf{u}^*\mathbf{B}\mathbf{y}^* \geq \mathbf{u}^*\mathbf{B}\mathbf{y} \text{ for all strategies } \mathbf{y}. \qquad (5)$$

In other words, it is not advantageous for any one player to depart from the status quo. This is definitely a *noncooperative* game concept because it uses the individual probability vectors $\mathbf{u}$ and $\mathbf{y}$ rather than the joint π_{ij}, and it is concerned with what happens when a single player changes his strategy.

It is sometimes convenient to refer to the pair of payoffs (α^*, β^*) in (5) as an equilibrium point, although one always wants to know the associated $\mathbf{u}^*$ and $\mathbf{y}^*$.

The reader can easily verify that a solution of a zero-sum game is an equilibrium point and an efficient point, and the payoff was shown to be unique in 4.4. In the present case a pair of equilibrium payoffs need be neither unique (Examples (a) and (e)) nor efficient (Examples (c) and (d)). However, every game with finite numbers of players and pure strategies has at least one equilibrium point in mixed strategies. Nash (1951) and the Lemke-Howson algorithm of 5.4 prove this for two players; in general see Debreu (1952) and Wilson (1971). If an equilibrium point is unique or if it dominates the others, it may be regarded as yielding the noncooperative solution $(\hat{\alpha}, \hat{\beta})$ of the game $\{I, J\}$.

According to (5), pure strategies k and l will define an equilibrium point provided

$$a_{kl} = \max_i a_{il} \qquad b_{kl} = \max_j b_{kj} \qquad (6)$$

that is, a_{kl} must be largest in its column l of $\mathbf{A}$ and b_{kl} must be largest in its row k of $\mathbf{B}$. For example, $k=l=1$ gives equilibrium points in Examples (a) to (c); $k=l=2$ gives a second (inefficient) equilibrium in Example (a). Thus we have the following.

Algorithm for Equilibrium Points in Pure Strategies

(i) In each column of $\mathbf{A}$, underline the largest element(s) in the column.

(ii) In each row of $\mathbf{B}$, underline the largest element(s) in the row. [Actually it would suffice to check the status of those (a_{ij}, b_{ij}) already underlined in (i).]

(iii) Those positions ij (if any) for which both a_{ij} and b_{ij} have been underlined are the equilibrium points in pure strategies. Equation (1) of 5.2 is an example.

Example (d): The Prisoner's Dilemma

(Attributed to Tucker and M. K. Flood.) Consider the game

	Altruistic	Egoistic	
Altruistic	(2,2)	(0,3)	(7)
Egoistic	(3,0)	(1,1)	

Here the second row of **A** and the second column of **B** are dominant; these strategies have been labeled "egoistic" to reflect the self-interest normally ascribed to the players of a game. More specifically, no matter what J does, I's choice of the egoistic strategy will yield 1 unit more to I and 2 units less to J; the position of J is entirely similar. The result is the paradox that individually dominant strategies produce jointly dominated payoffs.

Whether or not it is really true that each player hurts the other more than he helps himself, it is a fact that the egoistic strategies are mutually frustrating in this game. They produce the unique equilibrium point $(1,1)$, but this is not efficient, nor is the mixture $(3,0)/2+(0,3)/2=(1.5,1.5)$. The only symmetrical efficient solution is $(2,2)$, but to obtain it each player must be altruistic or trusting or daring enough to accept the risk of betrayal by the other player, or else they must give up or be deprived of the freedom (or necessity?) to pursue their self-interest. In the latter case there is the problem of finding a trustworthy authority, and the possibility that an attempt to compel powerful interests to "behave properly" may itself have very adverse effects on the payoffs.

In practice there is difficulty in defining the altruistic strategies (as in (16) of 5.2) and in measuring and comparing the payoffs of the two players. If the players' circumstances are different, should the payoffs be measured in absolute units such as tons or dollars, on a per capita basis, or as a fraction of the player's income or wealth? People's attitudes tend to be influenced by these considerations whereas the methods of this chapter generally are not. The given value of θ in 5.3 is a possible exception; the condition $a_{ij}=b_{ji}$ or $a_{ij}+b_{ij}=0$ is probably not, since the Nash solution of 5.3 supports the view that the underlying property is only hidden rather than destroyed by independent linear transformations of the a_{ij} and b_{ij}.

In the following interpretations one may wish to adjust the eight numbers in (7) by adding or subtracting a suitable constant.

(d_1) I and J are prisoners isolated from one another and charged with being jointly responsible for a crime. For each the altruistic strategy is to deny the charge while the egoistic strategy is to confess (truly or falsely), so as to benefit from the lenient treatment that has been promised to one who is alone in confessing and to escape the harsh treatment reserved for one who is alone in not confessing.

(d_2) I and J are nations or groups of people who share a common resource, such as a body of water, a petroleum reserve, a capacity for food production, a stream of revenue, a source of financial credit, or a tradition of mutual assistance. The altruistic strategy is to use the resource with restraint in order to limit its depreciation or depletion; the egoistic strategy is to try to exploit the resource rapidly before the other player does the same. The payoffs are estimated over a time period long enough to reflect the possible destruction of the resource.

(d_3) I and J are individuals or states engaged in a contest for power or prestige. The altruistic strategy is to accept the status quo, which has the value 2 in (7). The egoistic strategy is to undertake a prestige-building project. This is assumed to cost 1 unit more than its intrinsic worth, except that if only one player undertakes the project, he will gain 2 units at the expense of the other player. Thus the payoffs are calculated as $2-1=1, 2-2=0, 2+2-1=3$. ■

There is further discussion in 5.2.

Example (e): Chicken

This takes its name from the game that some teenagers play with oncoming automobiles. It arises from interpretations (d_2) and (d_3) above if the depletion of the resource has very grave consequences or the contest for prestige escalates into a mutually destructive conflict. The egoistic strategies are assumed to yield the lowest possible payoff to each player if both use them, but if only one player uses the egoistic strategy he is rewarded with the highest possible payoff as before. Thus the temptation to use the egoistic strategy tends to be less than in the Prisoner's Dilemma but still may be considerable as in the game

	Altruistic	Egoistic	
Altruistic	(19, 19)	(−42, 68)	(8)
Egoistic	(68, −42)	(−45, −45).	

Here no row or column is dominant but there are three equilibrium points $(68,-42)$, $(-42,68)$, and $(-2001,-2001)/52$, of which the first two are efficient and the third comes from $\mathbf{u}=\mathbf{y}=(3,49)/52$. The latter are given by

$$\mathbf{y}^*=\frac{(a_{22}-a_{12},\, a_{11}-a_{21})}{a_{11}+a_{22}-a_{12}-a_{21}}, \qquad \mathbf{u}^*=\frac{(b_{22}-b_{21},\, b_{11}-b_{12})}{b_{11}+b_{22}-b_{12}-b_{21}}$$

$$a^*=\frac{a_{11}a_{22}-a_{12}a_{21}}{a_{11}+a_{22}-a_{12}-a_{21}}, \qquad b^*=\frac{b_{11}b_{22}-b_{12}b_{21}}{b_{11}+b_{22}-b_{12}-b_{21}} \tag{9}$$

provided that all the components of $\mathbf{y}$ and $\mathbf{u}$ obtained thereby are nonnegative, or equivalently, neither row of $\mathbf{A}$ dominates the other and neither column of $\mathbf{B}$ dominates the other (with strict inequalities). In this case $\mathbf{A}$ determines $\mathbf{y}$ and $\mathbf{B}$ determines $\mathbf{u}$; each player has the minimax strategy determined by using the *other player's payoffs* in Exercise 9 of 4.2. The reader can easily verify that (9) satisfies the equalities in (5).

As before the altruistic solution (19, 19) is not an equilibrium point but it is symmetrical and efficient, and it is regarded as the cooperative solution. ■

Example (f)

A resource may take the form of a passing opportunity that will be wasted if it is not exploited, but it may be unprofitable for both players to attempt to do this. The reporters of Examples (b) and (c) will be in this position if the coverage of their event presents no obligation but is merely one of a number of opportunities. The same situation arises if two competing firms are thinking of entering a new market that can support the required initial investment by one firm but not by two. (Utilities supplying electricity, water, etc. are classic examples.) The payoffs might be

	Do nothing	Take action	
Do nothing	(0,0)	(0, 1)	(10)
Take action	(1,0)	(−1, −1)	

Here the equilibrium points are (1,0), (0,1), and (0,0), the latter arising not from $\mathbf{u}=\mathbf{y}=(1,0)$ but from $\mathbf{u}=\mathbf{y}=(.5,.5)$. The first two are efficient and their average (.5,.5) is an efficient symmetrical solution but not an equilibrium point. ■

EXERCISES 5.1

In 1 to 8 determine the efficient points and the equilibrium points (a_{ij}, b_{ij}) in the table, and (except in 7, 8) the mixed equilibrium strategies $\mathbf{u}^*, \mathbf{y}^*$, if any.

1. (0,0) (2,1)
(1,2) (3,3)

2. (0,0) (2,1)
(−2, −1) (3, −2)

3. (0,0) (3,2)
(2,3) (1,1)

4. (6,0) (3,5)
(0,6) (5,3)

5. (1,0) (−2, −1)
(0,1) (3,2)

6. (1,0) (−2, −1)
(0,1) (1,0)

7. (0,0) (0,5) (8,1)
(1,6) (−1,0) (0,0)
(7,0) (−2, −3) (0, −1)

8. (0,8) (6,7)
(2,9) (5,8)
(3,2) (7,0)

9. Show that the payoff points are collinear. Use this fact to solve the game.

$$\begin{bmatrix} (3,-1) & (2,1) \\ (-1,7) & (1,3) \\ (4,-3) & (-2,9) \end{bmatrix}$$

10. Each of two firms is going to open a store in one of three towns, of which the second is between the other two. If both choose the second town T_2, each will break even. If both choose T_1 or both choose T_3, each will lose 1 unit. If one chooses T_1 and one T_3, each will gain 1 unit. If one chooses T_2 and the other T_1 or T_3, the former will gain 2 units and the latter break even. Set up the bimatrix game and advise the firms.

11. I and J are candidates for the presidential nomination who must decide whether to campaign intensively in one or in both (strategy 3) of two states that are expected to be decisive. If each wins in one state the outcome is uncertain (payoff 1 for each), except that if they refrain from campaigning in the same state they will form a joint ticket with I nominated for president (worth 2) and J for vice-president (worth 1). If only one campaigns in a state, he will win it; otherwise I will win in state 1 and J will win in state 2. Losing the nomination after a campaign battle in the same state is worth -1. Set up the bimatrix game. How would you advise the candidates?

12. *The Marriage Game*. I and J are contemplating marriage or some other type of partnership. For each define four pure strategies, which may or may not be announced in advance, and plausible payoffs $a_{ij}=b_{ji}$.

5.2 ANALYSIS OF BIMATRIX GAMES

This section is concerned with some relations among the solution concepts of the preceding section, and the classification and more detailed discussion of 2 by 2 symmetric games, including the Prisoner's Dilemma and several extensions thereof.

The solution of the games $i \to j$ and $j \to i$ with ordered moves and pure strategies is nearly as simple for bimatrix games as for zero-sum games. Equation (1) illustrates this calculation as well as the determination of pure equilibria by underlining the largest a_{ij} in each column and the largest b_{ij} in each row.

$$\begin{array}{|ccc|l}
\hline
\underline{0},0 & 2,\underline{3} & 0,2 & {}^*2,3\ (i \to j) \\
\underline{0},\underline{1} & \underline{3},0 & \underline{1},\underline{1} & {}^*1,1 \\
\hline
0,1 & 3.0 & 1,1^* & (j \to i)
\end{array} \tag{1}$$

The points $(0,1)$ and $(1,1)$ are equilibria, both components being underlined; $(2,3)$ and $(1,1)$ are copied at the right because they have the largest b_{ij} in each row; the same is true of $(0,1)$, but it is passed over because it is dominated by $(1,1)$. At the bottom $(0,1)$ (which dominates $(0,0)$), $(3,0)$ and $(1,1)$ are copied because they have the largest a_{ij} in each column. The $j \to i$ solution $(1,1)$ has the largest $b_{ij}=1$ in its (added) row and dominates $(0,1)$. The $i \to j$ solution $(2,3)$ has the largest $a_{ij}=2$ in its (added) column.

Some bimatrix games resemble the zero-sum games in that I should prefer the version $j \to i$ in which he moves second. In other cases, such as (e) and (f) of 5.1, it is advantageous to move first. Equation (1) is an example in which *both* players should prefer that I move first, since $(2,3)$ dominates $(1,1)$.

We want to consider the logical relations among the following five statements about a bimatrix game:

P: There is a dominant *point* (a_{kl}, b_{kl}); it is the only efficient point. Thus

$$a_{kl} \geq a_{ij} \quad \text{and} \quad b_{kl} \geq b_{ij} \quad \text{for all } i \text{ and } j \tag{2}$$

S: There are dominant *strategies* represented by the kth row of $\mathbf{A}$ and the lth column of $\mathbf{B}$; that is,

$$a_{kj} \geq a_{ij} \quad \text{and} \quad b_{il} \geq b_{ij} \quad \text{for all } i \text{ and } j \tag{3}$$

Γ: (a_{kl}, b_{kl}) is the *cooperative* (Nash) solution $(\alpha^{\#}, \beta^{\#})$ defined in 5.3.
M: It is immaterial who moves first; $i \to j$ and $j \to i$ have the mutual solution (a_{kl}, b_{kl}).
Q: (a_{kl}, b_{kl}) is an *equilibrium* point, unique in that it dominates all other pure equilibrium points. Thus

$$a_{kl} \geq a_{il} \quad \text{and} \quad b_{kl} \geq b_{kj} \quad \text{for all } i \text{ and } j \tag{4}$$

and $a_{kl} \geq a_{pq}$ and $b_{kl} \geq b_{pq}$ when (a_{pq}, b_{pq}) is an equilibrium point. Then (a_{kl}, b_{kl}) is described as a solution of the (noncooperative) game $\{i, j\}$.

To avoid trivial complications and exceptions, we consider the case in which there are no ties for any maximum; this includes all the games having no two a_{ij} equal and no two b_{ij} equal. Then the following implications hold with a common kl:

$$\begin{array}{ccccc} P & \Longrightarrow & \Gamma & & \\ & \searrow & & & \\ S & \Longrightarrow & M & \Longrightarrow & Q \end{array} \tag{5}$$

The implications involving P are obvious. The proofs of the others are given in Exercises 17 and 18. Exercises 1 and 2 show that the implications cannot be reversed. Any of the Exercises 8 to 12 of 5.3 shows that a unique equilibrium point that is efficient is not necessarily the cooperative Nash solution.

Although (5) considers only pure strategies, it lends support to the idea of defining the *noncooperative solution* of a bimatrix game $\{I, J\}$ to be the equilibrium point $(\hat{\alpha}, \hat{\beta})$, pure or mixed, if such exists, that dominates all the other equilibrium points. The games of types C, C', C'' defined hereafter have no such solution. The following game shows domination of the pure equilibrium (2,2) by the mixed equilibrium (3,3) resulting from $\mathbf{u}=(.5,.5)$, $\mathbf{y}=(0,.5,.5)$.

$$\begin{array}{ccc} (2,2) & (4,0) & (2,1) \\ (0,0) & (2,6) & (4,5) \end{array} \tag{6}$$

Exercise 21 considers the $I \to J$ and $J \to I$ solutions of bimatrix games.

Classification of 2 by 2 Symmetric Games

A symmetric two-person game is one having $b_{ij}=a_{ji}$. (**B** is the transpose of **A**.) This says that if the two players exchange roles, by swapping their strategies i and j and their payoffs a and b, nothing is changed. Thus if each player has two pure strategies the game takes the form

$$\begin{matrix}(a_{11},a_{11}) & (a_{12},a_{21}) \\ (a_{21},a_{12}) & (a_{22},a_{22})\end{matrix} \qquad (7)$$

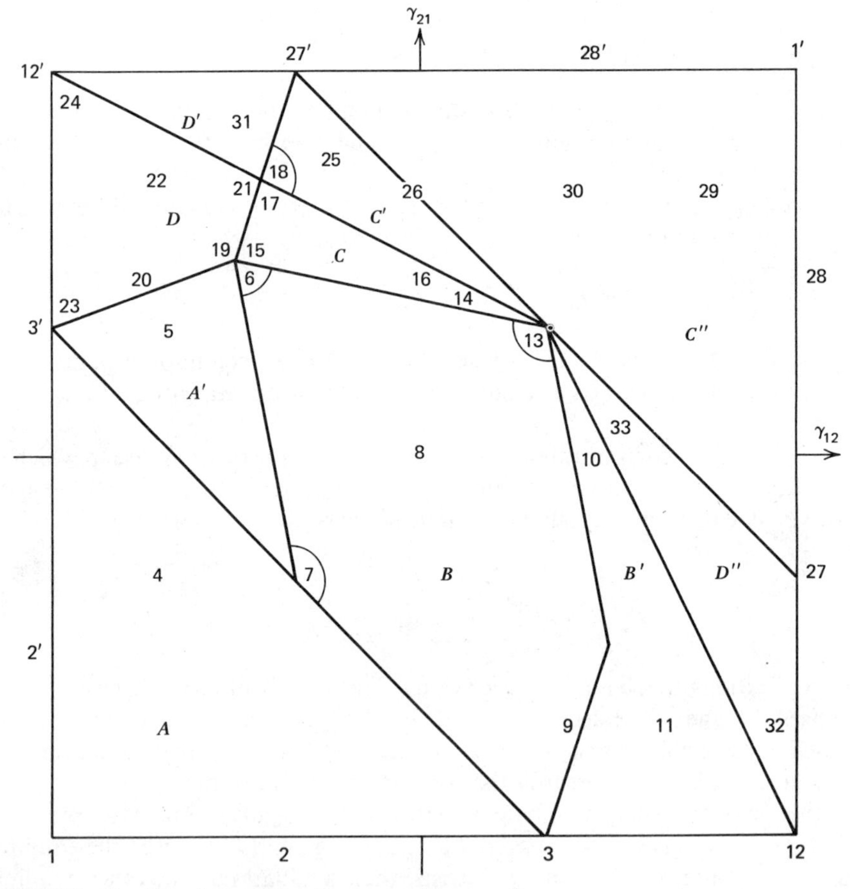

FIGURE 5.2 Classification of 2 by 2 games with $a_{ij}=b_{ji}=\gamma_{ij}$.
$\gamma_{11},\gamma_{22}=-(\gamma_{12}+\gamma_{21})/2\pm 1\mp\max(|\gamma_{12}|,|\gamma_{21}|)$.
$1\geq\gamma_{11}\geq\gamma_{22}\geq -1$. $|\gamma_{ij}|\leq 1=100\%$.
Labels 1 to 33 are circled in Table 5.2A; accented numbers require exchange of γ_{12} and γ_{21}.
Significance of the boundaries:
$2a_{11}=a_{12}+a_{21}$ on the line 12′-18-13-12.
$a_{11}=a_{21}$ on 3′-6-13-27. $a_{22}=a_{12}$ on 27′-6-7-3.
$a_{11}=a_{12}$ on 27′-13-3. $a_{22}=a_{21}$ on 3′-7.
$a_{11}=a_{22}$ on perimeter of square. Likewise for γ_{ij}.

By applying a linear transformation $\alpha' = c_1\alpha + c_0$ one can eliminate two of the parameters in (7). A convenient way to do this is to normalize the game by imposing some carefully chosen conditions on the payoffs, which then will be denoted by γ_{ij} rather than a_{ij}. The following conditions work well.

$$\gamma_{11} \geq \gamma_{22} \tag{8}$$

$$\gamma_{11} + \gamma_{12} + \gamma_{21} + \gamma_{22} = 0 \tag{9}$$

$$(\gamma_{11} - \gamma_{22})/2 + \max(|\gamma_{12}|, |\gamma_{21}|) = 1 \tag{10}$$

To reduce any 2 by 2 symmetric game to this standardized form one has only to renumber the pure strategies (thus interchanging the two rows and also the two columns) when necessary to make $a_{11} \geq a_{22}$; then subtract from each a_{ij} the average of all the a_{ij} to obtain a'_{ij} satisfying (9); and finally divide each a'_{ij} by $(a'_{11} - a'_{22})/2 + \max(|a'_{12}|, |a'_{21}|)$ to obtain γ_{ij} satisfying (10).

Equations (8) to (10) have the desirable properties of making all $|\gamma_{ij}| \leq 1$ and defining γ_{11} and γ_{22} as single-valued functions of γ_{12} and γ_{21}, thus:

$$\begin{aligned} \gamma_{11} &= 1 - \max(|\gamma_{12}|, |\gamma_{21}|) - (\gamma_{12} + \gamma_{21})/2 \\ \gamma_{22} &= -1 + \max(|\gamma_{12}|, |\gamma_{21}|) - (\gamma_{12} + \gamma_{21})/2. \end{aligned} \tag{11}$$

The γ_{ij} also satisfy

$$|\gamma_{12} - \gamma_{21}|/2 + \max(|\gamma_{11}|, |\gamma_{22}|) = 1 \tag{12}$$

so that with the slightly less convenient assumption $\gamma_{12} \geq \gamma_{21}$ one could use γ_{11} and γ_{22} as the independent parameters.

Thus each 2 by 2 symmetric game is represented by a point $(\gamma_{12}, \gamma_{21})$ in Figure 5.2. The various types of games correspond to 10 polygonal regions $A, A', B, B', C, D, C', C'', D', D''$. The major determinants of this classification are the properties listed in Table 5.2B. Table 5.2D defines five values $\alpha^{\#}$, $\hat{\alpha}$, α^{*}, α^{0} and α_0 associated with the games and includes these in the inequalities that give the analytic definitions of the interiors of the various regions. These conditions are valid for a_{ij}, $\alpha^{\#}$, etc. and for γ_{ij}, $\gamma^{\#}$, etc., before or after normalization.

Thus to classify any given 2 by 2 game having $a_{ij} = b_{ji}$, one may use Figure 5.2 after normalization or Tables 5.2B or D with or without any normalization. Table 5.2D is definitive.

The Nash cooperative value $\alpha^{\#}$, $\beta^{\#}$ or $\gamma^{\#}$ referred to hereafter is discussed in general terms in the following section. There Exercise 6 shows that for symmetric games as in this section, it is simply the unique efficient point that is symmetrical in the sense of having $\alpha^{\#} = \beta^{\#}$, as in (13) following.

TABLE 5.2A 33 Normalized Examples of 2 by 2 Games with $a_{ij}=b_{ji}=\gamma_{ij}$

Data ($\times 100$) for each example:

Nash cooperative values $=\alpha^{\#}$
Mixed (α^{*}) or dominant ($\hat{\alpha}$) eqlbm$=\alpha'$

$\alpha^{\#}$	③	γ_{11}	γ_{12}
α'	···	γ_{21}	γ_{22}

100	①	100	−100	67	②	67	−34	33	③	33	33
0	···	−100	100	4	A	−100	67	33	···	−100	33
83	④	83	−67	50	⑤	50	−67	50	⑥	50	−50
−4	A	−33	17	21	A′	33	−16	···	···	50	−50
100	⑦	100	−33	100	⑧	100	0	50	⑨	50	33
100	···	−33	−33	100	B	0	−100	50	B	−67	−16
30	⑩	30	50	33	⑪	33	67	0	⑫	0	100
30	B′	−10	−70	33	B′	−67	−33	0	B′	−100	0
33	⑬	33	33	33	⑭	33	29	48	⑮	48	−49
33	···	33	−100	33	C	35	−97	−25	C	51	−50
30	⑯	30	−10	19	⑰	19	−42	14	⑱	14	−42
20	C	50	−70	−38	C	68	−45	···	CD	70	−42
49	⑲	49	−51	41	⑳	41	−75	16	㉑	16	−45
−49	D	51	−49	−9	D	43	−9	−42	D	71	−42
25	㉒	25	−75	33	㉓	33	−99	1	㉔	1	−99
−25	D	75	−25	31	D	35	31	−1	D	99	−1
25	㉕	0	−25	33	㉖	0	0	33	㉗	−33	100
−19	C′	75	−50	0	C′C″	67	−67	···	···	−33	−33
67	㉘	−67	100	67	㉙	−34	67	50	㉚	−17	33
−4	C″	34	−67	4	C″	67	−100	4	C″	67	−83
20	㉛	−10	−50	12	㉜	−3	91	30	㉝	10	60
−30	D′	90	−30	−3	D″	−67	−21	10	D″	0	−70

The payoffs $(\gamma_{ij}, \gamma_{ji})$ are normalized so that $\gamma_{11} \geq \gamma_{22}$, $\gamma_{11}+\gamma_{12}+\gamma_{21}+\gamma_{22}=0$, $(\gamma_{11}-\gamma_{22})/2+\max(|\gamma_{12}|, |\gamma_{21}|)=1=100\%$. Plotted in Figure 5.2. Equilibria at $ij=12, 21$ iff $\gamma_{21} \geq \gamma_{11}$ and $\gamma_{12} \geq \gamma_{22}$.

Table 5.2A presents data (all multiplied by 100 and rounded to the nearest integer) for 33 normalized games identified by the numbers in Figure 5.2. The right half of each 2 by 4 array gives the γ_{ij}. The left column gives the Nash cooperative value, which for (7) is

$$\alpha^{\#} = \beta^{\#} = \max\left[a_{11}, a_{22}, (a_{12} + a_{21})/2\right] \tag{13}$$

and the mixed equilibrium value α^* of (9) of 5.1 (when it exists) or else the dominant equilibrium value $\hat{\alpha}$ (possessing property Q in (4)). In the 2 by 2 symmetric case $(\alpha^{\#}, \beta^{\#})$ is the only efficient point having $\alpha = \beta$. The three numbers in (13) are obtainable from the strategies $\mathbf{u} = \mathbf{y} = (1,0)$, $\mathbf{u} = \mathbf{y} = (0,1)$, and $\pi_{12} = \pi_{21} = 1/2$, respectively, in the notation of (2) of 5.1. The last of these is the joint mixed nonequilibrium strategy used cooperatively for the types C', C'', D', D''. Table 5.2C lists a dozen unnormalized examples for convenient reference.

Equations (11) and (13) show that γ_{12} and γ_{21} may be interchanged without changing the values of γ_{11}, γ_{22} or $\gamma^{\#}$. Nevertheless the resulting game is different, possibly of another type, except on the perimeter of the square where $\gamma_{11} = \gamma_{22}$. There the change is equivalent to a mere renumbering of the pure strategies $i = 1,2$ and $j = 1,2$. In this way the points $2', 3', 12', 27', 28'$ in Figure 5.2 are related to the corresponding unaccented numbers in Figure 5.2 and Table 5.2A, located on the perimeter, symmetrically with respect to the diagonal $\gamma_{12} = \gamma_{21}$. Thus D' and D'' (or

TABLE 5.2B Properties of 2 by 2 Games with $a_{ij} = b_{ji}$

Type of game	A, A'	B	B'	C	D	C', C''	D', D''
Nash solution (coop., efficient)	Pure equilibrium			Pure noneq.		Joint mixed noneq.	
Properties (2) to (3)	P	PS	S	$\cdots$	S	$\cdots$	S
†Efficient noneq.	0	0	2	1	3	0	2
†Efficient eq.	1	1	1	2	0	2	0
†Inefficient eq.	2*	0	0	1*	1	1*	1

†These lines give the number of points (a_{ij}, a_{ji}) of the three kinds indicated, except that the numbers 1* and 2* include one mixed equilibrium (eq.).
$\cdots$ indicates there is no dominant equilibrium.

TABLE 5.2C Examples (as in Table 5.2A but not normalized)

6	A	6	0	4	A'	4	0	4	B	4	2	5	B	5	3
$\frac{3}{2}$		0	2	2		3	1	4		2	0	5		0	0
3	B'	3	5	3	C	3	1	3	D	3	0	3	C'	2	1
3		0	0	2		4	0	1		4	1	$\frac{5}{4}$		5	0
$\frac{7}{2}$	C''	1	2	$\frac{7}{2}$	C''	1	5	$\frac{5}{2}$	D'	2	0	3	D''	2	6
$\frac{5}{3}$		5	0	$\frac{5}{3}$		2	0	1		5	1	2		0	0

TABLE 5.2D Solutions and Order Relations

$m_{12} \triangleq (a_{12}+a_{21})/2$. $a_{11} \geq a_{22}$ $\alpha=\beta$ for five solutions:

$\alpha^{\#}=\max(a_{11}, a_{22}, m_{12})=$Nash cooperative value.
$\hat{\alpha}$ =dominant equilibrium$=\{i, j\}$ solution (not in $CC'C''$).
$\alpha^{*}=(a_{11}a_{22}-a_{12}a_{21})/(a_{11}+a_{22}-a_{12}-a_{21})=$mixed equilibrium in $AA'CC'C''$; neither row of **A** dominates.
α^{0} =payoff when both players use α_0 strategies.
α_0 =maximin security value=value of zero-sum game **A** with mixed strategies.

A: $a_{11}=\alpha^{\#}=\hat{\alpha}>a_{22}>\alpha^{*}=\alpha^{0}=\alpha_0>\max(a_{12}, a_{21})$.
A': $a_{11}=\alpha^{\#}=\hat{\alpha}>a_{21}>\alpha^{*}>a_{22}=\alpha^{0}=\alpha_0>a_{12}$.
B: $a_{11}=\alpha^{\#}=\hat{\alpha}=\alpha^{0}>a_{12}=\alpha_0>a_{22}$; $a_{11}>a_{21}$.
B': $a_{12}>a_{11}=\alpha^{\#}=\hat{\alpha}=\alpha^{0}=\alpha_0>m_{12}>a_{21}$; $a_{11}>a_{22}$.
C: $a_{21}>a_{11}=\alpha^{\#}=\alpha^{0}>\alpha^{*}>\alpha_0=a_{12}>a_{22}$; $a_{11}>m_{12}$.
D: $a_{21}>a_{11}=\alpha^{\#}>a_{22}=\hat{\alpha}=\alpha^{0}=\alpha_0>a_{12}$; $a_{11}>m_{12}$.
C': $a_{21}>m_{12}=\alpha^{\#}>a_{11}=\alpha^{0}>\alpha^{*}>a_{12}=\alpha_0>a_{22}$.
C'': $m_{12}=\alpha^{\#}>\min(a_{12}, a_{21})>\alpha^{*}=\alpha^{0}=\alpha_0>a_{11}>a_{22}$.
D': $a_{21}>m_{12}=\alpha^{\#}>a_{11}>a_{22}=\hat{\alpha}=\alpha^{0}=\alpha_0>a_{12}$.
D'': $a_{12}>m_{12}=\alpha^{\#}>a_{11}=\hat{\alpha}=\alpha^{0}=\alpha_0>\max(a_{21}, a_{22})$.

Plausible Noncooperative Solutions

A, A', B, B': $\hat{\alpha}=\alpha^{\#}=a_{11}$. D'': $\hat{\alpha}=a_{11}$
C, C': α^{*} and D, D': $\hat{\alpha}=a_{22}$ (both $<$nonequilibrium a_{11}).
C'': $\alpha^{*}<a_{12}$ and a_{21} from efficient equilibria.

D and B') can be regarded as neighboring regions, and the whole square can be visualized as the surface of a sphere, or a double cone with vertices at $(-1,-1)$ and $(1,1)$. These vertices (labeled 1 and 1′) also represent essentially the same game (renumber the rows $i=1,2$ only).

Discussion of Examples

Games of types A, A', B, B' are distinguished by having identical cooperative and noncooperative solutions ($\alpha^{\#}=\hat{\alpha}=a_{11}$) and by having only one point, (a_{11}, a_{11}), that is both efficient and equilibrium. In B' even the ultrapessimistic maximin value α_0 equals $\alpha^{\#}$; the zero-sum case (12 or 12′ in Figure 5.2 and Table 5.2A), which has a saddle point, may be included in this type.

The other types C, D, C', C'', D', D'' possess none of these properties; the efficient cooperative Nash solution $(\alpha^{\#}, \alpha^{\#})$ is not an equilibrium nor is it obtained if each player uses the "egoistic" pure strategy that conceivably could give him the highest possible payoff. (This is strategy i or $j=2$, except in D'' and the adjacent half of C''.) In practice then there may be difficulty in realizing $\alpha^{\#}$ because of an inability of one or both players to communicate, or to enforce any agreement they may make, or to think and act rationally and cooperatively. Some of the noncooperative solutions that may result are listed at the bottom of Table 5.2D.

Types C and D are the games of Chicken and Prisoner's Dilemma as defined by Rapoport and Chammah (1966). They are distinguished from C', C'', D' and D'' in that pure (altruistic) strategies produce the Nash solution $(\alpha^{\#}, \alpha^{\#})=(a_{11}, a_{11})$, which is not an equilibrium point for any of the six types. In C only a one-sided betrayal (egotism, i or $j=2$) has the color of advantage, while in D the dominance of the egoistic strategies invites each player to betray no matter what the other does.

C, C' and C'' have a pair of efficient equilibria, which are efficient nonequilibria in D, D' and D''. In C, C' and C'' a change of strategy by a player dissatisfied with a pure equilibrium is bad for both players but a subsequent switch by the other player will be good for him and even better for the first player. In D, D' and D'' a departure from the dominant strategy of either player is always altruistic; it hurts the player who switches but helps the other more.

Types C', C'', D', D'' have a Nash solution that averages or oscillates between two efficient points. In C' and C'' these points are equilibria for which the two players have differing preferences. Oscillation is likely to occur with or without cooperation; the role of cooperation is to limit the oscillation to the efficient points. In D' and D'' there is a unique equilibrium, pure and inefficient; cooperation involves taking turns in leaving this false refuge in order to help the other player despite one's own (smaller) loss.

C'' is unique among the 10 types in having the efficient equilibria at $ij=12$ and 21 dominating $ij=11$ and 22 and dominating the mixed equilibrium. The result is a rapid depreciation in the desirability of the payoff a_{11} from C, where it is $\alpha^{\#}$, through C' to C'', where it is even less than α_0. In C'' one may have $a_{12}=a_{21}$ on the line joining 1′ to 13 (via 29) in Figure 5.2, so that the two dominant equilibria have identical payoffs. (A similar tie between $ij=11$ and 22 occurs between 1 and 3 or 3′ in Figure 5.2.) Then there is no dispute between the two players, although there may be a need for at least one-way communication between them to coordinate their choices and arrive at one of the dominant points. If a_{12} and a_{21} are nearly the same in C'', one or both players may feel that the difference is not worth fighting about, since the possible loss much exceeds the possible gain from insisting on the cooperative solution. Then either of (a_{12}, a_{21}) and (a_{21}, a_{12}) might be accepted. This is more true when $a_{21}>a_{12}$; then disagreement yields the smallest payoffs (a_{22}, a_{22}).

Every pair $(\mathbf{u}, \mathbf{y})$ is an equilibrium point for Example 6; each player determines the other's payoff but not his own. Examples 13, 18, 27 and 27′ also have infinite numbers of equilibrium points; in 13 these all make $\alpha=\beta=33$.

Displays (14) and (15) hereafter list the gains or (negative) losses for some examples of Chicken (C) and Prisoner's Dilemma (D) respectively, resulting from a change from the altruistic strategy 1 to the egoistic strategy 2. The first two differences give the egoist's gains, the third is the other player's loss $b_{21}-b_{11}$ when he remains an altruist, and the fourth is the amount each loses when both change from altruists to egoists.

(C)	Example	14	15	16	17	18	
Ego.	$a_{21}-a_{11}$	2	3	20	49	56	
Ego.	$a_{22}-a_{12}$	−126	−1	−60	−3	0	(14)
Alt.	$a_{12}-a_{11}$	−4	−97	−40	−61	−56	
Both	$a_{22}-a_{11}$	−130	−98	−100	−64	−56	

(D)	Example	19	20	21	22	23	24	
Ego.	$a_{21}-a_{11}$	2	2	55	50	2	98	
Ego.	$a_{22}-a_{12}$	2	66	3	50	130	98	(15)
Alt.	$a_{12}-a_{11}$	−100	−116	−61	−100	−132	−100	
Both	$a_{22}-a_{11}$	−98	−50	−58	−50	−2	−2	

In each of (14) and (15), the examples are numbered more or less in the order of increasing seductiveness of the egoistic strategy, which seems psychopathic in Examples 14, 15 and 19 with their small gains, but rather reasonable in Example 24 with its large gains of 98 and small joint loss of 2 units; the latter could be cheap insurance against a possible loss of 100. Perhaps the dilemma is most acute in Example 22, which is the normalized version of (d) of 5.1. Nevertheless in case D the egoistic strategy always has the potential for a moral dilemma in the sense that it is always relatively disastrous for the other player if he is an altruist.

The Examples (b) to (f) of 5.1 were of types B', D'', D, C and $C'-C''$ boundary, respectively. The remaining two examples are variants of the Prisoner's Dilemma that lie outside the classification scheme.

Example (g). The Prisoner's Trilemma

This is a symmetric 3 by 3 game that combines the features of Examples 19 and 24:

	Alt.	Prag.	Psy.	
Altruistic	(50,50)	(1,99)	(0,100)	
Pragmatic	(99,1)	(49,49)	(1,50)	(16)
Psychopathic	(100,0)	(50,1)	(2,2)	

The pragmatist expects the other player to be like himself, willing to accept a small loss rather than slaughter the other player, but also willing to take what he can on a one-for-one basis. ("Your loss is no more than my gain.") A sadistic or nihilistic strategy with nonpositive payoffs could be added to complete the psychological spectrum.

Example (h). A One-Sided Prisoner's Dilemma

J's belief→	I altruistic	I egoistic	
I altruistic	(2,3)	(0,2)	(17)
I egoistic	(3,0)	(1,1)	

Here (2,3) is the Nash solution of 5.3 and the $i \to j$ noncooperative solution while (1,1) is the unique equilibrium, the $j \to i$ or $\{I, J\}$ noncooperative solution, and the maximin solution.

The labels assigned to the strategies are appropriate for the $j \to i$ and $\{I, J\}$ version of the game. Only I has a dominant strategy ($i=2$), and the outcome of $j \to i$ depends only on I's ethics and on J's expectations concerning them, not on

J's own ethics. If J has to move first, $ij=22$ and the payoffs (1, 1) will result unless J believes that I will imitate J's altruism. On the other hand the $i \to j$ version involves no ethical problems at all, only the curiosity that it is advantageous for *both* players for I to move first, and in doing so, *I will reject his dominant strategy* $i=2$. This game is number 47 among the 78 games classified by Rapoport and Guyer (1966). ■

Metagames and Generalized Equilibria

Metagames begin with the observation (as at the end of 4.2) that even if a game is to be played only once, it may be reasonable to analyze it as if it were to be played many times. In the latter case, even if the moves are simultaneous in each play, they occur in sequence in the series of plays, and a player's current strategy may well be influenced by his past experience with his opponent. This was true in 4.3(iii) also, but now the need to cooperate in a variable-sum game leads to a different logic. A player may hope to influence the future behavior of his opponent; and with ordered moves it is no longer true that a dominant strategy is always preferable even for a confirmed egoist (see the preceding paragraph). On the contrary, as N. Howard (1971) p. 180 has noted, a player's use of a strictly dominant strategy (if he has one) makes *all* the equilibria of the game available for selection by his opponent(s); he renounces any threat that he may have.

The metagame theory and its variations use only pure strategies and the order relations (preferences) among each player's payoffs. In its simplest form it shares with Nash's definition of an equilibrium ((5) of 5.1) the property that each player needs to know only his own preferences on payoffs. This is relevant to *hypergames* (see Bennett et al., 1979, 1980), in which some players may fail to recognize the existence of certain strategies or be misinformed about the payoffs. Metagame theory was developed by N. Howard (1966–1974); also see Rapoport, Guyer and Gordon (1976).

Howard showed that it suffices to consider $n!$ "complete" metagames, one for each permutation of the n players. For the 2 by 2 Prisoner's Dilemma with the altruistic strategy c (cooperate) and the egoistic strategy d (defect) there are two of these metagames, in which one player has 4 strategies cc, cd, dc, dd and the other has 16 strategies $cccc, cccd, ccdc, \ldots, dddd$. The four letters in the latter indicate the strategy c or d chosen in response to cc, cd, dc, dd respectively, and the pairs of letters indicate the responses to c and d respectively. The reader can verify that each metagame has the original equilbrium resulting from $i=d, j=d$, and has acquired a new (meta) equilibrium resulting from $i=c$, $j=c$ in the original game and from $i'=cd$ and $j'=dcdd$ or $ccdd$ (or vice versa) in the metagame. Since cd is the second of the four pairs, the quadruples specify $j=c$ (the second of the four letters), which in turn specifies $i=c$, the first letter in $i'=cd$.

Since the new equilibrium dominates the old, one is tempted to say with Rapoport (1967) that a theoretical resolution for the Prisoner's Dilemma has been obtained; even with egoistic motivations, the inferior outcome dd is not logically inevitable, at least so long as the stakes are not too great. In general, however, the effect of metagame analysis is to supply (new) metaequilibria without necessarily determining a selection among them. (For example, in Chicken cc is obtained as a third equilibrium, which does not dominate the others.) In practice a similar result is obtained much more easily by weakening the definition of equilibrium so that more points can qualify. This is considered by N. Howard (1971) and Bennett, Dando and Sharp (1980); the latter thereby define an "axis" of "acceptable solutions."

A pair of pure strategies kl is *excluded from the axis* if and only if kl admits an "undeterrable improvement," one or more of (k) to (m) being true:

(k) There is a pl with $a_{pl} > a_{kl}$ and $a_{pj} \geq a_{kl}$ for all j.

(l) There is a kq with $b_{kq} > b_{kl}$ and $b_{iq} \geq b_{kl}$ for all i. (k) and (l) exclude j and i as deterrents.

(m) There is a pq strictly dominating kl such that $(a_{iq} > a_{pq}$ and $b_{iq} < b_{kl})$ for no i, and $(b_{pj} > b_{pq}$ and $a_{pj} < a_{kl})$ for no j. These conditions exclude defections (from the joint improvement pq) that would benefit the defector but injure the other player; in particular, if the dominating pq is an equilibrium we will have no $a_{iq} > a_{pq}$ and no $b_{pj} > b_{pq}$, and kl will not be in the axis.

In (16) the axis consists of the diagonal positions 11, 22, 33.

Utility Measurements

We are now concerned with changes more general than the linear transformations in 5.1(d). Let us take for granted the fact that in practice subjective factors may influence the estimates of what are in principle objective quantities. Then three overlapping types of payoff or utility may be considered:

(i) An objective utility is an amount of money, time or other desirable commodity or physical quantity, or the objective probability (with a relative-frequency interpretation) of a specified desirable event. This is usually required in zero-sum games, where the players share a common scale of utility, or more generally, scales that are linearly related.

(ii) A personal utility (usually called simply *utility*) is a personal (subjective) appraisal of one's personal situation. It is usually assumed to be measured by comparison with a hypothetical but objective probability scale as follows. Two utility values (for simplicity we will use 0 and 1) are assigned to two specific outcomes ω_0 and ω_1, of which ω_0 is very undesirable and ω_1 is very desirable. Then the utility α of an outcome ω that is better than ω_0 and worse than ω_1 is defined as the probability α such that the player has no preference between ω and a lottery in which he gets ω_0 with probability $1-\alpha$ and ω_1 with probability α. For example, as a rough but plausible approximation, it is sometimes assumed that the utility of accumulated wealth is proportional to the logarithm of its monetary value; to go from \$10,000 to \$20,000 is about as pleasant as to go from \$20,000 to \$40,000.

(iii) A semipersonal utility is a personal appraisal of a global situation (i.e., an outcome of the game), differing from (ii) in that it is influenced (in various possible ways) by the payoffs of the other players. At first sight this may appear to be the best definition; it considers all the facts and it leaves the door open to cooperation. However, there are difficulties. Both personal and semipersonal utilities invite misrepresentation. Semipersonal utilities are circular or implicit in their definition and so might require iterative calculations. By embedding ethical factors in the payoffs themselves, they make it hard to analyze such considerations in the game. Worst of all, they aggravate the drawbacks in being an altruist.

Consider the games

$$\begin{matrix} (4,4) & (0,6) \\ (6,0) & (2,2) \end{matrix} \quad (18) \qquad \begin{matrix} (4,4) & (3,6) \\ (3,0) & (2,2) \end{matrix} \quad (19)$$

(18) is the Prisoner's Dilemma, supposed to have objective or personal utilities. It becomes (19) after I has decided that J's welfare is as important as his own; thus he has replaced a_{ij} by $(a_{ij}+b_{ij})/2$. The trouble is that perhaps he should have replaced b_{ij} by $(a_{ij}+b_{ij})/2$ also, but under the rules for semipersonal utilities, that is J's prerogative, not I's.

The altruist I will choose the first row in both games. Now is is possible that J is decent enough to choose $j=1$ in (18), whereas after looking at (19) he may have no hesitancy about choosing $j=2$, because he thereby gains twice as much as I loses. (If I had used $a_{ij}+b_{ij}$, J could argue that (6,6) is a fairer division than (8,4)). The effect in (18) is to go from (4,4) to (0,6), as if J's welfare were at least twice as important as I's.

In short it can be unsatisfactory to make an interpersonal decision on a personal basis, whether the decision-maker is an altruist or an egoist.

EXERCISES 5.2

In 1 to 8 determine the $i \to j$ and $j \to i$ solutions and the presence (+) or absence (−) of the properties P, S, M, Q of (2) to (5).

1. $\begin{matrix} (0,0) & (0,3) & (0,0) \\ (3,0) & (1,1) & (0,0) \\ (0,0) & (0,0) & (2,2) \end{matrix}$ **2.** $\begin{matrix} (4,1) & (2,3) & (0,0) \\ (1,4) & (3,2) & (0,0) \\ (0,0) & (0,0) & (1,1) \end{matrix}$

3.–8. Use Exercises 3 to 8 of 5.1.

In 9 to 14 determine $\alpha^{\#}=\beta^{\#}$ (Nash cooperative value). Normalize and find the type $A, A', \ldots, D''$. (**A** is given; $\mathbf{B}=\mathbf{A}^T$.)

9. $\begin{matrix} 1 & -9 \\ 5 & 3 \end{matrix}$ **10.** $\begin{matrix} 9 & 0 \\ 11 & 6 \end{matrix}$ **11.** $\begin{matrix} 1 & 7 \\ 0 & 2 \end{matrix}$

12. $\begin{matrix} 5 & 6 \\ 7 & 0 \end{matrix}$ **13.** $\begin{matrix} 1 & 3 \\ 2 & 4 \end{matrix}$ **14.** $\begin{matrix} 21 & 13 \\ 25 & 1 \end{matrix}$

15.* Prove that every feasible (α_0, β_0) is dominated by some efficient (α, β).

16.* Prove that every efficient $(\alpha, \beta)=(a_{kl}, b_{kl})$ if and only if (a_{kl}, b_{kl}) dominates all the (a_{ij}, b_{ij}). (There is only one efficient (α, β) but the corresponding indices kl are not necessarily unique.)

Prove 17 to 19 in the absence of ties (see (2) to (5)).

17. S implies M. **18.*** M implies Q.

*Starred problems are more difficult (theoretical).

19:* S implies (a_{kl}, b_{kl}) is the only equilibrium, pure or mixed, barring ties.

20:* Matching pennies can be regarded as a bimatrix game not of the form (7) but still symmetric in the sense of (i) of 4.3. Show that it is the only normalized member of its family.

21:* Show how to find the $I \to J$ solution of (17). (See Table 4.1.)

22. Determine the axis for
(a) (17) of 5.2; (b) the Marriage Game, ex. 12 of 5.1, answer section;
(c) Chicken (8) of 5.1; (d) the Prisoner's Dilemma (7) of 5.1.

5.3 COOPERATIVE SOLUTIONS (NASH)

The problem of finding a cooperative or arbitrated solution (α, β) of a bimatrix game $(\mathbf{A}, \mathbf{B})$ can be stated and solved in four different ways, depending on which of the conditions (1) to (3) are imposed. In (a), (b) and (c) the parameter θ relating the utility scales is determined by fiat, experiment, or the use of transferable utility (money or some other valuable commodity); this permits one of the three conditions to be eliminated.

(a) uses (1) and (2) and ignores competition (3).
(b) uses (2) and (3); transfers of utility eliminate (1).
(c) uses (1) and (3); prohibition of transfers eliminates (2).
(d) uses (1), (2) and (3) to determine θ as well as (α, β). (d) is the most challenging problem; Nash (1953) gave two derivations of the solution, using a bargaining model and a set of axioms respectively. Only simultaneous moves ($\{I, J\}$ version) are treated here; ordered moves are in 10.7.

A basic requirement for a cooperative solution is that it yield a pair of payoffs (α, β) that is *efficient* in the sense of 5.1: even with perfect cooperation, neither of α and β can be increased without decreasing the other. This is symbolized by

$$(\alpha, \beta) \in \Omega' \tag{1}$$

where Ω' is the "northeast" boundary of the feasible region Ω.

In 5.1 and 5.2 there was no generally valid way to combine or to compare the payoffs α and β received by the two players; in effect each could measure his payoff on his own private scale, although considerations of collinearity or symmetry could make the use of a common payoff scale mathematically attractive in special cases. One can view Nash's solution (d) as a generalization of these special cases, a demonstration that *every bimatrix game* $(\mathbf{A}, \mathbf{B})$ *contains within itself the germ of a common utility scale*. This scale can be defined by relating the previously independent scales of measurement of α and β by a parameter θ satisfying $0 \le \theta \le 1$, such that $\theta\alpha$ and $(1-\theta)\beta$ can be regarded as measured in terms of the same unit. (If one merely compared $\theta\alpha$ and β, one would sometimes need to let $\theta \to \infty$.) The resulting payoffs are unique but θ need not be.

The introduction of θ appears to be necessary to choose among the points of Ω' in (1). It makes available another condition characteristic of cooperation, which is

$$\theta\alpha+(1-\theta)\beta=\max_{i,j}\left[\theta a_{ij}+(1-\theta)b_{ij}\right]\triangleq\sigma(\theta) \tag{2}$$

In fact, if kl is the (or a) maximizing value of ij, it is clear that $(\alpha,\beta)=(a_{kl},b_{kl})$ will satisfy (1) as well as (2), but (2) as it stands defines the straight line $\theta\alpha+(1-\theta)\beta=\sigma(\theta)$ (called a *supporting line*), not just the point (a_{kl},b_{kl}).

To represent the *competitive* aspect of the game we define the *minimax line*

$$\theta\alpha-(1-\theta)\beta=\min_{\mathbf{y}}\max_{\mathbf{u}}\mathbf{u}\left[\theta\mathbf{A}-(1-\theta)\mathbf{B}\right]\mathbf{y}\triangleq\delta(\theta) \tag{3}$$

which involves the value of the two-person, zero-sum game with mixed strategies $\mathbf{u}$ and $\mathbf{y}$ and payoffs $\theta a_{ij}-(1-\theta)b_{ij}$. Player I can try to increase the average of these payoffs by increasing his $\mathbf{uAy}$ or by decreasing his opponent's $\mathbf{uBy}$; the latter action is the *threat* (extortion, blackmail) aspect of the game. Similarly J wants to increase $\mathbf{uBy}$ or decrease $\mathbf{uAy}$.

Since I's attempts to increase $\mathbf{uAy}$ and decrease $\mathbf{uBy}$ are not independent but have to be accomplished by the choice of a single $\mathbf{u}$, Nash's form (3) is preferable to the expression

$$\min_{\mathbf{y}}\max_{\mathbf{u}}\theta\mathbf{uAy}-\max_{\eta}\min_{\omega}(1-\theta)\omega\mathbf{B}\eta \tag{3a}$$

as originally defined by Von Neumann and Morgenstern (1944) and Shapley (1953). However, in some important applications the zero-sum games $\mathbf{A}$ and $-\mathbf{B}$ have optimal strategies in common; then (3) and (3a) are the same. This is considered further in 5.7(b),(d).

The straight lines (2) and (3) have slopes of opposite signs but the same absolute value $\theta/(1-\theta)$. This relation between the slopes is evidently independent of scale changes in α and β; perpendicularity, which occurs only for $\theta=1/2$, does not have this independence. Thus using the same value of θ in (2) and (3) seems virtually inevitable, although Nash derived it from the maximization of $(\alpha-\alpha_3)(\beta-\beta_3)$ over the feasible region, where $\alpha_3=\mathbf{uAy}$ and $\beta_3=\mathbf{uBy}$ are determined by (3).

Suppose the payoff scales are changed so that one defines $a'_{ij}=c_1a_{ij}+c_0$ and $b'_{ij}=d_1b_{ij}+d_0$, where $c_1,d_1>0$. Then it can be verified that the corresponding solutions (α,β) and (α',β') for any of the problems (a) to (d) satisfy $(\alpha',\beta')=(c_1\alpha+c_0,d_1\beta+d_0)$ provided the new problem uses a θ' such that

$$\theta'/(1-\theta')=d_1\theta/c_1(1-\theta) \tag{4}$$

In particular the Nash solution (d) determines θ and θ' so that the requirement (4) is automatically satisfied, and in this desirable sense it is independent of scale changes (Exercise 16).

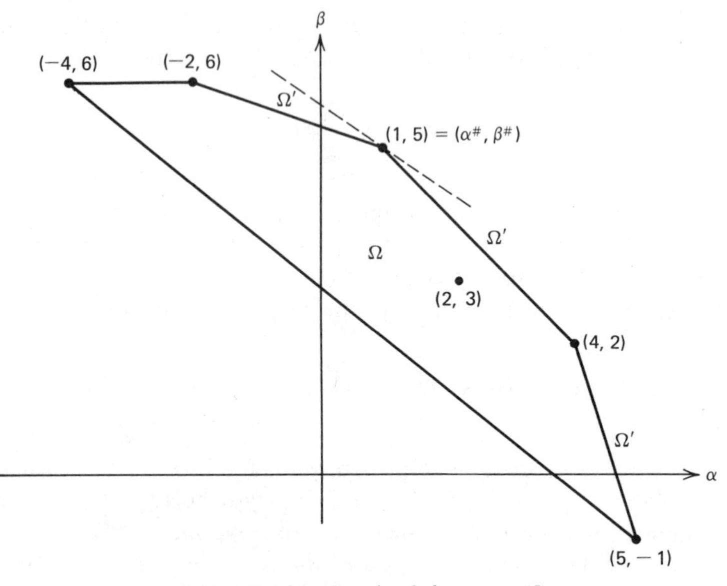

FIGURE 5.3A Graph of the game (5).

To illustrate the various problems and their solutions we shall use the example

$$\begin{matrix} (1, & 5) & (-4,6) & (\ \ 4,2) \\ (5,-1) & & (\ \ 2,3) & (-2,6) \end{matrix} \tag{5}$$

which is plotted in Figure 5.3A. The bow-shaped pentagon is the feasible region Ω, the convex hull of the points (a_{ij}, b_{ij}) in (5). Its three sides on the "northeast" between $(-2,6)$ and $(5,-1)$ are the set Ω' of efficient points.

(a) uses the given θ to *eliminate the competitive condition* (3) (the minimax line), except that in case of a tie the solution (most nearly) satisfying (3) may be preferred. In any case the solution (a_{kl}, b_{kl}) maximizes (2) and therefore satisfies (1), and

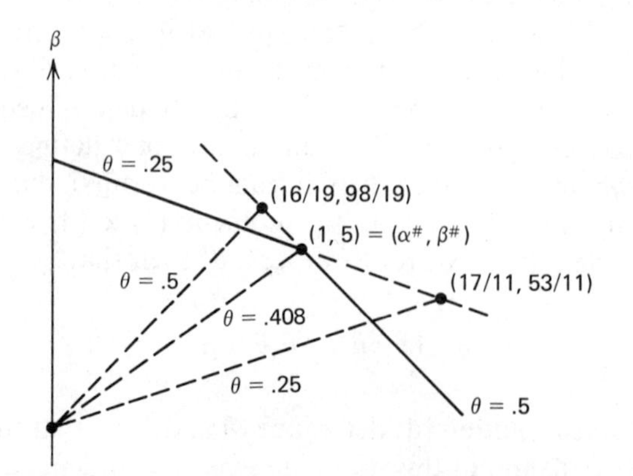

FIGURE 5.3B Solution (b) of (5) as a function of θ.

transferable utility is not needed even if it is available. The omission of (3) makes the matrix arrangement of the a_{ij} and b_{ij} irrelevant. The following are two possible applications:

(a_1) A solution is being imposed by an arbitrator who considers the relative (competitive) strengths of the two players to be immaterial.

(a_2) A single decision maker is choosing both i and j, which thus could be replaced by a single index. The payoffs a_{ij} and b_{ij} evaluate the alternative ij in terms of two different criteria such as performance by day and by night, or capacity and (negative) weight, or lifetime and (negative) cost, etc. The decision maker maximizes the weighted average $\theta a_{ij}+(1-\theta)b_{ij}$, which could easily be expanded to include additional criteria.

The solution (a) for (5) with $\theta=0.6$ is (4,2) since the maximum in (2) is $\sigma(0.6)=3.2=.6(4)+.4(2)$.

(b) uses the given θ to eliminate Ω' in (1) in favor of the supporting line (2). It is the only method allowing *side payments* from one player to another; these are called bribes when the rules of a larger game forbid them. These payments use *transferable utility*, which can be money or some other valuable commodity. If the latter also provides the common unit of measure of the a_{ij} and b_{ij}, then one has $\theta=1/2$.

Suppose for simplicity that $\theta=1/2$. Then if some point (α_0, β_0) on the line $\alpha+\beta=c$ is feasible, all the points on the line are feasible, because any such (α, β) can be obtained by choosing (α_0, β_0) and then transferring the amount $\alpha-\alpha_0=\beta_0-\beta$ from J to I. If the feasible region Ω is not already so defined, it is thus expanded from a finite polygon (the convex hull of the (a_{ij}, b_{ij})) to the narrowest strip (of slope $-\theta/(1-\theta)=-1$) that contains this polygon, as in Figure 5.3C. Thus (2) defines Ω' and so replaces (1). It is harmless if the strip is regarded as infinite in length; we never go far out on it anyway, because the minimax line (3) always intersects the convex hull of the (a_{ij}, b_{ij}) in the point $(\mathbf{uAy}, \mathbf{uBy})$.

Suppose that $\theta=0.6$ is specified with transferable utility. Then (3) and (5) give the zero-sum game

$$\begin{matrix} -1.4 & -4.8 & 1.6 \\ 3.4 & 0 & -3.6 \end{matrix}$$

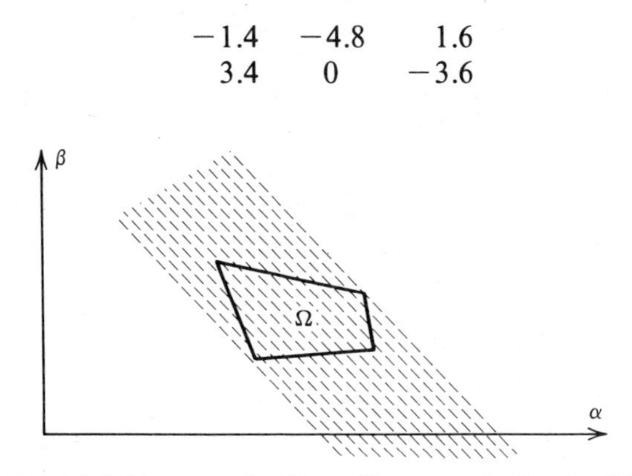

FIGURE 5.3C How transferable utility expands the feasible region.

where $-1.4=.6(1)-.4(5)$, etc. The first column can be omitted because it dominates the second and so Exercise 9 of 4.2 gives the optimal $\mathbf{u}=(.36,.64)$, $\mathbf{y}=(.52,.48)$, and the value $\delta(.6)=.6\alpha-.4\beta=-1.728$ in (3). The value of (2) is $\sigma(.6)=.6\alpha+.4\beta=3.2=.6(4)+.4(2)$. Solving these equations as in (7) gives $\alpha=3.68/3, \beta=6.16$. To obtain this point the players start at (4,2) and I gives up $8.32/3$ of his units. This is multiplied by $.6/.4$ to obtain the 4.16 that is added to J's payoff 2.

The intersection of (2) and (3) is illustrated in Figure 5.3B for three values of θ (those used in (d) below). The three dashed lines representing (3) are almost but not quite concurrent.

(c) uses the given θ to eliminate the supporting line (2) and the (possibly unacceptable) side payments it requires in (b), while retaining the efficient boundary Ω' in (1) and the minimax line (3). If the latter are inconsistent, their solution is defined to be the endpoint of Ω' that is closer to the line (3).

Suppose that $\theta=0.6$ is specified without transferable utility. Then (3) and (5) still give $.6\alpha-.4\beta=-1.728$ but the latter must be intersected with Ω'. Since $.6(1)-.4(5)=-1.4$ we expect the intersection with Ω' in Figure 5.3A will occur on the line segment joining (1,5) and (−2,6), whose equation is $\alpha+3\beta=16$. The solution is found to be $\alpha=6.08/11, \beta=56.65/11$. This is dominated by the solution (b), which has the larger feasible region, while both lie on the minimax line (3).

Nash's Cooperative Solution $(\alpha^{\#},\beta^{\#})$

(d) Nash's solution requires neither transferable utility nor a given value of θ. Instead it uses an *intrinsic value of* θ determined mathematically (from **A** and **B**) by imposing all three conditions (1), (2), (3). This is the value of θ for which problems (a), (b) and (c) have the same solution, so that the presence or absence of transferable utility or competition becomes immaterial.

The resulting $(\alpha^{\#},\beta^{\#})$ may fall on a corner or in the interior of a line segment in Ω'. In the latter case (2) will be satisfied only if $-\theta/(1-\theta)$ is the slope of the line segment in question. Thus the solution procedure involves using this slope to determine θ, using σ and δ in (2) and (3) to determine α and β, and observing whether $(\alpha,\beta)\in\Omega'$ as (1) requires. If it does not, the failure will suggest which line segment should be tested next, or else show that $(\alpha^{\#},\beta^{\#})$ lies at a specific corner. To begin one needs to find the endpoints of such a line segment, which will be extreme points of Ω that belong to Ω'. This may be done graphically or else numerically as follows.

Two such extreme points are the endpoints (α',β') and (α'',β'') of Ω'. The first is some (a_{kl}, b_{kl}) obtained, by maximizing b_{ij} and then (if there are ties) maximizing the associated a_{ij}; (α'',β'') is obtained by maximizing a_{ij} first and then b_{ij}. We determine θ' so that $-\theta'/(1-\theta')$ is the slope of the straight line joining these endpoints:

$$-\theta'/(1-\theta') \triangleq (\beta''-\beta')/(\alpha''-\alpha') \tag{6}$$

Then another corner of Ω', if there is one, is determined by putting $\theta=\theta'$ in (2); for this purpose (3) is not involved.

The process is repeated until two corners are found for which (2) does not yield a new corner between them; then a line segment of Ω' has been found. Equating its slope to $-\theta/(1-\theta)$ determines the first value of θ to be used in determining δ in (3). The number of terms in successive applications of (2) can be reduced by omitting those points that are dominated by efficient (a_{ij}, b_{ij}) already found, but the point (2,3) in Figure 5.3A shows that this procedure alone need not eliminate all the inefficient points.

The next step is to evaluate (2) and (3) and solve for α and β thus:

$$\alpha=(\sigma(\theta)+\delta(\theta))/2\theta \qquad \beta=(\sigma(\theta)-\delta(\theta))/2(1-\theta) \tag{7}$$

If (7) yields a point (α, β) in Ω' (in the line segment whose slope is $-\theta/(1-\theta)$), the solution has been found. Otherwise one moves along Ω' toward (α, β) to another line segment having a different value of θ, which increases or decreases according as α increases or decreases. As in the example below it may happen that two adjacent line segments in Ω' are found such that each (by the preceding sentence) seems to require a move to the other. In that case the corner (a_{ij}, b_{ij}) common to the two line segments is the Nash solution. The corresponding θ will lie between the θ's for the adjacent line segments. If it is wanted, it can be found by one of the methods illustrated in (10) to (13) below, but it may fail to be unique as in (15); $(\alpha^{\#}, \beta^{\#})$ is always unique.

If a line segment at one end of Ω' yields an (α, β) beyond the corresponding endpoint of Ω', that endpoint is thereby indicated as the Nash solution.

The value $\theta=0$ can be used when I cannot prevent J from getting $\beta'=\max_{ij} b_{ij}$, and $\theta=1$ can be used when I can get $\alpha''=\max_{ij} a_{ij}$. The requirement $(\alpha, \beta)\in\Omega'$ correctly selects an endpoint of Ω' even though one of the formulas (7) breaks down. A value $\theta=0$ or 1 could be *required* if one were considering a curved boundary; e.g., Ω' a quadrant of a circle having $(\alpha^{\#}, \beta^{\#})$ at one of its endpoints. This could arise from a continuous game in which the payoffs a and b are functions of two real numbers that index the pure strategies; see McKinsey (1952).

To solve the example (5) we may begin with $\theta=1/2$ so that $-\theta/(1-\theta)$ is equal to the slope -1 of the middle line segment of Ω' from (1,5) to (4,2). Then (3) gives the zero-sum game

$$\begin{bmatrix} -2 & -5 & 1 \\ 3 & -\frac{1}{2} & -4 \end{bmatrix} = (\mathbf{A}-\mathbf{B})/2$$

in which the first column can be deleted. Its value is

$$\delta(1/2)=(\alpha-\beta)/2=-41/19 \text{ while (2) gives}$$

$$\sigma(1/2)=(\alpha+\beta)/2=3=(1+5)/2=(4+2)/2$$

Solving as in (7) gives $\alpha=16/19, \beta=98/19$. This point is not on the line segment between (1,5) and (4,2) (whose slope determined $\theta=1/2$) but is more nearly between $(-2,6)$ and (1,5).

These points determine the next line segment to be tried, with $\theta=1/4$, the zero-sum game

$$\begin{matrix} -14/4 & -22/4 & -2/4 \\ 8/4 & -7/4 & -20/4 \end{matrix}$$

and its value $\delta(1/4)=-71/22$. Then $\sigma(1/4)=4$ and $(\alpha,\beta)=(17,53)/11$. This does not lie between $(-2,6)$ and $(1,5)$ but is more nearly on the previous (adjacent) line segment. We conclude that the Nash solution $(\alpha^{\#},\beta^{\#})=(1,5)$, the corner point common to the two line segments.

Problem (b) is a special case of (c) or (d) and so can be put in the form of the latter, although one wouldn't do this in practice. Define

$$a'=\min_{ij} a_{ij} \qquad \text{and} \qquad b''=\min_{ij} b_{ij} \tag{8}$$

Evidently (b) can never make $\alpha<a'$ nor $\beta<b''$. Thus if (a',b') and (a'',b'') satisfy (2) the line segment joining them will constitute Ω'. The solution of the zero-sum game (3) will be unchanged if one adjoins to the bimatrix game a row of identical points (a',b') and a column of (a'',b''), with (a',b'') at their intersection, because the new row and column are too unfavorable to be used. Thus with $\theta=0.6$ the solution of (5) with transferable utility is the same as the solution of (9) without transferable utility.

$$\begin{matrix} (1,5) & (-4,6) & (4,2) & (6,-1) \\ (5,-1) & (2,3) & (-2,6) & (6,-1) \\ (-4,14) & (-4,14) & (-4,14) & (-4,-1) \end{matrix} \tag{9}$$

Each of Exercises 8 to 12 has a unique and efficient equilibrium point that differs from the Nash solution. Thus we have seen that an equilibrium point needn't be unique; if it is unique, it still needn't be efficient; and if it is both unique and efficient, it may still differ from the Nash solution because it fails to reflect the threat capabilities of the players.

The remainder of this section is more mathematical and may be omitted if desired. It is concerned only with the Nash solution (d).

The Determination of θ at a Corner

If the value of θ is wanted in (d), we note that (3) must have the value $\theta-5(1-\theta)=6\theta-5$ because $(\alpha^{\#},\beta^{\#})=(1,5)$. Subtracting $6\theta-5$ from each element of $\theta\mathbf{A}-(1-\theta)\mathbf{B}$ gives the zero-sum game

$$\begin{matrix} 0 & -1-4\theta & 3 \\ 6-2\theta & 2-\theta & -1-2\theta \end{matrix} \tag{10}$$

which should have a value of zero. Since the kernel was found to consist of columns 2 and 3 when $\theta=1/2$ and $1/4$, this is our first choice to try now. If it is correct, then

(9) of 4.5 gives

$$\begin{vmatrix} -1-4\theta & 3 \\ 2-\theta & -1-2\theta \end{vmatrix} = 8\theta^2 + 9\theta - 5 = 0 \tag{11}$$

as the condition (equivalent to (3)) that (10) have the value zero. This makes $\theta^{\#} = (\sqrt{241} - 9)/16 = .4078$, and it only remains to verify that this does indeed make the game value of (10) equal to zero.

If the game were larger, an approximate determination of θ would probably be more practical. Let us try a value of θ between 1/2 and 1/4, say 0.4, which in (3) gives the game

$$\begin{matrix} -2.6 & -5.2 & 0.4 \\ 2.6 & -1.0 & -4.4 \end{matrix}$$

This has optimal $\mathbf{u} = (17,28)/45$ and $\mathbf{y} = (0,8,7)/15$ and value $\delta(.4) = -38.8/15 = -2.5867$ whereas the desired point $(1,5)$ requires $\delta(\theta) = (.4)1 - (.6)5 = -2.6$.

Mills (1956) implies that when the optimal strategies are unique one has the derivative

$$(d/d\theta)\delta(\theta) = \mathbf{u}(\mathbf{A}+\mathbf{B})\mathbf{y} \tag{12}$$

when θ has the value (0.4) for which $\mathbf{u}$ and $\mathbf{y}$ are optimal. For a proof note that $\mathbf{A}+\mathbf{B}$ is the coefficient of θ in (3) and the other two terms in the derivative are zero; for example,

$$\mathbf{u}\left[\theta\mathbf{A} - (1-\theta)\mathbf{B}\right] d\mathbf{y}/d\theta = \delta(\theta) \sum_j dy_j/d\theta = 0$$

This is true because $\Sigma y_j = 1$ implies $\Sigma\, dy_j/d\theta = 0$; by Theorem 3 of 4.5, for each j, either $y_j = 0$ (and remains so locally, so that $dy_j/d\theta = 0$) or else $\Sigma_i u_i[\theta a_{ij} - (1-\theta)b_{ij}] = \delta(\theta)$. A similar argument applies to the other term $(d\mathbf{u}/d\theta)[\theta\mathbf{A} - (1-\theta)\mathbf{B}]\mathbf{y} = 0$.

In the present case (12) equals $578/135 = 4.28$ while $(d/d\theta)[\theta - 5(1-\theta)] = 6$; these relate to the right and left members of (3) respectively. Then the Newton-Raphson tangent-line approximation to the discrepancy in (3) gives

$$\theta = 0.4 - (-2.6 + 2.5867)/(6 - 4.28) = 0.4078 \tag{13}$$

as a better approximation to $\theta^{\#}$.

Existence and Uniqueness Theorem

We first note that any two minimax lines (3) corresponding to different values of θ must intersect in Ω. Let δ_1 and δ_2 be the values of (3) corresponding to $\theta = \theta_1$ and θ_2, respectively. The absence of such an intersection would mean that the two minimax lines divided Ω into only three parts rather than four. Thus either for $(k, l) = (1,2)$ or else for $(k, l) = (2, 1)$.

$$\theta_k \alpha - (1-\theta_k)\beta \geq \delta_k, \qquad \theta_l \alpha - (1-\theta_l)\beta \leq \delta_l$$

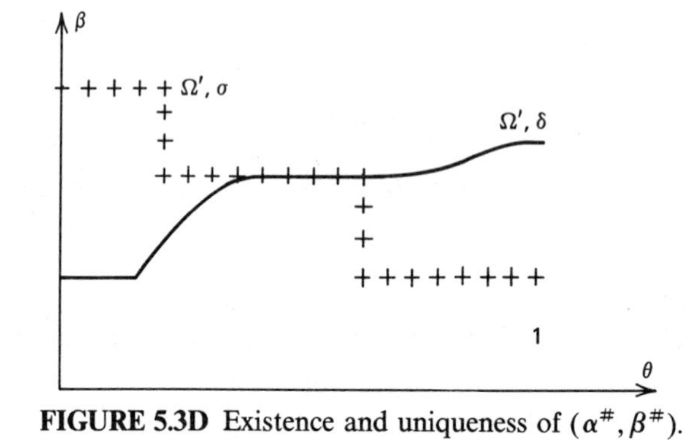

FIGURE 5.3D Existence and uniqueness of $(\alpha^{\#}, \beta^{\#})$.

would be impossible for any pair of mixed strategies $\mathbf{u}, \mathbf{y}$. But in fact one has

$$\mathbf{u}^k[\theta_k \mathbf{A} - (1-\theta_k)\mathbf{B}]\mathbf{y}^l \geq \delta_k$$

$$\mathbf{u}^k[\theta_l \mathbf{A} - (1-\theta_l)\mathbf{B}]\mathbf{y}^l \leq \delta_l \tag{14}$$

where $\mathbf{u}^k$ and $\mathbf{y}^k$ are the minimax mixed strategies producing δ_k $(k=1,2)$. The proof is completed by taking $\alpha = \mathbf{u}^k\mathbf{A}\mathbf{y}^l$ and $\beta = \mathbf{u}^k\mathbf{B}\mathbf{y}^l$.

The point $(\alpha^{\#}, \beta^{\#})$ will be determined as the common solution of problems (a) and (c) for a suitable value of θ. Both problems use Ω', whereon each of α and β is a continuous strictly decreasing function of the other. Thus it suffices to consider the relations between θ and β.

Problem (a) uses Ω' and σ to give the relation represented by the chain of plus signs in Figure 5.3D. Each horizontal portion corresponds to a corner of Ω', where the slope $-\theta/(1-\theta)$ of the supporting line can change while (α, β) remains fixed. Each vertical portion corresponds to a line segment in Ω', on which β can vary while the slope $-\theta/(1-\theta)$ is fixed.

Problem (c) uses Ω' and δ to give the solid line in Figure 5.3D; when the conditions would otherwise be inconsistent, the endpoint of Ω' closer to the minimax line (3) is used. The intersection property proved by (14) shows that β is a nondecreasing function of θ; it is also a continuous function of θ, because δ is (Exercise 8 of 5.6.). β can be constant while θ varies, as in the game

$$\begin{matrix} (\ \ 1,1) & (2,-2) \\ (-2,2) & (0,\ \ 0) \end{matrix} \tag{15}$$

for which $\alpha^{\#} = \beta^{\#} = 1$, $\pi_{11}^{\#} = 1$, and $.25 \leq \theta^{\#} \leq .75$.

θ varies from 0 to 1 along both graphs. Meanwhile on the broken line β traverses Ω' precisely while on the solid line β traverses all or part of Ω' in the reverse direction. We conclude that the two graphs always intersect and determine $\alpha^{\#}$ and $\beta^{\#}$ uniquely while $\theta^{\#}$ is either unique or confined to a closed interval.

Individual Rationality

It is desirable to show that the Nash solution is *individually rational*; that is, it gives each player at least as much as he could be certain of obtaining for himself without any cooperation. The latter values, called the *maximin* or the *security values*, are defined by

$$\alpha_0 = \max_{\mathbf{u}} \min_{\mathbf{y}} \mathbf{uAy} = \max_{\mathbf{u}} \min_{j} \sum_{i=1}^{m} u_i a_{ij} \tag{16}$$

$$\beta_0 = \max_{\mathbf{y}} \min_{\mathbf{u}} \mathbf{uBy} = \max_{\mathbf{y}} \min_{i} \sum_{j=1}^{n} b_{ij} y_j$$

the values of the zero-sum games $\mathbf{A}$ and $\mathbf{B}^T$, respectively. The assumptions are as pessimistic as possible; in calculating α_0, I assumes that J is indifferent to his own payoff and is interested solely in thwarting I.

Now suppose that $\theta = \theta^{\#}$ in (2) and (3) but the two pairs of strategies are replaced by a single pair, that is, I's security strategy $\mathbf{u}^0$ used in (16) and J's best response $\mathbf{y}'$ thereto in the game (3). Let I's resulting payoff (7) be denoted by $\alpha' = [\sigma'(\theta) + \delta'(\theta)]/2\theta$. Since the same pair of strategies is being used in both (2) and (3), this reduces to

$$\begin{aligned} \alpha' &= \{\mathbf{u}^0[\theta\mathbf{A} + (1-\theta)\mathbf{B}]\mathbf{y}' + \mathbf{u}^0[\theta\mathbf{A} - (1-\theta)\mathbf{B}]\mathbf{y}'\}/2\theta \\ &= \mathbf{u}^0\mathbf{A}\mathbf{y}' \geq \min_{\mathbf{y}} \mathbf{u}^0\mathbf{A}\mathbf{y} = \alpha_0 \end{aligned} \tag{17}$$

The departures from the optimal strategies in (2) and (3) make $\sigma'(\theta) \leq \sigma^{\#}(\theta)$ and $\delta'(\theta) \leq \delta^{\#}(\theta)$ so that $\alpha' \leq \alpha^{\#}$. Combining this with (17) makes $\alpha^{\#} \geq \alpha_0$ as desired. $\beta^{\#} \geq \beta_0$ is proved similarly.

Bonnardeaux, Dolait and Dyer (1976) reported an application to spacecraft trajectories. Owen (1971) allowed Ω in (d) to be any convex bounded set containing all (a_{ij}, b_{ij}). Shapley (1953) extended (b) (transferable utility) to n players; his results will be considered in 5.7 and 5.8. Harsanyi (1963, 1977), Shapley (1967, 1969), Owen (1972), Lemaire (1973) and others have extended the general case (d) to n players; the first two may require special devices to make the solution unique. Owen (1972) considered transfers of commodities having nonlinear utility functions.

EXERCISES 5.3

In 1 to 3 find the solutions (a) to (d) with $\theta = .6$ in (a) to (c). In 4 use the nongraphical method (6) to determine all the efficient (a_{ij}, b_{ij}) and the trial values of θ, for which (3) may have to be solved.

1. (2,3) (3,2) (0,0)
(4,0) (1,3) (0,4)

2. (3,4) (0,3) (4,0)
(2,1) (4,3) (1,5)

3. $(1,2)$ $(0,0)$ $(2,0)$
$(0,0)$ $(2,3)$ $(0,0)$
$(2,0)$ $(0,0)$ $(-1,6)$

4. $(0,9)$ $(8,2)$ $(7,3)$
$(4,7)$ $(2,8)$ $(9,1)$
$(10,-1)$ $(5,5)$ $(1,9)$

5. Show that the number of trial values of θ in (3) is always less than the number of efficient (a_{ij}, b_{ij}).

6. If $b_{ij}=a_{ji}$, show $\alpha^{\#}=\beta^{\#}=\max_{i,j}(a_{ij}+a_{ji})/2$, $\theta^{\#}=1/2$.

In 7 to 12, find $\alpha^{\#}$ and $\beta^{\#}$. These all have dominant strategies $i=2$ and $j=2$, but the Nash solution progresses from $i=2$ to $i=1$.

7. $(0,8)$ $(8,9)$
$(7,0)$ $(9,7)$

8. $(0,0)$ $(5,9)$
$(1,7)$ $(9,8)$

9. $(0,0)$ $(7,9)$
$(8,1)$ $(9,2)$

10. $(0,0)$ $(1,9)$
$(5,1)$ $(9,2)$

11. $(0,6)$ $(6,9)$
$(3,0)$ $(9,3)$

12. $(0,8)$ $(8,9)$
$(1,0)$ $(9,1)$

13, 14. In 1 and 2 find the maximin values (16).

15. Solve problems (a) to (d) in terms of θ, $a>0$, $b>0$.

$$\begin{matrix} (a,0) & (0,0) \\ (0,0) & (0,b) \end{matrix}$$

16. Prove that the Nash solution satisfies (4).

5.4 THE CALCULATION OF EQUILIBRIUM POINTS (LEMKE)

Nash (1951) gave the definition (1) for an equilibrium point and showed that every bimatrix game has at least one. However, in contrast to the zero-sum case it need not yield convex Cartesian product sets of equilibrium strategies nor unique values of the average payoffs $\mathbf{uAy}, \mathbf{uBy}$. The basic algorithm was given by Lemke and Howson (1964) and Lemke (1968).

As in (5) of 5.1, the probability vectors $\mathbf{u}$ and $\mathbf{y}$ for players I and J define an equilibrium point provided

$$\mathbf{uAy} \geq \boldsymbol{\omega}\mathbf{Ay} \qquad \text{and} \qquad \mathbf{uBy} \geq \mathbf{uB}\boldsymbol{\eta} \tag{1}$$

for all probability vectors $\boldsymbol{\omega}$ and $\boldsymbol{\eta}$. As in 4.2 and 4.4 it is permissible to restrict $\boldsymbol{\omega}$ and $\boldsymbol{\eta}$ (but not $\mathbf{u}$ and $\mathbf{y}$) to the pure strategies, so that each has one component equal to 1 and the other zero. Then the conditions (1) are replaced by

$$\alpha \triangleq \mathbf{uAy} \geq \sum_{j=1}^{n} a_{ij} y_j \qquad \text{for } i=1,\ldots,m \tag{2a}$$

$$\beta \triangleq \mathbf{uBy} \geq \sum_{i=1}^{m} u_i b_{ij} \qquad \text{for } j=1,\ldots,n \tag{2b}$$

$$\Sigma u_i = \Sigma y_j = 1 \qquad \text{all } u_i \geq 0 \qquad y_j \geq 0 \tag{3}$$

It is easy to show (as in Theorem 3 of 4.5 or (7) below) that equality must hold in (2a) when $u_i > 0$ and in (2b) when $y_j > 0$.

The analysis generalizes (1) to (5) of 4.6. As there, by adding a constant to each a_{ij} and another to each b_{ij}, we may determine the signs of **uAy** and **uBy** as we like. Todd (1974–1978) pointed out the advantages of assuming

$$\mathbf{uAy} > 0 \qquad \mathbf{uBy} > 0 \tag{4}$$

This is assured if some row (or positive combination of rows) of **A** is strictly positive and some column of **B** is strictly positive.

Thus we divide (2) by their (positive) left members, define

$$\mathbf{y}' = \mathbf{y}/\mathbf{uAy} \qquad \mathbf{u}' = \mathbf{u}/\mathbf{uBy} \tag{5}$$

and introduce slack x_i', v_j' to obtain

$$x_i' + \sum_{j=1}^{n} a_{ij} y_j' - 1 = 0 \qquad \text{for } i = 1, \ldots, m \tag{6a}$$

$$v_j' + \sum_{i=1}^{m} u_i' b_{ij} - 1 = 0 \qquad \text{for } j = 1, \ldots, n \tag{6b}$$

with all variables nonnegative. Multiplying (6a) by u_i, summing over i and using (5) gives $\mathbf{ux}' = 0$ or $\mathbf{u}'\mathbf{x}' = 0$; multiplying (6b) by y_j and summing gives $\mathbf{v}'\mathbf{y} = 0$ or $\mathbf{v}'\mathbf{y}' = 0$. Since the variables are nonnegative one has the *complementarity condition*

$$u_i' x_i' = 0 \qquad v_j' y_j' = 0 \qquad \text{for all } i, j \tag{7}$$

which is analogous to Theorem 3 of 4.5.

Equations (6) to (7) admit a trivial solution $\mathbf{u}' = \mathbf{0}$, $\mathbf{y}' = \mathbf{0}$. Any other solution will yield a solution of (2) to (3) via

$$\mathbf{y} = \mathbf{y}'/\Sigma y_j' \qquad \mathbf{u} = \mathbf{u}'/\Sigma u_i' \tag{8}$$

provided we can show that (5) is satisfied. Half the proof will suffice. Multiplying (6a) by u_i', summing over i and using (7) gives $\Sigma u_i' = \mathbf{u}'\mathbf{Ay}'$. Then (8) implies that $\mathbf{uAy} = \mathbf{u}'\mathbf{Ay}'/\Sigma u_i' \Sigma y_j' = 1/\Sigma y_j'$ so that (5) agrees with (8).

Equation (7) compensates somewhat for the absence of an objective function and avoids what otherwise would be a paradox, namely, the seeming determination of player J's strategy **y** solely by I's payoffs a_{ij} in (6a) and vice versa in (6b). However, this is indeed the situation after the identification of a kernel has been postulated as in (13) below; each player uses a minimax strategy determined by the other player's payoffs within an optimal kernel.

The Lemke-Howson-Todd Algorithm

With additive constants a and b included, the initial tableaus for (6) are

Row equations

$$\begin{array}{c|cc} 1 & y'_j & 1 \\ \hline x'_i & a_{ij}+a & -1 \end{array} \qquad (9a)$$

Column equations

$$\begin{array}{cc} 1 & v'_j \\ \hline u'_i & -b_{ij}-b \\ 1 & 1 \end{array} \qquad (9b)$$

with $\mathbf{x}, \mathbf{y}, \mathbf{u}, \mathbf{v} \geq \mathbf{0}$; $i=1,\ldots,m$; $j=1,\ldots,n$; at least one row of $a_{ij}+a$ and one column of $b_{ij}+b$ all positive. The negative transpose of (9b) could be used if row equations are preferred.

The notational question illustrated by (8) of 3.1 arises here. The use of the four letters x, y, u, v is convenient for many purposes except the description of the algorithm, where we use $(t_1,\ldots,t_{2m+2n})$ to represent

$$(y'_1,\ldots,y'_n,u'_1,\ldots,u'_m,v'_1,\ldots,v'_n,x'_1,\ldots,x'_m) \qquad (10)$$

Thus (7) becomes $t_h t_{h+m+n}=0$ for $h=1,\ldots,m+n$.

Each application of the algorithm yields one equilibrium point; repeated applications may be made in an effort to find new equilibria. Todd (1976) gives an example in which only one equilibrium (in pure strategies) can be found, although it is dominated by two mixed equilibria. The first application must begin with (9) but it offers a free choice of an initial index $k=1,\ldots,m+n$. After additional feasible complementary solutions have been found, there is a possibly valuable option to use them in place of (9) as the starting point, with a change in the value of k. This is one of the advantages of (4); it offers a hope of finding a mixed equilibrium even if all the rows and columns contain pure equilibria, which will preempt the applications starting at (9).

The index k is used to identify a nonbasic variable t_k that will be brought into the basis. This will be y'_k in (9a) if $1 \leq k \leq n$ and u'_{k-n} in (9b) if $n+1 \leq k \leq m+n$. As usual the preservation of feasibility determines the t_l that leaves the basis. The proof of validity requires that this determination be unique; this can be insured by the lexicographic method of 6.6 when it seems desirable.

Variables t_k and t_l are in the same tableau, (9a) or (9b). The dual partner of t_k, denoted by t_{k+m+n}, is in the other tableau; it remains basic while t_k becomes basic. Thus $t_k t_{k+m+n}$ is now likely to be positive rather than zero. This condition, with the *same basic pair*, will persist until the calculation is completed; the corresponding feasible basic solution is said to be *k-almost-complementary*. It also includes a dual pair, with variable indices, that are both nonbasic. After the first pivot step this nonbasic pair is t_l and t_{l-m-n}.

The pivot steps continue as follows:

(i) The pivot alternates from one tableau to the other (except perhaps when and if the algorithm is reapplied).

(ii) The variable entering the basis is one of those in the nonbasic pair—naturally, the one that has not just left the basis. Any other choice would either reverse the previous pivot or create another basic pair. This rule implies (i).

(iii) The variable leaving the basis is uniquely determined by (3) or (5) of 3.3 or its generalization in 6.6, so as to preserve feasibility. If it is a variable in the basic pair, a complementary feasible solution has been found; *stop or restart*. Otherwise a new nonbasic pair is created and one returns to (ii). A pivot of the required sign is always available because the feasible regions for (6) are both bounded; the simplicity of this argument, which determines the signs in the initial tableaus (9), is another advantage of (4).

Table 5.4 applies the algorithm to the example of 5.3. As in the Prisoner's Dilemma and other examples of 5.1 to 5.2, both players suffer somewhat from the absence of cooperation. Their equilibrium payoffs are (0, 30/7), whereas if each used his first pure strategy as in 5.3, each would get more, that is, (1, 5), but the latter is not an equilibrium.

Check columns are advisable as usual. The checking of the equilibrium point is suggested at the bottom of the table; it uses (3) and the following:

TABLE 5.4. Calculation of an Equilibrium Point

$$\mathbf{A}=\begin{bmatrix} 1 & -4 & 4 \\ 5 & 2 & -2 \end{bmatrix} \qquad \mathbf{B}=\begin{bmatrix} 5 & 6 & 2 \\ -1 & 3 & 6 \end{bmatrix}$$

Add 3 to each a_{ij}. Subtract each b_{ij} from 2.

1	y_1'	y_2'	y_3'	1
x_1'	4	−1	7	−1
x_2'	8	5*	1	−1

→

1	v_1'	v_2'	v_3'
u_1'	−3	−4	0
u_2'	3	−1	−4*
1	1	1	1

↙

5	y_1'	x_2'	y_3'	1
x_1'	28	1	36*	−6
y_2'	8	1	1	−1

→

$+4$	v_1'	v_2'	u_2'
u_1'	−12	−16*	0
v_3'	−3	+1	−1
1	+7	+3	+1

36	y_1'	x_2'	x_1'	1
y_3'	28	1	5	−6
y_2'	52	7	−1	−6

$+16$	v_1'	u_1'	u_2'
v_2'	+12	−4	0
v_3'	−15	+1	−4
1	+19	+3	+4

$\mathbf{y}^*=(0,6,6)/12$ $\qquad$ $\mathbf{u}^*=(3,4)/7$

$\mathbf{Ay}^*=(0,0)$ $\qquad$ $\mathbf{u}^*\mathbf{B}=(11,30,30)/7$

$\mathbf{u}^*\mathbf{Ay}^*=0=-3+36/12$ $\qquad$ (Checks by (11)) $\qquad$ $\mathbf{u}^*\mathbf{By}^*=30/7=2+16/7$

$v_2'y_2'=0$ only at the beginning and the end.

First pivotal column y_2' chosen arbitrarily, determines what follows.

Exchanges: y_2', x_2'; u_2', v_3'; y_3', x_1'; u_1', v_2', whose sequence is indicated by the arrows.

$$\mathbf{u^*Ay^*} = \max_i \sum_{j=1}^{n} a_{ij} y_j^* = 1/\sum y_j'$$

$$\mathbf{u^*By^*} = \max_j \sum_{i=1}^{m} u_i^* b_{ij} = 1/\sum u_i' \tag{11}$$

(The common values are the equilibrium payoffs α^* and β^*.) With (3), the first equality in each line, coming from (1) or (2), is enough for a check; as in Theorem 3 of 4.5, they imply (7). The second equalities come from (5).

If the algorithm starts with (9) and if the row or column determined by k contains an equilibrium in pure strategies, it is easily verified that the algorithm will find it and stop after one pivot step in each tableau. In general the finiteness of the number of pivot steps follows from the fact that for each k the number of k-almost-complementary (k-a.c.) basic solutions is finite and no solution can be encountered twice.

The latter is true because each k-a.c. solution is connected with only two others by means of a single pivot step, which brings one of the nonbasic pair into the basis. One of these pivots is used to arrive at the solution and the other to depart; a return to a previous solution would require a third link. Each feasible complementary solution is directly connected to only one k-a.c. solution, via the pivot that brings t_k into the basis. Thus a return to the initial solution would require a return to the first k-a.c. solution; the algorithm must end by producing a feasible complementary solution different from the initial one.

Fortunately we do not expect to have to generate all the k-a.c. solutions; many of them may be arranged in loops or paths that are not explored. The loops will not contain any complementary solution; each path will contain two, one at each endpoint. Since the trivial solution $\mathbf{u}'=\mathbf{0}$, $\mathbf{y}'=\mathbf{0}$ is complementary but not equilibrium, any one value of k suffices to show that in nondegenerate cases (when 6.6 is not required) the number of equilibrium points is odd. However, because of the dependence on links with the initial tableaus (9), the *discovery* of each additional equilibrium requires a *change* in k, perhaps to a value used previously.

The introduction of an artificial variable as in Lemke (1968) and Eaves (1971) produces a method that is probably even more flexible than the foregoing. One can begin by exploring the feasible solutions of one system, say (9a), in any manner one likes, while pivoting in the same positions in (9b) to keep it complementary. Whenever (9b) also is feasible but not in its initial form, an equilibrium point has been found.

When it is time to force the issue, a new row of arbitrary positive numbers $d_1, \ldots, d_n$ is introduced in (9b) as coefficients of a nonbasic artificial variable, say w. By pivoting on d_l where

$$-c_l/d_l = \max_j\{-c_j/d_j\colon c_j<0\} \tag{12}$$

one makes all $c_j \geq 0$ in the new tableau. At the same time w, which should be zero, becomes basic and positive, thus belying the apparent feasibility. There is no basic pair but there is a nonbasic pair of variables, one of which left the basis when w

entered. Next the other variable of the nonbasic pair is made basic, and the algorithm proceeds according to (i) to (iii) above. When w becomes nonbasic again, an equilibrium has been found. The procedure may be restarted as often as desired. In (9a) the artificial variable would be introduced with a column of negative coefficients.

Wilson (1971) gives a Lemke-type algorithm for any finite number of players. An existence proof for this case was given by Debreu (1952).

Kernels (All the Equilibrium Points)

To obtain all the optimal solutions or all the equilibrium points for a large system is a difficult problem that neither the simplex method nor the Lemke-type algorithms are well adapted to solve. However, this problem is particularly interesting in connection with bimatrix games because here the payoffs vary from one equilibrium point to another. Thus it may have to be solved, for example, if we want to find out whether there is an equilibrium that dominates all the others.

For any equilibrium strategies $\mathbf{y}^*, \mathbf{u}^*$ let

$$K=\{i: u_i^*>0\} \subset \{i: x_i^*=0\}$$

$$L=\{j: y_j^*>0\} \subset \{j: v_j^*=0\} \tag{13}$$

where the x_i and v_j are slack variables in (2a) and (2b), respectively. The inclusions follow from (7). To simplify the notation let $K=\{1,2,\ldots,k\}$, $L=\{1,2,\ldots,l\}$, $\alpha=\mathbf{uAy}$ and $\beta=\mathbf{uBy}$. Then by (2) and (3), $\mathbf{y}_L^*$, α^*, $\mathbf{u}_K^*$ and β^* satisfy

$$\begin{array}{cccccl} y_1 & \cdots & y_l & \alpha & 1 & \text{(rows)} \\ \hline a_{11} & \cdots & a_{1l} & -1 & 0 & \text{(14a)} \\ \vdots & & \vdots & \vdots & \vdots & \\ a_{k1} & \cdots & a_{kl} & -1 & 0 & \\ 1 & \cdots & 1 & 0 & -1 & \end{array} \qquad \begin{array}{c|ccccl} u_1 & b_{11} & \cdots & b_{1l} & 1 & \text{(columns)} \\ \vdots & \vdots & & \vdots & \vdots & \text{(14b)} \\ u_k & b_{k1} & \cdots & b_{kl} & 1 & \\ \beta & -1 & \cdots & -1 & 0 & \\ 1 & 0 & \cdots & 0 & -1 & \end{array}$$

with the other y_j and $u_i=0$. Unlike (9) of 4.5, (14) may fail to make $\mathbf{u}$ and $\mathbf{y}$ unique even if $k=l$. In any case *one must verify* $\mathbf{y}\geq\mathbf{0}$, $\mathbf{u}\geq\mathbf{0}$, *and the inequalities in* (2) *for* $i>k, j>l$.

THEOREM 1 Let $\mathbf{u}$ and $\mathbf{y}$ have k and l positive components, respectively. Then $\mathbf{y}, \alpha, \mathbf{u}, \beta$ is an equilibrium point for $\mathbf{A}, \mathbf{B}$ if and only if it is a *convex combination* of equilibria derived from *square* systems like (14) (each having its own $k=l$), which *share* the given $\mathbf{y}$ and α if $k\geq l$ or the given $\mathbf{u}$ and β if $k\leq l$. Thus *square kernels give all the extreme equilibria.*

Proof The "if" part is easily verified by (2) and (3), which are linear in $\mathbf{u}$ alone or in $\mathbf{y}$ alone. For the "only if," it suffices to consider the case $k<l$. We are given that the systems (14) have solutions $\mathbf{y}^*, \alpha^*$ and $\mathbf{u}^*, \beta^*$, respectively. It follows that the

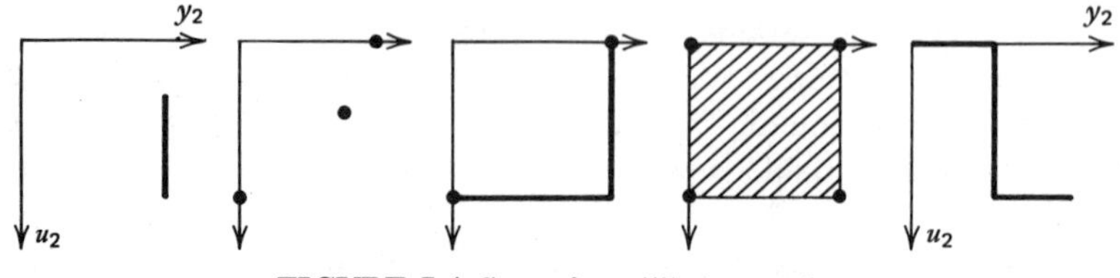

FIGURE 5.4 Sets of equilibrium points.

solution set Δ of (14a) is the nonempty intersection of a linear variety of rank at least $l-k$ with the region $\mathbf{x}\geq\mathbf{0}$, $\mathbf{y}\geq\mathbf{0}$ (Theorem 6 of 2.7). Since Δ is clearly bounded, any line through $\mathbf{y}^*$ will intersect Δ in a line segment, at each of whose endpoints at least one of $y_1,\ldots,y_l,x_{k+1},\ldots,x_m$ is zero. If it is a y_j, then l can be reduced at the endpoint; if it is an x_i, then k can be increased. The effect in (14b) is to relax the constraints; $\mathbf{u}^*,\beta^*$ remains a solution. $\mathbf{y}^*$ is a convex combination of the endpoints, which may coincide with $\mathbf{y}^*$. Repetition of this argument will bring k and l into equality and prove the theorem (cf. Theorem 4 of 2.2).

The truth of the theorem and the fact that the set of equilibrium points is not necessarily convex or a Cartesian product are illustrated by Figure 5.4, in which the heavy dots and lines and the shaded area depict the set of all equilibrium points. In the last of these games, note that the 2 by 2 kernel yields two extreme points with $\alpha=y_1=y_2=1/2$ and $\beta=u_1=0$ or 1.

0,3	1,2	2,3	1,4	1,1	0,2	1,1	0,1	1,1	0,1
3,0	1,2	4,1	0,0	2,0	0,0	1,0	0,0	0,0	1,0

When $k=l$, the square submatrices of $\mathbf{A}$ and $\mathbf{B}$ that appear in (14) are called *kernels*. Theorem 1 indicates how all the equilibrium points could be found by considering at most the $\binom{m+n}{m}-1$ pairs of kernels. This number can be reduced by using Theorem 2 following if nonpivotal (determinantal) methods of solution are employed for $m+n\leq 6$, or by considering only the solutions that are primal or dual feasible as in (12) above. The sufficiency of *square* kernels agrees with the fact that the usual methods of solution do not consider any others.

Equation (14) shows that (9) of 4.5 applies to bimatrix games provided the kernel is taken from $\mathbf{A}$ in calculating $\mathbf{y}$ and α and from $\mathbf{B}$ in calculating $\mathbf{u}$ and β. *I's payoffs determine J's strategies and vice versa, except for the essential verification (11).* In particular, if $\mathbf{A}$ and $\mathbf{B}$ (probably but not necessarily square) have *zero-sum* solutions $\mathbf{y}^*>0$ and a^* for $\mathbf{A}$ and $\mathbf{u}^*>0$ and b^* for $\mathbf{B}$ with all probabilities positive, then $\mathbf{y}^*,a^*,\mathbf{u}^*,b^*$ satisfy (14) with $k=m$ and $l=n$ and so form an equilibrium point for the bimatrix game.

Figure 5.4 illustrates the fact that equilibrium points may be lost if one discards the indices of rows dominated in $\mathbf{A}$ or columns dominated in $\mathbf{B}$. The Nash cooperative solution also may be lost; in Table 5.4 it is (1,5) in the first column, which is dominated in $\mathbf{B}$. However, the following is true with strict inequalities.

THEOREM 2 Let K and L be the sets of row and column indices in (14), which one may assume to be square by Theorem 1. If $a_{pj}>a_{qj}$ for all j in L, then row index

q may be omitted in forming sets K to pair with L. If $b_{ip} > b_{iq}$ for all i in K, then column index q may be omitted in forming sets L to pair with K. The dominant index p may represent a convex combination of rows or columns.

Proof The first case is typical. The difference of the row equations in (2a) with slack variables x_p and x_q is

$$\sum_{j \in L} (a_{pj} - a_{qj}) y_j + x_p = x_q$$

Since each term is nonnegative on the left and at least one $(a_{pj} - a_{qj}) y_j > 0$, one has $x_q > 0$ and therefore q should not be in K, by (13).

Theorem 2 shows that the first column may be omitted in Table 5.4 for the purpose of finding equilibrium points. Then (9) of 5.1 applied to columns 2 and 3 of **A** and **B** gives

$$y_2^* = \frac{4-(-2)}{4-(-2)+2-(-4)} = \frac{1}{2}, \qquad u_1^* = \frac{6-3}{6-3+6-2} = \frac{3}{7} \tag{15}$$

and so on. In general this must be checked by (3) and (11).

By placing distinct payoffs (it doesn't matter what they are) at random in the matrix or matrices, one can ask for the probability that a game does not have an equilibrium point in pure strategies. A moderate amount of enumeration gives the following probabilities that only mixed equilibria exist.

$m \times n$	2×2	2×3	3×3	$m \times \infty$	$\infty \times \infty$	
Zero sum	$\frac{1}{3}$	$\frac{1}{2}$	$\frac{7}{10}$	1	1	(16)
Bimatrix	$\frac{1}{8}$	$\frac{1}{6}$	$\frac{52}{243}$	$(1-1/m)^m$	$e^{-1} \doteq .368$	

(See the algorithm following (6) of 5.1.) Most bimatrix games have pure equilibria; most zero-sum games do not.

EXERCISES 5.4

1.–4. In the corresponding exercises of 5.3, find a mixed equilibrium (a) by the Lemke-Howson-Todd algorithm; (b) by kernels (14) as in (9) of 4.5.

Use kernels to find all the equilibria (y_2^*, u_2^*) as in Figure 5.4.

5.
$$\begin{matrix} 5,1 & 0,0 \\ 1,2 & 2,4 \end{matrix}$$

6.
$$\begin{matrix} 0,1 & 2,1 \\ 0,0 & 1,0 \end{matrix}$$

7.
$$\begin{matrix} 1,1 & 1,1 \\ 2,0 & 0,2 \end{matrix}$$

8.
$$\begin{matrix} 1,1 & 2,5 \\ 1,3 & 2,2 \end{matrix}$$

9. $\mathbf{A} = -\mathbf{B} =$
$$\begin{matrix} 3 & 5 & 2 \\ 5 & 3 & 6 \\ 2 & 6 & 0 \end{matrix}$$

10. For any one k, ignoring the feasibility requirement, find the largest possible number of k-a.c. solutions.

11. Use (2) and (3) to prove the "if" part of Theorem 1.

12. Show that if $\mathbf{A} = -\mathbf{B} > \mathbf{0}$, (5) of 4.6 is obtained instead of (6).

5.5 A SIMPLE PRODUCTION MODEL

This model has been used by Leontief (1936, 1966) in his studies of the national economy; it has also been applied within an industry or a single large firm. The model identifies m (classes of) goods with the indices $i=1,\ldots,m$; good i is produced by industry i and by no other. Census data or the like give the money value y_i^0 of the production of good i and the money value y_{ij}^0 of good i consumed by industry j in some time period. The difference $y_i^0-\Sigma_j y_{ij}^0 \triangleq x_i^0$, called the final demand for good i, represents the money value of good i used by households, government and foreign trade. (This use of x and y is more consistent with the rest of this book, but exchanges the roles they usually have in the literature.)

The input-output coefficients are defined by

$$a_{ij}=y_{ij}^0/y_j^0 \geq 0 \tag{1}$$

Since y_j^0 is a measure of the intensity with which industry j was operated, and y_{ij}^0 is the resulting consumption of good i, the usual linearity assumption implies that as the x's and y's vary, the a_{ij} will remain constant except insofar as the production technologies themselves change.

From the definitions of x_i^0 and a_{ij} it follows that the equations

$$y_j-a_{i1}y_1-\cdots-a_{im}y_m=x_i \tag{2}$$

for $i=1,\ldots,m$ are satisfied when $x_i=x_i^0$ and $y_j=y_i^0$. In matrix notation these equations take the form

$$(\mathbf{I}-\mathbf{A})\mathbf{y}=\mathbf{x} \tag{3}$$

In application they are solved to give $\mathbf{y}$ as a function of $\mathbf{x}$, thus:

$$\mathbf{y}=(\mathbf{I}-\mathbf{A})^{-1}\mathbf{x} \tag{4}$$

Equation (4) indicates how a change in the final demand $\mathbf{x}$ is expected to affect the industrial output $\mathbf{y}$. The theorem below shows that (4) exists and is nonnegative. Leontief reports that, considering its typically large size, the inversion of $\mathbf{I}-\mathbf{A}$ is not excessively vulnerable to round-off errors (see 7.5), and that (4) becomes less useful as more time elapses between the periods to which (1) and (4) respectively refer, as one should expect.

The above model, which has no objective function, is called a static Leontief model without substitution. It can be generalized in several ways that permit optimizing the growth rate or employment level or total profit, etc. The resulting models consider the relationships among successive time periods (in the dynamic models) and/or abandon the restriction to equal numbers of industries and goods (in models with substitution). Two of the simpler such models are considered in 5.6 below and in Dantzig (1955b) respectively. The latter requires only an abbreviated version of the simplex method. Some economic problems of linear-programming type that can be solved without using the simplex method at all are considered in 9.5, Ben-Israel et al. (1970), and Stanford (1976). A variety of models are described by Schumann (1968).

The following theorem has long been known; the proof is taken from Gale (1960) and Arrow. See also Koehler et al. (1975) and Seneta (1973).

THEOREM For any given $\mathbf{A} \geq \mathbf{0}$ and $\mathbf{x} > \mathbf{0}$ (all $x_i > 0$), a solution $\mathbf{y} \geq \mathbf{0}$ of $(\mathbf{I}-\mathbf{A})\mathbf{y} = \mathbf{x}$ exists if and only if $(\mathbf{I}-\mathbf{A})^{-1}$ exists and has nonnegative elements.

Proof Equation (4) proves the "if" part at once. For the other, suppose that $\mathbf{x}^0 > \mathbf{0}$, $\mathbf{y}^0 \geq \mathbf{0}$ is known to be a solution of (3). First we remark that in fact $\mathbf{y}^0 > \mathbf{0}$ since $y_j^0 = 0$ evidently makes $x_j^0 \leq 0$ in (3). Now let $\mathbf{x}, \mathbf{y}$ be any solution of (3) with $\mathbf{x} \geq \mathbf{0}$ and define

$$\theta = \max_j \{-y_j / y_j^0\} = -y_l / y_l^0 \tag{5}$$

If $\theta > 0$, then $\mathbf{x} + \theta \mathbf{x}^0 > \mathbf{0}$, $\mathbf{y} + \theta \mathbf{y}^0 \geq \mathbf{0}$ would be a solution of (3) having $y_l + \theta y_l^0 = 0$, which would contradict the preceding remark; hence we have $\theta \leq 0$ and the *lemma* that $\mathbf{x} \geq \mathbf{0}$ implies $\mathbf{y} \geq \mathbf{0}$. In particular, $\mathbf{x} = -\mathbf{x} = \mathbf{0}$ implies $\mathbf{y} \geq \mathbf{0}$ and $-\mathbf{y} \geq \mathbf{0}$, or $\mathbf{y} = \mathbf{0}$. This shows that the columns of $\mathbf{I}-\mathbf{A}$ are linearly independent and of rank m and hence $(\mathbf{I}-\mathbf{A})^{-1}$ exists by Theorem 2 of 7.1. Then setting $x_k = 1$ and all other $x_i = 0$ in (4) shows (by the lemma) that the kth column of $(\mathbf{I}-\mathbf{A})^{-1}$ is nonnegative, since it is the $\mathbf{y}$ in this case.

When $\mathbf{A} \geq \mathbf{0}$ and $(\mathbf{I}-\mathbf{A})^{-1} \geq \mathbf{0}$, Gale (1960) calls $\mathbf{A}$ a *productive* matrix. In that case it can be shown that the series

$$(\mathbf{I}-\mathbf{A})^{-1} = \mathbf{I} + \mathbf{A} + \mathbf{A}^2 + \mathbf{A}^3 + \cdots \tag{6}$$

converges. Thus to meet a net final demand $\mathbf{x} = \mathbf{I}\mathbf{x}$, one must have $\mathbf{y} = \mathbf{x} + \mathbf{A}\mathbf{x} + \mathbf{A}^2\mathbf{x} + \cdots$, where the primary additional consumption $\mathbf{A}\mathbf{x}$ implies a chain of secondary consumptions $\mathbf{A}^2\mathbf{x}$, $\mathbf{A}^3\mathbf{x}$, etc.

EXERCISES 5.5

In 1 to 5 determine whether the given $\mathbf{A}$ is productive and, if so, find the activity vector $\mathbf{y}$ that will precisely satisfy the final demand $x_i \equiv 1$.

1. $\begin{matrix} .1 & .2 \\ .3 & .1 \end{matrix}$ **2.** $\begin{matrix} 0 & .0 \\ 0 & .2 \end{matrix}$ **3.** $\begin{matrix} .8 & .3 \\ .9 & .2 \end{matrix}$

4. $\begin{matrix} .1 & .1 & .2 \\ .1 & .4 & .1 \\ .1 & .1 & .2 \end{matrix}$ **5.** $\begin{matrix} .9 & .3 & .3 \\ .0 & .9 & .0 \\ .0 & .0 & .9 \end{matrix}$

6. Use the theorem to show that $\mathbf{A} \geq \mathbf{0}$ is productive if $\Sigma_j a_{ij} < 1$ for all i or $\Sigma_i a_{ij} < 1$ for all j.

5.6 A BIMATRIX MODEL OF ECONOMIC GROWTH

The model of a self-contained economic system described in this section was derived by Kemeny, Morgenstern and Thompson (1956) and Thompson (1956) from a model of Von Neumann (1937). A very similar generalization was given independently by Gale (1956). Kemeny, Snell and Thompson (1957) and Gale (1960) also describe it; these books use $\mathbf{x}$ for the intensity vector and use the negative transpose of the matrix $\mu\mathbf{A}-\mathbf{B}$ defined here.

The growth of the economy is only a matter of scale, not qualitative changes. Thus it suffices to analyze it for one time period, during which the following definitions apply; $\mathbf{u}$, $\mathbf{y}$ and μ are to be determined.

u_i = price of good i ($=1,\ldots,m$). Goods can include capital equipment and personnel as well as goods consumed by people and machines.
y_j = intensity of process j ($=1,\ldots,n$).
$a_{ij}y_j$ = quantity of good i consumed by process j, required at the start of the time period. (Input.)
$b_{ij}y_j$ = quantity of good i produced by process j, available at the end of the same time period. (Output.)
μ', μ'' = growth factors for goods and for money.
$\mu\mathbf{A}-\mathbf{B}=(\mu a_{ij}-b_{ij})$ = a game matrix in which $\mu=\mu'=\mu''$ is determined so that the value is zero, and $\mathbf{u}/\Sigma u_i$ and $\mathbf{y}/\Sigma y_j$ are the optimal strategies.

These quantities are subject to the sign restrictions

$$\mathbf{u}\geq\mathbf{0},\qquad \mathbf{y}\geq\mathbf{0},\qquad \mathbf{A}\geq\mathbf{0},\qquad \mathbf{B}\geq\mathbf{0}\qquad (\text{none}=\mathbf{0})$$
$$\sum_i a_{ij}>0 \text{ for every } j,\qquad \sum_j b_{ij}>0 \text{ for every } i \tag{1}$$

E.g., this means all $a_{ij}\geq 0$ and some $a_{ij}>0$. Consequently the second line of (1) merely asserts that every column of $\mathbf{A}$ and every row of $\mathbf{B}$ has at least one positive component. Thus every process consumes at least one good and every good is produced by at least one process; nothing can be produced from nothing. This does not exclude the use of raw materials but it does preclude the consideration of any limits on the quantities available.

For each good, production in the current time period must be greater than or equal to consumption in the following time period, which is assumed to increase by a factor μ' because the y_j do so. Thus

$$b_{i1}y_1+\cdots+b_{in}y_n\geq\mu'(a_{i1}y_1+\cdots+a_{in}y_n)$$

for all $i=1,\ldots,m$. Furthermore when the strict inequality holds here, the corresponding goods are being produced in excess of demand; such goods are assumed to have prices $u_i=0$ as in 3.6. These conditions are summarized by

$$\mathbf{By}\geq\mu'\mathbf{Ay} \tag{2}$$
$$\mathbf{u}(\mu'\mathbf{A}-\mathbf{B})\mathbf{y}=0 \tag{3}$$

since (2) clearly implies that in (3) each term $u_i\Sigma_j(\mu' a_{ij}-b_{ij})y_j=0$.

The cost of the inputs to process j is $u_1a_{1j}+\cdots+u_ma_{mj}$ times the intensity y_j. If this money were invested at the interest rate $\mu''-1$ per period, the resulting total sum could not be less than the selling price of the outputs of process j, which is y_j times $u_1b_{1j}+\cdots+u_mb_{mj}$. (Note that the same prices u_i are used in both cases; they do not grow.) This is true because otherwise there would be a demand to borrow money to operate the especially profitable process j at a higher intensity; this demand (being otherwise unlimited) would cause μ'' to increase until j no longer yielded excess profits. (In this way inflation leads to high interest rates.) Thus we have

$$\mu''(u_1a_{1j}+\cdots+u_ma_{mj})\geq u_1b_{1j}+\cdots+u_mb_{mj}$$

for all $j=1,\ldots,n$. Furthermore when the strict inequality holds it is better to lend (at the interest rate $\mu''-1$) the money obtainable by reducing the intensity y_j of process j. This adjustment reduces μ'' as well as y_j and results either in restoring the equality above or in reducing y_j to zero. These conditions are summarized by

$$\mu''\mathbf{uA}\geq\mathbf{uB} \tag{4}$$

$$\mathbf{u}(\mu''\mathbf{A}-\mathbf{B})\mathbf{y}=0 \tag{5}$$

The final condition is that the economy should produce something of positive value. This requires $u_ib_{ij}y_j>0$ for some i and j or

$$\mathbf{uBy}>0 \tag{6}$$

since all the terms are nonnegative.

Equations (3), (5) and (6) assert that $\mu'\mathbf{uAy}=\mathbf{uBy}=\mu''\mathbf{uAy}\neq 0$ whence $\mu'=\mu''$; the common value will hereafter be denoted by μ. Then (2) and (4) become

$$(\mu\mathbf{A}-\mathbf{B})\mathbf{y}\leq 0,\qquad \mathbf{u}(\mu\mathbf{A}-\mathbf{B})\geq 0 \tag{7}$$

Equations (3) and (5) become identical consequences of (7) (multiply (7) on the left by $\mathbf{u}$ and on the right by $\mathbf{y}$); with (6) they imply $\mu>0$ and $\mathbf{uAy}>0$.

Equation (7) is satisfied if $\mathbf{u}$ and $\mathbf{y}$ are proportional to optimal probability vectors for the zero-sum game of matrix $\mu\mathbf{A}-\mathbf{B}$, whose value is zero. Theorems 2 and 3 below show that a μ satisfying (6) and (7) always exists and it is unique if all $a_{ij}+b_{ij}>0$. If (7) is multiplied by -1, it can be paraphrased by saying that the excess production $(\mathbf{B}-\mu\mathbf{A})\mathbf{y}$ of the various goods i must be nonnegative and the excess profits $\mathbf{u}(\mathbf{B}-\mu\mathbf{A})$ of the various processes j must be nonpositive. A good produced in excess has price $u_i=0$ and a process deficient in profitability has intensity $y_j=0$, so that the total monetary value $\mathbf{u}(\mathbf{B}-\mu\mathbf{A})\mathbf{y}$ of the excess is zero.

Application and Significance of the Model

One may ask whether money itself should not be included among the goods in the model, since it is produced by the banking system and passes from one owner to another like a commodity. The difference is that paper or credit money requires little in the way of inputs, thus violating (1). Not being subject to any significant technological restriction, it would be a free good, useless for value measurements, if its production were not controlled effectively. Since the model assumes that the prices u_i do not change from one period to the next, it is plausible that the money supply should have the same growth factor as the supply of other commodities. In any case the relative prices $u_i/\Sigma u_i$ in this model are not set by a government agency or a cartel but are generated by the interaction of demands and capabilities within the economy itself. The latter can be visualized as a giant computer that determines the most efficient allocation of resources.

Like capital equipment, skilled or unskilled labor is subject to production, as a result of training, biological reproduction or immigration, but both factors differ from other producible "goods" in being needed far in excess of their consumption by deterioration. Nevertheless they can be included in the model by resorting to the fiction that during the time period, each process consumes all the personnel and capital equipment required for its operation, but also produces them again, less attrition and depreciation; of course it also consumes the goods required by its personnel and equipment.

Properly understood, an ingeniously simplified model like this one can be of value in calling attention to factors it does not accommodate as well as those it does. One of the former is the variation in the risk or uncertainty of investment in various industries as well as in debt versus equity; in practice excess profits are required as insurance premiums on excess risk. Another is the finiteness of the planet on which we live; evidently the model has to be abandoned when some limitation on the availability of land, raw materials or waste disposal is encountered. Then presumably some processes might still expand while others would have to become stationary or contract.

Actually the finiteness of the earth would have stopped growth long ago in the absence of technological innovation. The latter, along with changes in consumer preferences and the standard of living, would require that $\mathbf{A}$ and $\mathbf{B}$ be functions of the time and also be enlarged to include new goods and processes. In today's rapidly changing world the qualitative equilibrium described by the game solution for any particular $\mathbf{A}(t_0)$ and $\mathbf{B}(t_0)$ does not have time to establish itself in all respects but it is still of interest. The model does reflect the possibility of increasing future consumption by decreasing current consumption $\mathbf{A}$ and thereby increasing the growth factor μ.

In speaking of Von Neumann's model in which all $a_{ij}+b_{ij}>0$ and hence μ is unique (Theorem 3 below), Dorfman, Samuelson and Solow (1958) compare it to a turnpike in the sense that it provides the fastest route between any two points (i.e., states of the economy) on the turnpike but will require secondary roads (other production patterns) to connect with other points.

Example 1

Since other factors beside the growth rate are important in evaluating an economic system, it is desirable to determine all the optimal $\mathbf{u}$ and $\mathbf{y}$ as in 4.5.

This is facilitated by the prior determination that the value is zero. We assume that $\Sigma u_i = \Sigma y_j = 1$.

Consider the system

$$\mathbf{A}=\begin{bmatrix} 2.0 & 0 & 0 \\ .4 & 1.0 & 0 \\ .2 & .2 & .4 \end{bmatrix} \qquad \mathbf{B}=\begin{bmatrix} 1.0 & 0 & 0 \\ .4 & 1.0 & 0 \\ .6 & .7 & 1.0 \end{bmatrix}$$

Here the matrices are square, b_{ii}/a_{ii} increases with i and the elements above the diagonal are zero. In such a case the allowable values of μ are found to be b_{ii}/a_{ii}, which make one of the diagonal elements equal to zero in $\mu\mathbf{A}-\mathbf{B}$. Thus this game matrix has three possible forms:

$\mu=0.5$			$\mu=1.0$			$\mu=2.5$		
0	0	0	1.0	0	0	4.0	0	0
−.2	−.5	0	0	0	0	.6	1.5	0
−.5	−.6	−.8	−.4	−.5	−.6	−.1	−.2	0

For $\mu=0.5$ the saddle points in row 1 show that the value is zero. The only optimal **u** is $(1,0,0)$ because every other yields a negative expectation in column 1 or 2. Every **y** is optimal because no row ever yields a positive expectation, but $y_1>0$ is needed to make $\mathbf{uBy}=u_1b_{11}y_1>0$. $\mu=.5$ results from insisting on using good 1 and process 1 even though process 1 (the first column) consumes more of good 1 (the first row) than it produces, and no other process produces any of good 1. The emphasis on the scarce good 1 makes the other prices zero, leaves producers indifferent as to whether they operate processes 2 and 3, and causes the economy to contract. If process 1 is truly indispensable, like eating or breathing, this deplorable state of affairs is the best that can be achieved within the model.

For $\mu=1$ there are four saddle points of value zero. We must have $y_1=0$ (see row 1) and $u_3=0$ (see columns 2 and 3) for optimality and $u_2y_2>0$ to make $\mathbf{uBy}>0$; otherwise **u** and **y** are arbitrary. It is required to use process 2, which produces no more of good 2 than it consumes. Hence $\mu=1$ even though process 3 also may be used.

For $\mu=2.5$ there are saddle points of value zero at $ij=13$ or 23. Row 2 requires $\mathbf{y}=(0,0,1)$. Inspection of kernels yields four basic (extreme) optimal vectors **u**, namely $(1,0,0)$, $(0,1,0)$, $(0,1,6)/7$ and $(3,80,600)/683$. Any convex combination having $u_3>0$ is acceptable. We don't want $u_3=0$ because good 3 is the only output (and input) of process 3, the only process being used. We don't want $u_2=0$ because if good 2 were free, process 2 would be more profitable (for the production of good 3) than process 3 is, but the resulting consumption of good 2 would raise its price. The high growth factor $\mu=2.5$ is made possible by the abandonment of goods 1 and 2 and the inefficient processes 1 and 2 that are the only means for producing them.

Theorem 3 following guarantees that there are no more allowable values of μ. Let us see what happens when $\mu=2$. For optimality the matrix

$$2\mathbf{A}-\mathbf{B}=\begin{bmatrix} 3.0 & 0 & 0 \\ .4 & 1.0 & 0 \\ -.2 & -.3 & -.2 \end{bmatrix}$$

must have $u_3=0$ (see column 3) and $\mathbf{y}=(0,0,1)$ (see row 2) so that $\mathbf{uBy}=0$ always. Thus process 3 is used exclusively, just as for $\mu=2.5$, but the zero price of good 3 leaves producers indifferent as to how much of good 3 is reinvested in process 3 and how much is piled up or destroyed as unwanted surplus. This indeterminacy in the model permits μ to be set at 2 (or any other value between .5 and 2.5) as the result of some other considerations. However, similar effects (with u_2 and $u_3>0$) could be obtained by adjusting **A** or **B**; for example, changing a_{33} from .4 to .5

For $\mu=.5$, 1.0 and 2.5, one has $u_i b_{ij} y_j>0$ (and $u_i a_{ij} y_j>0$) only for $ij=11,22,33$ respectively; the disjointness of these sets of values of i (and likewise j) and the occurrence of the zeros in **A** and **B** are asserted by Theorem 3 below. The corresponding signs of the $u_i y_j$ are

$$\mu=0.5:\ \begin{matrix} + & \oplus & \oplus \\ 0 & 0 & 0 \\ 0 & 0 & 0 \end{matrix} \qquad \mu=1.0:\ \begin{matrix} 0 & \oplus & \oplus \\ 0 & + & \oplus \\ 0 & 0 & 0 \end{matrix} \qquad \mu=2.5:\ \begin{matrix} 0 & 0 & \oplus \\ 0 & 0 & + \\ 0 & 0 & + \end{matrix}$$ ■

Example 2

$$\mathbf{A}=\begin{bmatrix} .1 & .1 & .2 \\ .1 & .4 & .1 \\ .1 & .1 & .2 \end{bmatrix} \qquad \mathbf{B}=\begin{bmatrix} 1 & 0 & 0 \\ 0 & 1 & 0 \\ 0 & 0 & 1 \end{bmatrix}$$

This type of problem (with $\mathbf{B}=\mathbf{I}$) differs from that of 5.5 (e.g., Exercise 4) only in that the excess production, if any, is reinvested rather than consumed. Since every $a_{ij}>0$, μ is unique by Theorem 3 below.

It is known (e.g., Karlin, 1966, p. 476) that if $\mathbf{B}=\mathbf{I}$ and all $a_{ij}>0$, then all u_i and y_j are positive and unique so that the whole matrix $\mu\mathbf{A}-\mathbf{B}$ is the kernel of the game. By (9) of 4.5, the value is zero only if the determinant is zero. It is convenient to let $1/\mu=\lambda=$ an eigenvalue of **A** and solve

$$\begin{vmatrix} .1-\lambda & .1 & .2 \\ .1 & .4-\lambda & .1 \\ .1 & .1 & .2-\lambda \end{vmatrix}=-\lambda(\lambda-.2)(\lambda-.5)=0 \tag{8}$$

The largest $\lambda=.5$ is the one that will give nonnegative probabilities. It makes $\mu=2$, $\mathbf{u}^*=(.2,.5,.3)$ and $\mathbf{y}^*=(.25,.50,.25)$. ■

Example 3

$$\mu\mathbf{A}-\mathbf{B}=\begin{bmatrix} \mu-3 & 3\mu-4 & 2\mu-3 \\ 2\mu-2 & \mu-3 & 2\mu-3 \end{bmatrix}$$

(a) Find the *eligible interval* in which μ must lie:

$$\left[\min_j \max_i (a_{ij}/b_{ij})\right]^{-1} \leq \mu \leq \left[\max_i \min_j (a_{ij}/b_{ij})\right]^{-1} \tag{9}$$

Values of ij having $a_{ij}=b_{ij}=0$ are to be ignored. If (9) were not true, one would have an all-negative column of $\mu a_{ij}-b_{ij}$ or an all-positive row, and the value would not be zero. In the example $3/2 \leq \mu \leq 3$.

(b) Find any values of μ that yield 0 as a saddle point. This can occur only if (9) reduces to a single point or $a_{ij}=b_{ij}=0$ for some ij (at the saddle point; see Exercise 9).

(c) (Optional) Find the value of the game for some μ in the middle of the eligible interval. If the value is positive (negative), discard the right (left) half of the eligible interval. $\mu=2$ gives the value .5. Thus $1.5 \leq \mu \leq 2$.

(d) Select a kernel (e.g., that found in (c)) and determine whether any μ in the eligible interval makes its determinant equal to zero. If so, see if that μ yields optimal strategies for the whole game (not just the kernel). If either test fails, repeat (c) or (d). Since (c) used the first two columns, set

$$\begin{vmatrix} \mu-3 & 3\mu-4 \\ 2\mu-2 & \mu-3 \end{vmatrix} = -5\mu^2+8\mu+1=0$$

whence $\mu=(\sqrt{21}-4)^{-1}=1.7165$. This does make the value zero because (11) of 4.2 is satisfied after Exercise 9 of 4.2 is used to determine $u_2^*=y_1^*=2-\sqrt{21}/3=.4725$, $u_1^*=y_2^*=-1+\sqrt{21}/3=.5275$, $y_3^*=0$. The third process is not used. Equation (12) of 5.3 is useful for an approximate determination of μ. ■

The remainder of this section, which is devoted to stating and proving the underlying theorems, may be omitted if desired.

THEOREM 1 If $\mathbf{C}$ is any real matrix, the system

$$\mathbf{uC}=\mathbf{v}\geq\mathbf{0} \qquad \mathbf{u}\geq\mathbf{0}$$

$$\mathbf{Cy}=-\mathbf{x}\leq\mathbf{0} \qquad \mathbf{y}\geq\mathbf{0} \tag{10}$$

possesses one or more solutions (not necessarily basic) such that

$$u_i+x_i>0 \qquad \text{and} \qquad v_j+y_j>0 \qquad \text{for all } i \text{ and } j \tag{11}$$

These are called *central solutions*. As usual, (10) implies $\mathbf{ux}+\mathbf{vy}=0$.

Remark. This is a stronger form of Theorem 5 of 3.3, implying that one can require each dual pair of variables to assume one zero and one positive value. Equation (10) differs from the canonical pair of linear programs (7) of 3.1 in having $\mathbf{b}=\mathbf{0}$, $\mathbf{c}=\mathbf{0}$, $d=0$, but this defines precisely the degenerate part (if any) of an optimal tableau (a submatrix $\mathbf{C}$ in $\mathbf{A}$, say). When a solution satisfying (11) has been found for the (homogeneous) rows and columns having $b_i^* = c_j^* = 0$, adding a sufficiently small positive multiple of this central solution to the basic optimal solution for the whole tableau will yield a nonbasic optimal solution for the latter satisfying (11). If the value $\operatorname{val}\mathbf{C}=0$, (10) determines optimal strategies $\mathbf{u}$ and $\mathbf{y}$ for the game $\mathbf{C}$. If $\operatorname{val}\mathbf{C}>0$, (10) can be satisfied by taking $\mathbf{u}=$an optimal strategy and $\mathbf{y}=\mathbf{0}$.

Proof (Tucker, 1963b) We first prove the theorem for the case in which $\mathbf{C}$ is a (square) skew-symmetric matrix $\mathbf{K}=-\mathbf{K}^T$ by considering the schema

$$
\begin{array}{c|ccl}
 & \mathbf{y} & \boldsymbol{\eta} & \\
\hline
\mathbf{u} & \mathbf{K} & \mathbf{K}+\mathbf{I} & =-\mathbf{x} \\
1 & \mathbf{0} & -\mathbf{e} & =z(\text{min.}) \\
\hline
 & \| & \geq & \\
 & \mathbf{v} & \mathbf{0} &
\end{array}
\qquad
\begin{array}{l}
\mathbf{u},\mathbf{v}\geq\mathbf{0} \\
\mathbf{x},\mathbf{y}\geq\mathbf{0} \\
\boldsymbol{\eta}\geq\mathbf{0}
\end{array}
\tag{12}
$$

in which the vector $-\mathbf{e}$ has every component $=-1$. As usual a vector may be either a row or a column depending on the context.

Premultiplying the row equations by $(\mathbf{y}+\boldsymbol{\eta})$ gives

$$\mathbf{y}\mathbf{x}+\mathbf{y}\boldsymbol{\eta}+\boldsymbol{\eta}\mathbf{x}+\boldsymbol{\eta}\boldsymbol{\eta}=0 \qquad \text{because } (\mathbf{y}+\boldsymbol{\eta})\mathbf{K}(\mathbf{y}+\boldsymbol{\eta})=0$$

Since all these terms are nonnegative we must have $\boldsymbol{\eta}\boldsymbol{\eta}=\Sigma\eta_j^2=0$ or $\boldsymbol{\eta}=\mathbf{0}$ whence $\min z=0$ and $\mathbf{K}\mathbf{y}=-\mathbf{x}$. Since the primal problem then has an optimal solution (with all variables zero), the dual has a feasible solution $\mathbf{u}\geq\mathbf{0}$ such that $\mathbf{u}\mathbf{K}=\mathbf{v}\geq\mathbf{0}$ and $\mathbf{u}(\mathbf{K}+\mathbf{I})=\mathbf{v}+\mathbf{u}\geq\mathbf{e}>\mathbf{0}$. Since $\mathbf{K}$ is skew symmetric, transposition gives $\mathbf{K}\mathbf{u}=-\mathbf{v}$ so that $\mathbf{y}=\mathbf{u}$ and $\mathbf{x}=\mathbf{v}$ satisfy the theorem when $\mathbf{C}=\mathbf{K}$. Replacing $\mathbf{u}=\mathbf{y}$, $\mathbf{v}=\mathbf{x}$ and $\mathbf{K}$ by $(\mathbf{u},\mathbf{y})$, $(\mathbf{x},\mathbf{v})$ and $\begin{bmatrix} \mathbf{0} & \mathbf{C} \\ -\mathbf{C}^T & \mathbf{0} \end{bmatrix}$ respectively gives the schema

$$
\begin{array}{c|ccl}
 & \mathbf{u} & \mathbf{y} & \\
\hline
\mathbf{u} & \mathbf{0} & \mathbf{C} & =-\mathbf{x} \\
\mathbf{y} & -\mathbf{C}^T & \mathbf{0} & =-\mathbf{v} \\
\hline
 & \| & \| & \\
 & \mathbf{x} & \mathbf{v} &
\end{array}
\qquad
\begin{array}{l}
\mathbf{u},\mathbf{v}\geq\mathbf{0} \\
\mathbf{x},\mathbf{y}\geq\mathbf{0}
\end{array}
\tag{13}
$$

with every component of $(\mathbf{u},\mathbf{y})+(\mathbf{x},\mathbf{v})$ positive. This completes the proof.

Balinski and Tucker (1969) give another proof, suggested by the lexicographic ordering of 6.6, that is probably more suitable for computational use. Other proofs, using mathematical induction, are given by Tucker (1956, 1968).

THEOREM 2 There exists some $\mu'=\mu''=\mu$ such that the game matrix $\mu\mathbf{A}-\mathbf{B}$ has the value zero. The optimal strategies $\mathbf{u}$ and $\mathbf{y}$ corresponding to the largest (or the smallest) such μ satisfy (6) and (7), hence also (2) to (5).

Proof The reduction to a game has already been given in (7) above. The existence of a μ making the value zero is a consequence of the following facts. The value of $-\mathbf{B}$ is $\min_j \Sigma_i u_i^*(-b_{ij})$ by (2a) of 4.5; this is negative because some $u_i^*>0$ and then some $b_{ij}>0$ by (1). Similarly the value $\operatorname{val}\mathbf{A}=\max_i \Sigma_j a_{ij} y_j^*$ is positive. These cases are approached as μ approaches 0 and ∞ respectively. Since $\operatorname{val}(\mu\mathbf{A}-\mathbf{B})$ is a continuous function of μ by Exercise 8, there is an intermediate (positive) value of μ for which $\operatorname{val}(\mu\mathbf{A}-\mathbf{B})=0$. An actual calculation of μ may involve some trial and error; (9) of 4.5 or (12) of 5.3 may be useful, with $\mathbf{A}+\mathbf{B}$ replaced by $\mathbf{A}$.

It only remains to show that (6) can be satisfied. Suppose on the contrary that μ makes $\operatorname{val}(\mu\mathbf{A}-\mathbf{B})=0$ but no optimal strategies $\mathbf{u}$ and $\mathbf{y}$ satisfy (6). This means that $u_i y_j>0$ implies $b_{ij}=0$ and also $a_{ij}=0$ by (5). Then the matrix

$$\mathbf{C}=\mu\mathbf{A}-\mathbf{B}=\begin{bmatrix} \mathbf{0} & \mathbf{P} \\ \mathbf{Q} & \mathbf{R} \end{bmatrix} \tag{14}$$

can be rearranged and partitioned so that $\mathbf{R}$ is the (largest) submatrix whose rows i and columns j have $u_i=y_j=0$ in every optimal solution. In particular this is true of the central solution $\mathbf{u}^*=(\mathbf{u}',\mathbf{0})$, $\mathbf{y}^*=(\mathbf{y}',\mathbf{0})$ which according to Theorem 1 also satisfies

$$(\mathbf{u}',\mathbf{0})\begin{bmatrix} \mathbf{0} & \mathbf{P} \\ \mathbf{Q} & \mathbf{R} \end{bmatrix}=(\mathbf{0},\mathbf{u}'\mathbf{P}) \qquad \text{with } \mathbf{u}'\mathbf{P}>\mathbf{0},\ \mathbf{u}'>\mathbf{0}$$

$$\begin{bmatrix} \mathbf{0} & \mathbf{P} \\ \mathbf{Q} & \mathbf{R} \end{bmatrix}\begin{bmatrix} \mathbf{y}' \\ \mathbf{0} \end{bmatrix}=\begin{bmatrix} \mathbf{0} \\ \mathbf{Q}\mathbf{y}' \end{bmatrix} \qquad \text{with } \mathbf{Q}\mathbf{y}'<\mathbf{0},\ \mathbf{y}'>\mathbf{0} \tag{15}$$

the strict inequalities holding for every component. Now the elements of $\mathbf{R}$ are irrelevant to the optimality of $\mathbf{u}^*$ and $\mathbf{y}^*$ and the zero block in (14) remains $\mathbf{0}$ for all values of μ. Thus within the limits such that $\mathbf{u}'\mathbf{P}\geq\mathbf{0}$ and $\mathbf{Q}\mathbf{y}'\leq\mathbf{0}$, (15) assures that μ can be increased and decreased without losing either $\operatorname{val}(\mu\mathbf{A}-\mathbf{B})=0$ or the optimality of $\mathbf{u}^*$ and $\mathbf{y}^*$.

THEOREM 3 Let $\mathbf{u}^0, \mathbf{y}^0, \mu^0$ and $\mathbf{u}^*, \mathbf{y}^*, \mu^*$ satisfy (6) and (7) with $\mu^0<\mu^*$. Then $\mathbf{u}^0, \mathbf{y}^*, \mu^0$ and $\mathbf{u}^0, \mathbf{y}^*, \mu^*$ satisfy (7) but not (6); instead $u_i^0 y_j^*>0$ implies $a_{ij}=b_{ij}=0$. Thus if all $a_{ij}+b_{ij}>0$, μ must be unique. More generally, $u_k^0 b_{kl} y_l^0>0$ and $u_p^* b_{pq} y_q^*>0$ implies $u_p^0=y_l^*=0$, $k\neq p$, and $l\neq q$, so that at most $\min(m,n)$ distinct values of μ satisfy (6) and (7).

Proof That $(\mathbf{u}^0,\mathbf{y}^*)$ is an optimal pair of strategies for both μ^0 and μ^* follows from (7), since $\mu^0\mathbf{A}\leq\mu^*\mathbf{A}$:

$$\mathbf{u}^0(\mu^*\mathbf{A}-\mathbf{B})\geq\mathbf{u}^0(\mu^0\mathbf{A}-\mathbf{B})\geq\mathbf{0}$$

$$(\mu^0\mathbf{A}-\mathbf{B})\mathbf{y}^*\leq(\mu^*\mathbf{A}-\mathbf{B})\mathbf{y}^*\leq\mathbf{0}$$

Then (5) requires $\mathbf{u}^0\mathbf{B}\mathbf{y}^* = \mu^0\mathbf{u}^0\mathbf{A}\mathbf{y}^* = \mu^*\mathbf{u}^0\mathbf{A}\mathbf{y}^*$ whence $\mathbf{u}^0\mathbf{A}\mathbf{y}^* = \mathbf{u}^0\mathbf{B}\mathbf{y}^* = 0$ and $u_i^0 y_j^* > 0$ implies $a_{ij} = b_{ij} = 0$. Finally $u_k^0 b_{kl} y_l^0 > 0$ and $u_p^* b_{pq} y_q^* > 0$ implies $u_k^0, b_{kl}, b_{pq}, y_q^* > 0$ while $\mathbf{u}^0\mathbf{B}y^* = 0$ implies $u_k^0 b_{kl} y_l^* = u_p^0 b_{pq} y_q^* = 0$. Thus $y_l^* = u_p^0 = 0$ and so $k \neq p$ and $l \neq q$; different values of μ satisfying (6) and (7) make $u_i b_{ij} y_j > 0$ (and $u_i a_{ij} y_j > 0$) in disjoint nonempty sets of i and disjoint nonempty sets of j. Example 1 above illustrates this theorem.

EXERCISES 5.6

Find the allowable $\mu, \mathbf{u}, \mathbf{y}$ for the given $\mu\mathbf{A} - \mathbf{B}$.

1.
$$\begin{array}{cc} \mu-3 & 2\mu-2 \\ 2\mu-2 & \mu-3 \end{array}$$

2.
$$\begin{array}{ccc} -2 & 2\mu-1 & \mu-3 \\ 3\mu-2 & \mu-3 & 2\mu-1 \end{array}$$

3.
$$\begin{array}{ccc} 2.5\mu-1 & 0 & 0 \\ .5\mu-.3 & \mu-1 & 0 \\ .1\mu-.6 & .3\mu-.8 & .5\mu-1 \end{array}$$

4.
$$\begin{array}{ccc} .1\mu-1 & .3\mu & .1\mu \\ .2\mu & .1\mu-1 & .2\mu \\ .4\mu & 0 & .1\mu-1 \end{array}$$

5. The transpose of 3.

6. List the processes for a primitive economy whose "goods" are grain, cows, milk, steers and people. The time period should be at least a year.

7. Describe an economy in which the goods are chickens and eggs and the processes are the laying and hatching of eggs (Kemeny, Snell and Thompson, 1957, p. 354).

8.* Prove that the absolute change in the value of a matrix game cannot exceed the maximum absolute change in the elements of the matrix.

9.
$$\begin{array}{ccc} 3\mu-5 & 2\mu & \mu-3 \\ 2\mu-1 & 3\mu-2 & 0 \\ -1 & \mu-1 & 3\mu-4 \end{array}$$

10.
$$\begin{array}{ccc} 3\mu-5 & \mu & 3\mu-2 \\ -1 & \mu-1 & \mu-2 \\ 4\mu-3 & 2\mu-1 & 2\mu-5 \end{array}$$

5.7 THE SHAPLEY VALUE FOR n-PERSON GAMES

The value in question is a vector $\boldsymbol{\phi}$ whose components are the calculated payoffs to the n players. One may think of it as realized in practice through cooperation or compulsory arbitration, or simply as a mathematical estimate of the strengths of the positions of the respective players. It exists in two versions that do not necessarily agree; of these (4) or (5) is a convenient stepping stone to the preferred (8). Both assume the existence of transferable utility, which makes (8) a generalization of 5.3(b). Shapley (1953) gave two derivations, which are discussed in (a) and in Theorem 1 of (d) following.

When it is necessary to refer to payoff vectors in general, other than the Shapley value $\boldsymbol{\phi}$, the literature often uses $\mathbf{u}$ (for *utility*). Since we have used $\mathbf{u}$ for a probability vector (a mixed strategy), we will use $(\alpha_1, \ldots, \alpha_n)$ for a typical payoff vector, generalizing the (α, β) of 5.1 to 5.4. Otherwise our notation is fairly standard. The existence of transferable utility (e.g., side payments of money) means that the feasible region is defined by

$$c \le \alpha_1 + \alpha_2 + \cdots + \alpha_n \le d \tag{1}$$

where c and d are constants. Thus the feasible region is an infinite slab between two parallel planes (or lines as in Figure 5.3C), each of which is perpendicular to the vector $(1,1,\ldots,1)$. Only some finite portion of this slab will actually be relevant.

(a) *The original Shapley formula* (4) depends on the characteristic function $v(S)$ of the game, which is the maximum total payoff that is surely obtainable by the set of players S (called a *coalition*) when they cooperate. Thus $v(S)$ is real valued and S is any of the 2^n subsets of the set $N=\{1,2,\ldots,n\}$ of the n players. In particular $v(N)=d$ in (1) and $v(\varnothing)=0$.

The derivation that is simplest (although not as convincing as Theorem 1 below) begins with a *permutation* $(h_1, h_2, \ldots, h_n)$ of the set N of all the players, who are identified by the integers $1,2,\ldots,n$. For any such permutation we can define a sequence of coalitions

$$S_0 = \varnothing, S_1 = \{h_1\}, \qquad S_2 = \{h_1, h_2\}, \ldots$$
$$S_{n-1} = \{h_1, \ldots, h_{n-1}\} = N - \{h_n\}$$
$$S_n = \{h_1, \ldots, h_{n-1}, h_n\} = \{1,2,\ldots,n\} = N \tag{2}$$

As his payoff each player is allotted the amount he adds to the total $v(S)$ when he joins the coalition, thus:

$$h_1 \text{ gets } v(S_1) - v(S_0) = v\{h_1\}$$
$$h_2 \text{ gets } v(S_2) - v(S_1) = v\{h_1, h_2\} - v\{h_1\}$$
$$h_3 \text{ gets } v(S_3) - v(S_2) = v\{h_1, h_2, h_3\} - v\{h_1, h_2\}$$
$$\cdots$$
$$h_n \text{ gets } v(S_n) - v(S_{n-1}) = v(N) - v(N - \{h_n\}) \tag{3}$$

In 4.1 to 4.2 the payoff could be a random quantity resulting from the random selection of pure strategies, but we studied only its expected value. Similarly the payoffs in (3) are random quantities because they depend on the permutation $\{h_1, \ldots, h_n\}$, which could be randomly selected. Since the $n!$ permutations are equally likely, the vector of expected values of the payoffs is the average of the $n!$ vectors in (3); naturally the components have to be rearranged in the standard order $1,2,\ldots,n$ before averaging.

A particular player of index h will have a chance of getting a payoff from each of the 2^{n-1} coalitions S of which he is a member. Let s be the number of players in S, where $1 \le s \le n$. Then the number of permutations of the n players that give h the payoff $v(S)-v(S-\{h\})$ is $(s-1)!(n-s)!$ This is obtained by permuting the $s-1$ members of S other than h (who must be the last member to join S), and also by permuting the $n-s$ nonmembers of S, who may enter the sequence in any order after h does. Adding these up and dividing by $n!$ gives Shapley's original formula

$$\phi_h' = \sum_{S \ni h} \frac{(s-1)!(n-s)!}{n!}\left[v(S)-v(S-\{h\})\right] \tag{4}$$

Here $s=|S|$ and the sum is over the 2^{n-1} coalitions S that contain h and are subsets of $\{1,2,\ldots,n\}$. For each s, (4) gets one nth of the average amount that h contributes when he joins a coalition of size $s-1$.

Evidently the terms subtracted in (4) correspond to the 2^{n-1} coalitions $S-\{h\}$ that do not contain h. Thus we may set $S-\{h\}=\overline{T}=N-T$, the complement of a coalition $T=\{h\}\cup\overline{S}$ containing $t=n-(s-1)$ elements. If we also set $s=n-t+1$ the subtracted terms become

$$-\sum_{T \ni h} \frac{(n-t)!(t-1)!}{n!} v(\overline{T})$$

Since the dummy variables T and t are summed out of the formula, they may be replaced by S and s, so that (4) acquires the form

$$\phi_h' = \sum_{S \ni h} \frac{(s-1)!(n-s)!}{n!}\left[v(S)-v(\overline{S})\right] \tag{5}$$

Only the pairing of the (formally) positive and negative terms has been changed.

(b) The *characteristic function* $v(S)$ was defined by Von Neumann and Morgenstern (1944) as the value of the zero-sum two-person game with S and $\overline{S}$ as the players I and J and payoffs equal to the total payoffs to S alone in the n-person game. As such it represents the minimum that S can be required to accept, an amount S can be certain of getting even if $\overline{S}$ is motivated to hurt S rather than to help himself; hence it is called the *security level* for S.

In the sequential model (2) to (3) the above definition seems appropriate just in terms of self-interest, without any assumption of malevolence. For example, consider the game between S_2 and $\overline{S}_2$. In the permutation being considered, (3) shows that the outcome of this game affects only the payments to h_2 and h_3, and only through the value of $v(S_2)$. The effect is such that h_2 wants a large $v(S_2)$ while h_3 wants a small $v(S_2)$; the value of $v(\overline{S}_2)$ is irrelevant at this point. This is the Von Neumann-Morgenstern definition of $v(S_2)$, which was also assumed by Shapley (1953).

Nevertheless (5) suggested to Harsanyi (1963) the desirability of redefining $v(S)-v(\overline{S})$, now denoted by $v(S/\overline{S})$, as the value of just one zero-sum game between S and $\overline{S}$. If the matrices of their total payoffs are denoted by $\mathbf{A}$ and $\mathbf{B}$, as if they were individual players as in 5.1 to 5.4, and $\operatorname{val}\mathbf{A}$ denotes the value of the zero-sum game of matrix **A**, then we have

$$\text{Old definition: } v(S)-v(\bar{S})=\text{val}\,\mathbf{A}-\text{val}\,\mathbf{B}^T$$
$$=\text{val}\,\mathbf{A}+\text{val}(-\mathbf{B}) \tag{6}$$

New definition: $v(S)-v(\bar{S})$ is replaced by

$$v(S/\bar{S})=-v(\bar{S}/S)=\text{val}(\mathbf{A}-\mathbf{B}) \tag{7}$$

These definitions may agree as in (17) and Theorem 3 or disagree as in (11) and (12). When (7) is substituted in (5) one obtains the generally accepted form

$$\phi_h''=\sum_{S\ni h}\frac{(s-1)!(n-s)!}{n!}v(S/\bar{S}) \tag{8}$$

which is called the (*modified*) *Shapley value*. Here there is one term for each of the 2^{n-1} subsets S (of N) that contain h, and $\bar{S}=N-S$. Since $\bar{N}$ is empty, $u(N/\bar{N})$ is the same as $v(N)$ (which equals d in (1)), and the old notation $v(N)$ may still be used.

If each coalition were permitted to choose for itself, it would generally choose the old definition because of the possible advantage of having two probability vectors at its disposal (one for **A** and one for $-\mathbf{B}$), regardless of what the opponent does, but such an option seems to have no counterpart in the real world (see (11) and (12) following). The new definition can still be related to permutations provided one does not consider $(h_1, h_2, \ldots, h_n)$ by itself, but only in combination with the reverse permutation $(h_n, \ldots, h_2, h_1)$, so that one gets a double pairing of terms as in (26) following.

With the new definition the calculation of a single game value for $\mathbf{A}-\mathbf{B}$ replaces val **A** and val($-\mathbf{B}$), and for $n=2$ it becomes consistent with Nash's solution 5.3(b), but its decisive advantage is that it gives better results in specific cases. The two examples (11) and (12) both have $n=2$,

$$v_{12}=v\{1,2\}=v(N)=8, \qquad v_{1-2}=v(\{1\}/\{2\})=\text{val}(\mathbf{A}-\mathbf{B})=4$$
$$(v_1, v_2)=(v\{1\}, v\{2\})=\text{the security values}=(0,0)$$
$$(\phi_1', \phi_2')=\text{the original Shapley value}=(4,4)$$
$$=(v_{12}+v_1-v_2,\, v_{12}-v_1+v_2)/2 \tag{9}$$
$$(\phi_1'', \phi_2'')=\text{the modified Shapley value (as in 5.3(b))}$$
$$=(v_{12}+v_{1-2},\, v_{12}-v_{1-2})/2=(6,2) \tag{10}$$

$$(a_{ij}, b_{ij}) \qquad\qquad \mathbf{A}-\mathbf{B}$$

$$\begin{bmatrix}(0,-8) & (1,1)\\(-1,-1) & (8,0)\end{bmatrix} \qquad \begin{bmatrix}8 & 0\\0 & 8\end{bmatrix} \tag{11}$$

$$\begin{bmatrix}(0,-4) & (8,0)\\(0,-4) & (8,0)\end{bmatrix} \qquad \begin{bmatrix}4 & 8\\4 & 8\end{bmatrix} \tag{12}$$

In (11) the original value (4,4) ignores the advantage that player 1 enjoys in the payoffs $(0,-8)$ and $(8,0)$; repeated noncooperative play would probably rotate clockwise among the four cells to give an average $(2,-2)$, or else remain at the unique equilibrium (8,0). In (12) the advantages are divided; player 1 has the better payoffs, but only player 2 has any control over the payoffs. However, the original value (4,4) ignores the fact that while player 2 can promote his interests by enforcing $\min \alpha_1 = 0$ or $\max \alpha_2 = 0$, he cannot do both at once; (0,0) is not attainable unless player 1 makes a gratuitous side payment. This makes (4,4) unrealistic, since it starts with (0,0) and gives half of the remaining eight units to each player.

Rosenthal (1976) considers examples similar to (12) and defends the original Shapley value (4,4) that was rejected above. He states that "our proposed solution is meant to apply to situations in which irrevocable self-commitments are possible, but the arbitrator is unable to know which player is in a better position to commit first," and concedes that "the harm to the threatener is never taken into account. Our model is therefore responsive to what may seem to be extremely irrational threats." (For an example, replace -4 by -1000 in (12).)

While they are relevant to the above considerations, Harsanyi (1963) and Chapters 11 and 12 of Harsanyi (1977) are mainly concerned with a bargaining model for the modified Shapley value and its extension to games without transferable utility, thus generalizing 5.3(d).

(c) *Examples* of n-person games having serious applications are given in Exercises 10 to 12, and more can readily be constructed along the lines of those in 5.1. The voting games considered in the following section 5.8 are sometimes regarded as applications. In a lighter vein let us consider the following three-person game, in which a verbal description of the payoffs is possible.

Three game theorists are going to a potluck supper. Player 1 is to bring spaghetti or seasoning, 2 is to bring hamburger or seasoning, and 3 is to bring hamburger or spaghetti. The required quantities of hamburger, spaghetti and seasoning cost 9, 5 and 3 units respectively. If they bring the three different ingredients, a platter of spaghetti and meatballs, worth 25 units, will be prepared and presented to the person who brought the hamburger; otherwise the incomplete ingredients will be discarded. Obviously, the three people had better cooperate and agree on what each is to bring, but how shall the person who gets the platter repay the other two?

Each player has two pure strategies, which may be identified by the corresponding cost 9, 5 or 3. Thus player 1 chooses $i=5$ or 3; 2 chooses $j=9$ or 3; and 3 chooses $k=9$ or 5. There are $2^3=8$ vectors of payoffs $(\alpha_1, \alpha_2, \alpha_3)$ to be specified, namely

ij	$k=9$	$k=5$
59	$(-5,-9,-9)$	$(-5,-9,-5)$
53	$(-5,-3,\ 16)$	$(-5,-3,-5)$
39	$(-3,-9,-9)$	$(-3,\ 16,-5)$
33	$(-3,-3,-9)$	$(-3,-3,-5)$

(13)

For example, the vector $(-5,-3,16)$ in row 53 and column 9 indicates that 1 has bought spaghetti at a cost of 5; 2 has bought seasoning at a cost of 3; and 3 has bought hamburger at a cost of 9 and has received spaghetti and meatballs worth 25, so that his payoff is $25-9=16$. This position (as well as $ijk=395$) determines $v(N)=v_{123}=\max(\alpha_1+\alpha_2+\alpha_3)=-5-3+16=8$.

To apply (8) we must solve three two-person (S and $\bar{S}$) zero-sum games with values $v(S/\bar{S})$, whose data are displayed in (14). Each of the two-person coalitions $S=\{2,3\}$, $\{1,3\}$, $\{1,2\}$ has $2^2=4$ (combinations of) pure strategies at its disposal; this requires some rearrangement in going from (13) to (14).

Players	$\{2,3\}$ vs. $\{1\}$	$\{1,3\}$ vs. $\{2\}$	$\{1,2\}$ vs. $\{3\}$
Payoffs	$\alpha_2+\alpha_3-\alpha_1$	$\alpha_1+\alpha_3-\alpha_2$	$\alpha_1+\alpha_2-\alpha_3$
	$(-13 \quad -15)$	$-5 \quad 14$	$(-5 \quad -9)$
Matrices	$-9 \quad 14$	$-1 \quad -7$	$(-24 \quad -3)$
	$18 \quad -9$	$(-3 \quad -9)$	$-3 \quad 18$
	$(-3 \quad -5)$	$(-24 \quad -5)$	$3 \quad -1$
$v(S/\bar{S})$	$v_{23-1}=3.42$	$v_{13-2}=-1.96$	$v_{12-3}=2.04$

(14)

The parenthesized rows are eliminated by domination as in 4.3(ii), and then the values of the remaining 2 by 2 games are calculated as in Exercise 9 of 4.2.

Since $v_{1-23}=-v_{23-1}$, a typical component of (8) for $n=3$ is

$$\phi_1''=-\tfrac{1}{3}v_{23-1}+\tfrac{1}{6}v_{12-3}+\tfrac{1}{6}v_{13-2}+\tfrac{1}{3}v_{123} \tag{15}$$

Substituting $v_{123}=8$ and the values from (14) gives

$$\phi''=(1.54,4.23,2.23) \tag{16}$$

If $ijk=539$ is used as the cooperative (joint) strategy, then (16) implies that player 3 should pay 6.54 to 1 and 7.23 to 2 in order to convert $(-5,-3,16)$ into (16). The payoff ϕ_2'' is largest because 2 has the best pair of strategies at his disposal—the inexpensive seasoning, and hamburger, which offers a chance to win the platter.

Another three-person game is the following.

ij	$k=1$	$k=2$	$i=1$	2	j or k	
11	(0,0,0)	(0,0,0)	(0	0)	(0	0)
12	(0,0,12)	(0,0,0)	12	0	(−12	0)
21	(0,0,0)	(0,12,0)	0	12	0*	12
22	(0,0,0)	(0,0,0)	(0	0)	(0	0)

(17)

$$v_{123}=12 \qquad v_{23-1}=6 \qquad v_{12-3}=v_{13-2}=0$$

Here (8) or (15) gives $\phi''=(2,5,5)$; this also is the value of ϕ' in (4), since $v_{123}=12$, $v_{23}=6$, and the other $v(S)$ are zero. It is what the players would obtain if 2 and 3 first divided v_{23} or $v_{23-1}=6$ between them to give $(0,3,3)$, and then all three players equally divided the remaining 6 units.

(d) The *uniqueness of the Shapley value* is established by Theorem 1, which is a variant of that given by Shapley (1953). Theorems 1 and 2 are valid for either definition of the characteristic function v, while Theorems 3 to 5 specify circumstances under which the two definitions are equivalent.

A *dummy* is a player who has no incentive to join any coalition because cooperation on his part yields no benefit to himself or anyone else. Inspection of the matrix form of the game (e.g., (13)) may reveal that some player is a dummy because his strategies have no effect on the others' payoffs and their strategies have no effect on his. However, for h to be a dummy it is sufficient that the characteristic function satisfy (18) or (19) for every S containing h.

$$v(S)=v(S-\{h\})+v\{h\} \tag{18}$$

$$v(S/\bar{S})=v(S-\{h\}/\bar{S}\cup\{h\})+2v\{h\} \tag{19}$$

Like Robinson Crusoe, the dummy may choose his own strategy so as to obtain his maximum $v\{h\}$. He is of interest to us because any reasonable evaluation should give him $v\{h\}$.

THEOREM 1 The coefficients or weights

$$c_s^n=\left[n\binom{n-1}{s-1}\right]^{-1}=\frac{(s-1)!(n-s)!}{n!} \tag{20}$$

in (4), (5) and (8) are uniquely determined by the formula's properties of symmetry, linearity (additivity) and efficiency, and its validity for dummy players.

Proof For simplicity we first consider $n=3$ and replace c_s^n and $v\{i, j\}$ by c_s and v_{ij}. We use the form (4) because it does not assume that $c_s=c_{n-s}$ but leaves this to be proved. For $n=3$ symmetry and linearity give

$$\phi_1=c_1v_1+c_2(v_{12}+v_{13})-d_2(v_2+v_3)+c_3v_{123}-d_3v_{23}$$

$$\phi_2=c_1v_2+c_2(v_{12}+v_{23})-d_2(v_1+v_3)+c_3v_{123}-d_3v_{13}$$

$$\phi_3=c_1v_3+c_2(v_{13}+v_{23})-d_2(v_1+v_2)+c_3v_{123}-d_3v_{12}$$

where c_1, c_2, d_2, c_3, d_3 are to be determined.

When player 1 is a dummy, (18) shows that the first of these equations should reduce to

$$v_1\equiv(c_1+2c_2+c_3)v_1+(c_2-d_2)(v_2+v_3)+(c_3-d_3)v_{23}$$

for all values of the v's, so that

$$1=c_1+2c_2+c_3, \qquad d_2=c_2, \qquad d_3=c_3$$

The first of these will be verified at the end of the proof. Now efficiency requires that $\phi_1+\phi_2+\phi_3\equiv v_{123}$ or

$$v_{123}\equiv(c_1-2c_2)(v_1+v_2+v_3)+(2c_2-c_3)(v_{12}+v_{13}+v_{23})+3c_3v_{123}$$

for all values of the v's, so that

$$1=3c_3, \qquad c_1-2c_2=0, \qquad 2c_2-c_3=0$$

or $\mathbf{c}=(1/3,1/6,1/3)$ as in (20) with $n=3$.

In the general case of n players, one obtains groups of $2\binom{n-1}{s-1}$ terms such as $(c_s-d_s)\Sigma v(S-\{p\})$ when p is dummy. (The sum is over coalitions S of size s that contain p.) These must be eliminated by setting $d_s=c_s$ for $2\le s\le n$. Then one has

$$\phi_h=c_1v_h+c_2\sum_{i\neq h}(v_{hi}-v_i)+c_3\sum_{i<j\neq h}(v_{hij}-v_{ij})+\cdots$$
$$\cdots+c_n\big[v(N)-v(N-\{h\})\big] \tag{21}$$

Summing (21) over h and using efficiency gives

$$\sum_{h=1}^{n}\phi_h=v(N)\equiv\sum_{s=1}^{n}\big[sc_s-(n-s)c_{s+1}\big]\sum_{S}v(S) \tag{22}$$

where the last sum is over the $\binom{n}{s}$ coalitions of size s. The coefficients s and $n-s$ can be inferred from the fact that in the c_s terms, each of the s members of S plays the role of h exactly once; and in the c_{s+1} terms, each of the $n-s$ nonmembers of S plays this role. Since the last term of (22) is just $nc_nv(N)$, we have $c_n=1/n$ and $c_s=(n-s)c_{s+1}/s$ for $s=n-1,n-2,\ldots,1$. This gives (20). If h is a dummy one can verify in (4), (8) or (21) that

$$v_h\equiv v_h\sum_{s=1}^{n}\binom{n-1}{s-1}c_s=v_h\sum_{s=1}^{n}\frac{1}{n}=v_h$$

A committee of three with majority voting (this is (29) below with $a=b=c=d=1$ and $\phi=(1,1,1)/3$) suffices to illustrate the fact that neither the Shapley value nor any other valuation method can be guaranteed to give every coalition as much as that coalition could obtain on its own, if it could be sure of remaining stable and not being superseded by some other coalition. However, such a guarantee can be given for the extreme cases of coalitions of size n (this is efficiency) and coalitions of size 1 (this is *individual rationality*). When both are true the payoff vector is said to be an *imputation*.

THEOREM 2 The Shapley value is an imputation, with either definition of the characteristic function.

Proof In (5) or (8), S admits s different values of h; thus the sum of (8) from $h=1$ to $h=n$ is

$$\sum_{h=1}^{n} \phi_h = \sum_{S} \frac{s!(n-s)!}{n!} v(S/\bar{S}) = v(N) \tag{23}$$

where S is nonempty. For the second equality note that $v(S/\bar{S}) = -v(\bar{S}/S)$ and both have the same coefficient $\binom{n}{s}^{-1} = \binom{n}{n-s}^{-1}$; also $\Sigma v(S/\bar{S}) = 0$ if $s = n/2$. Thus all the terms cancel except $v(N)$, which is obtained for $S=N$. This proves efficiency once more.

The old definition of v is superadditive, meaning

$$v(S) + v(T) \le v(S \cup T) \tag{24}$$

when S and T are disjoint. This is true because one of the strategies available to $S \cup T$ is the simultaneous use of the maximin strategies of the disjoint coalitions S and T. Then in particular $v(S) - v(S - \{h\}) \ge v\{h\}$ and (4) shows that

$$\phi'_h \ge \sum_{S \ni h} \frac{(s-1)!(n-s)!}{n!} v\{h\} = v\{h\} \sum_{s=1}^{n} \binom{n-1}{s-1} \frac{(s-1)!(n-s)!}{n!}$$

whence $\phi'_h \ge v\{h\} \sum_{s=1}^{n} \frac{1}{n} = v\{h\}$ (25)

With the new definition of v we need to combine the two methods of pairing terms in (4) and (8), thus:

$$\phi''_h = \sum_{R} \frac{(s-1)!(n-s)!}{n!} \left[v(R \cup \{h\}/T) + v(T \cup \{h\}/R) \right] \tag{26}$$

where R consists of $s-1$ players (player h being omitted), and the sum is over the $\binom{n-1}{s-1}/2$ partitions of N into disjoint subsets $R, \{h\}, T$, whose union is N. We shall show that

$$v(Rh/T) + v(Th/R) \ge v'(Rh/T) + v'(Th/R) \ge 2v\{h\} \tag{27}$$

Here Rh stands for $R \cup \{h\}$, and the two v' in the second member are evaluated by assuming that in both h uses his maximin strategy appropriate for $v\{h\}$, R uses a strategy optimal for $v(Th/R)$, and T uses a strategy optimal for $v(Rh/T)$. Thus the payoffs to R and T cancel out in the second member and we are left with twice the payoff to h, which cannot be less than $2v\{h\}$. The first inequality is true because the minimizing coalitions T and R use their optimal strategies in both members, while the maximizing coalitions Rh and Th use their optimal strategies in the first member v but not necessarily in the second member v'. The proof that $\phi''_h \ge v\{h\}$ is completed as in (25), by substituting (27) in (26).

THEOREM 3 The relation $v(S/\bar{S})=v(S)-v(\bar{S})$ is valid if any of (i) to (iii) holds. If this is true for all S, the two Shapley values (5) and (8) are obviously identical.

(i) The matrix games **A** and $-\mathbf{B}$ whose values are $v(S)$ and $-v(\bar{S})$ allow S to use a common optimal strategy in both; and $\bar{S}$ also has a common optimal strategy for **A** and $-\mathbf{B}$. In particular this is true if $\mathbf{B}=\mathbf{0}$ or all b_{ij} are equal (or all a_{ij} are equal).

(ii) The game in normal form is constant-sum; that is, $\Sigma_1^n \alpha_h = v(N)(=c=d$ in (1)) for all choices of strategies.

(iii) $v(S)$ is constant-sum in the sense that $v(S)+v(\bar{S})=v(N)$.

Proofs

(i) Apply (2c) of 4.5 to **A** and $-\mathbf{B}$ and add member to member. If $\mathbf{u}^*$ and $\mathbf{y}^*$ are the common optimal mixed strategies this gives

$$\mathbf{u}^*(\mathbf{A}-\mathbf{B})\mathbf{y} \geq \mathbf{u}^*(\mathbf{A}-\mathbf{B})\mathbf{y}^* \geq \mathbf{u}(\mathbf{A}-\mathbf{B})\mathbf{y}^*$$

for all probability vectors **u** and **y**. This shows that the middle member $v(S)-v(\bar{S})$ equals $v(S/\bar{S})$, the value of $\mathbf{A}-\mathbf{B}$. If $\mathbf{B}=\mathbf{0}$, all strategies are optimal for it, including those optimal for **A**.

(ii) Here $a_{ij}=v(N)-b_{ij}$; hypothesis (i) applies for all S, as does (iii).

(iii) causes (32) hereafter to reduce to $v(S)-v(\bar{S}) \leq v(S/\bar{S}) \leq v(S)-v(\bar{S})$.

Theorem 3 may be inapplicable even if all payoffs are 0 or 1, as in

$$\begin{bmatrix} 0,0 & 0,1 \\ 0,1 & 1,0 \end{bmatrix}$$

The characteristic function v gives a concise summary of the information needed to calculate the Shapley value (4) or (8) and to pursue various other investigations. Hence when $n \geq 3$ it is often taken as the starting point, usually in the original form $v(S)$, without any concern for what normal form (sets of pure strategies and the associated tables of payoffs as in (13) or (29)) the game might have. In general the normal form is not uniquely determined; for example, for any matrix game **A** the characteristic function tells us only that $v\{1\}=-v\{2\}=a^*$ or $v(1/2)=2a^*$ and $v\{1,2\}=0$. The following theorem defines a particular normal form such that any $v(S)$ is reproduced and also $v(S/\bar{S})=v(S)-v(\bar{S})$. It is stated in general terms, but it is simplified if every $v\{h\}=0$. This will be true if every player's payoff is measured from his security value; the effect is to subtract $\Sigma\{v\{h\}: h \in S\}$ from each $v(S)$ and to reduce (28) to $v(S_h)$.

THEOREM 4 Let the values of the characteristic function $v(S)$, superadditive as in (24) with $v(\varnothing)=0$, be given for all the subsets S of the set N of n players. In the corresponding normal form each player will have n pure strategies; when player h' chooses his hth strategy we say that *player* h′ *votes for player* h. Let S_h be the set of players who vote for player h. Then the game has the characteristic functions $v(S)$

and $v(S/\bar{S})=v(S)-v(\bar{S})$ provided the payoff to player h is

$$v\{h\}+v(S_h)-\sum\{v\{h'\}:h'\in S_h\} \tag{28}$$

This is simply $v\{h\}$ if h receives no votes or only one vote; and there is an option to reduce it to $v\{h\}$ whenever h does not vote for himself.

Proof When (28) is summed over all h in N the result is $\Sigma v(S_h)$, which is $\leq v(N)$ by repeated applications of (24), since the S_h form a partition of N. Furthermore the sum of the payoffs equals $v(N)$ if some $S_h=N$, all players voting for h. Thus $v(N)$ is correctly reproduced.

If all members of S vote for the same $h\in S$ and all members of $\bar{S}$ vote for the same $\bar{h}\in\bar{S}$, the sum of (28) over all $h'\in S=S_h$ is $v(S)$ and the sum over $\bar{S}$ is $v(\bar{S})$ as desired. It remains to see that $v(S)$ and $-v(\bar{S})$ are saddlepoint values for the matrices **A** and $-$**B** respectively. If some members of S change their votes (their joint strategy) while $\bar{S}$ does not, the effect in **A** is either limited to replacing h by some other member of S, which is immaterial, or else it replaces S by disjoint subsets thereof; then the sum of (28) over $h\in S$ is $\Sigma v(S_i)\leq v(S)$, where $\{S_i\}$ is a partition of S. The effect in $-$**B**, if any, is to enlarge $\bar{S}$ and/or to give it allies T_j. Then with $\bar{h}=1$ the effect is like that of replacing $-(v_{12}-v_2)$ by $-(v_{123}-v_2-v_3)-(v_{45}-v_4-v_5)$, which cannot be larger. If $\bar{S}$ is the one to change its strategy, the discussion is similar. The option to alter (28) has the effect of dissolving some new or inessential coalitions and replacing their $v(S_h)$ by $\Sigma\{v\{h'\}:h'\in S_h\}$. Finally Theorem 3(i) shows that $v(S/\bar{S})=v(S)-v(\bar{S})$.

Let $n=3$, $v_1=v_2=v_3=0$, $v_{23}=a$, $v_{13}=b$, $v_{12}=c$, $v_{123}=v(N)=d\geq 0$, and a, b, c lie in the interval $[0, d]$; then the altered version of (28) gives the normal form

i \ jk	11	12	13	21	22	23	31	32	33
1	$d00$	$c00$	$c00$	$b00$	$0a0$	000	$b00$	000	$00a$
2	000	000	000	$0c0$	$0d0$	$0c0$	000	000	$00a$
3	000	000	$00b$	000	$0a0$	$00b$	000	000	$00d$

(29)

Von Neumann and Morgenstern (1944), p. 243, obtained a normal form for the zero-sum case by having each player choose one of the 2^{n-1} subsets to which he belongs.

THEOREM 5 Arbitrary values of $v(S/\bar{S})$ can be represented as $v(S)-v(\bar{S})$, where $v(S)$ is superadditive as in (24). Then Theorem 4 provides a normal form for $v(S/\bar{S})$.

Proof Let r, s, t be the numbers of members in R, S, T. Define $v(\varnothing)=0$, $v(N)=v(N/\varnothing)$, $v\{h\}\equiv v_0$ (a constant to be determined), $v(S)=sv_0$ when $s<n/2$, $v(S)=sv_0$ also when $s=n/2$ and $v(S/\bar{S})\leq 0$ (in these two cases we shall say that S is "small"),

and

$$v(S)=v(S/\bar{S})+v(\bar{S})=v(S/\bar{S})+(n-s)v_0$$

when S is "large" (i.e., not small).

We have to show that $v(S\cup T)\geq v(S)+v(T)$ when S and T are disjoint. If $S\cup T$ is small, so are S and T, and each member is equal to $(s+t)v_0$. If $ST\triangleq S\cup T$ is large while S, T and $R\triangleq N-(S\cup T)$ are small, we need

$$v(ST/R)+rv_0\geq(s+t)v_0 \qquad \text{or} \qquad v(ST/R)/(n-2r)\geq v_0 \tag{30}$$

(If $r=s+t=n/2$, (30) reduces to $v(ST/R)\geq 0$, true by construction). If $S\cup T$ and T are large while R and S are small, we need

$$v(ST/R)+rv_0\geq sv_0+v(T/RS)+(r+s)v_0 \qquad \text{or}$$

$$\left[v(ST/R)-v(T/RS)\right]/2s\geq v_0 \tag{31}$$

It is impossible for both S and T to be large. If v_0 is chosen small enough, all instances of (30) and (31) can be satisfied and the theorem is proved.

The proof makes it clear that $v(S/\bar{S})$ does not determine $v(S)$ uniquely even when $v(S/\bar{S})=v(S)-v(\bar{S})$. Since the latter is untrue for many games, it also follows from Theorem 4 that $v(S)$ does not uniquely determine $v(S/\bar{S})$. (See Exercise 16 for examples.) So it appears that useful as $v(S)$ is, it does not serve to determine the Shapley value unless it happens to satisfy $v(S/\bar{S})=v(S)-v(\bar{S})$ (see (11) to (12)). To be sure, Theorem 5 permits a usable $v(S)$ to be determined, possibly different from the $v(S)$ as usually defined for the original game in normal form.

Equation (27) remains valid if R, h, T are replaced by S, $\bar{S}$, $\varnothing$ respectively to give $v(N)+v(\bar{S}/S)\geq 2v(\bar{S})$. Since $v(\bar{S}/S)=-v(S/\bar{S})$, interchanging S and $\bar{S}$ shows that

$$2v(S)-v(N)\leq v(S/\bar{S})\leq v(N)-2v(\bar{S}) \tag{32}$$

A game is called *inessential* if it satisfies

$$v(S)\equiv\sum\{v\{h\}:h\in S\}, \qquad v(S/\bar{S})\equiv v(S)-v(\bar{S}) \tag{33}$$

i.e., every player is a dummy (see (18)), and v is an additive function over N. In effect the Shapley value replaces the given game by an inessential game having $\phi(S)=\Sigma\{\phi_h: h\in S\}$ as its characteristic function.

(e) *Arbitrary probabilities for the* n! *permutations* in the construction of (2) to (4) still leave the Shapley value with its properties of linearity, efficiency and validity for dummy players as in Theorem 1; only the symmetry is lost. If one makes the desirable assumption that each permutation has the same probability as its reverse (as in (26) and Exercise 14 of 5.8), it is still possible to choose $n!/2-1$ of the probabilities independently. However, since the Shapley formulas then contain only $n(2^n-1)$ terms with $n(2^{n-2})$ different coefficients, it must be that different probabilities can lead to identical formulas.

THEOREM 6 Given that $h \in S$ and the terms $v(S)-v(S-\{h\})$ and $v(\bar{S}\cup\{h\})-v(\bar{S})$ have the same (nonnegative) coefficient in the (unsymmetrical) Shapley formula for ϕ_h, the number of independently variable coefficients (degrees of freedom) in ϕ is $(n-2)2^{n-2}$.

Proof There are $n(2^{n-2})$ distinct coefficients, denoted by $c_1, d_1, c_2, d_2, c_3, d_3$ in the case $n=3$, when we have

$$\begin{aligned}\phi_1 &= c_1(v_{12-3}+v_{13-2})+d_1(v_{123}-v_{23-1})\\ \phi_2 &= c_2(v_{23-1}+v_{12-3})+d_2(v_{123}-v_{13-2})\\ \phi_3 &= c_3(v_{13-2}+v_{23-1})+d_3(v_{123}-v_{12-3})\end{aligned} \tag{34}$$

The efficiency condition $\Sigma\phi_h \equiv v_N(=v_{123})$ leads to 2^{n-1} equations, one for each $v(S)-v(\bar{S})$ or $v(S/\bar{S})$, thus:

$$\begin{aligned} v_{23-1}&: -d_1+c_2+c_3=0\\ v_{13-2}&: -d_2+c_1+c_3=0\\ v_{12-3}&: -d_3+c_1+c_2=0\\ v_{123}&: \quad d_1+d_2+d_3=1\end{aligned} \tag{35}$$

We will show that these equations are independent, and that the n equations insuring validity for dummy players

$$2c_h+2d_h=1$$

are dependent on them. Thus there are $n(2^{n-2})-2^{n-1}=(n-2)2^{n-2}$ degrees of freedom by Theorem 6 of 2.7.

Suppose the 2^{n-1} equations (35) satisfied a homogeneous linear relation with coefficients $t(S/\bar{S})$, $t(N)$. Since (5) or (8) gives a solution of (35), the column of constants in the right members of (35) also satisfies the relation by Theorem 7 of 2.7. This implies $t(N)=0$; the relation does not involve the last equation $d_1+d_2+d_3=1$. For any n each coefficient c_h, d_h or whatever appears in exactly two of the equations (35), which may be represented by two nodes connected by an arc. The resulting graph is connected and guides the determination of the t's much as in the transportation problem of 11.2. Thus $t(N)=0$ implies all $t(N-\{h\}/\{h\})=0$, which implies all $t(N-\{h,h'\}/\{h,h'\})=0$, etc. The assumed linear relation vanishes.

If the player h is a dummy, he will have $\phi_h = v_h$ if and only if (twice) the sum of the coefficients in ϕ_h is 1. It suffices to make all the players dummies simultaneously, one particular $v_h = 1$ and the others zero. The effect is to make $v(S)=1$ or 0, $v(S/\bar{S})=1$ or -1 according as $h \in S$ or $\bar{S}$; also $\phi_h = 2c_h + 2d_h$ while the other equations (34) reduce to $0=0$. Thus we may form the sum $\Sigma \phi_{h'} = v(N)$ first and then make the substitution, which reduces it to $\phi_h = 1$. Thus $2c_h + 2d_h = 1$ as desired; equivalently, this can also be obtained by multiplying each equation (35) by the ± 1 value of its label $v(S/\bar{S})$ and adding. The proof is complete.

For $n \geq 3$ some of the new formulas can be obtained by the method of Owen (1971a), who represents the n players by n arbitrary fixed points on a circle or semicircle (or a sphere of dimension $n-2$ or less) according to their affinities. These points are regarded as distinct, but can approach coincidence in the limit. They determine the permutations of the players by the ordering of their distances from a variable point $\mathbf{x}$ on the circle. The sum of the arc lengths within which $\mathbf{x}$ determines a given permutation is proportional to its probability. For $n=3$, $\theta_1 + \theta_2 + \theta_3 = 1$ and $\theta_h \geq 0$ this is found to replace (15) by

$$2\phi_h = v_{h-ij} + v_{123} + \theta_h(v_{23-1} + v_{13-2} + v_{12-3} - v_{123}) \tag{36}$$

so that the desirability of a relatively large θ_h depends on the sign of $v_{23-1} + v_{13-2} + v_{12-3} - v_{123}$.

In case $n > 3$ Owen does not exhaust the possibilities because n points on an $(n-2)$-sphere account for only $n(n-2)$ parameters, of which $(n-1)(n-2)/2$ can be eliminated by using the measure and distance-preserving rotations of the sphere. This leaves at most $(n+1)(n-2)/2$ degrees of freedom, less than the $(n-2)2^{n-2}$ stated in Theorem 6.

Apparently the following are limiting cases of Owen's results. Let N consist of l disjoint blocs $N_1, \ldots, N_l$ of sizes $n_1, \ldots, n_l$ such that players in the same bloc must have consecutive positions in every permutation. Then only $l!n_1! \ldots n_l!$ different (equally probable) permutations are considered. It turns out that it can be disadvantageous to form a bloc. Thus if $n=3$, $v(N)=1$, and other $v(S)=0$ one has $\phi = (1,1,1)/3$; but if players 2 and 3 form a bloc (this means $\theta_1 = 0$ in (36)) the only permutations are 123, 132, 231, 321 which gives $\phi = (2,1,1)/4$. (Also see Table 5.8D and Exercise 20 in 5.8.) The following paragraph makes $\phi = (0,1,1)/2$.

For another way to get the probabilities, see the end of 5.8.

(f) Myerson's three papers give several more variations on the Shapley value; these also involve axioms and blocs of players, but dispense with permutations. Myerson (1977a) lets the players correspond to the nodes of a graph; if it is complete (each of the $n(n-1)/2$ pairs of players being connected by an arc) the original form of the Shapley value (4) or (5) is obtained. The Shapley formula also determines the values for incomplete graphs as follows. Let the graph be reduced to the nodes of S and the arcs incident with pairs of nodes in S; if this reduced graph is not connected, then $v(S)$ is to be replaced by the sum of the v's for the connected components of the reduced graph. Thus, there is *less* cohesion among the players than the Shapley value assumes, and the resulting solution may be inefficient if the whole graph with nodes in N is not connected. When v is superadditive as in (24), the addition of one arc to an incomplete graph never injures either of the two players whose nodes it connects. A dummy is simply a player all of whose incident arcs are irrelevant or nonexistent.

Myerson (1977b) uses Shapley's axioms unchanged but allows the characteristic function to depend on an *embedded* coalition, which means S together with a partition of $\bar{S}$. (For $n=3$ there are 10 of these.) A stronger axiom involving a decomposability property then holds. Curiously for $n=3$ a player gains by having a *small* characteristic function value associated with the three-part partition $\{1\}$, $\{2\}$, $\{3\}$. To relate the new value to the normal form of the game, Myerson (1978) uses an equilibrium point (not necessarily unique) in a game of threats involving 2^n-1 players, namely the coalitions of the original game.

Dubey, Neyman and Weber (1981) drop the assumption of efficiency in the Shapley value. Aubin (1981) considers "fuzzy" games in which each player has specified probabilities of joining the various coalitions.

EXERCISES 5.7

1. Review (8) and (15) and then proceed independently.
What and how many data are needed to get the Shapley value (8)?
How are these data defined or calculated?
How many terms does the formula for ϕ_h'' have? Why?
How are the coefficients obtained by counting permutations?
Write out the formula in the general form (8) and (without summations) the formulas for ϕ_1'' when $n=2,3,4$.

Find the (modified) Shapley value (8) in 2 to 12.

2. $n=4$; $v(N), v_{12-34}, v_{13-24}$, all $v_{pqr-t}=1$; others$=0$ (or -1).
3. $n=4$; $v(S/\bar{S})=|s-2|(s-2)/4$ where S has s members.
4. $n=3$; $v(N)=5, v_{23-1}=2, v_{13-2}=0, v_{12-3}=3$.
5. $n=3$; $v_{pq-r}=(p+q)^2-r^2$; $v(N)=42$.

6.

ij	$k=1$	$k=2$
11	$(1,1,1)$	$(2,2,-1)$
12	$(2,-1,2)$	$(-1,2,2)$
21	$(-1,2,2)$	$(2,-1,2)$
22	$(2,2,-1)$	$(1,1,1)$

7.

$k=1$	$k=2$
$(2,2,2)$	$(3,3,0)$
$(2,0,2)$	$(0,1,1)$
$(0,1,1)$	$(2,0,2)$
$(3,3,0)$	$(2,2,2)$

8. $ijk=$ 111 222 Other
$\alpha=(7,0,1)$ $(0,4,1)$ $(0,0,0)$
9. $\alpha(i,j,k)=(|2j-k|, ik, |2j-i|)$.
10. A small business, which is in difficulty, involves three players: (1) the manager, (2) the mortgagee (creditor), and (3) the owner. For these players the value of the status quo is (2,2,3). Two of the three must approve in order to change the status quo. The other alternatives are to liquidate the company, which has the

value $(0,8,2)$, and to replace the manager, which has the value $(0,5,8)$. (Use three pure strategies for each player; discuss the possibility of using four or five.)

11. Rework 10 with a change requiring unanimous approval.

12. In organizing a project three players 1,2,3 must assign themselves to three positions 1,2,2 (two are alike). Incompatible demands give $\boldsymbol{\alpha}=\mathbf{0}$; otherwise $\boldsymbol{\alpha}(1,2,2)=(3,1,2)$, $\boldsymbol{\alpha}(2,1,2)=(2,4,0)$, $\boldsymbol{\alpha}(2,2,1)=(1,2,5)$.

13. Which hypotheses of Theorem 1 are violated by $\alpha_h=v(N)/n$?

14. What normal form does Theorem 4 give when $n=2$?

15. Find a normal form having given v_{12} and $v_{1-2}(n=2)$.

16. Show that each of the games

$$\begin{bmatrix} (0,0) & (1,1) \\ (-1,-1) & (8,0) \end{bmatrix}, \quad \begin{bmatrix} (-1,-9) & (1,1) \\ (-1,-1) & (8,0) \end{bmatrix}$$

differs from (11) in one but not both of $v(S)$ and $v(S/\bar{S})$.

17. Use Theorem 5 to show that (29) gives a normal form for $v(S/\bar{S})$ for $n=3$ provided $a=v_{23-1}-v_0$, $b=v_{13-2}-v_0$, $c=v_{12-3}-v_0$, $d=v_{123}-3v_0$, and (ijk a permutation of 123) $v_0=\min[\min_k v_{ij-k}, (v_{123}-\max_k v_{ij-k})/2]$ is added to each of the resulting payoffs.

5.8 A PRIORI CALCULATION OF VOTING POWER

This section discusses the model of Shapley and Shubik (1954), which uses the Shapley value $\boldsymbol{\phi}$ of 5.7 but can be studied independently, and that of Banzhaf (1965), whose value $\boldsymbol{\beta}$ can be related to small departures from a neutral position as regards two alternatives. Some generalizations of $\boldsymbol{\beta}$ and alternatives to it are considered in (d), (e) and (f). Both $\boldsymbol{\phi}$ and $\boldsymbol{\beta}$ can be generalized to reflect a partitioning of the committee into blocs, or other measures of the affinities among its members.

(a) *Elementary algorithms for* $\boldsymbol{\phi}$ *and* $\boldsymbol{\beta}$. Consider a committee of four members (or four blocs of members who always vote together) having 8, 5, 2 and 1 votes respectively, with 10 votes required to approve a proposal; this is denoted by $[q;\mathbf{w}]=[10;8,5,2,1]$.

The *The Shapley-Shubik model* regards as equally likely the $4!=24$ permutations of the four members, that is,

8	5*	2	1	5	8*	2	1	2	8*	5	1	1	8	5*	2
8	5*	1	2	5	8*	1	2	2	8*	1	5	1	8	2*	5
8	2*	5	1	5	2	8*	1	2	5	8*	1	1	5	8*	2
8	2*	1	5	5	2	1	8*	2	5	1	8*	1	5	2	8*
8	1	5*	2	5	1	8*	2	2	1	8*	5	1	2	8*	5
8	1	2*	5	5	1	2	8*	2	1	5	8*	1	2	5	8*

(1)

In each permutation the votes are added from left to right and the first number that brings the total to 10 or more is starred. This number (or member) is called the *pivot* (a new meaning for an old word), because it marks the transition from a nonwinning to a winning coalition of members.

The Shapley value (a vector of $n=4$ components) is obtained by counting the number of times each member is pivotal and then dividing by the number of permutations. This gives

$$\boldsymbol{\phi}=(16,4,4,0)/24=(4,1,1,0)/6=(\phi_8,\phi_5,\phi_2,\phi_1)$$

Since $w_h=w_{h'}$ implies $\phi_h=\phi_{h'}$, it is convenient to use the components of w as subscripts for the components of $\boldsymbol{\phi}$; ϕ_8 is the Shapley-Shubik payoff to the member (or each of the members) having the weight 8. No accents are used since this value is consistent with both (4) and (8) of 5.7.

The *Banzhaf model* regards as equally likely the $2^4=16$ combinations of members (coalitions or subsets of N) that can support a given proposal. In the order 8, 5, 2, 1 these are defined by the plus signs in the 16 possible outcomes of the voting.

$$\begin{array}{cccc|cccc|cccc|cccc}
+* & + & + & + & +* & - & +* & + & - & + & + & + & - & - & + & + \\
+* & + & + & - & +* & - & +* & - & - & + & + & - & - & - & + & - \\
+* & +* & - & + & + & - & - & + & - & + & - & + & - & - & - & + \\
+* & +* & - & - & + & - & - & - & - & + & - & - & - & - & - & -
\end{array} \qquad (2)$$

Votes of Yes and No are denoted by $+$ and $-$, and starred $+$'s are pivotal (if any one of them is changed to $-$, the proposal is defeated). The Banzhaf value is obtained by counting the number of stars in each column and then dividing by the total number of stars. This gives $\boldsymbol{\beta}=(6,2,2,0)/10=(3,1,1,0)/5=(\beta_8,\beta_5,\beta_2,\beta_1)$. Other writers usually restrict the word "pivot" to the Shapley-Shubik model.

One could also mark pivotal Noes in (2), but that would simply double each of the counts. As it is, clearly it is sufficient to consider only the winning coalitions (capable of approving a proposal), namely, the first six combinations in (2). For some purposes it may be interesting not to require the Banzhaf values to add to 1, but to double the counts as just mentioned and then divide by 2^n, the number of combinations. In (2) this gives $(3,1,1,0)/4$.

Neither $\boldsymbol{\phi}$ nor $\boldsymbol{\beta}$ is proportional to $\mathbf{w}=(8,5,2,1)$ in this example. In particular the member having one vote is a *dummy* having $\phi_1=\beta_1=0$ because he is never pivotal. It is easy to see that $w\geq w'$ implies $\phi_w\geq\phi_{w'}$ and $\beta_w\geq\beta_{w'}$. Every permutation, but not every winning combination, must have exactly one pivot.

Pivots can be eliminated from the the Banzhaf value as follows. List the first six outcomes of (2), that is,

$$\begin{array}{cccc|cccc|cccc}
+ & + & + & + & + & + & - & + & + & - & + & + \\
+ & + & + & - & + & + & - & - & + & - & + & -
\end{array} \qquad (3)$$

which define all the winning coalitions W:

$$\begin{array}{ccc}
\{8,5,2,1\} & \{8,5,1\} & \{8,2,1\} \\
\{8,5,2\} & \{8,5\} & \{8,2\}
\end{array}$$

(If some W's contain no pivots, they must be listed nevertheless). Now in each column take the numbers of $+$'s (Yeses) minus the number of $-$'s (Noes); each member gets a reward of 1 when he is a member of a winning coalition of $+$'s and a penalty of 1 when he is a *non*member of such a coalition—a good way to keep score. This gives the same counts as before because (by the definition of a pivot) each nonpivotal $+$ can be replaced by $-$ to give another winning combination, and vice versa. Thus the $-$'s and the nonpivotal $+$'s cancel out, leaving only the pivotal $+$'s as before (starred in (2)). Pivots and permutations too can be eliminated from the Shapley-Shubik model as shown after (12) below.

Both values can be calculated from a single table having one row for each star in (2), and two columns.

5218*	6	528*1	2
85*21	2	518*2	2
58*21	2	815*2	2
82*51	2	812*5	2
28*51	2	218*5	2

(4)

Each winning coalition (to the left of the star, which marks the pivot) is listed once for every pivot it contains. As before the Banzhaf value is obtained by observing that the (member having) weight 8 has six stars, and each of the weights 5 and 2 has two.

The second column in (4) gives the corresponding number $(s-1)!(n-s)!$ of Shapley-Shubik permutations, which are listed individually in (1). Here s is the position of the pivot, $(s-1)!$ is the number of permutations of members preceding the pivot, and $(n-s)!$ is the number of permutations of members following the pivot. For example, in the first line in (4) there are six permutations of 5, 2, 1 to the left of the pivot, which leave it undisturbed; in the second line there are two permutations of 2, 1 on the right. As in (1) (but more compactly) the Shapley-Shubik value is obtained by observing that 8 is pivotal in $6+2+2+2+2+2=16$ permutations and each of 5 and 2 is pivotal in $2+2=4$. To guard against the omission of some possibilities one should confirm that the total number of permutations is $n!$ even if one is interested in $\boldsymbol{\beta}$ alone.

Weighted voting doesn't (always) work, according to the title and the text of Banzhaf (1965), especially with a smaller number of members or representatives of very different weights. The existence of various measures of voting power suggests that hair splitting is not worthwhile, but it is still true that all the criteria can agree in yielding powers that are far from being proportional to the weights; the resulting bias may either diminish or exaggerate the differences among the weights. For example, the member of weight 1 has one-third of the power in [51; 50, 50, 1], while such members have no power in [3; 3, 1, 1] or [4; 2, 2, 1]. In fact, apart from permutations (relabeling) of members, only one other power distribution (that for [3; 2, 1, 1]) is possible among a total of three members (Exercise 14). This limitation is overcome by having more members and (more nearly) equal weights.

Lucas (1974) should be consulted for a survey of the extensive literature concerning voting power and many applications. Brams (1975–1978) gives broader treatments of political questions.

(b) *An example with compact algorithms.* The United Nations Security Council has 5 permanent members—China, France, Great Britain, the Soviet Union and the United States—and 10 nonpermanent members having two-year terms. Except on procedural questions, approval of a proposal requires affirmative votes from all 5 permanent members and at least 4 nonpermanent members.

In this problem we do not want to list 5080 pivots as in (4) or even 848 winning combinations. Since there are only two kinds of members, represented by five M's and ten m's, we can divide the permutations or combinations into eight groups, one group for each line of Table 5.8A.

The total for column ϕ is $15!/5!10!=3003$; the 5 M's are regarded as indistinguishable and the 10 m's are indistinguishable. Thus no new permutations are generated to the right of the pivot while to the left the number of permutations (in column ϕ) runs from $(5+3)!/5!3!=56$ to $(10+4)!/10!4!=1001$. As before the pivot itself is not involved in these permutations. The proportion of pivots claimed by the 10 m's together is only $56/3003=.018648$ with the remainder of .981352 going to the M's. Thus on an individual basis one has $\phi_m=.001865$ and $\phi_M=.196270$.

The Banzhaf value is not concerned with permutations, but with the number of combinations in which the members to the right of the star vote No and those to the

TABLE 5.8A Calculation of Voting Powers ϕ and β in the UNSC

Prototype	s	ϕ	βW	βP
$M^5m^3m^*m^6$	9	56	210	840
$m^4M^4M^*m^6$	9	70		1050
$m^5M^4M^*m^5$	10	126	252	1260
$m^6M^4M^*m^4$	11	210	210	1050
$m^7M^4M^*m^3$	12	330	120	600
$m^8M^4M^*m^2$	13	495	45	225
$m^9M^4M^*m$	14	715	10	50
$m^{10}M^4M^*$	15	1001	1	5
Totals		3003	848	5080

Prototype permutation or combination of 5 M's and 10 m's with starred pivot. "Yes" votes to the left of the star.

Column s is the position of the pivot or the number of Yeses.

Column ϕ gives the number of permutations divided by 5!20! as if the 5 M's (also the 10 m's) were identical.

Column βW gives the number of winning combinations of 15 individuals with "No" votes to the right of the star. (In the ϕ case we don't care how these vote).

Column βP gives the number of Banzhaf pivots, 5 or (first line) 4 times column βW. It is the number of partitions of the 15 into the pivot, the set of other Yeses (on its left), and the set of Noes on its right.

Shapley-Shubik $\phi(m^{10})=56/3003$; Banzhaf $\beta(m^{10})=840/5080$.

left (including the pivot) vote Yes. Here the 10 m's are treated as distinguishable and we want the number of ways in which they can be divided into the indicated number of Yeses and Noes. In column βW these run from $(4+6)!/4!6!=210$ to $(10+0)!/10!0!=1$. Now the number of pivots per combination is 4 (the affirmative m's) in the first line and 5 (the affirmative M's) in the other lines. Thus column βP is obtained by multiplying βW by 4 or 5. Finally the proportion of pivots claimed by the 10 m's is $840/5080=.165354$, leaving .834646 for the five M's. The display (5) summarizes the results in percent.

	10 m	5 M	m	M
ϕ	1.86	98.14	0.19	19.63
β	16.54	83.46	1.65	16.69

(5)

The Banzhaf model gives the nonpermanent members m about 9 times the power that the Shapley-Shubik model does—apparently an unusual situation. The last line of Table 5.8A before the totals epitomizes the situation. One-third of the permutations allow an M to be pivotal in the last position, but all of these are contained in (without exhausting) just one of the 848 winning Banzhaf combinations, that in which every member votes Yes. From another point of view, consider that a typical permutation thoroughly mixes the 5 M's and the 10 m's; in each position the probability of having an M is $5/(5+10)$. Thus by the time all 5 M's have joined the winning coalition in the Shapley-Shubik model, most of the m's have very likely joined also, but less than 4 must have joined if an m is to be pivotal. In the Banzhaf model approval by the 5 M's does *not* change the probability of $1/2$ for approval by an m, and so there is a much larger probability of approval by only 4 m's, which makes them pivotal. This is discussed further in (c) and (f) below.

It is interesting but inessential to observe that the voting scheme of this example can be represented by

$$[5w+4; w,w,w,w,w,1,1,1,1,1,1,1,1,1,1] \tag{6}$$

with any $w>6$. Von Neumann and Morgenstern (1944) on pp. 466–470 gives examples of voting schemes with $n=6$ or 7 that cannot be represented as $[q;\mathbf{w}]$. One of these has the winning coalitions

$$124, 235, 346, 457, 561, 672, 713 \tag{7}$$

and every set that contains one or more of these. Since the authors assume that $N-S$ is winning if S is not, they do not give the example of Exercise 11 with $n=5$. The following theorem identifies a practically important area within which no representation $[q;\mathbf{w}]$ in terms of weighted voting exists.

THEOREM A bicameral legislature cannot be represented in the form $[q;\mathbf{w}]$ (i.e., as a single chamber or house with weighted voting) unless one of the houses is powerless or requires unanimous approval. (It is assumed that the two houses have no members in common; see Exercise 13.)

Proof Let m, n be the sizes of the two houses; k, l the numbers of votes required for approval; $\mathbf{u}, \mathbf{v}$ the nonnegative weights (to be shown nonexistent); and $u=\Sigma_1^m u_i/m$, $v=\Sigma_1^n v_j/n$ the average weights, respectively. We are assuming that $1 \le k \le m-1$ and $1 \le l \le n-1$ and either $k>m/2$ or $l>n/2$ or both.

In (8) to (10) the first column gives a typical inequality, the second column gives the number of such inequalities, and the third column is the average of the inequalities.

$$\sum_1^k u_i + \sum_1^l v_j \ge q \qquad \binom{m}{k}\binom{n}{l} \qquad ku+lv \ge q \tag{8}$$

$$mu + \sum_1^{l-1} v_j < q \qquad \binom{n}{l-1} \qquad mu+(l-1)v<q \tag{9}$$

$$\sum_1^{k-1} u_i + nv < q \qquad \binom{m}{k-1} \qquad (k-1)u+nv<q \tag{10}$$

Combining the averages as (10)+(9)−2(8) gives

$$(m-k-1)u+(n-l-1)v<0 \tag{11}$$

whereas the hypotheses imply the contrary (≥ 0). This completes the proof.

The $[q;\mathbf{w}]$ scheme would allow strong support in one house to compensate for lack of a majority in the other, but that cannot be permitted. Equations (6) to (11) confirm the uselessness of the weights w_i themselves as measures of power; they are usually not unique and they fail to exist in important cases.

In multicameral systems such as the UNSC it is usual for $\boldsymbol{\beta}$ to give relatively more power to the weaker members than $\boldsymbol{\phi}$ does.

(c) *Comparison of models*. Thus far this section has been developed independently of 5.7, but comparison of the two reveals at once the common assumption of equally probable permutations, from which $\boldsymbol{\phi}$ can be derived. In fact if one defines

$$v(S)=1 \text{ if } S \text{ is winning, } v(S)=0 \text{ otherwise,}$$

$$v(S/\bar{S})=v(S)-v(\bar{S}), \tag{12}$$

then it is easy to see that the Shapley values ((4) or (8) of 5.7) reduce to the Shapley-Shubik value $\boldsymbol{\phi}$ of this section. (The numerator of the coefficients $(s-1)!(n-s)!/n!$ is used in (4)). Then we can cite Theorem 4 of 5.7 to show that the Shapley-Shubik value can be derived from a game in normal form; each member is a player having n pure strategies, which can be interpreted as votes for the member himself or one of his colleagues.

The Shapley-Shubik value for a committee can also be obtained as a Banzhaf-type average over the following model. Each member has n pure strategies as in Theorem 4 of 5.7, but now strategy i consists in demanding to occupy position i in a permutation of the members. If every member chooses a different number, so that a permutation satisfying all their demands can be formed, the pivotal member receives a payoff 1; all other payoffs are 0. If the pure strategies are equally likely and independently chosen, the $n!$ permutations will be equally likely (each has the unconditional probability n^{-n}) and a Banzhaf-type average will simply count the pivots scored by each member as in (1). The normalization (to unit sum) of this average insures that it is efficient in the game-theoretic sense. The Shapley formula (5) or (8) of 5.7 applied to this model can yield a value different from either the standard $\boldsymbol{\phi}$ or $\boldsymbol{\beta}$.

This model can be modified by choosing for each member h an independently random real number t_h between 0 and 1 with a continuous uniform distribution. This number can be interpreted as the strength of the member's support for a given proposal. Arranging these numbers in decreasing order gives the Shapley permutations with equal probabilities as before; whereas interpreting $t_h > 1/2$ as Yes and $t_h \leq 1/2$ as No in the Banzhaf case yields the equal probabilities of $1/2$ for votes of Yes and No. Together with the independence of the members' votes, the latter is equivalent to the assumption that each combination of Yes-voters has the same probability 2^{-n} of occurring. This does not imply that the members are indifferent or uninformed, but rather that each has his own collection of attitudes and responsibilities, and a great variety of proposals are considered.

It seems unlikely that either $\boldsymbol{\phi}$ or $\boldsymbol{\beta}$ can be derived from an n-person game in *normal form* in which each member (player) has only the two pure strategies Yes and No, because of the need to attribute the payoffs to individuals, not just coalitions. Thus n pure strategies were used for $\boldsymbol{\phi}$ in (12) above and Theorem 4 of 5.7. For $\boldsymbol{\beta}$ two pure strategies for each member h are indeed sufficient, but they are the following: either retain the probability $p_h = 1/2$ for voting Yes as prescribed above, or else change it to $p_h = 1/2 + \varepsilon$. Here $|\varepsilon|$ is small ($\ll 1$) and independent of h, and for convenience $\varepsilon > 0$.

The origin of this game is described in (13) to (17) below, in which $\mathbf{p} = (p_1, \ldots, p_n)$ and $P(\mathbf{p})$ is the resulting probability that the proposal is approved. It arises when we do what is by far the simplest thing: regardless of what the other members do, when member h chooses $p_h = 1/2$ we give him 0, and when he chooses $p_h = 1/2 + \varepsilon$ we give him $\varepsilon \cdot \partial P / \partial p_h \geq 0$, the amount he contributes to the increase in $P(\mathbf{p})$ in (17). This is a most trivial game (inessential as in (33) of 5.7), in which each member naturally prefers $p_h = 1/2 + \varepsilon$. Nevertheless it suffices to yield the Banzhaf value, in that β_h turns out to be the proportion of the (small) total change in $P(\mathbf{p})$ that is contributed by member h or his strategy p_h. If ε is not small, one might prefer $\boldsymbol{\phi}$ or the nonadditive $\alpha(S)$ in (16) below, which is related to both $\boldsymbol{\beta}$ and $\boldsymbol{\phi}$.

The inessentiality of the game just defined for $\boldsymbol{\beta}$ has the effect that its theory, unlike that of ϕ, has no need for side payments from one player to another, but it is not clear that this has much significance for the applicability of the two values.

However, if the members compete with one another for bribes or other favors from outside, the resulting "market price" per unit of Δp_h would presumably be proportional to β_h for small departures from $p_h = 1/2$ and possibly proportional to $\alpha(S)$ in (16) for full support from the coalition S. In fact one might consider introducing a briber B as an $n+1$st player in the game such that S is winning if and only if $B \in S$ and $S - \{B\}$ was winning before. If B is an ordinary player, then the ratios among the β_h are unchanged (not so for $\boldsymbol{\phi}$). If he is a one-person subset on a par with a block consisting of all of N (see the end of 5.7), so that B must come first or last in each permutation, then the ratios among the ϕ_h are unchanged.

Dubey and Shapley (1979) give an axiomatic derivation of $\boldsymbol{\beta}$ and many other results and references. They show that if the quota q (votes required for approval) is a random variable uniformly distributed between $w/2$ and $w = \Sigma w_h$, then the average value of ϕ_h is simply w_h/w, the proportion of votes possessed by player h. The same is true of $\boldsymbol{\beta}$ only if the condition $\Sigma \beta_h = 1$ is abandoned, and the power calculated simply by dividing the number of pivots by 2^{n-1} as in (15) below.

Table 5.8B compares $\boldsymbol{\phi}$ and $\boldsymbol{\beta}$ in various respects, using the results above and in (d) following.

The remainder of this section is optional.

(d) *The probability gradient and internal structure.* Let each member h vote Yes with probability p_h and No with probability $1 - p_h$, independently of the other

Table 5.8B Comparison of Models of an n-Person Committee N

Shapley-Shubik $\boldsymbol{\phi}$	Banzhaf $\boldsymbol{\beta}$
Equally likely outcomes	
$n!$ permutations	2^n combinations
Committee's function	
General	Yes-or-no decisions
Logical basis, uniqueness	
Theorem 1 of 5.7 on general games	Gradient (14) of probability of committee's approval; or axioms
Significant values of $p_h \triangleq$ the probability that h votes Yes	
$0, 1$, possibly $1/2$ [in (16)]	$1/2, 1/2 + \varepsilon (\lvert\varepsilon\rvert \ll 1)$ (infinitesimal changes)
Characteristic function $v(S) =$	
0 or 1; or $\alpha(S)$ in (16)	$\varepsilon\, \Sigma\{\beta_h : h \in S\}$ (inessential)
Theory uses side payments within N?	
Yes	No
Division of power in a bicameral legislature	
Powers often equal (Ex. 14)	Larger house more powerful
Other characteristics	
Both are additive over the members h as in inessential games ((33) of 5.7). In an *average* sense $\boldsymbol{\phi}$ and ∇P (see (15) following) are both proportional to the weights $\mathbf{w}$.	

members. Then the probability that the proposal is approved is

$$P(\mathbf{p})=\sum_{W}\Pi\{p_h: h\in W\}\Pi\{1-p_{h'}: h'\notin W\} \tag{13}$$

where W runs through the winning coalitions. The gradient ∇P is the vector of partial derivatives

$$\begin{aligned}\partial P/\partial p_h = &P(p_1,\ldots,p_{h-1},1,p_{h+1},\ldots,p_n)\\ &-P(p_1,\ldots,p_{h-1},0,p_{h+1},\ldots,p_n)\end{aligned} \tag{14}$$

Since $P(\mathbf{p})$ is a linear function of each p_h individually, (14) is just the function that multiplies p_h in (13).

We will always set $\mathbf{p}=\bar{\mathbf{p}}\triangleq\{1/2,\ldots,1/2)$ in (14). Then we see that the right member is 2^{1-n} times the number of combinations in which member h is pivotal, or 2^{1-n} times the difference of the number of $+$'s (Yeses) and $-$'s (Noes) in column h of (3). In symbols this is

$$\partial P/\partial p_h = 2^{1-n}|\{W: h\in W\}|-2^{1-n}|\{W: h\notin W\}| \tag{15}$$

and $\boldsymbol{\beta}$ *is proportional to the probability gradient* $\nabla P(\bar{\mathbf{p}})$.

If the power of an individual member h is proportional to (14), then the power of the coalition S might be defined as

$$\begin{aligned}\alpha(S)&\triangleq P(\mathbf{p}^1)-P(\mathbf{p}^0)\\ &=2^{s-n}|\{W: S\subset W\}|-2^{s-n}|\{W: S\subset N-W\}|\end{aligned} \tag{16}$$

where $p_h^i=i$ ($=1$ or 0) for $h\in S$ and $p_h^i=1/2$ otherwise. Here $s=|S|=$the number of members in S; W's are the winning coalitions. According to (16), $\alpha(S)=1$ (the maximum possible) if and only if S is a winning coalition; for each of the nonwinning coalitions (16) requires a separate calculation.

The superadditivity of $v(S)$ (see (24) of 5.7) is needed in much of 5.7 but not in the basic Theorem 1. Hence in the Shapley formula (5) of 5.7 we may replace $v(S)$ by $\alpha(S)$ even though $\alpha(S\cup T)$ may be greater or less than $\alpha(S)+\alpha(T)$ when S and T are disjoint. The result is just the Shapley-Shubik value again, as when the simpler (12) is the starting point (Exercise 23). Since $\alpha(S)$ can be regarded as a nonadditive and global version of Banzhaf's value $\boldsymbol{\beta}$, this suggests that the localized and perturbational character of $\boldsymbol{\beta}$ (each p_h being near $1/2$), rather than its emphasis on binary choices, is its distinguishing feature and the basic reason why it often differs from $\boldsymbol{\phi}$. The reward for this restriction is the versatility of the linear expression (17) below; in particular an inessential game is obtained immediately without further calculation, as already noted. In accordance with Theorem 1 of 5.7, the Shapley value for the inessential game $\boldsymbol{\beta}$ is just $\boldsymbol{\beta}$ itself.

The remainder of this subsection (d) is based on linear approximations to the function $P(\mathbf{p})$ when $\mathbf{p}$ is near $\bar{\mathbf{p}}=(1/2,1/2,\ldots,1/2)$. These can be obtained by expressing P in terms of $\Delta p_h \triangleq p_h - 1/2$ for $h=1,\ldots,n$ and neglecting powers higher than the first, or by using the ideas of differentials or Taylor series from the calculus. The result is

$$\Delta P \triangleq P(p_1,\ldots,p_n) - P(\tfrac{1}{2},\ldots,\tfrac{1}{2}) \doteq \frac{\partial P}{\partial p_1}\Delta p_1 + \cdots + \frac{\partial P}{\partial p_n}\Delta p_n \tag{17}$$

where the coefficients $\partial P/\partial p_n$ are evaluated with all $p_h = 1/2$.

Equation (17) will not give a good approximation to (16), but when ε is small it does give

$$P\big(\bar{\mathbf{p}} + \varepsilon(\mathbf{p}^1 - \mathbf{p}^0)\big) - P(\bar{\mathbf{p}}) \doteq \varepsilon \sum_{h\in S} \frac{\partial P}{\partial p_h}$$

or

$$P(.51,.50,.51) - P(.5,.5,.5) \doteq .01\left(\frac{\partial P}{\partial p_1} + \frac{\partial P}{\partial p_3}\right)$$

for example. These small changes in P are proportional to the Banzhaf value

$$\beta(S) = \sum_{h\in S} \frac{\partial P}{\partial p_h} \Big/ \sum_{h\in N} \frac{\partial P}{\partial p_h} \tag{18}$$

Both (16) and (18) measure the power of S in terms of the effect of systematic changes in the p_h. In (16) these changes are the maximum possible within S and zero outside S, while in (18) they are uniform over N, but treated as if infinitesimal.

A more general approach is needed if the original assumption of independent reactions to all influences is to be maintained in whole or in part; that is, the Δp_h in (17) may be regarded as random variables having zero means, variances c_{hh} and covariances $c_{hh'}$. Then squaring (17) and averaging gives the variance of ΔP, which is the denominator of

$$\gamma(S) \triangleq \sum_{h\in S} \sum_{h'\in N} \frac{\partial P}{\partial p_h} c_{hh'} \frac{\partial P}{\partial p_{h'}} \Big/ \sum_{h\in N} \sum_{h'\in N} \frac{\partial P}{\partial p_h} c_{hh'} \frac{\partial P}{\partial p_{h'}} \tag{19}$$

In statistical terms the power $\gamma(S)$ is the regression coefficient of $\Sigma_S \Delta p_h \cdot \partial P/\partial p_h$ on $\Sigma_N \Delta p_{h'} \cdot \partial P/\partial p_{h'}$.

Like all the measures of power considered in this section (except (16)), (19) has the convenient property of being an additive function of S, which occurs only once in the right-hand member. Also it is the only measure that allows one or more members to have *negative powers*. This occurs when the members in question are mavericks whose reactions are more or less consistently opposed to those of the majority; the negative sign reflects the fact that support from these members is a bad omen. However, negative powers cannot occur if all $c_{hh'} \geq 0$, as they are in the three special cases mentioned in the next paragraph.

If all $c_{hh'}=1$, (19) reduces to (18) [$\gamma(S)=\beta(S)$]. The simplest new case is that of complete independence in which all $c_{hh}=1$ and all other $c_{hh'}=0$. Then we replace γ by δ and (19) becomes

$$\delta(S) \triangleq \sum_{h\in S}\left(\frac{\partial P}{\partial p_h}\right)^2 \Big/ \sum_{h'\in N}\left(\frac{\partial P}{\partial p_{h'}}\right)^2 \tag{20}$$

This could apply when S is a fixed group (e.g., the representatives from a given district) of diverse individualists. Evidently each $\partial P/\partial p_h$ in (19) and (20) can be replaced by β_h. Thus (20) gives

$$\delta_h=\beta_h^2 \Big/ \sum_{h'\in N}\beta_{h'}^2 \tag{21}$$

Another interesting special case, which still includes both β in (18) and δ in (20), is obtained when a partition $\{N_1, N_2, \dots\}$ of N is given, and $c_{hh'}=1$ if h and h' belong to the same N_i, and $c_{hh'}=0$ otherwise. Then if h is in N_i we have

$$\gamma_h=\beta_h \sum_{h'\in N_i}\beta_{h'} \Big/ \sum_j\left(\sum_{h''\in N_j}\beta_{h''}\right)^2 \tag{22}$$

This is considered further in (f) below and in Exercises 17 to 19.

The special case of Owen (1971a) described in the last paragraph of 5.7(e) yields a generalization of the Shapley-Shubik value ϕ that (like (22)) depends on partition of N. (See Exercise 20).

(e) *Asymptotic behavior*. Consider a system in which approval by one executive member M and q of the other $2q-1$ members m^{2q-1} is required. The calculations are based on Table 5.8C, which is like Table 5.8A.

TABLE 5.8C Voting Powers for 1M and q of $2q-1$ m's

Prototype	r	ϕ	βW	βP
$Mm^{q-1}m^*m^{q-1}$	$q+1$	q }	$\binom{2q-1}{q}$	$q\binom{2q-1}{q}$
$m^qM^*m^{q-1}$	$q+1$	1 }		$\binom{2q-1}{q}$
$m^{q+1}M^*m^{q-2}$	1	1	$\binom{2q-1}{q+1}$	$\binom{2q-1}{q+1}$
...	...	...	...	...
$m^{2q-1}M^*$	1	1	$\binom{2q-1}{2q-1}$	$\binom{2q-1}{2q-1}$
Totals		$2q$	2^{2q-2}	$q\binom{2q-1}{q}+2^{2q-2}$

According to the Banzhaf assumptions, the executive M and the committee of $2q-1$ other members m^{2q-1} each have probability $1/2$ for approving any proposal, which thus has probability $P=1/4$ for passing. Since all q of the m-pivots are in the first line, the column ϕ gives $\phi(M)=(2q-q)/2q=1/2$. The sum of the binomial coefficients in column βW is $2^{2q-2}=P\cdot 2^n$ because it includes exactly one of each pair of equal terms in the expansion of $(1+1)^{2q-1}$. Since there are q possible m's as pivots in the first line, we have

$$\beta(M)=2^{2q-2}\Big/\left[2^{2q-2}+q\binom{2q-1}{q}\right]\doteq(\pi/4q)^{1/2} \tag{23}$$

The approximate value is obtained from $\binom{2q-1}{q}=(2q-1)!/q!(q-1)!$ and Stirling's approximation

$$q!\doteq(2\pi q)^{1/2}q^q e^{-q} \text{ for large } q \tag{24}$$

Similarly (20) or (21) gives

$$\delta(M)=2^{4q-4}\Big/\left[2^{4q-4}+(2q-1)\left(\frac{q}{2q-1}\right)^2\binom{2q-1}{q}^2\right]$$

$$\to(1+2/\pi)^{-1}=.61102 \text{ as } q\to\infty \tag{25}$$

Thus the asymptotic behavior of four different estimates of the power of M in Table 5.8C are as follows:

$$\begin{matrix} \phi(M) & \delta(M) & \rho(M)\ [\text{see } (27)] & \beta(M) \\ 1/2 & \pi/(\pi+2) & \sqrt{\pi}/(\sqrt{\pi}+2) & (\pi/4q)^{1/2}\to 0 \end{matrix} \tag{26}$$

All but $\boldsymbol{\beta}$ give at least a roughly equal division of power between the executive M and the committee m^{2q-1}. Clearly $\boldsymbol{\beta}$ could not do this in (17) to (18), because an increase of Δp in each individual p_h will increase the joint approval probability of the m^{2q-1} by more than Δp, while the increase in p_M is still Δp.

The new value $\boldsymbol{\rho}$ was defined with the aim of obtaining something resembling $\boldsymbol{\beta}$ (more than $\boldsymbol{\phi}$ or $\boldsymbol{\delta}$ does), but with a value $\rho(M)$ not far from $1/2$. It is calculated by counting pivots as in $\boldsymbol{\beta}$ except that if r is the number of pivots occurring in a given coalition, then each pivot is counted as $1/\sqrt{r}$ rather than 1. Thus in Table 5.8C the first two entries in the last column are divided by $(q+1)^{1/2}$. This gives $\rho(M)$ equal to

$$\left[2^{2q-2}-\binom{2q-1}{q}+(q+1)^{-1/2}\binom{2q-1}{q}\right]\Big/\left[2^{2q-2}+(q+1)^{1/2}\binom{2q-1}{q}-\binom{2q-1}{q}\right]$$

$$\to 2^{2q-2}/2^{2q-2}(1+2/\sqrt{\pi})=.46984 \text{ as } q\to\infty \tag{27}$$

The remaining power is divided equally among the remaining $2q-1$ members. For small q the power $\rho(M)$ varies from .5 at $q=1$ to a minimum of .41283 at $q=6$.

For a moment let us generalize the model of Table 5.8C so that a proposal has to be approved by a specified number (more than half) of certain subcommittees; in a subcommittee of size $2q-1$, it suffices to have q votes in favor. If such a subcommittee has any Banzhaf pivots at all, it will have q; since this has a probability proportional to $2^{1-2q}\binom{2q-1}{q}$ or $1/\sqrt{q}$, the Banzhaf power of the subcommittee is nearly proportional to $q/\sqrt{q}=\sqrt{q}$ as in (23). To obtain approximately constant powers we need to introduce $1/\sqrt{q}$ as a weighting factor; since q may be hypothetical, in practice we use $1/\sqrt{r}$ as the weight of each of the r pivots occurring in a combination.

To relate ρ and δ, note that the β_h for each member of a subcommittee of size $2q-1$ is estimated to be proportional to $1/\sqrt{q}$. Thus the squaring of these values to get δ_h in (21) is another way to introduce a weighting factor that resembles $1/\sqrt{q}$ or $1/\sqrt{r}$.

The weight $1/\sqrt{r}$ also has a certain plausibility as a compromise between the Banzhaf weight 1, which measures the members' opportunities to change the outcome, and the weight $1/r$, which could be considered to measure their responsibilities for the outcome. At any rate the latter allows each equiprobable winning combination (that has pivots) to make a total contribution of $r/r=1$, which is divided among the r pivots.

Another asymptotic calculation, which (except for $\boldsymbol{\phi}$) gives results similar to (26), is obtained by replacing a member of weight w_h in $[q;\mathbf{w}]$ by w_h members of weight 1, where the w_h are integers. If these new members all vote independently, the result is simple; symmetry insures that $\boldsymbol{\phi}=\boldsymbol{\beta}=\boldsymbol{\rho}=\boldsymbol{\delta}=\mathbf{w}/\Sigma w_h$. It is more interesting to suppose that all the w_h votes in any one group are assigned to the side (Yes or No) that a majority of the w_h members favor. This *winner-takes-all* (WTA) procedure is followed within states in electing the U.S. president via the Electoral College; it also describes the effect of electing one representative from each district (under a two-party system) and giving him a weight proportional to the total size of the electorate in his district. If we further suppose that each w_h is odd, then the approval probability for a group of w_h is $1/2$, the same as for a single representative of weight w_h. This makes it plausible that one might wish to obtain the same distribution of power for winner-takes-all as for the original weighted voting scheme $[q;\mathbf{w}]$. In fact this doesn't happen with any precision, but $\boldsymbol{\rho}$ and $\boldsymbol{\delta}$ come closest as $l\to\infty$ in the scheme $[2l;2l-1,1^{2l}]$, which yields the following approximate values for the power of any one of the $2l$ unaffiliated members (for large l):

ϕ	δ	ρ	β	
$(3/4)^{3l-1/2}/l\sqrt{2}$	$2^{1-4l}\pi$	$2^{-2l-1}\sqrt{\pi}$	$(\pi/l)^{1/2}2^{-2l}$	(WTA)
$1/l(2l+1)$	2^{2-4l}	$2^{-2l-1/2}$	$(2^{2l-1}-1+2l)^{-1}$	$[q;\mathbf{w}]$
$\to 0$	$\pi/2$	$(\pi/2)^{1/2}$	$(\pi/4l)^{1/2}\to 0$	Ratio

(28)

The $[q;\mathbf{w}]$ case falls in the center of the squares in Figure 1 of Dubey and Shapley (1979), which finds $\boldsymbol{\beta}$ to be asymptotically more irregular than ϕ.

(f) *Comparison of power estimates in the UNSC.* One who likes the value 1/2 for the power of M in Table 5.8C can argue that $\lambda(m^{10})=.05160$ is the corresponding power for m^{10}, the 10 nonpermanent members of the UNSC in (b) above. In this example approval is equivalent to approval by each of two or more subcommittees. One of these is m^{10}; since at least four affirmative votes are required for approval, it has the approval probability $p_{10}=1-2^{-10}\left[1+\binom{10}{1}+\binom{10}{2}+\binom{10}{3}\right]=848/1024$, which can be inferred from Table 5.8A. The 5 permanent members can be regarded as five subcommittees, each having 1 member and approval probability $p_1=1/2$, or as one subcommittee of five requiring unanimous approval with probability $p_5=1/32$. Since p_5 should represent five times the power of p_1, the powers should be proportional to $\log p_h^{-1}$. On this basis the combined power of the 10 nonpermanent members is

$$\lambda(m^{10})=\log\frac{1024}{848}\bigg/\left(\log\frac{1024}{848}+5\log 2\right)=.05160 \tag{29}$$

However, it is not clear how this procedure could be generalized.

TABLE 5.8D Comparison of Voting Powers for m^{10} in the UNSC

	ϕ	δ	α	λ	ρ	β
Value	.01865	.01925	.03125	.05160	.13614	.16535
Reference	(b)	(20)	(16)	(29)	(27)	(b)

Dependence on Partitioning of M^5m^{10} into Blocs

Index k	Partition	$\phi^{(k)}$	$\gamma^{(k)}$
$1X$	$M^5, m, \ldots, m$	.36364	.00391
$5X$	$M, \ldots, M, m, \ldots, m$	$.01865=\phi$	$.01925=\delta$
11	M^5, m^{10}	.50000	.03777
51	M, M, M, M, M, m^{10}	.16667	.16405
1	M^5m^{10}	$.01865=\phi$	$.16535=\beta$
4	M, M, M, M^2m^{10}	.03788	.24802
Reference		5.7(e); ex. 20	(19), (22); ex. 19

The UNSC consists of 5 permanent members M^5 (all must approve) and 10 nonpermanent members m^{10} (at least 4 must approve).

The powers of m^{10} are maximized by $\phi^{(11)}$ and $\gamma^{(4)}$ respectively. $\phi^{(4)}(M^2m^{10})=1/4$, $\gamma^{(4)}(M^2m^{10})=.74882$. $\gamma(S)$ [but not necessarily $\phi(S)$] is increased by alliances within S and decreased by alliances within $N-S$ (Exercises 17, 18). Equation (16) and Exercise 22: $\alpha(m^l)=2^{-5}$ for $7\leq l\leq 10$. $\alpha(M^5)=848/1024=.82812$, $\alpha(m)=2^{-5}84/512=.00513$, $\alpha(M)=2^{-4}\alpha(M^5)=.05176$.

If $p_h\equiv p_m$ in m^{10} and $p_h\equiv p_M$ in M^5, (13) becomes $P(\mathbf{p})=p_M^5[1-(1-p_m)^7(1+7p_m+28p_m^2+84p_m^3)]$. Then if $p_m=p_M=p=(1/2)^{1/5}=.87055$, one obtains $P(\mathbf{p})=0.5$, $\partial P/\partial p_m=840p^8(1-p)^6=.001304$, $\partial P/\partial p_M=5p^4[1-(1-p)^7(1+7p+28p^2+84p^3)]=2.87175$, so that this version of $\boldsymbol{\beta}$ gives m^{10} only .00045.

The voting-power values considered in this section 5.8 are calculated in Table 5.8D for the 10 nonpermanent members. Six of these values are listed in the first three lines without reference to any particular affinities among the members. Then the values $\gamma^{(k)}$ in (19) and (22) (which generalize $\boldsymbol{\beta}$) and $\phi^{(k)}$ in 5.7(e) are tabulated for six different partitions of $N=M^5m^{10}$ into blocs. These latter calculations are considered in Exercises 19 and 20 [and 22 for $\alpha(S)$ in (16)].

It happens that $\beta(m^{10})$ is close to $\beta(M)$. The Banzhaf tendency to give more power to the larger of groups that otherwise seem to be equivalent is offset by the larger approval probability for m^{10}.

Frank and Shapley (1981) obtain unsymmetrical estimates of the power of the nine justices of the U. S. Supreme Court. A vector of three factor scores is determined for each justice by a principal components analysis of the 1978–79 voting records. The powers are proportional to areas on a sphere as in Owen (1971a), cited in 5.7(e) above. However, the permutation associated with an issue (a point on the sphere) is determined by the projections of the nine vectors (of various lengths) on the corresponding diameter, not by great circle distances on the sphere.

EXERCISES 5.8

In 1 to 11 calculate those of the values $\boldsymbol{\phi}, \boldsymbol{\beta}, \boldsymbol{\rho}, \boldsymbol{\delta}$ that you have studied.

1. $[9; 8, 5, 2, 1]$ **2.** $[16; 9, 7, 7, 6]$
3. $[5; 4, 2, 2, 1]$ **4.** $[5; 4, 2, 1, 1, 1]$
5. $[16; 9, 8, 4, 3, 2]$ **6.** $[w+2; w, w, 1, 1, 1]$ (any $w \geq 2$)
7. Two of M_1, M_2, M_3 and one of m_1, m_2 must approve.
8. Two of M_1, M_2, M_3 and one of m_1, m_2, m_3 must approve.
9. Two of M_1, M_2, M_3 and three of $m_1, \ldots, m_5$ must approve.
10. $[5; 3, 2, 1, 1]$
11. Two of M, m_1, m_2 and two of M, μ_1, μ_2 must approve. Show this has no $[q; \mathbf{w}]$.
12. Rework 7 with M_1 and m_1 being the same person. Find a $[q; \mathbf{w}]$ representation.
13. Explain how the proof of the theorem in (b) fails if the two legislative houses have some members in common.
14. Prove that the two houses of a legislature (without an executive) have equal Shapley-Shubik powers provided they have $n_1 = 2q_1 - 1, n_2 = 2q_2 - 1$. (Hint: What happens when the permutations are reversed?)
15. Determine all the possible vectors $\boldsymbol{\phi}$ and $\boldsymbol{\beta}$ for $n=3$ and so confirm the statement at the end of (a).
16. Use Exercises 3, 4, 7, 12 above to show that if two members merge their identities and acquire the sum of their previous weights, their resulting ϕ_h or β_h may be either more or less than the sum of their previous values, but not less than what either one had alone.
17. The optimal size for a coalition: Let n_i = the size of N_i in (22). Determine n_0 so as to maximize γ_h for $h \in N_0$ if (a) $n_i = 1$ for $i = 1, 2, \ldots, n - n_0$ (unorganized opposition). (b) $n_1 = n - n_0$; other $n_i = 0$ (organized opposition) ($\beta_h \equiv 1/n$).
18. In (22) show that if N_1 and N_2 are combined into a single bloc, $\gamma(N_1 \cup N_2)$ is not decreased. (It is increased if each of $N_1, N_2, \overline{N_1} \cap \overline{N_2}$ has positive power).
19. Give the formulas for the $\gamma^{(k)}$ in Table 5.8D in terms of $\beta = \beta(m^{10}) = .16535$ and $\bar{\beta} = \beta(M^5) = .83465$.

20. Derive the generalized Shapley values $\phi^{(k)}(m^{10})$ (as in the last paragraph of 5.7(e)) listed in Table 5.8D.

21. Show that for $n=10$ the number of possible models (e.g., sets of winning coalitions) exceeds the number of possible $\nabla P(1/2,\ldots,1/2)$, so that the latter cannot uniquely determine the former.

22.* Use (16) to show that 12 key values generate $\alpha(S)$ for 66 types of coalitions S in the UNSC in (b) and (f).

23.* Show that if $v(S)$ in (5) of 5.7 is replaced by $\alpha(S)$ from (16), the Shapley-Shubik value ϕ is still obtained, as it is from (12). (Hint: First prove the theorem when N is the only winning coalition. Then use the expendability of dummies of value zero and the linear dependence of ϕ on α or v).

REFERENCES (CHAPTER 5)

Arrow (1951)
Aubin (1981)
Aumann (1961)
Balinski-Cottle (1978)
Balinski-Tucker (1969)
Banzhaf (1965–1968)
Ben-Israel-Charnes-Hurter-Robers (1970)
Bennett-Dando (1979)
Bennett-Dando-Sharp (1980)
Bonnardeaux-Dolait-Dyer (1976)
Brams (1975–1978)
Cook-Seiford (1978)
Dantzig (1955b)
Debreu (1952)
Dixon (1968)
Dorfman-Samuelson-Solow (1958)
Dubey (1975)
Dubey-Neyman-Weber (1981)
Dubey-Shapley (1979)
Eaves (1971)
Frank-Shapley (1981)
Gale (1956, 1960)
Harsanyi (1963–1977)
Howard, N. (1966–1974)
Kakutani (1941)
Kemeny-Morgenstern-Thompson (1956)
Kemeny-Snell-Thompson (1957)
Koehler-Whinston-Wright (1975)
Kristol (1980)
Lemaire (1973)
Lemke (1965–1968)
Lemke-Howson (1964)
Leontief (1936–1966)
Lucas (1974)
Luce-Raiffa (1957)
Markowitz (1959)
McDonald (1975)
Mensch (1964)
Mills (1956)
Myerson (1977a, b; 1978)
Nash (1951–1953)
Owen (1968–1972)
Rapoport (1966–1974)
Rapoport-Chammah (1965–1966)
Rapoport-Guyer (1966)
Rapoport-Guyer-Gordon (1976)
Riker-Shapley (1968)
Rosenthal (1976)
Scarf (1973)
Schumann (1968)
Seneta (1973)
Shapley (1953 – 1974)
Shapley-Shubik (1954)
Shubik (1959–1980)
Stanford (1976)
Thompson (1956)
Todd (1974–1978)
Tucker (1956, 1963b, 1968)
Tucker-Luce (1959)
Von Neumann (1937)
Wilson (1971)
Young (1978)

CHAPTER SIX

The Two-Phase Simplex and Bounded Variables

The algorithms of 3.3 treated only one type of variable (nonnegative, excepting the objective functions) and required that a basic feasible solution be available initially. In this chapter these restrictions are dropped; artificial and unrestricted variables and equality constraints without slack variables are introduced. Sections 6.3 to 6.4 consider special techniques that sometimes reduce the number of iterations required. The introduction of bounded variables and their duals in 6.5 (which is independent of 6.3 to 6.4) results in a total of five types of variables. The convergence proof for the simplex method is perfected by the study of degeneracy and lexicographic ordering in 6.6.

6.1 EQUALITY CONSTRAINTS WITHOUT SLACK VARIABLES

The following is an example of the more general type of problem considered in this section:

$$\begin{aligned}
x_1 + 4x_2 - 3x_3 - x_4 - 2 &= 0\\
x_1 + x_2 - 4x_3 - 2x_4 + 3 &= 0\\
- x_3 + 4 &\le 0\\
-x_1 - 3x_4 + 20 &\ge 0\\
-9x_1 - 11x_2 + 28x_3 + 14x_4 + 1 &= z \text{ to be min.}\\
\text{All } x_i &\ge 0.
\end{aligned} \tag{1}$$

The algorithm of 3.3 is not immediately applicable because no feasible solution for (1) is readily apparent.

In (b) of 2.6 we used the tableau format to solve a system of linear equations by using zeros initially in place of basic variables and omitting the column of coefficients of these zeros after they became nonbasic. This same technique could be used to obtain slack variables for the equality constraints in (1); an additional row would be added to accommodate the objective function. However, in handling a linear program the following considerations emerge.

(a) It is undesirable to drop pivotal columns as in (b) of 2.6, because then the most important part of the dual solution is not obtained. Indeed the contrary is true in 7.2 to 7.6, where the *non*pivotal columns are dropped.

(b) In a linear program one wants first a *feasible* solution and finally an *optimal* solution, not just any or all solutions as in 2.6. Therefore it proves to be convenient, as in 3.2, to be able to interpret the column of constant terms in any tableau as the negative of the (not necessarily feasible) primal basic solution associated with the tableau. This requires that where the initial basic variables in the tableau would otherwise be replaced by zeros, the basic variables are now retained, but with the peculiarity that *their only feasible value is zero*—the most severe form of sign restriction. Because of this peculiarity they are called *artificial variables*; they will be denoted by x_i^0 in the primal problem and by u_j^0 in the dual if any should occur there.

TABLE 6.1 The Two-Phase Simplex Method

---

1. Convert each inequality into an equation by inserting a new nonnegative slack variable ($+x_i$ on the low side or $-x_i$ on the high side of the inequality).

2. Transpose all terms to the left side of each equation, and multiply by -1 where necessary to make the constant $b_i \leq 0$.

3. In the left side of every equation that does not already have a basic variable with coefficient $+1$ (necessarily feasible by step 2), insert an artificial basic x_i^0 with coefficient $+1$, where the index i is one not already in use. Thus all $x_i = -b_i \geq 0$. Set up the tableau. The coefficients c_j in the last row are those of z (to be minimized). Go to step 4 if there are artificial variables, and go to step 8 if not.

4. Define $z^0 = \Sigma_i |x_i^0| = \Sigma_i x_i^0$, including any $x_i^0 = 0$. Add a row labeled $-z^0$ with entries g_j for $j = 1, 2, \ldots, n, 0$: $g_j = -\Sigma_i \{a_{ij}: x_i^0 \text{ is artificial}\}$ where $a_{i0} \triangleq b_i$.

5. *Phase 1* pivots a_{kl}. If possible choose l by $\min_j \{g_j < 0: x_j \text{ not artificial}\} \triangleq g_l$ or better by $\min_j \{c_j/(-g_j): g_j < 0; x_j \text{ not artificial}\} \triangleq c_l/(-g_l)$ ((1) of 6.3). As usual, to keep $\mathbf{b} \leq \mathbf{0}$ one finds k from $\min_i \{-b_i/a_{il}: a_{il} > 0\} \triangleq -b_k/a_{kl} \geq 0$ with $a_{kl} > 0$.

6. Repeat step 5 until all $g_j \geq 0$, when Phase 1 ends.
 (i) If $\min z^0 > 0$, no feasible solution exists; *stop*.
 (ii) If $\min z^0 = 0$, the solution is feasible; go to step 7.

7. Every x_j having $g_j > 0$ is marked as artificial, even if not so originally. This is needed only if some $x_i^0 = 0$ are still basic. Pivots are never chosen in the columns of artificial variables and it is immaterial if their $c_j < 0$. Discard the row $-z^0$.

8. *Phase 2* pivots $a_{kl} > 0$. If possible choose l by $\min_j \{c_j < 0: x_j \text{ not artificial}\} \triangleq c_l$; row k as in step 5.

9. Repeat step 8 until Phase 2 ends thus:
 (i) If $c_l < 0$ and $a_{il} \leq 0$ for some x_l not artificial and for all i, $z^* = -\infty$; *stop*.
 (ii) If $c_j \geq 0$ for all x_j not artificial, an optimal solution has been found; *stop*.

Artificial variables are as convenient and respectable as the empty set or a constant function. By providing a measure of infeasibility, they supply an effective practical means for getting a feasible solution, and they are mathematically elegant in preserving the one-to-one correspondence between constraints and basic variables and in completing the class of five types of variables listed in (4) of 6.5.

The present problem is to find a tableau whose primal basic solution is feasible, in order to permit the use of the primal simplex algorithm of 3.3 in the subsequent Phase 2 of the calculation. The procedure is to construct a quantitative measure z^0 of the infeasibility of the existing basic solution and temporarily to adopt the minimization of z^0 rather than z as the objective of the program.

The *infeasibility form* z^0 is defined by

$$z^0 = \Sigma_i |x_i^0| = \Sigma_i x_i^0 \operatorname{sgn} x_i^0 = -\Sigma_i x_i^0 \operatorname{sgn} b_i \tag{2}$$

where $\operatorname{sgn} b_i = 1$ or -1 according as $b_i > 0$ or $b_i \leq 0$; the absolute value of x_i^0 is required in order that $z^0 = 0$ if and only if all $x_i^0 = 0$. Since x_i^0 is to be replaced by its value in terms of nonbasic variables, it is necessary to know the sign of $x_i^0 = -b_i$. Therefore the basic variables are defined so that the basic solution at all times is *quasifeasible*, meaning $x_i = -b_i \geq 0$ for all i, whether x_i is artificial or not. This implies that for every i, x_i is either feasible or artificial, and (2) reduces to

$$z^0 = \Sigma_i x_i^0 \qquad \text{where} \qquad x_i^0 = -\Sigma_j a_{ij} x_j - b_i \tag{3}$$

It follows that the coefficient g_j of x_j in z^0 (in the row to be labeled $-z^0$) is the *negative* of the sum over i of the coefficients a_{ij} of x_j in the rows labeled x_i^0:

$$g_j = -\Sigma_i \{a_{ij} \colon x_i^0 \text{ is artificial}\}. \tag{4}$$

Unlike ordinary slack variables, artificial variables can always be inserted with a coefficient chosen so that they are nonnegative in the basic solution.

Phase 1 is devoted to the minimization of z^0 while z has a secondary role, if any. Since $z^0 \geq 0$ by construction, it is never unbounded; the only question is whether $\min z^0 > 0$ (no feasible solution, when we may say $z^* = +\infty$) or $\min z^0 = 0$. In the latter case z^0 is dropped and z is minimized in Phase 2, with the usual two possibilities that z either has a finite optimum or is unbounded ($z^* = -\infty$).

More details are included in the description of the algorithm in Table 6.1 and in the discussion of the example hereafter. The flowchart in Figure 6.1 gives fewer details but indicates pictorially how one step leads into the next.

To apply this algorithm to the program (1), we first convert the two inequalities to equalities by inserting slack variables x_5 and $-x_9$ in the left members; these could be any subscripts not already in use, but 5 and 9 will turn out to be especially appropriate. Next we multiply the second, third, and fourth of the resulting equations throughout by -1 to make all the constant terms nonpositive (except in z); thus the resulting tableau will be quasifeasible as defined above. The equations

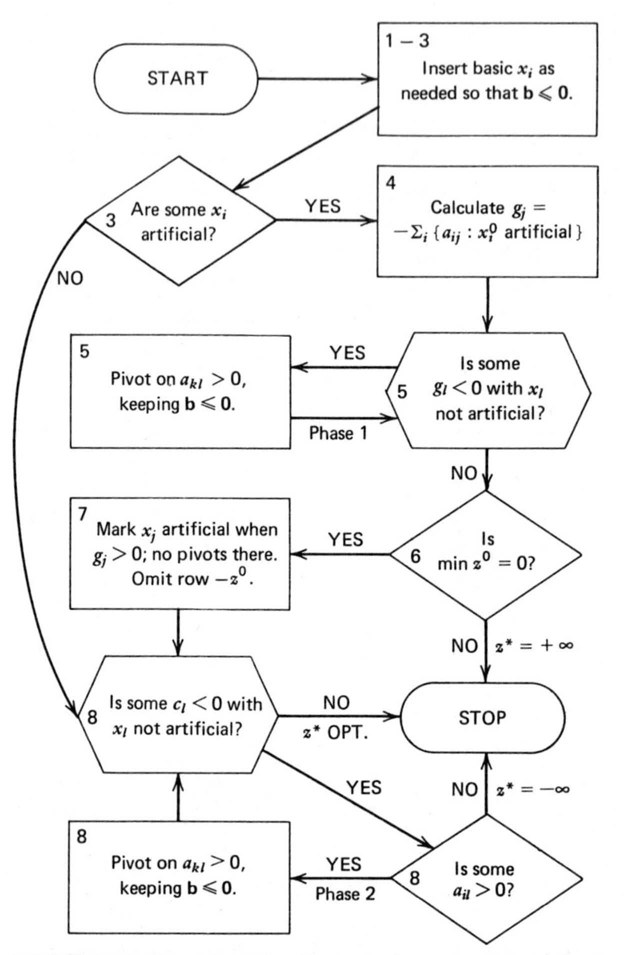

FIGURE 6.1 Flowchart for the two-phase simplex method.

now take the form (5), which is said to be in *standard form*.

$$
\begin{array}{rrrrrrl}
x_1 + & 4x_2 - & 3x_3 - & x_4 & & & - \ 2 = 0 \\
-x_1 - & x_2 + & 4x_3 + & 2x_4 & & & - \ 3 = 0 \\
& + & x_3 & & -x_5 & & - \ 4 = 0 \\
x_1 & & + & 3x_4 & & +x_9 & - 20 = 0 \\
-9x_1 - & 11x_2 + & 28x_3 + & 14x_4 & & & + \ 1 = z \min
\end{array} \tag{5}
$$

We recall that a primal basic variable must have a coefficient of $+1$ in one equation and 0 in all the other equations. Thus in (5), x_9 is basic (and so is $-z$, as usual) but no other variable is considered basic; x_5 is not, because its coefficient is negative. In order to form the usual tableau, every equation must contain a basic variable. Consequently artificial variables x_6^0, x_7^0, x_8^0 are added to the left members of the first three equations of (5). Then the tableau (6) can be formed.

Check	1	x_1	x_2	x_3	x_4	x_5	1
0	x_6^0	1	4*	−3	−1	0	−2
2	x_7^0	−1	−1	4	2	0	−3
−3	x_8^0	0	0	1	0	−1	−4
−15	x_9	1	0	0	3	0	−20
24	$-z$	−9	−11	28	14	0	1
5	$-z^0$	0	−3	−2	−1	1	9

(6)

All variables (with the possible exception of z) are required to be nonnegative in the optimal solution, and x_6^0, x_7^0, x_8^0 must be zero.

The infeasibility form is defined by

$$z^0 = x_6^0 + x_7^0 + x_8^0 \tag{7}$$

which implies that the last row of (6) is obtained as the negative of the sum of the three rows of (6) that have artificial variables as labels; negative, because the labels are $+x_i^0$ but $-z^0$. Thus the constant term in the row $-z^0$ is always nonnegative. The other coefficients in row $-z^0$ are denoted by g_j.

The use of the check values was explained in 2.5. Since we are not going to drop the columns of coefficients of artificial variables, we set these variables equal to 1 like all the others, including $-z$ and $-z^0$, in calculating check values; if such columns were to be dropped, then we would put all $x_i^0=0$ and $z^0=0$ in the check values. Thus we do not require that the check values satisfy (7), but instead have the following.

Check for row $-z^0$. In the initial tableau the sum of the check values for all the rows x_i^0 and $-z^0$ must equal the number of such rows, including $-z^0$. (Because of (4), the a_{ij} cancel out.)

In Phase 1, z^0 is used as the objective function while z may or may not have a role as indicated in step 5 of Table 6.1. Of course, neither the coefficients of z^0 nor of z are eligible to be pivots in either phase. The most negative coefficient of z^0 is -3 in column 2; this column also gives the minimum of the ratios

$$c_j/(-g_j) = -11/3, 28/2, 14/1$$

as in (1) of 6.3. The only positive pivot available in this column is $a_{62}=4$. Exchanging x_6^0 and x_2 gives

Check	4	x_1	x_6^0	x_3	x_4	x_5	1
0	x_2	1	1	−3	−1	0	−2
8	x_7^0	−3	1	13*	7	0	−14
−12	x_8^0	0	0	4	0	−4	−16
−60	x_9	4	0	0	12	0	−80
96	$-z$	−25	11	79	45	0	−18
20	$-z^0$	3	3	−17	−7	4	30

(8)

Now -17 is the most negative coefficient of z^0 and it also yields the minimum of $c_j/(-g_j)=79/17, 45/7$. Then the minimum of $-b_i/a_{i3}=14/13, 16/4$ gives $a_{73}=13$ (over 4) as the next pivot:

Check	13	x_1	x_6^0	x_7^0	x_4	x_5	1	
6	x_2	1	4	3	2	0	-17	
8	x_3	-3	1	4	7	0	-14	
-47	x_8^0	3*	-1	-4	-7	-13	-38	
-195	x_9	13	0	0	39	0	-260	
154	$-z$	-22	16	-79	8	0	218	
99	$-z^0$	-3	14	17	7	13	38	(9)

In column 1 there are three possible pivots, of which the 3 in row 8 gives the minimum ratio:

Check	3	x_8^0	x_6^0	x_7^0	x_4	x_5	1	
5	x_2	-1	1	1	1	1*	-1	
-9	x_3	3	0	0	0	-3	-12	
-47	x_1	13	-1	-4	-7	-13	-38	
2	x_9	-13	1	4	16	13	-22	
-44	$-z$	22	2	-25	-10	-22	-14	
12	$-z^0$	3	3	3	0	0	0	(10)

This completes Phase 1. The tableau is now truly primal feasible, and not merely quasifeasible, because the artificial variables are zero in the basic solution. However, *the termination criterion for Phase 1 is that the row* $-z^0$ *no longer contains any negative coefficients*; in fact it merely reproduces the definition (7) in the present case.

In Phase 2 we turn our attention to the objective function z. It has three negative coefficients, corresponding to the variables x_7^0, x_4, x_5. However, since the only feasible value of x_7^0 is zero, its negative coefficient does not afford any opportunity to reduce z, and it is therefore ignored. In general one has the following rule.

Artificial-Column Exclusion Rule. In Phase 1 or 2 one never chooses a pivot in a column j whose label is artificial, or whose entry g_j in row $-z^0$ was positive (not zero) at the end of Phase 1. Indeed, there is the possibility (though we have rejected it) of dropping these columns out of the tableau.

This rule is desirable in Phase 1, but in Phase 2 it is a necessary and sufficient condition that z^0 (which is assumed to be zero before any Phase 2 is attempted) is not inadvertently increased during Phase 2; it cannot decrease because the artificial variables (like the others) are kept nonnegative in Phase 2. This ban may apply to some nonartificial variables, in the event that some artificial variables are still basic, although necessarily of value zero. Such nonartificial variables must be distinguished in some way, such as by giving them a superscript zero, since they are like the artificial variables in having no feasible value other than zero. As soon as such provision is made for obeying the above rule, the row $-z^0$ is dropped at the end of Phase 1.

Returning to the example, we see that only the negative coefficients $-10/3$ and $-22/3$ in z can be considered; the latter, in column 5, is the more negative. Of the two positive pivots in this column, 1 is chosen because $1/1<22/13$. This pivot step yields the final tableau

Check	1	x_8^0	x_6^0	x_7^0	x_4	x_2	1
5	x_5	-1	1	1	1	3	-1
2	x_3	0	1	1	1	3	-5
6	x_1	0	4	3	2	13	-17
-21	x_9	0	-4	-3	1	-13	-3
22	$-z$	0	8	-1	4	22	-12

(11)

Thus the optimal feasible solution for (1), with all the x_h included, is $\mathbf{x}^*=(17,0,5,0,1,0,0,0,3)$, $z^*=-12$. The dual solution will be discussed in the following section; it is $\mathbf{u}^*=(0,22,0,4,0,8,-1,0,0)$. A solution in only two pivot steps is given in Table 6.4A.

EXERCISES 6.1

Find optimal solutions for the following, with all $x_h \geq 0$ and all x_i^0 artificial. Use check values.

1. Max $-5x_1+3x_2-x_3=-z$ subject to
$-3x_1+2x_2-x_3+1=0, \quad x_1-x_2-x_3\geq -5$

2. Min $x_3-x_4=z$ subject to
$3x_3+x_4=8, \quad x_3+2x_4\geq 1$

3. Max $x_1+3x_2-6x_3+4x_4$ subject to
$x_1-3x_2+x_3-6x_4+1=0, \quad x_1-x_2-x_3-3x_4=2$

4. Min $2x_3+7x_4-13x_5$ subject to
$2x_3-x_4+2x_5=1, \quad x_3+2x_4-3x_5=1$

5. Max $3x_4+2x_5+7x_6$ subject to $x_5+2x_6=4$,
$-x_4+2x_5+x_6\geq 2, \quad x_4-x_5+2x_6\geq 1$

6. Min $-7x_4+10x_5+x_6$ subject to $x_4-3x_5\leq -1$,
$x_4-x_5+2x_6=3, \quad x_5+2x_6\geq 1$

(Exercises 3 and 4 will be reconsidered in 6.3 and 5 and 6 in 6.4.)

7. Solve Example 5 of 3.4 with two pivot steps.

Exercises relating to special situations are given in 6.2.

6.2 UNRESTRICTED VARIABLES

These occur in the dual of the problem of 6.1. To determine its nature it is only necessary to observe that any equality constraint

$$a_{i1}x_1 + \cdots + a_{in}x_n + b_i = 0 \tag{1}$$

is equivalent to the pair of inequalities

$$\begin{aligned} a_{i1}x_1 + \cdots + a_{in}x_n + b_i &\leq 0 \\ -a_{i1}x_1 - \cdots - a_{in}x_n - b_i &\leq 0 \end{aligned} \tag{2}$$

Although we would not want to enlarge the problem in this way in practice, the substitution of (2) for (1) does allow us to apply the duality theory of 3.1 at once. To the inequalities (2) there corresponds a pair u_{i1}, u_{i2} of nonnegative dual variables that appear in the dual constraints and objective function in the form

$$\begin{array}{c} \cdots + a_{i1}(u_{i1} - u_{i2}) + \cdots \\ \cdots\cdots\cdots\cdots\cdots\cdots\cdots \\ \cdots + a_{in}(u_{i1} - u_{i2}) + \cdots \\ \cdots + b_i(u_{i1} - u_{i2}) + \cdots \end{array}$$

Since u_{i1} and u_{i2} occur always as the combination $u_{i1} - u_{i2} = u_i$, say, it is clear that u_{i1} and u_{i2} are never uniquely determined, and are restricted only to the extent that u_i is determined or restricted. Evidently the dual variable u_i corresponding to (1) is free of the specific nonnegativity restrictions that applied to u_{i1} and u_{i2}; at least until the other constraints are taken into account, it may be positive, negative, or zero (and in this sense is described as *unrestricted*). This is only to be expected, if one recalls the development of 3.1. There the dual multipliers u_i had to be nonnegative in order to preserve the sense of the inequalities or the signs of the slack variables they multiplied; but an equation remains an equation after multiplication by any number whatever. Thus the theorems of 3.3 remain valid in the present context.

The preceding argument is reversible. If one begins with a nonbasic variable that is unrestricted as to sign, it can be represented as the difference of two nonnegative variables, which in the dual correspond to a pair of inequalities such as (2) and hence to an equation such as (1). On the other hand, a constraint containing an unrestricted *basic* variable would serve only to define that variable and would not affect the problem otherwise.

It is easy to see that only one of the inequalities (2) would suffice, provided that the *sum* of the other inequalities, one from each pair, was included as a constraint.

When one wishes to emphasize that a variable such as u_i is not sign-restricted, it may be written as $\tilde{u}_i$ or $u_i^{\pm}$. The above arguments now permit us to form Table 6.2 in which the first four lines abstract the most important points in the remainder.

By including the dual variables, the initial and the final tableaus (6) and (11) of 6.1 may now be expanded into the schemas (3) and (4). In particular the dual

TABLE 6.2 Complementary Primal-Dual Pairs of Variables

Constraint	Primal or dual Basic variable	Dual or primal Nonbasic variable
Inequality	Nonnegative	Nonnegative
Equation	Artificial	Unrestricted

	Primal		Dual	
Constraint	Complementary Variables			Constraint
	Basic		Nonbasic	
$f_i' = 0$	x_i^0 Artificial		Unrestricted $u_i^{\pm}$	
$f_i' \leq 0$	x_i Nonnegative		Nonnegative u_i	
$f_i' = x_i^{\pm}$	$x_i^{\pm}$ Unrestricted	*	Artificial u_i^0	
	z Objective		Constants 1	
	Nonbasic		Basic	
	1 Constants		Objective $\overline{w}$	
	x_j^0 Artificial	*	Unrestricted $u_j^{\pm}$	$f_j = u_j^{\pm}$
	x_j Nonnegative		Nonnegative u_j	$f_j \geq 0$
	$x_j^{\pm}$ Unrestricted		Artificial u_j^0	$f_j = 0$

f_i' and f_j are nonhomogeneous linear functions of nonbasic variables only.

*The cases in these lines would usually be omitted in the initial tableau, and there is an option to drop them in other tableaus as well. Here the f_i' and f_j are not constrained, but merely define $x_i^{\pm}$ and $u_j^{\pm}$ respectively.

solution $\mathbf{u}^* = (0, 22, 0, 4, 0, 8, -1, 0, 0)$ is available for use in 3.5 and 3.6.

	x_1	x_2	x_3	x_4	x_5	1	1
$u_6^{\pm}$	1	4	−3	−1	0	−2	$= -x_6^0$
$u_7^{\pm}$	−1	−1	4	2	0	−3	$= -x_7^0$
$u_8^{\pm}$	0	0	1	0	−1	−4	$= -x_8^0$
u_9	1	0	0	3	0	−20	$= -x_9$
1	−9	−11	28	14	0	1	$= z(\min)$
	‖	‖	‖	‖	‖	‖	
1	u_1	u_2	u_3	u_4	u_5	$\overline{w}$ (max)	

(3)

	x_8^0	x_6^0	x_7^0	x_4	x_2	1	1
u_5	−1	1	1	1	3	−1	$= -x_5$
u_3	0	1	1	1	3	−5	$= -x_3$
u_1	0	4	3	2	13	−17	$= -x_1$
u_9	0	−4	−3	1	−13	−3	$= -x_9$
1	0	8	−1	4	22	−12	$= z(\min)$
	‖	‖	‖	‖	‖	‖	
1	$u_8^{\pm}$	$u_6^{\pm}$	$u_7^{\pm}$	u_4	u_2	$\overline{w}$ (max)	

(4)

When the dual variables are not all sign restricted, an all-negative column will not necessarily bar the existence of an optimal solution, nor can columns be eliminated by domination as in 4.3. On the other hand, a row equation that is either all nonpositive or all nonnegative and contains an artificial basic variable and a nonzero constant term obviously cannot be satisfied by nonnegative variables.

If there is linear dependence among the row equations before artificial variables are added, there is a corresponding nonuniqueness in the dual solution. That is, if $x_1^0,\ldots,x_\mu^0$ are artificial,

$$\sum_{i=1}^{\mu} \omega_i a_{ij} = \sum_{i=1}^{\mu} \omega_i b_i = 0 \qquad \text{for } j=1,\ldots,n \tag{5}$$

and $u_1^{\pm},\ldots,u_\mu^{\pm}$ are optimal values for the corresponding dual variables, then it is clear that $u_1^{\pm}+t\omega_1,\ldots,u_\mu^{\pm}+t\omega_\mu$ are also optimal values for any t, the other components of u remaining unchanged.

If (5) holds with not all $\omega_i=0$, then the relations (5) are mirrored in the homogeneous linear relations that will be found to exist among the x_i^0 alone, if one attempts to eliminate all the artificial variables from the basis. (See Theorem 7 of 2.7.) To reach this goal in a literal sense, it would be necessary to drop the rows of the tableau containing these homogeneous linear relations, but for practical purposes this is already accomplished by the rule prohibiting any pivot in the column of an artificial variable. If this rule is observed, artificial variables do not have to be distinguished in any other way in Phase 2 of 6.1.

1	x_5	x_6	x_7	x_8^0	1
x_1^0	2	−3	1*	1	0
x_2^0	−1	2	−1	−2	0
x_3^0	−1	1	0	−1	0
x_4	1	1	1	1	−4
$-z$	−1	−2	−3	−4	0
$-z^0$	0	0	0	2	0

3	x_4	x_3^0	x_1^0	x_8^0	1
x_7	1	5	2	−2	−4
x_2^0	0	3	3	−6	0
x_6	1	−1	−1	1	−4
x_5	1	−4	−1	4	−4
$-z$	6	9	3	−12	−24

(6)

In (6) the first tableau satisfies the requirements at the end of Phase 1. The variable x_8 is relabeled as x_8^0 because it has $g_8=2>0$; row $-z^0$ is dropped. After three Phase 2 pivots on $ij=17,36,45$, the optimal tableau is obtained. Row x_2^0 could have been dropped from the next to the last tableau.

It is not difficult to extend the algorithm of 6.1 to the case in which unrestricted variables are permitted in the primal problem as well as the dual. The result is an obligation *to make and to keep the unrestricted* $\mathrm{x}_{\mathrm{h}}^{\pm}$ *basic* as follows:

(a) Any nonbasic $x_l^{\pm}$ can be used to decrease z whenever $c_l\neq 0$, by giving $x_l^{\pm}$ the sign opposed to that of c_l; and if $c_l=0$, $x_l^{\pm}$ can be made basic without changing the other c_j. Thus $x_l^{\pm}$ probably belongs to some optimal basis or even to all of them. Consequently it should be made basic at the first opportunity. If this can be done by means of an exchange with a basic x_k^0, so much the better, but the preservation of quasifeasibility may prevent this; the method of 6.4 allows more freedom. (A vanishing pivot is the obstacle in (5) of 4.4). If $c_l > 0$, the usual rule for determining the pivotal row (or concluding that z is unbounded) will apply if $-x_l^{\pm}$ is used as the variable, with coefficients $-a_{il}$ and $-c_l$.

(b) Every basic $x_i^{\pm}$ must remain basic; no pivot is ever chosen in these rows. Indeed they could be dropped from the tableau, if desired, like the columns of x_j^0, since they merely define the values of the $x_i^{\pm}$ but impose no other restriction. (Recall that the pivotal row selection depends on the intent to keep the *restricted* basic variables nonnegative.)

EXERCISES 6.2

Find optimal solutions for the following, with all $x_h \geq 0$ and all x_i^0 artificial. They present some special situations.

1.

1	x_1	x_2	x_3	x_4	1
x_5^0	2	-1	1	3	-2
x_6^0	3	-1	-2	1	-5
$-z$	3	-1	-2	2	0

2.

1	x_1	x_2	x_3	1
x_4^0	-2	1	1	-1
x_5^0	1	1	-5	-2
$-z$	1	-2	1	0

3.

1	x_3	x_4	x_5	x_6	1
x_1^0	2	1	1	2	-3
x_2^0	4	-1	2	3	-6
$-z$	10	1	4	7	0

4.

1	x_3	x_4	x_5	1
x_1^0	3	-4	1	-2
x_2^0	-2	2	-1	-1
$-z$	1	1	1	0

5. In Figure 6.1, why is the existence of $a_{il} > 0$ questioned in Phase 2 but not in Phase 1?

6. If $\mathbf{x} \geq 0$, all $c_j \geq 0$ in the initial tableau, and there is a feasible primal solution, show that there are optimal primal and dual solutions.

7. What condition in the final tableau will guarantee that the optimal primal solution is unique? If not, how can other solutions be obtained? Apply to Exercise 1.

8. Use Theorem 4 of 3.3 and (2) of this section to prove *Farkas's lemma*: Every unrestricted $\mathbf{u}^{\pm}$ satisfying $\mathbf{u}^{\pm}\mathbf{A} \geq \mathbf{0}$ also satisfies $\mathbf{u}^{\pm}\bar{\mathbf{b}} \geq \mathbf{0}$, (if and) only if the latter inequality is a nonnegative linear combination of the former; i.e., only if there is a $\mathbf{y} \geq \mathbf{0}$ such that $\mathbf{A}\mathbf{y} = \bar{\mathbf{b}}$.

6.3 COLUMN SELECTION BY RATIO CRITERIA

The simple criterion $\min c_j$ for selecting the pivotal column in Phase 2 can be criticized because it is dependent on the scaling of the columns. That is, one can arrange to have *any* column for which $c_j < 0$ yield the $\min c_j$ by multiplying all the entries in the column by a sufficiently large $\sigma_j > 0$. Since this operation merely divides x_j by σ_j and multiplies u_j by σ_j, it does not change the optimal basis and so does nothing to make column j an efficient selection. Another criterion, $\max|\Delta z|$, completely avoids any dependence on scaling but is time consuming to apply to large problems.

Similar remarks apply to the criteria min g_j and max$|\Delta z^0|$ in Phase 1. However, in this case two independent and especially significant data are available for each column, namely c_j and g_j. If their ratio was used as a selection criterion, it would be independent of column scaling. To insure the termination of Phase 1, one wants to choose $g_j<0$, and since the relevant g_j are thus of constant sign and never zero, they are the logical numbers to put in the denominator. The minimization of z^0 now being assured, it is fitting to help with the minimization of z, which calls for algebraically small c_j. Thus we arrive at the *ratio-pricing* criterion for the selection of the pivotal column in Phase 1:

$$\min\{c_j/(-g_j): g_j<0\} \tag{1}$$

which Dantzig attributes to H. Markowitz. If $\omega_i = \operatorname{sgn} b_i$ when x_i is artificial or infeasible and $\omega_i=0$ otherwise, one has

$$g_j = \sum_i \omega_i a_{ij} \tag{2}$$

Evidently (1) can be described as maximizing the ratio of the rate of decrease $-c_j$ (which may be >0 or ≤ 0) of z to the rate of decrease $(-g_j>0)$ of z^0. Note that if all $c_j \geq 0$ and only one b_i is infeasible, (1) essentially reduces to the dual-simplex

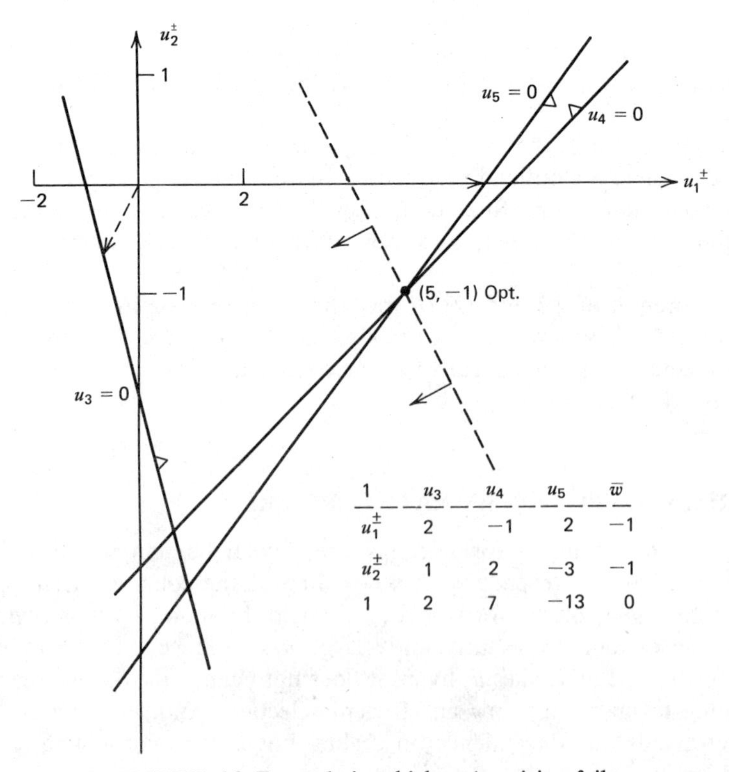

1	u_3	u_4	u_5	$\overline{w}$
$u_1^{\pm}$	2	−1	2	−1
$u_2^{\pm}$	1	2	−3	−1
1	2	7	−13	0

FIGURE 6.3 Example in which ratio-pricing fails.

criterion (5) of 3.3. However, the pivotal rows are different, and for the present we are assuming that all the infeasible x_i are artificial.

In the *dual* problem (1) relates to the line $u_i = \omega_i t$ ($\omega_i = 1, -1$, or 0 as in (2)), where i is in M so that u_i is nonbasic. As t increases from $-\infty$, each term in $\bar{w} = \mathbf{bu}_M$ increases (or remains at 0), so that the maximization of $\bar{w}$ is promoted. Along this line the column constraints take the form $c_j + tg_j \geq 0$, and (1) gives the largest value of t for which all constraints having $g_j < 0$ are satisfied. Those having $g_j \geq 0$ may or may not be satisfied; there is no guarantee that the line $u_i = \omega_i t$ will actually intersect the dual feasible region. This is illustrated in Figure 6.3, which relates to Exercise 4. Equation (1) takes u_3 out of the basis; this happens to be the wrong choice.

Beside its independence of column scaling, (1) has another invariance property. Note that when there are equality constraints without slack variables, there is a degree of arbitrariness in the c_j; any linear combination of the equalities could be added to z without changing the primal optimal solution. In particular, since z^0 is a sum of artificial variables only, any multiple tz^0 can be added to z, thus replacing c_j by $c_j + tg_j$. (The new problem is to minimize $z + tz^0$ subject to the same constraints as before.) In (1), the effect is merely to subtract t from every ratio, and hence the value or values of j that yield the minimum ratio are unchanged.

A consequence of this is that if (1) is used and the tableau is initially dual feasible (all $c_j \geq 0$), it will again be so at the end of Phase 1, though not necessarily during the intervening steps; thus no Phase 2 is required. To prove this, let $t = c_l/(-g_l) \geq 0$ be the minimum in (1); thus $c_j + tg_j \geq 0$ for all j but $c_l + tg_l = 0$. Then by using $z + tz^0$ in place of z and l as the pivotal column, dual feasibility will be preserved because $c_l + tg_l = 0$, and z^0 will be reduced because $g_l < 0$. A repetition of this procedure constitutes the *primal-dual method* given by Dantzig, Ford and Fulkerson (1956). The present discussion shows that if initially all $c_j \geq 0$, (1) will yield exactly the same pivot steps that the primal-dual method does, without requiring the calculation of any linear combinations $c_j + tg_j$. At the end of Phase 1, the addition or non-addition of a multiple of z^0 to z can no longer have any effect on the dual feasibility, which therefore must be restored.

Ratio pricing is often helpful in small problems. In eight large problems solved by Wolfe and Cutler (1963), in runs 6 and 21, p. 198, the number of iterations required by (1) varied from 55% to 162% of that required by min g_j, with a modest (geometric or logarithmic) average advantage of 6% for (1).

In solving small examples we let the following rules take precedence over (1) so long as the same column is not chosen repeatedly in different iterations:

(a) Suppose that for some indices k, l either

$$b_k > 0 \qquad a_{kl} < 0 \qquad \text{and} \qquad a_{kj} \geq 0 \text{ for all } j \neq l \tag{3}$$

or else x_k^0 is artificial and

$$b_k < 0 \qquad a_{kl} > 0 \qquad \text{and} \qquad a_{kj} \leq 0 \text{ for all } j \neq l \tag{4}$$

Then the equation in row k cannot be satisfied feasibly unless $x_l > 0$. Thus it is surely efficient to pivot in column l, as in Table 6.4A. The kth row is not necessarily pivotal nor is x_k necessarily absent from the optimal basis.

(b) Suppose that (a) does not apply, there are several $g_j<0$ but only one $c_l<0$, and either $g_l\leq 0$ or else $g_l>0$ and

$$-c_l/g_l>c_j/(-g_j) \qquad (g_j<0) \tag{5}$$

holds for several j. Then we may seek a positive pivot in column l. This differs from (1) only in the event that $g_l\geq 0$, so that the class of columns having $g_j<0$ and $c_j<0$ (which is preferred by (1)) is empty. Equation (5) is the condition that a column in this class be obtainable as a positive combination of columns l and j (Exercise 8), so that both z^0 and z can be decreased. Then the uniqueness of $c_l<0$ contrasted with the nonuniqueness of the $g_j<0$ satisfying (5) completes the evidence (not conclusive) in favor of l as the pivotal column. Exercise 4 is an example.

Column Selection in Phase 2

By pivoting in some rows and omitting others, we can ensure that all the basic x_i and nonbasic u_i are nonnegative rather than artificial or unrestricted. Consider a nonbasic x_j and basic u_j that also are nonnegative. The dual equation in column j is

$$\Sigma_i a_{ij}u_i+c_j=u_j\geq 0 \tag{6}$$

In seeking a pivotal column we shall require $c_j<0$ as usually to ensure convergence. If x_j is in an optimal basis one must have $u_j^*=0$. Thus when $c_j<0$, the problem according to (6) is to select j so that $\Sigma_i a_{ij}u_i/(-c_j)$ equals 1 rather than exceeds it, and hence to minimize this ratio. Since the optimal values of the u_i are only known to be nonnegative, we brashly replace them by 1 or $1/(-b_i)$ to obtain the criterion

$$\min_j\{g_j'/(-c_j): c_j<0;\ x_j \text{ not artificial}\} \tag{7}$$

Here g_j' might be defined as $\Sigma_i a_{ij}$ or else as

$$g_j'=\Sigma_i a_{ij}/\max(-b_i,\varepsilon) \tag{8}$$

The denominator is the larger of $-b_i$ and some small $\varepsilon>0$.

So long as all $-b_i\geq\varepsilon>0$, (7) and (8) are independent of both row and column scaling. By (3) of 3.3, the algebraically largest term in (8) is the reciprocal of the value $-b_i/a_{ij}$ of the new basic variable x_j (if $a_{ij}>0$). Thus if (8) is heuristically identified with its largest term, (7) and (8) tend to maximize the decrease $|\Delta z|=b_ic_j/a_{ij}$ in z as a result of pivoting in column j.

Division by $\max(-b_i,\varepsilon)$ could be used in calculating g_j provided these weights were left unchanged throughout Phase 1; updated values of b_i can be used in (8). Then (7) and (8) permit the example of Table 3.4, which required four iterations, to be solved in two iterations in Table 6.3. (The primal notation is used now.)

Kuhn and Quandt (1963) and Wolfe and Cutler (1963) obtained their fastest computer solutions by methods comparable with (7), obtaining g_j' or its square by

TABLE 6.3

Check	1	x_3	x_4	x_5	x_6	x_7	1
46	x_1	21	5	20	−1	1	−1
39	x_2	18	15	5	2*	−1	−1
−106	$-z$	−42	−27	−36	−2	0	0
$g_j'/(-c_j)$		$\frac{39}{42}$	$\frac{20}{27}$	$\frac{25}{36}$	$\frac{1}{2}$	...	
	2	x_3	x_4	x_5	x_2	x_7	1
131	x_1	60	25	45	1	1*	−3
39	x_6	18	15	5	1	−1	−1
−134	$-z$	−48	−24	−62	2	−2	−2
$3g_j'/(-c_j)$		$\frac{114}{48}$	$\frac{70}{24}$	$\frac{60}{62}$	...	$-\frac{2}{2}$	
	1	x_3	x_4	x_5	x_2	x_1	1
131	x_7	60	25	45	1	2	−3
85	x_6	39	20	25	1	1	−2
64	$-z$	36	13	14	2	2	−4

summing $\max(0, a_{ij})$ or $|a_{ij}|$ or a_{ij}^2 over i. Surprisingly, the last of these, which is

$$g_j' = \left[\Sigma_i a_{ij}^2\right]^{1/2} \tag{9}$$

resembles (2) and (8) in being adaptable (with a little more trouble) to the revised simplex method of 7.3, as shown by Goldfarb and Reid (1977). Previously Harris (1973) and Crowder and Hattingh (1975) had given approximations to (9) using only a limited number of rows of the current tableau **A**. In any case the number of iterations is reduced by as much as 50 percent. Also see Künzi et al. (1968) (duoplex algorithm).

To obtain a feeling for the effects of (8) and (9), consider four different situations involving choices between two different columns thus:

	$\mathbf{a}_{.1}$	$\mathbf{a}_{.2}$	$\mathbf{a}_{.1}$	$\mathbf{a}_{.2}$	$\mathbf{a}_{.1}$	$\mathbf{a}_{.2}$	$\mathbf{a}_{.1}$	$\mathbf{a}_{.2}$
$a_{3j}=$	1	2	1	1	1	2	2	4
$a_{4j}=$	2	4	−2	−2	−4	−2	−4	−3
$c_j=$	−1	−1	−2	−1	−1	−1	−1	−1
	*		*		*		*	
	(8)			(8)	(8)		(8)	
	(9)		(9)			(9)	(9)	

(10)

The b_i will be disregarded or assumed equal to -1.

In each pair of columns in (10) we have $a_{.1} \leq a_{.2}$, with strict inequality either for the c_j or for both a_{3j} and a_{4j}. Thus the column equations imply $u_2 > u_1 \geq 0$, and only the first column can belong to the optimal basis, as indicated by the asterisks in (10).

(This is the domination idea of 4.3.) As indicated, (8) and (9) each give the correct indication in three of the four cases. The same is true of $g'_j = \Sigma_i |a_{ij}|$, whereas

$$g'_j = \Sigma_i \max(0, a_{ij}) / \max(-b_i, \varepsilon) \tag{11}$$

with or without the denominator, is correct in every case. Note that changing from $a_{ij} < 0$ to $a_{ij} = 0$ should not make column j more likely to be pivotal, because it favors $x_j = 0$ and $u_i > 0$. This agrees with (8) and (11) but not with (9). However, it seems impractical to use (11) with the revised simplex method of 7.3.

Harris (1973) accepts small (e.g., $\leq .0005$) departures from primal feasibility because they may result from round-off errors; this allows more freedom in avoiding small pivots. She points out that the min c_j criterion favors columns that can supply very large pivots, but these must be offset by small pivots later; this leads to more roundoff error as well as more iterations.

EXERCISES 6.3

Use the methods of this section to find optimal solutions with all $x_h \geq 0$ and x_i^0 artificial.

1.

1	x_3	x_4	x_5	1
x_1	3	−2	1	−1
x_2	−4	5	1	−8
$-z$	−3	−5	−7	0

2.

1	x_3	x_4	x_5	1
x_1	3	1	2	−1
x_2	1	2	3	−3
$-z$	−1	−1	−1	0

3, 4. Work the corresponding problems of 6.1, using rules (a) and (b) of 6.3.

5.

1	x_3	x_4	x_5	x_6	1
x_1	1	1	−2	3	−1
x_2	0	3	5	−1	−2
$-z$	−1	−2	−1	−2	0

6.

1	x_1	x_2	x_3	1
x_4^0	2	−3	3	−3
x_5^0	3	2	−1	−3
$-z$	1	−7	8	0

7.

1	x_1	x_2	x_3	1
x_4^0	3	1	−3	−3
x_5^0	−1	2	2	−1
$-z$	1	−1	−2	0

8. Prove the assertion about (5).

6.4 FREER ROW SELECTION IN PHASES 1a AND 1b

The method of this section offers the saving of a few iterations and a reduction in the size of the problem by avoiding the introduction of artificial variables that are merely the absolute values of other variables already present in the problem. The method is related closely to the *composite methods* of Wolfe (1961, 1965), and more distantly to the method of Balinski and Gomory (1963), also in Balinski and Tucker (1969).

TABLE 6.4A

Check	1	x_1	x_2	x_3	x_4	1
0	x_6^0	1	4	−3	−1	−2
2	x_7^0	−1	−1	4*	2	−3
4	x_5	0	0	−1	0	4
−15	x_9	1	0	0	3	−20
24	$-z$	−9	−11	28	14	1
4	$-z^I$	0	−3	−2	−1	9
	4	x_1	x_2	x_7^0	x_4	1
6	x_6^0	1*	13	3	2	−17
2	x_3	−1	−1	1	2	−3
18	x_5	−1	−1	1	2	13
−60	x_9	4	0	0	12	−80
40	$-z$	−8	−16	−28	0	88
20	$-z^I$	−2	−14	2	0	30
	1	x_6^0	x_2	x_7^0	x_4	1
6	x_1	4	13	3	2	−17
2	x_3	1	3	1	1	−5
6	x_5	1	3	1	1	−1
−21	x_9	−4	−13	−3	1	−3
22	$-z$	8	22	−1	4	−12
(8	$-z^I$	2	3	2	1	−1)
	$-z^I$	1	0	1	0	0

The example of 6.1 can be solved in only two iterations rather than four, as shown in Table 6.4A. Of the four variables x_1, x_3, x_5, x_9 in the final basis, x_9 is initially basic and x_5 can be taken as initially basic; thus only two iterations, to bring x_1 and x_3 into the basis, are really required for optimization. One of the extra iterations occurred as a result of insisting on quasifeasibility; in effect, in 6.1 x_5 was forced to change from -4 to 1 via an intermediate stop at 0 rather than directly.

In the absence of quasifeasibility the infeasibility form is not derived solely from artificial variables and so is denoted by z^I instead of z^0:

$$z^I = \Sigma_i \{(-x_i)\,\mathrm{sgn}\, b_i \colon x_i \text{ is artificial or } b_i > 0\} \tag{1}$$

It has to be reduced to zero as before. In Table 6.4A we have $z^I = x_6^0 + x_7^0 - x_5$ initially. As regards column selection, one iteration is saved by applying the rule (a) of 6.3 to the initial tableau. Since $a_{53} = -1$ is the only negative entry in row 5 available to offset the infeasible $b_5 = 4$, one must have $x_3^* > 0$ and hence we pivot in column 3. In the second tableau (1) of 6.3 determines that column 1 is pivotal. The method of row selection will be considered shortly.

The final tableau illustrates one consequence of the abandonment of quasifeasibility. When x_5 changes from an infeasible $-13/4$ to a feasible $+1$, it must be dropped from the infeasibility form z^I; i.e., the definition of z^I is changed from $|x_5|+x_6^0+x_7^0=-x_5+x_6^0+x_7^0$ to $x_6^0+x_7^0$. Thus the next to the last row of Table 6.4A, obtained by the pivot rules and enclosed in parentheses, has to be corrected by subtracting the x_5 row, which gives the last row of the table. Alternatively, z^I could be calculated afresh from its definition. Basic artificial x_i^0 having zero values should be excluded from z^I because in any case they will be immediately removed from the basis by (2) below.

It remains to describe the method used in selecting pivotal rows. Obviously the number of iterations would be minimized if and only if at each iteration the variable entering the basis belonged to the optimal basis and the variable leaving the basis did not belong to the optimal basis. Since an artificial variable can seldom belong to the optimal basis, and an unrestricted variable nearly always belongs, it seems advisable to give these variables their ultimate status as quickly as possible. This part of Phase 1 will be called *Phase 1a*. The above example was special in that Phase 1a alone was enough to produce the optimal solution. Good selections of the variable to enter the basis are as desirable in Phase 1a as at any other time, and the methods of the previous sections are used first to select the pivotal column l.

Phase 1a Each pivotal row is selected with the following objects in view, in order of decreasing priority:

(a_1) Seek to remove an artificial variable from the basis.

(a_2) Make the new basic $x_l \geq 0$ if possible. (Both of these rules promote primal feasibility.)

(a_3) Make $|x_l|$ as small as possible, so as to minimize round-off errors and the amount of redefinition of z^I required by sign changes in basic variables.

Whenever possible we thus use

$$\min_i \{-b_i/a_{il} \geq 0\colon a_{il} \neq 0,\ x_i \text{ is artificial}\} \tag{2}$$

for row selection in Phase 1a. However, if z^I, as we assume for the sake of good column selection, includes the absolute value of any and all negative basic variables even if they are not artificial, it is possible that the set of rows i defined in (2) may be empty, and there is no certainty that any other column having $g_j<0$ would serve better.

In such a case it is suggested that the scope of (2) be enlarged thus.

$$\min_i \{|b_i/a_{il}|\colon a_{il} \neq 0,\ x_i \text{ is artificial}\} \tag{3}$$

The fact that (3) can make $x_l<0$ and z^I larger than before seems to be of secondary importance in view of the probability that the pivotal row and column are nevertheless well-selected. If (3) also fails because all $a_{il}=0$ for which x_i is artificial, the restriction to artificial x_i must be relaxed temporarily, barring a possibly vain search for another column. Then one uses the following procedure, which could be used for all of Phase 1, but is here recommended when x_l cannot remove an artificial variable.

Phase 1b Each pivotal row is selected according to the following principles:

(b_1) No feasible or artificial basic variable is permitted to change sign. Also the new basic $x_l \geq 0$.

(b_2) Subject to (b_1), the maximum number of basic variables are made feasible.

(b_3) At most two possible values of x_l now remain, corresponding to pivots of opposite signs. If there are two, choose a pivot that removes an artificial variable or whose sign is opposed to that of c_l. The latter chooses the smaller z and causes c_l to be replaced by a dual-feasible value.

If Phase 1a is not being used, so that some artificial x_i may be basic with $a_{il} \neq 0$, we assume as in 6.1 that they have $b_i \leq 0$; rule (b_1) preserves this condition. Then the above principles determine the following procedure for Phase 1b. Let

$$x_l = \xi_l \triangleq \min_i \{-b_i/a_{il} \geq 0 : a_{il} > 0\} \tag{4}$$

except as follows:

$$x_l = \max_i \{b_i/(-a_{il}) : a_{il} < 0\} \geq 0 \tag{5}$$

if ξ_l is undefined, and

$$x_l = \max_i \{b_i/(-a_{il}) \leq \xi_l : a_{il} < 0\} \geq 0 \tag{6}$$

if (6) is defined and nonnegative, $c_l \geq 0$, and (4) does not remove an artificial variable from the basis. Also in the event of a tie one prefers to remove an artificial variable from the basis. It is no longer true as in 6.3 that dual feasibility at the start implies dual feasibility at the end of Phase 1, but this seems to reflect simply a shortening of Phase 1. Computer codes will include a suitable restriction to prevent the acceptance of excessively large values in (5). The condition $g_l < 0$ implies that a pivot will be found in column l.

In solving exercises, students may find it helpful to study the graphical illustrations of formulas (2) to (6) in Figure 6.4, and to emphasize the learning of the plausible rules (a_1),...,(b_3), which are easier to remember and suffice to determine the computations. Values of $x_l = -b_i/a_{il}$ at which some x_i changes sign are designated according to the three possibilities for the prepivotal status of x_i, thus:

○ Artificial (treat like ● if present in rules (b_1), (b_2)).
● Feasible ($b_i \leq 0$) when $x_l = 0$; becoming infeasible.
I Infeasible ($b_i > 0$) when $x_l = 0$; becoming feasible.
↑ Origin where $x_l = 0$.
* Value $-b_i/a_{il}$ of the new basic variable x_l determined by (a_1) to (a_3), or one or two values determined by (b_1), (b_2).

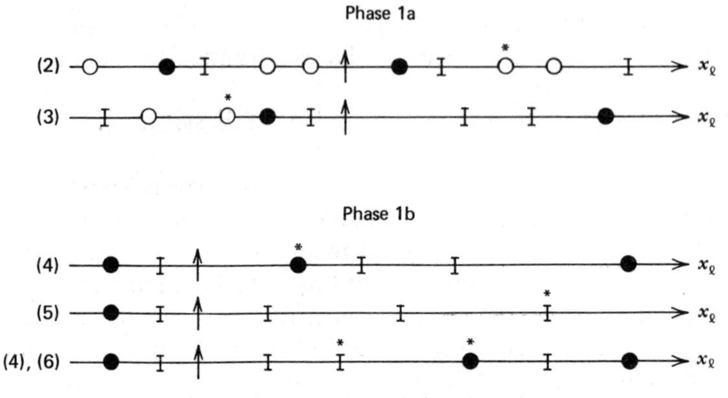

FIGURE 6.4 Selection of the pivotal row.

The *convergence proof* for Phases 1a and 1b depends on a lexicographic ordering among the vectors (α, β, z^I) where α=the number of artificial basic variables, β=the number of infeasible natural basic variables, and z^I is the value of the infeasibility form as usual. Then any iteration yields another vector $(\alpha', \beta', z^{I'})$ that is smaller than its predecessor in the sense that either $\alpha' < \alpha$, or $\alpha' = \alpha$ and $\beta' < \beta$, or $\alpha' = \alpha$ and $\beta' = \beta$ and $z^{I'} < z^I$. If all three components are unchanged, the argument continues as in 6.6.

Table 6.4B gives an example for Phase 1b.

TABLE 6.4B

1	y_1	y_2	y_3	y_4	1
x_1	−2	1	−1	0	5
x_2	0	−2	1	3*	−4
x_3	1	−2	1	−1	1
x_4	1	1	−6	1	−7
$-z$	1	5	−1	−5	0
$-z^I$	−1	−1	0	−1	6

3	y_1	y_2	y_3	x_2	1
x_1	−6	3	−3	0	15
y_4	0	−2	1	1	−4
x_3	3*	−8	4	1	−1
x_4	3	5	−19	−1	−17
$-z$	3	5	2	5	−20
$-z^I$	−6	3	−3	0	15

3	x_3	y_2	y_3	x_2	1
x_1	6	−13*	5	2	13
y_4	0	−2	1	1	−4
y_1	3	−8	4	1	−1
x_4	−3	13	−23	−2	−16
$-z$	−3	13	−2	4	−19
$-z^I$	6	−13	5	2	13

+13	x_3	x_1	y_3	x_2	1
y_2	−6	−3	−5	−2	−13
y_4	−4	−2	+1	+3	−26
y_1	−3	−8	+4	−1	−39
x_4	+13	+13	−78	0	−13
$-z$	+13	+13	+13	+26	−26

Thus $z^* = -2$, $\mathbf{x}^* = (0,0,0,1)$, $\mathbf{y}^* = (3,1,0,2)$.

If one supposes all $b_i \neq 0$ and enumerates the signs of the $-a_{il} \neq 0$ (or equivalently, the b_i) for which $-b_i/a_{il} > 0$, in the order of increasing values of $-b_i/a_{il}$, and assumes that these signs are independent and equally likely to be positive or negative, one obtains the following probabilities that this algorithm differs from the conventional procedure of 6.1 in a single preassigned pivot step:

Signs of b_i	$+-$	$++-$	$+++-$	$\ldots$		
$c_l<0$	$\frac{1}{4}$	$\frac{1}{8}$	$\frac{1}{16}$	$\ldots$	Total $=\frac{1}{2}$	
$c_l \geq 0$	0	$\frac{1}{8}$	$\frac{1}{16}$	$\ldots$	Total $=\frac{1}{4}$	(7)

Thus about 50 percent of the pivots are unconventional when $c_l<0$ and about 25 percent when $c_l \geq 0$. If Phase 1a was not used, the presence of artificial variables would make negative b_i more probable and so reduce these proportions. If ratio pricing is used, the case $c_l<0$ will tend to be the more probable while negative c_j remain.

Wolfe (1961) selects the pivotal row to minimize z^I; obviously this could serve as either an alternative or an auxiliary to (4) to (6) above. It involves a calculation to determine when g_l becomes positive as x_l increases: when $x_i=-b_i-a_{il}x_l$ changes sign (in either direction), add $|a_{il}|$ to the current value of g_l if x_i is sign-restricted, and add $2|a_{il}|$ if x_i is artificial. The experiments of Wolfe and Cutler (1963) ranked this procedure about equal to that of 6.1, both being superior to the other methods they tested.

EXERCISES 6.4

Use the methods of this section to find optimal solutions with all $x_h \geq 0$ and x_i^0 artificial.

1.

Chk	1	x_5	x_6	1
-1	x_1	2	1	-5
1	x_2	2	-1	-1
1	x_3	-1	-2	3
3	x_4	1	-1	2
3	$-z$	1	1	0

2.

Chk	1	x_5	x_6	1
-1	x_1^0	2	-1	-3
0	x_2	1	1	-3
0	x_3	-1	-2	2
1	x_4	-1	-3	4
4	$-z$	1	2	0

3, 4. Reverse the signs in row $-z$ above and solve.

5, 6. Apply the present methods to 5 and 6 of 6.1.

7, 8. Form the negative transposes of 1 and 3 and solve by 6.3 and/or 6.4.

9. Explain how the pivot kl in 3 is indicated merely by the signs of the tableau elements, the variables being nonnegative, and zero when nonbasic.

6.5 BOUNDED VARIABLES

A type of constraint that frequently occurs in practice is one that simply imposes an upper bound e_h on a variable x_h. (We assume $x_h \geq 0$ also, because if not, we could replace x_h by $e_h - y_h$, where y_h would be an ordinary nonnegative variable.) As a majority of the constraints could be of this type, it is fortunate that there is a technique due to Dantzig (1955a) whereby they need not appear explicitly in the tableau. In Table 6.5A the left-hand side presents the ordinary Phase 2 solution of a primal-feasible program using the four constraints required, while the right-hand side presents the corresponding compact solution involving only two explicit constraints. The required background will be developed before the example is discussed.

TABLE 6.5A Simplex Phase 2 with Bounded Variables

$$\begin{array}{lllll}
\text{Min } z = -2x_3 - x_4 \text{ s.t. all} & & & & x_h \geq 0,\ y_h \geq 0 \\
x_1 & & +x_3 & & +x_4 = 4 \\
 & x_2 & -x_3 & & -x_4 = 1 \\
 & x_2 + y_2 & & & = 4 \\
 & & x_3 + y_3 & & = 3
\end{array}$$

Original form

Check	1	x_2	x_3	x_4	1
-1	x_1	0	1	1	-4
-2	0	1*	-1	-1	-1
-2	y_2	1	0	0	-4
-1	y_3	0	1	0	-3
-2	$-z$	0	-2	-1	0

Check	1	x_3	x_4	1
-1	x_1	1	1	-4
-2	x_2	-1	-1	-1
0	y_2	1	1	-3
-1	y_3	1*	0	-3
-2	$-z$	-2	-1	0

Check	1	y_3	x_4	1
0	x_1	-1	1	-1
-3	x_2	1	-1	-4
1	y_2	-1	1*	0
-1	x_3	1	0	-3
-4	$-z$	2	-1	-6

Check	1	y_3	y_2	1
-1	x_1	0	-1	-1
-2	x_2	0	1	-4
1	x_4	-1	1	0
-1	x_3	1	0	-3
-3	$-z$	1	1	-6

Compact form

Check	1	$x_3^{3\#}$	x_4	1
-1	x_1	1	1	-4
-2	x_2^4	-1	-1	-1
-2	$-z$	-2	-1	0

Reflect x_3^3, which is blocked by its own bound $e_3 = 3$.

Check	1	y_3^3	x_4	1
0	x_1	-1	1	-1
-3	x_2^4	1	$-1^{\#}$	-4
-4	$-z$	2	-1	-6

Reflect x_2^4, which blocks x_4 by reaching its bound $e_2 = 4$.

Check	1	y_3^3	x_4	1
0	x_1	-1	1	-1
1	y_2^4	-1	1*	0
-4	$-z$	2	-1	-6

Check	1	y_3^3	y_2^4	1
-1	x_1	0	-1	-1
1	x_4	-1	1	0
-3	$-z$	1	1	-6

$\mathbf{x}^* = (1, 4, 3, 0)$, $z^* = -6$,
$\mathbf{u}^* = (0, -1, -1, 0)$

Bounded variables will be identified in general by a superscript e, which in numerical examples will be replaced by the specific value of the bound (a small positive integer). Of course this is consistent with our previous use of a superscript 0 for artificial variables. Computer programs sometimes use x_h^e/e_h as variables with bounds equal to 1, but Tomlin (1975) condemns this practice, referring to the experience of G. Hentges.

Each bounded variable x_h^e has a partner y_h^e such that $x_h^e+y_h^e=e_h>0$ and $x_h^e, y_h^e \geq 0$. Evidently at least one variable of each such pair must be basic, and the corresponding constraint (or one of them, for each pair) is omitted in the compact tableaus. Thus the latter each contain exactly one variable from each pair. However, any such x_h^e that does appear can be replaced by $e_h-y_h^e$ (or y_h^e replaced by $e_h-x_h^e$) whenever this is necessary to keep the nonbasic variables at a value of zero and to maintain the familiar signs for pivots, negative costs, etc. Such a replacement may be called a *reflection*; the "mirror" is at the midpoint $e_h/2$ of the bounded interval.

Prior to the omission of some of the rows as just described one has $\bar{w}^*=\mathbf{u}_M^*\mathbf{b}+d=\mathbf{c}\mathbf{x}_N^*+d=z^*=d^*$ as usual. After the omission (with no $-e_i$ any longer occurring as a component of $\mathbf{b}$) the relation takes the form

$$z^*=\mathbf{u}_M^*\mathbf{b}+d+\mathbf{u}^*\mathbf{x}^*=\mathbf{c}\mathbf{x}_N^*+d=d^* \tag{1}$$

Whenever y_h^e occurs in the final tableau one calculates $x_h^{e*}=e_h-y_h^{e*}$; if y_h^e is nonbasic, u_h^* is changed in sign so that $u_h^*=-c_h^*$. A term $u_h^*x_h^*$ in $\mathbf{u}^*\mathbf{x}^*$ is zero except when $x_h^*=e_h$; then it is the product of $-e_h$ as an omitted component of b and $-u_h^*$ as the dual variable corresponding to y_h. Other theorems in 3.3 may likewise require revision.

Note that in the original tableaus only the zero member (if any) of each pair x_h^*, y_h^* is associated with $u_h^* \neq 0$; thus in the second tableau on the left in Table 6.5A one has

$$z^*=\mathbf{u}_M^*\mathbf{b}+d=(0,0,1,1)(-4,-1,-3,-3)+0=-6$$

In a compact tableau $\mathbf{x}^*$ must be regarded as consisting of precisely those components x_h or y_h that actually appear in the tableau, and the signs of the u_h^* are chosen accordingly. Thus in the second tableau on the right of Table 6.5A one has

$$\begin{aligned} z^* &= \mathbf{u}_M^*\mathbf{b}+d+\mathbf{u}^*(x_1^*, x_2^*, y_3^*, x_4^*) \\ &= (0,-1)(-1,-4)-6+(0,-1,1,0)(1,4,0,0)=-6 \end{aligned}$$

The nonbasic variable x_l (or y_l) that we are attempting to increase will be called the *driving variable*. The first variable x_k (or y_k) that halts the increase in x_l (by threatening to overstep one of its bounds 0 or e_k) is called the *blocking variable*; as usual it need not be unique. There are *three types of blocking* according as x_k hits 0 or e_k, or x_l^e hits e_l. These cases will now be discussed in detail.

(α) Basic x_k, x_k^e or y_k^e hits 0 with $a_{kl}>0$. This is the type of blocking already familiar to us. The element $a_{kl}>0$ is starred and used as a *pivot*. The value of x_l is $-b_k/a_{kl}$. There is no reflection.

(β) Basic x_k^e (or y_k^e) hits its bound e_k with $a_{kl}<0$. The value of x_l is $(b_k+e_k)/(-a_{kl})=(\beta_k+e_k\delta)/(-\alpha_{kl})$, where $b_k=\beta_k/\delta$ and $a_{kl}=\alpha_{kl}/\delta$. There are two operations to be done:

(β_1) Mark a_{kl} with the sign # and from its row form a new row by *reflecting* x_k^e. Students should verify that this involves replacing a_{kj} by $-a_{kj}$ for each j, replacing b_k by $-(b_k+e_k)$ or β_k by $-(\beta_k+e_k\delta)$, and changing the row label from x_k^e to y_k^e or vice versa. This corresponds to having both x_k^e and y_k^e basic and replacing one of their rows by the other. In solving examples the new row can be written below the tableau without recopying the latter. The check value for the new row should be calculated by summing; the sum of the old and the new check values should be $(2-e_k)\delta$.

(β_2) In the new row, $-a_{kl}>0$ is starred and used as a *pivot*. The transform of the new row replaces the old row k, which is not transformed.

(γ) Nonbasic x_l^e (or y_l^e) hits e_l. Then we say $k=l$; the driving and blocking variables coincide. No pivot step is required but the column label x_l^e is marked with the sign # and x_l^e is *reflected*. This involves adding e_l times column l to the constants column, adding (e_l-2) times column l to the check column, changing the signs of all numbers in column l and changing its label from x_l^e to y_l^e or vice versa. This is equivalent to a pivot step exchanging x_l^e and y_l^e. Because of its simplicity this case is chosen whenever a multiplicity of possible blocking variables x_k gives a choice, unless a pivot step is needed to produce dual feasibility.

Note that a bounded basic x_i^e for which $a_{il}\neq 0$ is always a candidate to be blocking, in case (α) if $a_{il}>0$ and in case (β) if $a_{il}<0$. The reflected variable (if any) is always the blocking variable and it is reflected if and only if it reaches its upper bound. In Table 6.5A the two reflections are of types (γ) and (β) respectively; the first takes the place of the second pivot step on the left, while the second reconstructs one of the rows omitted from the tableau.

If the problem contains any unrestricted $x_h^{\pm}$, they can never be blocking but can be driving whenever $c_h\neq 0$. By replacing $x_j^{\pm}$ by $y_j^{\pm}=-x_j^{\pm}$ and changing the signs of all numbers in column j where necessary, we may assume that *a driving variable* x_1 *in this algorithm always has* $\mathrm{c}_1<0$ in Phase 2. It may be selected by $c_l=\min_j c_j$ or (7) of 6.3.

Thus x_3^3 is the first driving variable in Table 6.5A, with $c_3=-2$. Its increase has the following limitations (one for each of the above cases):

(α) x_1 blocks at $x_3^3=-b_1/a_{13}=4/1=4$.
(β) x_2^4 blocks at $x_3^3=(e_2\delta+\beta_2)/(-\alpha_{23})=(4-1)/1=3$.
(γ) x_3^3 blocks at its upper bound$=3$.

The smaller of these limits are the relevant ones, and the simpler case (γ) is chosen. Nonbasic x_3^3 is self-blocking and is reflected.

In the second tableau x_4 is the only possible driving variable. We have:

(α) x_1 blocks at $x_4=-b_1/a_{14}=1/1=1$.

TABLE 6.5B Simplex Phase 1 with Bounded Variables

Check	1	x_1	x_2^1	x_3	1
-2	x_4^4	$-1^{\#}$	-1	1	-2
-2	x_5^0	2	4	-1	-8
2	$-z$	-1	0	2	0
4	$-z^0$	-2	-4	1	8
0	y_4^4	1^*	1	-1	-2

(x_4^4 is replaced by $4-y_4^4$; then a pivot exchanges x_1 and y_4^4.)

Check	1	y_4^4	$x_2^{1\#}$	x_3	1
0	x_1	1	1	-1	-2
-2	x_5^0	-2	2	1	-4
2	$-z$	1	1	1	-2
4	$-z^0$	2	-2	-1	4

(x_2^1 is replaced by $1-y_2^1$.)

	1	y_4^4	y_2^1	x_3	1
-1	x_1	1	-1	-1	-1
-4	x_5^0	-2	-2	1^*	-2
1	$-z$	1	-1	1	-1
6	$-z^0$	2	2	-1	2

(Pivot exchanges x_3 and x_5^0.)

	1	y_4^4	y_2^1	x_5^0	1
-5	x_1	-1	-3	1	-3
-4	x_3	-2	-2	1	-2
5	$-z$	3	1	-1	1

$\mathbf{x}^*=(3,1,2,4,0)$, $z^*=1$, $\mathbf{u}^*=(0,-1,0,-3,-1)$.

When a reflection is required, this symbol is placed in the column of the driving variable and the row containing the symbol of the blocking variable; the latter is reflected.

(β) x_2^4 blocks at $x_4=(e_2\delta+\beta_2)/(-\alpha_{24})=(4-4)/1=0$.
(γ) x_4 has no upper bound ($e_4=+\infty$).

Case (β) yields the minimum, and so basic x_2^4 is reflected and then its replacement y_2^4 is exchanged with the driving variable x_4. This yields the optimal tableau.

If e_j is small, that is an argument against using x_j^e as a driving variable (if others are available), since it is then more likely to be ejected from the basis again.

If all primal infeasibilities are represented by artificial variables as in 6.1, the minimization of z^0 in Phase 1 is just like that of z in Phase 2; basic x_i^e that exceed their bounds e_i should be reflected as in (β_1) above before z^0 is formed. Table 6.5B illustrates the operation of Phase 1 in the presence of bounded variables; it happens that Phase 2 is not required. The composite method of 6.4 can also easily be used for Phase 1. In the optimal solution x_h^0 is generally nonbasic and the same tends to be true of x_h^e having small bounds e_h.

The Dual Simplex Method

This method is illustrated in Table 6.5C. Bounded variables require the addition of the following remarks as well as the ideas of Theorem 2″ of 3.3:

(i) If the cost coefficient c_j for the bounded nonbasic variable x_j^e is not already nonnegative, it will become so after x_j^e is reflected. Thus if all the nonbasic variables are bounded, the dual-simplex method is always applicable, though it is not necessarily as efficient as the two-phase method.

TABLE 6.5C Dual Simplex with Bounded Variables

The constants column has been doubled (added twice) in forming the check values, so that they are unchanged by reflections involving $e_h = 1$. Table 6.5B explains the symbol #.

Check	1	$x_1^{1\#}$	x_2^1	x_3	1
-2	x_4^1	-1	1	1	-2
9	x_5^0	-2	-3	-1	7
9	$-z$	1	4	3	0

($x_5^0 = -7$ is more infeasible than $x_4^1 = 2 > 1$. Min$\{1/2, 4/3, 3/1\}$ selects x_1^1, which increases to its bound 1. Let $x_1^1 = 1 - y_1^1$.)

Check	1	y_1^1	$x_2^{1\#}$	x_3	1
-2	x_4^1	1	1	1	-3
9	x_5^0	2	-3	-1	5
9	$-z$	-1	4	3	1

(Must continue to reduce $|x_5^0| = 5$. Min$\{4/3, 3/1\}$ selects x_2^1, which increases to its bound 1. Let $x_2^1 = 1 - y_2^1$.)

Check	1	y_1^1	y_2^1	x_3	1
-2	x_4^1	1	-1	1	-2
9	x_5^0	2	3	-1*	2
9	$-z$	-1	-4	3	5

(Pivot exchanges x_3 and x_5^0 and restores dual feasibility.)

Check	1	y_1^1	y_2^1	x_5^0	1
7	x_4^1	3	2	1	0
-9	x_3	-2	-3	-1	-2
36	$-z$	5	5	3	11

$\mathbf{x}^* = (1, 1, 2, 0, 0)$, $z^* = 11$, $\mathbf{u}^* = (-5, -5, 0, 0, 3)$.

(ii) The most infeasible row k may be taken as pivotal as usual. It is determined by the larger of

$$\max_i \{b_i > 0\} \qquad \text{and} \qquad \max_i \{-b_i - e_i > 0\colon x_i \text{ is bounded}\} \tag{2}$$

which maximizes the distance from the current value $x_i = -b_i$ to the closer endpoint of the feasible interval $[0, e_i]$. If $-b_k - e_k$ is the largest in (2), x_k should be reflected, so that $-b_k - e_k$ or $-(\beta_k + e_k\delta)/\delta > 0$ replaces b_k. Artificial variables x_i^0 are included in (2) with $e_i = 0$.

(iii) The driving variable is u_k and the blocking variable u_l is determined by

$$\min_j \{c_j/(-a_{kj})\colon a_{kj} < 0\} = c_l/(-a_{kl}) \tag{3}$$

as in 3.3. Since the new basic variables (primal and dual) are always feasible in 3.3, and a reflection is simpler than a pivot step, we allow x_l to be blocking if it is bounded. Thus if (3) would make $x_l > e_l$ (this is equivalent to $a_{kl}e_l + b_k > 0$) the pivot a_{kl} is abandoned and x_l merely reflected. To be sure, this will make $c_l \leq 0$, but any loss of dual feasibility will be restored after an acceptable pivot is finally found in row k by repeated applications of (3), as it must be if an optimal solution exists. In Table 6.5C two pivot steps are rejected in favor of nonbasic reflections before the pivot $a_{53} = -1$ is accepted in the third tableau.

Five Types of Variables

The following list differs from that of Orchard-Hays (1968) by the inclusion of the dual of a bounded or type 1 variable, called type 3, which will be discussed hereafter.

Type	Symbol	Name	Conditions for Feasibility	Conditions for Optimality
0	x_h^0	Artificial	$x_h^0=0$	No restriction
1	x_h^e, y_h^e	Bounded	$x_h^e+y_h^e=e_h>0$ $x_h^e\geq 0, y_h^e\geq 0$	(5) below
2	x_h	Nonnegative	$x_h\geq 0$	$c_h\geq 0$
3	$\check{x}_h^e, \check{y}_h^e$	Dual-bounded	(6) below	$0\leq c_h\leq \check{e}_h$
4	$x_h^{\pm}$	Unrestricted	No restriction	$c_h=0$

(4)

This numbering system has the property that if x_h is of type r, then its corresponding dual variable u_h is of type $4-r=4,3,2,1,0$, with the symbols $u_h^{\pm}$; $\check{u}_h^e, \check{v}_h^e$; u_h; u_h^e, v_h^e; u_h^0 corresponding to those in (4) in descending order.

To discuss the type 3 variables, consider $\check{u}_h^e$ and $\check{v}_h^e$, which are the duals of x_h^e and y_h^e, respectively. When x_h^e is basic, $\check{u}_h^e$ is zero as usual and when x_h^e is nonbasic, $\check{u}_h^e=c_h$, the cost coefficient for x_h^e in the current tableau. Feasibility for $\check{u}_h^e$ should be equivalent to optimality for x_h^e. Thus if x_h^e is at its lower bound 0, this means $\check{u}_h^e=c_h\geq 0$ as usual, so that z cannot decrease if x_h^e increases, while if x_h^e is at its upper bound e_h, then x_h^e can only decrease, and feasible $\check{u}_h^e\leq 0$.

Thus in the primal simplex algorithm the feasibility and optimality conditions for type 1 variables are the following:

$$\begin{array}{llll} \text{Basic} & 0\leq x_i^e\leq e_i & \text{with nonbasic} & \check{u}_i^e=0 \\ \text{Nonbasic} & x_j^e=0 & \text{with basic} & \check{u}_j^e=c_j\geq 0 \\ \text{Nonbasic} & x_j^e=e_j & \text{with basic} & \check{u}_j^e=c_j\leq 0 \end{array} \tag{5}$$

Taking the negative transpose of the tableau, which exchanges b, e, i, u with $-c, \check{e}, j, x$, respectively, gives the feasibility and optimality conditions for the type 3 variables $\check{x}_h^e$:

$$\begin{array}{ll} \text{Nonbasic}\ \ \check{x}_j^e=0 & \text{with}\ \ 0\leq u_j^e=c_j\leq \check{e}_j\ (\text{basic}) \\ \text{Basic}\ \ \check{x}_i^e=-b_i\geq 0 & \text{with nonbasic}\ \ u_i^e=0 \\ \text{Basic}\ \ \check{x}_i^e=-b_i\leq 0 & \text{with nonbasic}\ \ u_i^e=\check{e}_i \end{array} \tag{6}$$

where $\check{e}_h$ denotes the upper bound for u_h^e.

Since the existence of the pair x_h^e, y_h^e (when one is nonbasic) implies that the matrix **A** has a dispensable row that is a unit vector (all zero excepting one 1), dually the pair $\check{x}_h^e, \check{y}_h^e$ (when one is basic) implies that **A** has a dispensable column that is

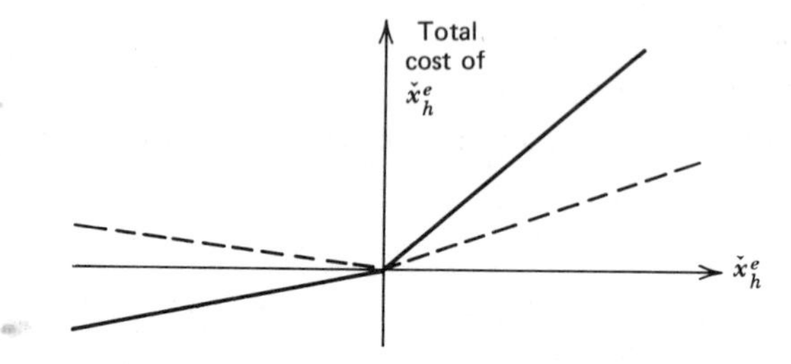

FIGURE 6.5 Cost functions for a type 3 variable.

the negative of a unit vector (all zero excepting one -1). Then one is moved to regard $\pm(\check{x}_h^e - \check{y}_h^e)$ as the basic variable in the row in which the -1 occurs; as the difference of two variables (each supposed to be nonnegative in the original problem), it would be unrestricted (type 4), except that one of its two parts, being truly basic, has a zero cost coefficient, while its other part (no matter which, if regarded as positive) is nonbasic with the cost $\check{e}_h > 0$.

The consequence is that a type 3 variable $\check{x}_h^e$ can be described as an unrestricted variable whose cost coefficient has two different values depending on the sign of $\check{x}_h^e$; the coefficient for $\check{x}_h^e \geq 0$ *exceeds* that for $\check{x}_h^e = -\check{y}_h^e \leq 0$ by the amount $\check{e}_h > 0$ when both are nonbasic. Figure 6.5 shows two ways in which the total cost of the unrestricted nonbasic $\check{x}_h^e$ might vary as a function of $\check{x}_h^e$. The dashed line shows a case in which $\check{x}_h^e = 0$ is optimal for the moment. If $\check{x}_h^e$ is the quantity of some commodity being bought or sold by one who is not a dealer or market maker, then $\check{e}_h$ is the spread between the bid and the asked market prices per unit of the commodity. The pair of variables y_3, y_4 in (1) of 3.7 thus corresponds to one primal variable of type 3.

The treatment of the $\check{x}_h^e = -\check{y}_h^e$ in the simplex algorithm may be inferred from the corresponding u_h^e. In particular, (iii) above indicates that in the primal simplex $\check{x}_k^e$ may block at 0 or u_k^e may block at $\check{e}_k$; in the latter case u_k^e is reflected. As a driving variable $\check{x}_l^e$ can increase when $c_l < 0$ and decrease when $c_l > \check{e}_l$; in the latter case one reflects u_l^e first. The reflection of a u_h^e is recorded by replacing the label $\check{x}_h^e$ by $\check{y}_h^e = -\check{x}_h^e$, but note that the latter substitution by itself is incorrect unless $\check{e}_h = 0$ so that one actually has $x_h^{\pm}$, of type 4.

EXERCISES 6.5

Find optimal solutions with all $x_h \geq 0$ and $0 \leq x_h^e \leq e$.

1. Max $x_3^1 + 3x_4 + 2x_5^1$ subject to
$x_1^4 = -2x_3^1 + x_4 - x_5^1 + 3, \qquad x_3^1 + x_4 + x_5^1 \leq 6$

2. Min $3x_3^3 + x_4^2 + 2x_5^1$ subject to
$2x_3^3 + 2x_4^2 + 3x_5^1 = 8, \qquad 2x_3^3 + 3x_4^2 + x_5^1 = 10$

3.

1	x_3^3	x_4^3	x_5	1
x_1	-2	-1	1	9
x_2	1	-1	-3	-1
$-z$	3	2	1	0

5.

1	x_3^3	x_4^3	x_5^3	x_6^3	1
x_1	1	-1	-3	-2	10
x_2^9	-2	-4	-3	-1	-1
$-z$	1	1	1	1	0

4.

1	x_3^1	x_4	x_5	1
x_1^2	2	-1	-1	-1
x_2	-1	4	-1	-4
$-z$	-3	-1	1	0

6.

1	x_3^2	x_4^2	x_5^2	x_6^2	1
x_1^0	3	-1	4	-2	-9
x_2^0	-1	5	-1	3	-10
$-z$	1	1	1	1	0

6.6 DEGENERACY AND LEXICOGRAPHY

The convergence of each phase of the simplex method of 6.1 depends on Theorem 3 of 3.3, whose proof was incomplete because the objective function z might remain forever at the same nonoptimal value. This phenomenon is called *circling* because, the number of different tableaus or bases being finite, it requires that the algorithm return again and again to the same tableaus. Two examples devised to illustrate this are quoted in Dantzig (1953), pp. 229–230. Computer codes generally disregard this possibility although Beale (1968, p. 18) reports that it has been observed in practice.

The avoidance of circling depends on making an appropriate choice whenever two or more rows are eligible to be pivotal as the result of a tie in the minimum ratio test (3) of 3.3. No matter what choice is made, however, the ensuing pivot step will make $b_i=0$ in the nonpivotal tied row(s). Any occurrence of $b_i=0$ is called (primal) degeneracy; it is frequently encountered in problems of all sizes. If some $b_i=0$, this means that the hyperplane defined by (basic) $x_i=0$ passes through the origin, which is already determined as the intersection of the hyperplanes on which the nonbasic variables are zero. Thus degeneracy might be described as involving "too many planes through the same point."

Dantzig (1963, p. 123) considered independent random choices among the tied rows when they occur. This could permit some circling, but the probability is zero that circling would continue indefinitely. Other methods prevent any circling whatever. The simplest, proposed and validated by Bland (1977), merely requires that the pivot a_{kl} be chosen so that k and l are the smallest eligible indices:

$$l=\min\{j\colon c_j<0\} \tag{1}$$

$$k=\min\{i\colon -b_i/a_{il}=\min_h(-b_h/a_{hl}\colon a_{hl}>0)\} \tag{2}$$

The results of Avis and Chvatal (1978) suggest that this method would not be very efficient for practical use. Fortunately (1) does not have to be used throughout the calculations, but only after iterations requiring (2).

Other methods exist whose concepts are needed for some other applications, such as (10) of 8.3 and the pivot step from (2) to (3) in 8.7, whether or not (1) and (2) are used. As applied to degeneracy, these methods exploit the fact that when the other coefficients are held constant, z^* and $\mathbf{x}^*$ are continuous functions of $\mathbf{b}$, so long as an optimal solution exists, though the dual $\mathbf{u}^*$ is not. Hence a way around degeneracy would seem to be to make suitable small changes in $\mathbf{b}$ such that ties no longer occur. This motivates a formal and exact procedure; the method is that of Dantzig, Orden and Wolfe (1955).

An easy way to insure that no ties can occur is to replace each b_i by $b_i - \varepsilon^i$, where $\varepsilon > 0$ is supposed to be so small that every multiple of the ith power ε^i occurring in the problem is greater than the higher powers that are subtracted from it. In general, a polynomial in ε, such as $\beta_0 + \beta_1\varepsilon + \cdots + \beta_m\varepsilon^m$, will be positive for sufficiently small $\varepsilon > 0$ if its coefficients β_j are not all zero, and the first nonzero coefficient (of lowest index) is positive. In this case the vector of coefficients is said to be lexicographically positive or *lexicopositive*, written

$$\boldsymbol{\beta} = (\beta_0, \beta_1, \ldots, \beta_m) \succ \mathbf{0} \tag{3}$$

Then $\boldsymbol{\alpha} \succ \boldsymbol{\beta}$ means $\boldsymbol{\alpha} - \boldsymbol{\beta} \succ \mathbf{0}$; this relation has many of the properties of the ordinary scalar inequality. By using this concept, one can avoid any reference to ε and its possible values.

Now the procedure is to replace each b_i in the initial tableau by the corresponding row of the matrix $[\mathbf{b}, -\mathbf{B}]$, where $\mathbf{B}$ is any invertible m by m matrix whose ith row is lexicopositive whenever $b_i = 0$. If $\mathbf{b} \leq 0$, this makes the rows of $[-\mathbf{b}, \mathbf{B}]$ lexicopositive. One must suppose that $\mathbf{B}$ is transformed at each pivot step like the rest of the tableau. If initially $\mathbf{B} = \mathbf{I}$, then at every iteration $\mathbf{B}$ will be part of the extended tableau $[\mathbf{I}, \mathbf{A}]$; in fact, precisely the part that is computed by the explicit-inverse method of 7.2 and there called $\mathbf{B}_r^{-1}$.

The convergence proof is now exactly the same as in the nondegenerate case, except that the lexicographic ordering of the vectors $\boldsymbol{\beta}_i$ (rows of $[-\mathbf{b}, \mathbf{B}]$) replaces or interpolates within the ordinary ordering of the first components. One may take $z = d = (d, 0, \ldots, 0)$ initially. Now the pivotal row k is selected to yield

$$\text{lexico-}\min_i \{\boldsymbol{\beta}_i / a_{il} : a_{il} > 0\} = \boldsymbol{\beta}_k / a_{kl} \tag{4}$$

The lexico-minimum is calculated by finding the set of all i that minimize the first component, then the subset of the set that minimize the second component, etc. There cannot be a tie, because that would require two rows of $\mathbf{B}$ to be proportional and rank $\mathbf{B} < m$, whereas $\mathbf{B}$ in any tableau is the product of two invertible matrices and thus of rank m (Theorem 2 of 7.1).

At each pivot step, (4) preserves the initial lexicopositivity of the $\boldsymbol{\beta}_i$, because

$$\boldsymbol{\beta}_i' = \boldsymbol{\beta}_i - \frac{a_{il}}{a_{kl}} \boldsymbol{\beta}_k = a_{il} \left(\frac{\boldsymbol{\beta}_i}{a_{il}} - \frac{\boldsymbol{\beta}_k}{a_{kl}} \right)$$

where $\boldsymbol{\beta}_i$ and $\boldsymbol{\beta}_k \succ 0$ and $a_{kl} > 0$. Then $\boldsymbol{\beta}_i' \succ \mathbf{0}$, by the first expression if $a_{il} \leq 0$, and by the second and (4) if $a_{il} > 0$. The change in z is $c_l \boldsymbol{\beta}_k / a_{kl} \prec \mathbf{0}$ since $c_l < 0$, $\boldsymbol{\beta}_k \succ 0$, and $a_{kl} > 0$. Thus z (in Phase 1, z^0) is decreasing in the lexicographic sense, no tableau can be repeated, and the algorithm must stop at or before the exhaustion of the finite collection of tableaus. The sequence of matrices $\mathbf{B}$ can be continued from Phase 1 into Phase 2 without restarting, despite the change in the objective function from z^0 to z.

The foregoing has used the vectors only as elements of vector spaces but the power series can also be multiplied together or divided one by another, forming what is called a nonarchimedean field. This field has all the properties of real numbers

TABLE 6.6 The Lexicographic Simplex Method

Check	1	x_5	x_6	x_7	1
−4	x_1	−2	1*	−3	−1
−7	x_2	−3	2	−4	−3
14	x_4	7	−2	9	−1
−5	x_4	−1	1	−4	−2
24	$-z$	9	−7	21	0

Check	1	x_5	x_1	x_7	1
−4	x_6	−2	1	−3	−1
1	x_2	1*	−2	2	−1
6	x_3	3	2	3	−3
−1	x_4	1	−1	−1	−1
−4	$-z$	−5	7	0	−7

First three components $(-b_i, a_{1i}, a_{2i})$ to find pivot a_{25}:
lexico-min$\{(1,-2,1)/1, (3,2,0)/3, (1,-1,0)/1\}=(1,-2,1)$.

	1	x_2	x_1	x_7	1
−2	x_6	2	−3	1	−3
1	x_5	1	−2	2	−1
3	x_3	−3	8	−3	0
−2	x_4	−1	1*	−3	0
1	$-z$	5	−3	10	−12

	1	x_2	x_4	x_7	1
−8	x_6	−1	3	−8	−3
−3	x_5	−1	2	−4	−1
19	x_3	5	−8	21	0
−2	x_1	−1	1	−3	0
−5	$-z$	2	3	1	−12

First three components $(-b_i, a_{1i}, a_{2i})$ determine pivot a_{41}:
lexico-min$\{(0,8,-3)/8, (0,1,-1)/1\}=(0,1,-1)$.
$\mathbf{x}^*=(0,0,0,0,1,3,0)$, $\mathbf{u}^*=(0,2,0,3,0,0,1)$, $z^*=-12$.

that are wanted here. Power-series products of vectors occur in the objective function when lexicography is used in quadratic programming. However, lexicography is perhaps most useful in integer programming. Table 6.6 illustrates lexicography in the Phase 2 simplex method and Exercise 2 does the same in Phase 1.

Since ε above is infinitesimal, it is plausible that it will suffice to store only the parameters (β_k, k) of the first nonzero term (the positive term of lowest order) in the polynomial $\beta_0+\beta_1\varepsilon+\cdots+\beta_m\varepsilon^m$ applicable to each row; Wolfe (1963) makes this idea simple and precise. Initially, the parameters (β^i, k_i) in row i are $(-b_i, 0)$. Every iteration begins by replacing each $\beta=0$ by 1 (or a random positive number) and replacing its k_i by k_i+1; the aim is to break the ties among the $\beta^i=0$. However, if all the (β^i, k_i) that yield $\max_i k_i \triangleq k^0$ are then ineligible to be pivotal because they have $a_{il}\le 0$, there will be no ties there, and the reverse operation is applied; (β^i, k^0) is replaced by $(0, k_i-1)$ in these rows. The result is that the only lexicographic parameters immediately affected by a pivot step are the β_i associated with $\max_i k_i$, to which the usual pivot rules apply.

EXERCISES 6.6

Use lexicography to optimize 1 and 2 with all $x_h\ge 0$ and x_i^0 artificial.

1.

Check	1	x_4	x_5	x_6	1
4	x_1	1	1	2	−1
2	x_2	0	2	1	−2
−1	x_3	−1	3	−1	−3
−3	$-z$	0	−3	−1	0

2.

Check	1	x_4	x_5	x_6	1
7	x_1^0	1	2	3	−1
9	x_2^0	1	1	4	−1
−3	x_3^0	1	−1	−2	−1
−3	$-z$	−3	0	−1	0

3. The real numbers have the archimedean property that by adding any positive number to itself a finite number of times, any other given real number may be exceeded. Show that the field of lexicographic vectors (ascending formal power series) does not have this property.

REFERENCES (CHAPTER 6)

Avis-Chvatal (1978)
Balinski-Gomory (1963)
Bland (1977)
Crowder-Hattingh (1975)
Dantzig (1951b, 1955a, 1963)
Dantzig-Ford-Fulkerson (1956)
Dantzig-Orden-Wolfe (1955)
Farkas (1902)
Goldfarb-Reid (1977)
Harris (1973)
Klee (1968)
Klee-Minty (1972)
Kuhn-Quandt (1963)
Künzi-Tzschach-Zehnder (1968)
Motzkin (1955)
Motzkin-Raiffa-Thompson-Thrall (1953)
Wolfe (1961–1965)
Wolfe-Cutler (1963)

CHAPTER SEVEN

The Revised Simplex; Matrices and Determinants

In Chapter 6 various procedures were considered for selecting the pivot steps that lead to the optimal solution, but the pivot steps thus selected were always carried out arithmetically by the simple rules of 2.4 or 2.5, with a few exceptions in 6.5. In the present chapter 7.2 to 7.6 are concerned with more efficient versions of the arithmetic of pivoting in the simplex method and in solving systems of linear equations.

Several major steps in the historical development may be distinguished. First the calculation of unneeded tableau elements was avoided as in 7.2. Then the explicit inverse $\mathbf{B}_r^{-1}$ was replaced by the product $\mathbf{E}^r \dots \mathbf{E}^1$ of more easily calculated elementary matrices, only one column $\boldsymbol{\eta}^r$ of each being stored, as in 7.3. More recently the triangular factorization $\mathbf{B}=\mathbf{LU}$ (7.4) has been recognized as still more advantageous. While all of this is helpful even with dense matrices, as indicated early in 7.5, the primary aim and reward has been to take advantage of the sparseness (small proportion of nonzeros) in the initial matrix $\mathbf{A}^0$ for most large practical problems (7.6).

The relevant elementary properties of matrices are reviewed and related to pivoting in 7.1, and the same is done for determinants in 7.7 and 7.8.

7.1 MATRIX ALGEBRA AND PIVOTING

Let the matrix $\mathbf{A}$ have m rows and n columns while $\mathbf{B}$ has n rows and p columns. Then products involving these matrices are defined as follows:

$$\mathbf{x}=-\mathbf{A}\mathbf{y} \quad \text{means} \quad x_i=-\sum_{j=1}^{n} a_{ij}y_j$$

$$\mathbf{v}=\mathbf{u}\mathbf{A} \quad \text{means} \quad v_j=\sum_{i=1}^{m} u_i a_{ij}$$

$$\mathbf{C}=\mathbf{A}\mathbf{B} \quad \text{means} \quad c_{ik}=\sum_{j=1}^{n} a_{ij}b_{jk} \tag{1}$$

for $i=1,\dots,m;\ j=1,\dots,n;\ k=1,\dots,p$.

The product **AB** can be formed if and only if **A** has as many columns as **B** has rows; it involves scalar products of rows of **A** with columns of **B**. In conformity with the latter statement, the vectors **x** and **y** in (1) are often regarded as columns and **u** and **v** as rows, although heretofore we have not insisted on that. What is essential is to make the distinction within each *matrix*, using rows or columns in forming scalar products according as the matrix is the left factor or the right. Note that (1) condenses (2) into (2′):

If $z_i=\Sigma_j a_{ij}y_j$ for all i and $y_j=\Sigma_k b_{jk}x_k$ for all j, then $z_i=\Sigma_j a_{ij}\Sigma_k b_{jk}x_k=\Sigma_k(\Sigma_j a_{ij}b_{jk})x_k$ for all i. (2)

If $\mathbf{z}=\mathbf{Ay}$ and $\mathbf{y}=\mathbf{Bx}$, then $\mathbf{z}=\mathbf{A}(\mathbf{Bx})=(\mathbf{AB})\mathbf{x}$. (2′)

Equation (2′) already exhibits one form of the *associative law* for matrix multiplication, which more generally is

$$\mathbf{D}=(\mathbf{AB})\mathbf{C}=\mathbf{A}(\mathbf{BC})=\mathbf{ABC}$$

meaning

$$d_{hk}=\Sigma_i\Sigma_j a_{hi}b_{ij}c_{jk} \tag{3}$$

Thus the grouping of the factors in a matrix product is immaterial provided the *order* is not changed; in general **AB** is different from **BA**. It may be verified that matrix multiplication would not be associative in general if it were defined in terms of scalar products of rows with rows or columns with columns.

By defining that

$$\mathbf{C}=\mathbf{A}+\mathbf{B} \quad \text{means} \quad c_{ij}=a_{ij}+b_{ij}$$

$$\mathbf{C}=t\mathbf{A} \quad \text{means} \quad c_{ij}=ta_{ij}$$

where t is any scalar, matrices of any given size m by n form a vector space that enjoys all the properties A1 to B5 of 2.1 as well as

$$(\mathbf{A}+\mathbf{B})\mathbf{C}=\mathbf{AC}+\mathbf{BC}$$

$$\mathbf{C}(\mathbf{A}+\mathbf{B})=\mathbf{CA}+\mathbf{CB} \tag{4}$$

$$t(\mathbf{AB})=(t\mathbf{A})\mathbf{B}=\mathbf{A}(t\mathbf{B}) \tag{5}$$

The zero matrix **0** has all of its elements equal to 0.

THEOREM 1 The rank ρ of the product of two or more matrices cannot exceed the rank of any of the factors:

$$\rho(\mathbf{A}_1\ldots\mathbf{A}_p)\leq\min\left[\rho(\mathbf{A}_1),\ldots,\rho(\mathbf{A}_p)\right]$$

Proof By Theorem 5 of 2.7, the rank of a matrix is equal to the minimum number of vectors that span its rows (i.e., generate them as linear combinations), and also is equal to the minimum number that span its columns. Then $\rho(\mathbf{A}_1\mathbf{A}_2) \le \rho(\mathbf{A}_2)$ because the rows of $\mathbf{A}_1\mathbf{A}_2$ are linear combinations of the rows of $\mathbf{A}_2$ and so cannot require any larger spanning set; and similarly $\rho(\mathbf{A}_1\mathbf{A}_2) \le \rho(\mathbf{A}_1)$ because the columns of $\rho(\mathbf{A}_1\mathbf{A}_2)$ are linear combinations of the columns of $\mathbf{A}_1$. This proves the theorem for $p=2$. The general result follows by induction since

$$\rho(\mathbf{A}_1 \dots \mathbf{A}_p) \le \min\left[\rho(\mathbf{A}_1 \dots \mathbf{A}_{p-1}), \rho(\mathbf{A}_p)\right]$$

$$\le \min\left\{\min\left[\rho(\mathbf{A}_1), \dots, \rho(\mathbf{A}_{p-1})\right], \rho(\mathbf{A}_p)\right\}$$

and the latter is $\min[\rho(\mathbf{A}_1), \dots, \rho(\mathbf{A}_p)]$.

A *unit matrix* $\mathbf{I}$ is a square matrix that has its diagonal elements (for which $i=j$) equal to 1, and the others zero. Evidently one has

$$\mathbf{I}\mathbf{A}_1 = \mathbf{A}_1 \qquad \text{and} \qquad \mathbf{A}_2\mathbf{I} = \mathbf{A}_2 \tag{6}$$

for any matrices $\mathbf{A}_1$ and $\mathbf{A}_2$, provided $\mathbf{I}$ has the appropriate number of rows and columns, which will be assumed.

THEOREM 2 An m by m matrix $\mathbf{B}$ has an *inverse* $\mathbf{B}^{-1}$ such that

$$\mathbf{B}\mathbf{B}^{-1} = \mathbf{B}^{-1}\mathbf{B} = \mathbf{I}$$

if and only if $\mathbf{B}$ has rank m. If the inverse exists, it is unique; it can be found by exchanging all the variables in a tableau of matrix $\mathbf{B}$, and rearranging the labels in their original order.

Proof It is necessary that $\mathbf{B}$ have rank m because $\mathbf{I}$ has rank m (see Theorem 1 above). To construct $\mathbf{B}^{-1}$, let us try to exchange all the variables in a tableau of matrix $\mathbf{B}$, thus:

$$\begin{array}{c|c} 1 & \mathbf{y} \\ \hline \mathbf{x} & \mathbf{B} \end{array} \qquad \begin{array}{c|c} 1 & \mathbf{x} \\ \hline \mathbf{y} & \mathbf{B}^{-1} \end{array} \qquad \text{or else} \qquad \begin{array}{c|cc} 1 & \mathbf{x}_K & \mathbf{y}_{\bar{L}} \\ \hline \mathbf{y}_L & \mathbf{B}_0 & \mathbf{B}_2 \\ \mathbf{x}_{\bar{K}} & \mathbf{B}_1 & \mathbf{0} \end{array} \tag{7}$$

where the components of $\mathbf{x}$ and $\mathbf{y}$ have the same order in both tableaus. If a complete exchange is possible, substituting from $\mathbf{x} = -\mathbf{B}\mathbf{y}$ into $\mathbf{y} = -\mathbf{B}^{-1}\mathbf{x}$ and vice versa as in (2′) gives $\mathbf{y} = \mathbf{B}^{-1}\mathbf{B}\mathbf{y}$ and $\mathbf{x} = \mathbf{B}\mathbf{B}^{-1}\mathbf{x}$, respectively. Since the tableaus show that the components of either $\mathbf{y}$ or $\mathbf{x}$ can be chosen arbitrarily, their coefficients must be equal in every equation, and so we find that $\mathbf{B}^{-1}\mathbf{B} = \mathbf{B}\mathbf{B}^{-1} = \mathbf{I}$.

The last tableau in (7) shows the only circumstance in which this calculation of $\mathbf{B}^{-1}$ can fail, that is, where the tableau contains a zero in every pivotal position that would otherwise permit additional variables to be exchanged. Then the tableau asserts that $\mathbf{x}_{\bar{K}}+\mathbf{B}_1\mathbf{x}_K=\mathbf{0}$ regardless of the value of $\mathbf{y}$. This linear relation therefore must hold also among the rows of $\mathbf{B}$ in the first tableau, and so the rank of $\mathbf{B}$ is less than m.

Finally, suppose that $\mathbf{B}$ had two inverses $\mathbf{B}_1$ and $\mathbf{B}_2$. By the preceding construction, we may assume that one of them is two sided (i.e., $\mathbf{BB}_1=\mathbf{B}_1\mathbf{B}=\mathbf{I}$), and hence that the two multiply $\mathbf{B}$ on opposite sides. Then

$$\mathbf{B}_1=\mathbf{B}_1\mathbf{I}=\mathbf{B}_1(\mathbf{BB}_2)=(\mathbf{B}_1\mathbf{B})\mathbf{B}_2=\mathbf{IB}_2=\mathbf{B}_2$$

Thus $\mathbf{B}_1=\mathbf{B}_2$ and the inverse is unique. This completes the proof.

A matrix that has an inverse is called *invertible* or *nonsingular*. Evidently $\mathbf{B}$ is the inverse of $\mathbf{B}^{-1}$.

THEOREM 3 The rank of a matrix is unchanged by multiplication by an invertible matrix.

Proof If $\mathbf{A}_2=\mathbf{BA}_1$, then $\operatorname{rank}\mathbf{A}_2\le\operatorname{rank}\mathbf{A}_1$ by Theorem 1. If $\mathbf{B}$ is invertible, then $\mathbf{A}_1=\mathbf{B}^{-1}\mathbf{A}_2$ and $\operatorname{rank}\mathbf{A}_1\le\operatorname{rank}\mathbf{A}_2$. Thus $\operatorname{rank}\mathbf{A}_1=\operatorname{rank}\mathbf{A}_2$.

Tableaus and Pivot Steps

In 7.1 to 7.3 the symbol $\mathbf{A}_N$ will be used in place of the former $\mathbf{A}$. Then the row equations $\mathbf{x}_M+\mathbf{A}_N\mathbf{x}_N+\mathbf{b}=\mathbf{0}$ and $z=\mathbf{cx}_N+d$ are represented by

$$\begin{array}{c|cc} 1 & \mathbf{x}_N & 1 \\ \hline \mathbf{x}_M & \mathbf{A}_N & \mathbf{b} \\ -z & \mathbf{c} & d \end{array} \qquad \text{or} \qquad \begin{array}{cccc} \mathbf{x}_M & -z & \mathbf{x}_N & 1 \\ \hline \mathbf{I} & \mathbf{0} & \mathbf{A}_N & \mathbf{b} \\ \mathbf{0} & 1 & \mathbf{c} & d \end{array} \tag{8}$$

according as the condensed or the extended tableau is used, as at the end of 2.4. Here $\mathbf{I}$ is the m by m unit matrix (in practice its rows become permuted), $\mathbf{x}_M$ is the vector of basic x_i, and $\mathbf{x}_N$ is the vector of nonbasic x_j.

Sometimes it is convenient to omit the column of $\mathbf{b}$ and/or the row of $\mathbf{c}$ (then they receive separate special treatment) or to absorb them into $\mathbf{A}_N$ (to simplify the formulas). In either case nothing is lost in the description of the arithmetic of pivoting. Thus the matrix of the extended tableau may be either

$$\mathbf{A}\triangleq\begin{bmatrix}\mathbf{I} & \mathbf{0} & \mathbf{A}_N & \mathbf{b}\\ \mathbf{0} & 1 & \mathbf{c} & d\end{bmatrix} \qquad \text{or} \qquad \mathbf{A}\triangleq[\mathbf{I},\mathbf{A}_N]$$

Similarly $\mathbf{I}$ and the square matrices $\mathbf{B}$, $\mathbf{B}^{-1}$ and $\mathbf{E}$ defined shortly may have $m+1$ rows instead of the m usually stated. In 7.1 to 7.3 one should have the matrix $\mathbf{A}$ of the extended tableau *in mind* even though the columns of $\mathbf{I}$ often will not be written out in displaying an example.

The advantage of the extended tableau is to permit a change of basis (pivot step) to be effected by matrix multiplication. Let us modify the example (11) and (12) of 2.4 by replacing the last column by $\mathbf{b}$ and adjoining a row and a column for $-z$.

Then the method of 7.3 performs the pivot on a_{13} as follows:

$$\begin{array}{c} \\ (x_3) \\ (x_2) \\ (-z) \end{array} \begin{array}{c} \begin{array}{ccc} (x_1) & (x_2) & (-z) \end{array} \\ \left[\begin{array}{ccc} 1/a_{13} & 0 & 0 \\ -a_{23}/a_{13} & 1 & 0 \\ -c_3/a_{13} & 0 & 1 \end{array}\right] \end{array} \cdot \begin{array}{c|cccccc} & -z & x_1 & x_2 & x_3 & x_4 & 1 \\ \hline (x_1) & 0 & 1 & 0 & a_{13}^* & a_{14} & b_1 \\ (x_2) & 0 & 0 & 1 & a_{23} & a_{24} & b_2 \\ (-z) & 1 & 0 & 0 & c_3 & c_4 & d \end{array}$$

$$= \begin{array}{c|cccccc} & -z & x_1 & x_2 & x_3 & x_4 & 1 \\ \hline (x_3) & 0 & \dfrac{1}{a_{13}} & 0 & 1 & \dfrac{a_{14}}{a_{13}} & \dfrac{b_1}{a_{13}} \\ (x_2) & 0 & -\dfrac{a_{23}}{a_{13}} & 1 & 0 & a_{24}-\dfrac{a_{14}a_{23}}{a_{13}} & b_2-\dfrac{a_{23}b_1}{a_{13}} \\ (-z) & 1 & -\dfrac{c_3}{a_{13}} & 0 & 0 & c_4-\dfrac{a_{14}c_3}{a_{13}} & d-\dfrac{b_1c_3}{a_{13}} \end{array} \tag{9}$$

or in brief, $\mathbf{EA}=\mathbf{A}'$. Note that the columns of $\mathbf{A}$ and $\mathbf{A}'$ are in the same order, while the ordered sets of the columns and the rows of $\mathbf{E}$ are identical with the ordered sets of the rows of $\mathbf{A}$ and $\mathbf{A}'$ respectively. The latter are themselves identical except for the replacement of $x_k = x_1$ by $x_l = x_3$.

To explain (9) let us recall from 2.4 that a pivot step has the effect of replacing the rows of $\mathbf{A}$ by certain linear combinations of these rows. This can be accomplished by multiplying $\mathbf{A}$ on the left by an m by m matrix $\mathbf{E}$ each of whose rows contains the coefficients of one of the linear combinations in question, so that $\mathbf{A}'=\mathbf{EA}$. For example, in the first matrix $\mathbf{E}$ in (9) above, its first row $1/a_{13},0,0$ accomplishes the division of the first row of $\mathbf{A}$ by the pivot a_{13}, and the other rows of $\mathbf{E}$ subtract the proper multiples of the first row of $\mathbf{A}$ from the other rows of $\mathbf{A}$.

Alternatively, note that $\mathbf{EI}=\mathbf{E}=\mathbf{I}'$, the result of applying the pivot rules of 2.4 to $\mathbf{I}$. Thus $\mathbf{E}$ is obtained by replacing the kth column of $\mathbf{I}$ by the transform of the pivotal column l in $\mathbf{A}$, because the latter acquires the label k after the usual exchange of labels. In this column of $\mathbf{E}$ (usually denoted by $\boldsymbol{\eta}$) one has

$$e_{lk} = 1/a_{kl} \quad \text{and} \quad e_{ik} = -a_{il}/a_{kl} \quad \text{for } i \neq l \tag{10}$$

by rules 2 and 4 of 2.4; the other elements of $\mathbf{I}$ are unchanged in $\mathbf{E}$. Despite the inequality of the indices k and l chosen from $\{1,\ldots,m+n\}$, the physical position of $e_{lk}=1/a_{kl}$ is always in the principal diagonal of $\mathbf{E}$. Whereas the $\mathbf{I}$ in $\mathbf{A}$ may be permuted, the $\mathbf{I}$ from which $\mathbf{E}$ is derived is not. $\mathbf{E}$ is called an *elementary matrix*.

A sequence of r pivot steps leads to a relation

$$\mathbf{A}^r = \mathbf{E}^r \ldots \mathbf{E}^1\mathbf{A}^0 \triangleq \mathbf{B}^{-1}\mathbf{A}^0 \tag{11}$$

where $\mathbf{A}^0$ is the matrix of the initial extended tableau, $\mathbf{A}^r$ is the extended matrix after r pivot steps, and the symbol $\mathbf{B}^{-1}$ can represent $\mathbf{E}^r \ldots \mathbf{E}^1$. The value of (11) is to show that once $\mathbf{B}^{-1}$ or the $\mathbf{E}$'s are known, there is no need to calculate all of $\mathbf{A}^r$, but only $\mathbf{b}^r$, the pivotal column, and the row $\mathbf{g}^r$ or $\mathbf{c}^r$.

In (11) the superscripts on the matrices indicate the pivot step to which they belong, while a subscript will indicate a selection of a group of columns corresponding to some basis such as M^0 or M^r. Since we have $\mathbf{A}^0_{M^0}=\mathbf{A}^r_{M^r}=\mathbf{I}$ by definition, (11) shows that $\mathbf{B}^{-1}=\mathbf{E}^r\ldots\mathbf{E}^1$ is given by

$$\mathbf{A}^r_{M^0}=\mathbf{B}^{-1}=\left(\mathbf{A}^0_{M^r}\right)^{-1} \tag{12}$$

(In (11) select columns M^0 to get the first equality in (12) and M^r to get the second.) Thus the r-step transforming matrix $\mathbf{A}^r_{M^0}=\mathbf{E}^r\ldots\mathbf{E}^1$ is the inverse of the matrix $\mathbf{A}^0_{M^r}$ (usually called simply the *basis* $\mathbf{B}$ or $\mathbf{B}_r$) consisting of those columns of $\mathbf{A}^0$ that are basic in $\mathbf{A}^r$. The column $\mathbf{b}^0$ is never part of any $\mathbf{B}$ but it has the same number of components (m or $m+1$) as a column of $\mathbf{B}$.

THEOREM 4 The components of any column vector $\mathbf{a}^r_{.j}$ of $\mathbf{A}^r$ are the coefficients occurring in the linear combination of the columns of $\mathbf{B}=\mathbf{A}^0_{M^r}$ that is equal to $\mathbf{a}^0_{.j}$. (The same is true if 0 is replaced by any other index r'.)

The proof follows by extracting the jth column of $\mathbf{B}\mathbf{A}^r=\mathbf{A}^0$, obtained by premultiplying (11) by $\mathbf{B}$. The theorem gives an elegant description of the arithmetic of pivoting, using the shortest possible vectors (of length $m+1$ or even m), and hence is widely used in the literature. Dantzig (1963, pp. 160–164) and Hadley (1962, pp. 158–162) discuss diagrams relating to the column space of $\mathbf{A}^0$. Hadley calls this the *requirements space* as contrasted with the *solution space* of vectors $\mathbf{x}$. Each of these has a dual counterpart.

We may now distinguish three related meanings of the word *basis*:

(i) The ordered *basic set* of x_j (or the ordered set M^r of their indices) that are dependent (solved for) in $\mathbf{A}^r$. The ordering is to agree with the ordering of the components of the column vectors.

(ii) The *basis* in the classic sense of Theorem 4, which is the ordered set of the column vectors of $\mathbf{A}^0$ (having indices in M^r) in terms of which $\mathbf{A}^r$ expresses the other columns of $\mathbf{A}^0$.

(iii) The square invertible *basis matrix* $\mathbf{B}=\mathbf{B}_r=\mathbf{A}^0_{M^r}$ whose columns are the basis (ii).

If $a^{r-1}_{k^r,l^r}$ is the pivot in $\mathbf{A}^{r-1}$ that takes it into $\mathbf{A}^r$, the row and column labels for the square matrices involved are the following:

$$\begin{array}{cc} & k^r\in M^{r-1} \\ l^r\in M^r & \boxed{\mathbf{E}^r} \end{array} \qquad \begin{array}{cc} & M^0 \\ M^r & \boxed{\mathbf{B}_r^{-1}=\mathbf{A}^r_{M^0}} \end{array} \qquad \begin{array}{cc} & M^r \\ M^0 & \boxed{\mathbf{B}_r=\mathbf{A}^0_{M^r}} \end{array} \tag{13}$$

In $\mathbf{E}^r$ the row labels M^r include l^r and the column labels M^{r-1} include k^r; otherwise they are the same. The column of $\mathbf{E}^r$ (that of index k^r) wherein it differs from $\mathbf{I}$ is denoted by $\boldsymbol{\eta}^r$.

It is possible to calculate the value of $\mathbf{c}^r$ or $\mathbf{g}^r$ for the rth basis even if the row for $-z$ or $-z^0$ has not been included in the matrix calculation (as $-z$ was in (9)). It suffices to show the procedure for z and c^r. The problem is this: Given the inverse of the m-rowed basis $\mathbf{A}^0_{M^r}=\mathbf{B}$, find the inverse of the $(m+1)$-rowed basis matrix

$$\begin{bmatrix} \mathbf{B} & \mathbf{0} \\ \mathbf{c}^0_B & 1 \end{bmatrix}$$

where $\mathbf{c}^0_B$ consists of those components of $\mathbf{c}^0$ (with indices in M^r) that go with $\mathbf{B}$. Evidently the inverse should have a similar form

$$\begin{bmatrix} \mathbf{B}^{-1} & \mathbf{0} \\ \boldsymbol{\pi} & 1 \end{bmatrix}$$

where $\boldsymbol{\pi}=\mathbf{c}^r_{M^0}$, the components of $\mathbf{c}^r$ with indices in M^0 that go with $\mathbf{B}^{-1}=\mathbf{A}^r_{M^0}$. To verify this one can multiply these partitioned matrices together almost as if they were 2 by 2 matrices of scalars, except that one must take care that the submatrices are multiplied in the same order as the large matrices. The result is

$$\begin{bmatrix} \mathbf{B}^{-1} & \mathbf{0} \\ \boldsymbol{\pi} & 1 \end{bmatrix}\begin{bmatrix} \mathbf{B} & \mathbf{0} \\ \mathbf{c}^0_B & 1 \end{bmatrix}=\begin{bmatrix} \mathbf{I}_m & \mathbf{0} \\ \boldsymbol{\pi}\mathbf{B}+\mathbf{c}^0_B & 1 \end{bmatrix}. \tag{14}$$

Since the right member is to be $\mathbf{I}_{m+1}$, one has $\boldsymbol{\pi}\mathbf{B}+\mathbf{c}^0_B=\mathbf{0}$ or $\boldsymbol{\pi}=-\mathbf{c}^0_B\mathbf{B}^{-1}$.

If the first matrix in (14) premultiplies $(\mathbf{b}^0, d^0)$ as well as an arbitrary (jth) column $(\mathbf{a}^0_{.j}, c^0_j)$ of $\mathbf{A}^0$, the result is

$$\begin{bmatrix} \mathbf{B}^{-1} & \mathbf{0} \\ \boldsymbol{\pi} & 1 \end{bmatrix}\begin{bmatrix} \mathbf{a}^0_{.j} & \mathbf{b}^0 \\ c^0_j & d^0 \end{bmatrix}=\begin{bmatrix} \mathbf{B}^{-1}\mathbf{a}^0_{.j} & \mathbf{B}^{-1}\mathbf{b}^0 \\ \boldsymbol{\pi}\mathbf{a}^0_{.j}+c^0_j & \boldsymbol{\pi}\mathbf{b}^0+d^0 \end{bmatrix}=\begin{bmatrix} \mathbf{a}^r_{.j} & \mathbf{b}^r \\ c^r_j & d^r \end{bmatrix}. \tag{15}$$

Then the principal formulas may be summarized as follows:

$$\mathbf{B}=\mathbf{A}^0_{M^r} \qquad \mathbf{B}^{-1}=\mathbf{A}^r_{M^0}$$

$$\mathbf{a}^r_{.j}=\mathbf{B}^{-1}\mathbf{a}^0_{.j} \qquad \mathbf{b}^r=\mathbf{B}^{-1}\mathbf{b}^0 \tag{16}$$

$$\boldsymbol{\pi}=-\mathbf{c}^0_B\mathbf{B}^{-1}=\mathbf{c}^r_{M^0} \tag{17}$$

$$\mathbf{c}^r=\boldsymbol{\pi}\mathbf{A}^0+\mathbf{c}^0 \tag{18}$$

$$d^r=\boldsymbol{\pi}\mathbf{b}^0+d^0=-\mathbf{c}^0_B\mathbf{b}^r+d^0$$

The last line results by treating $\mathbf{b}^0$ as a column of $\mathbf{A}^0$, and then reversing the roles of the 0th and the rth tableaus, or using the preceding relationships. The minus sign enters only when initial data premultiply current data to yield current data.

Because of their use in calculating $\mathbf{c}^r$ in (18) (this is called *pricing out*), the components of $\boldsymbol{\pi}=\mathbf{c}^r_{M^0}$ are called *simplex multipliers*. In the literature the sign of $\boldsymbol{\pi}$ is sometimes reversed; and the coefficient of $\mathbf{c}^0$ may be adjoined as an additional component $\pi_0=1$. The initial basis M^0 is sometimes called *logical* and the initial nonbasis N^0 called *structural.*

A derivation of (17) and (18) less dependent on matrix formalisms will be given following (9) of 7.3.

EXERCISES 7.1

1. Evaluate $\begin{bmatrix} 2 & -3 \\ -1 & 4 \\ 3 & 1 \end{bmatrix}\begin{bmatrix} 6 & 1 & -3 \\ 5 & 2 & 4 \end{bmatrix}$. By Theorem 1, what is the largest rank it could have? Is it invertible? Repeat with the factors in the opposite order.

2.–5. Verify (2) to (5) for general 2 by 2 matrices.

6. Evaluate $[5 \quad 1 \quad 2]\begin{bmatrix} 2 & -3 \\ -1 & 4 \\ 3 & 1 \end{bmatrix}\begin{bmatrix} 6 & 1 & -3 \\ 5 & 2 & 4 \end{bmatrix}$ in two different ways (see 1). Which is easier?

7. Let the matrices $\mathbf{A},\mathbf{B},\mathbf{C}$ have sizes m by n, n by p and p by q, respectively and all elements different from 0 or ± 1. How many multiplications are required to calculate **ABC**? In how many ways could **ABCD** be evaluated?

8. Pivot in the condensed tableau to show that

$$\begin{bmatrix} a_{11} & a_{12} \\ a_{21} & a_{22} \end{bmatrix}^{-1}=\begin{bmatrix} a_{22} & -a_{12} \\ -a_{21} & a_{11} \end{bmatrix}\div(a_{11}a_{22}-a_{12}a_{21})$$

if the denominator is not zero. Also confirm by multiplication.

Invert by using 8 and (14):

9.
$$\begin{matrix} 4 & 5 & 0 \\ 2 & 3 & 0 \\ 1 & 6 & 1 \end{matrix}$$

10.
$$\begin{matrix} -3 & 1 & 0 \\ 1 & 3 & 0 \\ 4 & -2 & 1 \end{matrix}$$

In 11 to 13 write out the initial extended tableau and use the given tableaus to read off $\mathbf{B}_1,\mathbf{B}_1^{-1}=\mathbf{E}^1,\mathbf{B}_2,\mathbf{B}_2^{-1},\mathbf{E}^2$. Write out the formulas that derive $\mathbf{A}^1$ from $\mathbf{A}^0$, $\mathbf{A}^2$ from $\mathbf{A}^1$, $\mathbf{A}^2$ from $\mathbf{A}^0$ and $\mathbf{A}^0$ from $\mathbf{A}^2$, and check a few of these numerically. Obtain $\mathbf{c}^2$ by (18).

11. Example 2 of 3.4. **12.** Table 6.3. **13.** Table 7.2 (omit z^0).

14. Use Theorem 1 and example $\mathbf{A}=\begin{bmatrix} 1 & 1 \\ 1 & 1 \end{bmatrix}$ to show that the effect of a pivot step cannot be reproduced by matrix multiplication of the condensed tableau alone.

7.2 THE EXPLICIT-INVERSE METHOD

This method is a halfway house between 2.4–2.5 and 7.3–7.4, not indispensable but possibly helpful to students. It avoids recalculating the whole tableau after each pivot step but retains the simplicity and inefficiency inherent in using the explicit inverse of the matrix of coefficients to solve a system of linear equations. Equations (1) and (2) of 7.5 give some idea of its computational efficiency.

For comparison, Table 7.2 gives the solution of the example of this section and the next by the method previously employed in 2.5 and 6.1. Table 7.2′ contains the primary description of the new method, which uses the second half of 7.1; it is applied to the example in Tables 7.2A and B. All of their data appear as a subset of Table 7.2, and it may be instructive for the reader to find them there. The text of this section is not self-sufficient but rather supplementary to Table 7.2′.

In Tables 7.2A and B the circled numbers indicate the sequence of the operations; when tagged they occur in both (A) and (B) (transfer of data from ①̀ to ①́, etc.) The superscript 0 marks x_7^0, x_8^0 and z^0 as artificial variables but marks other letters as having the values in the initial tableau. The rth cycle ($r=1,2,\dots$) is concerned with the transformation of the tableau or basis $r-1$ into r.

Table 7.2A begins with the initial tableau $\mathbf{A}_N^0$, which is used in the first application of step 1 to select the index of the pivotal column having $\min g_j<0$ or $\min c_j<0$. Thereafter Table 7.2B generates the columns of multipliers $\mathbf{g}_{M^0}^r=\boldsymbol{\pi}_{\mathrm{I}}^r$ in Phase 1 and/or $\mathbf{c}_{M^0}^r=\boldsymbol{\pi}_{\mathrm{II}}^r$ in Phase 2. These are transferred to the right side of Table 7.2A, where they are used in step 6 to generate the successive rows of $\mathbf{g}_{N^0}^r$ or $\mathbf{c}_{N^0}^r$ below $\mathbf{A}^0$ in Table 7.2A. The denominator δ^r adjoined to π^r comes from the unit vector for $-z^0$ or $-z$. A column $-2,4,11,-2,20,0,111,15$ obtained by summing over $\mathbf{A}_{N^0}$ and $\mathbf{b}^0$ (omitting $\mathbf{I}$) can be used to check the pricing out. For example, $111=(-2)(-9)+4(-7)+11(11)$ by pricing out and $111=3+10-9+107$ as a row sum. In returning to step 1, note that if M^0 contains nonartificial variables, the corresponding multipliers π_i^r (if negative) may determine the pivotal column.

Table 7.2B begins with the initial basic set M^0, inverse matrix $\mathbf{B}_0^{-1}=\mathbf{I}$, $\mathbf{b}^0$ and $\mathbf{g}_{M^0}^0$ or $c_{M^0}^0=\mathbf{0}$, to which $\mathbf{a}_{.l^1}^0$ is adjoined in step 2. This is written out only to display the logic of the method more clearly. In general the rth repetition of the pivot step 4 converts the previous subtableau $r-1$ in Table 7.2B into $\mathbf{A}_{M^0}^r$, the inverse of $\mathbf{B}_r=\mathbf{A}_{M^r}^0$, with the column $\mathbf{b}^r$ on the right, the row(s) $\mathbf{g}_{M^0}^r$ or $\mathbf{c}_{M^0}^r$ below (both are needed in changing from Phase 1 to Phase 2), and the denominator δ^r and the basic set M^r on the left. The pivotal columns $\mathbf{a}_{.l^{r+1}}^0$ (written as a row above the subtableau) and $\mathbf{a}_{.l^{r+1}}^r$ on the right are not entered until step 2 of the *following* cycle. The element $g_{l^{r+1}}^r$ or $c_{l^{r+1}}^r$ at the foot of $a_{.l^{r+1}}^r$ has to be transferred in step 1 rather than calculated in step 2 because the omission of the unit column for $-z^0$ or $-z$ in the inverse matrix results in the omission of the term $g_{l^{r+1}}^0$ or $c_{l^{r+1}}^0$. The nonpivotal column labels are omitted because they are always those in M^0. Hence if x_l is in M^0, $\mathbf{a}_{.l}^r$ will duplicate one of the columns of the inverse.

The pivot step 4 can be checked by writing $\mathbf{b}^0$ as a row above $\mathbf{B}_r^{-1}$ and verifying that its scalar products with the rows of $\mathbf{B}_r^{-1}$ yield the b_i^r, or check columns can be obtained by summing over M^0 and $\mathbf{b}^r$. In Table 7.2B this gives $(-7,-4,13)$, $(-16,-4,24)$, $(-16,-15,22,91)$ and $(-11,-15,-4)$ for $r=0,1,2,3$. For example, $-11=(1(-16)-(-7)(-15))/11$ by pivoting and $-11=1+1-13$ as a row sum. The checks in steps 5 and 6 are like 3.5 (a_3), (a_4).

TABLE 7.2 Original Simplex Solution (2.5 and 6.1)

1	x_1	x_2	x_3	x_4	x_5	1
x_6^0	3	−1	2	−1	3	−8
x_7^0	2	3	−1	1	4*	−5
$-z$	4	2	1	−1	5	0
$-z^0$	−5	−2	−1	0	−7	13

4	x_1	x_2	x_3	x_4	x_7^0	1
x_6^0	6	−13	11*	−7	−3	−17
x_5	2	3	−1	1	1	−5
$-z$	6	−7	9	−9	−5	25
$-z^0$	−6	13	−11	7	7	17

11	x_1	x_2	x_6^0	x_4	x_7^0	1
x_3	6	−13	4	−7	−3	−17
x_5	7	5	1	1*	2	−18
$-z$	3	10	−9	−9	−7	107

1	x_1	x_2	x_6^0	x_5	x_7^0	1
x_3	5	2	1	7	1	−13
x_4	7	5	1	11	2	−18
$-z$	6	5	0	9	1	−5

TABLE 7.2′ The Explicit-Inverse Method

Set up the initial tableau $\mathbf{A}^0_{N^0}$ and subtableau $\mathbf{B}_0^{-1}=\mathbf{I}$ in Tables 7.2A and 7.2B [referred to as (A), (B)], respectively. Let $r=1,2,\ldots$.

Step no.	1	2	3,4	5,6
$r=1$	②	③✓	④	⑤✓
$r=2$	⑥	⑦✓	⑧	⑨✓
$r=3$	⑩	⑪✓	⑫	⑬✓

1(A). The *pivotal column* $\mathbf{a}^0_{.l^r}$ is selected by $\min_h g_h^{r-1}$ or c_h^{r-1}; $h=1,\ldots,m+n$ with x_h not artificial. If $\min<0$, enter it at bottom of new column headed $\mathbf{a}^{r-1}_{.l^r}$ in (B). If $\min\geq 0$, *end the phase*. At end of Phase 1 go back to step 5 if $z^0=0$; *stop* if $\min z^0>0$.

2(B). The *current pivotal column* $\mathbf{a}^{r-1}_{.l^r}=\mathbf{B}_{r-1}^{-1}\mathbf{a}^0_{.l^r}$ is calculated by writing $\mathbf{a}^0_{.l^r}$ as a row above $\mathbf{B}_{r-1}^{-1}$ and multiplying row by row. The bottom component $g_{l^r}^{r-1}$ or $c_{l^r}^{r-1}$, entered in step 1, is *not* calculated in this way.

3(B). The *pivotal row* index k^r is found from $\min_i\{-b_i^{r-1}/a_{il^r}^{r-1}\colon a_{il^r}^{r-1}>0\}$. If $\mathbf{a}^{r-1}_{.l^r}\leq\mathbf{0}$, $z\to-\infty$; *stop*.

4(B). *Pivot* on $a^{r-1}_{k^r l^r}$ to get the new subtableau $\mathbf{A}^r_{M^0}=\mathbf{B}_r^{-1}$. Leave blank the positions of the pivotal column $\mathbf{a}^{r-1}_{.l^r}$. The common denominator in the algorithm changes from δ^{r-1} to δ^r.

5(B). The *multipliers* $\boldsymbol{\pi}^r=\mathbf{g}^r_{M^0}$ or $\mathbf{c}^r_{M^0}$ are normally obtained from the last row of the new subtableau (step 4). As a final check or to begin Phase 2, use $\boldsymbol{\pi}^r_{\mathrm{I}}=-\mathbf{g}^0_{M^r}\mathbf{B}_r^{-1}$, $\boldsymbol{\pi}^r_{\mathrm{II}}=-\mathbf{c}^0_{M^r}\mathbf{B}_r^{-1}$, $\mathbf{b}^0=\mathbf{B}_r\mathbf{b}^r$ (with the components of $\mathbf{g}^0,\mathbf{c}^0,\mathbf{b}^r$ properly rearranged), $\mathbf{c}^0\mathbf{x}^*_{N^0}=\mathbf{b}^0\mathbf{u}^*_{M^0}$.

6(A). *Pricing out*. $\boldsymbol{\pi}^r_{\mathrm{I}}\mathbf{A}^0_{N^0}+\mathbf{g}^0_{N^0}=\mathbf{g}^r_{N^0}$; $\boldsymbol{\pi}^r_{\mathrm{II}}\mathbf{A}^0_{N^0}+\mathbf{c}^0_{N^0}=\mathbf{c}^r_{N^0}$. Write $\boldsymbol{\pi}^r\delta^r$ as a column to the right of $\mathbf{A}^0$, insert its denominator δ^r below in line with $\mathbf{g}^0$ or $\mathbf{c}^0$, and multiply column by column to get $\mathbf{g}^r_{N^0}\delta^r$ or $\mathbf{c}^r_{N^0}\delta^r$. Check that g_j^r or $c_j^r=0$ for j in M^r. Return to step 1 with r increased by 1.

TABLE 7.2A Pricing Out in the Initial Tableau

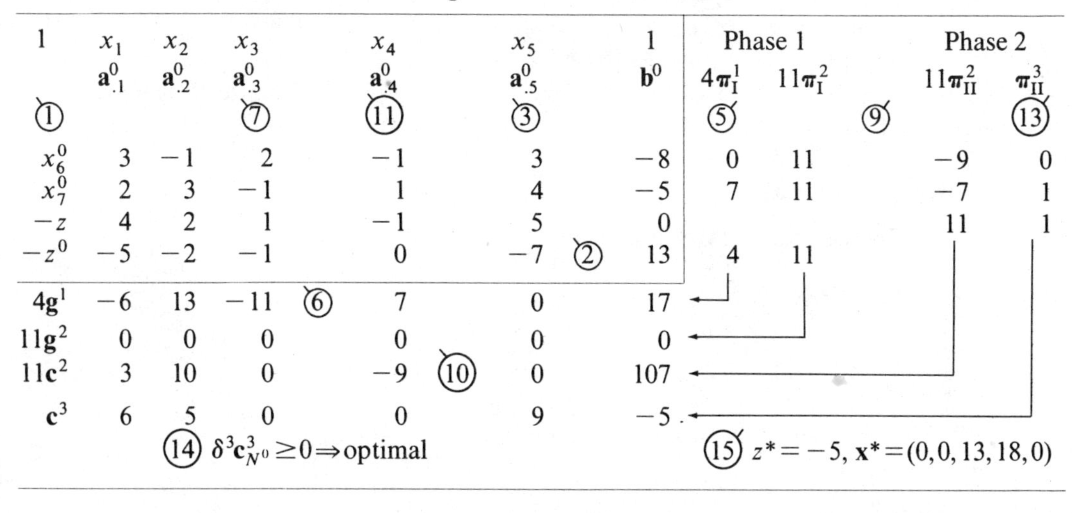

1	x_1	x_2	x_3		x_4		x_5		1	Phase 1			Phase 2	
	$\mathbf{a}^0_{.1}$	$\mathbf{a}^0_{.2}$	$\mathbf{a}^0_{.3}$		$\mathbf{a}^0_{.4}$		$\mathbf{a}^0_{.5}$		$\mathbf{b}^0$	$4\pi^1_{\mathrm{I}}$	$11\pi^2_{\mathrm{I}}$		$11\pi^2_{\mathrm{II}}$	π^3_{II}
(1)			(7)		(11)		(3)			(5)		(9)		(13)
x^0_6	3	−1	2		−1		3		−8	0	11		−9	0
x^0_7	2	3	−1		1		4		−5	7	11		−7	1
$-z$	4	2	1		−1		5		0				11	1
$-z^0$	−5	−2	−1		0		−7	(2)	13	4	11			
$4\mathbf{g}^1$	−6	13	−11	(6)	7		0		17					
$11\mathbf{g}^2$	0	0	0		0		0		0					
$11\mathbf{c}^2$	3	10	0		−9	(10)	0		107					
$\mathbf{c}^3$	6	5	0		0		9		−5					

(14) $\delta^3\mathbf{c}^3_{N^0} \geq 0 \Rightarrow$ optimal

(15) $z^* = -5$, $\mathbf{x}^* = (0, 0, 13, 18, 0)$

TABLE 7.2B. The Explicit Inverse Matrices

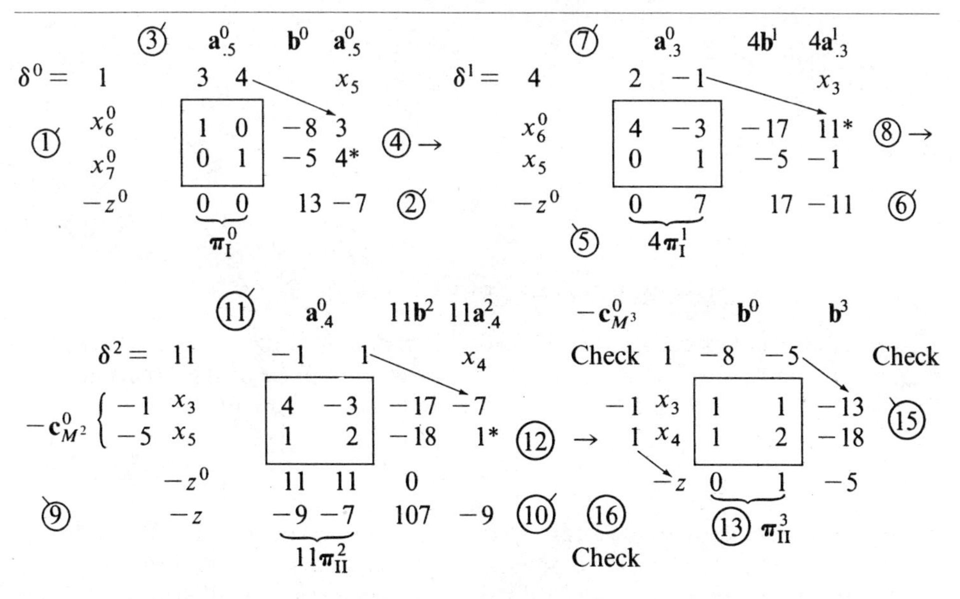

$\delta^0 = 1$

		(3) $\mathbf{a}^0_{.5}$: 3	4	$\mathbf{b}^0$	$\mathbf{a}^0_{.5}$ (x_5)	
(1)	x^0_6	1	0	−8	3	(4) →
	x^0_7	0	1	−5	4*	
	$-z^0$	0	0	13	−7	(2)
		π^0_{I}				

$\delta^1 = 4$

		(7) $\mathbf{a}^0_{.3}$: 2	−1	$4\mathbf{b}^1$	$4\mathbf{a}^1_{.3}$ (x_3)	
	x^0_6	4	−3	−17	11*	(8) →
	x_5	0	1	−5	−1	
	$-z^0$	0	7	17	−11	(6)
(5)		$4\pi^1_{\mathrm{I}}$				

$\delta^2 = 11$

		(11) $\mathbf{a}^0_{.4}$: −1	1	$11\mathbf{b}^2$	$11\mathbf{a}^2_{.4}$ (x_4)	
$-\mathbf{c}^0_{M^2}$: −1	x_3	4	−3	−17	−7	
−5	x_5	1	2	−18	1*	(12) →
	$-z^0$	11	11	0		
(9)	$-z$	−9	−7	107	−9	(10)
		$11\pi^2_{\mathrm{II}}$				

$-\mathbf{c}^0_{M^3}$				$\mathbf{b}^0$	$\mathbf{b}^3$	
Check		1	−8	−5		Check
−1	x_3	1	1	−13		(15)
1	x_4	1	2	−18		
	$-z$	0	1	−5		
(16) Check		(13) π^3_{II}				

The squares enclose $\mathbf{B}_r^{-1}\delta^r$ for $r = 0, 1, 2, 3$ with columns x^0_6, x^0_7.

EXERCISES 7.2

1. Find $\mathbf{c}^r$ and $z=d^r$ in 5 below if $\boldsymbol{\pi}_{\text{II}}^r=(11,9)/8$.
2. Find $\mathbf{g}^r$ and z^0 in 7 below if $\boldsymbol{\pi}_{\text{I}}^r=(14,8,0)/13$. (Find $\mathbf{g}^0$ first.)
3. Adjoin a vector $\mathbf{a}_{.8}^0=(2,1,-2)$ in Table 7.2A and optimize by extending Tables 7.2A, B by one more cycle (step 6 for the new column, steps 1–6, and finally step 1).
4. Rework 3 with $\mathbf{a}_{.8}^0=(3,2,-6)$.

Optimize by the explicit inverse method in 5 to 9.

5.

1	x_3	x_4	x_5	x_6	x_7	1
x_1	2	1	-1	3	-1	-1
x_2	0	2	2	-1	3	-2
$-z$	-3	-2	-1	-3	-2	0

6. (6) of 6.1. (Omit row x_9; this will not change the other x_h.)

7.

1	x_4	x_5	x_6	x_7	x_8	1
x_1^0	3	2	2	-1	1	-6
x_2^0	-2	-1	3	3	2	-7
x_3^0	1	2	-1	3	0	-7
$-z$	-2	-1	-6	2	-4	0

8. Table 6.3. **9.** Example 2 of 3.4.

10. Compare steps 2, 5 and 6 with 3.5 (b).

7.3 THE PRODUCT FORM OF INVERSE

This method, a compact form of the Jordan elimination of 2.6 (b), uses elementary factorization of the basis matrices $\mathbf{B}_r$ or their inverses in place of the explicit $\mathbf{B}_r^{-1}$ of 7.2. In principle the initial tableau $\mathbf{A}^0$ could be updated by the formula

$$\mathbf{A}^r=\mathbf{E}^r\ldots\mathbf{E}^2\mathbf{E}^1\mathbf{A}^0 \tag{1}$$

where the elementary matrix $\mathbf{E}^r$ corresponds to the rth pivot step and $\mathbf{E}^r\ldots\mathbf{E}^2\mathbf{E}^1=\mathbf{B}_r^{-1}=\mathbf{A}_{M^0}^r$ except for possible permutation of the columns. Each $\mathbf{E}^r$ is preserved by recording the column $\boldsymbol{\eta}^r$ in which $\mathbf{E}^r$ differs from $\mathbf{I}$ and its physical position p^r in $\mathbf{E}^r$. This $\boldsymbol{\eta}^r$ is derived from the pivotal column $\mathbf{a}_{.l^r}^{r-1}$ by rules 2 and 4 of 2.4: replace the pivot $a_{k^rl^r}^{r-1}$ in physical position p^r by its reciprocal and divide the other elements by $-a_{k^rl^r}^{r-1}$. Dantzig (1963) attributes the $\boldsymbol{\eta}$ vectors to Alex Orden. For a measure of their effectiveness, see (2) of 7.5.

The set M^r of indices of basic variables must be recorded as an ordered set to agree with the ordering of the components of the column vectors $\mathbf{b}^r$, $\mathbf{a}_{.l^{r+1}}^r$ and $\boldsymbol{\eta}^r$. M^0 is arranged in the order of the rows of the initial tableau $\mathbf{A}^0$. Then each M^r is determined from M^{r-1} in the obvious way, by removing the index k^r from M^{r-1}

and inserting l^r in the same physical position p^r without disturbing the remaining indices; p^r is a proxy for k^r, which is not remembered. This assures that the unit matrix **I** in **E** is not permuted but has all its 1's on the diagonal. It also implies that the ordering of M^r can be associated with $\boldsymbol{\eta}^{r+1}$ when necessary, as it is in Table 7.3B below.

We do not wish to calculate the whole product (1) but only to extract the columns $\mathbf{b}^r$ and $\mathbf{a}^r_{.l^{r+1}}$ and the row $\mathbf{g}^r$ or $\mathbf{c}^r$. One way to do this is to multiply on the right by suitable unit column vectors and on the left by a unit row vector $(0,\ldots,0,1)$. Multiplication being associative, the resulting products could be evaluated in various ways, but for computational efficiency it is essential to *begin the multiplication at the end having the vector as a factor*, so that all the partial products are vectors rather than matrices. This is the origin of the two kinds of transformations (2) and (8) (or (10) and (11)) below.

The multiplication of an elementary matrix **E** times a column vector **b** or $\mathbf{a}_{.l}$ proceeds thus:

$$\mathbf{Eb}=\begin{bmatrix}1 & \eta_1 & 0\\ 0 & \eta_2 & 0\\ 0 & \eta_3 & 1\end{bmatrix}\begin{bmatrix}b_1\\ b_2\\ b_3\end{bmatrix}=\begin{bmatrix}b_1+b_2\eta_1\\ b_2\eta_2\\ b_3+b_2\eta_3\end{bmatrix}=\mathbf{b}^{\#}+b_k\boldsymbol{\eta} \tag{2}$$

Here for convenience the row indices agree with the physical position. Thus the one-step transform of a column vector is obtained by using its pivotal component b_k in physical position p to multiply $\boldsymbol{\eta}$ and then adding the result to $\mathbf{b}^{\#}$, which is the original column with its pivotal component b_k replaced by 0. Note that this makes no change in the column **b** if and only if its component b_k in the pivotal row is zero.

A continued product such as

$$\mathbf{b}^r=\mathbf{E}^r\ldots\mathbf{E}^1\mathbf{b}^0 \qquad \text{or} \qquad \mathbf{a}^r_{.j}=\mathbf{E}^r\ldots\mathbf{E}^1\mathbf{a}^0_{.j} \tag{3}$$

is evaluated from right to left in the sequence $\mathbf{a}^0_{.j}$, $\mathbf{E}^1\mathbf{a}^0_{.j}$, $\mathbf{E}^2\mathbf{E}^1\mathbf{a}^0_{.j}$, etc. Since this uses the $\mathbf{E}^s$ or $\boldsymbol{\eta}^s$ in the order in which they were generated, (3) is called a *forward transformation* (FT). The $\mathbf{E}^s$ will generally be different if the order of the pivot steps is changed. (Exercise 10.)

The Backward Transformation

Multiplication of the type just described is impractical for computing more than a few components of $\mathbf{g}^r$ or $\mathbf{c}^r$. Ultimately the c^r_j are obtained by pricing out the initial tableau, thus:

$$\mathbf{c}^r=\mathbf{c}^0+\mathbf{c}^r_{M^0}\mathbf{A}^0=\mathbf{c}^0+\boldsymbol{\pi}^r\mathbf{A}^0 \tag{4}$$

If each $\mathbf{E}^r$ includes a row and a column for z, then $(\boldsymbol{\pi}^r,1)$ is simply the last row of $\mathbf{B}_r^{-1}$ and is extracted by premultiplying by the unit vector $\mathbf{i}_z$, a row having 1 in the z position and zeros elsewhere, thus:

$$(\boldsymbol{\pi}^r,1)=\mathbf{i}_z\mathbf{B}_r^{-1}=\mathbf{i}_z\mathbf{E}^r\mathbf{E}^{r-1}\ldots\mathbf{E}^1 \tag{5}$$

This is the usual procedure in computer codes, because it avoids the need to arrange the components of $\mathbf{c}^0$ in the proper order in the following.

A more general method is required in the event that any row has to be inserted or changed, as in some methods of integer and quadratic programming. If then the $\mathbf{E}^r$ do not include a row and a column for z, (17) of 7.1 gives

$$\boldsymbol{\pi}^r = \mathbf{c}^r_{M^0} = -\mathbf{c}^0_{M^r}\mathbf{B}_r^{-1} = -\mathbf{c}^0_{M^r}\mathbf{E}^r\mathbf{E}^{r-1}\ldots\mathbf{E}^1 \tag{6}$$

The ordering of the basic set M^r prescribes the ordering of the components of $-\mathbf{c}^0_{M^r}$. The example will use (6) for the sake of compactness, thus omitting the component $-(g^{r-1}_{l^r} \text{ or } c^{r-1}_{l^r})/a^{r-1}_{k^r l^r}$ in $\boldsymbol{\eta}^r$.

Equation (5) or (6) is efficiently evaluated from left to right by defining $\mathbf{c}^{rr} = \mathbf{i}_z$ or $-\mathbf{c}^0_{M^r}$ and

$$\mathbf{c}^{r,s-1} \triangleq \mathbf{c}^{rr}\mathbf{E}^r\mathbf{E}^{r-1}\ldots\mathbf{E}^s = \mathbf{c}^{rs}\mathbf{E}^s = \mathbf{c}^{rr}\mathbf{A}^r_{M^{s-1}} \tag{7}$$

After r postmultiplications the calculation ends with $\mathbf{c}^{r0} = \boldsymbol{\pi}^r$, but the components of the intermediate $\mathbf{c}^{rs}$ are not necessarily components of any $\mathbf{c}^{r'}$ or $-\mathbf{c}^{r'}$. Another method for using (7) is mentioned in (a) below.

TABLE 7.3 The Product-Inverse Method

(A), (B) refer to Tables 7.3A and 7.3B. $r=1,2,\ldots$.

1(A). The *pivotal column* $\mathbf{a}^0_{.l^r}$ is selected by $\min_h g^{r-1}_h$ or c^{r-1}_h; $h=1,\ldots,m+n$ with x_h not artificial. If $\min \geq 0$, *end the phase*. At end of Phase 1 go back to step 5 if $z^0=0$; *stop* if $\min z^0>0$.

2(B). Get the *current pivotal column* $\mathbf{a}^{r-1}_{.l^r}$ from $\mathbf{a}^0_{.l^r}$ by the *forward transformation* (FT) (3). (Use (2) or (10) $r-1$ times.)

3(B). The *pivotal row* index k^r in position p^r is found by $\min_i\{-b^{r-1}_i/a^{r-1}_{il^r}$: $a^{r-1}_{il^r}>0\}$. If $\mathbf{a}^{r-1}_{.l^r} \leq \mathbf{0}$, $z \to -\infty$; *stop*.

3′(B). Convert $\mathbf{a}^{r-1}_{.l^r}$ into $\boldsymbol{\eta}^r$ by replacing its pivot $a^{r-1}_{k^r l^r}$ in position p^r by its reciprocal (both starred in Table 7.3B) and dividing the other elements by $-a^{r-1}_{k^r l^r}$. [Optional; see (10) and (11) in (b) below.]

4(B). Convert $\mathbf{b}^{r-1}$ into $\mathbf{b}^r$ by one use of (2) or (10) (FT). Convert M^{r-1} into M^r by replacing k^r by l^r in position p^r.

5(A, B). *Multipliers* $\boldsymbol{\pi}^r$ (*backward transformation BT*). In Phase 1 take the components of $\mathbf{g}^{rr} = -\mathbf{g}^0_{M^r}$ from 7.3A and record them (in the order of M^r) in the first tier of rows above M^r in 7.3B. Calculate $\mathbf{g}^{r,r-1},\ldots,\mathbf{g}^{r0} = \mathbf{g}^r_{M^0} = \boldsymbol{\pi}^r_{\mathrm{I}}$ by (7) and (8) or (11). In Phase 2 replace $\mathbf{g}$ by $\mathbf{c}$. (Also see (a) below.)

6(A). *Pricing out.* $\boldsymbol{\pi}^r_{\mathrm{I}}\mathbf{A}^0_{N^0} + \mathbf{g}^0_{N^0} = \mathbf{g}^r_{N^0}$; $\boldsymbol{\pi}^r_{\mathrm{II}}A^0_{N^0} + \mathbf{c}^0_{N^0} = \mathbf{c}^r_{N^0}$. Write $\boldsymbol{\pi}^r = \mathbf{g}^r_{M^0}$ or $\mathbf{c}^r_{M^0}$ as a column to the right of $\mathbf{A}^0$, insert 1 or δ^r below in line with $\mathbf{g}^0$ or $\mathbf{c}^0$, and multiply column by column to get $\mathbf{g}^r_{N^0}$ or $\mathbf{c}^r_{N^0}$. Check that g^r_j or $c^r_j=0$ for j in M^r. Return to step 1 with r increased by 1.

$\boldsymbol{\eta}^r$, $\mathbf{b}^r$, $\mathbf{g}^r$ or $\mathbf{c}^r$, $\mathbf{a}^r_{.l^{r+1}}$ are found in that order in steps 3′, 4, 5 to 6, and 2. ①, ②,... in Tables 7.3A, B give $t+6(r-1)$ in step t of cycle $r=1,2,3$. Check the final $\mathbf{b}$ by substitution in $\mathbf{A}^0$.

An example of the product $\mathbf{c}^{rs}\mathbf{E}^s$ in (7) is

$$(c_1, c_2, c_3)\begin{bmatrix} 1 & \eta_1 & 0 \\ 0 & \eta_2 & 0 \\ 0 & \eta_3 & 1 \end{bmatrix} = (c_1, \Sigma c_i \eta_i, c_3) \tag{8}$$

which differs from $\mathbf{c}$ itself only in having the pth component (here c_2) replaced by the scalar product $\mathbf{c}\boldsymbol{\eta}$. The row transformation (5) to (7) using (8) is called the *backward transformation* (BT) because it uses the $\boldsymbol{\eta}^s$ in the reverse of the order in which they were generated.

To judge the symmetry of (6), substitute it in (4) and transpose one term to obtain

$$\mathbf{c}^0 = \mathbf{c}^r + \mathbf{c}^0_{M^r}\mathbf{B}_r^{-1}\mathbf{A}^0 = \mathbf{c}^r + \mathbf{c}^0_{M^r}\mathbf{A}^r \tag{9}$$

This is what one obtains by interchanging 0 and r in (4).

An easy derivation of (6) from first principles then goes as follows. From 2.4 we know that pivot steps are reversible and that repeated pivot steps change $\mathbf{c}$ by adding to it appropriate multiples of the other row vectors in $\mathbf{A}$. The latter is the operation of pricing out and it only remains to determine the multipliers. We begin by pricing out the rth tableau to obtain costs in the initial tableau: first the variables in M^r, to show that these multipliers must be $\mathbf{c}^0_{M^r}$; and then the variables in M^0, to obtain $\mathbf{c}^0_{M^r}\mathbf{B}_r^{-1} + \mathbf{c}^r_{M^0} = \mathbf{0}$, whose solution is (6). Finally, pricing out the variables M^0 in the initial tableau shows that the multipliers $\boldsymbol{\pi}^r$ there are $\mathbf{c}^r_{M^0}$ as stated in (6).

The reader may now consult the summary of the procedure in Table 7.3 and its application to the example of 7.2 in Tables 7.3A and 7.3B. Table 7.3A gives the initial tableau $\mathbf{A}^0_{N^0}$ with adjoined column vectors $\boldsymbol{\pi}^r_{\mathrm{I}} = \mathbf{g}^r_{M^0}$ (Phase 1) and $\boldsymbol{\pi}^r_{\mathrm{II}} = \mathbf{c}^r_{M^0}$ (Phase 2) and row vectors $\mathbf{g}^r_{N^0}$ and $\mathbf{c}^r_{N^0}$; it is identical with Table 7.2A except as regards the sequence numbers and the detachment of the denominators of the $\boldsymbol{\pi}^r$. Table 7.3B is concerned with the forward transformations (2) and (3) (progressing down) and the backward transformations (7) and (8) (progressing up), in which each $\boldsymbol{\eta}^r$ transforms the column vectors to its right. The original-simplex solution of the example was given in Table 7.2.

The sequence numbers ①, ②,..., have the value $t+6(r-1)$ for each step t in cycle $r=1,2,3$. When tagged they label data transferred between Tables 7.3A and 7.3B (e.g., from ⓪̸ to ⓪̸). As usual a superscript 0 marks x_h^0 and z^0 as artificial variables and marks other letters as having their values in the initial tableau. If preferred, the last seven columns of Table 7.3B (not B′) may be written in the order in which they are calculated, without segregating the forward and the backward transformations.

Variations and Refinements

(a) To generate any linear combination of the rows of $\mathbf{A}^r$ one has only to use (7) and define $\mathbf{c}^{rr}$ to be the row vector of the coefficients of the linear combination. (Multiply with this row vector on the left of (1).) In particular, if $\mathbf{c}^{rr} = \mathbf{i}_k$, the unit

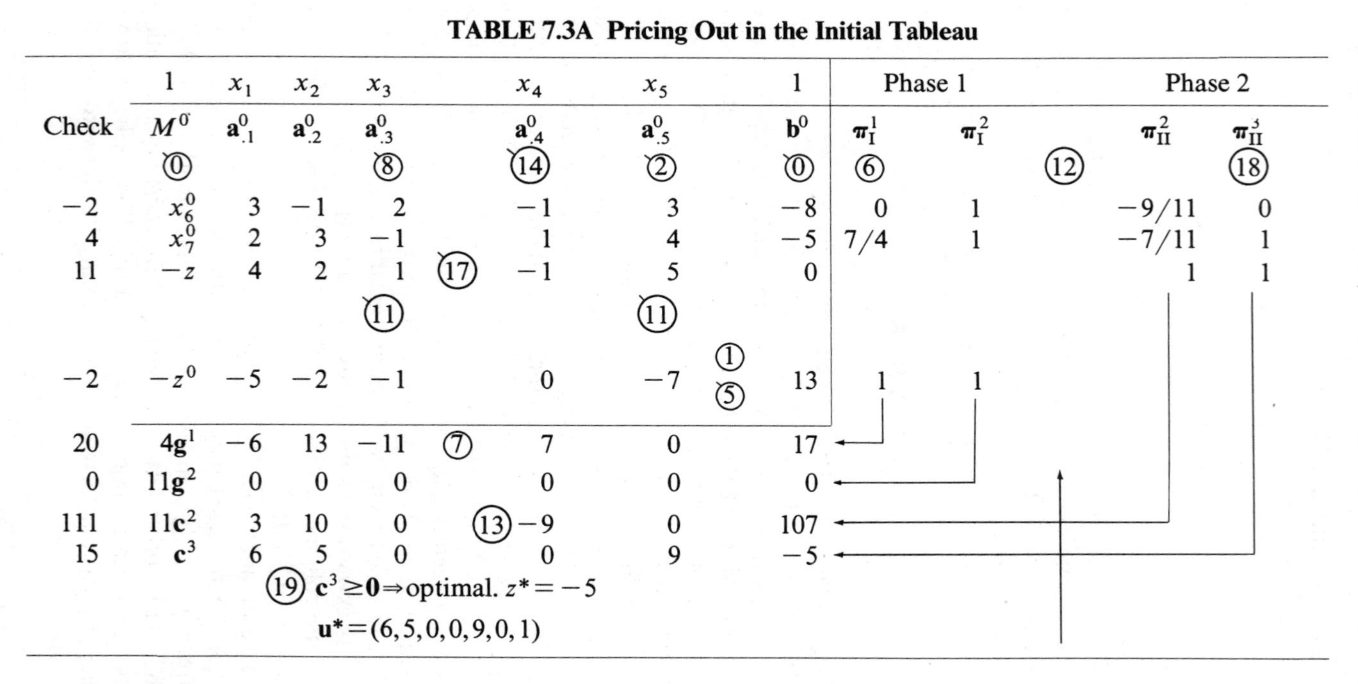

TABLE 7.3A Pricing Out in the Initial Tableau

	1	x_1	x_2	x_3		x_4	x_5		1	Phase 1			Phase 2	
Check	M^0	$\mathbf{a}^0_{.1}$	$\mathbf{a}^0_{.2}$	$\mathbf{a}^0_{.3}$		$\mathbf{a}^0_{.4}$	$\mathbf{a}^0_{.5}$		$\mathbf{b}^0$	$\boldsymbol{\pi}^1_{\mathrm{I}}$	$\boldsymbol{\pi}^2_{\mathrm{I}}$		$\boldsymbol{\pi}^2_{\mathrm{II}}$	$\boldsymbol{\pi}^3_{\mathrm{II}}$
	(0)			(8)		(14)	(2)		(0)	(6)		(12)		(18)
-2	x^0_6	3	-1	2		-1	3		-8	0	1		$-9/11$	0
4	x^0_7	2	3	-1		1	4		-5	$7/4$	1		$-7/11$	1
11	$-z$	4	2	1	(17)	-1	5		0				1	1
				(11)			(11)							
-2	$-z^0$	-5	-2	-1		0	-7	(1) (5)	13	1	1			
20	$4\mathbf{g}^1$	-6	13	-11	(7)	7	0		17					
0	$11\mathbf{g}^2$	0	0	0		0	0		0					
111	$11\mathbf{c}^2$	3	10	0	(13)	-9	0		107					
15	$\mathbf{c}^3$	6	5	0		0	9		-5					

(19) $\mathbf{c}^3 \geq \mathbf{0} \Rightarrow$ optimal. $z^* = -5$

$\mathbf{u}^* = (6,5,0,0,9,0,1)$

TABLE 7.3B Transformations

r	M^r		$\mathbf{b}^r$	$\boldsymbol{\eta}^{r+1}$		Forward			Backward					
						(2)	(8) ↓	(14)			↑			
0	x_6^0	(0)	-8	$-3/4$	(3)	3	2	-1	(5) 0	1		$-9/11$	0	
	x_7^0		-5	$1/4^*$	←	4^*	-1	1	7	7		-5	4	
1	x_6^0	(4)	$-17/4$	$4/11^*$	←		$11/4^*$	$-7/4$		1		-1	-1	
	x_5		$-5/4$	$1/11$	(9)		$-1/4$	$1/4$	$-\mathbf{g}_{M^1}^0$	7	(11)	-5	4	
2	x_3	(10)	$-17/11$	7		(15)		$-7/11$			↑		-1	(17)
	x_5		$-18/11$	11^*	←			$1/11^*$		$-\mathbf{g}_{M^2}^0$		$-\mathbf{c}_{M^2}^0$	1	
3	x_3	(16)	-13											
	x_4		-18			(20) $\mathbf{x}^* = (0, 0, 13, 18, 0, 0, 0)$							$-\mathbf{c}_{M^3}^0$	

Steps (11) through (13) are done first for **g** and then for **c**.

Examples: Check $111 = (-2)(-9) + 4(-7) + 11(11) = 3 + 10 - 9 + 107$.

$\mathbf{b}^2 \to \mathbf{b}^3$: $(-17, -18)/11 \to (-17/11, 0) - (18/11)(7, 11) = (-13, -18)$

$-\mathbf{c}_{M^3}^0 = (-1, 1) \to (-1, (-1, 1)(7, 11)) = (-1, 4)$

vector with its only nonzero component in the pivotal position, then (7) generates the *pivotal row* of $\mathbf{A}^r$ for use in the dual simplex method, or to permit an easy calculation of both $\mathbf{c}^r$ and $\mathbf{g}^r$ by adding the appropriate multiples of the pivotal row to $\mathbf{c}^{r-1}$ and $\mathbf{g}^{r-1}$, which are stored for this purpose (Zoutendijk, 1970). Since a large proportion of the computation time is consumed in step 6 (pricing out), one should thus avoid having to obtain more than one row by a full-scale pricing out. This method also permits the insertion of additional rows (constraints).

(b) m operations in each pivot step can be saved by dispensing with the calculation of $\boldsymbol{\eta}^s$ in step 3′ of Table 7.3; instead it suffices to store the corresponding *pivotal column* $\mathbf{a}^{s-1}_{.l^s} \triangleq \mathbf{a}^{s-1}$, which will be exemplified by (a_1, a_2, a_3) in this discussion. For the forward transformation the rules of 2.4 are applied to this column, to $\mathbf{b}^{s-1}$ (here represented by (b_1, b_2, b_3)) and to the columns of $\mathbf{A}^{s-1}$ to be transformed to $\mathbf{A}^s$. For example,

$$\begin{bmatrix} a_1 & b_1 \\ a_2{}^* & b_2 \\ a_3 & b_3 \end{bmatrix} \rightarrow \begin{bmatrix} b_1-(b_2/a_2)a_1 \\ b_2/a_2 \\ b_3-(b_2/a_2)a_3 \end{bmatrix} = \begin{bmatrix} b_1 \\ b_2 \\ b_3 \end{bmatrix} - \frac{b_2}{a_2}\begin{bmatrix} a_1 \\ a_2-1 \\ a_3 \end{bmatrix} \tag{10}$$

One step in the backward transformation gives

$$\mathbf{c}^{r,s-1} = \mathbf{c}^{rs}\mathbf{E}^s = (c_1, c_2, c_3)\begin{bmatrix} 1 & -a_1/a_2 & 0 \\ 0 & 1/a_2 & 0 \\ 0 & -a_3/a_2 & 1 \end{bmatrix}$$

$$= \left(c_1, \frac{-c_1a_1+c_2-c_3a_3}{a_2}, c_3\right) = \mathbf{c} - \frac{\mathbf{c}(a_1, a_2-1, a_3)}{a_2}\mathbf{i}_2 \tag{11}$$

where $\mathbf{i}_2 = (0,1,0)$, $c_i = c_i^{rs}$ and $a_i = a_{il^s}^{s-1}$. To avoid unnecessary divisions, a computer should store $1/a_k$ rather than a_k $(=a_2)$.

Equation (10) also facilitates the verification of the arithmetic in the tables by permitting the use of a common denominator as in 2.5. Thus in Table 7.3B we have $\mathbf{b}^2 = (-17, -18)/11$; the denominator 11 is the numerator of the previous pivot $11/4$; -17 is the numerator in the pivotal row of $\mathbf{b}^1$; and -18 is $[11(-5)-(-1)(-17)]/4$.

Tables 7.3A′ and B′ differ from 7.3A and B in using the $\mathbf{a}^s$ vectors in (10) and (11) rather than $\boldsymbol{\eta}^s$ in (2) and (8), and also in using detached denominators δ^s. To the left of the vertical line, the entries in the last four rows in 7.3A′ and in each pair of rows in 7.3B′ have the indicated common denominator 1, 4, 11 or 1 in accordance with 2.5. To the right of the vertical line, the entries in each *column* have the common denominator δ^r because $\mathbf{E}^r\mathbf{E}^{r-1}\ldots\mathbf{E}^s$ in (7) is simply $\mathbf{A}^r_{M^{s-1}}$ and the components of $\mathbf{c}^0$ and $\mathbf{A}^0$ are assumed to be integers.

Let the integer numerators be denoted by $\gamma_i^s = \delta^r c_i^{rs}$ and $\alpha_i^s = \delta^s a_{il^{s+1}}^s$ for $r \geq s \geq 0$. Then (11) shows that $\gamma_i^{s-1} = \gamma_i^s$ except that when $i = k^s \triangleq k$ one has

$$\gamma_k^{s-1} = \left(\gamma_k^s \delta^{s-1} - \sum_{i \neq k} \gamma_i^s \alpha_i^{s-1} \right) / \delta^s \qquad \text{for } r \geq s \geq 1 \tag{12}$$

Note that $\delta^s = \alpha_k^{s-1}$. For $\mathbf{c}^{rr} = -\mathbf{c}_{M^r}^0$ the common denominator δ^r is used for consistency although 1 would suffice. Note that in (10) the numerator b_2 is unchanged in the pivotal position but generally changed elsewhere, while in (12) the numerator changes only in the pivotal position. If $j \in M^0 \cap M^r$, then $-c_j^0 = c_j^r = 0$.

(c) As described thus far, the method has the disadvantage that the number of $\boldsymbol{\eta}^r$ or $\mathbf{a}^r$ will often come to exceed the size m of the explicit inverse; even before this occurs most of the $\boldsymbol{\eta}^r$ components will be nonzero even though $\mathbf{A}^0$ may have consisted mostly of zeros; and each of these effects will result in more computation and a more rapid accumulation of round-off errors. All these problems are made less serious by *reinverting* the basis $\mathbf{B}_r$ for one or more values of r; that is, occasionally expressing $\mathbf{B}_r^{-1}$ as the product of the *minimum number of factors* $\ldots \mathbf{E}^3\mathbf{E}^2\mathbf{E}^1$ (never more than m), in a fashion intended to *preserve as many zeros as possible* in the $\boldsymbol{\eta}^r$ or $\mathbf{a}^r$. Other versions of the simplex method also would require the equivalent of reinversions to reduce the accumulation of round-off errors in large problems. There is more on this in 7.4 and 7.6.

The term *crashing* refers to any of many techniques that aim to shorten the overall solution time by reducing the time per iteration even though the number of iterations may be increased. Typically they omit all backward transformations and make one pass through some or all of the nonbasic columns of $\mathbf{A}^0$, using some of the ideas of this subsection (c) and a composite simplex as in 6.4. A column can be brought into the basis if it is found to have $g_j^r < 0$ or $c_j^r < 0$ or eliminates an artificial variable or reduces the number of infeasible x_i. In particular a "reinversion" can be performed initially in the rows of the artificial variables and the columns having the smallest g_j^0 or c_j^0.

(d) The extended column vectors (including one c_j and/or g_j) make it possible to replace some of the backward row transformations by a few extra column transformations, which may involve less calculation. The pivot steps then fall into small groups called *passes*. (Only) at the beginning of each pass, all or part of the vector $\mathbf{g}^r$ or $\mathbf{c}^r$ is calculated; then several of the most promising columns are selected according to $\min g_j$ in Phase 1 and $\min c_j$ in Phase 2. The selected columns are brought up to date and the pivotal column is chosen from among them, perhaps by a more elaborate criterion as in 6.3, or $\max|\Delta z^0|$ or $\max|\Delta z|$, or a minimum number of entries above some positive threshold (see the end of 3.4). The pivot step is performed and the remainder of the selected columns, including their costs, is again brought up to date. If one of them again makes $|\Delta z|$ large enough, another pivot step is performed, and so forth; otherwise another pass is begun.

This *multiple column selection* technique also makes it easy to reject a column that yields an excessively small pivot. The term *suboptimization* is applicable if the pass

TABLE 7.3A′ Pricing Out in the Initial Tableau

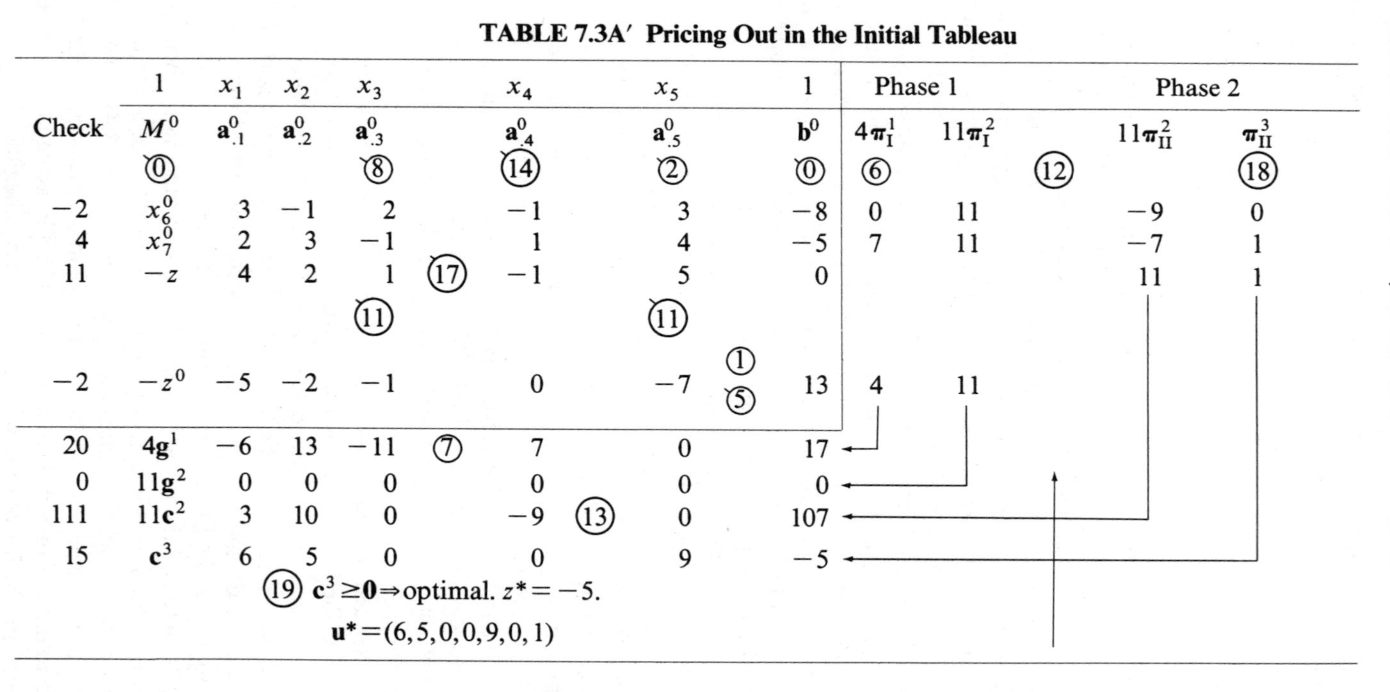

	1	x_1	x_2	x_3		x_4		x_5		1	Phase 1			Phase 2	
Check	M^0	$\mathbf{a}^0_{.1}$	$\mathbf{a}^0_{.2}$	$\mathbf{a}^0_{.3}$		$\mathbf{a}^0_{.4}$		$\mathbf{a}^0_{.5}$		$\mathbf{b}^0$	$4\pi^1_{\mathrm{I}}$	$11\pi^2_{\mathrm{I}}$		$11\pi^2_{\mathrm{II}}$	π^3_{II}
	(0)			(8)		(14)		(2)		(0)	(6)		(12)		(18)
−2	x^0_6	3	−1	2		−1		3		−8	0	11		−9	0
4	x^0_7	2	3	−1		1		4		−5	7	11		−7	1
11	$-z$	4	2	1	(17)	−1		5		0				11	1
				(11)				(11)							
−2	$-z^0$	−5	−2	−1		0		−7	(1) (5)	13	4	11			
20	$4\mathbf{g}^1$	−6	13	−11	(7)	7		0		17					
0	$11\mathbf{g}^2$	0	0	0		0		0		0					
111	$11\mathbf{c}^2$	3	10	0		−9	(13)	0		107					
15	$\mathbf{c}^3$	6	5	0		0		9		−5					

(19) $\mathbf{c}^3 \geq \mathbf{0} \Rightarrow$ optimal. $z^* = -5$.

$\mathbf{u}^* = (6, 5, 0, 0, 9, 0, 1)$

TABLE 7.3B′ Transformations

r	M^r		$\delta^r \mathbf{b}^r$	δ^r		Forward (FT)					Backward					
						②		⑧		⑭	⑤		↑			
0	x_6^0	⓪	−8		③	3		2		−1	0	11		−9	0	
	x_7^0		−5	1		4*		−1		1	28	77		−55	4	
1	x_6^0	④	−17	4		↓	⑨	11*		−7		11	⑪	−11	−1	
	x_5		−5					−1		1	$-\mathbf{g}_{M^1}^0$	77		−55	4	
2	x_3	⑩	−17						⑮	−7			↑		−1	⑰
	x_5		−18	11						1*		$-\mathbf{g}_{M^2}^0$		$-\mathbf{c}_{M^2}^0$	1	
3	x_3	⑯	−13	1								over			$-\mathbf{c}_{M^3}^0$	
	x_4		−18								4	11		11	1	$=\delta^r$
						⑳ $\mathbf{x}^* = (0,0,13,18,0,0,0)/1$										

δ^r = common denominator = $\pm|\mathbf{B}_r|$ = product of first r pivots. Column vectors $\delta^0\mathbf{a}^0, \delta^1\mathbf{a}^1, \delta^2\mathbf{a}^2$ (with starred pivots) replace $\boldsymbol{\eta}^1, \boldsymbol{\eta}^2, \boldsymbol{\eta}^3$

Calculation of $\delta^3\mathbf{b}^3$: $-13 = [1^*(-17) - (-7)(-18)]/11$; $-18 = -18$ above. FT likewise.

Calculation of last column, from bottom to top, using the $\delta^{r-1}\mathbf{a}^{r-1}$ at ⑮, ⑨, ③:

$1 = \delta_3$. $(-1, 1) = -\mathbf{c}_{M^3}^0 = -(c_3^0, c_4^0)$. $4 = (1(11) - (-1)(-7))/1$. $-1 = -1$ below. $4 = 4$ below. $0 = (-1(4) - 4(-1))/11$.

$1 = \delta_3$. $1 = (4(1) - 0(3))/4$. $0 = 0$ below. (See equations (11) and (12).)

persists until all the selected columns and also those they replace in the basis have acquired nonnegative costs, but this is liable to involve an excessive number of iterations.

(e) The use of Gaussian elimination (**LU** *triangular factorization*) and pivoting for sparseness is considered in 7.4 and 7.6. A clue to the possibilities here is offered by the fact that if a basis **B** is lower triangular (a_{ij} in **B** with $j>i$ implies $a_{ij}=0$), the $\mathbf{a}^r$ can be just the columns of **B** in the order $1,2,\ldots,m$; the sentence preceding (3) shows that with a triangular basis, the columns require no updating if the pivotal rows are taken in the order of increasing numbers of nonzero elements. One can do this in reinverting.

EXERCISES 7.3

1. Let the updated pivotal columns be $\mathbf{a}^0_{.1}=(3^*,-1,5)$, $\mathbf{a}^1_{.2}=(-2,4^*,1)/3$. With or without calculating $\boldsymbol{\eta}_1$ and $\boldsymbol{\eta}_2$, find the two-step forward and backward transforms of (a) $(1,2,3)$ (b) $(1,-1,1)$, (c) $(-1,4,1)$. Note that these vectors (and hence their transforms) satisfy (a)=2(b)+(c).
2. Suppose that $\mathbf{a}^0_{.5}=(1,0,3^*)$, $\mathbf{a}^0_{.6}=(4^*,1,-1)$, $\mathbf{a}^0_{.4}=(0,3^*,1)$ are the first three pivotal columns in $\mathbf{A}^0$, the pivotal positions being starred. Given that $\mathbf{b}^0=(-3,-1,-4)$, $\mathbf{c}^0_{4,5,6}=(-1,-2,-1)$, and $M^0=\{1,2,3\}$, calculate the equivalent of Table 7.3B and the $\boldsymbol{\pi}^r$.
3.–9. Use the product method to optimize 3 to 9 of 7.2. (Table (A) of the solution is essentially unchanged.)
10. Let $\mathbf{E}^1$ and $\mathbf{E}^2$ be elementary matrices with column vectors **e** and **f** whose components are indexed by their physical positions, of which p and q are pivotal ($p\neq q$). Show that $\mathbf{E}^1\mathbf{E}^2=\mathbf{E}^2\mathbf{E}^1$ if and only if $e_q=f_p=0$ or $\mathbf{E}^1=\mathbf{I}$ or $\mathbf{E}^2=\mathbf{I}$.
11. Let $p=q$ in 10 and find a single $\boldsymbol{\eta}$ equivalent to the consecutive pair $\boldsymbol{\eta}^1=\mathbf{e}$ and $\boldsymbol{\eta}^2=\mathbf{f}$.
12. What do $\boldsymbol{\eta}^1=(3^*,2)$ and $\boldsymbol{\eta}^2=(1,4^*)$ become if the order of the pivots is reversed?

7.4 GAUSSIAN ELIMINATION: B=LU

In 2.6 we briefly contrasted the Jordan and the Gaussian methods of solving systems of linear equations. Although the Jordan method is naturally associated with the basis changes of the simplex method, the Gaussian method requires only about two-thirds as many operations (see (2) of 7.5) and it can be made less vulnerable to round-off errors. It is applied to systems of linear equations and to the revised simplex method in this section and in 7.6.

For reasonable consistency with the linear programming notation used elsewhere in this book, the system of linear equations to be solved will be denoted by

$$\mathbf{Bx}=\bar{\mathbf{b}} \tag{1}$$

where $\mathbf{B}=(b_{ij})$ is an m by m invertible matrix.

The algorithm of 2.6(c) can be reduced to a very compact form, versions of which have been given by various writers including Doolittle (1878) and Crout (1941). As in Wilkinson (1963), we shall derive this form from the *triangular* (**LU**) *decomposition* of the matrix **B**. This means the determination of a lower-triangular matrix $\mathbf{L}=(l_{ij})$, for which $i<j$ implies $l_{ij}=0$, and an upper-triangular matrix $\mathbf{U}=(u_{ij})$, for which $i>j$ implies $u_{ij}=0$, such that

$$\mathbf{B}=\mathbf{LU} \tag{2}$$

L and **U** are not uniquely defined by these conditions; also they may fail to exist, even though **B** is invertible, until after its rows or columns have been permuted.

One or more columns $\bar{\mathbf{b}}$ can be adjointed to **B**, with the result that corresponding columns will be adjoined to **U**, without changing **L** or **U** itself. If $m=3$ and one such column is adjoined, (2) becomes

$$\left[\begin{array}{ccc|c} b_{11} & b_{12} & b_{13} & \bar{b}_1 \\ b_{21} & b_{22} & b_{23} & \bar{b}_2 \\ b_{31} & b_{32} & b_{33} & \bar{b}_3 \end{array}\right] \tag{3}$$

$$=\begin{bmatrix} l_{11} & 0 & 0 \\ l_{21} & l_{22} & 0 \\ l_{31} & l_{32} & l_{33} \end{bmatrix}\left[\begin{array}{ccc|c} u_{11} & u_{12} & u_{13} & u_{14} \\ 0 & u_{22} & u_{23} & u_{24} \\ 0 & 0 & u_{33} & u_{34} \end{array}\right]$$

$$=\left[\begin{array}{cccc|c} l_{11}u_{11} & l_{11}u_{12} & l_{11}u_{13} & & l_{11}u_{14} \\ l_{21}u_{11} & l_{21}u_{12}+l_{22}u_{22} & l_{21}u_{13}+l_{22}u_{23} & & l_{21}u_{14}+l_{22}u_{24} \\ l_{31}u_{11} & l_{31}u_{12}+l_{32}u_{22} & l_{31}u_{13}+l_{32}u_{23} & & l_{31}u_{14}+l_{32}u_{24} \\ & & +l_{33}u_{33} & & +l_{33}u_{34} \end{array}\right]$$

The square matrices yield $m^2=9$ equations to be satisfied by $m(m+1)/2=6$ of the l_{ij} and a like number of u_{ij}; the adjoined column involves $m=3$ additional equations and u_{ij}.

For purposes of explanation and hand calculation it is simplest to suppose that (3) is solved in the following traditional sequence. It is assumed that none of the pivots $l_{ss}u_{ss}$ is zero.

1. Choose l_{11}, u_{11} so that $l_{11}u_{11}=b_{11}$; then calculate

$$u_{12}=b_{12}/l_{11} \qquad u_{13}=b_{13}/l_{11} \qquad u_{14}=\bar{b}_1/l_{11}$$

$$l_{21}=b_{21}/u_{11} \qquad l_{31}=b_{31}/u_{11}$$

Equality now holds in the first row and column.

2. Choose l_{22}, u_{22} so that $l_{22}u_{22}=b_{22}-l_{21}u_{12}$; then calculate

$$u_{23}=(b_{23}-l_{21}u_{13})/l_{22} \qquad u_{24}=(\bar{b}_2-l_{21}u_{14})/l_{22}$$
$$l_{32}=(b_{32}-l_{31}u_{12})/u_{22}$$

Equality now holds in the first two rows and columns.

3. Choose $l_{33}u_{33}=b_{33}-l_{31}u_{13}-l_{32}u_{23}$; then calculate

$$u_{34}=(\bar{b}_3-l_{31}u_{14}-l_{32}u_{24})/l_{33}$$

Equation (3) has now been solved; the sequence is indicated pictorially in (13) below.

In general the equations

$$b_{ij}=\sum_{h=1}^{\min(i,j)} l_{ih}u_{hj} \tag{4}$$

(for $i=1,\ldots,m$ and $j=1,\ldots,m$ or more) are solved in m cycles. In cycle $s(=1,\ldots,m)$ one chooses l_{ss} and u_{ss} so that

$$l_{ss}u_{ss}=b_{ss}-\sum_{h=1}^{s-1} l_{sh}u_{hs} \tag{5}$$

and then calculates

$$u_{sj}=l_{ss}^{-1}\left(b_{sj}-\sum_{h=1}^{s-1} l_{sh}u_{hj}\right) \qquad \text{for } j=s+1,\ldots,m \text{ or more} \tag{6}$$

$$l_{is}=u_{ss}^{-1}\left(b_{is}-\sum_{h=1}^{s-1} l_{ih}u_{hs}\right) \qquad \text{for } i=s+1,\ldots,m \tag{7}$$

This completes the factorization $\mathbf{B}=\mathbf{LU}$ and reduces (1) to

$$\mathbf{LUx}=\mathbf{Ly}=\bar{\mathbf{b}} \qquad \text{where } \mathbf{y}=\mathbf{Ux} \tag{8}$$

Thus one may first solve the system $\mathbf{Ly}=\bar{\mathbf{b}}$ for $\mathbf{y}$ and then solve $\mathbf{Ux}=\mathbf{y}$ for $\mathbf{x}$. The triangular nature of $\mathbf{L}$ and $\mathbf{U}$ makes these systems very easy to solve. There is one equation containing only one unknown; when it is solved and the value substituted, there is another equation in only one unknown, etc. This is the back-substitution procedure already mentioned in 2.6(c). The solution procedures are given by

$$y_i=l_{ii}^{-1}\left(\bar{b}_i-\sum_{j=1}^{i-1} l_{ij}y_j\right) \qquad \text{for } i=1,\ldots,m \tag{9}$$

$$x_i=u_{ii}^{-1}\left(y_i-\sum_{j=i+1}^{m} u_{ij}x_j\right) \qquad \text{for } i=m,\ldots,1 \tag{10}$$

If $\bar{\mathbf{b}}$ is adjoined to $\mathbf{B}$ as in (3), the corresponding column $\mathbf{y}$ is adjoined to $\mathbf{U}$, since (3) [with $\mathbf{y}=(u_{14}, u_{24}, u_{34})$] shows that it satisfies $\mathbf{Ly}=\bar{\mathbf{b}}$. In fact the algorithm is equivalent to premultiplying the system $\mathbf{Bx}=\bar{\mathbf{b}}$ by $\mathbf{L}^{-1}$ (another lower-triangular matrix, not actually computed) to obtain $\mathbf{Ux}=\mathbf{L}^{-1}\bar{\mathbf{b}}=\mathbf{y}$. Thus in this case only the system $\mathbf{Ux}=\mathbf{y}$ has to be solved by back substitution, but the same arithmetic operations are performed in either event.

The determinant of a triangular matrix is evidently equal to the product of its diagonal elements. Then Theorem 6 of 7.7 shows that the determinant of $\mathbf{B}$ is

$$|\mathbf{B}|=|\mathbf{L}|\cdot|\mathbf{U}|=\prod_{s=1}^{m}(l_{ss}u_{ss}) \tag{11}$$

If $\mathbf{B}$ is symmetric ($b_{is}=b_{si}$), (5) is replaced by

$$l_{ss}=u_{ss}=\left(b_{ss}-\sum_{h=1}^{s-1}l_{sh}u_{hs}\right)^{1/2} \tag{12}$$

especially if $\mathbf{B}$ is positive-definite. Otherwise pure imaginary numbers will occur, but none with both real and imaginary parts. A worse drawback is the possible occurrence of pivots that are zero or nearly so. An inductive proof using (6) and (7) with $j=i$ obviously shows that $u_{si}=l_{is}$; $\mathbf{L}$ and $\mathbf{U}$ are transposes of one another, and only one of them has to be calculated and recorded. (See Exercise 5.)

If $\mathbf{B}$ is not symmetric, it is customary to set each l_{ss} equal to 1 or to that power of 2 or 10 that is closest to (12) in absolute value; (5) to (7) then determine the other elements of $\mathbf{L}$ and $\mathbf{U}$. The approximation to (12) scales the matrices in a nearly symmetrical fashion, but we shall take $l_{ss}=1$ for simplicity. Then the triangular row equations in $\mathbf{U}$ are given explicitly in (13) below, and a slight reduction in round-off error is perhaps achieved by dividing by u_{ii} in (10) rather than by l_{ii} (earlier) in (6) or (9). If $\mathbf{B}$ is symmetric, $l_{ss}=1$ requires writing or storing both $\mathbf{L}$ and $\mathbf{U}$ but no more calculation than (12) because $l_{is}=u_{si}/u_{ss}$ ($i>s$) can be used in place of (7). This is still dependent on the assumption that the pivotal positions are on the diagonal.

The procedure is summarized in Tables 7.4A and C and in Table 7.4B is used to obtain the solution $x=(18.6, -3.6, 11.7, -1.4)$ for the system

$$\begin{aligned}
2x_1+x_2-2x_3+3x_4&=6\\
x_1+3x_2\qquad\quad+2x_4&=5\\
-3x_1-x_2+4x_3-x_4&=-4\\
-x_1-x_2+2x_3+x_4&=7
\end{aligned}$$

Check columns are used as in 2.6; the l_{ij} are not included in the row sums but they are checked via their effect on the other entries. The back substitution must be checked separately, either by using the final check column as $\mathbf{y}$ in (10) to obtain x_i+1 or by substituting the x_i in the original equations.

TABLE 7.4A Compact Gaussian Solution of $\mathbf{Bx}=\bar{\mathbf{b}}$

Define $\mathbf{B}=\mathbf{LU}$, $\mathbf{Ux}=\mathbf{y}$ where $l_{ij}=0$ $(i<j)$, $u_{ij}=0$ $(i>j)$. If the sums in (14) are accumulated in a calculator, nothing need be written down except the m^2+m elements of $\mathbf{B}$ and $\bar{\mathbf{b}}$ and the m^2+2m nonzero elements of $\mathbf{L}-\mathbf{I}$, $\mathbf{U}$, $\mathbf{y}$ and $\mathbf{x}$ in (13) in sequence ①, ②,..., ⑪ (for $m=4$). No division in ①, ③, ⑤, ⑦.

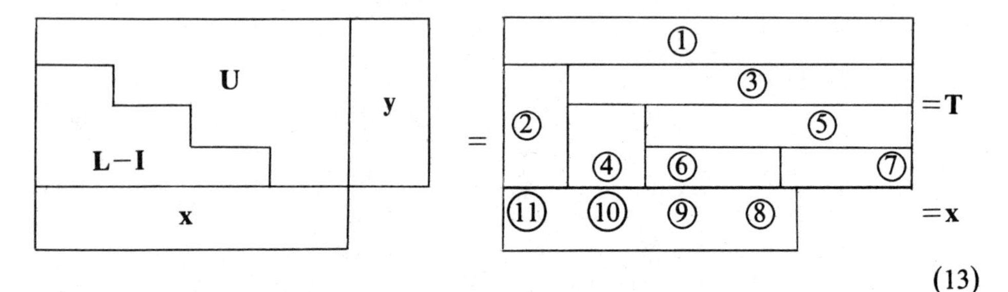

(13)

The elements t_{ij} of $\mathbf{T}$ and x_i of $\mathbf{x}$ are calculated by

$$t_{ij}=\left(b_{ij}-\sum_{h=1}^{\min(i,j)-1} t_{ih}t_{hj}\right)\div\begin{bmatrix}1 & \text{if } i\leq j\\ t_{jj} & \text{if } i>j\end{bmatrix} \tag{14}$$

$$x_i=u_{ii}^{-1}\left(y_i-\sum_{j=i+1}^{m} u_{ij}x_j\right) \qquad \text{for } i=m,\ldots,1 \tag{15}$$

The positions of the t_{ih} and t_{hj} for $h=1,2,3$ in finding t_{45} and t_{54} by (14) are shown in (16); the pivots are starred and $\mathbf{y}$ is omitted.

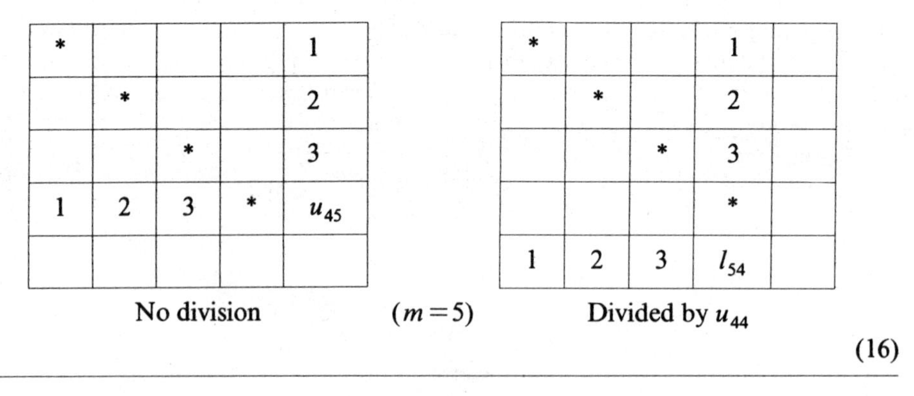

(16)

TABLE 7.4B Example of $\mathbf{Bx}=\mathbf{LUx}=\mathbf{Ly}=\bar{\mathbf{b}}$

B				$\bar{\mathbf{b}}$		**U**				**y**	
2	1	−2	3	6	→	2.0*	1.0	−2.0	3.0	6.0	①
1	3	0	2	5		.5	2.5*	1.0	.5	2.0	③=T
−3	−1	4	−1	−4	L−I	−1.5	.2	.8*	3.4	4.6	⑤
−1	−1	2	1	7		−.5	−.2	1.5	−2.5*	3.5	⑦
						②	④	⑥			
						⑪	⑩	⑨	⑧		
(Data for $\mathbf{Bx}=\bar{\mathbf{b}}$)						18.6	−3.6	11.7	−1.4	=**x**	

For example, $l_{43}=1.5=[2-(-.5)(-2.0)-(-.2)(1.0)]/.8,\quad x_4=3.5/(-2.5)$

$$y_4=u_{45}=3.5=7-(-.5)(6.0)-(-.2)(2.0)-(1.5)(4.6)$$

$$x_1=18.6=[6.0-1.0(-3.6)-(-2.0)(11.7)-(3.0)(-1.4)]/2.0$$

Check columns are (10, 11, −5, 8), (10, 6, 8.8, 1). For example, in row 3:

$$-5=-3-1+4-1-4;\quad 8.8=.8+3.4+4.6=-5-(-1.5)(10)-(.2)6.$$

TABLE 7.4C t_{ij} as Elements b_{ij}^s of $\mathbf{B}^s$ (after pivots $b_{11}^0,\ldots,b_{ss}^{s-1}$)

$$u_{ij}=b_{ij}^{i-1}\quad (i\le j),\qquad l_{ij}=\bar{b}_{ij}^{j}=-b_{ij}^{j}=b_{ij}^{j-1}/b_{jj}^{j-1}\quad (j<i)$$

b_{11}^0	b_{12}^0	b_{13}^0	b_{14}^0
$\bar{b}_{21}^1=\dfrac{b_{21}^0}{b_{11}^0}$	$b_{22}^1=b_{22}^0-\bar{b}_{21}^0 b_{12}^0$	$b_{23}^1=b_{23}^0-\bar{b}_{21}^1 b_{13}^0$	$b_{24}^1=b_{24}^0-\bar{b}_{21}^1 b_{14}^0$
$\bar{b}_{31}^1=\dfrac{b_{31}^0}{b_{11}^0}$	$\bar{b}_{32}^2=\dfrac{b_{32}^0-\bar{b}_{31}^1 b_{12}^0}{b_{22}^1}$	$b_{33}^2=b_{33}^0-\bar{b}_{31}^1 b_{13}^0 - \bar{b}_{32}^2 b_{23}^1$	$b_{34}^2=b_{34}^0-\bar{b}_{31}^1 b_{14}^0 - \bar{b}_{32}^2 b_{24}^1$
$\bar{b}_{41}^1=\dfrac{b_{41}^0}{b_{11}^0}$	$\bar{b}_{42}^2=\dfrac{b_{42}^0-\bar{b}_{41}^1 b_{12}^0}{b_{22}^1}$	$\bar{b}_{43}^3=(b_{43}^0-\bar{b}_{41}^1 b_{13}^0 - \bar{b}_{42}^2 b_{23}^1)/b_{33}^2$	$b_{44}^3=b_{44}^0-\bar{b}_{41}^1 b_{14}^0 - \bar{b}_{42}^2 b_{24}^1-\bar{b}_{43}^3 b_{34}^2$

TABLE 7.4D Computer Calculation of Fourth Row of T Above (B, T stored by rows). ( ) = content of same register j in preceding row.

b_{41}^0	b_{42}^0	b_{43}^0	b_{44}^0
$\bar{b}_{41}^1=b_{41}^0/b_{11}^0\to$	$b_{42}^0-\bar{b}_{41}^1 b_{12}^0$	$b_{43}^0-\bar{b}_{41}^1 b_{13}^0$	$b_{44}^0-\bar{b}_{41}^1 b_{14}^0$
$\bar{b}_{41}^1$	$\bar{b}_{42}^2=(\ )/b_{22}^1\to$	$(\ )-\bar{b}_{42}^2 b_{23}^1$	$(\ )-\bar{b}_{42}^2 b_{24}^1$
$\bar{b}_{41}^1$	$\bar{b}_{42}^2$	$\bar{b}_{43}^3=(\ )/b_{33}^2\to$	$b_{44}^3=(\ )-\bar{b}_{43}^3 b_{34}^2$

Let the m by m matrix obtained from $\mathbf{B} \triangleq \mathbf{B}^0$ by pivoting in positions $11, 22, \ldots, ss$ be denoted by $\mathbf{B}^s$ (not to be confused with the basis $\mathbf{B}_r = \mathbf{A}^0_{M^r}$ in 7.1 to 7.3, which is an instance of $\mathbf{B}^0$). Then Table 7.4C uses the elements b^s_{ij} of $\mathbf{B}^s$ or their negatives $\bar{b}^s_{ij}$ to express the elements of $\mathbf{T} = [\mathbf{L}\backslash\mathbf{U}]$ and their method of calculation in the sequence (13), which requires only one (possibly double-precision) register to accumulate products. Table 7.4C depicts $\mathbf{T}$ and so remains valid if each b^s_{ij} or $\bar{b}^s_{ij}$ in the table is replaced by t_{ij}.

Table 7.4C shows how the compact Gaussian procedure could be derived from the pivot rules of 2.4 instead of from $\mathbf{B} = \mathbf{LU}$. For this we need the relation $l_{ij} = \bar{b}^j_{ij} = b^{j-1}_{ij}/b^{j-1}_{jj}$ $(j<i)$. Then, for example,

$$\begin{aligned}\bar{b}^3_{43} = t_{43} = -b^3_{43} &= \left(b^0_{43} - \bar{b}^1_{41} b^0_{13} - \bar{b}^2_{42} b^1_{23}\right)/b^2_{33} \\ &= \left(b^0_{43} - b^0_{41} b^0_{13}/b^0_{11} - b^1_{42} b^1_{23}/b^1_{22}\right)/b^2_{33} \qquad (17)\end{aligned}$$

which reproduces exactly the effect of the pivots in positions 11, 22 and 33. From the point of view of the all-integer method of 2.5, Table 7.4C shows how the various common denominators relate to $\mathbf{T}$; that for t_{ij} is δ_s, where the index $s = \min(i-1, j)$. Here $\delta_0 = 1$, $\delta_1 = b_{11}$, $\delta_2 = b_{11}b_{22} - b_{12}b_{21}$, etc.

A computer usually prefers to calculate by rows or by columns (not both), which requires m registers, one for each column in Table 7.4D. This table illustrates the calculation of the last row of $\mathbf{T}$ in Table 7.4C when $\mathbf{B}$ and $\mathbf{T}$ are stored by rows. Each row in 7.4D can use data to the left in the same row; beyond this the four rows require data only from the fourth row of $\mathbf{B} = \mathbf{B}^0$ and the first, second and third rows of $\mathbf{T}$, respectively. The calculation is completed in the last row of Table 7.4D, which is the same as the last row of $\mathbf{T}$ in Table 7.4C. Then the back substitution (15) goes as before; it is a triangular form of the backward transformation (repeated use of (11) of 7.3). However, if the equations involved in the back substitution cut across the stored vectors, the calculation is similar to that in Table 7.4D; it is a triangular form of the forward transformation (repeated use of (10) of 7.3), in which the value of each x_j is inserted into *all* the remaining equations as soon as it is found.

In the *revised simplex* of 7.3, $\mathbf{B}$ is the current basis. One may say that the purposes of the forward and the backward transformations are to find the solutions $\mathbf{x}_{M^r} = -\mathbf{b}^r$ and $\mathbf{u}_{M^0} = \boldsymbol{\pi}^r$ for the equations

$$\mathbf{B}\mathbf{x}_{M^r} + \mathbf{b}^0 = \mathbf{0} \qquad \text{and} \qquad \mathbf{u}_{M^0}\mathbf{B} + \mathbf{c}^0_{M^r} = \mathbf{0} \qquad (18)$$

respectively as in (16) and (14) of 7.1. Thus the factorization $\mathbf{B} = \mathbf{LU}$ provides an alternate way to perform these transformations, at least up to the most recent reinversion (7.3(c)), by solving (18).

The computer solution of (18) is essentially like that in Table 7.4D, but works with columns rather than rows. In 7.3 the basis $\mathbf{B}$ or its inverse was defined in terms of column vectors $\mathbf{a}^r$ or $\boldsymbol{\eta}^r$ occurring in elementary factor matrices. The same thing can be done for $\mathbf{LU}$ *without any additional calculation*. The $\mathbf{a}^r$ are simply the columns of $\mathbf{L}$ taken in order of decreasing length followed by the columns of $\mathbf{U}$ in order of decreasing length because by 2.4 or (2) of 7.3, a column requires no updating if it has zero components in all the preceding pivotal rows. The pivots, such as they are, are on the diagonals of **L** and **U**, or what would be the diagonals if they were not permuted. Unit column vectors are omitted as usual.

The result is that there are twice as many $\mathbf{a}^r$ as in 7.3 but on the average each has only half as many nonzero elements (or less; see 7.6). According to (2) of 7.5, this requires only about two-thirds as many operations as 7.3. Table 7.4C also indicates the advantage. A corresponding table for 7.3 or Jordan elimination would have b_{ij}^{j-1} in *every* position, not just as numerators of the l_{ij}; the larger superscripts in the upper triangle imply that more work is done in 7.3. This is necessary while the simplex rule $\min_i\{-b_i/a_{il}\colon a_{il}>0\}$ is operating and the reuse of pivotal rows may be required, but it is not necessary during reinversion.

In 7.6 the assumption that the pivotal positions are $ij=11,\ldots,mm$ will be dropped.

EXERCISES 7.4

Find **L** (with all $l_{ss}=1$), **U**, **y** and **x** with diagonal pivots in 1 to 4. **B** and $\bar{\mathbf{b}}$ are given.

1.
$$\begin{array}{rrrr|r} 3 & -2 & 0 & 2 & -1 \\ 0 & 1 & 1 & -3 & 0 \\ -1 & 0 & -1 & 2 & 2 \\ 2 & 1 & -2 & 0 & -3 \end{array}$$

2.
$$\begin{array}{rrr|r} 5 & -4 & 1 & 11 \\ -2 & 1 & 3 & 6 \\ 3 & 2 & 1 & 10 \end{array}$$

3.
$$\begin{array}{rrrr|r} 1 & 0 & 3 & 2 & 5 \\ -1 & 2 & 1 & 0 & -5 \\ 3 & 1 & 0 & 1 & 11 \\ 2 & -1 & -2 & 3 & -2 \end{array}$$

4.
$$\begin{array}{rrrr|r} -2 & 0 & -1 & 1 & 2 \\ 1 & 3 & 2 & -1 & 1 \\ -2 & -1 & 1/2 & 2 & -1 \\ 1 & 1 & -1 & 0 & 4 \end{array}$$

5.* Use least squares to express z approximately as a third-degree polynomial in v, given the points

$$\begin{array}{lrrrrrr} v= & -4 & -3 & -1 & 2 & 3 & 5 \\ z= & -11 & -8 & 1 & -2 & -1 & 9 \end{array}$$

Hint: Write down six equations for the four unknown coefficients. Reduce to a system of four equations by multiplying by the transpose of the matrix of coefficients, without simplifying any equation by multiplying or dividing. Pivot down the diagonal, using a calculator and making $l_{ij}=u_{ji}$, since **B** is symmetric with $b_{ij}=\Sigma_v v^{i+j-2}$.

6. Write out analogs of Table 7.4D for storage by columns.

7.5 SPEED AND PRECISION

The size of problem that can be handled successfully is limited by the number of operations with its associated time requirement and the accumulation of round-off errors. This section briefly considers the effect of these factors on the comparative performance of three methods for solving linear programs or systems of linear equations.

The Number of Operations

The count will be limited to multiplications and divisions because additions and subtractions can be performed more rapidly and usually occur about the same number of times as multiplications and divisions. Then the pivot rules of 2.4 (with rule 5 in the form $a'_{ij}=a_{ij}-a'_{lj}a_{il}$) show that exactly one operation is required for each element of a new tableau obtained by pivoting. This remains true for the singly contracting tableaus in Jordan elimination in 2.6(b) but not quite for the doubly contracting tableaus in Gaussian elimination in 2.6(c), because for the latter the a'_{lj} are not included in the new tableau. Computer time may be saved by replacing repeated divisions such as $a'_{lj}=a_{kj}/a_{kl}$ $(j=1,2,\ldots)$ by multiplications by a_{kl}^{-1}.

(a_1) In *linear programming* the original simplex method using 2.4 requires $(m+1)(n+1)$ operations per iteration. In the explicit-inverse of 7.2 most of the operations come from steps 2 (updating the pivotal column), 4 (updating the inverse by pivoting), and 6 (pricing out). The numbers of operations may be represented by φm^2, $(m+1)^2$ and φmn, respectively, where φ is the density (proportion of nonzeros) in $\mathbf{A}^0$. Then to a rough approximation the explicit-inverse will require fewer operations provided

$$\varphi m^2+m^2+\varphi mn<mn$$

or

$$\varphi<(n-m)/(n+m) \tag{1}$$

as shown in Dantzig (1963, pp. 216–217). (His n is replaced by $m+n$ to agree with the notation of this book.)

Most large practical problems easily satisfy (1). Although (1) denies any possible advantage to the explicit-inverse method when $\varphi=1$, this is no longer true when the amount of pricing out is cut back as in 7.3(d). Nevertheless much more efficient methods are available.

(a_2) For *systems of linear equations* the following table gives the approximate number of operations required to obtain $\mathbf{B}^{-1}$ or something equivalent thereto, and then the number required to process each column b (or one entering the basis) by any of the methods considered. The "square" case has all $b_{ij}\neq 0$ while the triangular has $b_{ij}=0$ for $i>j$, say.

		Square	Triangular
Explicit inverse $\mathbf{B}^{-1}$		m^3	$m^3/6$
Jordan elimination	$\boldsymbol{\eta}$	$m^3/2$	$m^2/2$
(Revised simplex 7.3)	$\mathbf{a}$	$m^3/2$	0
Gaussian elimination (**LU**)		$m^3/3$	0
Each additional column $\mathbf{b}$		m^2	$m^2/2$
Number of storage slots		m^2	$m^2/2$

(2)

The last two lines apply to all the methods.

The zeros in this table reflect the fact that if the matrix of coefficients of a system of linear equations is elementary ((10) to (11) of 7.3) or triangular ((10) of 7.4), any sort of inversion in solving the equations is wasted effort; with or without it, the next to the last line of the table applies because the equations can be solved at once by back substitution. Thus the name of the game is not matrix inversion but matrix factorization, and the *triangular factorization* **LU** is the better of the two. The data in (2) will now be derived.

(a_3) *Jordan elimination* as in 2.6(b) does not transform the pivotal columns. (It does keep them in storage so that $m(n+1)$ storage slots are occupied throughout.) Hence if the matrix of coefficients is m by m, it has to be transformed $m-1$ times into tableaus of $(m-1)m,(m-2)m,\ldots,1\cdot m$ elements respectively for a total operation count of

$$m\sum_{1}^{m-1} r=m^2(m-1)/2 \tag{3}$$

Each of the $n-m+1$ additional columns requires m^2 operations. The following qualifications apply.

(i) If one is obliged to find the inverse $\mathbf{B}^{-1}$ explicitly, one should not use tableaus contracting from $2m$ columns to the m columns in $\mathbf{B}^{-1}$, but only m columns of m elements for each of m pivot steps as in 2.6(a) or Theorem 2 of 7.1; this requires m^3 operations.

(ii) Equation (3) holds for the revised-simplex format also if (10) and (11) of 7.3 are used; otherwise (in the notation used there) the calculation of $\eta=(-a_{1l}/a_{kl},\ldots,1/a_{kl},\ldots,-a_{ml}/a_{kl})$ for each of m steps will increase (3) by m^2 operations. If the pivotal positions are the same, 2.6(b) and (10) to (11) of 7.3 prescribe exactly the same arithmetic operations but in different orders. In 2.6(b) the rth pivotal column is used to effect one-step transformations of all the remaining $n+1-r$ columns (not yet pivotal), while in 7.3(b) the first r pivotal columns are used to effect an r-step transformation of the $r+1$st pivotal column. Efficiency demands that the latter procedure be used when one has $r\leq m+1<n+1$ for all r and the pivotal columns are not known in advance.

(iii) If an element of **B** is left untransformed only if its row *and* column have been pivoted, as in (10) of 4.5, the square m by m requires $m^3-\Sigma_1^m r^2=(4m^3-3m^2-m)/6$ operations, plus $m(2m+1)$ for **b**, **c** and d.

(a_4) *Gaussian elimination* begins with the triangular decomposition $\mathbf{B}=\mathbf{LU}$. For $m=4$ and **B** not symmetrical the numbers of operations required for the decomposition is

$$\begin{matrix} 0 & 0 & 0 & 0 \\ 1 & 1 & 1 & 1 \\ 1 & 2 & 2 & 2 \\ 1 & 2 & 3 & 3 \end{matrix}$$

arranged as in (13) of 7.4. In general the total of these numbers is

$$2\sum_{r=1}^{m-1} r(m-r) = \sum_{r=1}^{m-1} r(r+1) = (m^3-m)/3. \tag{4}$$

This is about two-thirds of the number required by Jordan elimination, and according to Klyuyev and Kokovkin-Shcherbak (1965) it is the best possible for a general $\mathbf{B}$. Then to solve $\mathbf{Bx}=\bar{\mathbf{b}}$ requires $m(m-1)/2$ operations to compute $\mathbf{y}=\mathbf{L}^{-1}\bar{\mathbf{b}}$ and $m(m+1)/2$ to compute $\mathbf{x}$ by back-substitution, or a total of m^2 additional operations for every $\bar{\mathbf{b}}$ that is processed, just as for Jordan elimination. To find $\mathbf{B}^{-1}$ by the formula $\mathbf{U}^{-1}\mathbf{L}^{-1}$ requires m^3 operations as before (Exercise 3).

(a_5) A *determinant* has a value equal to the product of the m pivots required in Gaussian elimination by Theorem 11 of 7.8. This gives (4) plus $m-1$ for its evaluation in the absence of zeros. However, one operation can be saved near the end by evaluating the 2 by 2 determinant directly instead of performing the $m-1$st pivot. The following table compares this with (3) and an alternative method using Theorem 5 of 7.7, which is best for algebraic work.

	m	2	3	4	5	6
Gaussian	$(4)+m-2$	2	9	22	43	74
Jordan	$(3)+m-2$	2	10	26	53	94
Minors	(6)	2	9	28	75	186

(5)

The numbers in the last row of (5) are given by

$$\sum_{2}^{m} j\binom{m}{j} = m(2^{m-1}-1) \tag{6}$$

where j is the number of multiplications (one per term) in the expansion of a j-rowed minor and $\binom{m}{j}$ is the number of such minors in the first j columns of $\mathbf{B}$. (See Gentleman and Johnson, 1974.) All such minors are evaluated for $j=2,\ldots,m$.

Control of Round-Off Errors

If a number given as $x+\varepsilon$ has a true value $x\neq 0$ and an absolute error $|\varepsilon|$, its *relative error* is defined to be $|\varepsilon/x|$ and the number of significant figures is approximately $\log|x/\varepsilon|$, where the logarithm is to the base 10 or whatever is the base of the number system. It is easy to show that after a series of arithmetic operations consisting of addition, multiplication, division, or extraction of roots, all applied to positive numbers and carried out with sufficient precision, the relative error of the result will not much exceed $p+1$ times the largest relative error of any input, where p is the number of multiplications plus the number of divisions. If the input errors cancel one another to any extent, as they tend to do, the relative error of the output will be less than the indicated limit.

The *subtraction* of positive numbers, or the addition of numbers of unspecified signs, is conspicuously absent from the preceding list of operations. Compare

$$\begin{bmatrix} 100 & 99 \\ 100 & 100 \end{bmatrix}^{-1} = \begin{bmatrix} 1 & -.99 \\ -1 & 1 \end{bmatrix} \quad \text{with} \quad \begin{bmatrix} 100 & 99 \\ 101 & 100 \end{bmatrix}^{-1} = \begin{bmatrix} 100 & -99 \\ -101 & 100 \end{bmatrix} \tag{7}$$

reflecting the change in the determinant from 100 to $1=(100)(100)-(99)(101)$. Although the elements of the original matrices differ by at most 1%, the subtraction required in evaluating the determinant has reduced the number of significant figures to zero. Some suggestions for reducing round-off errors will now be mentioned.

(b_1) Since addition or subtraction is the operation wherein relative precision is lost, many calculators, large or small, are designed to accumulate sums of products of two factors each with *double precision* (retaining twice the usual number of digits) without any loss of time or special programming. The final sum is then cut to the normal length. This device is fully effective (e.g. in (7)) only if the input data are exact but it is beneficial in general. One of the attractions of the compact Gaussian algorithm of 7.4 is that it takes the maximum advantage of this feature. Programs are also available to perform *all* operations with double precision, but this takes about three times as long.

(b_2) Another useful principle of linear computation is to *prefer pivots that are large in absolute value* when a choice is available, as in the pivotal positioning of 7.4 discussed in 7.6. A small pivot is likely to have arisen as a difference of larger quantities as in (7) and also is likely to produce such a situation again.

To illustrate this, consider the following two sequences of pivots for inverting the matrix by Theorem 2 of 7.1:

$$\begin{bmatrix} 10^* & 9 & -9 \\ 9 & 8 & 1 \\ -9 & 1 & 1 \end{bmatrix} \to \begin{bmatrix} 1 & 9 & -9 \\ -9 & -1^* & 91 \\ 9 & 91 & -71 \end{bmatrix} \div 10 \to \begin{bmatrix} -8 & 9 & 81 \\ 9 & -10 & -91 \\ -81 & 91 & 821^* \end{bmatrix} \tag{8}$$

$$\downarrow \qquad\qquad\qquad\qquad \downarrow$$

$$\begin{bmatrix} 1 & 9 & -9 \\ -9 & -1 & 91 \\ 9 & 91 & -71^* \end{bmatrix} \div 10 \to \begin{bmatrix} -1 & -18 & -9 \\ 18 & 821^* & 91 \\ -9 & -91 & -10 \end{bmatrix} \div 71 \to \begin{bmatrix} -7 & 18 & -81 \\ 18 & 71 & 91 \\ -81 & 91 & 1 \end{bmatrix} \div 821$$

These data are exact, but the last two matrices on the right, following the pivot -1, are related by $-8\cdot 821+81^2=7$, which involves a loss of three significant figures. When the lower sequence involving the larger pivot -71 is used, no more than two significant figures are lost in any one calculation. This is the best that can be done, since one cannot avoid evaluating the combination $10\cdot 8-9\cdot 9=-1$.

By Theorem 11 of 7.8, the product of all the pivots is always equal to the determinant, so that early success in finding large pivots may require the acceptance of small pivots later, but at least fewer iterations are thereby adversely affected. Suitable scaling should improve the effectiveness of selecting the largest pivot; there is more on this in Harris (1973) and Tomlin (1975).

(b_3) In the simplex method the accumulated errors may be increased by extraneous pivot steps (hence the need for reinversion as in 7.3(c)). Consider the initial and final tableaus

1	x_3	x_4	x_5	1
x_1	1	2	102	-1
x_2	2	1	199*	-1
$-z$	-1	-1	-100	0

3	x_5	x_1	x_2	1
x_4	5	2	-1	-1
x_3	296	-1	2	-1
$-z$	1	1	1	-2

(9)

If one avoids pivoting in column x_5, perhaps by using (7) of 6.3 to obtain independence of column scaling, the loss of nearly two significant digits is confined to that column. Otherwise the loss occurs in a column that is pivotal thereafter and so affects the whole tableau. The row selection rule $\min_i\{-b_i/a_{il}: a_{il}>0\}$ in the simplex method has somewhat the same effect as partial positioning; *after* l is selected, it is independent of row and column scaling.

(b_4) *To improve the precision* of a solution $\mathbf{x}^1$ obtained by any method, the following procedure may be repeated several times, for $s=1,2,\ldots$ until the sequence of solutions $\mathbf{x}^s$ appears to have converged satisfactorily. For each s, the vector of *residuals* $\bar{\mathbf{b}}-\mathbf{B}\mathbf{x}^s$ is computed with *extra precision*; this requires m^2 multiplications for each value of s. Then the equations

$$\mathbf{B}(\mathbf{x}^{s+1}-\mathbf{x}^s)=\bar{\mathbf{b}}-\mathbf{B}\mathbf{x}^s \tag{10}$$

are solved for the corrections $\mathbf{x}^{s+1}-\mathbf{x}^s$, which are to be added to $\mathbf{x}^s$ to obtain the improved solution $\mathbf{x}^{s+1}$. This requires only m^2 additional multiplications or divisions ((2) above) since normally one can use the same matrices $\mathbf{L},\mathbf{U}$ (or vectors in the product form) that were used to produce the first solution $\mathbf{x}^1$. It is a notable fact that an acceptable convergence of the $\mathbf{x}^s$ may be obtained without any accompanying decrease in the residuals. Thus in Figure 7.5 $\mathbf{x}^2$ is closer to the intersection but no closer to the lines.

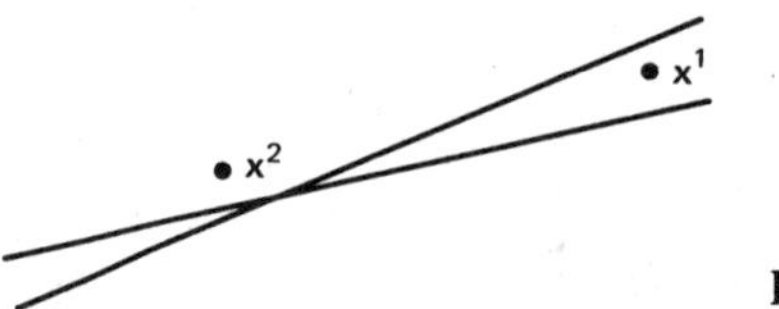

FIGURE 7.5

The Gauss-Seidel or relaxation type of method also improves the precision of $\mathbf{x}^s$ iteratively. However, to be useful it requires a dominant diagonal in $\mathbf{B}$, which is not typical of linear programming problems.

Round-off errors in linear systems have been studied extensively by Wilkinson (1961, 1963, 1965); Forsythe and Moler (1967) is a good introduction.

EXERCISES 7.5

1. How many operations are required to solve the linear equations of 2 and 5 of 7.4?
2. Show that to verify that $\mathbf{B}\mathbf{B}^{-1}=\mathbf{I}$ or $\mathbf{L}\mathbf{U}=\mathbf{B}$ requires the same number of operations as the original determination of $\mathbf{B}^{-1}$ or $\mathbf{L}$ and $\mathbf{U}$.

3. Show that the calculation of $\mathbf{U}^{-1}\mathbf{L}^{-1}$, like other methods of getting $\mathbf{B}^{-1}$, requires m^3 operations.
4. If the optimal basis $M = \bar{K} \cup L$ is known in advance, how many operations are required to find z^* and the complete vectors $\mathbf{x}^*$ and $\mathbf{u}^*$ in a linear program? Ignore the presence of zeros except those in unit vectors in the basis, and consult (8) of 4.5.
5. Show that the relative error in the sum of two positive numbers cannot exceed the larger of the relative errors of the two numbers. Extend to a sum of p positive numbers.
6. Show that the relative error in a product or quotient of two numbers is at worst approximately equal to the sum of the relative errors of the two numbers.
7. Extend 6 to show that at worst the relative error in $x_1^{p_1} x_2^{p_2} \cdots x_k^{p_k}$ (where the p_i are integers) is approximately $|p_1|\theta_1 + |p_2|\theta_2 + \cdots + |p_k|\theta_k$, where θ_i is the relative error in x_i. (Logarithms and the calculus give the same result for any real p_i.)

7.6 PIVOTING FOR SPARSENESS OR SIZE

$$\mathbf{B} = \begin{bmatrix} 3 & 5 & 0 & 0 \\ 2 & 0 & 6 & 0 \\ 0 & 4 & 0 & 9 \\ 0 & 0 & 7 & 8 \end{bmatrix} = 54 \begin{bmatrix} -192 & 315 & 240 & -270 \\ 126 & -189 & -144 & 162 \\ 64 & -96 & -80 & 90 \\ -56 & 84 & 70 & -72 \end{bmatrix}^{-1} \tag{1}$$

$$= \begin{bmatrix} 1 & 0 & 0 & 0 \\ 2/3 & 1 & 0 & 0 \\ 0 & -6/5 & 1 & 0 \\ 0 & 0 & 35/36 & 1 \end{bmatrix} \begin{bmatrix} 3 & 5 & 0 & 0 \\ 0 & -10/3 & 6 & 0 \\ 0 & 0 & 36/5 & 9 \\ 0 & 0 & 0 & -3/4 \end{bmatrix} \tag{2}$$

Corresponding pivotal columns (read left to right):

Pivots at 11, 22, 33, 44				Pivots at 11, 44, 22, 33			
3*	5/3	3	−15/4	3*	0	5/3	3
2	−10/3*	−9/5	9/4	2	0	−10/3*	−9/5
0	4	36/5*	5/4	0	9	4	−27/40*
0	0	7	−3/4*	0	8*	0	7/8

(3)

Another example for **LU**:

$$\begin{bmatrix} 2 & 5 & 7 \\ 3 & 6 & 0 \\ 4 & 0 & 0 \end{bmatrix} \begin{bmatrix} 1 & 0 & 0 \\ 0 & 1 & 0 \\ 0 & 0 & 1 \end{bmatrix} = \begin{bmatrix} 1 & 0 & 0 \\ 3/2 & 1 & 0 \\ 2 & 20/3 & 1 \end{bmatrix} \begin{bmatrix} 2 & 5 & 7 \\ 0 & -3/2 & -21/2 \\ 0 & 0 & 56 \end{bmatrix} \tag{4}$$

7.6 PIVOTING FOR SPARSENESS OR SIZE

We have said little so far about a very important factor, the *sparseness* of the initial tableaus $\mathbf{A}^0$ encountered in practice. Beale (1975) remarked that even in very large programs, a typical variable x_j occurs in only about six equations; its other coefficients are zero. This implies a corresponding reduction in the number of operations, to the extent that the sparseness can be carried over to the factored basis. Since arithmetic operations are the cause of losses of sparseness (as well as losses of time and precision) it is not surprising that versions of the methods found to require the fewest operations in 7.5 when applied to a full matrix are also relatively effective in preserving sparseness.

Table 7.6A shows a matrix $\mathbf{B}$ with 8 nontrivial elements ($\neq 0$ or 1), $\mathbf{B}^{-1}$ with 16 and $\mathbf{LU}$ with 10. (3) shows the pivotal columns (whose transforms are the $\boldsymbol{\eta}^r$), arranged in left-to-right order; these contain 13 and 11 nontrivial elements for two different pivot sequences. Equation (4) shows two different $\mathbf{LU}$ factorizations of another matrix having 6 and 9 nontrivial elements.

Reports of experience with large problems confirm the desirability of the $\mathbf{LU}$ factorization and the dependence of the sparseness on the choice of pivots during reinversion or rather refactorization. To determine the pivots one may arrange $\mathbf{B}$ so that it becomes

$$\mathbf{B}=\begin{bmatrix}\mathbf{B}_{11} & \mathbf{0} & \mathbf{0}\\ \mathbf{B}_{21} & \mathbf{B}_{22} & \mathbf{0}\\ \mathbf{B}_{31} & \mathbf{B}_{32} & \mathbf{B}_{33}\end{bmatrix} \tag{5}$$

with $\mathbf{B}_{11}$ and $\mathbf{B}_{33}$ lower triangular (or empty). Beale suggested factoring the second of the obvious three factors of (5) to separate $\mathbf{B}_{22}$ and $\mathbf{B}_{32}$ and obtain

$$\mathbf{B}=\begin{bmatrix}\mathbf{B}_{11} & \mathbf{0} & \mathbf{0}\\ \mathbf{B}_{21} & \mathbf{I} & \mathbf{0}\\ \mathbf{B}_{31} & \mathbf{0} & \mathbf{I}\end{bmatrix}\begin{bmatrix}\mathbf{I} & \mathbf{0} & \mathbf{0}\\ \mathbf{0} & \mathbf{B}_{22} & \mathbf{0}\\ \mathbf{0} & \mathbf{0} & \mathbf{I}\end{bmatrix}\begin{bmatrix}\mathbf{I} & \mathbf{0} & \mathbf{0}\\ \mathbf{0} & \mathbf{I} & \mathbf{0}\\ \mathbf{0} & \mathbf{B}_{32} & \mathbf{I}\end{bmatrix}\begin{bmatrix}\mathbf{I} & \mathbf{0} & \mathbf{0}\\ \mathbf{0} & \mathbf{I} & \mathbf{0}\\ \mathbf{0} & \mathbf{0} & \mathbf{B}_{33}\end{bmatrix} \tag{6}$$

Here only the second factor, which involves $\mathbf{B}_{22}$ alone, fails to be lower triangular. It may be possible to recognize additional (interior) triangular factors in $\mathbf{B}_{22}$.

A small example of Beale's factorization is

$$\begin{bmatrix}a_1 & b_1 & 0\\ a_2 & b_2 & 0\\ a_3 & b_3 & 1\end{bmatrix}=\begin{bmatrix}a_1 & 0 & 0\\ a_2 & 1 & 0\\ a_3 & 0 & 1\end{bmatrix}\begin{bmatrix}1 & b_1/a_1 & 0\\ 0 & b_2-a_2b_1/a_1 & 0\\ 0 & b_3-a_3b_1/a_1 & 1\end{bmatrix} \tag{7}$$

$$=\begin{bmatrix}a_1 & 0 & 0\\ a_2 & 1 & 0\\ 0 & 0 & 1\end{bmatrix}\begin{bmatrix}1 & b_1/a_1 & 0\\ 0 & b_2-a_2b_1/a_1 & 0\\ 0 & 0 & 1\end{bmatrix}\begin{bmatrix}1 & 0 & 0\\ 0 & 1 & 0\\ a_3 & 0 & 1\end{bmatrix}\begin{bmatrix}1 & 0 & 0\\ 0 & 1 & 0\\ 0 & b_3 & 1\end{bmatrix}$$

In general each form involves 6 nontrivial elements, but if $b_3 = 0$, there are 5, 6 and 5 respectively.

Methods for factoring $\mathbf{B}_{22}$ in (5) and (6) with the aim of preserving sparseness may be classified according as they do or do not determine all the pivotal positions in advance of the actual pivoting. In the first category is the early suggestion to choose each pivot to obtain (one of) the smallest available value(s) of $(n_i - 1)(m_j - 1)$, where n_i is the number of nonzeros in row i of $\mathbf{B}_{22}$ and m_j is the number in column j. This is derived from rules like those of 2.4, which show that no more than $(n_i - 1)(m_j - 1)$ nonzeros can be created by a pivot on b_{ij}. More elaborate methods in this category are given by Hellerman and Rarick (1971, 1972). In the second category is the method recommended by Dantzig (1969), which minimizes n_i in the *updated* rows and columns not yet pivoted (i.e., in the doubly contracting tableau), and then minimizes m_j. No pivot is accepted whose absolute value is less than some prescribed tolerance. Lin and Mah (1977) also give a method in the first category.

Pivotal Positioning in LU Factorization

Pivotal positioning involves specifying two permutations of $1, 2, \ldots, m$, one giving the row indices and the other the column indices in the order of pivoting. In 7.4 both of these were assumed to be identities (the natural order of the integers). In a small dense problem one can usually assume that *one* of these is the identity, obtaining what is called partial positioning. In a linear programming reinversion this would mean taking the basic columns as they are stored in $\mathbf{A}^0$ and determining the pivotal row index in each column, perhaps to maximize the absolute value of the pivot.

This is illustrated in Table 7.6B, which treats the problem of Table 7.4B. Here $\mathbf{U}$ consists of the (starred) pivots in $\mathbf{T}$ and the elements to their right; the other t_{ij} are elements of $\mathbf{L}$. As usual the value of p gives the physical position of the pivot in the corresponding column. The terms subtracted from b_{ij} in calculating t_{ij} continue to be the scalar product of vectors in the row and column of t_{ij}, but now the components of the column vector, say, must be rearranged as explained in Table 7.6B. This explanation continues to hold even if both pivotal permutations are arbitrary. The rows are used in the reverse order of their pivots in the back substitution. The initial check column contains the row sums of $[\mathbf{B}, \mathbf{b}]$; the final check column contains the row sums of $\mathbf{U}$ (including $\mathbf{y}$).

Pivoting by columns is shown in Table 7.6B because that is immediately applicable to the revised simplex method of 7.3 and in any case involves no inconvenience for a computer. However, as noted in the table, on paper the numerators of the l_{ij} have to be recorded and maximized separately (given that $l_{ii} = 1$); in $\mathbf{T}$ itself the pivoting for size is indicated only by the fact that all $|l_{ij}| \leq 1$. In pivoting for size by rows as in Exercise 2, the pivot $\max_j |u_{ij}|$ in row $i (i = 1, \ldots, m)$ can be found directly in $\mathbf{T}$ without any additional writing. Then $\mathbf{L}$ consists of the elements of $\mathbf{T}$ below the starred pivots and in the back substitution the x_j are determined in the reverse order of pivoting.

Pivoting for size by columns or rows is presumably more effective if distortion is avoided by combining it with scaling of rows or columns respectively as in Tomlin (1975b).

TABLE 7.6B Pivoting for Size in Columns 1, 2, 3, 4

Data for $\mathbf{Bx}=\bar{\mathbf{b}}$						$p=3$	$p=2$	$p=4$	$p=1$			
	$\mathbf{B}$			$\bar{\mathbf{b}}$		②	④	⑥	⑧	$\mathbf{y}$		
						$\mathbf{L}$						
2	1	−2	3	6		−2/3	1/8	1/2	5/4*	−7/4	⑨	⑩
1	3	0	2	5		−1/3	8/3*	4/3	5/3	11/3	⑤	⑫
									$\mathbf{U}$			
−3	−1	4	−1	−4	①	−3*	−1	4	−1	−4	③	⑬
−1	−1	2	1	7		1/3	−1/4	1*	7/4	37/4	⑦	⑪
Sequence						②	④	⑥				
①, ②,...						$x_1=18.6$	−3.6	11.7	$-1.4=x_4$			
all $\lvert l_{ij}\rvert\le 1$						⑬	⑫	⑪	⑩			

First pivot -3^* is absolutely largest in first column of **B**. Divide rest of column by -3 and copy the rest of third row.

$\text{Max}\{1-(-2/3)(-1), 3-(-1/3)(-1), |-1-(1/3)(-1)|\}=8/3^*$. Divide the others by pivot $8/3^*$ to get $l_{12}=1/8, l_{42}=-1/4$.

$\text{Max}\{-2-(-2/3)4-(1/8)(4/3), 2-(1/3)4-(-1/4)(4/3)\}=1^*$. As in rule 5′ of 2.5, the factors of each term $-t_{il}t_{kj}$ lie at opposite corners of the rectangle whose other corners are the t_{ij} being evaluated and a *previous* pivot t_{kl}. Then $x_1=[-4-(-1)(-1.4)-4(11.7)-(-1)(-3.6)]/(-3)=18.6$ (in row 3). Check columns are $(10, 11, -5, 8)$ and $(-1/2, 28/3, -5, 12)$. For example, $-1/2=10-(-2/3)(-5)-(1/8)(28/3)-(1/2)(12)=(5/4)-(7/4)$.

The preservation of sparseness requires complete freedom in pivoting. One procedure is to rearrange the stored columns of **B** explicitly in the desired order of pivoting and then to handle the permutation of row indices implicitly as in Table 7.6B. This is similar to what the product method does in 7.3.

As an example of arbitrary pivoting without initial rearrangement of columns, pivoting in the positions α, β, γ takes

$$\mathbf{B}=\begin{bmatrix} 6 & 7^{\gamma} & 2 \\ 1 & 5 & 9^{\alpha} \\ 8^{\beta} & 3 & 4 \end{bmatrix} \quad \text{into} \quad \mathbf{T}=\begin{bmatrix} \frac{13}{7} & \frac{90}{17}^{\gamma} & \frac{2}{9} \\ 1 & 5 & 9^{\alpha} \\ \frac{68}{9}^{\beta} & \frac{7}{9} & \frac{4}{9} \end{bmatrix} \tag{8}$$

where the elements of $\mathbf{L}-\mathbf{I}$ are $13/7$, $2/9$ and $4/9$. This gives the factorization

$$\mathbf{B}=\begin{bmatrix} 1^{\gamma} & \frac{2}{9} & \frac{13}{17} \\ 0 & 1^{\alpha} & 0 \\ 0 & \frac{4}{9} & 1^{\beta} \end{bmatrix}\begin{bmatrix} 0 & \frac{90}{17}^{\gamma} & 0 \\ 1 & 5 & 9^{\alpha} \\ \frac{68}{9}^{\beta} & \frac{7}{9} & 0 \end{bmatrix}=\mathbf{LU} \tag{9}$$

in which the columns of **L** are permuted as compared with **T**.

In forming the **a** or $\boldsymbol{\eta}$ vectors, however, the columns of **L** are taken first in order of decreasing length (numbers of potential nonzeros) and then the columns of **U** in order of decreasing length; for **U** this is the opposite of the pivoting order in (8). This corresponds to reading the following matrix product from *right to left*.

$$\begin{bmatrix} 1 & 0 & 0 \\ 0 & \frac{1}{9}^{\alpha} & 0 \\ 0 & 0 & 1 \end{bmatrix}\begin{bmatrix} 1 & 0 & 0 \\ 0 & 1 & -\frac{9}{68} \\ 0 & 0 & \frac{9}{68}^{\beta} \end{bmatrix}\begin{bmatrix} \frac{17}{90}^{\gamma} & 0 & 0 \\ -\frac{85}{90} & 1 & 0 \\ -\frac{119}{810} & 0 & 1 \end{bmatrix}$$

$$\times\begin{bmatrix} 1 & 0 & -\frac{13}{17} \\ 0 & 1 & 0 \\ 0 & 0 & 1^{\beta} \end{bmatrix}\begin{bmatrix} 1 & -\frac{2}{9} & 0 \\ 0 & 1^{\alpha} & 0 \\ 0 & -\frac{4}{9} & 1 \end{bmatrix}\mathbf{B}=\begin{bmatrix} 0 & 1 & 0 \\ 0 & 0 & 1 \\ 1 & 0 & 0 \end{bmatrix} \qquad (10)$$

This uses the elementary matrices **E** containing the $\boldsymbol{\eta}$ columns and produces the permutation of **I** that occurs in the updated extended tableau. In practice the corresponding **a** vectors $(2/9,1^{\alpha},4/9),\ldots,(0,9^{\alpha},0)$ (columns of (9)) could be used instead as in (10) and (11) of 7.3.

As in 7.3, the permutation of rows is not explicit in (10) but is accomplished in effect by the indication of the pivotal position in each vector. However, explicit permutation of rows as well as columns may facilitate the determination of the pivotal positions in reinversion to preserve sparseness as in Hellerman and Rarick (1971, 1972), or the storage of dense columns for which only the triangular foreshortening is considered relevant for closer packing. This foreshortening is offset by the doubling of the number of vectors but the better preservation of sparseness and the reduction in the number of operations in reinversion ((2) of 7.5) remain as net gains.

Equation (10) can perfectly well be combined with the full-length vectors produced by the simplex iterations as in 7.3. However, some of the papers mentioned at the end of this section consider methods for updating the **LU** factorization itself after each simplex iteration.

The Factorization B=QR and Updating

Triangular matrices are not the only ones that make a system of linear equations easy to solve. An *orthogonal matrix* **Q** also does this because by definition it has $\mathbf{Q}^{-1}=\mathbf{Q}^T$. While possibly less convenient in some respects, it does promote numerical stability (the control of round-off errors).

Suppose that $\mathbf{B}=\mathbf{QR}$ where **Q** is orthogonal and **R** is right (i.e., upper) triangular. Then $\mathbf{Bx}+\mathbf{b}=\mathbf{0}$ is equivalent to $\mathbf{QRx}+\mathbf{b}=\mathbf{0}$ or $\mathbf{Rx}+\mathbf{Q}^T\mathbf{b}=\mathbf{0}$, which can be solved by back substitution. $\mathbf{uB}+\mathbf{c}=\mathbf{0}$ is equivalent to $\mathbf{uQR}+\mathbf{c}=\mathbf{0}$, which can be solved for $\mathbf{uQ}\triangleq\mathbf{v}$ by back substitution and then $\mathbf{u}=\mathbf{vQ}^T$. One also has $\mathbf{B}^T\mathbf{B}=\mathbf{R}^T\mathbf{Q}^T\mathbf{QR}=\mathbf{R}^T\mathbf{R}$, which is called the *Cholesky* (triangular) factorization of the symmetric positive-definite matrix $\mathbf{B}^T\mathbf{B}$. The latter would never be calculated in linear programming but usually is in linear regression, where **B** is replaced by a matrix having $m>n$ (see Beale, 1970).

Q may be determined by the modified Gram-Schmidt orthogonalization (Rice, 1966) or represented as a product of more elementary orthogonal matrices. Two examples of the latter are the Householder (1958) matrices $\mathbf{I}-2\mathbf{q}\mathbf{q}^T$ with $\mathbf{q}^T\mathbf{q}=1$, and the Givens matrices, which differ from $\mathbf{I}$ in having $q_{kk}=-q_{ll}=\cos\theta$ and $q_{kl}=q_{lk}=\sin\theta$ for some particular k and l. Both of these have $\mathbf{Q}=\mathbf{Q}^T=\mathbf{Q}^{-1}$ individually and as products. Some of the papers discussing the **QR** factorization or methods for updating the **LU** or the **QR** factorization after each simplex iteration are Golub and Saunders (1970), Bartels, Golub and Saunders (1970), Bartels (1971), Forrest and Tomlin (1972), Tomlin (1972), Gill, Golub, Murray, and Saunders (1974), Gill and Murray (1974), and Saunders (1975). The first two of these also consider least squares and quadratic programming; the last suggests that the **LU** updating procedure of Bartels and Golub (1969) need be no more difficult to implement on a computer than the less stable method of Forrest and Tomlin (1972).

EXERCISES 7.6

$$\text{Let } \mathbf{B}= \begin{matrix} 0 & 1 & -2 & -1 \\ 3 & -1 & 1 & 2 \\ -1 & 2 & -2 & 1/2 \\ 1 & 0 & 1 & -1 \end{matrix} \qquad \bar{\mathbf{b}}= \begin{matrix} 2 \\ 1 \\ -1 \\ 4 \end{matrix} \text{ in 1 and 2.}$$

1. Solve $\mathbf{B}\mathbf{x}=\bar{\mathbf{b}}$ as in Table 7.6B.
2. Solve $\mathbf{B}\mathbf{x}=\bar{\mathbf{b}}$ by pivoting for size in rows 1, 2, 3, 4.
3. Factor as in (6): $\begin{matrix} 2 & 0 & 3 & 4 \\ 4 & 5 & 6 & 7 \\ 0 & 0 & 0 & 8 \\ 8 & 0 & 9 & 2 \end{matrix}$
4. Recompute the factorization (10) for **B** in (8), using the pivotal positions $ij=21, 13, 32$.

7.7 INTRODUCTION TO DETERMINANTS

Most readers will already be acquainted with determinants but may be interested in seeing how they are related to pivot steps in a matrix in 7.8. The only prerequisites in this book for 7.7 and 7.8 are Chapter 2 and 7.1 through Theorem 2.

Theorem 2 of 7.1 provides the context within which the concept of a determinant naturally arises. One can imagine the passage from $\mathbf{x}=\mathbf{B}\mathbf{y}$ to $\mathbf{y}=\mathbf{B}^{-1}\mathbf{x}$ actually being carried out in terms of the symbols b_{ij}, either by pivoting or some similar procedure. The common denominator obtained for the values of the y_j in terms of the symbols b_{ij} is (except for a possible change of sign) precisely what is meant by $\det\mathbf{B}$, $|\mathbf{B}|$, or B. This indicates the principal uses for the determinant: its nonvanishing supplies the concrete condition necessary and sufficient for **B** to have an inverse and to be of full rank m; and it permits one to write down formal expressions for inverse matrices, tableau elements, solutions of linear equations, and the like. The determinant also has a geometrical interpretation; its absolute value gives the volume of the m-dimensional parallelopiped whose edges are the row vectors (or the column vectors) of **B**.

From the above we see that $|\mathbf{B}|$ must be some kind of polynomial in the b_{ij}. Furthermore it must be a homogeneous function (of degree $p>0$) of the elements of each row or column; otherwise a set of vectors could be made linearly dependent merely by multiplying one of them by a nonzero scalar, which is impossible. To illustrate this, suppose that $b_{11}^2+b_{22}^2-b_{12}^2-b_{21}^2$ was proposed as the definition of $|\mathbf{B}|$ when $m=2$. Then multiplying the first row by t yields the expression $(b_{11}^2-b_{12}^2)t^2+b_{22}^2-b_{12}^2$, which may vanish for a suitable value of t. Since $|\mathbf{B}|$ is a polynomial, p must be an integer, and the nature of the process of solving linear equations (no squaring or repeated use of the same denominator) suggests that $p=1$, as indeed it is.

So we expect that each term in the polynomial $|\mathbf{B}|$ contains as a factor exactly one b_{ij} from each row and exactly one from each column; every value of i occurs once and every value of j occurs once. The total degree of each term is then m as an m-dimensional volume requires. Since the order of the factors doesn't matter, they can be arranged according to increasing values of i; then there are m values of j to put in the first position, $m-1$ values remaining to put in the second position, etc., making $m(m-1)\cdots 2\cdot 1=m!$ possible terms. Since the rows and columns of $\mathbf{B}$ can be rearranged without changing its rank, we expect all these terms to appear and perhaps (by symmetry) to have the same coefficient. However, if two rows or two columns are the same, the rank is reduced and the determinant must be zero. This would not be true in general if all the coefficients were identical.

Thus the closest that we can come to complete symmetry is to give half of the $m!$ terms a coefficient of $+1$ and the other half a coefficient of -1. For the whole determinant, the sign is chosen so that the diagonal term $b_{11}b_{22}\cdots b_{kk}\cdots b_{mm}$ has coefficient $+1$. Then the sign affecting each of the other terms is $+$ or $-$ according as an even or an odd number of interchanges of pairs of columns is required to bring the term into the diagonal position.

The formal *definition of the determinant* $|\mathbf{B}|=B$ is

$$\begin{vmatrix} b_{11} & \cdots & b_{1m} \\ \vdots & & \vdots \\ b_{m1} & \cdots & b_{mm} \end{vmatrix} = \sum_{\pi} (-1)^{q(\pi)} b_{1\pi(1)} b_{2\pi(2)} \cdots b_{m\pi(m)} \tag{1}$$

Here $\pi(1),\pi(2),\ldots,\pi(m)$ is a permutation (arrangement) of $1,2,\ldots,m$; the sum is over the $m!$ possible permutations π; and $q(\pi)$ is the number of *inversions* in π. An inversion occurs whenever $h<i$ but $\pi(h)>\pi(i)$, and the total number is the sum of the number of positive integers less than $\pi(1)$, the number $<\pi(2)$ and following it in the permutation (thus $\neq\pi(1)$), the number $<\pi(3)$ and following it (thus $\neq\pi(1)$ or $\pi(2)$), etc.

The definition shows that

$$|b_{11}|=b_{11} \qquad \begin{vmatrix} b_{11} & b_{12} \\ b_{21} & b_{22} \end{vmatrix} = b_{11}b_{22}-b_{12}b_{21}$$

$$\begin{vmatrix} b_{11} & b_{12} & b_{13} \\ b_{21} & b_{22} & b_{23} \\ b_{31} & b_{32} & b_{33} \end{vmatrix} = b_{11}b_{22}b_{33}+b_{12}b_{23}b_{31}+b_{13}b_{21}b_{32}$$

$$-b_{13}b_{22}b_{31}-b_{12}b_{21}b_{33}-b_{11}b_{23}b_{32}$$

THEOREM 1 The definition (1) is unchanged if the factors are written in an arbitrary order, so that $|\mathbf{B}|$ becomes

$$\sum_{\mu \text{ or } \nu} (-1)^{q(\mu)+q(\nu)} b_{\mu(1),\nu(1)} b_{\mu(2),\nu(2)} \cdots b_{\mu(m),\nu(m)} \tag{2}$$

Thus $q(\pi)$ is replaced by the sum of the number of inversions $q(\mu)$ among the row subscripts $\mu(1),\ldots,\mu(m)$ plus the number $q(\nu)$ among the column subscripts $\nu(1),\ldots,\nu(m)$. In the summation, either of the permutations μ or ν is held fixed while the other assumes the $m!$ possibilities.

Proof If $\mu(i)=i$ for $i=1,\ldots,m$ and $\nu=\pi$, then (2) reduces to (1). In any case the same $m!$ terms are produced, and only the signs remain to be considered. It is easy to see that one can go from (1) to (2) or vice versa by a series of interchanges of pairs of adjacent factors. Such an interchange introduces or eliminates only one inversion among the $\mu(i)$—that involving the two row subscripts of the adjacent factors, and so changes $q(\mu)$ by one unit; similarly it changes $q(\nu)$ by one unit. Thus the parity (evenness or oddness) of $q(\mu)+q(\nu)$ is unchanged, and (2)=(1).

THEOREM 2 Transposition of a square matrix (converting rows to columns and vice versa without change of order) leaves its determinant unchanged: $|\mathbf{B}^T|=|\mathbf{B}|$.

Proof Transposition makes it necessary to write the factors in (1) in the order $b_{\mu(1),1},\ldots,b_{\mu(m),m}$, where μ is another permutation called the inverse of π. The result is obtained by setting $\nu(j)=j$ for all j in Theorem 1.

Theorem 2 is used frequently in the rest of this chapter, without citation, usually by way of asserting a theorem for rows or columns, but giving the proof for only one of these two cases.

THEOREM 3 If two rows or two columns are interchanged, the sign of the determinant is changed. If two rows or two columns are identical, the determinant is zero.

Proof The theorem is true if two *adjacent* rows are interchanged, because exactly one inversion is gained or lost in (1). If rows i and k are interchanged, $q(\pi)$ changes by an odd integer, because $|i-k|$ interchanges of adjacent rows are required to carry row i across row k, and then $|i-k|-1$ are required to take row k back to position i. If two identical rows or columns are interchanged, this shows that $|\mathbf{B}|=-|\mathbf{B}|$ or $|\mathbf{B}|=0$.

THEOREM 4 If some multiple of row k is added to another row i, the value of the determinant is unchanged (similarly for columns).

Proof This follows from the linearity in the elements of the altered row or column, and Theorem 3. For example

$$\begin{vmatrix} b_{11} & tb_{11}+b_{12} \\ b_{21} & tb_{21}+b_{22} \end{vmatrix} = \begin{vmatrix} b_{11} & tb_{11} \\ b_{21} & tb_{21} \end{vmatrix} + \begin{vmatrix} b_{11} & b_{12} \\ b_{21} & b_{22} \end{vmatrix}$$

$$= t\begin{vmatrix} b_{11} & b_{11} \\ b_{21} & b_{21} \end{vmatrix} + \begin{vmatrix} b_{11} & b_{12} \\ b_{21} & b_{22} \end{vmatrix} = \begin{vmatrix} b_{11} & b_{12} \\ b_{21} & b_{22} \end{vmatrix}$$

The determinant B_{kl} obtained by deleting row k and column l is called a (first) *minor* of $|\mathbf{B}|$. When multiplied by $(-1)^{k+l}$ it is called the *cofactor* of b_{kl} in $|\mathbf{B}|$ because of the following theorem.

THEOREM 5 (Expansion by minors of the kth row or lth column). $|\mathbf{B}|$ equals

$$\sum_{j=1}^{m} (-1)^{k+j} b_{kj} B_{kj} \qquad \text{or} \qquad \sum_{i=1}^{m} (-1)^{i+l} b_{il} B_{il} \tag{3}$$

Proof It is clear that either formula provides exactly the required $m!$ terms; only the signs have to be verified. For the first formula, apply Theorem 1 with the factor b_{kj} occurring first in each term. Then the removal of this factor removes exactly $k-1$ inversions from $q(\mu)$ and $j-1$ from $q(\nu)$, so that a factor $(-1)^{k+j-2}=(-1)^{k+j}$, which does not appear in terms of B_{kj}, is needed in the corresponding terms of $|\mathbf{B}|$.

Repeated application of Theorem 5 provides a means for evaluating determinants. Before each application it is often desirable to make repeated application of Theorem 4 so as to obtain a row or column containing at most one nonzero element, so that (3) contains only one minor.

THEOREM 6 Determinants of the same size can be multiplied by multiplying their (square) matrices in either order, with or without transposition:

$$|\mathbf{A}|\cdot|\mathbf{B}| = |\mathbf{AB}| = |\mathbf{BA}| = |\mathbf{A}^T\mathbf{B}| = |\mathbf{AB}^T|, \text{ etc.}$$

Proof Apply linearity and Theorem 3 to $|\mathbf{AB}|$, to obtain multiples of $|\mathbf{A}|$ that combine to give $|\mathbf{A}|\cdot|\mathbf{B}|$. Thus for $m=2$:

$$\begin{vmatrix} a_{11}b_{11}+a_{12}b_{21} & a_{11}b_{12}+a_{12}b_{22} \\ a_{21}b_{11}+a_{22}b_{21} & a_{21}b_{12}+a_{22}b_{22} \end{vmatrix}$$

$$= b_{11}\begin{vmatrix} a_{11} & a_{11}b_{12}+a_{12}b_{22} \\ a_{21} & a_{21}b_{12}+a_{22}b_{22} \end{vmatrix} + b_{21}\begin{vmatrix} a_{12} & a_{11}b_{12}+a_{12}b_{22} \\ a_{22} & a_{21}b_{12}+a_{22}b_{22} \end{vmatrix}$$

$$= b_{11}b_{12}\begin{vmatrix} a_{11} & a_{11} \\ a_{21} & a_{21} \end{vmatrix} + b_{11}b_{22}\begin{vmatrix} a_{11} & a_{12} \\ a_{21} & a_{22} \end{vmatrix} + b_{21}b_{12}\begin{vmatrix} a_{12} & a_{11} \\ a_{22} & a_{21} \end{vmatrix} + b_{21}b_{22}\begin{vmatrix} a_{12} & a_{12} \\ a_{22} & a_{22} \end{vmatrix}$$

$$= 0 + (b_{11}b_{22} - b_{21}b_{12})\begin{vmatrix} a_{11} & a_{12} \\ a_{21} & a_{22} \end{vmatrix} + 0 = \begin{vmatrix} b_{11} & b_{12} \\ b_{21} & b_{22} \end{vmatrix} \cdot \begin{vmatrix} a_{11} & a_{12} \\ a_{21} & a_{22} \end{vmatrix}$$

Theorem 3 produces the sign changes in the terms of $|\mathbf{B}|$ and ensures that no two b_{ij} in the same row or column are multiplied together. Another proof is suggested in Exercise 11.

THEOREM 7 The square matrix $\mathbf{B}$ has an inverse if and only if $|\mathbf{B}| \neq 0$. In that case the inverse matrix has the element $(-1)^{i+j}B_{ji}/|\mathbf{B}|$ in row i and column j; that is, it is obtained from the *transpose* of the matrix of cofactors by dividing each cofactor by $|\mathbf{B}|$.

Proof If $|\mathbf{B}| \neq 0$, use Theorem 5 to condense the elements of $\mathbf{B}\mathbf{B}^{-1}$ or $\mathbf{B}^{-1}\mathbf{B}$. On the diagonal one has $|\mathbf{B}|/|\mathbf{B}| = 1$; elsewhere the numerator reduces to a determinant that is zero because it has two identical rows or columns, respectively. If $\mathbf{B}^{-1}$ exists, Theorem 6 shows that $|\mathbf{B}| \cdot |\mathbf{B}^{-1}| = |\mathbf{B}\mathbf{B}^{-1}| = |\mathbf{I}| = 1$, so that $|\mathbf{B}| \neq 0$.

THEOREM 8 (Cramer's Rule for Solving Linear Equations) If $\mathbf{Bx} = \bar{\mathbf{b}}$ and $|\mathbf{B}| \neq 0$, then the unique solution is $x_j = |\mathbf{B}_j|/|\mathbf{B}|$ for $j = 1, \ldots, m$, where $\mathbf{B}_j$ is the matrix obtained by replacing the jth column of $\mathbf{B}$ by the column of constants $\bar{b}$.

Proof To derive the result and prove uniqueness, multiply $\mathbf{Bx} = \mathbf{b}$ on the left by $\mathbf{B}^{-1}$. This gives $\mathbf{x} = \mathbf{B}^{-1}\bar{\mathbf{b}}$, which by Theorems 7 and 5 has the components $|\mathbf{B}_j|/|\mathbf{B}|$ as stated. Multiplying $\mathbf{x} = \mathbf{B}^{-1}\bar{\mathbf{b}}$ on the left by $\mathbf{B}$ shows that it is indeed a solution.

THEOREM 9 A determinant is zero if and only if its rows and/or its columns are linearly dependent.

Proof Suppose that the rows are linearly dependent. Then for some row k and some constants c_i one has

$$b_{kj} = \sum_{i \neq k} c_i b_{ij}$$

for all j, and the kth row can be reduced to zero by subtracting the multiples c_i of the other rows. By Theorem 4 this does not change the value of the determinant, which is thus seen to be zero.

Conversely, suppose that the determinant is zero. If all its elements are zero, its rows and columns satisfy every possible linear relation. Otherwise there is an integer ρ (the rank of the matrix), satisfying $1 \leq \rho \leq m-1$, such that some ρ-rowed minor determinant Λ is not zero but every $\rho+1$-rowed minor is zero. Then by using Theorem 5 to expand suitably chosen $\rho+1$-rowed minors containing Λ, it is easy to see that every row of the determinant is linearly dependent on the rows in which Λ lies, and every column is dependent on the columns in which Λ lies. This type of argument also establishes the following.

THEOREM 10 Any matrix $\mathbf{A}$ is of rank ρ if and only if it contains a nonzero ρ-rowed minor Λ, and every $\rho+1$-rowed minor (or every such minor containing Λ) is zero.

EXERCISES 7.7

1. Count the inversions and find the sign to be applied to the term $b_{31}b_{24}b_{12}b_{43}$ in $|\mathbf{B}|$.
2. By inspection show without expanding that

$$\begin{vmatrix} 1 & 5 & 6 \\ 4 & 2 & 3 \\ -3 & 0 & 1 \end{vmatrix} = \begin{vmatrix} 6 & 3 & 1 \\ 1 & 4 & -3 \\ 5 & 2 & 0 \end{vmatrix} = \begin{vmatrix} 9 & 3 & 1 \\ 5 & 4 & -3 \\ 7 & 2 & 0 \end{vmatrix}$$

3. By inspection show that none of these is invertible:

$$\begin{bmatrix} 6 & -9 & 7 \\ 2 & -3 & 5 \\ -4 & 6 & -1 \end{bmatrix}, \quad \begin{bmatrix} a_{11} & a_{12} & a_{13} \\ a_{21} & 0 & 0 \\ a_{31} & 0 & 0 \end{bmatrix}, \quad \begin{bmatrix} 0 & -1 & 1 \\ 1 & 0 & -1 \\ -1 & 1 & 0 \end{bmatrix}$$

4. Use Theorem 5 repeatedly to evaluate

$$\begin{vmatrix} 0 & a & 0 & b \\ d & 0 & 0 & 0 \\ 0 & e & 0 & f \\ 0 & 0 & c & 0 \end{vmatrix}, \quad \begin{vmatrix} 1 & 0 & 0 & 1 \\ 0 & 1 & 0 & 1 \\ 0 & 0 & 1 & 1 \\ 1 & 1 & 1 & 0 \end{vmatrix}, \quad \begin{vmatrix} a_{11} & 0 & 0 & 0 \\ a_{21} & a_{22} & 0 & 0 \\ a_{31} & a_{32} & a_{33} & 0 \\ a_{41} & a_{42} & a_{43} & a_{44} \end{vmatrix}$$

5. Express the following products as single determinants, multiplying row times column:

$$\begin{vmatrix} 2 & 3 & 1 \\ -3 & 1 & 2 \\ 1 & 4 & -1 \end{vmatrix} \cdot \begin{vmatrix} -1 & 4 & -1 \\ 1 & -2 & 3 \\ 1 & 1 & 5 \end{vmatrix}, \quad \begin{vmatrix} 1 & 1 & -1 \\ 1 & 2 & 0 \\ -1 & 0 & 3 \end{vmatrix} \cdot \begin{vmatrix} 1 & 3 & 2 \\ 3 & 1 & 1 \\ 2 & 1 & 1 \end{vmatrix}$$

6. Use Theorem 7 to invert

$$\begin{bmatrix} a & f & e \\ 0 & b & d \\ 0 & 0 & c \end{bmatrix}, \quad \begin{bmatrix} -1 & 4 & -1 \\ 1 & -2 & 3 \\ 1 & 1 & 5 \end{bmatrix}, \quad \begin{bmatrix} a & b & 0 & 0 \\ c & d & 0 & 0 \\ 0 & 0 & u & v \\ 0 & 0 & x & y \end{bmatrix}$$

7. Solve Exercise 2 of 7.4 by Cramer's rule.
8. Simplify and expand

$$\begin{vmatrix} x_1+2y & x_1+y & x_1-3y \\ x_2-y & x_2+y & x_2 \\ x_3+3y & x_3-2y & x_3+y \end{vmatrix}$$

9. Find the area of the triangle whose vertices are (1,2), (5,4), (3,7).

10. Find the volume of the tetrahedron whose vertices are (0,0,0), (2,1,0), (3,0,1), (0,2,3).

11. Prove Theorem 6 by using the lemma that if one polynomial (in $2m^2$ variables) is always zero whenever another (irreducible) polynomial is zero, then the first polynomial is divisible by the second.

7.8 DETERMINANTS AND PIVOTING

This section extends the numbered sequence of theorems begun in 7.7 as it explores the relationships between determinants and pivot steps. The *block pivot* is the matrix $\mathbf{\Lambda}$ or its determinant Λ formed by the a_{ij} in the original tableau lying in those rows and columns whose labels have been exchanged (an odd number of times).

THEOREM 11 The determinants of the block pivot Λ and the basis B are equal to ± 1 times the product of any equivalent sequence of individual pivots. For Λ the sign is $(-1)^q$ where q is the number of interchanges of pairs of rows or columns required to bring the pivotal positions into the principal diagonal of Λ.

Proof Consider first a Λ lying in rows i and k and columns j and l. Any of the four entries (if not zero) can be used as the first pivot, after which the second is uniquely determined. In each case rule 5 of 2.4 shows that the product of the two pivots is $\pm(a_{ij}a_{kl}-a_{il}a_{kj})$, which is Λ if the pivots lie in the principal diagonal of Λ and $-\Lambda$ if they do not. In the latter case $q=1$ since an interchange of rows i and k will put them in the principal diagonal. Repeated applications of this result suffice to relocate any sequence of pivots so that they fall in order in the diagonal of any Λ; thus $(-1)^q$ times the product of any sequence of nonzero pivots, one in each row and each column of a given Λ, always has the same sign.

To identify the product of the pivots with the determinant Λ, one uses induction on the order λ of Λ. If no two pivots lie in the same row or column, the inductive step from $\lambda-1$ to λ is well illustrated by the following transformation of Λ in the case $\lambda=3$:

$$\begin{vmatrix} a_{11} & a_{12} & a_{13} \\ a_{21} & a_{22} & a_{23} \\ a_{31} & a_{32} & a_{33} \end{vmatrix} = \begin{vmatrix} a_{11} & a_{12} & a_{13} \\ 0 & a_{22}-a_{12}a_{21}/a_{11} & a_{23}-a_{13}a_{21}/a_{11} \\ 0 & a_{32}-a_{12}a_{31}/a_{11} & a_{33}-a_{13}a_{31}/a_{11} \end{vmatrix}$$

$$= a_{11}\begin{vmatrix} a_{22}-a_{12}a_{21}/a_{11} & a_{23}-a_{13}a_{21}/a_{11} \\ a_{32}-a_{12}a_{31}/a_{11} & a_{33}-a_{13}a_{31}/a_{11} \end{vmatrix} \tag{1}$$

The first equality uses Theorem 4; $-a_{21}/a_{11}$ times the first row is added to the second, and $-a_{31}/a_{11}$ times the first row is added to the third. The second equality uses Theorem 5. The resulting determinant of $\lambda-1=2$ rows is exactly that obtained by pivoting on a_{11} and then deleting its row and column. By the inductive hypothesis the smaller determinant is already known to be equal to $\pm$ the product of the remaining pivots (other than a_{11}), and the theorem is proved in this case.

A more general proof, allowing superfluous pivots, would pivot on an entry in (3) below; when this is multiplied by Λ (assumed equal to $\pm$ the product of the previous pivots), the new block pivot is obtained, after a possible adjustment in sign.

Finally, if the rows and columns are suitably rearranged one has

$$B=\begin{vmatrix} \mathbf{\Lambda} & \mathbf{0} \\ \mathbf{P} & \mathbf{I} \end{vmatrix} \tag{2}$$

Using Theorem 5 to eliminate the unit columns on the right shows that $B=\pm\Lambda$.

THEOREM 12 If the first λ basic variables x_k in a tableau of matrix **A** have been exchanged with the first λ nonbasic variables y_l without any change of order, then the typical entries in the resulting tableau are

1	x_k exchanged	y_j not exchanged
y_l exchanged	$(-1)^{k+l}\Lambda_{kl}/\Lambda$	$(-1)^{l+\lambda}\Lambda^{\cdot j}_{\cdot l}/\Lambda$
x_i not exchanged	$(-1)^{k+\lambda+1}\Lambda^{i\cdot}_{k\cdot}/\Lambda$	Λ^{ij}/Λ

(3)

where $1\le k\le\lambda$; $1\le l\le\lambda$; $i, j\ge\lambda+1$. The block pivot Λ is the λ by λ determinant in the upper left corner of A. Λ_{kl} is obtained from Λ by deleting row k and column l. $\Lambda^{i\cdot}_{k\cdot}$ is obtained from Λ by deleting row k and adjoining row i *at the bottom*. $\Lambda^{\cdot j}_{\cdot l}$ is obtained by deleting column l and adjoining column j *on the right*. Λ^{ij} is a $\lambda+1$ by $\lambda+1$ determinant obtained by adjoining row i at the bottom and column j on the right.

An equivalent statement is obtained by writing the initial and final tableaus as

1	$\mathbf{y}_L$	$\mathbf{y}_{\bar{L}}$
$\mathbf{x}_K$	$\mathbf{A}_{11}$	$\mathbf{A}_{12}$
$\mathbf{x}_{\bar{K}}$	$\mathbf{A}_{21}$	$\mathbf{A}_{22}$

and

1	$\mathbf{x}_K$	$\mathbf{y}_{\bar{L}}$
$\mathbf{y}_L$	$\mathbf{A}_{11}^{-1}$	$\mathbf{A}_{11}^{-1}\mathbf{A}_{12}$
$\mathbf{x}_{\bar{K}}$	$-\mathbf{A}_{21}\mathbf{A}_{11}^{-1}$	$\mathbf{A}_{22}-\mathbf{A}_{21}\mathbf{A}_{11}^{-1}\mathbf{A}_{12}$

(4)

with $\mathbf{A}_{11}=\mathbf{\Lambda}$ having λ rows and columns (Tucker, 1960). Note how this generalizes the pivot rules of 2.4.

Proof The entry $(-1)^{k+l}\Lambda_{kl}/\Lambda$ in row l and column k of (3) follows from Theorem 2 of 7.1 and Theorem 7 above. For the rest, exchange *one* more pair of variables x_i and y_j in (3) (an additional row or column may be adjoined to **A** if necessary). Then apply the same two theorems to show that the elements of $(\Lambda^{ij})^{-1}$ are

1	x_k	x_i
y_l	$(-1)^{k+l}\Lambda^{ij}_{kl}/\Lambda^{ij}$	$(-1)^{l+\lambda+1}\Lambda^{\cdot j}_{\cdot l}/\Lambda^{ij}$
y_j	$(-1)^{k+\lambda+1}\Lambda^{i\cdot}_{k\cdot}/\Lambda^{ij}$	Λ/Λ^{ij}

(5)

where $1 \le k \le \lambda$ and $1 \le l \le \lambda$ as before but $i, j \ge \lambda+1$ are regarded as fixed and in the last position $\lambda+1$ in Λ^{ij}.

Now by making a (reverse) pivot on Λ/Λ^{ij} in (5), one obtains the remaining elements of (3) in the form stated, while for the (l, k) position one obtains

$$(-1)^{k+l}(\Lambda\Lambda^{ij}_{kl} - \Lambda^{i}_{k.}\Lambda^{\cdot j}_{\cdot l})/\Lambda\Lambda^{ij} \tag{6}$$

Since this must agree with the $(-1)^{k+l}\Lambda_{kl}/\Lambda$ already obtained, we have proved that if $i>k$ and $j>l$, then

$$\Lambda\Lambda^{ij}_{kl} - \Lambda^{i}_{k.}\Lambda^{\cdot j}_{\cdot l} = \Lambda_{kl}\Lambda^{ij} \tag{7}$$

or with B replacing Λ^{ij}

$$BB_{ik,jl} = B_{ij}B_{kl} - B_{il}B_{kj} = \begin{vmatrix} B_{kl} & B_{kj} \\ B_{il} & B_{ij} \end{vmatrix} \tag{8}$$

Since (7) is an identity between two polynomials, it continues to hold even if $\Lambda=0$ or $\Lambda^{ij}=0$.

Equation (4) is obtained by matrix algebra analogous to that of 2.4: Premultiply the first equation by $\mathbf{A}_{11}^{-1}$ to get its replacement; then premultiply the result by $\mathbf{A}_{21}$ and subtract from the second equation. Comparing Λ^{ij}/Λ in (3) with the (i, j) element of $\mathbf{A}_{22} - \mathbf{A}_{21}\mathbf{A}_{11}^{-1}\mathbf{A}^{12}$ in (4) gives (after multiplication by $\Lambda = A_{11}$) what is called the Cauchy expansion of Λ^{ij} by the elements of row i and column j.

THEOREM 13 A matrix is of rank λ if and only if its λ-rowed minors are proportional and not all zero.

Proof If (K, L) denotes the minor determinant in rows K and columns L, the condition is that

$$(K_1, L_1)(K_2, L_2) = (K_1, L_2)(K_2, L_1) \tag{9}$$

for all K_i, L_j of size λ. If (9) is true, (8) shows that every $\lambda+1$-rowed minor B is zero and the rank is λ. (If any B were not zero, it would have to contain a nonzero $B_{ik,jl}$, by two applications of Theorem 5, but (8) and (9) forbid that.)

Conversely, (9) is trivial unless one of the minors in it is nonzero, say $(K_1, L_1) = A_{11} \neq 0$. Since $|\mathbf{A}_{11}^{-1}| = 1/A_{11}$, (9) follows from Theorem 6 and the inference

$$\mathbf{A}_{22} = \mathbf{A}_{21}\mathbf{A}_{11}^{-1}\mathbf{A}_{12} \tag{10}$$

that one may draw from (4) when A_{11} is a minor of maximum rank λ. If (10) were false, an additional pair of variables could be exchanged in (4) and the rank would exceed λ by Theorem 4 of 2.7. If the minors in (9) have rows $K_1 \cap K_2$ or columns $L_1 \cap L_2$ in common, these may be duplicated to permit the minors in (10) or (4) to be disjoint.

THEOREM 14 (Tucker, 1960) Let L and L' be the sets of row and column labels in a tableau of matrix **A** that are exchanged by a series of pivot steps yielding a matrix **B**. Then there is a one-to-one correspondence between the minor determinants $A(G,G')$ in rows G and columns G' of **A** and the $B(H',H)$ of **B** such that their absolute values are proportional:

$$B(H',H) = \pm A(G,G')/A(L,L')$$

$$= \pm A(G,G')B(L',L) \tag{11}$$

where

$$G = (H'-L') \cup (L-H) \tag{12}$$

$$G' = (H-L) \cup (L'-H')$$

$$H' = (G-L) \cup (L'-G')$$

$$H = (G'-L') \cup (L-G)$$

Since no index is used twice (for a row and a column) in the same tableau, all intersections are empty except $H \cap L \triangleq HL, H'L', GL, G'L', GH', G'H$.

Equation (12) asserts that the set G (or G') of the row (or the column) indices of the minor $A(G,G')$ consists of those row (or column) indices of the minor $B(H',H)$ that lie outside the block pivot $B(L',L)$, and the *column* (or the *row*) indices of $B(L',L)$ that lie outside $B(H',H)$. The change of wording in the latter phrase corresponds to the exchange of labels. Exercise 8 is a simple example.

The result is that each row or column label belongs to an even number (0 or 2) of the sets $G \cup G'$, $H \cup H'$ and $L \cup L'$. In more detail, a row or column outside the block pivot intersects both minors $A(G,G')$ and $B(H',H)$ or else neither; a label inside the pivot appears in exactly one of these minors. The block pivots $A(L,L')$ and $B(L',L)$ are reciprocals, and each corresponds to an empty determinant (no rows or columns) of value 1 in the other tableau. This determines the constant of proportionality.

Proof (T. D. Parsons, 1967) Using the column equations, compose the linear transformations

$$\begin{array}{c|ccc|} \mathbf{x} & \mathbf{A}_{11} & \mathbf{A}_{12} & \mathbf{0} \\ \mathbf{y} & \mathbf{A}_{21} & \mathbf{A}_{22} & \mathbf{0} \\ \mathbf{z} & \mathbf{A}_{31} & \mathbf{A}_{32} & \mathbf{I} \\ \hline & \| & \| & \| \\ & \mathbf{u} & \mathbf{v} & \mathbf{z} \end{array} \quad \text{and} \quad \begin{array}{c|ccc|} \mathbf{u} & \mathbf{I} & \mathbf{B}_{12} & \mathbf{B}_{13} \\ \mathbf{v} & \mathbf{0} & \mathbf{B}_{22} & \mathbf{B}_{23} \\ \mathbf{z} & \mathbf{0} & \mathbf{B}_{32} & \mathbf{B}_{33} \\ \hline & \| & \| & \| \\ & \mathbf{u} & \mathbf{y} & \mathbf{w} \end{array} \tag{13}$$

On the common solution set of

$$\begin{array}{c|ccc|} & & & \\ \mathbf{x} & \mathbf{A}_{11} & \mathbf{A}_{12} & \mathbf{A}_{13} \\ \mathbf{y} & \mathbf{A}_{21} & \mathbf{A}_{22} & \mathbf{A}_{23} \\ \mathbf{z} & \mathbf{A}_{31} & \mathbf{A}_{32} & \mathbf{A}_{33} \\ & \| & \| & \| \\ & \mathbf{u} & \mathbf{v} & \mathbf{w} \end{array} \quad \text{and} \quad \begin{array}{c|ccc|} & & & \\ \mathbf{u} & \mathbf{B}_{11} & \mathbf{B}_{12} & \mathbf{B}_{13} \\ \mathbf{v} & \mathbf{B}_{21} & \mathbf{B}_{22} & \mathbf{B}_{23} \\ \mathbf{z} & \mathbf{B}_{31} & \mathbf{B}_{32} & \mathbf{B}_{33} \\ & \| & \| & \| \\ & \mathbf{x} & \mathbf{y} & \mathbf{w} \end{array}$$

the result must reduce to

$$\begin{array}{c|ccc|} \mathbf{x} & \mathbf{A}_{11} & \mathbf{0} & \mathbf{A}_{13} \\ \mathbf{y} & \mathbf{A}_{21} & \mathbf{I} & \mathbf{A}_{23} \\ \mathbf{z} & \mathbf{A}_{31} & \mathbf{0} & \mathbf{A}_{33} \\ & \| & \| & \| \\ & \mathbf{u} & \mathbf{y} & \mathbf{w} \end{array}$$

Since $\mathbf{x},\mathbf{y},\mathbf{z}$ are completely unrestricted (independent), the last matrix must equal the product of those in (13). Taking the determinants by Theorem 6 and applying Theorem 5 to the columns containing $\mathbf{I}$ gives

$$\begin{array}{c} \mathbf{x} \\ \mathbf{y} \\ \end{array}\!\begin{array}{|cc|} \mathbf{A}_{11} & \mathbf{A}_{12} \\ \mathbf{A}_{21} & \mathbf{A}_{22} \\ \hline \mathbf{u} & \mathbf{v} \end{array}\!\begin{array}{c} \mathbf{v} \\ \mathbf{z} \\ \end{array}\!\begin{array}{|cc|} \mathbf{B}_{22} & \mathbf{B}_{23} \\ \mathbf{B}_{32} & \mathbf{B}_{33} \\ \hline \mathbf{y} & \mathbf{w} \end{array} = \pm \begin{array}{|cc|} \mathbf{A}_{11} & \mathbf{A}_{13} \\ \mathbf{A}_{31} & \mathbf{A}_{33} \\ \hline \mathbf{u} & \mathbf{w} \end{array}\!\begin{array}{c} \mathbf{x} \\ \mathbf{z} \\ \end{array}$$

with each label $\mathbf{u},\mathbf{v},\mathbf{w},\mathbf{x},\mathbf{y},\mathbf{z}$ occurring twice. This is

$$A(L,L')\cdot B(H',H) = \pm A(G,G')$$

as required (see Figure 7.8A). A negative sign is possible if $\mathbf{A}_{11}$ is not square and hence $\mathbf{I}$ is not centered on the diagonal. This completes the proof.

Theorem 14 generalizes Theorem 12 and has two corollaries in common with the following.

THEOREM 15 (Sylvester, 1851) Let $C=|\mathbf{C}|$ be a γ-rowed minor of the α-rowed determinant $A=|\mathbf{A}|$. Let $D_\beta = D_\beta(\mathbf{A},\mathbf{C})$ be the determinant of order $\binom{\alpha-\gamma}{\alpha-\beta} = \binom{\alpha-\gamma}{\beta-\gamma}$ whose elements are all the β-rowed minors of $\mathbf{A}$ that contain $\mathbf{C}$. Here $\alpha \geq \beta \geq \gamma$ and minors in the same row (column) of D_β lie in the same rows (columns) of A. Then

$$D_\beta = \pm A^{\binom{\alpha-\gamma-1}{\alpha-\beta}} C^{\binom{\alpha-\gamma-1}{\alpha-\beta-1}} \tag{14}$$

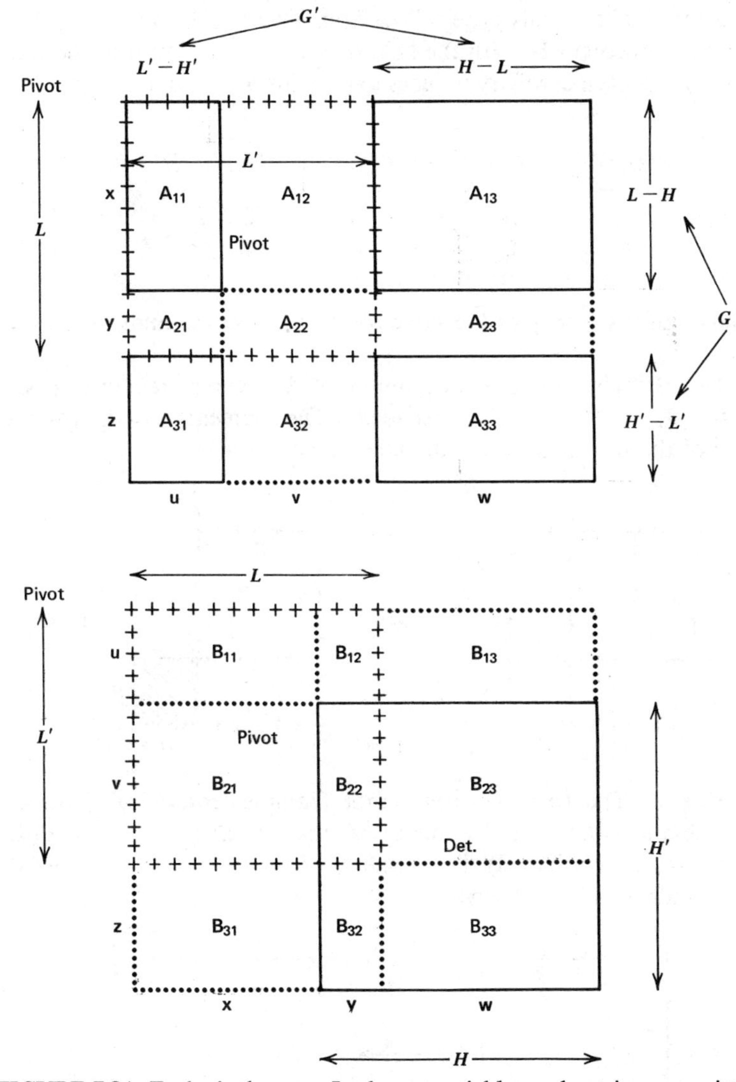

FIGURE 7.8A Tucker's theorem. Irrelevant variables and entries are omitted.

Gröbner (1956, p. 117) gives a proof. One could let **A** be a diagonal matrix to show that the exponents in (14) are determined by (vector times binomial coefficient)

$$(\gamma,\beta)\binom{\alpha-\gamma}{\alpha-\beta}=(\gamma,\alpha)\binom{\alpha-\gamma-1}{\alpha-\beta}+(\gamma,\gamma)\binom{\alpha-\gamma-1}{\alpha-\beta-1} \tag{15}$$

D_β is a minor of the βth *adjugate* of **A**, which is the determinant of *all* the β-rowed minors of **A**. Equation (14) gives its value when $\gamma=0$ and $C=1$. It rank is $\binom{\rho}{\beta}$ where ρ is the rank of **A**.

The two corollaries on this page follow from Theorem 15 by setting $\beta=\alpha-1$ or $\gamma+1$, and from Theorem 14 with the help of the data displayed in the figure. If in addition $\alpha-\gamma=2$, each corollary reduces to (8) with $B=A$ and $B_{ik,jl}=C$.

Cor.	β	$A(G,G')$	$A(L,L')$	$B(H',H)$	H	H'	D_β	λ
1	$\alpha-1$	C	A	$\pm C/A$	$L-G$	$L'-G'$	$\pm A^{\alpha-\gamma-1}C$	α
2	$\gamma+1$	A	C	$\pm A/C$	$G'-L'$	$G-L$	$\pm AC^{\alpha-\gamma-1}$	γ

Use the plus sign if **C** occupies the same rows as it does columns in **A**.

COROLLARY 1 The $(\alpha-\gamma)$-rowed minor of $\mathbf{A}^{-1}$ complementary to **C** in **A** is $B(H',H)=\pm D_{\alpha-1}/A^{\alpha-\gamma}=\pm C/A$ (shaded). The elements of $D_{\alpha-1}=\pm A^{\alpha-\gamma-1}C$ are $(\alpha-\gamma)^2$ of the Λ_{kl} in (3) inside the block pivot.

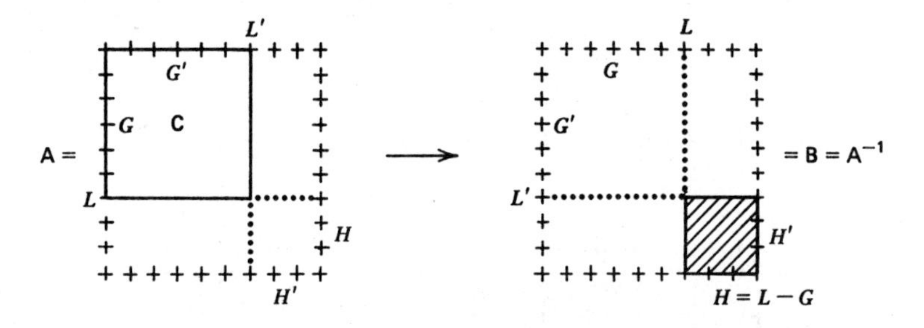

COROLLARY 2 The $(\alpha-\gamma)$-rowed minor (complementary to **C** in **A**) in the tableau **B** obtained from **A** by inverting the block pivot **C** is $B(H',H)=\pm D_{\gamma+1}/C^{\alpha-\gamma}=\pm A/C$ (shaded). The elements of $D_{\gamma+1}=\pm AC^{\alpha-\gamma-1}$ are $(\alpha-\gamma)^2$ of the Λ^{ij} in (3) outside the block pivot.

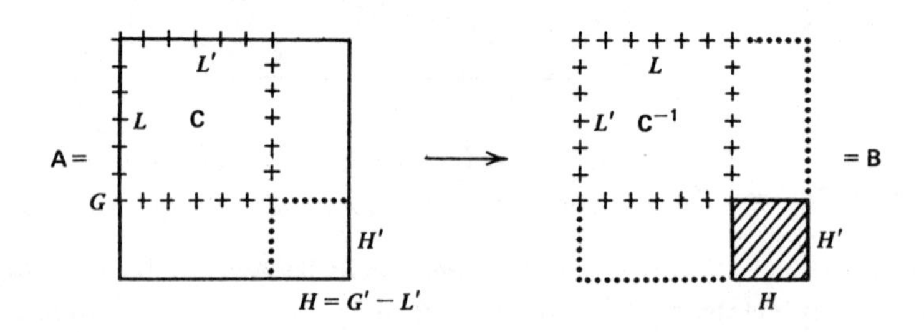

In addition to its use in solving exercises in 2.5, the following theorem will be applied in 8.5 with "integer" interpreted as an element of the ring of polynomials in the parameter t, with or without integer coefficients.

THEOREM 16 Suppose that a tableau of integers **A** over a common denominator δ is to be pivoted as in 2.5. If possible, select as a pivot any one nonzero element ε of **A** such that δ and ε have no factor in common. Then if rule 5′ of 2.5 produces integer quotients in the new tableau **B** (over ε), the same will be true of any and all additional pivot steps made according to 2.5. (If a common factor is divided out as in (2) of 2.5, the test must be reapplied.)

Proof If $\delta=1$, any one pivot step will produce an integer tableau as required, and the following proof is much simplified. In general consider any other equivalent tableau obtained by the rules of 2.5. Induction on λ, by pivoting on some element $\Lambda^{ij}/\delta^{\lambda}$ in (16), will be possible if we prove that no fractions occur when (3) is written in the form

$\Lambda/\delta^{\lambda-1}$	x_k exchanged	y_j not exchanged
y_l exchanged	$(-1)^{k+l}\Lambda_{kl}/\delta^{\lambda-2}$	$(-1)^{l+\lambda}\Lambda_{\cdot l}^{\cdot j}/\delta^{\lambda-1}$
x_i not exchanged	$(-1)^{k+\lambda+1}\Lambda_{k.}^{i.}/\delta^{\lambda-1}$	$\Lambda^{ij}/\delta^{\lambda}$

(16)

with the common denominator $\Lambda/\delta^{\lambda-1}$. Thus the theorem is equivalent to showing that every λ-rowed minor of **A** (the various Λ's in (16)) is divisible by $\delta^{\lambda-1}$. (This is obvious if $\delta=1$.) By hypothesis every μ-rowed minor of **B** is an integer, which by Theorem 14 is equal to $\pm\varepsilon^{\mu-1}/\delta^{\lambda-1}$ times a λ-rowed minor of **A**. Here $\mu=\lambda+1$, λ or $\lambda-1$ according as the minor of **A** involves neither, one or both of the indices of the pivot $\varepsilon/\delta=A(L, L')$. Since δ and ε are assumed to have no common factor, the theorem is proved.

EXERCISES 7.8

In 1 and 2 express the specified elements in the final tableau in terms of determinants with elements in the initial tableau.

1. $a_{51}^2=-3/4$, $a_{53}^2=-5/4$, $c_1^2=9/4$, $c_3^2=19/4$ in Example 2 of 3.4. (Superscript 2 denotes final tableau.)
2. $a_{62}^2=1$, $a_{63}^2=39$, $c_2^2=2$, $c_3^2=36$ in Table 6.3.
3. Interpret (4) in terms of Table 7.2′.
4. Evaluate the determinants of the cofactors of the 1, 4, 9, and 16 elements in the lower right corner of the determinant.

$$\begin{vmatrix} 3 & -2 & 0 & 2 \\ 0 & 1 & 1 & -3 \\ -1 & 0 & -1 & 2 \\ 2 & 1 & -2 & 0 \end{vmatrix} = 3$$

5. Suppose the m by m matrix **A** is of rank $m-1$. Show how to use cofactors of **A** to solve $\mathbf{Ax}=\mathbf{0}$.
6. Describe how Figure 7.8A is specialized to yield the diagrams on p. 312.
7. Use Theorem 16 to explain why pivot steps in the matrix will never produce terms in t^2 or t^3.

$$\begin{bmatrix} 2-3t & -1+4t & 3-2t \\ -4+t & 2+t & -6+5t \\ 2t & -3t & t \end{bmatrix} \text{ over } t$$

8. Use Theorem 14 to derive $|x_j|$ in Theorem 8.
9. Show that the tableau fails to satisfy the conclusion of Theorem 16. Explain why.

4	x_3	x_4	x_5
x_1	4*	2	2
x_2	2	3	4

REFERENCES (CHAPTER 7)

Bartels (1971)
Bartels-Golub (1969)
Bartels-Golub-Saunders (1970)
Beale (1970, 1975)
Bodewig (1959)
Buchet (1971)
Crout (1941)
Dantzig (1963a, 1968, 1969)
Doolittle (1878)
Fieldhouse (1980)
Forrest-Tomlin (1972)
Forsythe-Moler (1967)
Fulkerson-Wolfe (1962)
Gacs-Lovasz (1979)
Gentleman-Johnson (1974)
Gill-Golub-Murray-Saunders (1974)
Gill-Murray (1970, 1974)
Gill-Murray-Wright (1981)
Goldfarb-Reid (1977)
Golub-Saunders (1970)
Gröbner (1956)
Hellerman-Rarick (1971–1972)
Householder (1958–1964)
Khachian (1979)
Klyuyev-Kokovkin-Shcherbak (1965)
Lin-Mah (1977)
Markowitz (1957)
Ralston (1965)
Reid (1971)
Rice (1966)
Rose-Willoughby (1972)
Saunders (1975)
Smith-Orchard-Hays (1963)
Sylvester (1851)
Tomlin (1972, 1975a, b)
Tucker (1960a)
Wilkinson (1961–1965)
Willoughby (1969)
Wolfe (1980)
Zoutendijk (1970)

CHAPTER EIGHT

Parameters and Multiple Objectives

In practical applications the user of linear programming will not necessarily be satisfied merely to know the $\mathbf{x}^*, \mathbf{u}^*, \mathbf{z}^*$ corresponding to one $\mathbf{A}^0, \mathbf{b}^0, \mathbf{c}^0$. He should also know the effect of changes in these quantities and the effect of departures from the optimal $\mathbf{x}^*$. The first two sections are concerned with discrete alterations or the small variations that are possible without changing the optimal basis. This is called *postoptimal analysis* or *sensitivity analysis* or *ranging*. The other sections are concerned with generating a family of optimal solutions that depend on one or more parameters t, which is called *parametric programming*; or with generating all the points that are efficient with respect to a set of *multiple objectives*.

The parameter itself may be subject to optimization by the programmer but more frequently its value is determined by outside influences that he cannot control. Examples are the market price of an input or output, the interest rate on borrowed money, the quantity or quality of a key raw material, the rate of operation of a related industry, the outside temperature or humidity, the speed of an airplane or spacecraft, or the probability of one of two alternative situations. Specific applications are considered in 8.1–8.3 and 8.9.

8.1 ADD, DROP OR CHANGE A COLUMN OR A ROW

The tableau in which this operation is specified is referred to as the initial tableau, as indeed it normally is, and its entries are denoted by $\mathbf{A}^0, \mathbf{b}^0, \mathbf{c}^0, d^0$. The corresponding alteration in the final tableau has to be determined for $\mathbf{b}, \mathbf{c}, d$ at least; if these no longer satisfy the optimality conditions, then some of the a_{ij} may have to be revised also to permit reoptimization. Table 8.1 lists seven useful formulas.

Equations (1) and (5) appear in (4) of 3.5(b), and (1), (6), (7) are (16) to (18) of 7.1. (1) and (7) merely form the appropriate linear combinations of rows of $\mathbf{b}^0$ and $\mathbf{A}^0$, respectively. (3) is obvious and shows that (4) and (7) are equivalent. (7) implies (5) and (6) when the roles of $\mathbf{A}^0, \mathbf{c}^0$ and $\mathbf{A}, \mathbf{c}$ are interchanged. The remaining formula (2) can be obtained by using the initial row equations to express $\mathbf{b}^0_{M^0}$ in terms of $\mathbf{x} = -\mathbf{b} = (-\mathbf{b}_M, \mathbf{0}_N)$ thus:

$$\mathbf{b}^0_{M^0} = \left[\mathbf{I}_{M^0}, \mathbf{A}^0_{N^0}\right]\mathbf{b} = \mathbf{b}_{M^0} + \mathbf{A}^0_{N^0}\mathbf{b}_{N^0}$$

Solving for $\mathbf{b}_{M^0}$ gives (2); it is used only in 8.1 and 8.2.

TABLE 8.1 Formulas for Updating $\mathbf{A}^0=[\mathbf{I}_{M^0},\mathbf{A}^0_{N^0},\mathbf{b}^0]$

Use (1) or (2) for column $\mathbf{b}$ or $\mathbf{a}_{Ml}$; (3)–(7) for row $\mathbf{c}$ or $\mathbf{a}_{kN}$. Use (3) to (4) or (5) to (7) according as the row is or is not regarded as part of $\mathbf{A}^0$ and part of the basis $\mathbf{B}=\mathbf{A}^{-1}_{M^0}=\mathbf{A}^0_M$ (the columns of the initial matrix $\mathbf{A}^0$ having indices in the current basic set M).

$$\mathbf{b}_M=\mathbf{A}_{M^0}\mathbf{b}^0_{M^0} \quad \text{and} \quad \mathbf{a}_{Ml}=\mathbf{A}_{M^0}\mathbf{a}^0_{M^0l}=\mathbf{B}^{-1}\mathbf{a}^0_{M^0l} \tag{1}$$

(forward transformation); $\mathbf{a}_{Ml}=l$th column of $\mathbf{A}$.

$$\mathbf{b}_{M^0}=\mathbf{b}^0_{M^0}-\mathbf{A}^0_{N^0}\mathbf{b}_{N^0} \quad \text{and} \quad \mathbf{a}_{M^0l}=\mathbf{a}^0_{M^0l}-\mathbf{A}^0_{N^0}\mathbf{a}_{N^0l} \tag{2}$$

$$\mathbf{c}_{M^0}=\mathbf{i}_z\mathbf{A}_{M^0} \quad \text{and} \quad \mathbf{a}_{kM^0}=\mathbf{i}_k\mathbf{A}_{M^0} \tag{3}$$

(the unit vector $\mathbf{i}_z$ or $\mathbf{i}_k$ selects the desired row of $\mathbf{A}$),

$$\mathbf{c}_{N^0}=\mathbf{c}_{M^0}\mathbf{A}^0_{N^0} \quad \text{and} \quad \mathbf{a}_{kN^0}=\mathbf{a}_{kM^0}\mathbf{A}^0_{N^0} \tag{4}$$

$$\mathbf{c}_N=\mathbf{c}^0_N-\mathbf{c}^0_M\mathbf{A}_N=-\mathbf{c}^0_{N^0}\mathbf{C}^{-1} \tag{5}$$

and

$$\mathbf{a}_{kN}=\mathbf{a}^0_{kN}-\mathbf{a}^0_{kM}\mathbf{A}_N=-\mathbf{a}^0_{kN^0}\mathbf{C}^{-1}$$

where $\mathbf{C}^{-1}$ consists of the rows of $\begin{bmatrix}\mathbf{A}_N\\ -\mathbf{I}_N\end{bmatrix}$ that have indices in N_0 ($\mathbf{C}^{-1}$=dual inverse basis).

$$\mathbf{c}_{M^0}=-\mathbf{c}^0_M\mathbf{A}_{M^0} \quad \text{and} \quad \mathbf{a}_{kM^0}=-\mathbf{a}^0_{kM}\mathbf{A}_{M^0} \tag{6}$$

$$\mathbf{c}_{N^0}=\mathbf{c}^0_{N^0}+\mathbf{c}_{M^0}\mathbf{A}^0_{N^0} \quad \text{and} \quad \mathbf{a}_{kN^0}=\mathbf{a}^0_{kN^0}+\mathbf{a}_{kM^0}\mathbf{A}^0_{N^0} \tag{7}$$

Equation (3) or (6) is equivalent to the backward transformation of 7.3. Formula (4) or (7) describes pricing out; here N^0 can be replaced by $M^0\cup N^0=M\cup N$ (all the columns of the extended tableau). Equations (1) and (2) are duals of (5) and (7), which are variants of (6) and (4) respectively.

The case of a change in $\mathbf{b}^0$ and/or $\mathbf{c}^0$ is simplified by the fact that it changes nothing but d, $\mathbf{b}$ and/or $\mathbf{c}$ prior to reoptimization. This case has already been treated in 3.5(b) and 7.1 to 7.3. The main ideas of the remaining cases are summarized in (a) to (c) as follows and more details are given thereafter in (a′) to (c′).

(a) To *add* a column q or a row p requires that at least its c_q or b_p be updated to the previously optimal basis. If optimality is found to be lost, the whole vector must be updated by (1) or (6) and (7) or the equivalent in 7.3, and then a pivot determined in it.

(b) To *drop* a column q or a row p is equivalent to replacing x_q, u_q by x^0_q, $u^\pm_q$ or x_p, u_p by $x^\pm_p$, u^0_p in the notation of Table 6.2. Reoptimization is not required if the index q or p has not been exchanged, or if $b_q=0$ or $c_p=0$. The dropping of the new variables is optional after reoptimization, if any.

(c) To *change* a column or a row is equivalent to adding the new one and dropping the old.

(a′) *To add a column* of coefficients $\mathbf{a}^0_{M^0q}$, c^0_q for the nonbasic x_q, the current c_q is calculated as the qth component of (7):

$$c_q = c^0_q + \mathbf{c}_{M^0}\mathbf{a}^0_{M^0q} \tag{8}$$

Then if $c_q \geq 0$ (or more generally, if c_q satisfies the appropriate optimality condition in (4) of 6.5), the previous solution (with $x^*_q=0$, $u^*_q=c_q$ appended) is still optimal and nothing more need be done. Otherwise the primal simplex algorithm is used to optimize the solution after the new column has been brought up to date by (1) above or a forward transformation in 7.3.

(a″) *To add a row* with basic x_p and entries $\mathbf{a}^0_{pN^0}$, b^0_p, the current b_p is calculated as the pth component of (2):

$$b_p = b^0_p - \mathbf{a}^0_{pN^0}\mathbf{b}_{N^0} \tag{9}$$

Then if $x_p = -b_p$ is feasible, the previous solution (with $x^*_p = -b_p$, $u^*_p=0$ appended) is still optimal and nothing more need be done. If $x_p = -b_p$ is not feasible, the new row is brought up to date by (5) or (6) and (7) above, or by a backward transformation and pricing out in 7.3, as if it defined an objective function.

The additional row does not require a reinversion in the revised simplex. Inspection of (2) or (10) of 7.3 shows that the forward transformations in the product algorithm will work as usual if the *updated* row is adjoined to the *initial* extended tableau and a *zero* component is adjoined in the corresponding position in all the vectors $\boldsymbol{\eta}^r$ previously calculated. The extended $\boldsymbol{\eta}^r$ have no effect on the new updated row, which is as it should be, and the updating of the other rows is also unaffected. Since the forward and backward transformations are equivalent (e.g., if used to calculate all of **A**), the backward transformation also works.

To make the new solution feasible one may use the dual simplex method of 3.3 or (1) of 6.3 together with 6.4. In the latter case one has $z^I=|x_p|$ until x_p becomes feasible or it is shown that x_p has no feasible value. (x_p may or may not be artificial.)

(b′) *To drop the column* of x_q is equivalent to setting $x_q=0$. Hence if x_q is still nonbasic in the final tableau, one has only to drop its column in the final tableau; the remainder of the optimal solution is unchanged. If x_q has changed from nonbasic to basic, it must be treated as artificial. The resulting reoptimization is similar to that in the preceding paragraph.

(b″) *To drop the row* of x_p is equivalent to dropping its sign-restriction. Hence if x_p is still basic in the final tableau, one has only to drop its row in the final tableau; the remainder of the optimal solution is unchanged.

If x_p has changed from basic to nonbasic and is sign restricted, then it has $c_p \geq 0$ in the optimal tableau, and the interesting values of x_p are the previously forbidden negative ones. Thus x_p is replaced by $-x^{\pm}_p$, but the minus sign is not entered in the label; instead it is used to change the signs of the a_{ip} and c_p. Then the primal simplex method is used to restore $\mathbf{c} \geq \mathbf{0}$. When $x_p^{\pm}$ enters the basis, it may be dropped. The corresponding component of the $\boldsymbol{\eta}$ vectors in 7.3 cannot be dropped until after the next reinversion; until then one merely requires $i \neq p$ in $\min_i\{-b_i/a_{il}: a_{il} > 0\}$.

If x_p is artificial and its column has been retained in the tableau, the above procedure is unchanged except that if $c_p < 0$, x_p^0 is replaced by $x_p^{\pm}$ without changing any signs in the column. If the column of x_p^0 has not been retained, it may be necessary to resolve the problem from the beginning after dropping the corresponding row equation from the initial tableau.

(c′) *To change a column or a row* whose label x_h has the same status (in or out) with respect to the initial and the final bases, one simply drops the corresponding column or row in the initial and the final tableaus and calculates the new c_h or b_h. If it is not feasible, reoptimization is done as in (a′) or (a″).

If the label has changed from basic to nonbasic or vice versa, the new column or row, say with label y_h, should be added by (a′) or (a″) above. In this procedure any sign that an optimal solution is lacking must be judged in the light of the new type (artificial as in (b′) or unrestricted as in (b″)) assigned to x_h. If the pivot is not too small and the change in the row or column is slight, one might at once eliminate x_h by exchanging it with y_h in the final tableau; it is possible that the resulting solution is still optimal. Otherwise one of the algorithms of 3.3 or Chapter 6 is used to reoptimize. The details will depend on whether primal or dual feasibility or both are lost.

The effect of changing a single element a_{pq}^0, b_p^0, or c_q^0 will be discussed further in the following section.

As an example, consider the following initial and final tableaus with $\mathbf{x} \geq \mathbf{0}$:

1	x_3	x_4	x_5	1
x_1	1	−2	−1	−1
x_2	−2	1	−1*	1
$-z$	3	3	1	0

(10)

1	x_3	x_4	x_2	1	y_5
x_1	3	−3	−1	−2	0
x_5	2*	−1	−1	−1	−1
$-z$	1	4	1	1	−1

(11)

Here (5) of 3.3 or (1) of 6.3 (with $z^I = -x_2$) has chosen column x_5 as pivotal in the initial tableau.

Now suppose that column x_5 is to be replaced by $y_5 \geq 0$ with coefficients, 1, 1, −2. The latter are updated by

$$\begin{bmatrix} 1 & -1 & 0 \\ 0 & -1 & 0 \\ 0 & 1 & 1 \end{bmatrix} \begin{bmatrix} 1 \\ 1 \\ -2 \end{bmatrix} = \begin{bmatrix} 0 \\ -1 \\ -1 \end{bmatrix}$$

where the matrix $\mathbf{B}^{-1}$ consists of those columns of the final extended tableau (11) whose labels were basic in the initial tableau (10). The new column is adjoined to the final tableau as indicated. It would imply that z is unbounded but for the fact that

x_5 is now artificial (and u_5 unrestricted). However, primal and dual feasibility have both been lost.

Now $z^0 = x_5^0$, and only x_3 can reduce x_5^0 to 0. Thus reoptimization begins by exchanging x_3 and x_5^0:

2	x_5^0	x_4	x_2	y_5	1
x_1	-3	-3	1	3*	-1
x_3	1	-1	-1	-1	-1
$-z$	-1	9	3	-1	3

(12)

3	x_4	x_2	x_1	1
y_5	-3	1	2	-1
x_3	-3	-1	1	-2
$-z$	12	5	1	4

(13)

(13) gives the new optimal solution.

(d) *Application to electric power production*. Suppose it is desired to add 10 million kilowatts (10 gigawatts) to the capacity existing in an area. (For comparison total production in all of the United States is equivalent to nearly 300 gigawatts of continuous operation.) Of these 10, y_1 are produced from "cheap coal," y_2 from "clean coal," y_3 from oil, and y_4 from some nonpolluting alternative source such as solar, geothermal or windmill. (While nuclear power should be evaluated too, it is not included because its special complexities would strain the credibility of a simple model. We assume that no unexploited water power is available.) Because of the balance-of-payments problems associated with imported oil and the probable time required for further development of the alternative technology, it is stipulated that at least 5 of the 10 gigawatts must be produced from coal. The two kinds of coal may differ in their original sulfur content or the efficiency with which their exhaust gases are cleaned or both.

Although air pollution takes various forms (particles, sulfur oxides, carbon monoxide, etc.), only one form will be considered for simplicity, and an upper limit of 13 units will be imposed. An upper limit might also be set on the total construction cost of the project, but this has not been done because depreciation and interest or dividend charges on this cost already must be included in the operating cost z, which is to be minimized. Here $z = \mathbf{cy}$ is measured in kilodollars ($1000) per hour, and each c_j is measured in mills (tenths of a cent) per kilowatt hour. Not to make the problem too easy, it is assumed the cleaner the fuel j, the higher its cost c_j.

The assumed initial tableau and the resulting optimal tableau for this problem are the following, with all $x_i, y_j \geq 0$. (Clearly the minimization of z will exclude any excess production of power, so that $x_6^* = 0$ and x_6 could be regarded as artificial. The index 5 is reserved for later use.)

1	y_1	y_2	y_3	y_4	1
x_6	-1	-1	-1	-1	10
x_7	-1	-1	0	0	5
x_8	4	2	1	0	-13
$-z$	20	30	42	50	0

(14)

2	x_8	y_1	y_3	x_6	1
y_4	-1	-2	1	-2	-7
y_2	1	4	1	0	-13
x_7	1	2	1	0	-3
$-z$	20	20	4	100	740

(15)

This says that $y_2=13/2=6.5$ gigawatts (the maximum permitted by the antipollution standards) will be produced from clean coal at a cost of 195, and $y_4=7/2=3.5$ gigawatts will be produced by the nonpolluting technology at a cost of 175. The minimum total cost is $740/2=370$ (i.e., \$370,000 per hour of full-scale operation). Whenever convenient, the notation x_j could be used in place of y_j ($j=1,2,3,4$).

(d′) Additional technologies, or revisions in the parameters of the four already considered, are likely to be of interest. Hence we consider a new column vector $\mathbf{a}^0_{.5}=(-1, a^0_{75}, a^0_{85}, c^0_5)$ of coefficients of a new variable y_5; here a^0_{75} is -1 or 0 or possibly in between. Equation (1) gives $\mathbf{B}^{-1}\mathbf{a}^0_{.5}=\mathbf{a}_{.5}$ or

$$\begin{bmatrix} -1 & 0 & -\frac{1}{2} & 0 \\ 0 & 0 & \frac{1}{2} & 0 \\ 0 & 1 & \frac{1}{2} & 0 \\ 50 & 0 & 10 & 1 \end{bmatrix} \begin{bmatrix} -1 \\ a^0_{75} \\ a^0_{85} \\ c^0_5 \end{bmatrix} = \begin{bmatrix} -a^0_{85}/2+1 \\ a^0_{85}/2 \\ a^0_{75}+a^0_{85}/2 \\ -50+10a^0_{85}+c^0_5 \end{bmatrix} = \mathbf{a}_{.5} \tag{16}$$

If the last component $c_5 \triangleq -50+10a^0_{85}+c^0_5 \geq 0$, the new column cannot reduce z and so it need not be considered further (unless other changes are made in the model). Note that this condition can be obtained simply by substituting the optimal dual solution $\mathbf{u}^*=(50,0,10)$ into the new column constraint $-u_6+a^0_{75}u_7+a^0_{85}u_8+c^0_5 \geq 0$.

For the other cases, suppose someone believes a better coal-burning technology can be developed having $a^0_{85}=2.4$ and $c^0_5=22$; since it is not immediately available, it is assigned $a^0_{75}=-.5$. Then (16) gives $\mathbf{a}_{.5}=(-.2, 1.2, .7, -4)=(-4,24,14,-80)/20$. Adjoining this column to the optimal tableau (15) (with denominator 20) and pivoting on 14/20 to exchange y_5 and x_7 gives (17). Thus

14	x_8	y_1	y_3	x_6	x_7	1
y_4	-5	-10	9	-14	4	-55
y_2	-5	4*	-5	0	-24	-55
y_5	10	20	10	0	20	-30
$-z$	180	220	68	700	80	5060

(17)

$\mathbf{y}^*=(0,55,0,55,30)/14$ if all five technologies are available; the cost is reduced from 370 to 361.4.

For the remaining possibilities we suppose that all five technologies are not available; instead y_5 is a replacement for y_1 or y_2. If it replaces y_1, (17) is still optimal, because it already makes $y_1=0$. If it replaces y_2, the latter must be regarded as artificial ($=0$) in (17). The only way to make $y_2=0$ is to increase y_1 to $55/4$ (pivot on 4/14), but this makes $y_5=-245/14$; there is no feasible solution in this case.

Finally we suppose that the initial constants column $\mathbf{b}^0$ is reduced from $(10,5,-13,0)$ to $(9,4,-20,0)$ in order to get a larger feasible region. Matrix multiplication as in (16) gives the column $\mathbf{b}=(1,-10,-6,250)=(2,-20,-12,500)/2$, which is to replace the last column in (15). Then a dual simplex pivot step exchanging y_4 and y_1 gives the optimal tableau

2	x_8	y_4	y_3	x_6	1
y_1	1	−2	−1	2	−2
y_2	−1	4	3	−4	−16
x_7	0	2	2	−2	−10
$-z$	10	20	14	80	520

(18)

The reduction in $\mathbf{b}^0$ has enlarged the feasible region, permitting the use of the two cheapest (coal-burning) processes ($y_1^* = 1, y_2^* = 8$) and reducing the cost from 370 to 260.

(d″) Additional constraints may be needed to limit additional types of pollution or to modify the given constraint in row 8. Hence we consider the constraint

$$x_9 + a_{91}^0 y_1 + \cdots + a_{94}^0 y_4 + b_9^0 = 0 \tag{19}$$

with $x_9 \geq 0$ and $b_9^0 < 0$. Equation (5) gives $\mathbf{a}_{9.} = -\mathbf{a}_{9.}^0 \mathbf{C}^{-1}$ or

$$\begin{bmatrix} a_{91}^0 & a_{92}^0 & a_{93}^0 & a_{94}^0 & b_9^0 \end{bmatrix} \begin{bmatrix} 0 & 1 & 0 & 0 & 0 \\ -\frac{1}{2} & -2 & -\frac{1}{2} & 0 & \frac{13}{2} \\ 0 & 0 & 1 & 0 & 0 \\ \frac{1}{2} & 1 & -\frac{1}{2} & 1 & \frac{7}{2} \\ 0 & 0 & 0 & 0 & 1 \end{bmatrix} \tag{20}$$

Note that the last row of the matrix is a unit vector corresponding to the primal label 1 or the dual label $\bar{w}$ (which are never exchanged) in the last column. If the last component of (20) is nonpositive, that is,

$$b_9 \triangleq 6.5a_{92}^0 + 3.5a_{94}^0 + b_9^0 \leq 0 \tag{21}$$

the new row is feasible and the optimal solution is unchanged. Note that (21) can be obtained simply by substituting the optimal solution $\mathbf{y}^* = (0, 6.5, 0, 3.5)$ in (19) and requiring $x_9 \geq 0$.

For the other cases, suppose that someone doubts the advisability of planning to produce more than 6 gigawatts from sources other than cheap coal. Then (19) becomes $y_2 + y_3 + y_4 \leq 6$, $\mathbf{a}_{9.}^0$ is $(0, 1, 1, 1, -6)$, (20) gives $\mathbf{a}_{9.} = (0, -1, 0, 1, 4)$, and the corresponding constraint is $x_9 - y_1 + x_6 + 4 = 0$. This could be made feasible by making $y_1 \geq 4$, but (15) shows this would make both y_2 and x_7 negative. Thus the new constraint or one of the old ones must be abandoned or modified. Since $x_7^* = 1.5$, its constraint already has slack and its abandonment would not suffice by itself. The total abandonment of the first row of (14) would be rather drastic, resulting in a total production of only 5 gigawatts as specified by the second row of (14). Instead we might modify both of these constraints as follows. Increase y_1 as much as (15) permits, to 13/4, with x_7 unrestricted; then decrease x_6 from 0 to $-3/4$, just enough to make $x_9=0$. This gives $\mathbf{y}=(3.25,0,0,6)$ with total production reduced from 10 to 9.25, and heavy reliance on the expensive alternative technology. Alternatively, if x_7 is reduced to zero but not made negative we get $\mathbf{y}=(1.5,3.5,0,2.5)$ with total production reduced to 7.5.

Finally we suppose the initial cost coefficients $(\mathbf{c}^0,\mathbf{d}^0)$ are changed from $(20,30,42,50,0)$ to $(18,30,38,60,0)$. Then matrix multiplication as in (20) gives $\mathbf{c}=(15,18,-7,60,405)=(30,36,-14,120,810)/2$. Using this as the last row of (15) and exchanging x_7 and y_3 gives the new optimal tableau

1	x_8	y_1	x_7	x_6	1
y_4	-1	-2	-1	-1	-2
y_2	0	1	-1	0	-5
y_3	1	2	2	0	-3
$-z$	22	25	14	60	384

(22)

Here $\mathbf{y}=(0,5,3,2)$.

EXERCISES 8.1

Use the methods of this section to revise the optimal solutions of the following, supposing that in the initial tableau (a) every entry in the middle column of $\mathbf{A}^0$ and $\mathbf{c}^0$ is decreased by 1, or (b) every entry in the top row is increased by 1.

1, 2. Exercises 1, 2 of 6.1 (both tableaus are in 6.1 answers).

3.

1	x_4	x_5	x_6	1
x_1	-2	1	-3	-1
x_2	1	-3	-1	6
x_3	-4	-1	1	5
$-z$	3	1	8	0

13	x_3	x_2	x_6	1
x_1	-5	6	-50	-2
x_5	-1	-4	3	-29
x_4	-3	1	-4	-9
$-z$	10	1	113	56

4.

1	x_4	x_5	x_6	1
x_1	4	0	-1	-4
x_2	-3	2	1	-1
x_3	1	1	-2	-5
$-z$	-1	-2	1	0

8	x_1	x_2	x_6	1
x_4	2	0	-2	-8
x_5	3	4	1	-16
x_3	-5	-4	-15	-16
$-z$	8	8	8	-40

8.2 RANGES FOR INDIVIDUAL SCALARS

Ranging is discussed by W. O. Blattner in 12-4 of Dantzig (1963) and by Orchard-Hays (1968). It is a type of parametric programming in which one or more elements a_{ij}^0, b_i^0 or c_j^0 of the initial tableau are considered as parameters *one at a time*, while the others remain fixed at their specified values, called base values, and the optimal basis M remains unchanged. Some x_h may also be considered as a parameter; in this case *one nonbasic* x_j varies in the optimal tableau and thereby causes the basic x_i to vary. For simplicity the base values of the elements of the optimal tableau **A** with basis M will be denoted by a_{ij}, b_i, c_j, whereas a_{ij}^*, b_i^*, c_j^* will denote the variable values of these same optimal elements considered as functions of the parameter. The following three questions will be considered for each kind of parameter:

(i) Within what range (interval) can the parameter vary without changing the optimal basis? Besides putting a convenient limit on the scope of the investigation, the restriction to a single optimal basis avoids the expense of starting and stopping certain activities in the physical system described by the linear program, and it permits (iii) to give exact results for parameters other than a_{kl}^0. Furthermore, in combination with the practice of specifying the range for the *change* Δs in the parameter s, it causes the range to depend only on the optimal base values a_{ij}, b_i, c_j, the optimal basis M, and the information $(i \in M^0, j \notin M^0)$ about the initial basis M^0 that is contained in the notation a_{ij}^0, b_i^0 or c_j^0 for the parameter. To be sure, many of these advantages would be retained, and more territory might be covered, if one chose the base value of some parameter s so that it admitted two optimal bases, one valid for $\Delta s \geq 0$ and the other valid for $\Delta s \leq 0$; or this situation may occur by chance.

To simplify matters and to reduce the chance of getting a range of zero length, it will be assumed that as far as possible *all artificial variables have been pivoted out of the optimal basis*. However, it should be noted that nonartificial variables that were marked as artificial at the end of Phase 1 (because their only feasible value was zero) may acquire other feasible values during ranging.

(ii) What are the corresponding limits for the change Δz in z^*? It will turn out that each limit (other than 0 or $\pm\infty$) has the form $-b_{i^z}c_{j^z}/a_{i^z j^z}$ in terms of certain elements of the optimal tableau.

(iii) Within the limits that (i) imposes on Δs, what is the multiplier that converts Δs into Δz in (ii)? This question is meaningful because the relation between s (other than a_{kl}^0) and z is *linear*, as long as M remains the optimal basis. The multiplier is the slope of the linear graph; it is denoted by the calculus notation $\partial z^*/\partial s$.

(a) *Parameter* $\mathrm{x_q}$ *or* $\mathrm{x_p}$. This type of variation could result from imposing an upper or a lower bound on the variable such as to make the original optimum infeasible.

If $x_q \geq 0$ is not in the optimal basis, then $x_q = 0$ and

$$\partial z^*/\partial x_q = c_q \tag{1'}$$

wherein x_q can only increase, not decrease. As in the primal-simplex pivot step in (7) of 3.2 (but not requiring $c_q < 0$), primal feasibility requires

$$0 \leq x_q \leq \min_i\{-b_i/a_{iq}: a_{iq} > 0\} \triangleq -b_{p'}/a_{p'q} \tag{1''}$$

if a change of basis is to be avoided. The product of (1′) and (1″) gives

$$0 \leq \Delta z = c_q x_q \leq -b_{p'}c_q/a_{p'q} \tag{1}$$

If $x_p \geq 0$ is in the optimal basis M, its variation must be induced by varying one of the nonbasic nonartificial x_j, and the latter is chosen to increase z^* as slowly as possible. Now

$$\frac{\partial z^*}{\partial x_p} = \frac{\partial z^*}{\partial x_j} \Big/ \frac{\partial x_p}{\partial x_j} = \frac{c_j}{-a_{pj}}$$

where the nonbasic variables other than x_j are held constant. Thus we choose

$$\frac{\partial z^*}{\partial x_p} = -\min_{j \in N'} \{c_j/a_{pj}: a_{pj} > 0\} \triangleq -c_{q'}/a_{pq'} \leq 0$$

or

$$\min_{j \in N'} \{c_j/(-a_{pj}): a_{pj} < 0\} \triangleq -c_{q''}/a_{pq''} \geq 0 \tag{2'}$$

according as x_p is decreasing or increasing. In each case the absolute value is minimized, all j for which x_j is artificial are omitted in N', and a minimum over an empty set is interpreted as $+\infty$. (In (2′) that would mean x_p could not change in that direction.)

Now the range for $\Delta x_p = x_p + b_p$ is the closed interval from

$$-a_{pq'} \min_i \{-b_i/a_{iq'}: a_{iq'} > 0\}$$

to

$$-a_{pq''} \min_i \{-b_i/a_{iq''}: a_{iq''} > 0\} \tag{2''}$$

As in (1″) the minima are the largest allowable values for the nonbasic $x_{q'}$ and $x_{q''}$; multiplication by the negatives of their coefficients in row p gives the corresponding limits on Δx_p. If the minima in (2″) are denoted by $-b_{p'}/a_{p'q'}$ and $-b_{p''}/a_{p''q''}$, multiplication by $c_{q'}$ and $c_{q''}$, respectively, shows that

$$0 \leq \Delta z \leq -b_{p'}c_{q'}/a_{p'q'} \quad \text{or} \quad -b_{p''}c_{q''}/a_{p''q''} \tag{2}$$

according as x_p is decreasing or increasing. It may be desirable to change q' and/or q'' so as to lengthen the interval (2″) (especially if it contains only the value zero for Δx_p) even though (2) also is lengthened and $\partial z^*/\partial x_p$ no longer minimized.

Although we do not expect to reduce x_p to zero by pivoting in row p, it is true that the second formula in (2′) agrees with (5) of 3.3, used in selecting a dual-simplex pivot. It is associated with increasing x_p, which makes $x_p + b_p > 0$ as in the dual simplex. The first formula in (2′) is similar except that an unconventional *positive* sign is specified for the pivot.

Parameters in $\mathbf{b}^0$, $\mathbf{c}^0$ or $\mathbf{A}^0$

Pairs of pivots (both primal or both dual) of opposite signs as in (2′) are encountered repeatedly in this section. This is true because (except in (1″) and (2″)) the increment (in the parameter), which replaces the usual nonbasic variable, may be either positive or negative. These pivot steps are not executed in full, but only carried far enough to yield the desired information.

In the above x_p was in the optimal basis and x_q was not. In the remainder of this section we denote the row of the initial tableau in which the parameter occurs by p if x_p remains basic in the optimal tableau, and by k if x_k changes from basic to nonbasic. Similarly x_q remains nonbasic while x_l changes from nonbasic to basic. This agrees with our use of a_{kl} to denote a pivot.

The derivation of the remaining formulas will be standardized by relying on the following observation. In the initial *extended* tableau $\mathbf{A}^0$ the replacement of any one entry a_{ij}^0 (we may suppose $b_i^0 = a_{i0}^0$, $c_j^0 = a_{0j}^0$) by $a_{ij}^0 + t$ can be accomplished by postmultiplication (on the *right*) by an elementary matrix $\mathbf{E}$ that differs from the unit matrix $\mathbf{I}_{m+n+1}$ only in having t instead of 0 in the (unique) position ij. This is true because the only effect is to alter column j by adding to it t times the unit vector in column i. Now the optimal extended tableau $\mathbf{A}$ is converted to $\mathbf{A}^*$ (depending on t) in just the same way, although it may have to be followed by a pivot step. This is a consequence of Theorem 4 of 7.1, or the associativity of matrix multiplication, since pivot steps can be accomplished by multiplication on the *left* as in (9) of 7.1 or (1) of 8.1; thus we have $\mathbf{AE} = (\mathbf{B}^{-1}\mathbf{A}^0)\mathbf{E} = \mathbf{B}^{-1}(\mathbf{A}^0\mathbf{E})$. Obviously $\mathbf{E}$ could be generalized as in 7.1 to contain and to generate an arbitrary column of numbers or parameters. Table 8.2A illustrates the conversion of $\mathbf{A}$ in (10) into $\mathbf{A}^*$ in (11).

For alternative derivations, references to (1) through (7) of 8.1 are also given; such a formula is not used in toto but only incrementally, to supply the appropriate vector or scalar multiplier for Δb_h^0, Δc_h^0 or Δa_{ij}^0.

(b) *Parameter* b_h^0 *or* c_h^0. Here Δb_h^0 prescribes the multiple of column h that is added to the constants column. If $h = p$ remains basic, the only effect in $\mathbf{A}^*$ is to replace b_p by $b_p + \Delta b_p^0 \leq 0$, so that

$$\Delta z = 0 \qquad \text{for} \qquad \Delta b_p^0 \leq -b_p \tag{3}$$

If $h = k$ becomes nonbasic, one obtains

$$z^* = d^* = c_k \Delta b_k^0 + d, \qquad \partial z^* / \partial b_k^0 = c_k \tag{4'}$$

$$b_i^* = a_{ik} \Delta b_k^0 + b_i \leq 0 \qquad \text{for all } i \text{ in } M$$

which also follow from (1) of 8.1. Then

$$-\min_i \{ b_i / a_{ik} : a_{ik} < 0 \} \leq \Delta b_k^0 \leq \min_i \{ -b_i / a_{ik} : a_{ik} > 0 \} \tag{4''}$$

The two limits may be described as relating to sign reversed and conventional primal-simplex pivot steps in column k of $\mathbf{A}$, but Δb_k^0 replaces the nonbasic variable x_k.

Δc_h^0 prescribes the multiple of column $(-z)$ (a unit vector) that is added to column h. The only effect in $\mathbf{A}^*$ is to replace c_h by $c_h^* = c_h + \Delta c_h^0 \geq 0$. If $h = q$ has remained nonbasic,

$$\Delta z = 0 \qquad \text{for} \qquad \Delta c_q^0 \geq -c_q \tag{5}$$

If $h = l$ has become basic in $\mathbf{A}^*$, it is necessary to reduce the cost $c_l^* = \Delta c_l^0$ to zero again, by multiplying row l by $-\Delta c_l^0$ and adding to row $(-z)$. This gives

$$z^* = d^* = -b_l \Delta c_l^0 + d, \qquad \partial z^* / \partial c_l^0 = -b_l \tag{6'}$$

$c_j^* = -a_{lj} \Delta c_l^0 + c_j \geq 0$ for all j in N' (nonartificial), which also follow from (5) of 8.1. Then

$$-\min_{j \in N'} \{c_j / (-a_{lj}) : a_{lj} < 0\} \leq \Delta c_l^0 \leq \min_{j \in N'} \{c_j / a_{lj} : a_{lj} > 0\} \tag{6''}$$

Here $-\Delta c_l^0$ replaces u_l in the dual-simplex criteria in row l.

Recall that the indices $1, 2, \ldots, m+n$ are divided into four categories with typical members k, l, p, q thus:

		l	q		k	q
i	k	Initial		l	Optimal	
	p	tableau $\mathbf{A}^0$		p	tableau $\mathbf{A}$	.

In the initial tableau x_i and b_i^0 occur only in row i as the sum $x_i + b_i^0$, so that $\Delta b_i^0 = \Delta x_i$ (any increment $\Delta \ldots$ can be associated with either x_i or b_i^0 as one chooses). Similarly $\Delta c_j^0 = -\Delta u_j$ because adding $\Delta \ldots$ to c_j^0 in the jth column equation is equivalent to subtracting it from u_j (here $j = l$ or q). This equivalence involves all the b_i^0 and c_j^0 but only the initially basic x_i and u_j. Since the base values of b_i^0 and x_i come from different tableaus, they are generally different.

In the optimal tableau $x_k = \Delta x_k$ and $u_l = \Delta u_l$ are nonbasic with zero base values. Applying the above results to them permits (4″) and (6″) to be derived from the familiar (1″) and its dual counterpart, as already suggested above, provided $x_k < 0$ and $u_l < 0$ are permitted. The other variables involved in the equivalence are x_p and u_q, which are basic in the optimal tableau. When they are caused to increase as in (2′) and (2″) and Exercise 8, the effect is to extend the results (3) for b_p^0 and (5) for c_q^0 into an adjacent interval with a new optimal basis—a result that otherwise would be left for 8.3. It permits $\Delta z \neq 0$.

(c) *Parameter* a_{ij}^0. Here Δa_{ij}^0 prescribes the multiple of column i that is added to column j in the extended tableau.

Δa_{pq}^0. If p remains basic and q remains nonbasic in $\mathbf{A}$, the only effect is to replace a_{pq} by $a_{pq}^* = a_{pq} + \Delta a_{pq}$. Thus

$$\Delta z = 0 \qquad \text{for} \qquad -\infty < a_{pq}^0 < \infty \tag{7}$$

Δa_{kq}^0. If k becomes nonbasic in **A** and q remains nonbasic, the only effect is to replace the qth column $\mathbf{a}_{Mq}$ by $\mathbf{a}_{Mq}^* = \mathbf{a}_{Mq} + \mathbf{a}_{Mk}\Delta a_{kq}^0$, which also follows from (1) of 8.1. The only element of interest is $c_q^* = c_q + c_k \Delta a_{kq}^0 \geq 0$, which also follows from the qth component of the first formula (7) of 8.1. Thus

$$\Delta z = 0 \qquad \text{for} \qquad \Delta a_{kq}^0 \geq -c_q/c_k(c_k > 0) \tag{8'}$$

$$\Delta z = 0 \qquad \text{for} \qquad \Delta a_{kq}^0 \leq -c_q/c_k \qquad (c_k < 0,\ x_k \text{ artificial}) \tag{8''}$$

Equation (7) applies if $c_k = 0$.

Δa_{pl}^0. If p remains basic and l becomes basic in **A**, the only effect is to replace a zero in **A** by $a_{pl}^* = \Delta a_{pl}^0$. Since l is basic in **A**, this must be reduced to zero again by multiplying row l by $-\Delta a_{pl}^0$ and adding to row p to obtain $\mathbf{a}_{pN}^* = \mathbf{a}_{pN} - \mathbf{a}_{lN}\Delta a_{pl}^0$. This also follows from (5) of 8.1 with $k = p$. The only element of interest is $b_p^* = b_p - b_l \Delta a_{pl}^0 \leq 0$, which also follows from (2) of 8.1. Thus

$$\Delta z = 0 \qquad \text{for} \qquad \Delta a_{pl}^0 \leq b_p/b_l \qquad (\text{use (7) if } b_l = 0) \tag{9}$$

$\Delta a_{kl}^0 \triangleq t$. If k and l are both exchanged, a_{kl}^0 being an element of the block pivot $\mathbf{\Lambda}$, the matrix postmultiplication that changes a_{kl}^0 to $a_{kl}^0 + t$ in $\mathbf{A}^0$ is applied to **A** in (10) of Table 8.2A. Here x_i represents any basic variable other than x_l and $-z$, and x_j is any nonbasic variable other than x_k. To restore the unit vector in the basic column l in (11), it is necessary to pivot on $1 + a_{lk}t$ in (11) to obtain (12), in which all the basic unit-vector columns are omitted and τ represents $t/(1 + a_{lk}t)$. Its coefficient is the rank-one matrix $-\mathbf{a}_{Mk} \otimes \mathbf{a}_{lN}$ (column k times row l of **A**).

For (12) to be optimal, τ must satisfy the four inequalities in (13). They define the same two intervals as in (4″) and (6″). The only differences result from the replacement of the middle members Δb_k^0 and Δc_l^0 by $-b_l\tau$ and $c_k\tau$, respectively. This implies that the two intervals, which formerly applied independently to two different parameters, now apply simultaneously to the single parameter τ. Intersecting the two intervals (after division by $-b_l$ and c_k, if these are not zero) gives some interval $\tau_1 \leq \tau \leq \tau_2$, which includes $\tau = 0$. This usually translates into the interval (14) for $t \triangleq \Delta a_{kl}^0$.

In deriving (14) it is assumed that $1 + ta_{lk} > 0$. This is indeed true if $b_l \neq 0$ or if $c_k \neq 0$ and x_k is not artificial, because (12) shows that $c_k \geq 0$ and $b_l \leq 0$ are divided by $1 + a_{lk}t$ to get $c_k^* \geq 0$ and $b_l^* \leq 0$. On the other hand, if $b_l = c_k = 0$ (and hence $x_l^* = u_k^* = 0$), (13) imposes no restrictions, (14) is irrelevant, and it is clear that a_{kl}^0 can vary from $-\infty$ to ∞ without changing the values of z^*, $\mathbf{x}^*$ and $\mathbf{u}^*$. In this case one hardly need be concerned with the fact that the given optimal basis M^* does not exist when $t = -1/a_{lk}$, which corresponds to $\tau = \pm\infty$.

In the remaining case when $b_l = 0$, $c_k \neq 0$, and x_k is artificial, negative values of $1 + a_{lk}t$ may be theoretically admissible. Then (14) is replaced by

$$t \leq t_2 \triangleq \tau_2/(1 - a_{lk}\tau_2) \qquad \text{or} \qquad t_1 \triangleq \tau_1/(1 - a_{lk}\tau_1) \leq t \tag{15}$$

TABLE 8.2A Parameter $t \triangleq \Delta a_{kl}^0$ (k and l exchanged)

$$\begin{array}{c} \\ (x_i) \\ (x_l) \\ (-z) \end{array} \begin{array}{c} \begin{array}{ccccc} x_i & x_l & x_k & x_j & 1 \end{array} \\ \begin{bmatrix} 1 & 0 & a_{ik} & a_{ij} & b_i \\ 0 & 1 & a_{lk} & a_{lj} & b_l \\ 0 & 0 & c_k & c_j & d \end{bmatrix} \end{array} \begin{bmatrix} 1 & 0 & 0 & 0 & 0 \\ 0 & 1 & 0 & 0 & 0 \\ 0 & t & 1 & 0 & 0 \\ 0 & 0 & 0 & 1 & 0 \\ 0 & 0 & 0 & 0 & 1 \end{bmatrix} \tag{10}$$

$$=\mathbf{AE}= \begin{array}{c} (x_i) \\ (x_l) \\ (-z) \end{array} \begin{array}{c} \begin{array}{ccccc} x_i & x_l & x_k & x_j & 1 \end{array} \\ \begin{bmatrix} 1 & a_{ik}t & a_{ik} & a_{ij} & b_i \\ 0 & (1+a_{lk}t)^* & a_{lk} & a_{lj} & b_l \\ 0 & c_k t & c_k & c_j & d \end{bmatrix} \end{array} \tag{11}$$

while $\mathbf{A}^0\mathbf{E}$ differs from $\mathbf{A}^0$ only in having a_{kl}^0 increased by $t=\Delta a_{kl}^0$. Pivoting on $1+a_{lk}t$ to make x_l basic again gives (12), the optimal condensed tableau $\mathbf{A}^*$, in which $\tau=t/(1+a_{lk}t)$.

$$\begin{array}{l|lll} 1 & x_k & x_j & 1 \\ \hline x_l & a_{lk}-\tau a_{lk}^2 & a_{lj}-\tau a_{lk}a_{lj} & b_l-\tau b_l a_{lk}=b_l/(1+a_{lk}t) \\ x_i & a_{ik}-\tau a_{ik}a_{lk} & a_{ij}-\tau a_{ik}a_{lj} & b_i-\tau b_l a_{ik}=b_i^* \le 0 \\ -z & c_k-\tau c_k a_{lk} & c_j-\tau c_k a_{lj} & d-\tau b_l c_k=d+\Delta z \\ & =c_k/(1+a_{lk}t) & =c_j^* \ge 0 & \end{array} \tag{12}$$

$$-\min_{i\in M}\{b_i/a_{ik}\colon a_{ik}<0\}\le -b_l\tau \le \min_{i\in M}\{-b_i/a_{ik}\colon a_{ik}>0\} \text{ and}$$

$$-\min_{j\in N'}\{c_j/(-a_{lj})\colon a_{lj}<0\}\le c_k\tau\le \min_{j\in N'}\{c_j/a_{lj}\colon a_{lj}>0\} \tag{13}$$

are derived from $\mathbf{b}^*\le\mathbf{0}$ and $\mathbf{c}^*\ge\mathbf{0}$. Artificial x_j are omitted from N'. If (13) makes $\tau_1\le\tau\le\tau_2$, then usually

$$\tau_1/(1-a_{lk}\tau_1)\le\Delta a_{kl}^0 \triangleq t\le\tau_2/(1-a_{lk}\tau_2)$$

$$\partial z^*/\partial a_{kl}^0=-b_l c_k \text{ at } t=0;\ \Delta z=-b_l c_k\tau \tag{14}$$

with $t\neq -1/a_{lk}$. In this case the failure of (14) is indicated by the fact that $t_2<t_1$ and $t_1t_2>0$. The following is a simple illustration.

Initial tableau, $\mathbf{x}\ge\mathbf{0}$; not optimal for any t.

$$\begin{array}{c|ccc} 3 & x_l & x_m & 1 \\ \hline x_k^0 & 1+3t^* & 1 & 0 \\ -z & -1 & 2 & 0 \end{array} \tag{16}$$

Optimal when $t \le -1/2$ or $-1/3 < t$ by (15), so that $(1+2t)/(1+3t) \ge 0$. (13) gives $-\infty < \tau \le 1$.

$1+3t$	x_k^0	x_m	1
x_l	3	1*	0
$-z$	1	$1+2t$	0

(17)

Optimal when $t \le -1/2$

1	x_k^0	x_l	1
x_m	3	$1+3t$	0
$-z$	-2	$-1-2t$	0

(18)

In Table 8.2A we revert to the condensed tableau in (12) for the sake of compactness, but if $a_{lk} \ne 0$, condensed tableaus could have been used throughout. The procedure is to make a (reverse) pivot on a_{lk} to get $1/a_{lk}$, add t to the latter, and pivot again on $t+1/a_{lk}$.

(d) *Examples and details.* Table 8.2B gives an example with the initial and the final tableaus and some ranging calculations. The organization of more comprehensive data is illustrated in Table 8.2C, which applies to (20) (electric power production). The first and last columns identify the appropriate formula in Table 8.2D. The base values of the b_i^0, c_j^0 and a_{ij}^0 are taken from the initial tableau, as the superscript zeros indicate; the base values of the x_i and y_j and all the other data are derived from the optimal tableau (20). The a_{6j}^0 and a_{7j}^0 are not ranged as they are unlikely to vary.

The range of Δs (the parameter minus its base value) is the interval within which the optimal basis contains the same indices M. (An extension for formulas (3) and (5) is mentioned below.) This range can be transformed by the addition of the number on its left or (except a_{kl}^0 in (14)) by multiplication by the number on its right. For example, consider the range $[-1.5, 0]$, which means $-1.5 \le \Delta y_2 \le 0$, in the first of the lines for y_2. Addition of the base value $y_2 = 6.5$ gives $5 \le y_2 \le 6.5$, and

TABLE 8.2B An Example (Initial and Optimal Tableaus)

1	y_1	y_2	y_3	1
x_4	1	1	-2	-1
x_5	-2*	1	1	1
x_6	1	2	1	-2
$-z$	3	-2	3	0

3	x_5	x_4	y_3	1
y_2	1	2	-3	-1
y_1	-1	1	-3	-2
x_6	-1	-5	12	-2
$-z$	5	1	12	4

(19)

Range $b_5^0 = -1$, $c_1^0 = 3$, $a_{51}^0 = -2$; keep $M = \{2, 1, 6\}$. $\mathbf{x} \ge \mathbf{0}$, $\mathbf{y} \ge \mathbf{0}$.

Since x_5 changes from basic to nonbasic it is denoted by x_k; since y_1 changes from nonbasic to basic it corresponds to x_l. (The other possibilities are x_p basic initially and finally, and x_q always nonbasic.) (4″) gives $-2 \le \Delta b_5^0 \le 1$ with $i=1$ and 2 minimizing $|b_i/a_{i5}|$; (6″) gives $-4 \le \Delta c_1^0 \le 1$ with $j=3$ and 4 minimizing $|c_j/a_{1j}|$ in the final tableau. Multiplying these ranges by $c_5 = 5/3$ and $-b_1 = 2/3$, respectively, gives $-10/3 \le \Delta z \le 5/3$ and $-8/3 \le \Delta z \le 2/3$. The latter is also the intersection of the two and hence shows the effect of a_{51}^0. Dividing it by $-b_1 c_5 = 10/9$ gives $\tau_1 = -12/5$ and $\tau_2 = 3/5$. In (14) with $a_{15} = -1/3$ these give $-12 \le \Delta a_{51}^0 \le 1/2$.

TABLE 8.2C Electric Power Production (14)–(15) of 8.1

Initial Tableau

1	y_1	y_2	y_3	y_4	1
x_6	-1	-1	-1	-1	10
x_7	-1	-1	0	0	5
x_8	4	2	1	0	-13
$-z$	20	30	42	50	0

Optimal Tableau

2	x_8	y_1	y_3	x_6	1
y_4	-1	-2	1	-2	-7
y_2	1	4	1	0	-13
x_7	1	2	1	0	-3
$-z$	20	20	4	100	740

(20)

Extreme Values of $\Delta z = (\partial z^*/\partial s)\Delta s = -b_i c_j / a_{ij}\ (ij = ij^z)$

Parameter s	Base Value	Range of Δs	$\frac{\partial z^*}{\partial s}$	Range of Δz	$i'j'$	Pivot $i''j''$	Equations
$y_1 = x_q$	0	$[0, 1.5]$	10	$[0, 15]$	—	71^z	(1)
$y_2 = x_l$	6.5	$[-1.5, 0]$	-4	$[0, 6]$	23	73^z	(2)
	6.5	$[-3, 0]$	-5	$[0, 15]$	21	71^z	(2)
$y_3 = x_q$	0	$[0, 3]$	2	$[0, 6]$	—	73^z	(1)
$y_4 = x_l$	3.5	$[-1.5, 0]$	-4	$[0, 6]$	43	73^z	(2)
	3.5	$[0, 1.5]$	10	$[0, 15]$	41	71^z	(2)
$b_6^0 = b_k^0$	10	$[-3.5, \infty)$	50	$[-175, \infty)$	46^z	41	(4)
$b_7^0 = b_p^0$	5	$(-\infty, 1.5]$	0	0	—	—	(3)
$b_8^0 = b_k^0$	-13	$[-7, 3]$	10	$[-70, 30]$	$48, 78^z$	41, —	(4)
$c_1^0 = c_q^0$	20	$(-\infty, -10]$	1.5	$(-\infty, 0]$	71	—	—
	20	$[-10, \infty)$	0	0	—	—	(5)
$c_2^0 = c_l^0$	30	$(-\infty, 4]$	6.5	$(-\infty, 26]$	23^z	73	(6)
$c_3^0 = c_q^0$	42	$(-\infty, -2]$	3	$(-\infty, 0]$	73	—	—
	42	$[-2, \infty)$	0	0	—	—	(5)
$c_4^0 = c_l^0$	50	$[-10, 4]$	3.5	$[-35, 14]$	$41, 43^z$	71, 73	(6)
$u_7 = u_p$	0	$[0, \infty)$	-3	$(-\infty, 0]$	—	—	—
$a_{81=kq}^0$	4	$[-1, \infty)$	a_{lk}	0	—	71	(8′)
$a_{82=kl}^0$	2	$[-.7, .5]$	0.5	$[-70, 26]$	$48, 23^z$	41, 73	(14)
$a_{83=kq}^0$	1	$[-.2, \infty)$		0	—	73	(8′)
$a_{84=kl}^0$	0	$[-2, 1/3]$	-0.5	$[-35, 14]$	$41, 43^z$	21, 73	(14)

a_{82}^0 has $-14/13 \leq \tau \leq 0.4$. a_{84}^0 has $-1 \leq \tau \leq 0.4$. Model has $a_{84}^0 \geq 0$.
$x_k = x_6$, x_8 leave the basis; $x_l = y_2$, y_4 enter. $x_p = x_7$ remains basic;
$x_q = y_1$, y_3 remain nonbasic. $\Delta b_i^0 = \Delta x_i$; $\Delta c_j^0 = -\Delta u_j$.

TABLE 8.2D Formulas for Ranging

All the input data are in the optimal tableau **A**.

(a) $\partial z^*/\partial x_q = c_q$ ($q \in N'$, not artificial). (1′)

$$0 \le x_q \le \min_i \{-b_i/a_{iq}\colon a_{iq} > 0\} \qquad (1'')$$

$$\partial z^*/\partial x_p = -\min_{j \in N'} \{c_j/a_{pj}\colon a_{pj} > 0\} \triangleq -c_{q'}/a_{pq'} \le 0$$

or

$$\min_{j \in N'} \{c_j/(-a_{pj})\colon a_{pj} < 0\} \triangleq -c_{q''}/a_{pq''} \ge 0 \qquad (2')$$

according as $x_p < -b_p$ or $x_p > -b_p$ ($p \in M$). Let $q = q'$ or q'': Limits on Δx_p are $-a_{pq}\min_i\{-b_i/a_{iq}\colon a_{iq} > 0\}$.

In (b) and (c) the row or the column index of the parameter is denoted by p or q, respectively, if it is not exchanged, and by k or l if exchanged to make x_k nonbasic or x_l basic.

(b) $\Delta z = 0$ for $\Delta b_p^0 \le -b_p$ or $\Delta c_q^0 \ge -c_q$. (3), (5)

$$\partial z^*/\partial b_k^0 = c_k, \quad \partial z^*/\partial c_l^0 = -b_l \text{ (both} \ge 0) \qquad (4'), (6')$$

$$-\min_i\{b_i/a_{ik}\colon a_{ik} < 0\} \le \Delta b_k^0 \le \min_i\{-b_i/a_{ik}\colon a_{ik} > 0\} \qquad (4'')$$

$$-\min_{j \in N'}\{c_j/(-a_{lj})\colon a_{lj} < 0\} \le \Delta c_l^0 \le \min_{j \in N'}\{c_j/a_{lj}\colon a_{lj} > 0\} \qquad (6'')$$

(c) $\Delta z = 0$ for $-\infty < a_{pq}^0 < \infty$. (7)

$$\Delta z = 0 \text{ for } \Delta a_{kq}^0 \ge -c_q/c_k \qquad (c_k > 0) \qquad (8')$$

$$\Delta z = 0 \text{ for } \Delta a_{kq}^0 \le -c_q/c_k \qquad (c_k < 0;\ x_k \text{ artificial}) \qquad (8'')$$

$$\Delta z = 0 \text{ for } \Delta a_{pl}^0 \le b_p/b_l \qquad (\text{use (7) if } b_l = 0 \text{ or } c_k = 0) \qquad (9)$$

$$\Delta z = -b_l c_k \tau, \qquad \tau_1/(1 - a_{lk}\tau_1) \le \Delta a_{kl}^0 \le \tau_2/(1 - a_{lk}\tau_2) \qquad (14)$$

where $\tau_1 \le \tau \le \tau_2$ is the interval satisfying both (4″) and (6″) with middle members $-b_l\tau$ and $c_k\tau$, respectively. Multiply by c_k and $-b_l$, respectively (if nonzero), intersect to get the interval for $-b_l c_k \tau = \Delta z$, divide by $-b_l c_k$ to get $[\tau_1, \tau_2]$, then use (14). Thus (4), (6) and (14) form a computational unit, as do (1) and (2).

multiplication by $\partial z^*/\partial y_2 = -4$ gives $0 \le \Delta z \le 6$ or $370 \le z^* \le 376$, since 740/2 is the base value of z. The extreme value max $\Delta z = 6$ can also be calculated as $-b_7c_3/a_{73} = (3/2)(4/2)/(1/2)$ in (20) by using the indices $i^z j^z = 73$.

The calculations are like those treated in more detail in 8.3 and 8.5, but here we are usually concerned with only one optimal tableau. (Exceptions involving b_p^0, c_q^0, a_{kl}^0 are noted below.) In most cases the calculation begins with a minimum-ratio criterion that determines an extreme value of Δs (positive, negative or zero) and a pair of indices $i'j'$. One of the latter agrees with the (or an) index of the parameter; the other identifies a b_i^* or c_j^* that has become zero and will become infeasible if Δs moves farther from zero. If this is unique (ties are discussed in 8.3), a primal or dual simplex pivot $a_{i''j''}$ in the corresponding row or column will restore optimality in some adjacent interval of the parameter. In contrast with $i'j'$, the usual sign restrictions apply during the determination of $i''j''$. Of course it has at least one index in common with $i'j'$, and its determination involves another minimum ratio in a direction perpendicular to the first. The corresponding pivot step is not actually performed. As indicated by the superscript z in Table 8.2C, one of the pairs $i'j'$ and $i''j''$ is the $i^z j^z$ such that $-b_{i^z}c_{j^z}/a_{i^z j^z}$ is the extreme value of Δz. Equation (4″) determines i' or (2′) or (6″) determines j'.

For example, consider the parameters y_2 (first of the two lines) and c_2^0, which have the same pairs $i'j' = 23$ and $i''j'' = 73$. The method of 3.5(b) or the argument leading to (6″) above shows that the last row of the optimal tableau (20) is $(20 - \Delta c_2^0, 20 - 4\Delta c_2^0, 4 - \Delta c_2^0, 100, 740 + 13\Delta c_2^0)/2$ as a function of Δc_2^0, whose coefficients are $-a_{2j}$, $-b_2$ (the negatives of the entries in row y_2). Thus the upper limit max Δc_2^0 for Δc_2^0 is $\min(20/1, 20/4, 4/1) = 4 = c_3/a_{23} = c_{j'}/a_{i'j'}$, the dual simplex type of calculation in (6″). This also is the minimum of $|\partial z^*/\partial y_2|$ in (2′). In both cases the pivot $a_{i''j''}$ is determined by the primal simplex calculation max $y_3 = \min(7/1, 13/1, 3/1) = 3 = -b_7/a_{73} = -b_{i''}/a_{i''j''}$ as in (1″). For the parameter c_2^0 one has max $\Delta z = -b_2$ max $\Delta c_2^0 = (13/2)4 = 26$, coming entirely from $i'j'$, because the pivot on a_{73} has $c_3 = (4 - \max \Delta c_2^0)/2 = 0$. For the parameter y_2 one has instead max $\Delta z = -b_7c_3/a_{73} = 6$, coming entirely from the pivot on $a_{i''j''}$, because the determination of min $|\partial z^*/\partial y_2| = 4$ changes neither z nor c_2.

The relations $\Delta b_i^0 = \Delta x_i$ and $\Delta c_j^0 = -\Delta u_j$ discussed at the end of (b) make some of the parameters superfluous, regardless of the intended application. In doing the calculating it may be convenient to think in terms of $\mathbf{x}$ and $\mathbf{u}$, and to range the nonbasic variables in the optimal tableau first, so that their data are available for use in the subsequent ranging of basic variables. In reporting the results in an applied problem, however, statements in terms of b_k^0, c_l^0, b_p^0, c_q^0 seem to have more immediate interpretations. At most it suffices to consider $m+n$ parameters x_q, b_k^0, b_p^0, x_l, which are arranged in the order of their indices in Table 8.2C; another $m+n$ parameters u_p, c_l^0, c_q^0, u_k, of which the (dual basic) u_k have been omitted in the table; and as many a_{ij}^0 as one is interested in.

The x_l, which change from nonbasic to basic, are of particular interest because they represent the positive levels of activity in the problem. They are not equivalent to any b_i^0. One may want to decrease some x_l either because it has turned out to be inconveniently large, or because it is close enough to zero that one suspects its

TABLE 8.2E

s	Base	Δs	$\partial z^*/\partial s$	Δz	$i'j'$	$i''j''$	Ref.
b_6^0	-2	$(-\infty, \frac{2}{3}]$	0	0	—	64	(3)
x_6	$\frac{2}{3}$	$[0, \frac{5}{6}]$	$\frac{1}{5}$	$\frac{1}{6}$	64	24^z	(2)
b_6^0	-2	$(-\infty, \frac{2}{3}, \frac{3}{2}]$	$0, \frac{1}{5}$	$[0, \frac{1}{6}]$	64	24^z	(3), (2)
c_3^0	3	$[-4, \infty)$	0	0	—	63	(4)
u_3	4	$[0, \frac{4}{5}]$	$-\frac{1}{6}$	$-\frac{2}{15}$	63	64^z	Ex. 8
c_3^0	3	$[-\frac{24}{5}, -4, \infty)$	$\frac{1}{6}, 0$	$[-\frac{2}{15}, 0]$	63	64^z	Ex. 8

positivity contributes little to the optimization. Economies of scale (implying a smaller c_l^0) could motivate an increase in x_l.

Two adjacent intervals for Δc_q^0 ($q=1$ or 3) are listed in Table 8.2C. For $-\infty<\Delta c_1^0 \le -10$ one has $\partial z^*/\partial c_1^0 = 1.5$ and $-\infty < \Delta z \le 0$; for $-10 \le \Delta c_1^0 < \infty$ one has $\partial z^*/\partial c_1^0 = 0$ and $\Delta z = 0$. The second interval comes from (5) and the first from Exercise 8 and $\Delta c_q^0 = -\Delta u_q$ (see the end of (b)). The second but not the first is necessarily semi-infinite. To illustrate this consider Table 8.2E, which treats the two cases $p=6$ and $q=3$ in (19); the last line in each group of three combines the information in the first two lines.

In this application x_p or u_q is increasing and all sign restrictions are observed. In (19) we have $\partial z^*/\partial u_3 = -\min_i\{-b_i/a_{i3}\colon a_{i3}>0\} = -2/12$ for $i=i'=6$. This $a_{i'j'} = a_{63} = \frac{12}{3}$ is a pivot that preserves optimality as c_3^0 decreases by a little more than 4, but it is not a pivot when u_6 increases and thereby increases u_3. In the latter case the pivot $a_{i''j''} = a_{64} = -5/3$ is determined by $\min_j\{c_j/(-a_{6j})\colon a_{6j}<0\} = \min\{5/1, 1/5\} = 1/5$. (This distinction explains why ranging x_p or u_q over the given optimal basis M is equivalent to ranging b_p^0 or c_q^0 over an adjacent interval and basis.) Multiplying by $b_6 = -2/3$ gives $\Delta z = -2/15$, and dividing by $\partial z^*/\partial u_3 = -1/6$ gives $\max \Delta u_3 = 4/5$. This is subtracted from $-u_3 = -c_3 = -12/3$ to give $-24/5$, the second critical value of Δc_3^0. The last line of Table 8.2E now asserts that $\Delta z = 0$ for $-4 \le \Delta c_3^0 < \infty$, and Δz changes linearly to the value $-2/15 = (1/6)(-24/5+4)$ when $\Delta c_3^0 = -24/5$.

To return to Table 8.2C, note that the basic y_2 and y_4 ($=x_l$) each get two lines, for different reasons. In the case of y_4 one line is for $\Delta y_4 < 0$ and one for $\Delta y_4 > 0$. On the other hand $\Delta y_2 > 0$ is impossible, but two different ways of decreasing y_2 are considered. They show that y_2 (power produced from clean coal) can be reduced more cheaply ($|\partial z/\partial y_3| = 4$) in the first line by increasing the nonbasic y_3 (power from oil) than by increasing y_1 (power from cheap coal). This conclusion seems wrong until one notes that the pollution restriction requires less power y_4 from the expensive nonpolluting technology in the first case, and more in the second. Nevertheless, if it were worth the cost, the second line shows that the substitution of cheap coal for some of the clean coal would permit a greater reduction in the use of the latter (e.g., if it was in short supply). The two resulting y vectors are $(0,5,3,2)$ and $(1.5,3.5,0,5)$ with $z^*=376$ and 385, as compared with the base values $(0,6.5,0,3.5)$ and 370.

In (8′), (8″) and (9) the denominator c_k or b_l has the index that has been exchanged; the sense of the inequality reflects the fact that $x_q^*=0$ permits $a_{kq}^0 \to +\infty$, and $u_p^*=0$ permits $a_{pl}^0 \to -\infty$; and an upper limit is ≥ 0, a lower limit ≤ 0. To determine $i''j''$ in the cases a_{kl}^0 it is necessary to calculate some elements of (12) with the appropriate end-point value of τ inserted. These are the elements of $\mathbf{b}^*$, $\mathbf{c}^*$ and the row or column in which the pivot $a_{i''j''}$ is to be found.

For the parameters a_{ij}^0 in (7) to (14) the values of $\partial z^*/\partial s$ have been omitted in Table 8.2C because they are either zero or variable. In particular, when (14) is used, the $\partial z^*/\partial s$ slot is allotted to the important constant a_{lk}, and τ_1 and τ_2 should also be recorded in these cases. It is then possible to do most of the intervening arithmetic mentally, while keeping the optimal tableau and the formulas of Table 8.2D in view. The range $[t_1, t_2]$ of Δa_k^0 can be checked by a reverse calculation of $[\tau_1, \tau_2]$ via $\tau = t/(1+a_{lk}t)$.

Now consider the parameter $a_{kl}^0 = a_{82}^0$; this uses (14) because both indices are exchanged in the optimal solution. First $-70 \leq \Delta z \leq 26$ is obtained by intersecting the corresponding Δz ranges for b_8^0 (from (4)) and c_2^0 (from (6)), already listed in the table. Then $-14/13 \leq \tau \leq 0.4$ is obtained by dividing by $-b_l c_k = (13/2)(20/2) = 65$, which is the product of the $\partial z^*/\partial s$ values for b_8^0 and c_2^0. Finally the formula $t = \tau/(1 - a_{lk}\tau)$ with $a_{28} = 1/2$ converts the τ-interval into $-0.7 \leq \Delta a_{82}^0 \leq 0.5$. This can be checked by a reverse calculation of $-14/13 \leq \tau \leq 0.4$ via $\tau = t/(1+a_{lk}t)$—the same as the inverse formula except for the sign preceding a_{lk}. If $-b_l$ or $c_k = 0$, it could be represented by a small $\varepsilon \geq 0$.

(e) *Ranging to eliminate infeasibility*. If the primal problem is found to have no feasible solution, and no error can be found to explain this result, one may try to make the final basis feasible by changing one or more of the b_i^0; if x_i is not an artificial variable, only a decrease in b_i^0 can be useful. Similarly, if the dual problem has no feasible solution (the objective is unbounded), one may try to make the final basis optimal by increasing one or more of the c_j^0. As elsewhere in this section, we vary only one parameter at a time; otherwise the methods of the following sections apply. Success depends on the choices of the parameter and of the final basis.

With some changes, formulas (3) to (6″) of (b) give the desired results. If z is unbounded, (6″) is applicable to the last tableau (which now is not optimal), except when there is a j such that $c_j < 0$, $a_{lj} = 0$, and x_j is not artificial; in the latter case *no* value of Δc_l^0 will work (make the final basis optimal). Otherwise a minimum over an empty set is taken to be $+\infty$ as usual. The second case in which no value of Δc_l^0 will work is that in which the left member (the lower bound) in (6″) exceeds the right member (the upper bound), which can happen if some $c_j < 0$.

To have the best chance of eliminating primal infeasibility by varying parameters it is desirable to pivot all the artificial variables out of the basis, even though feasibility is not thereby obtained. Then (4′) to (4″) remain valid except that no value of Δb_k^0 will work if there is an i such that $b_i > 0$ and $a_{ik} = 0$, or if the left member of (4″) exceeds the right member. For example, Exercise 4 of 3.4 has $\mathbf{x} \geq \mathbf{0}$, the initial basis $\{1,2,3\}$, and the final constraints

$$\begin{array}{llll} 6x_5 & & & -11x_4 - x_1 - 4x_2 - 9 = 0 \\ & 6x_6 & & -7x_4 + x_1 - 2x_2 - 3 = 0 \\ & & 6x_3 & +13x_4 + 5x_1 + 8x_2 + 27 = 0 \end{array} \tag{21}$$

Then (4″) gives $-9 \leq \Delta b_1^0 \leq \min\{3/1, -27/5\} = -5.4$ and $-\min\{9/4, 3/2\} = -1.5 \leq \Delta b_2^0 \leq -27/8$ as conditions either of which will make $\{5,6,3\}$ an optimal basis. However, the second condition must be rejected because it requires $-1.5 \leq -27/8$, which is impossible. So we are left with $-9 \leq \Delta b_1^0 \leq -5.4$, or else $\Delta b_3^0 \leq -27/6$, which comes from (3). Naturally these intervals cannot now contain the original values $\Delta b_i^0 = 0$, as they did previously in (b).

The most useful remaining case is that in which only one artificial x_p remains in the basis and its $b_p \neq 0$ is the only infeasibility. Then (3) is replaced by $b_p + \Delta b_p^0 = 0$ or $\Delta b_p^0 = -b_p$ as a sufficient condition for feasibility. Furthermore if $b_i < 0$ for all $i \neq p$ and $a_{pl} \neq 0$ for some x_l not artificial, then feasibility can be obtained when $\Delta b_p^0 = -b_p - \varepsilon \operatorname{sgn} a_{pl}$, for sufficiently small $\varepsilon \leq 0$, by exchanging x_p and x_l. Finally consider the condition $b_p^* = a_{pk} \Delta b_k^0 + b_p = 0$. If $a_{pk} = 0$, this excludes all values of Δb_k^0 if $b_p > 0$ and excludes none if $b_p \leq 0$. If $a_{pk} \neq 0$, $\Delta b_k^0 = -b_p / a_{pk}$ is the only value that can make row p feasible, and it may have to be discarded if it makes some other row infeasible.

For example, Exercise 4 of 6.2 has $\mathbf{x} \geq \mathbf{0}$, the initial basis $\{1,2\}$, and the final constraints

$$\begin{array}{ll} 3x_3 & +x_1^0 - 4x_4 + x_5 - 2 = 0 \\ & 3x_2^0 + 2x_1^0 - 2x_4 - x_5 - 7 = 0 \end{array} \tag{22}$$

Here $b_p = -7/3$ and $\Delta b_2^0 = -b_2 = 7/3$ will produce feasibility; also any larger value can be accommodated by exchanging x_2^0 and x_4. The feasible $x_2^0 = -b_2^* = 0$ also results from Δb_1^0 or $\Delta x_1 = 7/2$, but this must be rejected because it makes $x_3 = -1/2$.

EXERCISES 8.2

Find the ranges of the specified parameters and the corresponding ranges for Δz; use (19) in 1 to 6.

1. $x_1, \ldots, x_6$

2. b_1^0, b_2^0, b_3^0

3. c_4^0, c_5^0, c_6^0

4. a_{36}^0, d^0

5. a_{16}^0, a_{35}^0

6. $a_{14}^0, a_{25}^0, a_{15}^0$

7. Use the optimal tableau for the applied Exercise 2 of 3.7; the initial basis was x_0, x_1^0, x_2, x_3. Parameters $(h=1,2)$: $y_h, y_3, b_h^0, c_h^0, a_{1h}^0, a_{3h}^0, a_{01}^0$.

17	x_1^0	y_2	x_3	1
x_0	-1	3	-2	-1
y_1	-6	1	5	-6
x_2	-8	-27	18	-25
y_3	7	13	-3	-10
$-z$	-29.3	14.8	0.9	111.8

Note that a_{lk} is in row y_l and column x_k; $\mathbf{x},\mathbf{y}\geq\mathbf{0}$.

8. Derive results like those in (1) and (2) for parameters u_p and u_q.

9. Derive (12) to (15) from (11), $\mathbf{b}^*\leq\mathbf{0}$, $\mathbf{c}^*\geq\mathbf{0}$, $\tau_1\leq\tau\leq\tau_2$.

8.3 ONE PARAMETER IN b AND c ($T_0^*\neq\varnothing$)

The problem of this section and the next is to find $\mathbf{x}^*$, $\mathbf{u}^*$, and z^* as functions of the scalar parameter t, which is assumed to enter the data only via the definitions

$$b_i=b_{i0}+b_{i1}t$$

$$c_j=c_{0j}+c_{1j}t \tag{1}$$

and in d. It is easy to see that if (1) holds for the initial data, relations of the same form will also hold after any number of pivot steps on various a_{ij}, which never depend on t. In general d (the value of z) will be a quadratic polynomial in t. The problem is a special case of that treated by Graves (1963) for polynomials, and has also been discussed by Dantzig (1963) and Orchard-Hays (1968). An application has already been given in 4.7(e), where the parameter was the Bayes loss α.

Step 1: To Find T_0^*

Here we determine the values of t, if any, for which a given tableau or basic solution is primal or dual feasible or optimal. The procedure may be applied to the initial tableau or to one that is known to be optimal for some $t=t^0$, perhaps because that case has been optimized previously. The latter alternative is often available and preferable in applied problems in which the range of permitted parametric variation is rather limited; it only requires that the b_{i1} and/or c_{1j} be updated by the methods of 3.5(b), 7.1 to 7.3, or Table 8.1.

Let us suppose that the artificial x_i^0, if any, have become nonbasic and their columns are ignored. (With artificial variables in the basis there is little chance of having primal feasibility.) If $b_{i1}=0$, then $b_i=b_{i0}$ is feasible for all values of t or for none according as $b_{i0}\leq 0$ or $b_{i0}>0$. Otherwise we have for each i the equivalent feasibility conditions

$$b_{i0} + b_{i1}t \leq 0 \tag{2}$$

$$t \geq -b_{i0}/b_{i1} \text{ if } b_{i1} < 0 \qquad \text{and} \qquad t \leq -b_{i0}/b_{i1} \text{ if } b_{i1} > 0 \tag{3}$$

By forming the intersection of all the semi-infinite intervals (3), one finds that the *conditions for primal feasibility* are the following:

$$\begin{aligned}
&\text{For } no\ i \text{ is } b_{i1} = 0 \text{ and } b_{i0} > 0; \text{ and}\\
&\theta_1' \leq t \leq \theta_1'', \text{ where}\\
&\theta_1' = \max_i\{-b_{i0}/b_{i1}\colon b_{i1} < 0\} \ (= -\infty \text{ if no } b_{i1} < 0)\\
&\theta_1'' = \min_i\{-b_{i0}/b_{i1}\colon b_{i1} > 0\} \ (= +\infty \text{ if no } b_{i1} > 0)
\end{aligned} \tag{4}$$

If some $b_{i1} = 0$ and $b_{i0} > 0$, or $\theta_1'' < \theta_1'$, there is no value of t making the given basis primal feasible.

Similarly, equivalent *conditions for dual feasibility* are:

$$c_{0j} + c_{1j}t \geq 0 \text{ for each } j \tag{5}$$

$$\begin{aligned}
&\text{For } no\ j \text{ is } c_{1j} = 0 \text{ and } c_{0j} < 0; \text{ and}\\
&\theta_2' \leq t \leq \theta_2'', \text{ where}\\
&\theta_2' = \max_j\{-c_{0j}/c_{1j}\colon c_{1j} > 0\} \ (= -\infty \text{ if no } c_{1j} > 0)\\
&\theta_2'' = \min_j\{-c_{0j}/c_{1j}\colon c_{1j} < 0\} \ (= +\infty \text{ if no } c_{1j} < 0)
\end{aligned} \tag{6}$$

If some $c_{1j} = 0$ and $c_{0j} < 0$, or $\theta_2'' < \theta_2'$, there is no value of t making the given basis dual feasible.

For *optimality*, (4) and (6) must both hold; that is, since only nonnegative variables are involved, it is necessary and sufficient that

$$\begin{aligned}
&\theta' \triangleq \max(\theta_1', \theta_2') \leq t \leq \min(\theta_1'', \theta_2'') \triangleq \theta''\\
&b_{i1} = 0 \text{ implies } b_{i0} \leq 0\\
&c_{1j} = 0 \text{ implies } c_{0j} \geq 0
\end{aligned} \tag{7}$$

If $\theta' \leq \theta''$ and the other two conditions are met, the interval $[\theta', \theta'']$ is denoted by T_r^* $(r = 0, 1, 2, \ldots)$. In this section we are assuming that T_0^* is not empty; i.e., that the given tableau is optimal for at least one value of t.

In solving exercises the novice should no doubt begin with (2) and (5), and possibly use (4) and (6) for checking. However, the latter can perhaps be remembered by noting that each ratio includes a minus sign and is a value of t for which some b_i or c_j is zero, and the three remaining features are associated thus (in alphabetical order in each line):

Feasible signs ($b_{i1}<0$ or $c_{1j}>0$)	Infeasible signs ($b_{i1}>0$ or $c_{1j}<0$)
Maximization	Minimization
Lower limits	Upper limits

As usual the sign of the *denominator* (the coefficient of t) is critical.

The above results show that for a particular basis, the values of t that make it optimal form a single interval, which may be empty, finite, semiinfinite, or all inclusive. Next we show that the same conclusion holds for all possible bases collectively.

THEOREM If the parameter t enters the linear program only as in (1), then the set T^* of values of t for which the program has an optimal solution is a convex set (a single interval in this case), regardless of the types and status of the x_h.

Proof Let $\mathbf{x}^h, \mathbf{u}^h$ be optimal solutions for $t=t_h$ $(h=0,1)$. Since these are also feasible solutions, substitution (or Theorem 2 of 2.2, with t included among the variables) shows that $(1-s)\mathbf{x}^0+s\mathbf{x}^1$ and $(1-s)\mathbf{u}^0+s\mathbf{u}^1$ are feasible solutions for $t=(1-s)t_0+st_1$ and $0\le s\le 1$. The latter will not be optimal unless their scalar product is zero, but Theorem 4 of 3.3 assures us that optimal $\mathbf{x}$ and $\mathbf{u}$ do exist for $t=(1-s)t_0+st_1$. This proves T^* is convex.

Step 2: To Find T_r^* ($r>0$)

In this section 8.3 we are assuming that (7) yields a nonempty interval of optimality T_0^* for the given tableau. In many cases T_0^* will overlap the interval of interest T. If it does not, then one should try to optimize the tableau after setting $t=t^0$, the end-point of T that is closest to T_0^*. If this optimum is lacking, the preceding theorem shows it will be lacking for all t in T; in any case the methods of 3.3 or 6.1 can be used. However, if T is not sharply defined, it may be more informative to use the method about to be described, as if one were already in T, and so cause t to move toward t^0 or T.

Subject to remaining in or moving toward T, the procedure is to try to find other basic solutions that are optimal in t intervals adjoining those already found (T_0^* was the first). This requires one or more pivot steps for each new interval. The union of these intervals known at any time is denoted by $[t', t'']$. It is convenient to change t consistently in one direction as long as necessary before returning to the first optimal basic solution and changing t in the reverse direction (unless one began outside T or with $t=\pm\infty$). For most of the discussion it will suffice to consider $t>t''$.

Neither the new basis (optimal for some $t>t''$) nor the interval of optimality is immediately known. However, if the new basis exists at all, it will be optimal for some $t=t''+\varepsilon$, where the letter ε represents a sufficiently small positive number. The situation is clarified by substituting $t=t''+\varepsilon$ in (1) to obtain

$$b_i = b_{i0} + b_{i1}t'' + b_{i1}\varepsilon \triangleq b_i'' + b_{i1}\varepsilon \tag{8}$$

$$c_j = c_{0j} + c_{1j}t'' + c_{1j}\varepsilon \triangleq c_j'' + c_{1j}\varepsilon \tag{9}$$

In the case $t<t'$ one substitutes $t=t'-\varepsilon$ to (1) to obtain

$$b_i = b_{i0} + b_{i1}t' - b_{i1}\varepsilon \triangleq b_i' - b_{i1}\varepsilon \tag{8'}$$

$$c_j = c_{0j} + c_{1j}t' - c_{1j}\varepsilon \triangleq c_j' - c_{1j}\varepsilon \tag{9'}$$

Aside from the replacement of double accents by single, the effect is to replace b_{i1} and c_{1j} by $-b_{i1}$ and $-c_{1j}$, because t is decreasing.

By the definition of t'', at least one b_i'' or c_j'' is zero, with $b_{i1}>0$ or $c_{1j}<0$ (infeasible signs). If this is true of many b_i and c_j, it may be necessary to resort to a full-fledged application of the two-phase methods of 6.1 or 6.4 to extend the interval of optimality or to prove that cannot be done (see Exercise 4). However, it is usual that only a few b_i or c_j become infeasible, and then a single pivot step will often suffice to restore optimality, as follows.

(a) Primal feasibility persists for $t>t''$ if and only if for each i in (8) one has either $b_i''<0$ or else $b_i''=0$ and $b_{i1}\leq 0$. In this case the primal simplex method is used with some $c_l=c_{1l}\varepsilon<0$. In determining $\min_i\{-b_i/a_{il}: a_{il}>0\}=-b_k/a_{kl}$, one first sets $t=t''$ to obtain $b_i=b_i''$. If this does not determine k uniquely, one must try to break the tie by using (8), which indicates that b_i should be replaced by b_{il} in the tied cases. This is a rudimentary form of lexico-minimum; although its origin is not the same as in 6.6, the latter (in a subordinate role) could be combined with it to give the vectors

$$(-b_i'', -b_{i1}, a_{ij'}, a_{ij''}, \dots) \tag{10}$$

for use in lexicographic ordering. Here $M^0=\{j', j'', \dots\}$.

In the present case it is preferable not to define $c_l=\min c_j$. If it is possible to optimize in a single pivot step, it is obvious that choosing l to yield $\max_l|\Delta z|= b_k c_l/a_{kl}$ will always find the appropriate pivot, with or without a parameter, but the chance of optimizing in one step is much greater here. Equations (8) and (9) with $c_l''=0$ give

$$|\Delta z|/\varepsilon = b_k'' c_{1l}/a_{kl} + \varepsilon b_{k1} c_{1l}/a_{kl} \tag{11}$$

Here k depends on l, and l is chosen first to maximize $b_k'' c_{1l}/a_{kl}$; then one must try to break a tie (if any) by maximizing $b_{k1}c_{1l}/a_{kl}$. Of course $b_k''\leq 0$, $c_{1l}<0$, $a_{kl}>0$.

(b) Dual feasibility persists for $t>t''$ if and only if for each j in (9) one has either $c_j''>0$ or else $c_j''=0$ and $c_{1j}\geq 0$. In this case the dual simplex is used with some

$b_k = b_{k1}\varepsilon > 0$. In determining $\min_j\{c_j/(-a_{kj}): a_{kj} < 0\} \triangleq c_l/(-a_{kl})$, one first sets $t = t''$ to obtain $c_j = c_j''$, and then breaks ties (if any) by replacing c_j by c_{1j}. Thus l depends on k in the expression

$$\Delta z/\varepsilon = b_{k1}c_l''/(-a_{kl}) + \varepsilon b_{k1}c_{1l}/(-a_{kl}) \tag{12}$$

Here $b_{k1} > 0, c_l'' \geq 0, a_{kl} < 0$; k is chosen first to maximize $b_{k1}c_l''/(-a_{kl})$ and then to maximize $b_{k1}c_{1l}/(-a_{kl})$.

(c) If there is only one $c_l < 0$ for $t = t'' + \varepsilon$ and no other $c_j = 0$ for $t = t''$, dual feasibility is obtained for some $t > t''$ by pivoting on any $a_{kl} > 0$. In fact this replaces c_l by $-c_l/a_{kl} > 0$ and all other c_j by $c_j - a_{kj}c_l/a_{kl} = c_j'' + c_{1j}\varepsilon - a_{kj}c_{1l}\varepsilon/a_{kl}$. Since all these $c_j'' > 0$ by hypothesis, one has dual feasibility for ε sufficiently small.

Instead of choosing k arbitrarily, for the sake of progress toward primal feasibility we generalize the primal simplex method (a) thus:

$$\min_i\{|b_i|/a_{il}: a_{il} > 0\} \triangleq |b_k|/a_{kl} \tag{13}$$

where only $b_i \leq 0$ are considered unless this excludes all i. If $a_{il} \leq 0$ for all i, no feasible dual solution exists for any $t > t''$; otherwise (13) produces either optimality or case (b).

Similarly if there is only one $b_p > 0$ for $t = t'' + \varepsilon$ and no other $b_i = 0$ for $t = t''$, primal feasibility is obtained for some $t > t''$ by pivoting on $a_{pq} < 0$. The dual simplex method (b) is generalized thus:

$$\min_j\{|c_j|/(-a_{pj}): a_{pj} < 0\} \triangleq |c_q|/(-a_{pq}) \tag{14}$$

where only $c_j \geq 0$ are considered unless this excludes all j. If $a_{pj} \geq 0$ for all j, no feasible primal solution exists for any $t > t''$; otherwise (14) produces either optimality or case (a).

Evidently the second component of (10)–(12) can come into play in breaking ties in (a)–(c) only if t occurs in both **b** and **c** as in Example 2 below.

(d) Suppose that the pairs of conditions preceding (13) and (14) are both satisfied and $a_{pl} \neq 0$. Then optimality is restored by one pivot step using (13) if $a_{pl} < 0$ or (14) if $a_{pl} > 0$, provided a pivot of the required sign (opposed to that of a_{pl}) can be found; otherwise there is no optimal solution for $t > t''$. In any case immediate optimality could not result from pivoting on a_{pl} itself.

Suppose that only one b_i and/or one c_j becomes infeasible for $t = t'' + \varepsilon$. Then (c) and (d) show that if reoptimization is possible at all, it can usually be accomplished in one pivot step. The exception is the case $a_{pl} = 0$ in (d).

(e) If $t > t''$ makes $b_k > 0$ in a row in which all $a_{kj} \geq 0$ or makes $c_q < 0$ in a column in which all $a_{iq} \leq 0$, the optimal solution cannot be extended to any $t > t''$. The same remark applies also to $t < t'$. When further extension (within the interval of interest T) is found to be impossible in both directions, the calculation stops.

Tableau

The following is a specimen of the tableau that will be used in these problems:

1	x_1	x_2	x_3	1	t
x_4	a_{41}	a_{42}	a_{43}	b_{40}	b_{41}
x_5	a_{51}	a_{52}	a_{53}	b_{50}	b_{51}
$\{1$	c_{01}	c_{02}	c_{03}	d_{00}	d_{01}
$\{t$	c_{11}	c_{12}	c_{13}	d_{10}	d_{11}

(15)

The first two rows are read as usual. The last two rows taken together specify that one minimizes $z=(c_{01}+c_{11}t)x_1+(c_{02}+c_{12}t)x_2+(c_{03}+c_{13}t)x_3+(d_{00}+d_{10}t)+(d_{01}+d_{11}t)t$, in which the last two groups of terms reduce to

$$d_{00}+(d_{10}+d_{01})t+d_{11}t^2$$

However, in pivoting it is convenient to proceed as if each of the entries d_{10} and d_{01} had independent significance. The four d_{ij} are segregated so that one does not inadvertently confuse them with the b_{ij} or c_{ij}. Additional rows and columns may be added for z^0, for checking (put $t=1$ and pretend that every row has its own basic variable of value 1), and to give the numerical values of the b_i and/or c_j for some particular value of t.

In the dual problem the first three columns of (15) are read as usual ($u_1=a_{41}u_4+a_{51}u_5+c_{01}+c_{11}t$, etc.), while the last two columns taken together specify that one maximizes $\bar{w}=(b_{40}+b_{41}t)u_4+(b_{50}+b_{51}t)u_5+(d_{00}+d_{01}t)+(d_{10}+d_{11}t)t$.

Example 1

Optimize (4) of 8.1 for all convex combinations of the original $\mathbf{b}=(-1,1)$ and the new $\mathbf{b}=(2,-1)$. This makes $\mathbf{b}=(-1+3t,1-2t)$, which is not feasible for any t. However, since the original problem showed that the basis x_1, x_5 was optimal for $t=0$, we begin with that. It is only necessary to update the column $(3,-2,0)$ of coefficients of t as in 3.5(b) or (1) of 8.1 to obtain the first tableau below.

1	x_3	x_4	x_2	1	t	
x_1	3	−3	−1*	−2	5	←
x_5	2	−1	−1	−1	2	(16)
$-z$	1	4	1	1	−2	

$z^*=1-2t$ for $t\leq .4$

1	x_3	x_4	x_1	1	t	
x_2	−3	3	−1	2	−5	
x_5	−1	2	−1	1	−3	(17)
$-z$	4	1	1	−1	3	

$z^*=-1+3t$ for $.4\leq t$

Since it is already dual feasible, optimality merely requires $b_1=-2+5t\leq 0$ and $b_5=-1+2t\leq 0$. Solving for t gives $t\leq .4$ and $t\leq .5$; the intersection $t\leq .4$ is the interval of optimality. When $t>.4$, b_1 becomes positive; the dual-simplex algorithm selects a negative pivot in the first row (marked by an arrow) by the criterion $\min_j\{c_j/(-a_{1j}): a_{1j}<0\}=\min\{4/3, 1/1\}=1$. The resulting tableau is optimal for the remaining values of t, namely $t\geq.4$. Since the problem specified convex combinations, only $t\in[0,1]$ is considered.

This example and the others assume $\mathbf{x}\geq\mathbf{0}$. ■

Example 2

1	x_3	x_4	1	t
x_1	3	2*	−3	0
x_2	4	1	−6	1
{ 1	6	4	0	2
{ t	−3	−2	0	1

(18)

$z^*=2t+t^2$ for $t\leq2$

2	x_3	x_1	1	t
x_4	3	1	−3	0
x_2	5	−1	−9	2
{ 1	0	−4	12	4
{ t	0	2	−6	2

(19)

$z^*=6-t+t^2$ for $2\leq t\leq4.5$

The object here is to illustrate the use of the b_{i1} and c_{1j} in breaking ties in (a) (and similarly in (b)). The primal simplex is used with $t=2+\varepsilon$ to eliminate the negative $c_3=-3\varepsilon$ and $c_4=-2\varepsilon$. The indicated pivot a_{14} is chosen because its $|\Delta z|/\varepsilon=(-3)(-2)/2=3$ is larger than the $|\Delta z|/\varepsilon=(-6+2+\varepsilon)(-3)/4=3-3\varepsilon/4$ produced by pivoting on $a_{23}=4$. In column x_3 the pivot a_{23} would be chosen because $(6-2-\varepsilon)/4<3/3$. ■

Example 3

Optimize for all $t\leq1.45$ with $\mathbf{x}\geq\mathbf{0}$.

Check	1	x_1	x_2	x_3	1	t	
−1	x_4	−2*	1	−1	−3	3	←
−3	x_5	−2	−1	2	−1	−2	
8	{ 1	1	2	4	0	0	
−2	{ t	1	−2	−2	0	0	
(t=1)	$-z$	2	0	2			

We should first determine as in (2) to (7) for what values of t, if any, primal or dual feasibility holds in (20). Primal feasibility requires $-3+3t\leq0$ and $-1-2t\leq 0$, or $-1/2\leq t\leq1$. Dual feasibility requires $1+t$, $2-2t$, $4-2t\geq0$, or $-1\leq t\leq1$. Intersecting these two intervals gives $-0.5\leq t\leq1$ as the optimality condition for (20); in step 1 we have found $T_0^*=[-0.5,1]$.

Since T_0^* does not contain $T=(-\infty,1.45]$, we try to optimize for $t>1$. This makes $b_4>0$ and $c_2<0$ the two infeasible entries, and case (d) of step 2 applies. Since $a_{pl}=a_{42}=1>0$, (14) requires a negative pivot in row x_4 as in the dual-simplex method. The values of the c_j are calculated at the transition point $t=1$

and min$\{2/2,2/1\}$ selects x_1 to enter the basis. Performing the pivot step gives:

Check	2	x_4	x_2	x_3	1	t	$(t=1.4)$
1	x_1	-1	-1	1^*	3	-3	-1.2
-4	x_5	-2	-4	6	4	-10	-10.0
15	{ 1	1	5	7	-3	3	
-5	{ t	1	-3	-5	-3	3	

(21)

This is optimal for $1 \leq t \leq 1.4$. For $t > 1.4$, c_3 becomes infeasible, and the values of the b_i for $t=1.4$ are tabulated on the right. Min$\{1.2/1, 10/6\}$ selects x_1 to leave the basis. The primal-simplex pivot step of case (a) gives:

Check	1	x_4	x_2	x_1	1	t
1	x_3	-1	-1	2	3	-3
-5	x_5	2	1	-6	-7	4
4	{ 1	4	6	-7	-12	12
0	{ t	-2	-4	5	6	-6

(22)

This is optimal for $1.4 \leq t \leq 1.5$. Since a solution was required only for $t \leq 1.45$, there is no need to increase t further. However, we do need to return to (20) and consider $t < -.5$. This makes b_5 infeasible. The dual-simplex step of case (b) exchanges x_5 and x_1 and gives:

Check	2	x_5	x_2	x_3	1	t
4	x_4	-2	4	-6	-4	10
3	x_1	-1	1	-2	1	2
13	{ 1	1	3	10	-1	-2
-7	{ t	1	-5	-2	-1	-2

(23)

This is optimal for $-1 \leq t \leq -.5$. For $t < -1$, c_5 becomes negative, and since all $a_{i5} \leq 0$, z is unbounded for $t < -1$ as in case (e). This accounts for all t in $T = (-\infty, 1.45]$ and the calculation is ended.

Because the optimal solutions jump from one extreme point to another at transitional values of t, they are usually discontinuous functions of t. The graph of z^* versus t is continuous but has corners at which its slope is discontinuous, as illustrated in this graph for this example in Figure 8.3.

$$
\begin{aligned}
z^* &= -.5 - 1.5t - t^2 && (-1 \leq t \leq -.5) \\
&= 0 && (-.5 \leq t \leq 1) \\
&= -1.5 + 1.5t^2 && (1 \leq t \leq 1.4) \\
&= -12 + 18t - 6t^2 && (1.4 \leq t \leq 1.5)
\end{aligned}
\tag{24}
$$

■

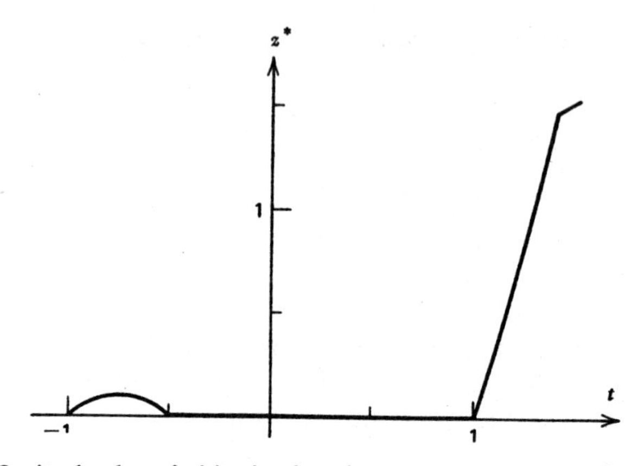

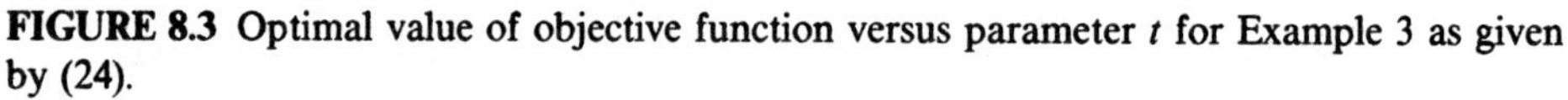
FIGURE 8.3 Optimal value of objective function versus parameter t for Example 3 as given by (24).

Example 4

This is a parametric version of the problem of 8.1(d). We begin by assuming that the new power plants are to be built to meet an expected increase in demand over some time period, such as the next five years. Since the amount of the increase is not certain, the power requirements 10 and 5 million kilowatts will be replaced by $10t$ and $5t$; thus t is proportional to the increase in demand, or the growth rate. If the allowable total pollution (13 units) were also replaced by $13t$, the effect would be merely to multiply the last column by t in every tableau. Hence we shall see what happens when $-b_8^0=13$ remains independent of $t\geq 0$.

Presumably the costs c_j^0 in row $-z$ will all be increasing functions of t, but c_3 may increase most rapidly because of the oil shortage. Since the ratios of these costs determine everything except a factor applied to z, we assume that $c_3^0=32+10t$ while $c_1^0=20, c_2^0=30, c_4^0=50$ as before. The resulting initial tableau, optimal tableau for $.8\leq t\leq 1.3$, optimal tableau for $0\leq t\leq .325$, and other data are shown in Table 8.3.

As in 8.1 or 3.5, by updating $\mathbf{b}^0$ and $\mathbf{c}^0$ one can convert (15) of 8.1 into (26) without repeating the pivot steps that produced the former. Row x_7 in (26) shows there is no feasible solution for $t>1.3$. The smaller t (or the growth rate) is, the cheaper the fuels that can be used. If $0<t<0.325$, the only positive y_j is $y_1=10t$; "cheap coal" is used exclusively. ■

EXERCISES 8.3

Optimize the following for all possible values of t ($\mathbf{x}\geq\mathbf{0}$).

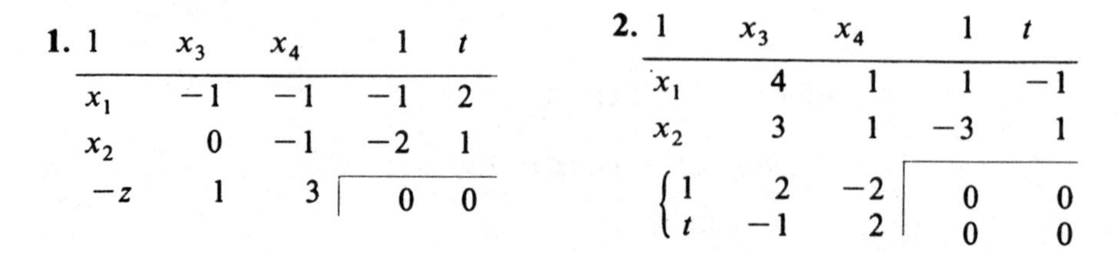

1.

1	x_3	x_4	1	t
x_1	-1	-1	-1	2
x_2	0	-1	-2	1
$-z$	1	3	0	0

2.

1	x_3	x_4	1	t
x_1	4	1	1	-1
x_2	3	1	-3	1
{ 1	2	-2	0	0
{ t	-1	2	0	0

TABLE 8.3 Electric Power Production Versus Growth Rate t

	Check	1	y_1	y_2	y_3	y_4	1	t	
	7	x_6	−1	−1	−1	−1	0	10	
Initial:	4	x_7	−1	−1	0	0	0	5	
	−5	x_8	4	2	1	0	−13	0	(25)
	133	{1	20	30	32	50	0	0	
	11	t	0	0	10	0	0	0	

		2	x_8	y_1	y_3	x_6	1	t	
	−9	y_4	−1	−2	1*	−2	13	−20	
Opt.	−5	y_2	1	4	1	0	−13	0	
$t \in$	3	x_7	1	2	1	0	−13	10	(26)
[.8, 1.3]	866	{1	20	20	−16	100	−260	1000	
⋮	22	t	0	0	20	0	0	0	

		1	y_2	y_3	y_4	x_6	1	t	
	−7	y_1	1	1	1	−1	0	−10	
Opt.	23	x_8	−2	−3	−4	4	−13	40	
$t \in$	−3	x_7	0	1	1	−1	0	−5	(27)
[0, .325]	273	{1	10	12	30	20	0	200	
	11	t	0	10	0	0	0	0	

Critical t_c	0		.325		.65		.80		1.3	
$z^*(t_c)$	0		65		195		270		520	
$z^*(t)$		$200t$		$400t-65$		$200t^2+210t-26$		$500t-130$		
Opt. basis		187		127		327		427		(28)
y_2^*		0		$20t-6.5$		$13-10t$		6.5		
Other $y_j^*>0$		$10t$		$6.5-10t$		$20t-13$		$10t-6.5$		

3. Example of 8.1(c) for linear combinations of the original $(\mathbf{c}, d)=(3,3,1,0)$ (for $t=0$) and $(-1,4,3,0)$ (for $t=1$).

4.

1	x_3	x_4	x_5	1	t
x_1	2	3	−2	−2	1
x_2	3	−1	−1	−2	1
{1	2	2	1	0	0
t	−1	−1	0	0	0

5.

1	x_3	x_4	1	t
x_1	3	−1	0	−1
x_2	2	4	−2	1
{1	0	3	0	0
t	1	−2	0	0

8.4 ONE PARAMETER IN b AND c ($T_0^* = \varnothing$)

The methods of 8.3 suffice whenever some value of t, say t^0, is known to permit the linear program to have an optimal solution. If necessary the program can be optimized for $t=t^0$ and the resulting tableau used to start the procedure.

The present section considers the case in which the set T_0^* determined in step 1 of 8.3 is empty; this is the set of values of t for which the given tableau and its basis are optimal. The starting point is Theorem 4 of 3.3, which asserts that a linear program (for any given value of t, in the present case) has an optimal solution if and only if it has feasible primal and dual solutions; thus the latter may be investigated separately if necessary and the resulting two intervals in t intersected.

The symbolism $\text{val}\,\mathbf{A}<0$ will designate the case in which feasible primal solutions $\mathbf{X}$ exist for *all* $\mathbf{b}$ and $\mathbf{c}$ (not just those occurring in the given problem or generated by varying t), and $\text{val}\,\mathbf{A}>0$ will mean that feasible dual solutions $\mathbf{u}$ exist for all $\mathbf{b}$ and $\mathbf{c}$. These cases are mutually exclusive because they can give opposite signs to $\mathbf{u}_M\mathbf{A}\mathbf{x}_N$. There is also a residual case $\text{val}\,\mathbf{A}=0$.

In Exercise 9 those readers who are acquainted with the game theory of 4.4 and 4.5 are asked to show that $\text{val}\,\mathbf{A}$ can be interpreted as the value of the game whose payoff matrix is $\mathbf{A}$. Here $\mathbf{A}$ is the matrix of coefficients of the x_j (independent of t) in any equivalent condensed tableau from which all but the nonnegative x_h have been eliminated. The precise value of the game (if $\neq 0$) varies from one tableau to another; it remains unknown and of no interest in the present context.

The determination of a trial value t^0 of t depends on the values $\theta_1', \theta_1'', \theta_2', \theta_2''$ of t defined in 8.3. Although we are considering the case in which no t makes the given tableau optimal, it may still be possible to choose t^0 to satisfy one or even both of the conditions

$$\theta_1' \le t^0 \le \theta_1'' \qquad (1) \qquad \theta_2' \le t^0 \le \theta_2'' \qquad (2)$$

perhaps thereby achieving primal or dual feasibility. Note that (1) is more or less irrelevant when artificial variables are in the basis, and nonbasic artificial variables are ignored in computing (2). If neither (1) nor (2) is possible and the sign of $\text{val}\,\mathbf{A}$ is unknown, it is suggested that t^0 be (approximately) the median of the θ's, half being $\le t^0$ and half $\ge t^0$. Then the simplex method is applied to try to find an optimal solution for $t=t^0$. If it succeeds, the methods of 8.3 are then employed. If it fails, the nature of the failure (in Phase 1 or 2) indicates that $\text{val}\,\mathbf{A}\ge 0$ or $\text{val}\,\mathbf{A}\le 0$ respectively.

If $\text{val}\,\mathbf{A}\ge 0$ one may ignore $\mathbf{c}$ temporarily and seek a primal-feasible solution by minimizing the infeasibility form z^0 or z^I over $\mathbf{x}$ and t jointly. If $\text{val}\,\mathbf{A}\le 0$ one may ignore $\mathbf{b}$ temporarily and seek a dual-feasible solution by minimizing w^I, the sum of the infeasible basic $-u_j$, over $\mathbf{u}$ and t. In these minimizations t may be regarded as one of the x_h or u_h, but of the appropriate type (unrestricted, nonnegative or bounded). t may or may not be allowed to become basic, but if so, it must be made nonbasic again (but not necessarily zero) before the second phase of the optimization begins.

If primal-feasible solutions are nonexistent for some t and dual-feasible solutions are nonexistent for some t, $\text{val}\,\mathbf{A}=0$ is indicated. Then it may be necessary to minimize z^I over **x** and t (ignoring **c**) *and* to minimize w^I over **u** and t (ignoring **b**). If both minima are zero, one uses (a) or (b) of 8.3 to try to extend one or both of these feasible solutions to other values of t for which both primal and dual-feasible (and therefore optimal) solutions exist.

Of course it is particularly easy to substitute the values $t=0$ or $\pm\infty$. In the latter cases $-b_i$ and c_j are replaced by $\mp b_{i1}$ and $\pm c_{1j}$, or more precisely (for breaking ties) by the vectors

$$(\mp b_{i1}, -b_{i0}, a_{ip}, a_{iq}, \dots) \tag{3}$$

$$(\pm c_{1j}, c_{0j}, -a_{rj}, -a_{sj}, \dots) \tag{4}$$

with the lexicographic ordering. The first two components are essential.

Convergence follows from the use of lexicographically ordered vectors (if necessary) and the fact that no basic solution is generated more than once (or twice, if val $\mathbf{A}=0$).

Example 1

Optimize for all possible t with $\mathbf{x}\geq\mathbf{0}$ and x_1^0 artificial.

(5)

1	x_3	x_4	x_5	1	t
x_1^0	-2	1	-3^*	0	1
x_2	-3	2	-3	2	1
1	3	0	2	0	0
t	-1	1	-1	0	0
$t=1$	2	1	1		

(6)

3	x_3	x_4	x_1^0	1	t
x_5	2	-1	-1	0	-1
x_2	-3^*	3	-3	6	0
1	5	2	2	0	2
t	-1	2	-1	0	-1

Since x_1^0 is artificial, (5) is never primal feasible; $\{0\}\cap(-\infty,-2]=\varnothing$. However, it is dual feasible for t in $(-\infty,3]\cap[0,\infty)\cap(-\infty,2]=[0,2]$. Hence we set $t=1$ and use the dual simplex method to eliminate x_1^0 from the basis. It is ignored thereafter except that its column provides values of the unrestricted $u_1^{\pm}=c_{01}+c_{11}t$.

The tableau (6) is again never primal feasible but it is dual feasible for t in $(-\infty,5]\cap[-1,\infty)=[-1,5]$. We might set $t=5$ to contribute to primal feasibility but in any case $a_{23}=-3/3$ is the only possible pivot.

(7)

3	x_2	x_4	x_1^0	1	t
x_5	2^*	1	-3	4	-1
x_3	-3	-3	3	-6	0
1	5	7	-3	10	2
t	-1	1	0	-2	-1

$z^*=(10-t^2)/3$ for $4\leq t\leq 5$

(8)

2	x_5	x_4	x_1^0	1	t
x_2	3	1	-3	4	-1
x_3	3	-1	-1	0	-1
1	-5	3	3	0	3
t	1	1	-1	0	-1

$z^*=(3t-t^2)/2$ for $5\leq t$

The resulting tableau (7) is optimal for t in $[4,\infty)\cap(-\infty,5]\cap[-7,\infty)=[4,5]$, and row x_5 shows there can be no primal feasible solution for $t<4$. For $t>5$, c_2 becomes infeasible and a primal simplex step gives (8). ■

Example 2

Optimize for all possible t and $\mathbf{x}\geq\mathbf{0}$.

Check	1	x_1	x_2	x_3	1	t	$(t=3)$
12	x_4	-1	1	3	14	-6	-4
31	x_5	-3	2	8	39	-16	-9
-19	x_6	2	-2	-5	-26	11	7
-47	$\{1$	19	-16	-51	0	0	
15	t	-6	5	15	0	0	
$(t=3)$		1	-1	-6			

(9)

$\theta_1'=\max\{14/6,39/16\}=2.44$, $\theta_1''=\min\{26/11\}=2.36$, $\theta_2'=\max\{16/5,51/15\}=3.40$, $\theta_2''=\min\{19/6\}=3.17$. Since the lower limits exceed the upper limits, (9) is not primal or dual feasible for any t. Hence we let $t^0=3$, an integer median that exceeds 2.44 and 2.36 and is exceeded by 3.40 and 3.17. Then the two-phase simplex gives the optimal tableau.

Check	1	x_4	x_5	x_6	1	t
3	x_1	6	-1	2	-7	2
0	x_2	1	-1	-1	1	-1
5	x_3	2	0	1	2	-1
145	$\{1$	0	3	-3	247	-103
-42	t	1	-1	2	-77	32

(10)

Thus $z^*=247-180t+32t^2$ for $2\leq t\leq 3$. Row x_3 and column x_5 show there are no optimal solutions outside this interval.

Evidently we were fortunate to obtain (10) so easily, especially since $\operatorname{val}\mathbf{A}=0$ ($a_{35}=0$ is a saddle point in (10)). A choice of $t^0>3$ would not have led to an optimal tableau. For the record we may note that the basis 126 is primal feasible for $2\leq t\leq 2.75$ and inconsistent for $t<2$, and 526 is PF for $2.75\leq t$; thus PF is possible if and only if $2\leq t$. Also 126 is dual feasible for $t\leq 1.5$ and inconsistent for $3<t$, 136 is DF for $1.5\leq t\leq 2$, and 135 is DF for $2\leq t\leq 3$; thus DF is possible if and only if $t\leq 3$. Intersection $2\leq t$ and $t\leq 3$ shows that optimal solutions exist if and only if $2\leq t\leq 3$. However, one should proceed to obtain (10) as soon as one has found even one value t^0 for which an optimal solution exists, and even the precise result $2\leq t\leq 3$ does not require every one of the bases just mentioned. ■

EXERCISES 8.4

Optimize for all possible $t, \mathbf{x} \geq \mathbf{0}$, x_i^0 artificial.

1.

1	x_1	x_2	x_3	1	t
x_4	2	−1	3	−2	−1
x_5	1	2	−1	1	−2
{1	−1	1	−1	0	0
{t	1	−1	0	0	0

2.

1	x_3	x_4	1	t
x_1	2	−1	0	−1
x_2	−1	1	1	0
{1	0	1	0	0
{t	−1	0	0	0

3.

1	x_3	x_4	x_5	1	t
x_1	1	1	−1	−3	2
x_2	−2	1	3	2	−1
{1	−2	−1	1	0	0
{t	1	3	−1	0	0

4.

1	x_1	x_2	1	t
x_3	−2	1	−1	1
x_4	−3	−2	1	0
{1	1	−1	0	0
{t	1	2	0	0

5.

1	x_4	x_5	x_6	1	t
x_1^0	−1	−1	4	1	−1
x_2^0	1	0	1	−4	1
x_3	2	1	1	−4	0
{1	2	−4	5	0	0
{t	−1	1	0	0	0

6.

1	x_1	x_2	1
x_3	−4	2	−3
x_4	2	−1	−4
x_5	3	−3	−5
{1	−2	1	0
{t	1	−2	0

7.

1	x_3	x_4	x_5	x_6	1	t
x_1^0	2	−1	3	−1	−10	1
x_2^0	−1	3	2	−2	−2	−1
{1	−3	6	−3	0	0	0
{t	1	−3	0	2	0	0

8.

1	x_3	x_4	1
x_1	−1	−4	2
x_2	−3	−1	1
{1	−1	−2	0
{t	1	2	0

9. Prove that $\mathbf{Ax}+\mathbf{b} \leq \mathbf{0}$ has a feasible solution $\mathbf{x} \geq \mathbf{0}$ for every $\mathbf{b}$ if and only if the value of the game $\mathbf{A}$ is negative.

8.5 ONE PARAMETER IN A COLUMN OR ROW OF A

The most general form of tableau encountered here is

$\delta_0 + \delta_1 t$	x_j	tx_j	1	t
x_i	a_{ij}	α_{ij}	b_{i0}	b_{i1}
$-z$	c_{0j}	c_{1j}	d_0	d_1

(1)

in which the former constant a_{ij} or a_{ij}/δ has been replaced by

$$a_{ij}(t)=(a_{ij}+\alpha_{ij}t)/(\delta_0+\delta_1 t) \tag{2}$$

However, this section differs from the others in that pivoting in the initial tableau usually produces something more complicated. In order that the latter remain of the form (1), it is generally necessary in practice that in the *initial* tableau one has all the coefficients of t $(\alpha_{ij}, b_{i1}, c_{1j}, \delta_1)$ equal to zero except for $(\boldsymbol{\alpha}^0_{.q}, c^0_{1q})$ in one column q or $(\boldsymbol{\alpha}^0_{p.}, b^0_{p1})$ in one row p. The sufficiency of each condition follows from (3) of 7.8 and the definition of the determinant in (1) of 7.7, or from (4) and (5) below. Exercise 7 generalizes these conditions.

Such problems have been discussed by Orchard-Hays (1968). In many respects the algorithm is like that of 8.3 and 8.4. However, (17) of 8.2 and Exercises 2(a) and 6 of this section show that the optimizable values of t need not form a convex set. This implies that more searching for optimal tableaus may be required.

Tables 8.5A and B give examples for the column and the row cases respectively; most of the discussion is concerned with the former. In Table 8.5A the column

TABLE 8.5A Optimize for All t (in column x_2)

Check	1	x_1	x_2	tx_2	1
0	x_3	1	-2	-1	1
-1	x_4	-1^*	1	-3	1
0	x_5	-2	-1	1	1
7	$-z$	1	1	1	3

Check	1	x_4	x_2	tx_2	1
-1	x_3	1	-1^*	-4^*	2
1	x_1	-1	-1	3	-1
2	x_5	-2	-3	7	-1
6	$-z$	1	2	-2	4

Chk 1	$1+4t$	x_4	tx_4	x_3	tx_3	1	t		$\omega(tx_2)$	Chk 0
1	x_2	-1	0	-1	0	-2	0		4^*	1
3	x_1	-2	-1	-1	3	-3	2		7	2
6	x_5	-5^*	-1^*	-3	7	-7	10	←	19	5
30	$-z$	3	2	2	-2	8	12		-10	4

$z^*=(8+12t)/(1+4t)$ for $-.25<t\leq .70$ $\quad 12=-[4(-4^*)-2(-2)]/1$

$30=-[6(-1^*-4^*)-(-1)(2-2)]=1+4+3+2+2-2+8+12$ (Chk 1)

$4=-[6(-1^*)-(-1)2] \quad =1+3+2+8-10$ (Chk 0)

Chk 1	$5+t$	x_5	tx_5	x_3	tx_3	1	t	$\omega(tx_2)$	Chk 0
0	x_2	-1	0	-2	0	-3	0	1^*	0
0	x_1	-2	-1	1	-1	-1	-2	-3	0
-6	x_4	-1	-4	3	-7	7	-10	-19	-5
42	$-z$	3	2	1	3	19	8	7	35

$z^*=(19+8t)/(5+t)$ for $.70\leq t$. All signs changed. Use pivot -5 and divisor 1 to find columns $x_5, x_3, 1, \omega(tx_2)$, Chk 0.
Use pivot $-5-1=-6$ and divisor $1+4=5$ for Chk 1. For tx_3, t: $8=-[12(-1^*)-10(2)]/4$ or (in last tableau) $[19(1^*)-(-3)(7)]/5$.

headed tx_2 in the initial tableau contains the vector $\alpha^0_{.q}$ and $t=1$ is used in calculating check values. If the problem could be optimized while keeping $x_q=x_2$ nonbasic, t would remain confined to this column and the resulting basis would be optimal for all values of t such that $c^*_q(t)\geq 0$. The first pivot $a_{41}=-1$ was selected in an attempt to do this. However, row x_3 in the second tableau shows that no feasible solution exists having $x_2=0$. Therefore the next pivot on $-1-4t$ exchanges x_3 and x_2.

Before discussing the mechanics of computing and checking the third and fourth tableaus, we note that the third is optimal for $-.25<t\leq .70$. This calculation differs from (2) to (7) of 8.3 in that the sign of the denominator $1+4t\neq 0$ must also be considered. Thus the crucial requirements are $-2/(1+4t)\leq 0$ and $(-7+10t)/(1+4t)\leq 0$ in rows x_2 and x_5. Row x_2 shows there is no feasible solution for $t\leq -.25$. When $t\geq .70$, $b_5>0$ is infeasible. In this row there is only one pivot $a_{54}(t)=(-5-t)/(1+4t)$ that is negative as required when $t=.70$. It produces the last tableau, which is optimal for $.70\leq t$.

TABLE 8.5B Optimize for All t (in row x_2)

Chk 1	1	x_3	tx_3	x_1	tx_1	x_5	tx_5	1	t	Chk 0
2	x_4	−1	0	1	0	2	0	−1	0	2
0	x_2	1*	4*	1	−3	3	−7	−2	2	4
−3	$-z$	−2	0	1	0	1	0	−4	0	−3
5	$\omega(-tu_2)$	−4		3		7		−2		5

Chk 1	$1+4t$	x_2	tx_2	x_1	tx_1	x_5	tx_5	1	t	Chk 0
10	x_4	1	0	2	1	5*	1*	−3	−2	6
0	x_3	1	0	1	−3	3	−7	−2	2	4
−15	$-z$	2	0	3	−2	7	−10	−8	−12	5
25	$\omega(-tu_2)$	4*		7		19		−10		21

$z^*=(-8-12t)/(1+4t)$ for $-.25<t\leq .70$ $\quad -12=4^*(-4)-(-2)2$

$-15=(-3)(1^*+4^*)-0(-2+0)=1+4+2+0+3-2+7-10-8-12$ (Chk 1)

$5=(-3)(1^*)-4(-2) \quad =1+2+3+7-8$ (Chk 0)

Chk 1	$5+t$	x_2	tx_2	x_1	tx_1	x_4	tx_4	1	t	Chk 0
10	x_5	1	0	2	1	1	4	−3	−2	6
8	x_3	2	0	−1	1	−3	7	−1	−3	2
−12	$-z$	3	0	1	2	−7	10	−19	−8	−17
−8	$\omega(-tu_2)$	1*		−3		−19		7		−9

$z^*=(-19-8t)/(5+t)$ for $.70\leq t$. No optimum for $t\leq -.25$.
Use pivot 5 and divisor 1 to find columns $x_2, x_1, x_4, 1$, Chk 0.
Use pivot $5+1=6$ and divisor $1+4=5$ for Chk 1. Others:
$-8=[(-12)(1^*)-(-10)(-2)]/4$ or (in last tableau) $[(-19)(1^*)-(3)(7)]/5$.

In performing the last pivot (on $-5-t$) the exact-division principle of Theorem 16 of 7.8 applies to the polynomials in t. In the notation of (1) of 2.5, which has its own use for the letter α, we can therefore assert that

$$\frac{(\pi_0+\pi_1 t)(\alpha_0+\alpha_1 t)-(\beta_0+\beta_1 t)(\gamma_0+\gamma_1 t)}{\delta_0+\delta_1 t}$$

$$=\frac{\pi_0\alpha_0-\beta_0\gamma_0}{\delta_0}+\frac{\pi_1\alpha_1-\beta_1\gamma_1}{\delta_1}t \qquad \text{if } \delta_0\delta_1\neq 0 \tag{3}$$

For example, one can write $(10+29t+21t^2)/(2+3t)=10/2+21t^2/3t=5+7t$ *if* one knows the quotient is a binomial. It is immaterial whether the numerical coefficients are integers, but as usual they will remain so if they are all integers in the initial tableau.

If the denominator is a constant or monomial, the first member of (3) is evaluated directly without short cuts; otherwise $(\delta_1\delta_2\neq 0)$ *it is as if one were performing two independent pivot steps, one on the constant terms and one on the coefficients of* t. Nevertheless the presence of t has doubled the number of arithmetic operations; this can largely be avoided as follows, when t occurs only in column q in the initial tableau.

Suppose that the column $\mathbf{a}^0_{\cdot q}$ belongs to the current basis, so that its updated version is the unit vector $\mathbf{i}_q$. Then the updated version of $\mathbf{a}^0_{\cdot q}(t)=\mathbf{a}^0_{\cdot q}+\boldsymbol{\alpha}^0_{\cdot q}t \triangleq \mathbf{a}^0_{\cdot q}+\omega^0 t$ is

$$\mathbf{B}^{-1}\left(\mathbf{a}^0_{\cdot q}+\omega^0 t\right)=\mathbf{i}_q+\mathbf{B}^{-1}\omega^0 t \triangleq \mathbf{i}_q+\omega t \tag{4}$$

where $\mathbf{B}^{-1}$ is the inverse basis matrix as in 7.1. To bring this column into the basis in place of $\mathbf{i}_q$ requires a pivot on its qth element, which is $1+\mathbf{a}_{qM^0}\omega^0 t=1+\omega_q t$. This replaces the constant a_{ij} by

$$a_{ij}(t)=a_{ij}-a_{qj}\omega_i t/(1+\omega_q t) \qquad (\text{when } \delta_0=1)$$

With an arbitrary $\delta_0\neq 0$ in (1) (used to avoid fractions in hand calculations) this becomes

$$a_{ij}(t)=(a_{ij}+\alpha_{ij}t)/(\delta_0+\omega_q t) \text{ where}$$

$$\alpha_{ij}=(\omega_q a_{ij}-\omega_i a_{qj})/\delta_0 \tag{5}$$

Equation (5) is equivalent to the application of rule 5′ of 2.5; the pivot is ω_q, usually not in the current pivotal row k. Indeed, (5) is unusual in that both members really refer to the same tableau. Its great value is to permit the calculation of coefficients of t to be limited to those required to determine the next pivot; that is, the b_{i1} and c_{1j} and then the α_{ij} in the pivotal column l (primal simplex) or pivotal row k (dual simplex). Before this is done one must update the a_{ij} and ω_i by either of the available methods. Equation (3) or 2.5 gives, for example,

$$\omega_i' = (a_{kl}\omega_i - a_{il}\omega_k)/\delta_0 \tag{6}$$

This ω_i' then is used as ω_i in (5). If the methods of 7.3 are employed instead of (6), it will be necessary to update only the columns $\mathbf{b}_{.0}$, ω and $\mathbf{a}_{.l}$, and the rows $\mathbf{c}_{0.}$, $\mathbf{a}_{q.}$ and $\mathbf{a}_{k.}$.

The last two tableaus in Table 8.5A show two different check columns, which are distinguished by the value 0 or 1 that seems to be assigned to t (except as regards ω). Actually both set $t=1$ but Chk 0 allows only the column $\mathbf{a}_{.q}$, not $\mathbf{a}_{.q}+t\mathbf{a}_{.q}$, in the basis. Chk 0 is used with (5) and (6), is updated as in (6), and for each i should agree with $\delta_0+\Sigma_j a_{ij}+b_{i0}+\omega_i$. Chk 1 is used with (3), gives values of $s_i \triangleq \delta_0+\delta_1+\Sigma_j(a_{ij}+\alpha_{ij})+b_{i0}+b_{i1}$, and is updated by

$$s_i' = [(a_{kl}+\alpha_{kl})s_i - (a_{il}+\alpha_{il})s_k]/(\delta_0+\delta_1) \tag{7}$$

If $\delta_0+\delta_1=0$, some other value of t (not 1 or 0) is called for, as in Exercise 2(b).

If t occurs initially in a row, forming the negative transpose of the tableau would convert it to the column case already considered, but this would be inconvenient in the context of Chapter 6 or 7. Hence the row case is illustrated in Table 8.5B, which gives the negative transposes of the last three tableaus in Table 8.5A. However, since the notation and checking in both tables emphasizes the row equations, the second table is not altogether redundant. Note that the auxiliary row labeled $\omega(-tu_2)$ begins as the *negative* of the coefficients of t in row x_2, which are shifted one column to the left. Thus the ω vectors and the common denominators are the only entries that do not even change sign in going from Table 8.5A to B.

EXERCISES 8.5

1. Why is the problem more tractable when t occurs independently in a column *and* a row (**b** and **c**) in 8.3 than in 8.5?
2. How is the solution in Table 8.5A changed (a) if x_2 is not sign restricted; (b) if x_4 is not sign restricted?

Optimize 3 to 6 for all possible t with $\mathbf{x}\geq\mathbf{0}$, x_i^0 artificial.

3.

1	x_4	x_5	tx_5	1
x_1	2	1	−1	−1
x_2	1	1	1	−2
x_3	3	−1	2	−3
$-z$	−1	−1	1	0

4.

1	x_4	tx_4	x_5	tx_5	1	t
x_1	2	1	1	−1	−1	2
x_2	−1	0	3	0	−2	0
x_3	1	0	2	0	−3	0
$-z$	−1	0	−2	0	0	0

5.

1	x_3	x_4	x_5	tx_5	1
x_1^0	3	−1	1	−1	−3
x_2^0	−1	2	1	1	−4
$-z$	−1	−1	−1	0	0

6.

1	x_3	tx_3	x_4	tx_4	1	t
x_1	2	1	−3	−1	−1	−1
x_2	−1	−1	2	1	−3	1
$-z$		−1	1	1	0	1

7. Use (3) of 7.8 to show that if each $a_{ij}^0(t)=a_{ij}^0+\alpha_{ij}^0 t$, then the highest power of t occurring in other tableaus cannot exceed the rank ρ of the matrix of α_{ij}^0. (The case $\rho=1$ is a slight generalization of this section. Values **b** and **c** are to be included in **A**.)

8.6 SEVERAL PARAMETERS IN b AND c

The theory and practice of this section are essentially the same as in 8.3 and 8.4 except for the nature of the convex set T_r^* (in the parameter space) within which the rth basic solution is optimal. T_r^* is no longer an interval but rather a convex polygonal set in two or more dimensions. Each set can have many adjoining neighbors, each of which can usually be reached by a single pivot stop. Thus it is more difficult to insure that the whole parameter space T is mapped out exactly once, although a graphical presentation works well with two parameters. As in 8.3, the union of all the T_r^* is also convex. We shall limit the discussion to a pair of examples.

The first example, which has one parameter t_1 in **b** and another t_2 in **c**, is shown in Table 8.6A and Figure 8.6A. There are four regions having optimal solutions and one region having none. As in 8.3 to 8.5, each pivot selection is related to the vanishing (becoming infeasible) of some b_k or c_l as indicated by the arrowheads.

The second example has both parameters t_1 and t_2 in **b** alone. The optimality conditions will not be written out since they always consist of $b_i \leq 0$ for the three values of i in the basis.

Check	1	x_4	x_5	1	t_1	t_2		
−4	x_1	−1	−3*	0	−1	0	←	(1)
−2	x_2	−1	−1	0	0	−1		
−1	x_3	−2	−1	−1	1	1		
4	$-z$	1	2	0	0	0		

Check	3	x_4	x_1	1	t_1	t_2		
4	x_5	1	−1	0	1	0		(2)
−2	x_2	−2*	−1	0	1	−3	←	
1	x_3	−5	−1	−3	4	3		
4	$-z$	1	2	0	−2	0		

	2	x_2	x_1	1	t_1	t_2		
2	x_5	1	−1	0	1	−1		
2	x_4	−3	1	0	−1	3		(3)
4	x_3	−5*	1	−2	1	7	←	
2	$-z$	1	1	0	−1	−1		

	5	x_3	x_1	1	t_1	t_2		
7	x_5	1	−2*	−1	3	1	←	
−1	x_4	−3	1	3	−4	−3		(4)
−4	x_2	−2	−1	2	−1	−7		
7	$-z$	1	3	−1	−2	1		

	2	x_3	x_5	1	t_1	t_2		
−7	x_1	−1	−5	1	−3	−1		(5)
1	x_4	−1	1	1	−1	−1		
−3	x_2	−1*	−1	1	−1	−3	←	
7	$-z$	1	3	−1	1	1		

	1	x_2	x_5	1	t_1	t_2		
−2	x_1	−1	−2	0	−1	1		(6)
2	x_4	−1	1	0	0	1		
3	x_3	−2	1	−1	1	3		
2	$-z$	1	1	0	0	−1		

The corresponding regions of optimality are depicted in Figure 8.6B. Most of them have three sides, one for each of $b_1, b_2, b_3 \leq 0$. However, in (3) and (4) the condition on $2-t_1-7t_2$ is redundant (implied by the other two). Consequently their regions of optimality do not have a common boundary although the tableaus are related by a single pivot step. In the style of the answer section, Figure 8.6B would be partially described by noting that four regions having optimal bases 123, 124, 524, 523 (in counterclockwise order) surround $(t_1, t_2) = (0,1)$; 123, 523, 543, 143 surround $(0,0)$; and 123, 143, 142 surround $(1,0)$.

For more than two parameters see Van de Panne (1975a).

TABLE 8.6A Linear Programming with Two Parameters

1	x_3	x_4	1	t_1	
x_1	-1	1	-1	0	(a)
x_2	2	-1^*	0	1	←
{1	0	1	0	0	
t_2	1	0	0	0	

Opt. for $t_1 \leq 0, 0 \leq t_2$.

1	x_3	x_2	1	t_1	
x_1	1	1	-1	1	(b)
x_4	-2^*	-1	0	-1	←
{1	2	1	0	1	
t_2	1	0	0	0	

Opt. for $0 \leq t_1 \leq 1; -2 \leq t_2$.

2	x_2	x_4	1	t_1	
x_1	1	1^*	-2	1	(c)
x_3	1	-1	0	1	
{1	0	2	0	0	
t_2	-1	1	0	-1	
		↑			

Opt. for $t_1 \leq 0; -2 \leq t_2 \leq 0$.

1	x_2	x_1	1	t_1	
x_4	1	2	-2	1	(d)
x_3	1	1	-1	1	
{1	-1	-2	2	-1	
t_2	-1	-1	1	-1	

Opt. for $t_1 \leq 1; t_2 \leq -2$.

In (d), row x_3 shows there is no feasible solution for $1 \leq t_1$, and the four entries in the lower right corner show $z^* = 2 - t_1 + t_2 - t_1 t_2$.

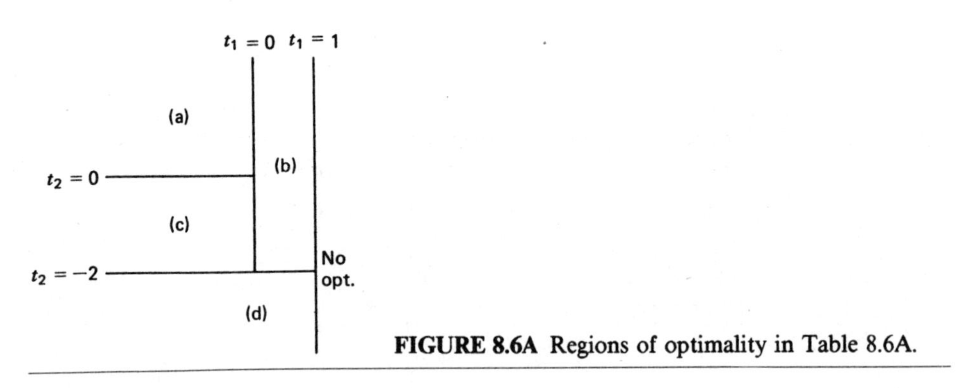

FIGURE 8.6A Regions of optimality in Table 8.6A.

EXERCISES 8.6

1. Show that the proof of the theorem of 8.3 holds in this section.
2. When $m = n = 2$ and there are two parameters, show that the boundaries of the T_r^* consist of one or two sets of lines, each set being either concurrent or parallel.
3. State T_r^*, M_r^*, z^* for (1) to (4) and Figure 8.6A when $t_1 = t_2 = t$.
   Optimize 4 to 8 for all possible values of t_1 and t_2.
4. Check

	1	x_3	x_4	1	t_1
4	x_1	1	2	-1	1
3	x_2	3	1	-1	-1
1	{1	1	-1	0	0
0	t_2	-2	1	0	0

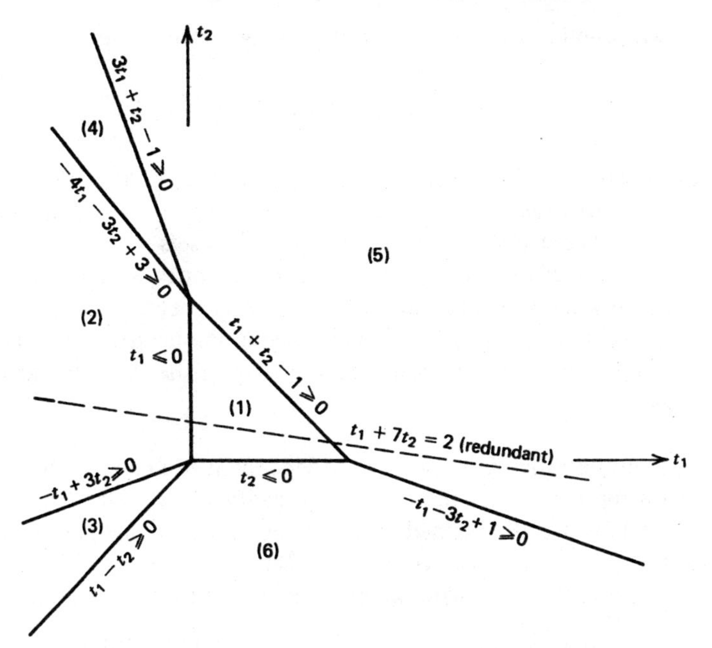

FIGURE 8.6B Regions of optimality for Tableaus (1)–(6).

5.

1	x_3	x_4	1	t_1	t_2
x_1	1	1	1	−2	1
x_2	2	3	1	1	−3
$-z$	−2	−1	0	0	0

6.

1	x_3	x_4	1	t_1	t_2
x_1	−2	−1	−1	1	0
x_2	−1	−3	−1	0	1
$-z$	1	1	0	0	0

7.

1	x_3	x_4	1	t_1	t_2
x_1	1	−1	2	0	−2
x_2	−1	1	0	−1	0
1	−2	0	0	0	0
t_1	2	0	0	0	0
t_2	0	−1	0	0	0

8.

1	x_3	x_4	1	t_1	t_2
x_1	2	3	−5	1	2
x_2	1	2	−1	−1	1
1	−2	−1	0	0	0
t_1	1	0	0	0	0
t_2	0	1	0	0	0

8.7 NONLINEAR PARAMETERS IN b AND c

Although the examples will consider only one scalar parameter t, in the preliminary theory it is as easy to consider a vector **t** of parameters occurring in **b** or **c**. If the b_i or $-c_j$ are not linear functions of **t**, the next best property that they can possess is the following.

A real-valued function $f(\mathbf{t})$ defined for all $\mathbf{t}$ in some convex set C is called *convex* if

$$(1-s)f(\mathbf{t}^0)+sf(\mathbf{t}^1)\geq f[(1-s)\mathbf{t}^0+s\mathbf{t}^1] \tag{1}$$

for all s satisfying $0\leq s\leq 1$ and every pair of points $\mathbf{t}^0, \mathbf{t}^1$ in C. If $-f(\mathbf{t})$ is convex, $f(\mathbf{t})$ is called *concave*. Obviously a linear function such as $\mathbf{ct}+d$ is both convex and concave. The word "concave" is not usually applied to sets.

The graph of a convex or a concave function (of one or two scalar variables) looks convex or concave when viewed from *below*. Evidently such a function will define a convex set T when it is used in an inequality having the proper sense, as illustrated in Figure 8.7A and proved in Exercise 1. This motivates the following generalization of the theorem of 8.3.

THEOREM Suppose that $\mathbf{A}$ is constant (independent of t) and each $b_i(\mathbf{t})$ and $c_j(\mathbf{t})$ is linear when it represents ($\pm$) the value of an artificial variable. (These conditions and those that follow are all assumed to be satisfied in some one particular tableau.) Then if every $b_i(\mathbf{t})$ and $-c_j(\mathbf{t})$ is convex, the set T^* of optimizable values of t is convex. This assertion (a) and two others are tabulated thus:

If every $b_i(\mathbf{t})$ is	every $c_j(\mathbf{t})$ is	and $d(\mathbf{t})$ is	then (conclusion)
(a) convex	concave	any	T^* is convex
(b) convex	constant	convex	$z^*(\mathbf{t})$ is convex
(c) constant	concave	concave	$z^*(\mathbf{t})$ is concave

Proof The artificial variables can be eliminated by pivoting or as in (1) of 6.2, while the other conditions continue to hold. Now let $\mathbf{y}^0$ and $\mathbf{y}^1$ be optimal values for the nonbasic components of $\mathbf{x}$ corresponding to $\mathbf{t}^0$ and $\mathbf{t}^1$, respectively. Let $0\leq s\leq 1$ and

$$\mathbf{y}^s=(1-s)\mathbf{y}^0+s\mathbf{y}^1 \qquad \mathbf{t}^s=(1-s)\mathbf{t}^0+s\mathbf{t}^1$$

Since optimal solutions are feasible,

$$\mathbf{0}\geq\mathbf{A}\mathbf{y}^0+\mathbf{b}(\mathbf{t}^0) \qquad \mathbf{0}\geq\mathbf{A}\mathbf{y}^1+\mathbf{b}(\mathbf{t}^1)$$

Then multiplying by $(1-s)$ and s, respectively, adding, and using the convexity of each $b_i(\mathbf{t})$ gives

$$\mathbf{0}\geq\mathbf{A}\mathbf{y}^s+(1-s)\mathbf{b}(\mathbf{t}^0)+s\mathbf{b}(\mathbf{t}^1)\geq\mathbf{A}\mathbf{y}^s+\mathbf{b}(\mathbf{t}^s)$$

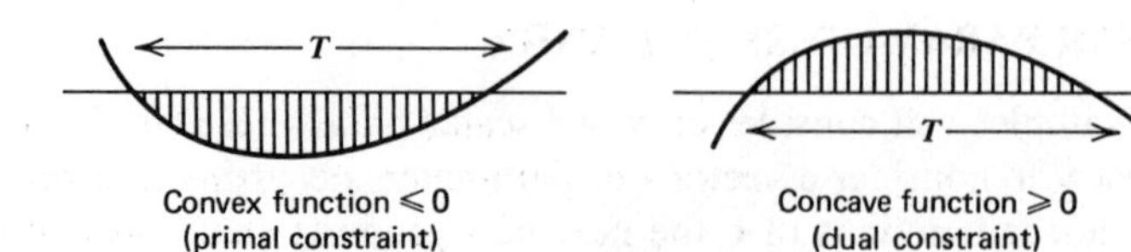

FIGURE 8.7A

This shows that $\mathbf{y}^s$ is a feasible (primal) solution corresponding to $\mathbf{t}^s$. Similarly we can construct a feasible dual solution $\mathbf{u}^s$ corresponding to $\mathbf{t}^s$. Then by Theorem 4 of 3.3 there is an optimal solution for $\mathbf{t}^s$, and so T^* is convex. Finally by the minimality of z^* and the convexity of d,

$$z^*(\mathbf{t}^s) \leq \mathbf{c}\mathbf{y}^s + d(\mathbf{t}^s) \leq (1-s)\left[\mathbf{c}\mathbf{y}^0 + d(\mathbf{t}^0)\right] + s\left[\mathbf{c}\mathbf{y}^1 + d(\mathbf{t}^1)\right]$$

Since the last member is $(1-s)z^*(\mathbf{t}^0) + sz^*(\mathbf{t}^1)$, this shows that z^* is a convex function of $\mathbf{t}$. Part (c) is proved similarly, or by taking the negative transpose.

Figure 8.3 shows that $z^*(t)$ as a whole need be neither convex nor concave when t occurs linearly in both $\mathbf{b}$ and $\mathbf{c}$, although each piece of the graph (parabolic in 8.3 and hyperbolic in 8.5) is either convex or concave. The convexity of $z^*(\mathbf{t})$ is associated with the possible occurrence of $z^* = +\infty$ when $\mathbf{b}(\mathbf{t})$ becomes large enough to exclude feasible solutions.

When the conditions of the theorem are not known to be satisfied for some basis, the generalization of 8.3 requires the recognition that the sets T^* and $\cup T_r^*$ no longer need be convex (in one dimension, intervals). Thus there is less incentive to get an optimal solution for *some* $t = t^0$ as soon as possible, as it may still be necessary to map out all of the relevant set T, piece by piece, and some of the pieces for which optimal solutions exist may be isolated points.

Nonlinear Polynomials

We now suppose that the b_i and c_j are polynomials in a scalar parameter t, whose superscripts will be exponents. Such problems have been discussed by Graves (1963). Consider the tableau

Check	1	x_1	1	t	t^2	t^3
2	x_2	1*	0	0	1	−1
2	x_3	1	−2	5	−3	0
2	{ 1	1	0	0	0	0
−1	{ t	−2	0	0	0	0
		↑				

(2)

The first row shows that no feasible solution can exist when $t^2 - t^3 > 0$; that is, when $t < 0$ or $0 < t < 1$. However, (2) is optimal for the isolated value $t = 0$, and it is primal feasible for $t \geq 1$. To verify the latter and to prepare for optimization, we may set $t_1 = t - 1$ and calculate

$$-b_2 = t_1 + 2t_1^2 + t_1^3$$

$$-b_3 = t_1 + 3t_1^2$$

For $t \geq 1$ or $t_1 \geq 0$, c_1 is negative and a pivot is required in the column of x_1. $\mathrm{Min}(-b_2/1, -b_3/1)$ will depend on t_1, but as in 8.3, values of t_1 close to zero are to be considered first, and so $-b_2/1$ is chosen because $-b_3/1 + b_2/1 = t_1^2 - t_1^3$ has its first nonzero coefficient positive. This is a lexicographic ordering like that of 6.6; it takes priority over the latter *except* at an isolated point of optimality such as $t=0$ in (2), where it is irrelevant, a small departure from $t=0$ not being possible at all. In place of the coefficients of t_1^h one may use the hth derivatives with respect to t evaluated at $t=t_1$, which differ only by the constant factor $h!$

Pivoting on a_{21} gives

Check	1	x_2	1	t	t^2	t^3	
2	x_1	1	0	0	1	-1	(3)
0	x_3	-1^*	-2	5	-4	1	←
0	$\{1$	-1	0	0	-1	1	
3	$\{t$	2	0	0	2	-2	

Since $-b_3=(t-1)^2(2-t)$, this is optimal for t in $[1,2]$, the intersection of $\{0\}\cup[1,\infty)$, $(-\infty,2]$, and $[1/2,\infty)$. When $t>2$ we can use the dual simplex method, taking the negative pivot in row x_3:

Check	1	x_3	1	t	t^2	t^3	
2	x_1	1	-2	5	-3	0	(4)
0	x_2	-1	2	-5	4	-1	
0	$\{1$	-1	2	-5	3	0	
3	$\{t$	2	-4	10	-6	0	

This is optimal for $t\geq 2$ since $-b_1=(3t-2)(t-1)$. The corner elements assert that in this tableau

$$z^*=2-5t+3t^2+t(-4+10t-6t^2)=2-9t+13t^2-6t^3$$

In general in this type of problem the determination of the appropriate pivots and intervals of optimality will require the calculation of numerical approximations to irrational roots of certain algebraic equations, the selection of the smallest of these, and its substitution into the appropriate polynomials. Although these roots thus determine the t-intervals and the selection of the pivots, they will not enter into the resulting tableaus so long as the latter are expressed in terms of the original parameter t.

Convergence depends on the convergence of the simplex method and the finiteness of the numbers of different tableaus and the intervals in t that they determine.

Piecewise Linear Functions

More generality can be obtained by not insisting that each nonlinear function be represented by the same analytic expression throughout its whole domain. It might be thought that the effect would be to split the problem into independent parts, one for each interval in t for which each function is again represented by a single expression. However, this point of view is often impractical because of the number of expressions involved and the fact that the nonlinear composite functions will usually still be continuous, so that the optimal bases in adjacent intervals may still differ by very few pivot steps, and the ordering of the values of the parameter t remains significant.

Thus the principal effect of using different analytic expressions for the same function (b_i or c_j) in different intervals of t is to introduce the transition points between such consecutive intervals into the problem in addition to those determined by the vanishing of some b_i or c_j. However, if the approximations are not specified in advance, they may be determined as the solution progresses, so as to take advantage of points at which some b_i or c_j vanishes.

Piecewise-linear interpolation is the simplest of these procedures and ties in well with linear programming. It uses straight line segments connecting adjacent selected points on the graphs of the function. In Figure 8.7B the dotted line shows a crude approximation by linear interpolation to the function $f(t)=t\sqrt{3-3t}$ in the interval $0 \le t \le 1$.

For example, suppose that a linear program is to be solved for $0 \le t \le 3$, with the b_i and c_j in the initial tableau determined by linear interpolation between consecutive values in the following table:

$t=$	0		1		2		3
$b_1=$	-1		0		2		4
$b'_1=$		1		2		2	
$b_2=$	-1		-1		-2		-4
$b'_2=$		0		-1		-2	
$c_3=$	1		0		-2		-3
$c'_3=$		-1		-2		-1	
$c_4=$	1		-1		-2		-4
$c'_4=$		-2		-1		-2	

(5)

The alternate rows labeled b'_1, b'_2, c'_3, c'_4 give the slopes of the linear approximations to the b_i and c_j in the respective intervals $0 \le t \le 1$, etc. These slopes are equal to the

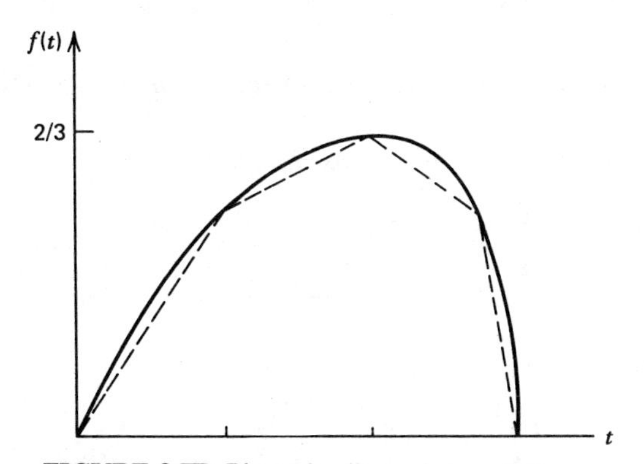

FIGURE 8.7B Piecewise-linear interpolation.

changes in the b_i and c_j divided by the change in t. In the interval $\tau \le t \le \tau+1$, the auxiliary parameter $t_\tau = t-\tau$ will be used.

The initial tableau is given as

1	x_3	x_4	1	t
x_1	3	−2	−1	1
x_2	−1	1*	−1	0
{ 1	1	1	0	0
{ t	−1	−2	0	0
		↑		

(6)

with all $x_h \ge 0$. The row and column labeled 1 contain the values for $t=0$ in the first column of (5). The row and column labeled t contain the values of the slopes for $0 \le t \le 1$ from the second column of (5). Thus (6) specifies the problem for $0 \le t \le 1$; it is optimal for $0 \le t \le 0.5$.

For $t > .5$, c_4 becomes negative. The pivot a_{24} gives

1	x_3	x_2	1	t
x_1	1*	2	−3	1
x_4	−1	1	−1	0
{ 1	2	−1	1	0
{ t	−3	2	−2	0
	↑			

(7)

This is optimal for $0.5 \le t \le 2/3$. For $t > 2/3$, c_3 becomes negative. The pivot a_{13} gives

1	x_1	x_2	1	t
x_3	1	2	−3	1
x_4	1	3	−4	1
{ 1	−2	−5	7	−2
{ t	3	8	−11	3

(8)

This is optimal only for $2/3 \le t \le 1$, because the linear interpolations are valid only for $0 \le t \le 1$.

The tableau is now rewritten with new b_i, c_j and d thus:

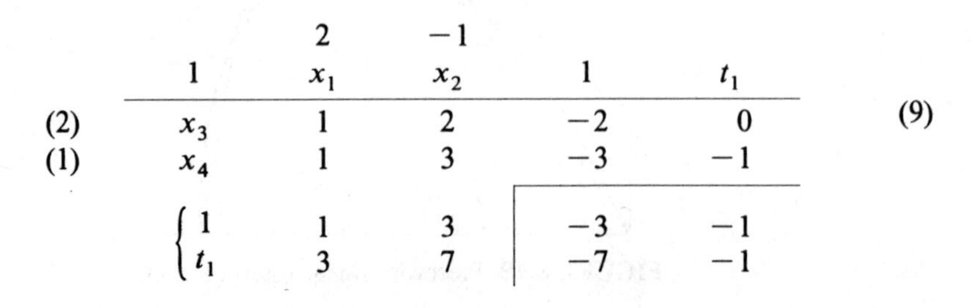

		2	−1		
	1	x_1	x_2	1	t_1
(2)	x_3	1	2	−2	0
(1)	x_4	1	3	−3	−1
	{ 1	1	3	−3	−1
	{ t_1	3	7	−7	−1

(9)

Here, $t_1 = t-1$ and the a_{ij} are the same as before. Since the b_i, c_j and d are continuous functions of t, their values at $t_1 = 0$ (the entries in the row and column labeled 1) are equal to their values at $t=1$ in the previous tableau; of course they differ from the values at $t=1$ in (5) because of the change of basis. In particular the constant term in d, namely $d_{00} = -3$, is $7-(2+11)\cdot 1+3\cdot 1^2$.

The entries in the row and column labeled t_1 in (9) are obtained by updating (as in 3.5(b) or 7.3) the slopes corresponding to $1 \le t \le 2$ in (5). The latter have been adjoined to the tableau as in 3.5(b), those for c_3 and c_4 being reversed in sign and parenthesized. Then for example

$$b_{31} = 2\cdot 1 - 1\cdot 2 = 0 \qquad d_{01} = 2\cdot 1 - 1\cdot 3 = -1$$

$$c_{12} = 2\cdot 2 + 1\cdot 3 = 7 \qquad d_{10} = 2(-2) + 1(-3) = -7$$

$$d_{11} = 2\cdot 0 + 1(-1) = 2\cdot 3 + (-1)7 = -1$$

in the notation of (15) of 8.3. Note that in these calculations the row or the column labeled 1 is involved either in its entirety (excepting the common element $d_{00} = -3$) or not at all. This will prevent the mistake of entering d_{11} where d_{01} or d_{10} belongs.

The above tableau is optimal for $0 \le t_1 \le 1$, which is equivalent to its whole range of validity $1 \le t \le 2$. Setting $t_2 = t_1 - 1$ and updating as before produces the following tableau, which is optimal for $0 \le t_2 \le 1$ (equivalent to $2 \le t \le 3$). This completes the calculation.

		2	-2		
	1	x_1	x_2	1	t_2
(1)	x_3	1	2	-2	-2
(2)	x_4	1	3	-4	-4
	{ 1	4	10	-12	-12
	{ t_2	3	8	-10	-10

(10)

If updating and pivoting both appear to be required at the same transition value of t, the updating must be done first, as it may eliminate or modify the need to pivot.

EXERCISES 8.7

1. Show that if $s_1, s_2 \ge 0$ and $f_1(\mathbf{t})$ and $f_2(\mathbf{t})$ are both convex or both concave, then $s_1 f_1(\mathbf{t}) + s_2 f_2(\mathbf{t})$ has the same property; and if $f(\mathbf{t})$ is convex, the set of $\mathbf{t}$ satisfying $f(\mathbf{t}) \le 0$ is convex.
2. Reoptimize the second example [(5) and (6)] with

$$\mathbf{A}^0 = \begin{bmatrix} -1 & 3 \\ 2 & -1 \end{bmatrix} \text{ instead of } \begin{bmatrix} 3 & -2 \\ -1 & 1 \end{bmatrix}$$

Optimize 3 to 6 for all possible values of t.

3.

1	x_3	x_4	1	t	t^2
x_1	1	-1	0	2	-1
x_2	1	2	-2	3	-1
1	-1	0	0	0	0
t	2	0	0	0	0
t^2	-1	1	0	0	0

4.

1	x_3	x_4	1	t
x_1	-1	2	-1	1
x_2	2	1	-1	-1
1	-3	-1	0	0
t	-4	2	0	0
t^2	1	-1	0	0

5.

1	x_3	x_4	1	t	t^2	t^3
x_1	2	3	0	2	-1	0
x_2	-1	-2	3	-1	-3	1
$-z$	1	1	0	0	0	0

6.

1	x_3	x_4	1
x_1	-4	1	$2\sinh s$
x_2	-1	-2	$2\cosh s$
$-z$	1	1	0

(Let $t = e^s > 0$)

8.8 MULTIPLE-OBJECTIVE LINEAR PROGRAMMING

In most of this book, as in the literature, a scalar objective function is taken for granted because of the basic importance and tractability of the problems that result. Nevertheless there are many situations that involve a multiplicity of more or less conflicting objectives such as cost, time and quality; short-term, intermediate and long-term consequences; profitability, risk, public favor and the general welfare, etc.

The games against nature in 4.7 could be regarded as a special case in which (ordinarily) there are no constraints and the various objectives (one for each state of nature) happen to be measured in terms of a common unit, although they may have points of reference (origins) ε_i that are different and unknown. It seems less useful to regard n-person games, with one objective for each player, as a special case. Although one may speak of "competing" objectives, in this section this competition takes place only in the mind(s) of one or more decision makers; while the latter might be considered as arbitrators, they are really the players of the game. (We assume that there is (at least effectively) only one decision maker.) Neither would one want to assume that the objectives have to be treated symmetrically (or even linearly) as the n players were in most of 5.7.

Nevertheless the theories of multiple objectives and of n-person games do share a special interest in the set of efficient points within the feasible region (cf. 5.3(a$_2$)). In 5.1 an efficient point was defined as one where no player could get more unless at least one player got less. In symbols, a feasible $\boldsymbol{\alpha}^0$ is efficient if $\alpha_h^0 \le \alpha_h$ for all h, with $\boldsymbol{\alpha}$ feasible, implies that $\boldsymbol{\alpha}^0 = \boldsymbol{\alpha}$. This assumes that each player h wants to maximize his payoff α_h, whereas our simplex tableau minimizes in the primal problem.

To reconcile these conventions when the objective h replaces the payoff to player h, we change the tableau only to the extent of replacing the label $-z$ or $-z_h$ by $\bar{z}_h = -\mathbf{c}_{h\cdot}\mathbf{x}_N - d_h$, and reporting the values of the latter. Then the desirable (but not necessarily demanded) maximization of $\bar{z}_h$ is consistent with our usual minimization of $\mathbf{cx}_N$. The hth objective row in the tableau reads $\bar{z}_h + \mathbf{c}_{hN}\mathbf{x}_N + d_h = 0$ and the whole tableau (in the form of (1) of 3.3 without artificial variables) is

$$\begin{array}{c|cc} 1 & \mathbf{x}_N & 1 \\ \hline \mathbf{x}_M & \mathbf{A} & \mathbf{b} \\ \bar{\mathbf{z}} & \mathbf{C} & \mathbf{d} \end{array} \qquad \begin{array}{c} \mathbf{x} \geq \mathbf{0} \\ \text{"max"}\,\bar{\mathbf{z}} \end{array} \tag{1}$$

Here the vector $\mathbf{c}$ of cost coefficients has been replaced by the μ by n matrix $\mathbf{C}$, and the scalars $-z$ and d have been replaced by vectors with μ components.

Now a feasible vector $\bar{\mathbf{z}}^0$ is *efficient* if $\bar{\mathbf{z}}^0 \leq \bar{\mathbf{z}}$ (feasible) implies $\bar{\mathbf{z}}^0 = \bar{\mathbf{z}}$. Note that the objectives are serving as constraints in the condition $\bar{\mathbf{z}}^0 \leq \bar{\mathbf{z}}$.

Some main points of the theory will now be sketched. More rigorous treatments of some of the topics are available for example in Stoer and Witzgall (1970).

Like any linear program, (1) may fail to have any feasible solutions or some or all of the objectives may be unbounded. In addition (1) may be an unsatisfactory model in yet another way, exemplified by $\bar{z}_1 = x_1$, $\bar{z}_2 = -x_1$ (so that $\bar{z}_1 + \bar{z}_2 \equiv 0$), or a little more subtly by $\bar{z}_1 = x_1$, $\bar{z}_2 = -x_1 + x_2 + 3$, $\bar{z}_3 = -x_1 - x_2$, so that $2\bar{z}_1 + \bar{z}_2 + \bar{z}_3 \equiv 3$. Since the coefficients 2, 1, 1 are all positive, it is clear that one $\bar{z}_h$ can increase only if one or more others decrease, and so every feasible point is efficient, no matter what the constraints may be. This proves half of Theorem 1. It should be noted that if $\boldsymbol{\lambda}\bar{\mathbf{z}}$ is constant for one choice of nonbasic variables $\mathbf{x}_N$, it is constant for all N.

THEOREM 1 Given that the feasible region of (1) has an interior point, the point is efficient if and only if there exists a $\boldsymbol{\lambda} > \mathbf{0}$ such that $\boldsymbol{\lambda}\mathbf{C} = \mathbf{0}$ (or $\boldsymbol{\lambda}\bar{\mathbf{z}} \equiv$ constant).

Proof The "if" part was proved above. Both parts follow by applying Theorem 4 of 3.3 to the schema

$$\begin{array}{c|ccc|l} & \boldsymbol{\zeta} & \boldsymbol{\xi}^{\pm} & 1 & \\ \hline \boldsymbol{\lambda} & \mathbf{I} & \mathbf{C} & \mathbf{0} & = \mathbf{0} \\ 1 & -\mathbf{e} & \mathbf{0} & \mathbf{0} & = -\zeta_0 \\ \hline & \geq & \| & \| & ((-\zeta_0) \text{ to be min.}) \\ & \mathbf{0} & \mathbf{0} & \bar{w} & \end{array} \tag{2}$$

The "only if" part may be stated with the hypothesis that $-\mathbf{C}\boldsymbol{\xi} \geq \mathbf{0}$ implies $\mathbf{C}\boldsymbol{\xi} = \mathbf{0}$ for all $\boldsymbol{\xi}$; and the conclusion that there exists a $\boldsymbol{\lambda} \geq \mathbf{e} \triangleq (1, \ldots, 1)$ such that $\boldsymbol{\lambda}\mathbf{C} = \mathbf{0}$. For the purposes of this proof $\boldsymbol{\xi}$ is not sign restricted (and hence written as $\boldsymbol{\xi}^{\pm}$ in (2) with dual variables replaced by zeros) because it represents a (small) departure from an *interior* point of the feasible region. Now in the rows of (2) we have $-\mathbf{C}\boldsymbol{\xi} = \boldsymbol{\zeta} \geq \mathbf{0}$, which implies $\mathbf{C}\boldsymbol{\xi} = \mathbf{0}$, $\boldsymbol{\zeta} = \mathbf{0}$ and $\zeta_0 = 0$; thus an optimal solution is $\boldsymbol{\xi} = \mathbf{0}$, $\boldsymbol{\zeta} = \mathbf{0}$, $\zeta_0 = 0$. Then by Theorem 4 of 3.3, the column equations of (2) also have an optimal (feasible) solution, whose existence is the desired conclusion.

The more useful models are those in which the two equivalent conditions in the theorem do not hold, so that all the efficient points are on the boundary of the feasible region. Among these the *efficient extreme points* are of particular interest for the following reasons.

(a) By Theorem 8 of 3.3, every extreme point is a basic solution that can be generated by suitable pivot steps, and conversely. (In (1) we could arrange to have $\bar{\mathbf{z}} \geq \mathbf{0}$ if necessary.)

(b) All the efficient extreme points can be enumerated by pivoting from one to the other, without any inefficient extreme points intervening.

(c) Every efficient point is a convex combination of efficient extreme points (but not conversely), if all the objectives are bounded. Theorems 9 and 4 below state and prove (b) and (c).

THEOREM 2 $\bar{\mathbf{z}}^0$ is an efficient point for (1) if and only if it is obtainable from max $\boldsymbol{\lambda}\bar{\mathbf{z}}$ with some $\boldsymbol{\lambda} > \mathbf{0}$.

Proof The "if" part is obvious in the contrapositive form; if $\bar{\mathbf{z}}^0$ were not efficient one would have a feasible $\bar{\mathbf{z}}^1 \geq \bar{\mathbf{z}}^0$ with some $\lambda_h \bar{z}_h^1 > \lambda_h \bar{z}_h^0$, so that $\boldsymbol{\lambda}\bar{\mathbf{z}}^0$ would not be the maximum. The "only if" part differs from that in Theorem 1 only by allowing some $\xi_j \geq 0$, so that the conclusion is $\boldsymbol{\lambda}^0\mathbf{C} \geq \mathbf{0}$ for some $\boldsymbol{\lambda}^0 > \mathbf{0}$. This is enough to insure that $\boldsymbol{\lambda}^0\bar{\mathbf{z}}$ is maximized.

THEOREM 3 Let $\bar{\mathbf{z}}^0$ lie in the (relative) interior of a face Q, defined by restricting the feasible region by requiring one or more x_i or $x_j = 0$. (In particular this is true if $\bar{\mathbf{z}}^0$ is expressible as a positive convex combination of all the extreme points of Q.) Then if $\bar{\mathbf{z}}^0$ is efficient, every $\bar{\mathbf{z}}$ in Q (including its boundary) is efficient.

Proof By Theorem 2, there is a $\boldsymbol{\lambda} > \mathbf{0}$ such that $\boldsymbol{\lambda}\bar{\mathbf{z}}^0 \geq$ all feasible $\lambda\bar{z}$. Now there are only two possibilities:

(i) $\boldsymbol{\lambda}\bar{\mathbf{z}}^Q = \boldsymbol{\lambda}\bar{\mathbf{z}}^0$ for every $\bar{z}^Q$ in Q, so that $\boldsymbol{\lambda}\bar{\mathbf{z}}^Q \geq$ all feasible $\boldsymbol{\lambda}\bar{\mathbf{z}}$, and $\bar{\mathbf{z}}^Q$ is efficient by Theorem 2.

(ii) There are two points $\bar{\mathbf{z}}^1$ and $\bar{\mathbf{z}}^2$ in Q, collinear with $\bar{\mathbf{z}}^0$ and on opposite sides of it, such that $\boldsymbol{\lambda}\bar{\mathbf{z}}^1 < \boldsymbol{\lambda}\bar{\mathbf{z}}^0 < \boldsymbol{\lambda}\bar{\mathbf{z}}^2$. This would be true because $\bar{\mathbf{z}}^0$ is an interior point of Q, and $\boldsymbol{\lambda}\bar{\mathbf{z}}$ is a linear function of $\bar{\mathbf{z}}$. However, the hypothesis excludes $\boldsymbol{\lambda}\bar{\mathbf{z}}^0 < \boldsymbol{\lambda}\bar{\mathbf{z}}^2$, and so the proof is complete.

THEOREM 4 If all the objectives are bounded above, every efficient point $\bar{\mathbf{z}}$ of (1) is a convex combination of efficient extreme points.

Proof Every feasible point satisfies (without slack) some maximum number of independent constraints $x_j = 0$. These equalities together with the remaining inequalities determine a feasible convex subset Q_r containing the given $\bar{\mathbf{z}}$ and contained in a linear variety of minimum dimension $r \leq n$. The proof is by induction on r, using the fact that convex combinations of convex combinations are again convex combinations as in Theorem 4 of 2.2.

If $r = 0$, $\bar{\mathbf{z}}$ itself is extreme and the theorem is certainly true. If $r > 0$, $\bar{\mathbf{z}}$ will lie in the interior of Q_r. Any straight line in Q_r through $\bar{\mathbf{z}}$ will intersect the boundary of Q_r

in two new points $\bar{\mathbf{z}}^1$ and $\bar{\mathbf{z}}^2$, because otherwise some $\bar{z}_h$ would be able to increase indefinitely. (Convexity precludes more than two points.) By construction, $\bar{\mathbf{z}}$ is a convex combination of $\bar{\mathbf{z}}^1$ and $\bar{\mathbf{z}}^2$, so that the latter must be efficient as in Theorem 3. As boundary points of Q_r, each of them satisfies an additional equality and so is contained in some Q_{r-1}, for which the theorem has already been proved. This completes the proof.

The converse of Theorem 4 is not necessarily true even for a pair of efficient extreme points separated by only one pivot step. This is illustrated by

1	x_1	x_2	1
x_3	1	1	-1
$\bar{z}_1$	-3	-2	0
$\bar{z}_2$	3	1	-3

$$(\mathbf{x}\geq\mathbf{0},\ \text{“max”}\,\bar{\mathbf{z}}) \tag{3}$$

Here $\bar{\mathbf{z}}=(3,0)$, $(2,2)$, $(0,3)$ are efficient extreme points corresponding to the basic $x_i=x_1, x_2, x_3$, respectively, but convex combinations of $(3,0)$ and $(0,3)$ are not efficient. Theorem 8 following will repair the damage.

If a primal basic solution is degenerate, pivoting in one of the rows having $b_i=0$ will change the designation of the variables as basic or nonbasic without changing the values of the primal variables, but the values and the feasibility of the dual solution may change. The result is that efficiency criteria stated as in Theorem 2 apply to feasible sets of basic indices rather than to basic values of $\mathbf{x}$, and a feasible basic $\mathbf{x}$ is efficient if and only if it has at least one basis for which it is revealed as efficient. With an unfortunate choice of basis, $\mathbf{C}$ may indicate an apparent opportunity to improve the solution (as if it were inefficient) when, in fact, the effort will be completely frustrated by the presence of the constraints having $b_i=0$. (Figure 11.4 pictures an example with a single objective function.)

The search for a suitable degenerate basis can be combined with the rest of the algorithm as follows. The matrix $\mathbf{C}'$ is formed by adjoining to $\mathbf{C}$ those rows of $\mathbf{A}$ that have $b_i=0$ as in Evans and Steuer (1973) and Ecker and Kouada (1978); then the criteria of Theorems 5 and 6 are obtained for inefficiency and for efficiency. The notation is the same as in (2), but now $\boldsymbol{\xi}=\mathbf{x}_N\geq\mathbf{0}$, $\mathbf{C}$ is replaced by $\mathbf{C}'$, and $\mathbf{I}$ is enlarged correspondingly, but the additional components of $\mathbf{e}$ are *zeros*, not ones. In (2) thus expanded the existence of optimal dual and primal solutions is asserted by the equivalent conditions (e_0) and (e_2) in Theorem 6 below.

THEOREM 5 Each of (d_1) and (d_2) is equivalent to the *inefficiency* (or deficiency) of the basic solution $\mathbf{x}=(\mathbf{0},-\mathbf{b})\geq\mathbf{0}$ in (1):

(d_1) There exists a $\boldsymbol{\xi}\geq\mathbf{0}$ such that $\mathbf{C}'\boldsymbol{\xi}\leq\mathbf{0}$ and $\mathbf{c}_{h\cdot}\boldsymbol{\xi}<0$ for the row of some $\bar{z}_h$ in $\mathbf{C}$.

(d_2) Max $\zeta_0=+\infty$ where $\zeta_0=\Sigma_h(\bar{z}_h+d_h)=-\Sigma_j(\Sigma_h c_{hj})\xi_j$, $\boldsymbol{\xi}\geq\mathbf{0}$, $\mathbf{C}'\boldsymbol{\xi}\leq\mathbf{0}$. In the final tableau derived from $\mathbf{C}'$ there is a nonpositive column with a negative entry in the last row ζ_0. Nonbasic variables can include $\bar{z}'_h=\bar{z}_h+d_h\geq0$.

(d_3) Val $\mathbf{C}'<0$ is sufficient; val $\mathbf{C}'\leq 0$ is necessary. Here val $\mathbf{C}'$ is the value of the game of matrix $\mathbf{C}'$.

Proof Criterion (d_1) says that one can choose $\boldsymbol{\xi} \triangleq \mathbf{x}_N \geq \mathbf{0}$ so as to increase one or more $\bar{z}_h$ without decreasing any, and without immediately violating any constraint.

(d_1) implies $\mathbf{eC}\boldsymbol{\xi}<0$ and $\zeta_0=-\mathbf{eC}(t\boldsymbol{\xi})\rightarrow+\infty$ as $t\rightarrow\infty$. (When all the rows of $\mathbf{A}$ are considered in (1), $\max\zeta_0$ may well be finite but will still be positive.) This proves (d_2).

(d_2) implies there exists a $\boldsymbol{\xi}\geq\mathbf{0}$ such that $\mathbf{eC}\boldsymbol{\xi}<0$ and $\mathbf{C}'\boldsymbol{\xi}\leq 0$. Hence some $\mathbf{c}_{h\cdot}\boldsymbol{\xi}<0$ as required in (d_1).

(d_3) follows from (1) and (1′) of 4.5 when $\mathbf{C}'\boldsymbol{\xi}$ replaces $\mathbf{Ay}$.

THEOREM 6 Each of (e_0), (e_1) and (e_2) is equivalent to the *efficiency* of the basic solution $\mathbf{x}=(\mathbf{0},-\mathbf{b})\geq\mathbf{0}$ in (1):

(e_0) There exists a $\boldsymbol{\lambda}\geq\mathbf{0}$ such that $\boldsymbol{\lambda}\mathbf{C}'\geq\mathbf{0}$ and $\lambda_h>0$ for every $\bar{z}_h$ (i.e., for every row of $\mathbf{C}$).

(e_1) For every $\boldsymbol{\xi}\geq\mathbf{0}$, $\mathbf{C}'\boldsymbol{\xi}\leq\mathbf{0}$ implies $\mathbf{C}\boldsymbol{\xi}=\mathbf{0}$.

(e_2) $\operatorname{Max}\zeta_0=0$ where $\zeta_0=\Sigma_h(\bar{z}_h+d_h)=-\Sigma_j(\Sigma_h c_{hj})\xi_j$, $\boldsymbol{\xi}\geq\mathbf{0}$, $\mathbf{C}'\boldsymbol{\xi}\leq\mathbf{0}$. In the final tableau derived from $\mathbf{C}'$, the (last) row ζ_0 has nonnegative coefficients. Nonbasic variables can include $\bar{z}'_h=\bar{z}_h+d_h\geq 0$. $\zeta_0=-\mathbf{eC}\boldsymbol{\xi}$.

(e_3) Val $\mathbf{C}'>0$ or $\boldsymbol{\lambda}\geq\mathbf{0}$ with $\boldsymbol{\lambda}\mathbf{C}'>\mathbf{0}$ is sufficient; val $\mathbf{C}'\geq 0$ is necessary.

Proof (e_1) to (e_3) are the negations of (d_1) to (d_3), respectively. For (e_2), note that the equations are homogeneous, the $(\mathbf{b},\mathbf{d})$ column having been replaced by zeros; thus 0 and $+\infty$ are the only possibilities for $\max\zeta_0$.

(e_2) implies (e_0); as λ_h we may take the (positive) coefficient of $\bar{z}'_h$ in the row ζ_0 of the final tableau, after ζ_0 has been replaced by its equivalent $\Sigma_h\bar{z}'_h=\Sigma_h(\bar{z}_h+d_h)$. The omission of the nonnegative terms involving the ξ_j makes $\boldsymbol{\lambda}\mathbf{C}'\geq\mathbf{0}$ rather than $=\mathbf{0}$. Criterion (e_0) implies $\boldsymbol{\lambda}\mathbf{C}'\boldsymbol{\xi}\geq 0$ while the hypothesis $\mathbf{C}'\boldsymbol{\xi}\leq\mathbf{0}$ of (e_1) implies $\boldsymbol{\lambda}\mathbf{C}'\boldsymbol{\xi}\leq 0$. Thus $\boldsymbol{\lambda}\mathbf{C}'\boldsymbol{\xi}=0$ and each nonpositive term $\lambda_h\mathbf{c}_{h\cdot}\boldsymbol{\xi}=0$. Since $\lambda_h>0$ for every row h in $\mathbf{C}$, we must have $\mathbf{C}\boldsymbol{\xi}=\mathbf{0}$, the conclusion of ($e_1$). This shows that ($e_0$) implies ($e_1$).

In the actual generation of efficient extreme points, (d_1) and (e_0) are useful in small examples, the existence of $\boldsymbol{\xi}$ or $\boldsymbol{\lambda}$ being confirmed by inspection. The criterion (e_0) is also suitable for a general algorithm as Theorem 8 will show. Although (d_3) and (e_3) fail to be definitive in the special case val $C'=0$, they are very easy to remember, because the necessary condition val $\mathbf{C}'\geq 0$ for efficiency is analogous to the necessary and sufficient condition $\mathbf{c}\geq\mathbf{0}$ or $\min c_j\geq 0$ for the optimization of a single objective. These conditions immediately indicate that an all-positive row in $\mathbf{C}'$ guarantees efficiency and an all-negative column guarantees inefficiency. However, (d_1) with $\xi_l>0$ and the other $\xi_j=0$ shows that a nonpositive column l in C' implies inefficiency if the column contains at least one negative element in a row of $\mathbf{C}$.

Domination among columns of $\mathbf{C}'$ or among rows of $\mathbf{A}$ may permit some to be disregarded as usual, but a row of $\mathbf{C}$ must not be eliminated by domination. For example, if $\mathbf{C}'=\mathbf{C}$ is given by (4), the third row is responsible for the inefficiency, but it is dominated by the sum of the first two rows. The problem arises because in (d_1) the "column player" chooses not only the column weights ξ_j but also the row h.

$$\begin{matrix} 2 & -2 \\ -2 & 2 \\ -1 & -1 \end{matrix} \tag{4}$$

The MOLP Algorithm

The flowchart in Figure 8.8A gives an overview. The six boxes from 1 to 3b will be called *phases* since in general each consists of a sequence of pivot steps, and Phases 1 and 2 are indeed the familiar simplex phases of 6.1, which try to maximize $\mathbf{e}\bar{\mathbf{z}}=\Sigma_h \bar{z}_h$. ($\mathbf{C}$ may be updated simultaneously or at the end of Phase 2.) If this attempt is successful, which it is likely to be in a practical problem, Phases 2a and 2b are not needed. (On the other hand, 2a and 2b could always replace Phase 2 if desired.) In any case Phases 3a and 3b may be repeated once for each of the efficient extreme points. Phase 3a determines where more of these points will be found (on maximal efficient faces), and Phase 3b actually calculates them.

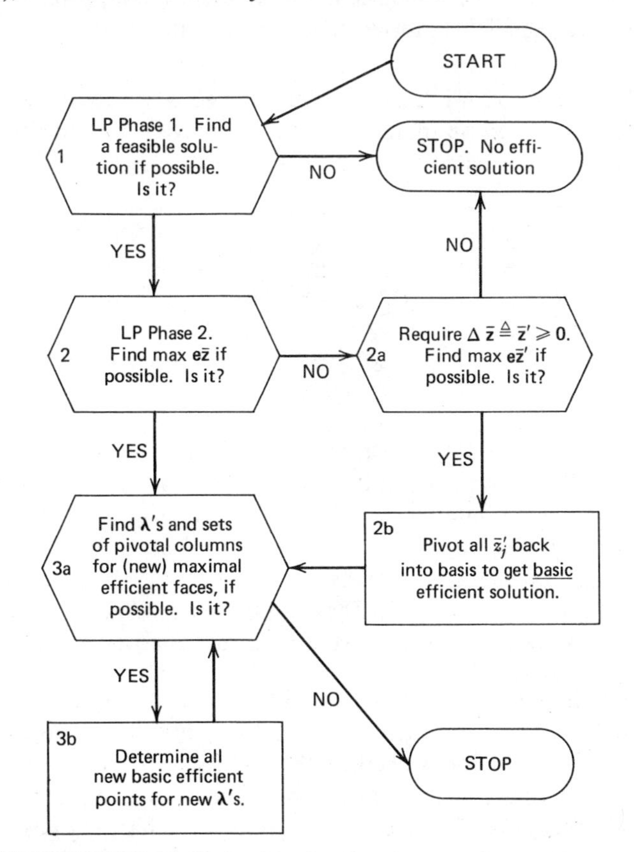

FIGURE 8.8A Multiple objective linear programming algorithm.

Phase 2a is needed if Phase 1 yields a feasible solution but $\mathbf{e}\bar{\mathbf{z}}$ is unbounded above. This is illustrated by the problem

$$x_1+2x_2\leq 3 \qquad x_1+3x_2\leq 4 \qquad \text{“max”}\,\bar{\mathbf{z}}=(x_1,x_2) \tag{5}$$

Contrary to the usual assumption, x_1 and x_2 are here to be *unrestricted in sign*, but they could represent differences of nonnegative variables as in 6.2. Then $\bar{z}_1+\bar{z}_2$, $\bar{z}_1$ and $\bar{z}_2$ are all unbounded but the extreme point $\bar{\mathbf{z}}=\mathbf{x}=(1,1)$ and all the other points on the boundary are efficient. By Theorem 2 *some* $\boldsymbol{\lambda}\bar{\mathbf{z}}$ (e.g., $2\bar{z}_1+5\bar{z}_2$) must have a maximum, but one does not know in advance what $\boldsymbol{\lambda}$ (if any) will work.

According to Theorem 7 below (as in Ecker and Kouada, 1975), Phase 2a proceeds as follows:

(f_1) Begin with some primal-feasible tableau.

(f_2) Replace each label $\bar{z}_h$ by $\bar{z}'_h \triangleq \bar{z}_h+d_h$ and each entry d_h by zero. (This change can be reversed at the end of Phase 2b.)

(f_3) Require that $\bar{\mathbf{z}}'\geq\mathbf{0}$ and try to maximize $\mathbf{e}\bar{\mathbf{z}}'=\mathbf{e}\bar{\mathbf{z}}+\mathbf{ed}$. If it is unbounded, stop; there is no efficient solution. Otherwise an efficient solution has been found. However, it will be nonbasic (nonextreme) because some of the $\bar{z}'_h$ will become nonbasic; this leaves fewer nonbasic x_h than a basic solution requires.

THEOREM 7 In the following LP tableau let $\mathbf{x}\geq\mathbf{0}$, $\bar{\mathbf{z}}'\geq\mathbf{0}$, $\mathbf{b}\leq\mathbf{0}$; $\mathbf{e}\bar{\mathbf{z}}'=\Sigma\bar{z}'_h$ is to be maximized. Then it has an optimum solution iff the MOLP (“max”$\bar{\mathbf{z}}$ or $\bar{\mathbf{z}}'$, without $\bar{\mathbf{z}}'\geq\mathbf{0}$, as in (1)) has an efficient solution.

1	$\mathbf{x}_N$	1
$\mathbf{x}_M$	$\mathbf{A}$	$\mathbf{b}$
$\bar{\mathbf{z}}'$	$\mathbf{C}$	$\mathbf{0}$
$\mathbf{e}\bar{\mathbf{z}}'$	$\mathbf{eC}$	0

Proof The “only if” part is obvious by contradiction; if the purported optimum solution of the LP was not efficient, $\Sigma\bar{z}'_h$ could surely be increased and so the solution would not be optimum after all. For the converse suppose that $\bar{\mathbf{z}}'=\bar{\mathbf{z}}^e$ is an efficient solution. By Theorem 2 there exists a $\boldsymbol{\lambda}>\mathbf{0}$ such that $\bar{\mathbf{z}}^e$ maximizes $\boldsymbol{\lambda}\bar{\mathbf{z}}'$. Since $\bar{\mathbf{z}}'\geq\mathbf{0}$ we then have $\lambda_h\bar{z}'_h\leq\boldsymbol{\lambda}\bar{\mathbf{z}}'\leq\boldsymbol{\lambda}\bar{\mathbf{z}}^e$ or $\bar{z}'_h\leq\boldsymbol{\lambda}\bar{\mathbf{z}}^e/\lambda_h$ for all h. Thus $\Sigma\bar{z}'_h\leq\boldsymbol{\lambda}\bar{\mathbf{z}}^e\cdot\Sigma\lambda_h^{-1}$ and $\max\mathbf{e}\bar{\mathbf{z}}'$ exists in the LP. The assumption $\mathbf{x}\geq\mathbf{0}$ goes with $\mathbf{b}\leq\mathbf{0}$ (primal feasibility) but otherwise is merely conventional, not essential.

Phase 2b obtains a basic (extreme) solution that is still efficient, by exchanging nonbasic $\bar{z}'_h$ (no longer sign restricted) with basic x_i while preserving the feasibility of $\mathbf{x}=-\mathbf{b}$. There is no need for concern with dual feasibility or the $\bar{z}'_h$ as objectives, although they are retained in the tableau and their coefficients updated. Ecker and Hegner (1978) suggest that first priority be given to exchanging basic $x_k=-b_k=0$ with some nonbasic $\bar{z}'_l$, whether the pivot be positive or negative. Next the usual formula $\min_i\{-b_i/a_{il}\colon a_{il}>0,\ x_i \text{ basic}\}$ is used, where the column and row labels are $\bar{z}_l'$ and x_i respectively. Finally if no positive pivot exists, the formula $\min_i\{b_i/a_{il}: a_{il}\neq 0, x_i \text{ basic}\}$ determines a negative pivot, which is acceptable since $\bar{z}_l'$ is allowed to become negative if necessary.

It is easy to see that primal feasibility is preserved and that every $\bar{z}_h'$ can be made basic; if all the x_i coefficients in some $\bar{z}_h'$ column were zero, then $\bar{z}_h'$ could vary while $\mathbf{x}$ was unchanged, whereas $\bar{z}_h'$ depended on $\mathbf{x}$ initially and hence always. As for efficiency, the x_j that were nonbasic at the start of Phase 2b remain nonbasic. Since their zero values define the face on which the solution lies, the solution remains on this face while moving to an extreme point. Thus it remains efficient by Theorem 3.

Phase 3a determines which constraints can be removed (which nonbasic x_j can become basic or $\to\infty$) in order to go from a given efficient extreme point to a maximal efficient face having the point on its boundary. A convex combination of efficient points is again efficient if and only if they lie on such a face (including its boundary and vertices); thus the problem of (3) and the invalid converse of Theorem 4 is surmounted. An example at the end of Section 8.9 illustrates the possible importance of nonextreme efficient points obtained as convex combinations.

A special case of the following theorem is required. Yu and Zeleny (1975) and Ecker, Hegner and Kouada (1980) have given similar theorems and algorithms for Phases 3a and 3b. The former begins the Phase 3a enumeration with facets (faces having the fewest constraints, usually one) and uses an incidence matrix relating extreme points to facets. Other duality theorems have been given by Isermann (1977, 1978).

THEOREM 8 Let a dual solution $\boldsymbol{\lambda}>\mathbf{0}$ (μ components), $\mathbf{u}\geq\mathbf{0}$ ($m+n$ components) satisfying $\mathbf{u}_N=\mathbf{u}_M\mathbf{A}+\boldsymbol{\lambda}\mathbf{C}$ (in some tableau (1) of primal basis M) be called *feasible*. Then the *convex hull* Q of a set of feasible points $\mathbf{x}^p$ ($p=1,\ldots,q$) is *efficient* (i.e., it corresponds to efficient points $\bar{\mathbf{z}}$) if and only if there exists a feasible $(\boldsymbol{\lambda},\mathbf{u})$ such that $\mathbf{u}\mathbf{x}^p=\mathbf{0}$ for all p. (Theorem 6 (e_0) is a special case with $q=1$.)

Proof Let the efficient point $\mathbf{x}^e$ be the given point $\mathbf{x}^1$ if $q=1$, and a point in the (relative) interior of the convex hull Q if $q>1$. By Theorem 2, $\mathbf{x}^e$ determines a $\boldsymbol{\lambda}>\mathbf{0}$ such that $\boldsymbol{\lambda}\mathbf{C}\mathbf{x}_N^e=\min\boldsymbol{\lambda}\mathbf{C}\mathbf{x}_N$ over feasible $\mathbf{x}_N$, and $\boldsymbol{\lambda}\mathbf{C}\mathbf{x}=\boldsymbol{\lambda}\mathbf{C}\mathbf{x}^e$ for all $\mathbf{x}$ in Q (see the proof of Theorem 3). Thus all the $\mathbf{x}^p$ are optimal solutions for the linear program with the single objective $\boldsymbol{\lambda}\mathbf{C}\mathbf{x}_N$, and Theorems 4 and 7* of 3.3 assure the existence of the required feasible (and optimal) $\mathbf{u}$.

Conversely any convex combination $\mathbf{x}$ of feasible $\mathbf{x}^p$ will be feasible and satisfy $\mathbf{u}\mathbf{x}=\mathbf{0}$. Since $\mathbf{u}$ is feasible, Theorem 7* of 3.3 shows that $\mathbf{u}$ and $\mathbf{x}$ are optimal for the linear program with the objective $\boldsymbol{\lambda}\mathbf{C}\mathbf{x}_N$. Since $\boldsymbol{\lambda}>\mathbf{0}$ by hypothesis, $\mathbf{x}$ is efficient by Theorem 2.

In practice the $\mathbf{x}^p$ will be extreme points and Q will be an efficient face of the feasible region of points $\mathbf{x}$, preferably maximal. There is a corresponding face of points $\bar{\mathbf{z}}$, which can be of lower dimensionality. (The image of a basis for the $\mathbf{x}$ plane as a linear variety must contain a basis for the $\bar{\mathbf{z}}$ plane). Extreme points for $\mathbf{x}$ correspond to extreme points for $\bar{\mathbf{z}}$, but some of the latter may coincide. In three dimensions $\boldsymbol{\lambda}$ can be visualized as a vector perpendicular to the efficient face of points $\bar{\mathbf{z}}$.

Phase 3a begins with a tableau produced in Phase 2, 2b or 3b, whose basic solution $\mathbf{x}^e$ is efficient. The constants column $\mathbf{b}$, $\mathbf{d}$ and the rows of $\mathbf{A}$ having $b_i \neq 0$ are omitted to obtain the dual of (2) described just before Theorem 5. Theorem 8 confirms the appropriateness of this; we now have one of the $\mathbf{x}_N^p = \mathbf{x}_N^e = \mathbf{0}$, so that $\boldsymbol{\lambda}\mathbf{C}\mathbf{x}_N^e = 0 = \mathbf{b}\mathbf{u}_M$, and $u_i = 0$ unless $b_i = 0$. The object is to find a *new value of* $\boldsymbol{\lambda}$ in Theorem 6(e_0) or 8. If it exists, the maximization of the new $\boldsymbol{\lambda}\bar{\mathbf{z}}$ may yield new efficient points in Phase 3b, in addition to reproducing the original one.

The condition $\lambda_h > 0$ for h in the objective rows H is insured by replacing $\boldsymbol{\lambda}$ by $\boldsymbol{\lambda}' + \mathbf{e}$ where $\boldsymbol{\lambda}' \geq \mathbf{0}$ and $\mathbf{e}$ has unit components in H and zeros elsewhere. This involves no loss of generality because $(\boldsymbol{\lambda}, \mathbf{v})$ can always be multiplied by a positive factor so that the smallest λ_h equals 1. The substitution produces the familiar vector $\mathbf{eC}$ in the last row of the tableau, in which only the column equations are of interest:

$$\begin{array}{cc|l} 1 & v_j & j \in N \\ \hline \lambda_i' & a_{ij} & i \in M_0 \ (b_i = 0) \\ \lambda_h' & c_{hj} & h \in H \text{ (objectives)} \\ 1 & \mathbf{e}c_{Hj} & \mathbf{e} = (1, 1, \ldots, 1) \text{ (components in } H) \\ & & \boldsymbol{\lambda}' \geq \mathbf{0}, \ \mathbf{v} \geq \mathbf{0} \end{array} \qquad (6)$$

Here the $\lambda_i = \lambda_i'$ as well as the v_i are some of the components of $\mathbf{u}$ in Theorem 8; the other u_i are always zero, because they have $b_i \neq 0$.

Since the last row $\mathbf{eC}$ gives the values of the (nonnegative) basic variables in the first row, the first requirement is to make the last row nonnegative; if necessary a dual form of the Phase 1 of 6.1 or 6.4 can be used, with the dual infeasibility form in an additional column. In the context of the algorithm this effort should always succeed, because the initial point has already been proved efficient. Thereafter this nonnegativity is preserved by using dual simplex steps as in (5) of 3.3; one may choose any row other than $\mathbf{eC}$, provided it contains at least one negative entry for a pivot and does not lead to unnecessary repetition of previous tableaus.

Since $\mathbf{v} = (\boldsymbol{\lambda}' + \mathbf{e})\mathbf{C}' = \boldsymbol{\lambda}\mathbf{C}' =$ the coefficients for $\boldsymbol{\lambda}\bar{\mathbf{z}}$ (perhaps updated by the constraints in M_0), every net exchange of variables in (6) reduces one of these coefficients to zero or keeps it at zero. Other v_j may become zero while remaining basic. Thus any column labels v_j that acquire zero values in (6) *identify the columns in the full primal tableau (1) in which primal-simplex pivots will give other extreme points on the same efficient face* determined by $\boldsymbol{\lambda}$ in Theorem 8. (If no positive pivot exists in such a column, it determines a semi-infinite efficient edge). Because $\boldsymbol{\lambda}\mathbf{C}'$ has zero components in these columns, pivots therein will not change $\boldsymbol{\lambda}\mathbf{C}'$ and its required nonnegativity. The other condition $\mathbf{u}\mathbf{x}^p = 0$ in Theorem 8 is also satisfied since all the x_j that become basic and probably positive are paired with $u_j = v_j = 0$.

Each maximal efficient face incident with the initial $\mathbf{x}^e$ corresponds to a set of v_j that can be reduced to zero; this set is maximal in the sense of not being contained in any other such set. Thus it is unique if all n of the v_j can be reduced to zero (some can remain basic); then every point is efficient by Theorem 1. A smaller maximal set may be unique or there may be several, such as $\{1,2\}$, $\{2,3\}$, $\{3,4\}$.

Phase 3b performs primal simplex pivot steps in the full tableau (1) in the rows of **A** and in those columns j belonging to a maximal set discovered in Phase 3a. The object is to find all the extreme points on the corresponding maximal efficient face; some or all of these may have been encountered previously on other faces. Subfaces of the face are perhaps of secondary interest, but each is easily recognized and defined as the convex hull of a maximal set of extreme **x** having $x_j=0$ when j belongs to a specified set of indices, including those applicable to the maximal efficient face itself. The following theorem assures that all the efficient extreme points can be found by the algorithm.

THEOREM 9 Every pair of efficient extreme points of (1) is connected by a chain of pivot steps, each of which defines an efficient edge (or identical efficient extreme points, in degenerate cases).

Proof Theorem 2 permits the use (at least conceptually) of a parametric programming procedure like that of 8.6 with parameters $\lambda_1,\ldots,\lambda_{\mu-1}>0$ and $\lambda_1+\cdots+\lambda_{\mu-1}<1$. The theorem of 8.3, whose proof is equally valid for $\mu-1$ parameters, shows that those $\boldsymbol{\lambda}$ yielding an optimal (efficient) solution form a convex region, which is divided into convex subregions corresponding to the various efficient extreme points. (See Figures 8.6B and 8.9.) Connecting two such subregions by a straight line or other continuous curve shows which subregions (or efficient extreme points) should be traversed by the desired chain of pivot steps. The boundary point where the curve crosses from one subregion to another determines a common $\boldsymbol{\lambda}$, which makes the edge efficient by Theorem 8. Lexicographic methods may be required as in 6.6 or 8.3.

Enumeration of Vertices Many algorithms have been published for the determination of all the vertices of an n-dimensional polyhedron or polytope (extreme points of a feasible region), without any objectives or efficiency concept at all. Before mentioning some of these, one should note that they can be used in both Phase 3a and 3b, and they confront the most difficult aspect of the problem, the huge amount of enumeration required by all but the smallest programs. In Phase 3a one can find all the vertices for the column equations of (6), and then discard those that do not yield maximal sets of $v_j=0$ (nonbasic or basic); the generation of many of the unwanted vertices is unavoidable in any case. In Phase 3b one wants all the vertices of certain faces, which are polyhedra of lower dimensionalities, already determined (in Phase 3a) to be entirely efficient.

Any two vertices of a polyhedron are connected by a sequence of pivot steps, which can be determined by the primal simplex method of 3.3. The objective is to minimize (reduce to zero) the linear function $-\Sigma_i f_i(\mathbf{x})$, where $f_i(\mathbf{x})\leq 0$ are all the inequalities that are satisfied as equalities $f_i(\mathbf{x})=0$ at the terminal vertex. (This sum defines a supporting plane such that the optimum is unique).

The faces of a polyhedron (of various dimensionalities) are elements of the Boolean *lattice* of subsets consisting of all the possible unions of various faces. The individual faces form a smaller lattice that is non-Boolean because the union is replaced by the least upper bound, the smallest face containing the given faces. It is a little more concise to use index sets of zero variables (among the $m+n$ components of $\mathbf{x}$) rather than basic or nonzero variables, although this reverses the relation of containment in going from faces to index sets. Thus the intersection of faces corresponds to the least upper bound of the index sets, and the least upper bound of faces corresponds to the intersection of the index sets. The whole polyhedron may have an empty index set, and the empty face has the all-inclusive index set.

For example, the faces (in the general sense) of a square pyramid whose base is $x_5=0$ have the following index sets:

Square pyramid	∅
(Two-dimensional) faces	1,2,3,4;5 (base)
Edges	12,23,34,14; 15,25,35,45
Vertices	1234,125,235,345,145
Empty set	12345

The dimensionalities here decrease from 3 to -1, the dimensionality of the empty set! Opposite edges of the base, such as 15 and 35, have an empty intersection; this is reflected in the fact that the union 135 of their index sets is not contained in the index set for any vertex, but only in that for the empty set (12345, meaning $\{1,2,3,4,5\}$). In practice it is desirable to keep the indices in a standard (increasing) order.

In small exercises no special method of enumeration is needed; we simply record all the relevant tableaus as usual and verify that no additional tableaus can be produced. (See the largest example of the following section in Tables 8.9B, C.) In larger problems this quickly becomes impractical even for a computer. In particular the examples of Brown (1961) show that it may be impossible to enumerate all of the q vertices in the ideal number $(q-1)$ of pivot steps.

The three methods cited in the following paragraphs all use the sets of indices N of nonbasic variables to identify the vertices, but Dyer-Proll and Mattheiss also have special devices to determine more quickly whether a potentially new vertex has already been listed. All the methods use the tableau or its 7.3 equivalent to list permanently the index set and the coordinates $\mathbf{x}=-\mathbf{b}$ for the current basic solution, and also to list temporarily the index sets for those adjacent vertices not yet listed permanently. The latter list grows at first but it is called temporary because it is empty at the end of the algorithm. This list is essential to assure that no vertices are overlooked, and to provide index sets that are known to yield feasible basic solutions. It could easily include coordinates $\mathbf{x}$ also, but that would not obviate the necessity for getting the corresponding tableaus eventually, to see if *they* have any new adjacent vertices. Manas-Nedoma and Mattheiss store exactly one tableau at any one time; it may be small compared to the data on the vertices. Manas-Nedoma and Dyer-Proll use the number of pivot steps as a measure of distance or path length, and Mattheiss could do so too.

Manas and Nedoma (1968) gives a method so simple and basic that it has almost been described in the preceding paragraph. It remains to say that if and when a new tableau cannot be produced in a single pivot step, then several steps (the smallest number) are performed to arrive at a closest index set (vertex) on the temporary list.

The index set in question is then transferred from the temporary to the permanent list, as it is also when a single pivot step happens to produce a vertex on the temporary list.

Dyer and Proll (1977) arranges the vertices in a tree of shortest paths (from the initial vertex) generated as in 9.3. This results from the fact that the first vertex in the temporary list is always the next to be listed permanently; the permanent list preserves the order of the temporary. No tableau is stored, but with each vertex there is stored the pivotal column in the tableau for the preceding vertex in the tree. To determine the vertices adjacent to a given vertex, many of the columns of the tableau have to be updated from the initial tableau by the forward transformation (10) of 7.3. However, there is no need to update those columns whose exchange with any row index would give an index set already listed permanently.

Mattheiss and Rubin (1980) is an excellent review of many methods; it includes computational experience favorable to Mattheiss (1973). The latter introduces a new independent variable $y \geq 0$ whose coefficient in the hth inequality (including $-x_j \leq 0$) is $(\Sigma_j a_{hj}^2)^{1/2}$ initially. It is assumed that the feasible region X is bounded and has a nonempty interior so that $0 < \max y < \infty$ and Y defined below is not empty.

Thus the original polyhedron X (where $y=0$) becomes a face of a polyhedron $X \cup Y$, where $y>0$ on Y. In Figure 8.8B, X has the vertices X_1 to X_6 in the plane of the paper and Y has the vertices Y_1 to Y_4 above the paper, with max y attained at Y_4. If the value of y is ignored so that the Y_p are projected onto the paper ($y=0$), the Y_p appear as interior points of X, and X_pY_p and $X_{p+3}Y_p$ are bisectors of the angles at X_p and X_{p+3} ($p=1,2,3$).

Pivoting in the y column shows that every X_p is adjacent to some Y_q. Hence it suffices to calculate tableaus only for the Y_q, which are nearly always less numerous than the X_p (see Mattheiss and Schmidt, 1980); then only the **b** column has to be calculated at the X_p. In Figure 8.8B there are six X_p and only four Y_q, but five full-scale pivot steps in both cases, connecting the vertices in sequences such as $X_1X_4X_2X_5X_3X_6$ and $X_1Y_1Y_4Y_2Y_4Y_3$. This assumes that the index sets of the neighbors are listed when a given tableau is first produced. Each is delisted after *its* tableau and neighbors have been determined; only the index sets and coordinates of the X_p are listed permanently. When it is necessary to make a fresh start, Mattheiss goes to the largest value of y remaining in his temporary list.

The enumeration of all the Y_q evidently includes the maximization of y, which requires pivotal columns having negative coefficients in the row labeled y; the same

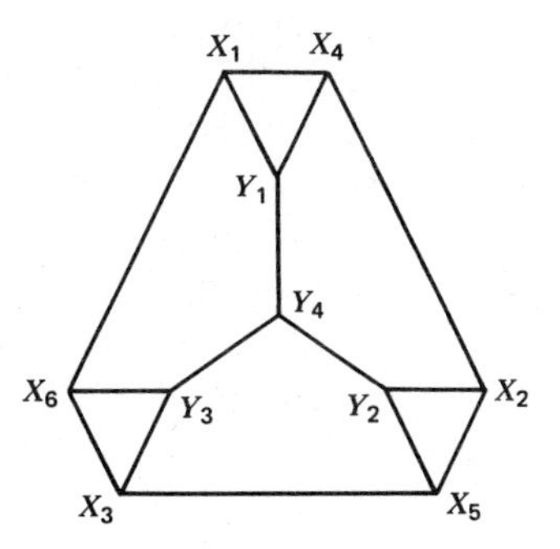

FIGURE 8.8B

is often the case when a new tableau is to be produced to match an index set taken from the list. However, the opposite is true in listing the neighboring vertices of a tableau, when one has to consider only pivotal columns (most of them not actually pivoted) having nonnegative coefficients in row y, so that y is nonincreasing. This results from a conceptual reversal of the primal-simplex sequence of pivot steps leading from any X_p, where $y=0$, to Y^*, where y is maximized; in this sequence y is nondecreasing.

8.9 MULTIPLE OBJECTIVES: EXAMPLES AND VARIATIONS

Table 8.9A presents the tableaus for the small example in (3) of 8.8, which has a triangular feasible region. All three vertices but only two sides (edges) are efficient. Table 8.9D gives two examples of Phase 3a in degenerate cases in which $b_4=0$ and $\operatorname{val}\mathbf{C}'=0$, so that the efficiency tests as stated in Theorems 5(d_3) and 6(e_3) of 8.8 are inconclusive. Recall that $\operatorname{val}\mathbf{C}'$ is the value of the game whose matrix consists of the objective function coefficients $\mathbf{C}$ together with the row(s) of $\mathbf{A}$ having $b_i=0$.

For the principal example the primal tableaus of Phases 2 and 3b are given in Table 8.9B and the dual tableaus of Phase 3a are in Table 8.9C. These are generated in the sequence (D_0), (E_5), (4), (E_4), (5), (6), (E_3), (E_2), (E_1). The feasible region has the facial lattice (incidence relations) of a triangular prism; one four-sided face and one additional edge are efficient. These are depicted in Figure 8.9 along with a diagram (similar to Figure 8.6B) relating them to all the possible values of $\boldsymbol{\lambda}>\mathbf{0}$.

The fact that all the extreme points have been found in Table 8.9B permits Phase 3a to be shortened. E_1, E_2 and E_3 need not be used as initial points because no extreme points are available to form additional maximal efficient faces with them; E_1 to E_3 already lie on one such face, E_4 and E_5 have already been investigated, and D_0 is inefficient. If one more extreme point were available, it would be used as an initial point for Phase 3a.

In principle $\boldsymbol{\lambda}$ is not unique; e.g., the edge E_4E_5 has the direction $(2,3,-1)$, and any positive combination of $(1,0,2)$ and $(0,1,3)$ could serve as its $\boldsymbol{\lambda}$. Nevertheless the two Phase 3a calculations in (4) and (5) of Table 8.9C both give $\boldsymbol{\lambda}=(1,1,5)$, thus confirming that the same edge is involved in both cases. The algorithm has eliminated the ambiguity by maximizing the number of (smallest) $\lambda_i=1$.

Our last example, which has no constraints, touches on the analogy between multiple objectives and the states of nature in 4.7, and on the difficult matter of measuring utilities in practical applications.

The decision maker J has a fairly secure job that pays \$20,000 per year after taxes, but also has an intriguing opportunity to go into business for herself. Thus she must decide among the status quo $(j=1)$, changing to the new activity $(j=2)$, or attempting to do both for a while $(j=3)$. States of nature might include her future health and that of the businesses she is involved with; she decides to condense these to good times $(i=1)$ and hard times $(i=2)$. As objectives she uses the dimensions of

TABLE 8.9A Simple Example with Two Objectives

The feasible region is a triangle whose vertices correspond to the three tableaus in (1).

Row Equations in Phase 3b

1	x_2	x_3	1
x_1	1	1	-1
$\bar{z}_1$	1	3	-3
$\bar{z}_2$	-2	-3	0
	0	3	#
	-3	-3	←

1	x_1	x_3	1
x_2	$1^{\#}$	1^{*}	-1
$\bar{z}_1$	-1	2	-2
$\bar{z}_2$	2	-1	-2
	0	3	
	3	0	

1	x_1	x_2	1
x_3	1	1	-1
$\bar{z}_1$	-3	-2	0
$\bar{z}_2$	3	1	-3
*	-3	-3	
→	3	0	

(1)

Column Equations in Phase 3a

1	λ'_1	v_3
v_1	-1	-2
λ'_2	2	3
eC	1	3

1	v_1	v_3
λ'_1	$-1^{\#}$	2
λ'_2	2	-1^{*}
eC	1	1

1	v_1	λ'_2
λ'_1	3	2
v_3	-2	-1
eC	3	1

(2)

The middle tableaus in (1) and (2) are associated, the latter being a dual tableau like (6) of 8.8. The other tableaus are obtained by the # pivots in going to the left and * in going to the right. In the first tableau of (2) the basic dual solution is $\boldsymbol{\lambda}'=(1,0)$, whence $\boldsymbol{\lambda}=(1,0)+(1,1)=(2,1)$; the last tableau gives $\boldsymbol{\lambda}=(0,1)+(1,1)=(1,2)$. The two unlabeled rows below (1) give the coefficients of $\boldsymbol{\lambda}\mathbf{z}$ for these two solutions, namely $2z_1+z_2$ and z_1+2z_2. These are nonnegative row vectors in four of the six cases, and we conclude as follows:

Efficient vertices $\bar{\mathbf{z}}$	$V_1(3,0)$		$V_2(2,2)$		$V_3(0,3)$
Efficient edges		V_1V_2		V_2V_3	
$\boldsymbol{\lambda}$		$(2,1)$		$(1,2)$	

(3)

The edge V_1V_3 is not efficient because there is no $\boldsymbol{\lambda}>\mathbf{0}$ making $\boldsymbol{\lambda}\mathbf{C}\geq 0$ for both V_1 and V_3 (here $\mathbf{C}'=\mathbf{C}$).

TABLE 8.9B Extreme Points for MOLP

Phase 2

1	x_1	x_2	x_3	1
x_4	2	2	1	-4
x_5	1*	-1	-1	-1
$\bar{z}_1$	-1	-1	-2	0
$\bar{z}_2$	-1	-2	-1	0
$\bar{z}_3$	-10	11	20	0
$\mathbf{e}\bar{\mathbf{z}}$	-12	8	17	0

(D_0)

Inefficient by column x_1

1	x_5	x_2	x_3	1
x_4	-2	4*	3	-2
x_1	1	-1	-1	-1
$\bar{z}_1$	1	-2	-3	-1
$\bar{z}_2$	1	-3	-2	-1
$\bar{z}_3$	10	1	10	-10
$\mathbf{e}\bar{\mathbf{z}}$	12	-4	5	-12

(E_5)

Efficient by max $\bar{z}_3$ (unique)

Phase 3b

4	x_5	x_4	x_3	1
x_2	-2	1	3*	-2
x_1	2	1	-1	-6
$\bar{z}_1$	0	2	-6	-8
$\bar{z}_2$	-2	3	1	-10
$\bar{z}_3$	42	-1	37	-38
$\mathbf{e}\bar{\mathbf{z}}$	40	4	32	-56

(E_4)

Efficient by max $\mathbf{e}\bar{\mathbf{z}}$ or (4)

3	x_5	x_4	x_2	1
x_3	-2	1	4	-2
x_1	1*	1	1	-5
$\bar{z}_1$	-3	3	6	-9
$\bar{z}_2$	-1	2	-1	-7
$\bar{z}_3$	50	-10	-37	-10
$\mathbf{e}\bar{\mathbf{z}}$	46	-5	-32	-26

(E_3)

Efficient by (6) in Table 8.9C.

1	x_1	x_4	x_2	1
x_3	2	1	2*	-4
x_5	3	1	1	-5
$\bar{z}_1$	3	2	3	-8
$\bar{z}_2$	1	1	0	-4
$\bar{z}_3$	-50	-20	-29	80
$\mathbf{e}\bar{\mathbf{z}}$	-46	-17	-26	68

(E_2)

2	x_1	x_4	x_3	1
x_2	2	1	1	-4
x_5	4	1	-1	-6
$\bar{z}_1$	0	1	-3	-4
$\bar{z}_2$	2	2	0	-8
$\bar{z}_3$	-42	-11	29	44
$\mathbf{e}\bar{\mathbf{z}}$	-40	-8	26	32

(E_1)

Efficient by (6) in Table 8.9C. Also see Figure 8.9.

money (m) and time (t), hoping that to some extent they can reflect the many interactions between her personal values and the activities, people, things and ideas involved in her decision.

To help with the comparison of money and time, J makes a list of pairs of values (m, t) that seem about equally desirable, thus:

Income m in \$1000/year	9	10	20	30	50	100
Work t in hours/week	0	10	30	40	50	60
$1-0.35\log_{10} m$	.666	.650	.545	.483	.405	.300
$2-\log_{10}(100-t)$	0	.046	.155	.222	.301	.398

(9)

TABLE 8.9C Phase 3a

Efficient Edge at E_5

1	v_5	v_2	v_3		1	v_5	λ_3'	v_3	
λ_1'	1	-2	-3		λ_1'	21	2	17	
λ_2'	1	-3	-2		λ_2'	31	3	28	(4)
λ_3'	10	1*	10		v_2	10	1	10	
eC	12	-4	5		**eC**	52	4	45	

No other pivot. $\boldsymbol{\lambda}=(0,0,4)+(1,1,1)=(1,1,5)$ makes $\boldsymbol{\lambda}\mathbf{c}_{.2}=0$.
Pivoting in this column 2 in (E_5) of Table 8.9B gives efficient vertex E_4 and edge E_5E_4.

Efficient Edge and Face at E_4

4	v_5	v_4	v_3		2	λ_2'	v_4	v_3	
λ_1'	0	2	-6		λ_1'	0	1	-3*	
λ_2'	-2*	3	1	$\overset{*}{\rightarrow}$	v_5	-4	-3	-1	
λ_3'	42	$-1^{\#}$	37		λ_3'	42	31	29	
eC	40	4	32	$\overset{\#}{\downarrow}$	**eC**	40	32	26	$\overset{*}{\downarrow}$

1	v_5	λ_3'	v_3		3	λ_2'	v_4	λ_1'	
λ_1'	21	2	17		v_3	0	-1	-2	
λ_2'	31	3	28	(5)	v_5	-6	-5	-1	(6)
v_4	-42	-4	-37		λ_3'	63	61	29	
eC	52	4	45		**eC**	60	61	26	

$\boldsymbol{\lambda}=(0,0,4)+(1,1,1)=(1,1,5)$ shows edge E_4E_5 is efficient as in (4). See Figure 8.9.

$\boldsymbol{\lambda}=(26,60,0)/3+(1,1,1)=(29/3,\ 21,\ 1)$ makes $\boldsymbol{\lambda}\mathbf{c}_{.3}=\boldsymbol{\lambda}\mathbf{c}_{.5}=0$. Pivots in these columns in (E_4) of Table 8.9B give efficient vertices E_3, E_2, E_1 and face $E_4E_3E_2E_1$. No more $\boldsymbol{\lambda}$'s exist.

Since (9) defines an indifference curve, the utility or disutility of each point (m,t) should be the same; influenced by 4.7, J uses disutility. She finds that the sum of the functions in the last two lines is nearly constant, varying from .666 to .706. This confirms the consistency of her original estimates and also provides measures of the disutilities of other situations. Note that the disutility will approach infinity (though not very rapidly) as $m\rightarrow 0$ or $t\rightarrow 100$, reflecting J's extreme distaste for having no income or for working 100 hours/week; these values define asymptotes for the indifference curve.

TABLE 8.9D Efficiency Tests with $b_4=0$, val $C'=0$ (Phase 3a)

Inefficient

1	v_1	v_2	v_3	
λ_4	5*	0	−3	
λ'_1	2	3	−1	
λ'_2	−4	−5	2	
λ'_3	−3	1	1	
eC	−5	−1	2	

5	λ_4	v_2	v_3	
v_1	1	0	−3	
λ'_1	−2	15	1*	
λ'_2	4	−25	−2	
λ'_3	3	5	−4	
eC	5	−5	−5	

1	λ_4	v_2	λ'_1	
v_1	−1	9	3	
v_3	−2	15	5	
λ'_2	0	1	2	(7)
λ'_3	−1	13	4	
eC	−1	14	5	

Nonpositive column makes it impossible to get $\mathbf{eC}\geq 0$ or $\boldsymbol{\lambda}'\geq\mathbf{0}$. (Theorem 5($d_2$) of 8.8)) Primal labels in (7) show that $(x_1, x_3, \Delta\bar{z}_3)=(1,2,1)x_4$ (increasing) while $x_2, \bar{z}_1, \bar{z}_2$ are unchanged. Inefficient.

Efficient

1	v_0	v_1	v_2	v_3	
λ_4	2	−1	−5	1	
λ'_1	1	−1	−1	−3	
λ'_2	−1	−3	13	5	
λ'_3	−1	3*	−5	2	
eC	−1	−1	7	4	

3	v_0	λ'_3	v_2	v_3	
λ_4	5	1	−20	5	
λ'_1	2*	1	−8	−7	
λ'_2	−6	3	24	21	
v_1	−1	1	−5	2	
eC	−4	1	16	14	

2	λ'_1	λ'_3	v_2	v_3	
λ_4	−5	−1	0	15	
v_0	3	1	−8	−7	
λ'_2	6	4	0	0	(8)
v_1	1	1	−6	−1	
eC	4	2	0	0	

Nonpositive column immaterial since $\mathbf{eC}\geq\mathbf{0}$. $\boldsymbol{\lambda}=(4,0,2,0)/2+(1,1,1,0)=(3,1,2,0)$. Since $\mathbf{v}\neq\mathbf{0}$, $\boldsymbol{\lambda}\mathbf{C}=\mathbf{0}$ and every feasible point is efficient (Theorem 1 of 8.8). λ_4 goes with a constraint having $b_4=0$.

Next J estimates the values of (m,t) for each combination of i and j, and then translates these into disutilities as in (9):

	$j=1$	$j=2$	$j=3$	Disutilities			
$i=1$	(20,40)	(30,40)	(40,70)	.767	.705	.962	(10)
$i=2$	(16,40)	(10,50)	(5,40)	.801	.951	.977	
Average with $u=(2,1)/3$				.778	.787	.967	

Taking $2/3$ as the probability of good times, J finds that her present job ($j=1$) has the least disutility (.778) by a small margin. She also calculates that if the probability of good times exceeded $(951-801)/(951-801+767-705)=.708$, the new job ($j=2$)

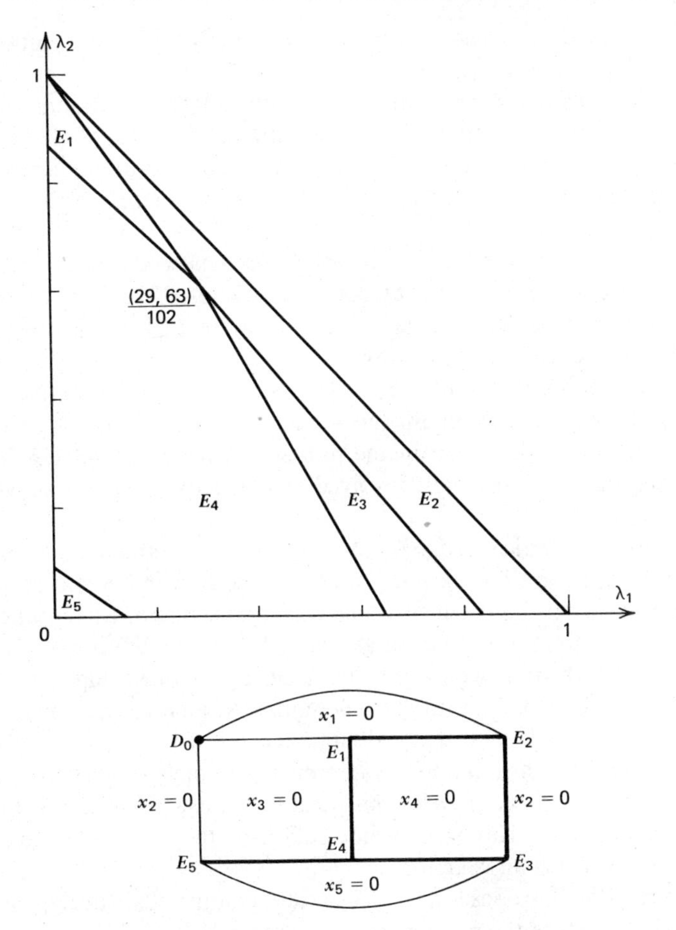

Equations clockwise from lower left:

$$2.3\lambda_1 + 3.3\lambda_2 = 0.3 \qquad \lambda_1 + \lambda_2 = 1$$

$$12.6\lambda_1 + 14.6\lambda_2 = 12.6 \qquad 18\lambda_1 + 16\lambda_2 = 15$$

$$11.7\lambda_1 + 8.7\lambda_2 = 8.7 \qquad 17.1\lambda_1 + 10.1\lambda_2 = 11.1$$

Geometrical configuration of efficient vertices, edges and face: The face $E_1E_2E_3E_4$ corresponds to the point $\boldsymbol{\lambda} = (29, 63, 3)/102$. The decision maker's task is reduced to choosing a convex combination (weighted average) of $E_5(1,1,10)$ and $E_4(4,5,19)/2$, or a convex combination of E_4 with $E_3(9,7,10)/3$, $E_2(8,4,-80)$ and $E_1(2,4,-22)$. The feasible region resembles a triangular prism with bases $x_1 = 0$ and $x_5 = 0$.

FIGURE 8.9

would be preferable. Here J has used the Bayes solution 4.7(a). The method of 4.7(e) also could be used.

The foregoing examples have illustrated both categories of multiple-objective problems, those with constraints and those without. In both of them interaction between the decision-maker and the computer is an important consideration in practice, as in Steuer (1977). Some approaches to constrained multiple objectives are the following.

(g_1) Probably the simplest and most widely used procedure is to replace some (often all but one) of the objectives by constraints whose bounds b_h represent goals to be achieved. For example, all three of the constraints in (14) of 8.1 (electric power production) could be viewed in this way.

(g_2) The weights λ_h, the bounds b_h or the values $\bar{z}_h$ of the objectives, if not too numerous, can be regarded as parameters so that 8.3 or 8.6 applies. A more concise use of the λ_h has been described in the present sections 8.8 and 8.9. The Bayes-constrained minimax in 4.7(e) really involved two objectives z and α, of which α was treated as a parameter.

(g_3) The efficient vertices, edges and faces can be enumerated and tested by means of pivot steps. This has been our principal concern in 8.8 and 8.9.

(g_4) A very specialized procedure, which can replace max $\mathbf{e}\bar{\mathbf{z}}$ as a means of getting an initial efficient solution, is to arrange the $\bar{z}_h$ in order of decreasing importance and then maximize them lexicographically. That is, the most important, say $\bar{z}_1$, is maximized first; if this does not uniquely determine $\mathbf{x}$, then the next most important is maximized while $\bar{z}_1$ is held at its maximum, and so on.

(g_5) A feasible $\bar{\mathbf{z}}$ may be obtained by minimizing (subject to the constraints) some measure of its distance from an infeasible ideal point, usually obtained by maximizing each $\bar{z}_h$ individually. This is a generalization of the minimax regret in 4.7(c). Benayoun et al. (1971) is an example.

(g_6) Stewart (1981) proposes the use of multivariate statistical methods (e.g., factor analysis) to reduce the number of objectives.

The following approaches are usually described with reference to unconstrained multiple objectives, as in (9) and (10) above. Nevertheless they can be made available in constrained problems too by the device of restricting the decision maker to a finite set of selected efficient points, extreme or otherwise, such as are produced by the methods (g_3) of this section.

(h_1) A linear or nonlinear function may be constructed with the aim of mapping the vector of objective values into a scalar objective that can be optimized as in (9) and (10) above. The reader may consult Keeney and Raiffa (1976) and Farquhar (1977) in this area.

(h_2) The data analyzed may be limited to qualitative preferences rather than quantitative measures of the objectives. Section 3 of Roy (1971) is an example of this. The ideas of Cook and Seiford (1978) on preferential voting may also be relevant here.

In conclusion we may recall several of the less obvious ways in which 8.8 and 8.9 differ from the preceding sections and chapters. First, in contrast to 8.6, this section does not fully explore the space of parameter values $\boldsymbol{\lambda}$ or t but only obtains those particular values that are needed for efficiency tests. Second, in contrast to the case of one scalar objective, the set of efficient points is not necessarily convex. Hence one should obtain not only the efficient extreme points (basic solutions), but also (via Theorem 8) the structure (edges and faces of dimensions $1,2,\dots,\mu-1$) that indicates how these extreme points determine other efficient points.

To illustrate the importance of efficient points that are not extreme, suppose there are only two efficient extreme points, $\bar{\mathbf{z}}=(3,1)$ and $(1,3)$. Then by Theorem 9 the edge joining them must be efficient, and the common $\boldsymbol{\lambda}$ in Theorem 8 will be proportional to $(1,1)$. Now if the decision maker's indifference curves have the form $\bar{z}_1\bar{z}_2=$constant, his optimal solution will be the intermediate efficient point $(2,2)$, which is not extreme. However, an inefficient point, extreme or not, obviously cannot maximize a utility function $f(\bar{\mathbf{z}})$ that is an increasing function of each $\bar{z}_h$ (e.g., each $\partial f/\partial \bar{z}_h>0$).

EXERCISES 8.9

Find the efficient extreme points and the efficient edges and faces.

1.

1	x_1	x_2	1
x_3	1	2	−1
$\bar{z}_1$	−1	−3	0
$\bar{z}_2$	−1	−1	0

2.

1	x_1	x_2	1
x_3	1	−1	−1
$\bar{z}_1$	−3	1	0
$\bar{z}_2$	−1	2	0

3.

1	x_1	x_2	x_3	1
x_4	1	2	3	−1
$\bar{z}_1$	1	−1	−1	0
$\bar{z}_2$	−1	3	−1	0
$\bar{z}_3$	−1	−2	4	0

4.

1	x_4	x_5	x_6	1
x_1	1	−1	1	−1
x_2	−1	2	3	−1
x_3	1	4	−1	−1
$\bar{z}_1$	−2	−1	−1	0
$\bar{z}_2$	−1	3	−1	0
$\bar{z}_3$	−1	−1	−4	0

5.

1	x_4	x_5	x_6	1
x_1	2	1	1	−1
x_2	1	3	1	−1
x_3	1	1	4	−1
$\bar{z}_1$	−1	1	−1	0
$\bar{z}_2$	1	−2	−3	0
$\bar{z}_3$	−1	−4	1	0

REFERENCES (CHAPTER 8)

Arrow (1951)
Bell-Keeney-Raiffa (1977)
Benayoun-Montgolfier-Tergny-Laritchev (1971)
Blin (1973, 1977)
Brown (1961)
Cochrane-Zeleny (1973)
Cook-Seiford (1978)
Dantzig (1963)
Dinkelbach (1969)
Dyer-Proll (1977)
Ecker-Hegner (1978)
Ecker-Hegner-Kouada (1980)
Ecker-Kouada (1975–1978)
Evans-Steuer (1973a, b)
Farquhar (1977)
Geoffrion (1968)
Graves (1963)
Gray-Sutherland (1980)
Isermann (1977, 1978)

CHAPTER NINE

Optimal Paths and Networks

Chapters 9 to 11 (other than 10.5 to 10.7) are largely independent of what has gone before even though a form of the simplex method is used in Chapter 11. A few ideas from graph theory are introduced in 9.1. Then 9.2 to 9.7 are concerned with the determination of optimal trees and paths in a network whose arcs are characterized by additive lengths, costs or durations denoted by c_{ij}. In 9.8 it is shown that analogues of these same algorithms determine paths of maximum capacity, although capacities in series are combined by taking their minimum rather than their sum.

9.1 GRAPHS, TREES AND NETWORKS

The general idea of a network is familiar enough; it is a collection of locations or objects that are tied together in some way, by mechanical connections, transportation or communication links, or more abstract relations. There are networks of highways, airline routes, pipelines, television stations, telephone channels, and wholesale or retail distribution points, as well as electrical networks.

In mathematical programming the term *network* usually implies the presence of various numerical parameters such as the quantity of some commodity to be shipped from or to each point; the cost or distance or time lapse c_{ij} associated with shipment from i to j; the maximum capacity or rate of shipment on each route; and occasionally the probability that a particular route is usable or selected for use. These parameters may or may not be the same for travel or shipment in opposite directions on the same route.

Most of the remainder of this volume will be concerned with or interpretable in terms of the quantitative analysis of networks. This section briefly introduces the purely structural or qualitative characteristics of networks when the numerical parameters are lacking or ignored; then we speak of *graphs* rather than networks.

(Undirected) Graphs

Figure 9.1A shows an example of a graph having six nodes 12, 13, 23, 34, 24, 14 and twelve arcs. Some possible interpretations of this graph are the following:

(a_1) The nodes are airports and the arcs indicate which pairs are connected by nonstop flights in both directions.

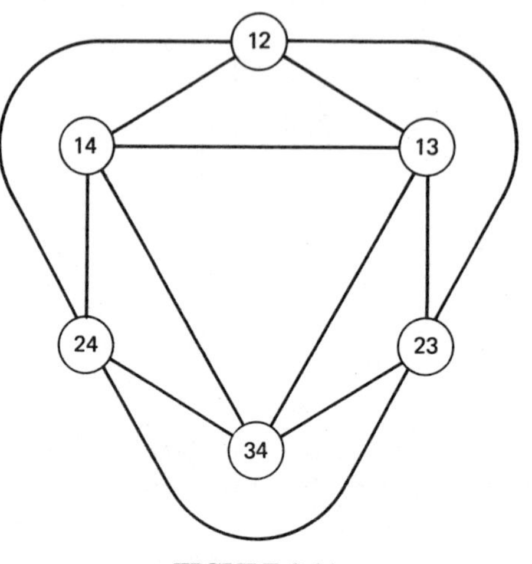

FIGURE 9.1A

(a_2) Node kl is a bilingual person who knows languages k and l. Arcs connect persons who can converse directly.

(a_3) Node kl is the basic set $\{x_k, x_l\}$ in a linear program having $m=n=2$. The arcs represent pivot steps. (See Exercise 1.)

(a_4) The nodes are six carbon atoms forming a hypothetical molecule; the arcs are valence bonds.

(a_5) The nodes are the six vertices of a regular octahedron, the arcs are the 12 edges, and the eight faces are the areas in the figure. (The unbounded exterior region must be counted as one of these.)

(a_6) The six nodes are the edges of a tetrahedron or the faces of a cube; the arcs indicate the relations of adjacency.

An (undirected) *graph* in its simplest form may be defined as a finite set of p *nodes* (or points or vertices) ($p \geq 1$) together with a set of q unordered pairs $\{i, j\}$ of distinct nodes ($0 \leq q \leq p(p-1)/2$). Each pair of nodes in the latter set (and no others) is connected by an *arc* (or line or edge) that does not meet any intervening node. The parenthesized terms are often encountered in the literature but *node* and *arc* have the advantage of being least often needed for use in any other sense.

The two endpoints of any arc are called *adjacent nodes*; each is *incident* with the arc and vice versa. Two arcs that are incident with the same node are called *adjacent arcs*. In discussing the theory it is convenient to use the integers $1, \ldots, p$ to identify the nodes; in solving exercises the use of capital letters such as $A, B, C, \ldots$ avoids possible confusion with other numerical data.

Figure 9.1B shows some more examples of graphs, with their values of (p, q): In the first row, the two graphs labeled (5, 5) are regarded as identical (as are those labeled (4, 1)), despite their different appearances, but each of the other graphs is different from any other in the figure.

If there is no arc connecting nodes i and j, it may be possible to get from one to the other by traversing several arcs. An (undirected) *path* connecting the nodes i_0

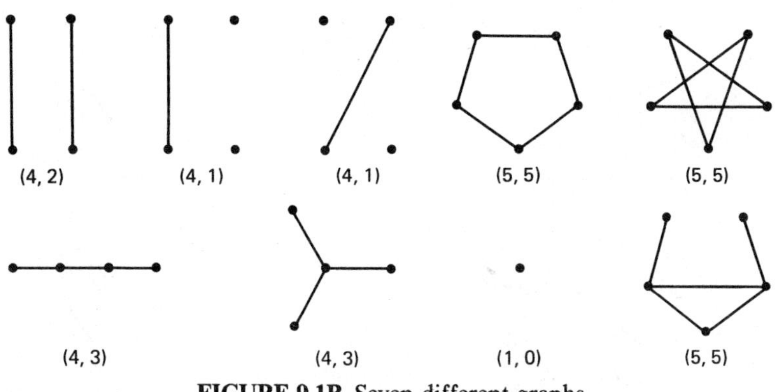

FIGURE 9.1B Seven different graphs.

and i_m is a sequence $(i_0, i_1, \ldots, i_m)$ of distinct nodes such that every consecutive pair $\{i_{k-1}, i_k\}$ is connected by an arc, for $k=1,2,\ldots,m$. If $i_0=i_m$ and $m\geq 3$ are required, the nodes still being distinct otherwise, the sequence $(i_0, i_1, \ldots, i_m)$ defines a *loop*. An undirected graph or network is said to be *connected* if every pair of its nodes can be connected by a path. In Figure 9.1B all but the first three graphs are connected; only the three labeled (5,5) contain loops.

Directed Graphs (Digraphs)

If a direction is assigned to each arc in a graph, one obtains a *directed* graph, whose arcs are ordered pairs (i, j) of distinct nodes i and j, directed from i to j $(i\to j)$. With p nodes, there may be as many as $p(p-1)$ directed arcs. If there are no nodes i and j such that (i, j) and (j, i) are both arcs, the directed graph is said to be *oriented*; then the number of arcs cannot exceed $p(p-1)/2$, as in the undirected case.

The following list shows some of the ways in which directed arcs may arise in practice.

(b_1) Transportation or communication may be possible or useful in only one direction along an arc, as in shipping from a production or storage point to a consumer (11.1), or broadcasting a television program to a receiving set, or drifting in a river or airstream.

(b_2) The arc (i, j) may indicate that i is an event that must precede event j in time, as in 9.4, 9.5, Chapter 10, and the diagram in the preface.

(b_3) Formula (2) of 2.2 illustrates the use of directed arcs to represent relations of implication between statements or containment between sets.

(b_4) The nodes i, j may be objects or courses of action and (i, j) or $i\to j$ means that i is preferred to j. A troublesome phenomenon (the voters' paradox) is the fact that if the three members of a committee or electorate have preferences $A\to B\to C$, $B\to C\to A$, $C\to A\to B$, then the committee by majority vote will have the inconsistent preferences $A\to B\to C\to A$.

In a directed graph or network it must be made clear whether or not all the arcs in a path or loop are required to be directed in the same sense, so that they meet head to tail. The following definitions aim to accomplish this.

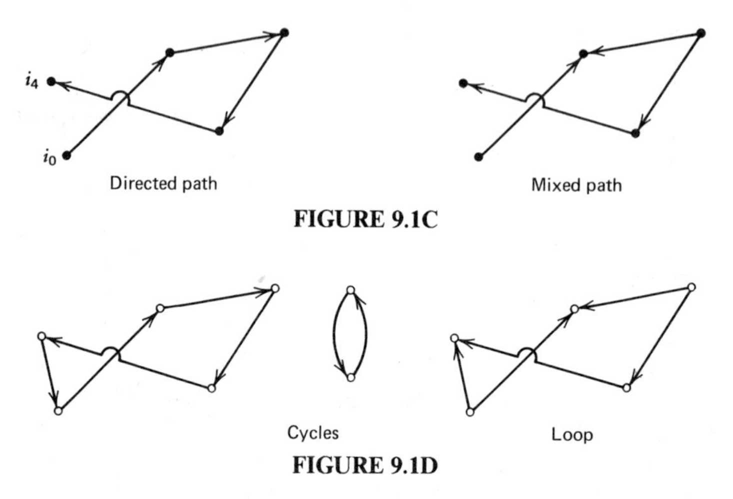

FIGURE 9.1C

FIGURE 9.1D

A *directed path* or *dipath* from i_0 to node i_m is a sequence $(i_0, i_1, \ldots, i_m)$ of distinct nodes such that each consecutive ordered pair (i_{k-1}, i_k) is a directed arc from i_{k-1} to i_k, for all $k=1,2,\ldots,m$. A *mixed path* connecting node i_0 and node i_m is a sequence $(i_0, i_1, \ldots, i_m)$ of distinct nodes such that for each $k=1,2,\ldots,m$ at least one of the ordered pairs (i_{k-1}, i_k) and (i_k, i_{k-1}) is a directed arc (Figure 9.1C). If these definitions are altered by requiring $i_0 = i_m$, one obtains a (directed) *cycle* and a (mixed) *loop*, respectively (Figure 9.1D). A directed graph containing no directed cycles is called *acyclic*.

Thus a directed path or a cycle is a mixed path or a loop in which two adjacent arcs (arrows) always meet head to tail. Only a directed path from i_0 to i_m is useful for communication from i_0 to i_m. However, in transportation this is not always true because a decrease in the amount shipped from i to j is equivalent to an incremental shipment in the reverse direction. An acyclic graph is always oriented, because two nodes and the two directed arcs between them would form a cycle. In an undirected graph a loop must contain at least three nodes and three arcs.

A directed graph or network will be called bilaterally connected or *biconnected* if it has a directed path from every node i to every node j (and hence also from j to i), and unilaterally connected or *uniconnected* if for every pair of its nodes i and j there is a directed path either from i to j or from j to i. It may be called *mixoconnected* if every pair of nodes is connected by a mixed path.

Trees

A *tree* may be defined as a graph or network of p nodes ($p \geq 1$), either undirected or else oriented, having any two of the following three properties.

(c_1) It has $q = p-1$ arcs.

(c_2) It is connected or mixoconnected.

(c_3) It contains no loop. (In the directed case, loops having only two arcs are already excluded by the restriction to oriented graphs.)

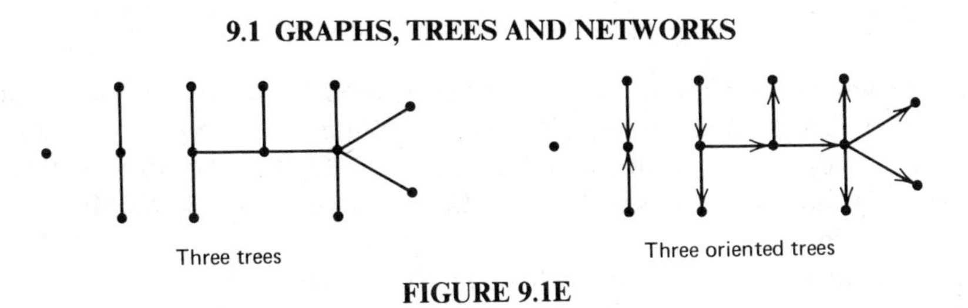

FIGURE 9.1E

The consistency of this definition results from the following slightly stronger statement.

THEOREM Each of the following is a necessary and sufficient condition for a graph G having p nodes and q arcs to be a tree.

(d_1) G is connected and has no loops.
(d_2) G is connected and $p \geq q+1$.
(d_3) G has no loops and $p \leq q+1$.

Proof The conditions are satisfied when $p=1, q=0$ (an isolated node). Now suppose (d_1) holds and $q>0$. By extending a path as far as one can, a (terminal) node incident with only one arc can be found. After this node and its incident arc have been deleted, (d_1) will still hold and the value of $p-q$ will be unchanged. Repetition of this procedure reduces q to 0, and then connectedness requires $p=1$. Thus $p-q=1$ initially as well as finally, and (d_2) and (d_3) are proved.

Conversely, suppose (d_1) does not hold. If G is neither connected nor free of loops, (d_2) and (d_3) are both false. If G lacks only connectedness, (d_2) is false and G can be converted into a connected graph G' satisfying (d_1) by inserting an appropriate number of new connecting arcs. Then $p-1=q'>q$ and (d_3) also is false. If G is connected but has one or more loops, (d_3) is false and G can be converted into G'' satisfying (d_1) by deleting one arc in each loop. Then $p-1=q''<q$ and (d_2) also is false.

A *rooted tree* is obtained by selecting one node (say 0) to be called the *root*, and orienting each arc so that there is a directed path from 0 to each of the other nodes. (In some trees all these orientations may be reversed.) The last tree of Figure 9.1E has a root in the upper left corner. Any connected directed graph may have a root in this sense, and rooted acyclic graphs are useful generalizations of rooted trees (e.g., in Chapter 10).

Unrooted trees will be encountered as optimal networks in 9.2 and as basic solutions of transportation problems in 11.2. Rooted trees will occur in the shortest-path algorithms of 9.3 to 9.6 and the decision and game networks of 10.3 to 10.7.

EXERCISES 9.1

In 1 to 13 sketch and describe the graph.

1. Let the nodes be the basic solutions of (3) of 2.5 and the arcs be the pivot steps.
2. Let the nodes be the first 12 positive integers and let (i, j) be a directed arc iff j/i is a prime integer >1.

3. The knight in chess must move two squares in one direction and one square in the other. Let the nodes be the nine squares of a 3 by 3 chessboard, and let two nodes be adjacent iff a knight can go from one to the other in a single move.
4. The nodes are the subsets of $\{1,2,3\}$. Let (i, j) be a directed arc iff subset i is contained in subset j.
5. Let a linear program in nonnegative variables be given and let $\mathbf{x}$ and $\mathbf{u}$ be any primal and dual solutions. The four nodes represent the following properties. P_1: $\mathbf{u}$ and $\mathbf{x}$ are optimal. P_2: $\mathbf{u}$ and $\mathbf{x}$ are feasible. P_3: $\mathbf{ux}=0, \mathbf{u}\geq\mathbf{0}, \mathbf{x}\geq\mathbf{0}$. P_4: $\mathbf{u}\geq\mathbf{0}, \mathbf{x}\geq\mathbf{0}$. (i, j) is a directed arc iff P_i implies P_j.
6. The three nodes represent the following properties of certain real numbers t. P_1: $t^3\geq 0$. P_2: $e^t\geq 1$. P_3: $\sqrt{t}$ is real. (i, j) is a directed arc iff P_i implies P_j for each t.

7 to 13 have four nodes. The given values of $c_{ij}=5$, 6, 7 or 8 merely indicate which arcs are present; also the arcs are undirected if and only if $c_{ij}=c_{ji}$ for all i and j.

7.

$i\diagdown j$	1	2	3	4
1	—	7	6	5

8.

$i\diagdown j$	1	2	3	4
1	—	7	6	5
2	7	—	—	—

9.

$i\diagdown j$	1	2	3	4
1	—	7	6	5
2	7	—	—	—
3	6	—	—	—
4	5	—	—	—

10.

$i\diagdown j$	1	2	3	4
1	—	7	—	5
2	—	—	—	—
3	6	—	—	—
4	—	—	—	—

11.

$i\diagdown j$	1	2	3	4
1	—	7	—	—
2	—	—	—	—
3	—	6	—	5
4	—	—	—	—

12.

$i\diagdown j$	1	2	3	4
1	—	7	—	—
2	—	—	6	—
3	—	—	—	5
4	8	—	—	—

13.

$i\diagdown j$	1	2	3	4
1	—	7	—	—
2	—	—	6	—
3	—	—	—	5
4	—	—	—	—

14. Show that a graph is a directed path if and only if it is a uniconnected oriented tree.
15. Show that a graph is a directed cycle if and only if it is mixoconnected and every node is incident with exactly one incoming and one outgoing directed arc.
16. Show that an undirected graph is a loop if and only if it is connected and every node is incident with exactly two arcs.
17. Show that the set of connected components of a graph is unique; that is, if $G_k \cap G_l \neq \varnothing$ then $G_k = G_l$.

9.2 SHORTEST SPANNING TREES ($c_{ij}=c_{ji}$)

Most of the problems in Chapters 9 and 11 have the minimization of transportation costs as a basic application. This section is concerned with optimizing the construction of a new transportation or communication network rather than the use of an existing facility.

Suppose then that the network is required to connect (*span*) the p given nodes, and let the cost of an undirected arc between nodes i and j be $c_{ij}=c_{ji}$ if it is constructed and 0 otherwise. It is desired to minimize the sum of the c_{ij} over those arcs that are actually constructed; this will be the shortest spanning network if the costs are used as the measures of distance or vice versa. This interpretation suggests that all $c_{ij}\geq 0$; then the shortest network will be a *tree*, because any loop could be eliminated by deleting one of its arcs without increasing the cost. If some $c_{ij}<0$, we shall add the requirement that the network be a tree; then there must be exactly $p-1$ arcs. The identity of the shortest tree will be unaffected if each c_{ij} is replaced by $c_{ij}+c$, where c is any constant, but there is no need to eliminate any negative values in this manner.

This problem was discussed by Kruskal (1956) and Dijkstra (1959) but most of the present treatment derives from the longer paper of Prim (1957). The idea is delightfully simple. One begins with a subtree consisting of only one arbitrarily selected node, which will not affect the length of the final tree; this subtree is expanded by adding one arc at a time (with its incident node) until all p nodes have been connected by means of $p-1$ arcs. *At each stage the added arc is an arc of shortest length* c_{ij} *connecting a node* i *in the subtree with a node* j *not yet in the subtree.* We will show that this produces a shortest spanning tree if $c_{ij}=c_{ji}$ and the arcs having finite c_{ij} form a connected graph. In the literature this algorithm has been described as greedy (brooking no delay in gratification) or myopic (not looking ahead), and related to matroids (Lawler, 1976 or Welsh, 1976).

The algorithm can be applied by inspection of a small network such as Figure 9.2A, in which the heavy lines indicate the resulting shortest spanning tree of length 37. If one starts at node A one obtains the arc lengths

$$c_{AC}=-10=\min(c_{AB}=19, c_{AD}=27, c_{AC}=-10)$$

$$c_{CF}=\quad 4=\min(19,27,18,36,4)$$

$$c_{FG}=\quad 11=\min(19,27,18,36,35,24,11,26)$$

and so on to GK, KH, CB, BD and DE. Some negative "lengths" are included to show that the algorithm can handle them. The resulting undirected tree may be

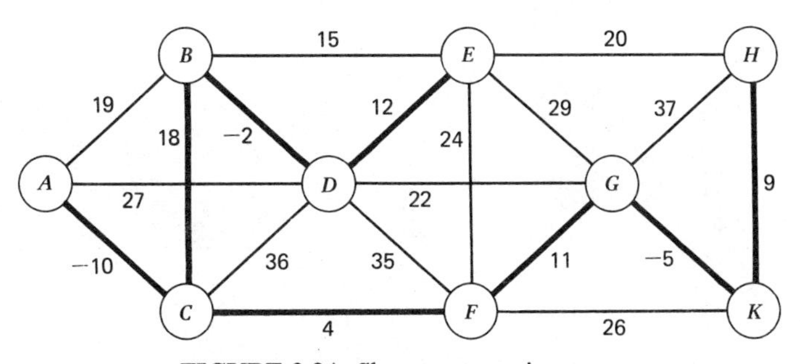

FIGURE 9.2A Shortest spanning tree.

recorded compactly as

$$\begin{matrix} EDBCFGKH \\ A \end{matrix}$$

It is equally easy to carry out the calculation by using only the square table of the c_{ij} as in Table 9.2A, which also states the rules of the procedure. The column headed r records the sequence in which the rows are marked. This may be useful for reviewing the work; otherwise it would suffice to circle the letter labels, A, C, etc. The set of r circled row labels (the nodes of the subtree) is denoted by S_r. Although the arcs are undirected, each $c_{ij}=c_{ji}$ appears twice in the table, once for each direction, and the rules must restrict the selection of each $\min c_{ij}$ to arcs *from* S_r (in marked rows) *to* its complement $\bar{S}_r$ (in columns with nothing struck out); otherwise loops could be formed. Table 9.2A shows the effect of starting at A, striking out its column of c_{iA}, marking its row (A) and $r=1$, circling $c_{AC}=-10=\min(19,-10,27)$ (the shortest arc out of A), striking out the rest of the column of c_{iC}, and marking the row (C) and $r=2$.

TABLE 9.2A Shortest Spanning Tree Algorithm I

0. The initial node is immaterial; choose A. Strike out its column of c_{iA} and mark its row by circling A or entering $r=1$.

1. If some c_{ij} in the marked rows are not struck out or circled, circle a smallest one of these (call it c_{kl}); otherwise stop.

2. Strike out the column l (c_{kl} excepted). This eliminates redundant data and prevents the formation of loops.

3. Mark row l and return to step 1.

The last two columns in the following table are shown as of the completion of the algorithm; the rest of the table is shown after the first completion of step 3. The next cycle will result in circling $c_{CF}=4=\min(19,27,18,36,4)$, striking out the other elements in column F, and marking row F with $r=3$.

$i\setminus j=$	A	B	C	D	E	F	G	H	K	r	SST Arcs
(A)	—	19	(−10)	27						1	AC
B	19	—	18	−2	15					7	BD
(C)	−10	18	—	36		4				2	CF, CB
D	27	−2	36	—	12	35	22			8	DE
E		15		12	—	24	29	20		9	
F			4	35	24	—	11		26	3	FG
G				22	29	11		37	−5	4	GK
H					20		37	—	9	6	
K						26	−5	9	—	5	KH

A partly filled table such as 9.2A can be stored compactly (in nine rows and six columns in this case) by using $i=1,\ldots,p$ as the labels of rows in which are listed the values of c_{ij} (perhaps in a nondecreasing order) and the corresponding indices j. There are no column labels.

Algorithm II: Recursive Minimization

The method of Table 9.2A involves a minimum of writing and is convenient for small exercises. For large problems one should exploit the existence of many common elements in the successive minimizations. This involves the use of recursion within each minimization as well as in the successive selection of shortest arcs.

As before, the initial node is selected arbitrarily; it is the sole member of S_1. In general let S_r denote the set of r nodes in the subtree and $\bar{S}_r$ the complementary set of nodes. Then the rth arc and the $r+1$st node $[r+1]$ to be added to the subtree are determined by the minimizing values of i and j in

$$\min_{j\in\bar{S}_r}\left(\min_{i\in S_r} c_{ij}\right) \tag{1}$$

c_{ij} that are undefined are ignored or set equal to $+\infty$.

For each r and each j the value of the bracketed expression is calculated as the minimum of only (one or) two quantities, namely, the corresponding expression for the preceding value of r and a single value of c_{ij} using the latest value of i, namely $[r]$. Thus for $j\in\bar{S}_r$

$$\min_{i\in S_r} c_{ij}=\min\left(\min_{i\in S_{r-1}} c_{ij}, c_{[r]j}\right) \tag{2}$$

the shortest single arc from the subtree S_r to node j in $\bar{S}_r$. Here $[r]$ is the rth node to be added to the subtree, thereby converting S_{r-1} into S_r. Then (2), the minimum over the subtree, has to be minimized over j, which runs through the nodes not yet in the subtree.

Table 9.2B gives the details involved in applying this algorithm to the previous example, whose data are now presented in Table 9.2C. In Table 9.2A and C there is exactly one circled c_{ij} in each column except the first, because each node after the first is reached exactly once via the tree.

Proof of Validity

The final network and all the partial networks obviously are connected and the number of arcs is one less than the number of nodes; thus they are trees by (d_2) of 9.1. If the given network is connected, the final tree T will span it (contain all the nodes). To show that T is a shortest spanning tree T_* we prove the following.

LEMMA Given any subset S containing from 1 to $p-1$ nodes, T_* must contain (one of) the shortest arc(s) connecting a node in S with a node not in S.

TABLE 9.2B Shortest Spanning Tree Algorithm II (Rules)

The first table in 9.2C shows the given $c_{ij}=c_{ji}$.

The second table in 9.2C shows the results of the first minimization (2) (recursive over i in S_r) for each r and j. The column headed $[r]$ lists the nodes in the order in which they are added to the tree. The r dashes in row r exclude arcs that would lead back to the subtree and form loops. ④ is the length of CF (smallest in its row) in the tree but ⑱ is the length of CB, not HB; one must *take the label* C *from the row in which the length* 18 *first occurs.*

Construction of the second table in 9.2C:

0. The rows are labeled $r=1,\ldots,p$. The initial column A is filled with dashes. In the first row write $[r]=A$ and copy the values of c_{Aj} from row A of the first table. Set $r=2$.
1. Circle a smallest entry in the latest row $r-1$ (if none, stop).
2. Fill the rest of its column $j=[r]$ with $p-r+1$ dashes.
3. Enter the letter symbol for node $j=[r]$ in the next row r.
4. In each column j in row r enter the smaller of the preceding entry in row $r-1$ and the value of $c_{[r]j}$ from the first table, except where neither number is defined or a dash has been entered. Increase r by 1 and return to step 1. For example, in row 2 replace the preceding larger 19 by $c_{CB}=18$; repeat the preceding $27<36=c_{CD}$; enter the solely available $c_{CF}=4$.

Proof Let $\{l,\bar{l}\}$ be such a shortest arc with l in S and $\bar{l}$ in its complement $\bar{S}$. If $\{l,\bar{l}\}$ is not in T_*, let it be added to T_*. This will form a loop that includes $\{l,\bar{l}\}$ and lies in $T_*\cup\{l,\bar{l}\}$. The loop must also include at least one other arc $\{k,\bar{k}\}\neq\{l,\bar{l}\}$ with k in S and $\bar{k}$ in $\bar{S}$. If the loop is broken by removing $\{k,\bar{k}\}$, the resulting network $T_0 \triangleq T_*\cup\{l,\bar{l}\}-\{k,\bar{k}\}$ is also a spanning tree. (The assumption $c_{ij}=c_{ji}$ is needed for this; see Exercise 15.) Since T_0 cannot be shorter than T_*, this implies $c_{l\bar{l}}\geq c_{k\bar{k}}$, but we also have $c_{l\bar{l}}\leq c_{k\bar{k}}$ by the minimality that defined $\{l,\bar{l}\}$. Thus $c_{k\bar{k}}=c_{l\bar{l}}$ and $\{k,\bar{k}\}$ is an arc in T_* of the sort that is asserted to exist.

The algorithm is obviously derived from the lemma. To complete the validation of the algorithm, suppose for the moment that all the c_{ij} are distinct real numbers so that ties cannot occur. Then the algorithm uniquely determines the final tree T and guarantees that no other spanning tree could satisfy the necessary conditions given by the lemma. The number of spanning trees is finite, being less than the number of selections of $p-1$ arcs out of $p(p-1)/2$. If the network is connected, it contains at least one spanning tree. (Delete one arc in each loop in the network until all loops are eliminated.) Thus there is at least one spanning tree, and T must be the one and only shortest spanning tree T_* in this case.

If the c_{ij} are not all distinct they can be made so by adding to them suitable nonnegative fractional parts of any chosen $\varepsilon>0$. Ties can thereby be broken in any desired way, and if ε is less than the minimum difference between pairs of distinct c_{ij}, the other choices in the algorithm will be unaffected. Thus the algorithm can still

TABLE 9.2C Shortest Spanning Tree Algorithm II (Data)
Table of Arc Lengths $c_{ij}=c_{ji}$

i \ $=j$	A	B	C	D	E	F	G	H	K
A		19	−10	27					
B	19		18	−2	15				
C	−10	18		36		4			
D	27	−2	36		12	35	22		
E		15		12		24	29	20	
F			4	35	24		11		26
G				22	29	11		37	−5
H					20		37		9
K						26	−5	9	

Values of (2): $\min_{i \in S_r} c_{ij} = \min(\min_{i \in S_{r-1}} c_{ij}, c_{[r]j})$

r	A	B	C	D	E	F	G	H	K	$[r]$	SST Arcs
1	—	19	(−10)	27						A	AC
2	—	18	—	27		(4)				C	CF
3	—	18	—	27	24	—	(11)		26	F	FG
4	—	18	—	22	24	—	—	37	(−5)	G	GK
5	—	18	—	22	24	—	—	(9)	—	K	KH
6	—	(18)	—	22	20	—	—	—	—	H	CB
7	—	—	—	(−2)	15	—	—	—	—	B	BD
8	—	—	—	—	(12)	—	—	—	—	D	DE
9	—	—	—	—	—	—	—	—	—	E	
	A	B	C	D	E	F	G	H	K		

produce the same T and it will be shortest for $\varepsilon>0$. If $z(\varepsilon)$ and $z^*(\varepsilon)$ are the lengths of T and T_* as functions of ε, one has

$$z(0) \leq z(\varepsilon) \leq z(0) + (p-1)\varepsilon$$

$$z^*(0) \leq z^*(\varepsilon) \leq z^*(0) + (p-1)\varepsilon \tag{3}$$

and $z(\varepsilon)=z^*(\varepsilon)$ for $\varepsilon>0$. This implies that $z(0)$ and $z^*(0)$ both lie between $z(\varepsilon)-(p-1)\varepsilon$ and $z(\varepsilon)$, and so $z(0)=z^*(0)$ as desired because ε can be chosen arbitrarily small.

Extensions

(a) The proof above shows that the tree T_* produced by the algorithm will minimize any symmetric nondecreasing function of the $p-1$ values of c_{ij} in the tree (as in (4) below) as well as their sum. The effect of the symmetry of the function and of $c_{ij}=c_{ji}$ is to make immaterial the order in which the arcs are chosen and hence to obviate the heed for foresight. To maximize a symmetric nondecreasing function of the c_{ij} as in (5) below one need only choose the largest eligible c_{ij} at each step. From this point of view only two interesting spanning trees (the shortest T_* and the longest T^*) are defined by a symmetric matrix of numbers c_{ij}.

For example, let c_{ij} denote the *probability* of loss of fluid, voltage, pressure, security or the like as the result of the failure of the arc $\{i, j\}$ within a specified time. Then if the various arcs are assumed to fail independently, $T=T_*$ will minimize the probability

$$1-\prod_{\{i,j\}\in T}(1-c_{ij}) \tag{4}$$

that one or more failures occur in the tree. It is an accidental circumstance that an equivalent result can be obtained by replacing c_{ij} by $|\log(1-c_{ij})|$ and minimizing the sum of the latter.

As another example, let c_{ij} denote the *capacity* of an arc, which is the maximum quantity of something that it can transport per unit time. A conservative definition of the capacity of a tree T is

$$\min\{c_{ij}:\{i,j\}\in T\} \tag{5}$$

the minimum of the capacities of its arcs. This is again a nondecreasing function of the c_{ij}. Although it cannot be reduced to a sum as (4) can, it is nevertheless maximized by the same tree T^* that maximizes the total "length" $\Sigma\{c_{ij}:\{i,j\}\in T\}$.

Reliabilities and capacities will be reconsidered in 9.3 and 9.8.

(b) A *forest* is defined to consist of one tree or of several trees that are not connected to one another; that is, it is a graph without loops. The determination of one or more *shortest spanning forests* F^* consisting of specified numbers q of trees is of interest as a means for dividing the set of nodes into clusters in terms of proximity (see Gower and Ross, 1969; Zahn, 1971; Kershenbaum and Van Slyke, 1972).

One way to do this is simply to delete the $q-1$ longest arcs in the shortest spanning tree. Equivalently one may select the $p-q$ shortest available arcs, rejecting those that form loops with arcs already chosen. The rejected arcs are those both of whose endpoints are currently listed as belonging to the same connected set of nodes (a subtree).

The lemma now assumes that F^* must consist of q separate subtrees and that the set of p nodes is partitioned into more than q disjoint sets S_h. It is asserted that F^* contains one of the shortest of the arcs connecting pairs of nodes not both in the same S_h. If F^* does not contain such a shortest arc $\{l, \bar{l}\}$, the latter is added to F^*. If this creates a loop, the loop is broken as before, with S denoting either of the distinct sets to which l and $\bar{l}$ belong. If no loop is formed, note that one or more of the q

subtrees of F^* must contain nodes from at least two distinct sets $S_{h'}$, and $S_{h''}$, since there are more than q of the S_h. In this subtree one can find $c_{k\bar{k}} \leq c_{l\bar{l}}$ by the minimality of F^*, but $c_{k\bar{k}} \geq c_{l\bar{l}}$ by the minimality of $\{l, \bar{l}\}$. Here $k \in S_{h'}$ and $\bar{k} \in S_{h''}$.

(c) After the elimination of any loops among them, some arcs may be required to belong to the optimal spanning tree or forest because these facilities already exist and do not require construction. This reduces to the standard case after these existing arcs are given zero or arbitrary c_{ij} values smaller than any other, or their actual c_{ij} are circled wherever they occur to indicate they are to be treated as if small. Similarly m particular nodes may be designated as subcenters, no two of which may belong to the same subtree. This reduces to (b) above (with q replaced by $q-m+1$) after $m-1$ arcs connecting the m subcenters have been given minimal c_{ij} values, thus excluding any other connections among them.

(d) One may ask for the shortest tree connecting some subset of the given nodes, the connection status of the remaining nodes not being specified. If one knew which of the latter were in fact connected, the algorithms above could still be used to determine the tree. If only two nodes i and j have to be connected, one has the shortest-path problem discussed in the remainder of this chapter; d_{ij} denotes the length of the path. If three nodes j, k, l have to be connected, the shortest-tree length is the smallest of the four numbers

$$d_{jk}+d_{kl},\ d_{jk}+d_{jl},\ d_{jl}+d_{kl}, \qquad \min_i(d_{ij}+d_{ik}+d_{il}) \tag{6}$$

In general the connection of m nodes may use as many as $q=m-2$ junction points like i in (6), but no more. In Figure 9.2B the junction points are the open circles, and the arcs represent nonoverlapping shortest paths in the original network. Given this use of shortest path lengths d_{ij}, a junction point will be useful only if it meets at least three arcs, while each of the given m nodes will meet at least one arc. To connect the $m+q$ nodes requires $m+q-1$ arcs, each of which has exactly two

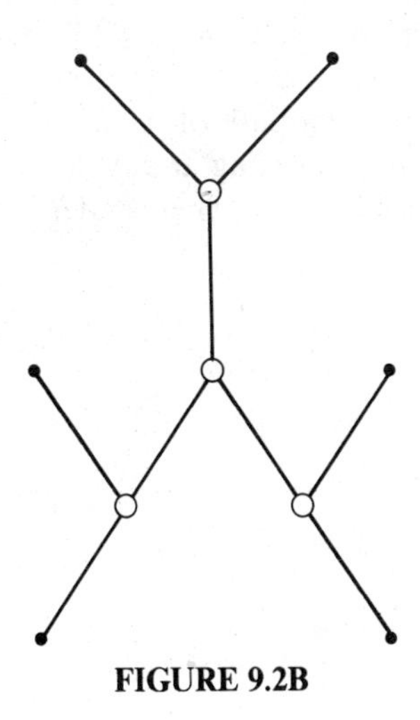

FIGURE 9.2B

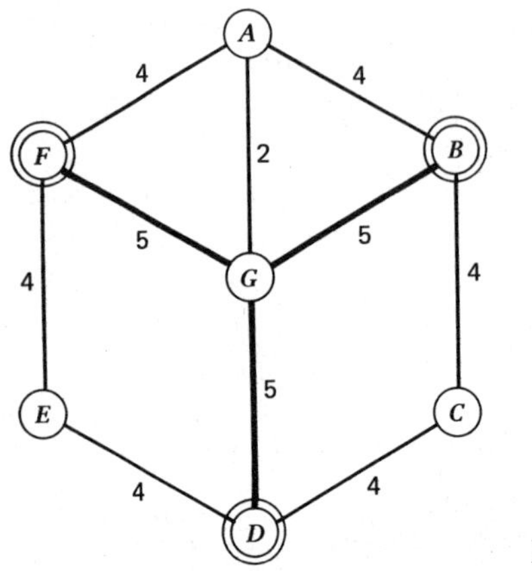

FIGURE 9.2C

ends. Thus one has

$$2(m+q-1)\geq 3q+m \qquad \text{or} \qquad m-2\geq q \tag{7}$$

Figure 9.2C is an example in which B, D and F are most efficiently connected by way of G, but this tree has no arcs in common with either the shortest spanning tree for the whole network or the shortest paths connecting any two of the nodes B, D and F. (d) is the Steiner network problem (Lawler, 1976).

(e) If $c_{ij}\neq c_{ji}$ the algorithm will produce the shortest oriented spanning tree provided the smaller of c_{ij} and c_{ji} is used as their common value. There may be more interest in taking $c_{ij}+c_{ji}$ as the common value, but this does not cause the algorithm to produce the shortest biconnected spanning network (a cycle of length 7) in Figure 9.2D, and it fails to give any spanning network if the given network is a one-way cycle. To find the shortest cycle connecting all the nodes is the difficult traveling salesman problem. In Figure 9.2E the shortest spanning tree rooted at A is ACB of length 3 but the algorithm chooses ABC of length 4. See Edmonds (1967) and Karp (1972).

(f) The algorithm will generate a path of $p-1$ arcs if i is restricted to be one of the two endpoints of the subpath existing at any stage $r\geq 3$. However, the resulting path $ACBD$ in Figure 9.2F has length 6 whereas $ABCD$ has length 5.

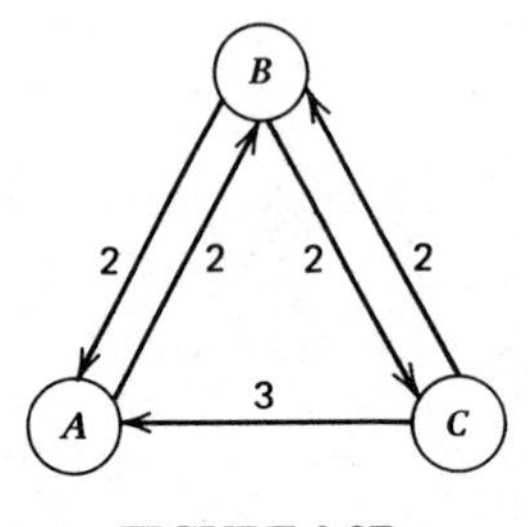

FIGURE 9.2D

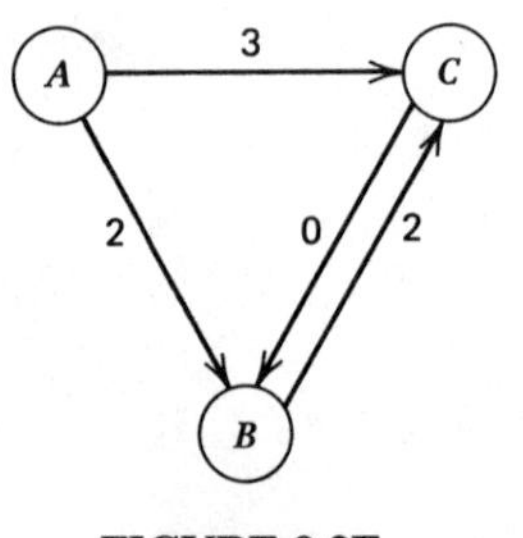

FIGURE 9.2E

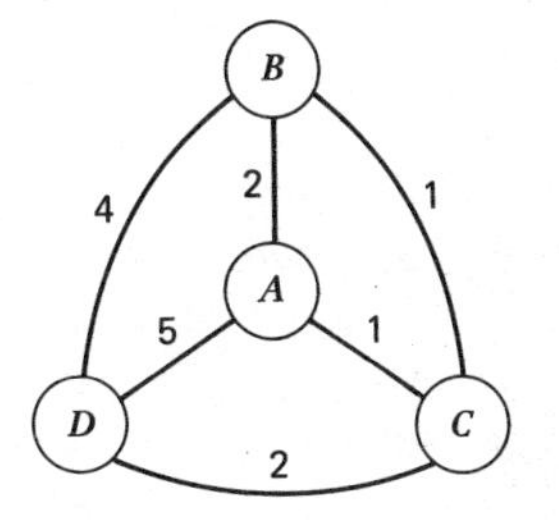

FIGURE 9.2F

EXERCISES 9.2

Find the shortest spanning trees in 1 to 8 by Algorithm I or II. Show your work.

1. Verify the solution given in Table 9.2A.
2. $=c_{ij}$

	A	B	C	D	E	F	G
A	—	17	19	6	22	5	21
B	17	—	5	22	29	13	25
C	19	5	—	17	23	26	19
D	6	22	17	—	20	8	9
E	22	29	23	20	—	4	3
F	5	13	26	8	4	—	18
G	21	25	19	9	3	18	—

3. Like 2 but replace c_{ij} by $30-c_{ij}$.
4. Like 2 but replace c_{ij} by the sum of its digits.
5. Like 2 but add 10 to odd values of c_{ij}.
6. Like Table 9.2C but replace c_{ij} by $30-c_{ij}$.
7. Let $c_{ij}=c_i+c_j$ for $i\neq j$ and $c_i\leq c_{i+1}$ for $1\leq i\leq p-1$.
8. Let $c_{ij}=|c_i-c_j|$ for $i\neq j$ and $c_i\leq c_{i+1}$ for $1\leq i\leq p-1$.
9. The following are independent probabilities of failure.

$$c_{AB}=.03 \quad c_{BC}=.01 \quad c_{CD}=.05 \quad c_{DE}=.03 \quad c_{EF}=.04$$
$$c_{AC}=.02 \quad c_{BD}=.04 \quad c_{CE}=.02 \quad c_{DF}=.01 \quad (c_{ji}=c_{ij})$$

Find the most reliable spanning tree and its reliability.

For 10 to 13 use the data of Table 9.2C and Exercise 2 of 9.2.

10, 11. Divide the nodes into two subsets in terms of proximity.
12, 13. Find the *SST* containing arcs *CD* and *DG* and all nodes.
14. If (6) (for $m=3$) is said to involve four cases, how many cases are required for $m=4$?
15. Explain how attempts to extend the proof of validity of the algorithm to Figures 9.2E and F fail.
16. Suppose that $c_{kl}<c_{ij}$ and $c_{pq}<c_{ij}$ for all $\{i,j\}\neq\{k,l\}$ or $\{p,q\}$. Prove that the arcs $\{k,l\}$ and $\{p,q\}$ belong to the shortest spanning tree T_*.
17. How is the lemma changed if one is concerned with shortest oriented trees D_* with node 1 as the root?

9.3 SHORTEST PATHS FROM ONE NODE ($c_{ij} \geq 0$)

The *total length or cost or elapsed time* for a path traversing the (possibly directed) arcs (A, B),(B, C),...,(F, G),(G, H) is given by $c_{AB} + c_{BC} + \cdots + c_{FG} + c_{GH}$. For fixed endpoints A, H and any number $0, 1, 2, \ldots$ of variable intermediate nodes $B, C, \ldots, F, G$ the minimum value of this expression is denoted by d_{AH}. Determining the shortest or cheapest or fastest paths is obviously a problem of fundamental importance in network theory and its applications. Various algorithms for this purpose will be considered in 9.3, 9.4, 9.6, and 9.7 but none of these determines only one shortest path and its subpaths. (Obviously the subpaths must be shortest paths also.)

This section finds the shortest paths from (or to) one particular node to (or from) some or all of the other nodes in an order of nondecreasing length. The assumption $c_{ij} \geq 0$ implies there is no advantage in allowing loops or cycles to be formed. Thus if only one shortest path is accepted for each of the other nodes, it is clear that all the shortest paths from one node form a tree. In this and many other respects, including the existence of analogous versions of each, this algorithm resembles that of 9.2. The principal differences are the following.

(i) The tree may be oriented (rooted at the initial node) because the present assumption is $c_{ij} \geq 0$ rather than $c_{ij} = c_{ji}$. Accordingly the proof of validity is different.

(ii) Apart from ties, only one minimizing tree was defined in 9.2. Here there could be p or $2p$ of them, one or two for each initial node, which may be either an origin or a destination if $c_{ij} \neq c_{ji}$. If most of these are required, Section 9.7 is an alternative.

(iii) Shortest paths involve minimizations over $d_{Ai} + c_{ij}$ rather than c_{ij} alone. This is the only change that needs to be kept in mind in doing exercises (Step 1 in Table 9.3A). It is responsible for the phenomenon illustrated by (10) below.

According to the valuable review by Dreyfus (1969), the best version of this algorithm for large problems was apparently first published by Dijkstra (1959). However, it is convenient to begin with a *single-labeling procedure* proposed by various authors at about that same time.

Let d_{Ai} denote the shortest path length from some fixed node A to node i; let node $A = [1]$ be the sole member of S_1, and $d_{AA} = 0$. In general let S_r consist of the r nodes $i = [1], \ldots, [r]$ whose shortest path lengths d_{Ai} from A have been determined at stage r, and let $\bar{S}_r$ consist of the remaining nodes. The shortest paths from A to the other nodes of S_r, if unique, form a subtree. Then the rth arc to be added to this subtree is the (i, j) that minimizes

$$d_{A[r+1]} = \min_{i \in S_r} \left(d_{Ai} + \min_{j \in \bar{S}_r} c_{ij} \right) \qquad (r = 1, \ldots, p-1) \tag{1}$$

The added node $[r+1]$ is the j that minimizes (1). Values of (i, j) for which c_{ij} is undefined are ignored or else such c_{ij} are set equal to $+\infty$. To determine the shortest path lengths d_{iA} *to A* one uses

$$d_{[r+1]A} = \min_{j \in S_r} \left(\min_{i \in \bar{S}_r} c_{ij} + d_{jA} \right) \qquad (r = 1, \ldots, p-1) \tag{2}$$

If a small network diagram is given, (1) permits the successive $d_{A[r]}$ to be written down by inspection. In Figure 9.3 they are included within the circle representing the node, while each c_{ij} labels its arc on the right side near its initial node i. The shortest paths are indicated by heavy lines and arrowheads. Where necessary a small black square indicates a nonexistent directed arc (or infinite cost).

First A is labeled with $d_{AA} = 0$ because it is the given node. The next two labels (for D and C) are determined thus: $d_{AD} = 6 = d_{AA} + \min(c_{AB}, c_{AD}, c_{AC}) = 0 + \min(9, 6, 7)$. $d_{AC} = 7 = \min(0+7, 6+2) =$ the smaller of $d_{AA} + \min(c_{AB}, c_{AC})$ and $d_{AD} + \min(c_{DB}, c_{DE}, c_{DG}, c_{DF}, c_{DC})$. There is a tie for the next minimum and hence B and F can be labeled in either order or simultaneously: $d_{AB} = d_{AF} = 8 = \min(0+9, 6+2, 7+1)$ where $0+9 = d_{AA} + c_{AB}$, $6+2 = d_{AD} + \min(c_{DB}, c_{DE}, c_{DG}, c_{DF}) = d_{AD} + c_{DB}$ and $7+1 = d_{AC} = c_{CF}$.

The remaining labels $d_{AG} = 11$, $d_{AK} = 17$, $d_{AE} = 18$ and $d_{AH} = 20$ are determined in that order. The resulting oriented tree of shortest paths with root at A may be recorded compactly as

$$\begin{matrix} & E & \\ BDA'CFKH. \\ & G & \end{matrix}$$

It is even easier to make the calculation when the c_{ij} are given in a square table as in Table 9.3A, which also gives a stepwise description of the procedure. The set of r circled row labels (the nodes of the subtree) is denoted by S_r. Whether or not the arcs are directed, it is necessary to restrict the attention to arcs going in the appropriate direction—in this case, leading *from* S_r (in marked rows) *to* its complement $\bar{S}_r$ (the nodes not yet evaluated). Thus columns are struck out as they enter S_r to prevent the generation of new (erroneous) values of d_{Ai} greater than those already obtained. The successive minima obtained in step 1 are the shortest path lengths recorded in the column d_{Ai}. Table 9.3A shows the effect of starting at A, striking out its column of c_{iA}, marking its row Ⓐ and $d_{Ai} = 0$, circling $c_{AD} = 6$, which yields

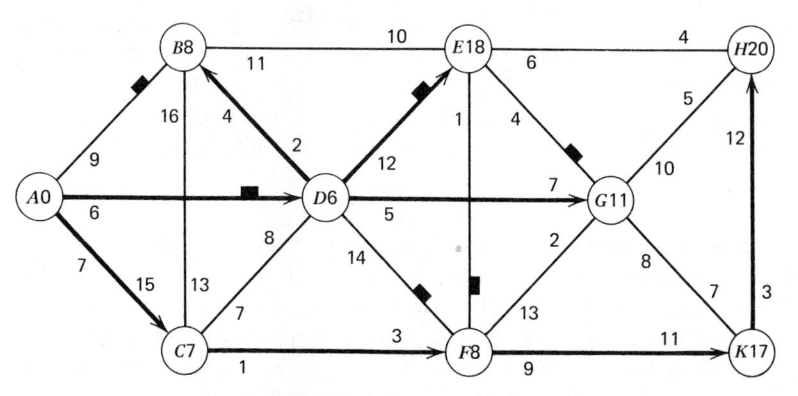

FIGURE 9.3 Shortest paths from *A*.

TABLE 9.3A Shortest Paths from Node A: Algorithm I

0. Suppose that shortest paths from A are required. Strike out column A, mark row A as Ⓐ, and enter $d_{Ai}=0$.
1. Circle a c_{kl} that yields $\min_{ij}(c_{ij}+d_{Ai})$, taken over marked rows i and c_{ij} not struck out or circled; if none such, stop.
2. Strike out the column l (c_{kl} excepted) as irrelevant.
3. Mark row l as ⓛ, enter $d_{Al}=\min$ in step 1, return to 1.

The last two columns in the table below are shown as of the completion of the algorithm; the rest of the table is shown after the first completion of step 3. Next step 1 circles $c_{AC}=7$ because it yields the smaller of $\min(9,7)+0$ and $\min(2,8,12,14,5)+6$. Strike out the rest of column C; in row C enter Ⓒ and $d_{AC}=7+0=+7$; etc.

$i \diagdown j=$	A	B	C	D	E	F	G	H	K		d_{Ai}	Arcs
Ⓐ	—	9	7	⑥						+	0	AD, AC
B		—	16	4	11					+	8	
C	15	13	—	7		1				+	7	CF
Ⓓ		2	8	—	12	14	5			+	6	DB, DG, DE
E		10			—	1	4	6		+	18	
F			3			—	13		9	+	8	FK
G				7		2	—	10	8	+	11	
H					4		5	—	12	+	20	
K						11	7	3	—	+	17	KH

$\min(9,7,6)+0$ (the shortest arc out of A), striking out the rest of the column D, and marking the row Ⓓ and $d_{AD}=6+0=+6$.

If all $c_{ij}=c_{ji}$ where defined, the problem is the same whether the fixed node is an origin or a destination, and so the format of Table 9.3A is always applicable. However, if some $c_{ij}\neq c_{ji}$ and a fixed destination is specified, rows and columns must be interchanged either in the table or in the rules but not both. If the table of c_{ij} is left unchanged, the description of the beginning of Algorithm I can be generalized as follows so as to apply to both cases. (Table 9.3D illustrates the other case.)

Orientation of Algorithm I. Origins (tails of arrows) are represented by rows in order that their subscripts i may precede the column subscripts j, which represent destinations (heads of arrows). The fixed node (A or whatever) has an assigned status of origin or designation, the other status being excluded. The fixed node is *marked in its assigned status* by entering Ⓐ and $d_{Ai}=0$ as labels of row A if A is an origin, or entering Ⓐ and $d_{jA}=0$ as labels of column A if A is a destination. *The* *excluded status of* A *determines the crossing out* of a column of c_{iA} or a row of c_{Aj}, respectively. The remainder of the algorithm follows the orientation thus established.

A partly filled table such as 9.3A can be stored compactly (in nine rows and five columns in this case) by using $i=A, B, \ldots$ as labels of rows in which are listed the values of c_{ij} (perhaps in a nondecreasing order) and the corresponding indices j. There are no column labels.

Algorithm II: Recursive Minimization

The preceding method (single labeling) involves a minimum of writing and is convenient for small exercises. The method given by Dijkstra (1959) is more efficient for computer-sized problems. Like the corresponding method in 9.2, it uses recursion within each minimization to take advantage of the many common elements in successive minimizations.

To do this one reverses the order of the minimizations in (1) to obtain

$$d_{A[r+1]} = \min_{j \in \bar{S}_r} \left[\min_{i \in S_r} \left(d_{Ai} + c_{ij} \right) \right] = \min_{j \in \bar{S}_r} d^r_{Aj} \tag{3}$$

for $r=1,2,\ldots, p-1$. Here we define $d^1_{Aj} = c_{Aj}$ and

$$d^r_{Aj} = \min_{i \in S_r} \left(d_{Ai} + c_{ij} \right) = \min\left(d^{r-1}_{Aj}, d_{A[r]} + c_{[r]j} \right) \tag{4}$$

for $r=2,\ldots, p-1$ and j in $\bar{S}_r$. The last equality is obvious because $d^{r-1}_{Aj} = \min_{i \in S_{r-1}}(d_{Ai} + c_{ij})$; this includes everything to be minimized in (4) except the term for $i=[r]$, the node whose addition converts S_{r-1} into S_r. At the same time $[r]$ disappears as a possible value for j.

Evidently d^r_{Aj} is the shortest distance from A to j when the nodes preceding j are required to belong to S_r. In particular $d^1_{Aj} = c_{Aj}$, only the node A being available to precede j.

Tables 9.3B and 9.3C give a stepwise description of the algorithm and apply it to the previous example. In Tables 9.3A and C there is exactly one circled c_{ij} in each column except A, because each node other than A is reached via exactly one shortest path from A.

Proof and Extensions

It will be shown inductively that (1) or (3) determines the $r+1$st smallest of the shortest path lengths. Since all $c_{ij} \geq 0$, (one of) the smallest possible of these is certainly $d_{AA} = 0$, and the next is $\min_{j \neq A} c_{Aj}$ as (1) asserts. Now suppose that

$$0 = d_{AA} \leq d_{A[2]} \leq \cdots \leq d_{A[r]} \tag{5}$$

are the r smallest of the shortest path lengths d_{Ai} determined by (1). When the shortest path from A to any node k in $\bar{S}_r$ is traced out, let i be the last node of S_r that is encountered, and let it be followed by j in $\bar{S}_r$. Then the shortest path from A to k (of length d^*_{Ak}, say) must follow a shortest path from A to i, whose length d_{Ai} in (5) is already known, then the arc from i to j of length c_{ij}, and possibly additional arcs

TABLE 9.3C Shortest Paths from A: Algorithm II (Data)
Table of Arc Lengths $c_{ij} \geq 0$

$i \backslash j=$	A	B	C	D	E	F	G	H	K		d_{Ai}
A		9	7	6						+	0
B			16	4	11					+	8
C	15	13		7		1				+	7
D		2	8		12	14	5			+	6
E		10				1	4	6		+	18
F			3				13		9	+	8
G				7		2		10	8	+	11
H					4		5		12	+	20
K						11	7	3		+	17

Values of (4): $d_{Aj}^{r} = \min(d_{Aj}^{r-1}, c_{[r]j} + d_{A[r]})$

r	A	B	C	D	E	F	G	H	K	[r]	Arcs
1	—	9	7	⑥						A	AD
2	—	8	⑦	—	18	20	11			D	AC
3	—	⑧	—	—	18	8	11			C	DB
4	—	—	—	—	18	⑧	11			B	CF
5	—	—	—	—	18	—	⑪		17	F	DG
6	—	—	—	—	18	—	—	21	⑰	G	FK
7	—	—	—	—	⑱	—	—	20	—	K	DE
8	—	—	—	—	—	—	—	⑳	—	E	KH
9	—	—	—	—	—	—	—	—	—	H	
	A	B	C	D	E	F	G	H	K		

of total length $d_{jk}^{*} \geq 0$. Thus

$$d_{Ak}^{*} = d_{Ai} + c_{ij} + d_{jk}^{*} \geq d_{Ai} + c_{ij} \geq d_{A[r+1]} \tag{6}$$

the second inequality following from (1). Since (1) shows that $d_{A[r+1]}$ is the length of some path from A to a node of $\bar{S}_r$, (6) shows that is the smallest of all such lengths. Also $d_{A[r]} \leq d_{A[r+1]}$ by the assumption accompanying (5). This completes the proof.

(a) The proof remains valid if S_1 *consists of a subset of distinguished nodes* rather than a single node; then the shortest paths from or to S_1 form a forest with one tree (possibly without arcs) for each node in S_1.

In the previous example, let us suppose that hospitals are located at C and H and it is desired to find the length d_{j*} of the shortest path to a hospital from each of the other nodes. Unless the unsymmetrical table of c_{ij} is transposed, one must label the *columns* C and H with $d_{CC} = 0$ and $d_{HH} = 0$ and strike out the rows of C and H since arcs leaving C or H cannot be used. (We are using Algorithm I.)

There is a tie for the shortest arc into C or H, namely FC and KH of length 3. These entries are circled and the remainder of rows F and K struck out. Columns F and K receive the labels $d_{FC} = d_{KH} = 0 + 3 = 3$. Table 9.3D now applies.

TABLE 9.3B Shortest Paths from A: Algorithm II (Rules)

The first table in 9.3C shows the given $c_{ij} \geq 0$ and a column of shortest path lengths d_{Ai} (circled in the second table).

The second table in 9.3C shows the results d^r_{Aj} of the first minimization (4) (recursive over i in S_r) and the nodes $[r]$ in the order in which they are added to the tree S_r of shortest paths. The r dashes in row r exclude nodes already in the tree. The value ⑧ in column B is d_{AB}; its last arc is DB, not CB. One must *take the label* D *from the row in which the length* 8 *first occurs*.

Construction of the second table in 9.3C:

0. The rows are labeled $r = 1, \ldots, p$. The initial column A is filled with dashes. In the first row write $[r] = A$ and copy the values of c_{Aj} from row A of the first table. Set $r = 2$.
1. Circle a smallest entry $d_{A[r]}$ in the latest row $r-1$ (if none, stop).
2. Fill the rest of column $j = [r]$ with $p - r + 1$ dashes.
3. Enter the letter symbol for node $j = [r]$ in the next row r.
4. Copy the circled $d_{A[r]}$ in row $[r]$ and column d_{Ai} of the first table.
5. In each column j in row r enter the smaller of the preceding entry d^{r-1}_{Aj} in row $r-1$ and the sum of the numbers $c_{[r]j}$ and $d_{A[r]}$ of the first table, except where neither d^{r-1}_{Aj} nor $c_{[r]j}$ is defined or a dash has been entered. Increase r by 1 and return to step 1. For example, in row 2 repeat the preceding $7 < 8 + 6$; elsewhere enter the smaller $c_{Dj} + d_{AD} = 2 + 6$, etc.

TABLE 9.3D

$i \diagdown j$ =	A	B	Ⓒ	D	E	Ⓕ	G	Ⓗ	Ⓚ
A	—	9	7	6					
B		—	16	4	11				
C	15	13	—	7		1			
D		2	8	—	12	14	5		
E		10			—	1	4	6	
F			③			—	13		9
G				7		2	—	10	8
H					4		5	—	12
K						11	7	③	—
	+	+	+	+	+	+	+	+	+
d_{j*}	7	12	0	8	4	3	5	0	3
Arcs			FC	BD		EF		KH	
			AC			GF			
			DC						

According to (2) the next arc is determined by the minimum of $c_{ij}+d_{j*}$ with $i=A, B, D, E, G$ and $j=C, F, H, F$. This is $c_{EF}+d_{F*}=1+3=4=d_{E*}$. Proceeding in this way one obtains the path lengths d_{j*} and the trees $BDC\overset{A}{\underset{G}{F}}E$ and KH oriented toward C and H.

An equivalent procedure, valid in 9.6 and 9.7 also, is to set $c_{ij}=0$ for all i and j in S_1, and then start at any node of S_1.

(b) For simplicity let the lengths of the arcs in a path be denoted by $c_1, c_2, \ldots, c_r$ and let the object be to minimize $f_r(c_1, \ldots, c_r)$ over all possible paths with given endpoints. If the functions f_r have the properties

$$f_r(c_1, \ldots, c_r) \leq f_{r+1}(c_1, \ldots, c_r, c_{r+1}) \tag{7}$$

$$f_{r+1}(c_1, \ldots, c_r, c_{r+1}) = f_2(f_r(c_1, \ldots, c_r), c_{r+1}) \tag{8}$$

then (1) or (3) is replaced by

$$d_{1[r+1]} = \min_{i \in S_r} \min_{j \in \bar{S}_r} f_2(d_{1i}, c_{ij}) \tag{9}$$

Thus far we have had $f_r(c_1, \ldots, c_r) = c_1 + \cdots + c_r$ and the condition $c_{ij} \geq 0$ was required to satisfy (7).

As in (a) of 9.2, one application involves letting q_{ij} = the probability of a failure in arc (i, j) and perhaps setting $c_{ij} = \log(1-q_{ij})^{-1}$ since f_r can be defined as a sum of these logarithms. The present case differs from 9.2(a) in that the change from q_{ij} to c_{ij} (equivalent to a change in f_2) can result in the selection of different arcs if the failure probabilities are high. Thus if

$$c_{12} = c_{23} = .4 \qquad c_{13} = .7 \tag{10}$$

the path $(1, 3)$ has the shorter length but the same numbers used as q_{ij} give $(1, 2, 3)$ as the more reliable path because

$$2|\log(1-.4)| < |\log(1-.7)| \qquad \text{or} \qquad (1-.4)^2 > 1-.7$$

Another application is obtained by letting c_{ij} = the negative of the capacity of arc (i, j) and $f_r(c_1, \ldots, c_r) = \max\{c_1, \ldots, c_r\}$. This is discussed in some detail in 9.8.

(c) A further generalization allows the function f_2 in (8) to depend not only on the length $c_{r+1} = c_{ij}$ of the last arc but also on its indices i and j. In terms of (6) this means that c_{ij} can be a function of d_{Ai} and d^*_{jk} can be a (nonnegative) function of $d_{Ai} + c_{ij}(d_{Ai})$. Hence Dreyfus (1969) noted that Dijkstra's algorithm can be used to study traffic problems in which the travel time depends on the traffic density, which depends on the time of day and therefore on the previous travel times. Then (1) and

(2) respectively are replaced by

$$d_{A[r+1]} = \min_{i \in S_r} \left[d_{Ai} + \min_{j \in \bar{S}_r} c_{ij}(d_{Ai}) \right] \tag{11}$$

$$d_{[r+1]A} = \min_{j \in S_r} \left[\min_{i \in \bar{S}_r} c_{ij}(d_{jA}) + d_{jA} \right] \tag{12}$$

for $r = 1, \ldots, p-1$.

EXERCISES 9.3

Find the shortest paths by Algorithm I or II. Show your work.

1. Complete the solution begun in Table 9.3A.
2. From node K in Table 9.3A.
3. From node E in Table 9.3A.
4. To node E in Table 9.3A.
5. From node C in Exercise 2 of 9.2.
6. To node G in Exercise 2 of 9.2.
7. From node C or H in Table 9.3D.
8. Show that these algorithms fail for d_{12} in a network having $c_{12} = 1, c_{13} = 2, c_{32} = -2$.

9.4 CRITICAL PATHS IN ACYCLIC SCHEDULING NETWORKS

In the event that the network is oriented and has no cycles, it admits a shortest-path algorithm even simpler than that of the preceding section. For simplicity we assume that the nodes are numbered so that $i<j$ whenever (i, j) is an arc (c_{ij} is defined).

A method for doing this, together with many applications, is described by Fulkerson (1966). Note that if every node had one or more outgoing arcs, then any directed path could be extended until it returned to one of its previous nodes and formed a cycle. Hence at least one node has no outgoing arc and a similar proof (extending the path backward) shows that at least one node has no incoming arc. Then one may begin by assigning the index 1 to one of the latter nodes; after this node and all its arcs are deleted, another node having no incoming arc is given the index 2, and so on until all are indexed.

If the nodes are arranged in this order, the table of the c_{ij} for an acyclic network will be upper triangular, with no c_{ij} defined for $i>j$ (below the diagonal). However, there is no need to rewrite the table for this purpose. In solving exercises as in Table 9.4, letters $A, B, C, \ldots$ will be assigned to the nodes arbitrarily, and then the index numbers will be recorded adjacent to the letters as they are determined. A node having no incoming arc corresponds to a column j in which no c_{ij} is defined (in uncovered rows), and the deletion of such nodes after they are indexed corresponds to the temporary deletion of the corresponding rows, perhaps by covering them with pencils or strips of cardboard.

This indexing assures that the indices of the nodes encountered along a directed path are always increasing. Hence if $d_{hh}=0$ as usual, the other shortest path lengths d_{hj} from a particular node h can be calculated (or shown not to exist) in order of increasing values of $j>h$ by

$$d_{hj} = \min_{h \leq i < j} (d_{hi} + c_{ij}) \tag{1}$$

Only those values of i are used for which d_{hi} has been calculated and c_{ij} is defined; if no such i exists, then no path exists and d_{hj} is undefined. The subpath from h to i must itself be optimal and the indexing of the nodes assures that only i satisfying $h \leq i < j$ have to be considered.

There is no sign restriction on the c_{ij}. However, for the application that will be considered in this section, one has $c_{ij} \geq 0$ provided one seeks the *longest* path and hence replaces the min in (1) by *max*. This convention will be used here and in the first part of Section 9.5.

This section considers the scheduling of the component activities that make up a large project to avoid unnecessary loss of time and money. For example, the establishment of a new factory or shopping center involves such activities as a feasibility study, site selection and zoning approval, design of the facility, financing, signing contracts and hiring personnel for construction and later for operation, construction of needed public utilities, landscaping, furnishing, advertising to attract clientele, etc. Some of these tasks can be performed simultaneously, but others cannot be begun until their prerequisite tasks have been completed.

PERT (Project Evaluation and Review Technique) and CPM (Critical-Path Method) are two of the better known acronyms for procedures relating to such problems. They may take account of the variation, controllable or random, in the times required to perform the various tasks as well as their relation to monetary costs and available resources. However, we consider only what can be said about specified times without regard to other factors.

The temporal structure of a project is represented by a network in which *each directed arc or arrow represents a task* (or part of a task) whose duration is the length c_{ij} of the arc, and *each node represents an event*. An *event* is a time point or date at which a portion of the project arrives at a specified state, such that one can begin the tasks corresponding to the arcs leaving that node. These networks must be *oriented and acyclic* (without cycles), as otherwise one would have the absurdity of an event or a task following itself.

One may wish to allow more than one arc to connect the same ordered pair of nodes (i, j), but the longest such arc is the significant one. Some arcs (often of zero length) may represent necessary temporal relations not associated with any actual task. For example, in Figure 9.4C below the arc having length b_2 cannot be eliminated even if $b_2=0$.

By extending any directed path forward and backward as far as possible, one concludes that there is at least one node, called a *start event*, that has no incoming arcs, and at least one node, called a *terminal* or *end event*, that has no outgoing arcs. If either of these events is not unique, it can be made so by introducing a new node to occupy the role uniquely, and connecting it with the previous candidates by arcs of zero length. (A unique start event is a root as defined in 9.1.) In describing the procedure, but not in solving the example below, we adopt this convention of uniqueness whenever it simplifies the language and describe the resulting mixoconnected network as *birooted*. It has a directed path from start to end via any specified intermediate node.

The practical problem involves tradeoffs between consumption of money and time. Since we are ignoring costs for the sake of simplicity, this reduces to minimizing the time requirements by making the c_{ij} as small as possible and beginning each task as soon as possible. Then it remains (1) to calculate the resulting project duration t^* and (2) to determine which tasks (i, j) deserve the most attention because their durations c_{ij} have the most effect on t^*.

Arcs that occur in series along a directed path represent tasks that must be performed in the corresponding sequence, with their durations adding together; two such sequences (parallel paths) that have no arcs or nodes in common except their endpoints may (and for efficiency usually must) be performed simultaneously. Thus it is clear that, given good management, the project duration t^* is the length of a *longest path* in the network, which is called a *critical path*. It is also clear that the tasks (i, j) on this path are especially important because any increase in their c_{ij} will result in a corresponding increase in t^*; for other tasks some delay (up to a limit called the float) may be harmless.

The methods of 9.3 cannot be used here because the replacement of the maximization by minimization would result in negative arc lengths. To be sure, one can use the fact that $c_{ij} \geq 0$ here to show that an acceptable ordering of the nodes (in the sense that every arc (i, j) has $i<j$) can be found such that $j<k$ implies $s_j \leq s_k$ in (2) below, but there seems to be no incentive to do this.

Let s_i = the longest path length from the start event 1 to event i. Then s_i is the earliest time (measured from $s_1 = 0$) at which event i can occur. Equation (1) is replaced by

$$s_j = \max_{i<j} \{s_i + c_{ij}, 0\} \qquad (j = 1, 2, \ldots, p) \tag{2}$$

which describes the *forward pass*. Only those $i<j$ are used for which c_{ij} is defined; s_i is always defined for $i<j$. The 0 in (2) has the effect of setting $s_i = 0$ for (all) the start event(s). If there is more than one, (2) automatically maximizes the other path lengths over the choice of the start event also. In words, s_j is the largest sum obtained by adding the lengths c_{ij} of the arrows with heads at j to the s_i $(i<j)$ already determined at the tails of the arrows. The $s_i + c_{ij}$ corresponding to nonexistent arcs (in particular for $i \geq j$) are simply omitted; otherwise one could put such $c_{ij} = -\infty$.

At this point one knows the duration of the project, which is $t^* = \max s_i$. Before attempting to find the critical arcs one should calculate the latest event times, t_i, which are the smallest numbers obtainable by diminishing t^* by the longest path lengths to (any or all of) the end or terminal event(s). It is convenient to calculate these differences directly by the *backward pass*

$$t_i = \min_{j>i} \{t_j - c_{ij}, t^*\} \qquad (i=p, p-1, \ldots, 1) \tag{3}$$

The appearance of t^* insures that every end event has $t_i = t^*$ even though some end events may have small indices i. In words, t_i is the smallest of the differences obtained by subtracting the lengths c_{ij} of the arrows with tails at i from the t_j $(j>i)$ already determined at the heads of the arrows.

Although (2) maximizes and (3) minimizes, there is a tendency in either case to use large values of c_{ij}. To preserve the form of (2) one could define $\bar{t}_j = t^* - t_j$ and replace (3) by

$$\bar{t}_i = \max_{j>i} \{c_{ij} + \bar{t}_j, 0\} \qquad (i=p, p-1, \ldots, 1) \tag{4}$$

Now consider any typical arc (k, l) of length $c_{kl} \geq 0$. The associated times may be represented as in Figure 9.4A. They satisfy the inequalities

$$\begin{aligned} s_k + c_{kl} &\leq s_l \qquad \leq t_l \\ s_k + c_{kl} &\leq t_k + c_{kl} \leq t_l \end{aligned} \tag{5}$$

of which the first and the last follow from (2) and (3). All of them relate to the definition of s_l as the longest path length from 1 to l and $t^* - t_l$ as the longest from l to p. For example, since t^* is the longest path length from 1 to p we have $t^* \geq s_l + (t^* - t_l)$ or $s_l \leq t_l$, in agreement with their interpretation as earliest and latest times.

The difference $t_l - s_l \geq 0$ is called the *slack* for event l. If one or more arcs into l (or one or more out of l) were lengthened by this amount while the other c_{ij} were unchanged, the project would not be delayed, because the longest path length through l would not exceed $s_l + (t_l - s_l) + (t^* - t_l) = t^*$.

The difference $t_l - s_k - c_{kl} \geq 0$ is called the (*total*) *float* for the task (k, l). It specifies the amount by which c_{kl} could be increased without delaying the project, provided no other c_{ij} is changed; if $c_{kl} = t_l - s_k$, the length $s_k + c_{kl} + (t^* - t_l)$ of the longest path through (k, l) reduces to t^*. Project managers often list their tasks in the order of increasing values of the float.

Critical tasks are those for which the float $t_l - s_k - c_{kl}$ is zero, which is equivalent to saying that they lie on some critical path. *Critical events* k are those for which the slack $t_k - s_k$ is zero; they also lie on critical paths. The critical tasks in toto need not be confined to a path or even a tree, but *may form any acyclic network* (Exercise 8). The inequalities (5) show that if task (k, l) is critical, so are events k and l, but the converse is not true (see *BF* in Exercise 7).

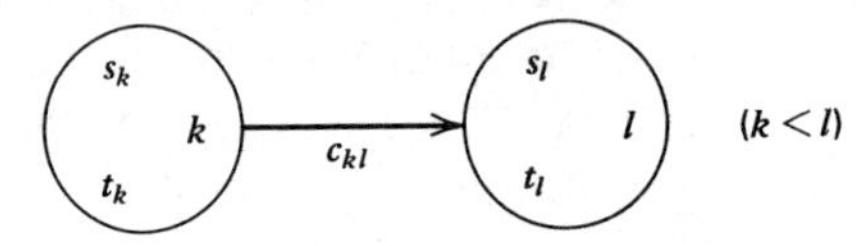

FIGURE 9.4A Scheduling arc and two nodes.

The difference $s_l - s_k - c_{kl} \geq 0$ is called the *free float*. If the other c_{ij} are unchanged, c_{kl} could be increased by this amount without detracting from the floats available for the following tasks, because no s_i is changed and no t_j with $j \geq l$ is changed.

Finally, $\max(s_l - t_k - c_{kl}, 0)$ is called the *independent float* for the task (k, l). If it is positive (5) becomes

$$s_k + c_{kl} \leq t_k + c_{kl} < s_l \leq t_l \tag{6}$$

and it is clear that c_{kl} could be increased by $s_l - t_k - c_{kl}$ without changing any s_i or t_j. Thus if the other c_{ij} are unchanged, one could decide, before the project is begun, to increase c_{kl} by this amount without reducing the floats available for any other task.

Figure 9.4B and Table 9.4 solve an example in which C and G are start events and A and K are end events. See Figure 9.4A for the notation.

Fulkerson (1961) and Kelley (1961) discuss critical-path analysis when the cost of each task is a linear function of the time allotted to it.

Sequencing *n* Jobs Through Several Operations

Bellman formulated the following problem. Suppose that one has n jobs, each of which requires the use of machine A for a_i time units and then (after a delay of at least b_i) machine C for c_i time units ($i = 1, \ldots, n$). Each machine can handle only one job at a time, but the times b_i are allowed to overlap, unless one has either $\min a_i \geq \max b_i$ or $\min c_i \geq \max b_i$, which makes overlapping b_i unnecessary. For three jobs done in the sequence 1, 2, 3, this gives the network in Figure 9.4C. The previous methods are more efficient, but obviously there are only $n = 3$ possible paths, and so the critical path length has the formula

$$t = a_1 + \max(b_1 + c_1 + c_2, a_2 + b_2 + c_2, a_2 + a_3 + b_3) + c_3 \tag{7}$$

The nontrivial part of the problem is to determine a sequence for the jobs that will minimize t. Johnson (1954, 1959) and Mitten (1959) showed that this is accomplished by using the following sequence on both machines. *First start those jobs having* $a_i < c_i$, *in order of increasing values of* $d_i = a_i + b_i$; *then start the remaining jobs, having* $a_i \geq c_i$, *in order of decreasing values of* $d_i = b_i + c_i$. (In general, $d_i = b_i + \min(a_i, c_i)$). Jackson (1956) extended this result to the case in which some jobs require only one machine (these are scheduled in the middle section of that

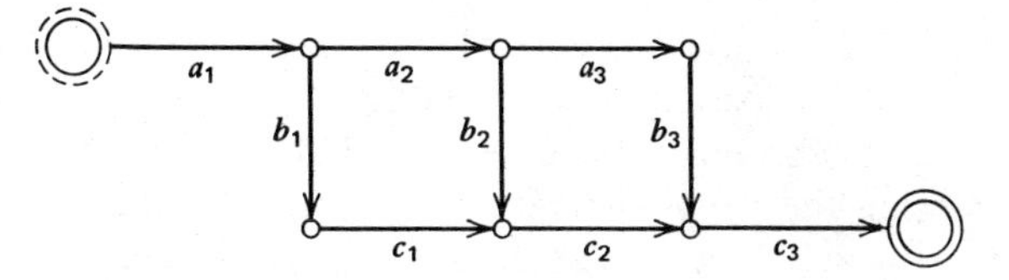

FIGURE 9.4C Three jobs through two operations.

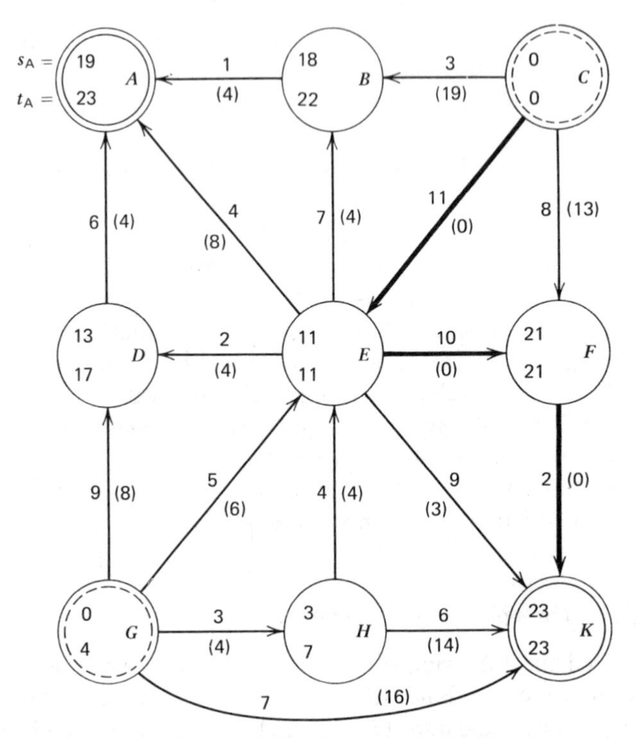

FIGURE 9.4B Critical path CEFK in an acyclic network. Earliest times s_i in order CG, H, E, BDF, AK; then the latest times t_j in the reverse order. Start with $s_C = s_G = 0$, $s_H = 3+0$, $s_E = \max(11+0, 5+0, 4+3) = 11, \ldots, t_A = t_K = t^* = 23$, $t_F = 23-2, \ldots, t_H = \min(11-4, 23-6) = 7, \ldots$. The value of the total float $t_j - s_i - c_{ij}$ is in parentheses; it is zero if and only if the arc is critical.

machine's sequence) and some require that machine C precede A (these are the first jobs done by C and the last done by A).

Scheduling is considered in Chapter 9 of Müller-Merbach (1970) and is the sole topic in Conway, Maxwell, and Miller (1967), O'Brien (1969), Baker (1974) and Coffman (1976).

EXERCISES 9.4

1. A man is to be dressed as fast as possible, with undershirt, shorts, shirt, tie, trousers, jacket, socks, boots (too large to pass through the trouser legs), and wallet, which must go in the trousers after they are put on. Draw a network, assuming that five aides can work simultaneously and the man either stands on two feet or sits as he is dressed.

2. A couple must move as quickly as possible from city I to a nearby city II. Both persons are needed for loading and unloading the truck and agreeing on a new apartment. Other tasks, which either can do alone, are packing, cleaning both apartments, renting the truck and returning it to I, driving the truck and their own car, and viewing prospective apartments. Draw a plausible network, leaving such things as meals and sleep to be improvised.

TABLE 9.4. Critical Path CEFK in an Acyclic Network

i	j	8 A	5 B	1 C	6 D	4 E	7 F	2 G	3 H	9 K	Earliest s_i
8	A	—									19
5	B	1	—						(c_{ij})		18
1	C		3	—		11	8				0
6	D	6			—						13
4	E	4	7		2	—	10			9	11
7	F						—			2	21
2	G				9	5		—	3	7	0
3	H					4			—	6	3
9	K									—	23
Latest	t_j	23	22	0	17	11	21	4	7	23	

1. Index the nodes $C=1$, $G=2$, etc. as at the beginning of 9.4.
2. Let $s_1 = s_C = 0$. Let $t^* = \max s_i$ after using

$$s_j = \max_{i<j} \{s_i + c_{ij}, 0\} \qquad \text{for} \qquad j=2,3,\ldots, p. \tag{2}$$

For example, $s_8 = s_A = \max(1+18, 6+13, 4+11, 0) = 19$.

3. Let $t_p = t_K = t^* = 23$ and use

$$t_i = \min_{j>i} \{t_j - c_{ij}, t^*\} \qquad \text{for} \qquad i=p-1,\ldots,2,1. \tag{3}$$

Thus $t_2 = t_G = \min(17-9, 11-5, 7-3, 23-7, 23) = 4$.

4. The set Z of critical nodes is $\{i\colon s_i = t_i\} = \{C, E, F, K\}$.
5. Critical arcs (i, j) have $i \in z$, $j \in z$, $t_j = s_i + c_{ij}$. CE, EF, FK are critical; thus the critical path is $CEFK$.

Find the s_i and t_j in 3 to 7.

	AB	AC	AD	BC	BD	BE	CD	CE	DE
3. $c_{ij}=$	1	3	3	1	4	3	2	5	1
4. $c_{ij}=$	4	5	3	2	5	4	1	2	3

5.

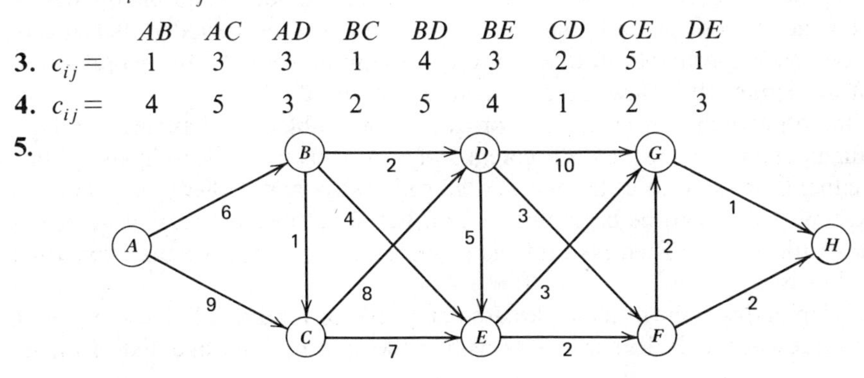

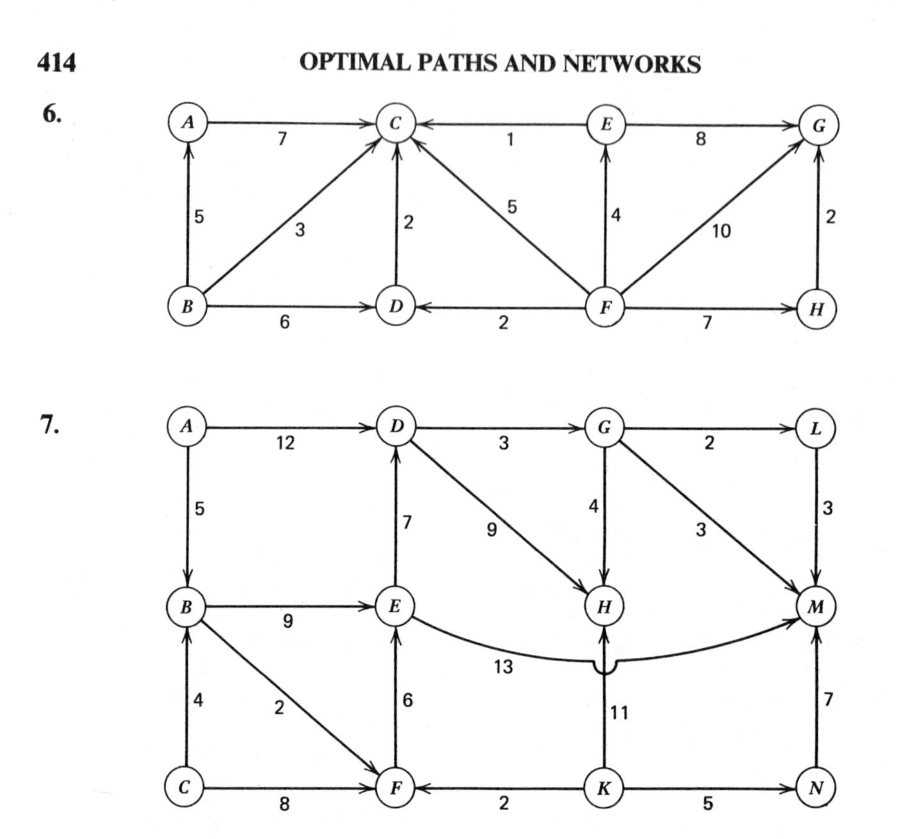

8. Show that with suitable values for the c_{ij} in any acyclic network, every one of its arcs can be critical.

9.5 THE WAREHOUSE, REPLACEMENT, ASSORTMENT

(a) The *warehouse problem* was proposed by Cahn (1948). A dealer assumes that he knows for $r=1,\ldots,n$ the price b_r at which he can buy one unit of a commodity at the beginning of time period r and store it until the beginning of time period $r+1$; the cost c_r merely to carry (store) one unit during time period r; and the price d_r at which he can dispose of (sell) one unit at the end of time period r. Because of restrictions on his available storage space or capital, he cannot store more than a units of the commodity. How can he maximize his profit?

We first observe that this is a linear programming problem, and as such will have its optimum at an extreme point (Theorem 9 of 3.3). Thus the dealer will always buy or sell either 0 or as much as he can; and he will always have either 0 or a units on hand, except perhaps at the beginning or end. Let us assume at first that he begins and ends with 0 units. Then his profit is proportional to a, and we may take $a=1$ without loss of generality.

After eliminating some of the nodes used on p. 140 of Ford and Fulkerson (1962), one can represent the dealer's transactions by flows in a network like that of Figure

9.5A for $n=3$:

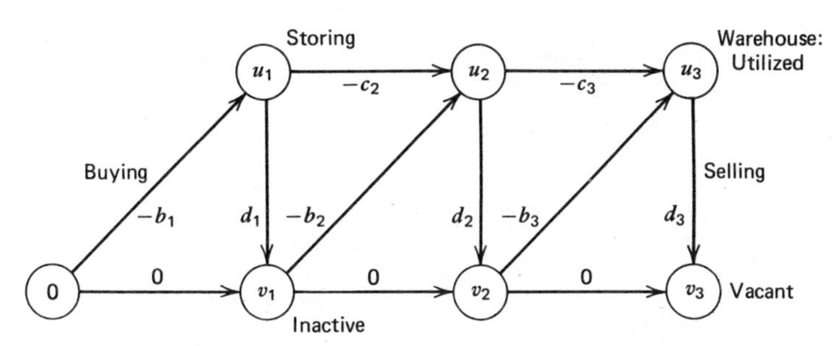

FIGURE 9.5A Network for the warehouse problem.

This is an acyclic network like those of 9.4, but now both positive (d_r) and negative ($-b_r, -c_r$) arc lengths occur. The nodes are labeled with their longest path lengths u_r, v_r from the starting node 0; the end node is the one labeled $v_3 = v_n$.

Each node defines a particular state for the dealer's business. Nodes in the lower row, labeled 0 or v_r, correspond to 0 units on hand, while nodes in the upper row, labeled u_r, correspond to $a=1$ unit on hand. The two arcs leaving each node ($r=0,1,2$) correspond to the alternatives the dealer has to choose between in the corresponding state, and the label on the arc is the increment of revenue the dealer receives when he chooses that alternative. Arcs labeled 0 correspond to "doing nothing," which is a legitimate alternative.

It is now apparent that there is a one-to-one correspondence between the possible histories of the dealer's activities in the three time periods and the paths from node 0 to node v_3 that are consistent with the arrows. Furthermore, the length of such a path is equal to the dealer's profit (for $a=1$). As in 9.4 the problem is simply to find the longest path. In the present problem (2) of 9.4 takes the form

$$u_r = \max(u_{r-1} - c_r, v_{r-1} - b_r)$$

$$v_r = \max(v_{r-1}, d_r + u_r) \tag{1}$$

for $r=1,\dots,n$; one begins with $u_0 = -\infty$, $v_0 = 0$, $u_1 = -b_1$. Table 9.5 applies these formulas to an example, both in the network in Figure 9.5B and in tabular form.

If the warehouse contains a fraction θ_0 of its capacity initially and a fraction θ_n at the end, it is only necessary to change the lengths of four arcs, thus:

Arc	$(0, u_1)$	$(0, v_1)$	(v_{n-1}, v_n)	(u_n, v_n)	
Old length	$-b_1$	0	0	d_n	(2)
New length	$-b_1(1-\theta_0) - c_1\theta_0$	$d_0\theta_0$	$-b_n\theta_n$	$d_n(1-\theta_n)$	

Equation (1) still holds as before except that

$$u_1 = -b_1(1-\theta_0) - c_1\theta_0, \qquad v_1 = \max(d_0\theta_0, u_1 + d_1)$$

$$v_n = \max\left[v_{r-1} - b_n\theta_n, \quad u_n + d_n(1-\theta_n)\right] \tag{3}$$

TABLE 9.5 The Warehouse Problem

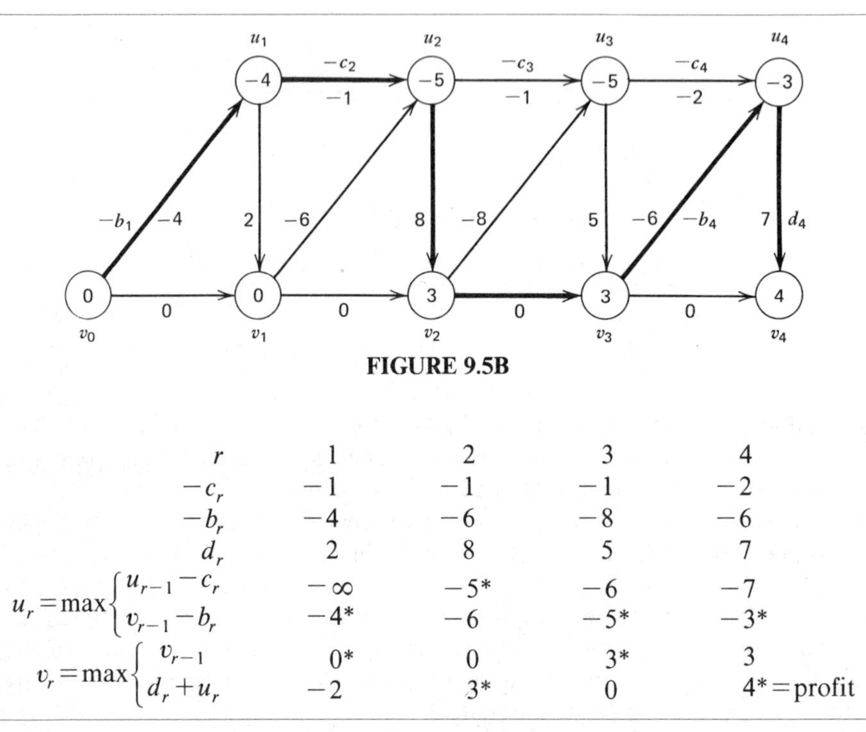

FIGURE 9.5B

		1	2	3	4
	r	1	2	3	4
	$-c_r$	-1	-1	-1	-2
	$-b_r$	-4	-6	-8	-6
	d_r	2	8	5	7
$u_r=\max$	$u_{r-1}-c_r$	$-\infty$	-5^*	-6	-7
	$v_{r-1}-b_r$	-4^*	-6	-5^*	-3^*
$v_r=\max$	v_{r-1}	0^*	0	3^*	3
	d_r+u_r	-2	3^*	0	$4^*=\text{profit}$

The stars identify the values of u_r and v_r (the larger of the two entries), calculated in the order $u_1, v_1, \ldots, u_4, v_4$. Then the longest path is traced in reverse from $v_4=4^*$ to $v_0=0$ by noting the positions of the stars. Since $4^*=d_4+u_4$, the preceding arc length is d_4 and the node is u_4; $u_4=-3^*=v_3-b_4$ supplies a length $-b_4$ and a node v_3; $v_3=3^*=v_2$ supplies a length 0 and a node v_2, etc. The optimal decisions are put on record by writing $v_4=4=-b_1-c_2+d_2+0-b_4+d_4$.

If one wanted to use θ_0 as a variable parameter, one would use a backward pass as in (4) of 9.4, beginning with $v_4=0$. (See Exercise 7.)

The problem becomes even simpler if the quantity to be sold in each period is specified, because then the arcs of length d_r do not affect the decisions. Then all $v_r=0$ and one wishes to minimize the cost $\bar{u}_r$ of supplying one unit of the commodity at the end of period r. Equation (1) reduces to

$$\bar{u}_r=\min(\bar{u}_{r-1}+c_r, b_r) \tag{4}$$

for $r=2,\ldots,n$, with $\bar{u}_1=b_1$. The profit is $d_r-\bar{u}_r$. The network is essentially like that of Figure 9.4C with all $a_i=0$ and maximization replaced by minimization.

(b) *Equipment replacement*. The owner of a piece of equipment, such as an automobile or a production machine, must decide from time to time whether to continue to operate it despite its increasing inefficiency or to accept the cost of replacing it by a new machine. Suppose first that the age i of the machine is the only significant variable; it is measured in terms of the constant time interval between successive decisions as the unit. For $i \geq 1$ let $c_{i-1,i}$ be the cost of maintaining the machine from age $i-1$ to age i, and let c_{i0} be the cost of replacing a machine of age i by a new machine; the other c_{ij} are undefined. In practice some finite number of nodes will suffice (one more than the greatest age considered).

This directed graph contains one cycle having each possible number of arcs $2, 3, \ldots$; the cycle of $m+1$ arcs has the cost $c_{01} + \cdots + c_{m-1,m} + c_{m0}$. One should minimize the average cost per time period:

$$\min_{m} \left(c_{01} + c_{12} + \cdots + c_{m-1,m} + c_{m0} \right) / m \tag{5}$$

The replacement itself is represented by an arc but it does not occupy a time period.

Because money can earn interest, (5) should be modified when substantial periods of time are involved. If the (compound) interest rate is denoted by $(\lambda^{-1} - 1) \cdot 100$ percent per period, $\lambda^m c_{m0}$ dollars in hand now will become c_{m0} dollars in m periods and so suffice to cover the cost of the next replacement, for example. Division by m is no longer exactly correct, the periods no longer being quite equivalent. Instead one should minimize the total future discounted cost z_m. If no particular limit is set on the duration of the process this is (Bellman and Dreyfus, 1962)

$$\begin{aligned} z_m &= \sum_{r=0}^{\infty} \lambda^{mr} \left(c_{01} + \lambda c_{12} + \cdots + \lambda^{m-1} c_{m-1,m} + \lambda^m c_{m0} \right) \\ &= c_{01} + \lambda c_{12} + \cdots + \lambda^{m-1} c_{m-1,m} + \lambda^m \left(c_{m0} + z_m \right) \end{aligned}$$

Solving for z_m shows that m should minimize

$$z_m = \left(c_{01} + \lambda c_{12} + \cdots + \lambda^{m-1} c_{m-1,m} + \lambda^m c_{m0} \right) / (1 - \lambda^m) \tag{6}$$

In practice the costs are likely to depend on the calendar time as well as the age of the existing machine. Then one can define c_{ij} $(i<j)$ to be the cost of maintaining a machine from its acquisition at time i until j and then replacing it at time j.

(c) *Assortment*. In this problem, treated by White (1969), a manufacturer has to decide in how many sizes (or quality levels) he will offer his product. If this parameter is higher than customers require, the product will be acceptable but the cost of manufacturing and stocking many items may become excessive. Examples are the size of a briefcase, automobile or house; the quantity of serum in a one-shot package; the purity of a chemical; the maximum temperature a material can withstand, etc.

Let 0 be an initial node and let the nodes $1, 2, \ldots, n$ correspond to the sizes that are demanded in increasing order. Let c_{ij} be the total cost of using the size j to service all the demands h satisfying $i < h \leq j$. One may assume $c_{ij} = \infty$ if $i > j$, so that the network is acyclic. Then the shortest path from 0 to n solves the assortment problem; its nodes (other than 0) correspond to the sizes that should be produced and its length d_{0n} is the minimum cost. Since c_{ij} and d_{0j} are presumably nondecreasing functions of j, the algorithms of 9.3 and 9.4 (1) are identical in this case; $[r] = r - 1$ in 9.3.

EXERCISES 9.5

Find the maximum profit for each of the warehouse problems 1 to 3, beginning and ending with an empty warehouse, (a) without and (b) with the assumption that a unit amount of the commodity is to be sold in each time period.

1.

c_r	1	2	0	1
b_r	2	3	4	2
d_r	7	2	5	6

2.

1	0	1	2	0
4	3	6	7	4
6	10	5	9	2

3.

c_r	1	2	0	2	1	0
b_r	10	6	12	14	10	6
d_r	13	8	9	19	11	5

In 4 to 5 use the following costs for equipment maintenance and replacement. When should the machine be replaced? ($i=1$ to 8)

$c_{i-1,i}$	4	5	7	10	13	16	20	25
c_{i0}	25	40	50	59	66	70	73	75

4. Without discounting.
5. With $\lambda=.9$.
6. What peculiarities of Table 9.5 may require special attention?
7. What do (1) and (3) become if a backward pass is used as in (4) of 9.4, beginning with $\bar{v}_n=0$?

9.6 SHORTEST PATHS FROM ONE NODE (GENERAL)

The algorithm of this section requires more computation than those of 9.3 and 9.4 but the c_{ij} are not sign restricted and cycles are permitted as long as their lengths are nonnegative (when minimizing). This slight restriction excludes any attempt to improve a path by including a cycle within it. The algorithm is usually attributed to Moore (1957) but it was also proposed by Ford (1956) and Bellman (1958), and studied by Bellman, Cooke and Lockett (1970).

Let $d_{AA}=0$ as usual. Then if shortest path lengths d_{Aj} from node A to node j exist, they satisfy

$$d_{Aj}=\min_{i\neq j}\left(d_{Ai}+c_{ij}\right) \qquad \text{for } j\neq A \tag{1}$$

because j does have some preceding node i and the path from A to i must itself be optimal. Equation (1) differs from (1) of 9.4 only in having more values of i to be considered and hence requiring the use of successive approximations for its solution.

If a better approximation is not available we may begin with

$$d_{AA}^1=0,\ d_{Aj}^1=c_{Aj} \qquad (=+\infty \text{ if arc } (A,j) \text{ is absent and } j\neq A) \tag{2}$$

and proceed with iterative substitutions as in the theorem below. First we define a *negative cycle* as a cycle of $m \geq 2$ nodes and arcs such that

$$\begin{aligned} &c_{ij}+c_{ji}<0 && (m=2), \\ &c_{ij}+c_{jk}+c_{ki}<0 && (m=3), \text{ etc.} \end{aligned} \tag{3}$$

THEOREM Let the c_{ij} for $i \neq j$ be arbitrary real numbers or $+\infty$ and define all $c_{jj}=0$. Let $d_{Aj}^1=c_{Aj}$ and

$$d_{Aj}^{r+1}=\min_{\text{all } i}\left(d_{Ai}^r+c_{ij}\right) \tag{4}$$

for all nodes $j=1,\ldots,p$ and for $r=1,\ldots,p-1$ or less. Then:

(i) If there is no negative cycle, (4) converges to the shortest path lengths d_{Aj} in at most $p-2$ iterations ($d_{Aj}^{r+1}=d_{Aj}^r=d_{Aj}$ for some $r \leq p-1$ and all j).
(ii) If there is a negative cycle, (4) will never converge.
(iii) There is a negative cycle if and only if $d_{Aj}^p \neq d_{Aj}^{p-1}$ for some j.

Proof Note that (2) and (4) insure that d_{Aj}^r is the length of the shortest generalized path from A to j containing exactly r generalized arcs. The word *generalized* refers to

TABLE 9.6A Shortest Paths from *A*
(The table is transposed for typographical convenience.)

j \ i	*A*	*B*	*C*	*D*	*E*	*F*	*G*	*H*	*K*	$=i$
A	(0)		15							
B	9	(0)	13	2	10					
C	7	16	(0)	8		3				
D	6	4	7	(0)			7			
E		11		12	(0)			4		c_{ij}
F			1	14	1	(0)	2		11	
G				5	4	13	(0)	5	7	
H					6		10	(0)	3	
K						9	8	12	(0)	
	+	+	+	+	+	+	+	+	+	
$d_{Ai}^1=$	0	9	7	6						$=c_{Ai}$
$d_{Ai}^2=$	0	8	7	6	18	8	11	(omit if (b) is used)		
$d_{Ai}^3=$	0	8	7	6	18	8	11	21	17	
$d_{Ai}^4=$	0	8	7	6	18	8	11	20	17	$=d_{Ai}$
Arc		*DB*	*AC*	*AD*	*DE*	*CF*	*DG*	*KH*	*FK*	

No further changes. Tree is $B\,\overset{E}{\underset{G}{D}}\,A'CFKH$ (root at A).

Method of calculation: $d_{Aj}^{r+1}=\min_i(c_{ij}+d_{Ai}^r)$ (4)

$$d_{AH}^4=\min(6+18,10+11,0+21,3+17)=20$$

$=c_{KH}+d_{AK}$. Hence arc KH is in the shortest path.

the fact that (4) allows any given node to occur repeatedly; the resulting "path" may differ from the ordinary path (without repeated nodes) by containing one or more "arcs" (i, i) or cycles.

Arcs (i, i) and cycles of zero length will not delay the convergence and can be eliminated if they occur in a shortest path. Cycles of positive length cannot occur because their elimination would produce a shorter path. Thus in the absence of negative cycles, d_{Aj}^r is the shortest path length from A to j containing *at most r* arcs and no repeated nodes (if $j \neq A$). Since $p-1$ is the maximum number of arcs in a path, the truth of (i) is clear. Statement (ii) is true because repeated traverses of a negative cycle will cause all d_{Aj}^r to approach $-\infty$ as r increases. (A special rule banning such repetition would produce a more difficult problem.) Statement (iii) follows from (i) and (ii).

Instead of defining $c_{jj}=0$, one can define $d_{AA}^r=0$ and $i \neq j$ in (4), provided there are no negative cycles.

The tables display two examples of the algorithm corresponding to parts (i) and (iii) of the theorem respectively. A value of $i \neq j$ that yields the minimum in (4) determines the next to the last node in the shortest path. This is recorded only for the final $r=4$ in Table 9.6A but (because of its possible usefulness at $r=4$) for every $r=1,\ldots,7$ in Table 9.6B.

TABLE 9.6B Shortest Paths from A Not Found
(because EGF is a negative cycle: $-2+1+0<0$)

	A	B	C	D	E	F	G	$=i$
A	(0)	2	1	2				
B	1	(0)	2	−1		3		
C	4	3	(0)	5	6			
j D	3	4	3	(0)	1	4	3	c_{ij}
E			2	2	(0)	0	3	
F		5		2	1	(0)	1	
G				−1	−2	2	(0)	
	+	+	+	+	+	+	+	
$d_{Ai}^1=$	0	$A1$	$A4$	$A3$				$=c_{Ai}$
$d_{Ai}^2=$	0	$A1$	$A4$	$A3$	$D5$	$D5$	$D2$	
$d_{Ai}^3=$	0	$A1$	$A4$	$A3$	$D5$	$G3$	$D2$	
$d_{Ai}^4=$	0	$A1$	$A4$	$A3$	$F3$	$G3$	$E1$	((b) used)
$d_{Ai}^5=$	0	$A1$	$A4$	$A3$	$F3$	$G2$	$E1$	
$d_{Ai}^6=$	0	$A1$	$A4$	$A3$	$F2$	$G2$	$E1$	
$d_{Ai}^7=$	0	$A1$	$A4$	$A3$	$F2$	$G2$	$E0$	

For example, $d_{AF}^2=\min(5+1,2+3)=5=c_{FD}+d_{AD}^1$. Stop after d_{Ai}^7 ($r=7=$number of nodes). One could stop after d_{Ai}^4 because its predecessor nodes F, G, E in columns E, F, G include no indices outside the set of column indices $\{E, F, G\}$; this indicates a negative cycle.

Short Cuts

This algorithm is rather laborious when it is necessary to carry out all $p-2$ iterations, each involving as many as $p(p-1)$ comparisons to determine the minima in (4). Fortunately, the situation is often much better than this, for the following reasons.

(a) Nonexistent arcs (having $c_{ij}=+\infty$) may be omitted in (4).

(b) One may replace d^r_{Ai} by d^{r+1}_{Ai} in (4) for those values of i for which d^{r+1}_{Ai} is already known. By allowing more freedom in the number of arcs to be used, this may permit d^r_{Aj} to decrease to its limit more rapidly. It may be helpful to index the nodes more or less in the order of increasing values of d_{Aj}, if there is any a priori information about this. In the special cases in 9.4 and 9.5, (a) and (b) were used so effectively that the final values of the d_{Aj} were found immediately in a single iteration.

(c) Let S_r be the set of j for which $d^r_{Aj}<d^{r-1}_{Aj}$; then its complement $\bar{S}_r$ is the set of j for which $d^r_{Aj}=d^{r-1}_{Aj}$. By (4), d^{r+1}_{Aj} is the smaller of

$$\min_{i\in S_r}\left(d^r_{Ai}+c_{ij}\right) \quad \text{and} \quad \min_{i\in \bar{S}_r}\left(d^r_{Ai}+c_{ij}\right)=\min_{i\in \bar{S}_r}\left(d^{r-1}_{Ai}+c_{ij}\right)$$

Since the latter is $\geq d^r_{Aj}\geq d^{r+1}_{Aj}$, we conclude that

$$d^{r+1}_{Aj}=\min\left[d^r_{Aj}, \min_{i\in S_r}\left(d^r_{Ai}+c_{ij}\right)\right] \tag{5}$$

(Only those $d^r_{Ai}<d^{r-1}_{Ai}$ can make $d^{r+1}_{Aj}<d^r_{Aj}$.) In the absence of negative cycles the number of elements in S_r cannot exceed $p-r+1$ (Exercise 9), and so (5) reduces the number of operations by about 50 percent or more. In general S_{r+1} need not be a subset of S_r, but it is (for $r=2,\ldots,p-2$) when the d_{Aj} can be calculated in a single iteration as in 9.4.

(d) Dreyfus (1969) may be consulted for still more efficient but more complicated methods due to J. Y. Yen and to Dantzig, Blattner and Rao.

For the effect of negative cycles see 5.4 of Müller-Merbach (1970), for example.

EXERCISES 9.6

Apply the method of 9.6 to the table of the c_{ij}. Transposition is optional. Show your work.

	A	B	C	D	E	F	G	$=j$
A	(0)	3	-1	2				
B	4	(0)	1	5		-2		
C	2	3	(0)	1	6			
$i=D$	1	2	4	(0)	3	5	-1	
E			1	6	(0)	2	3	
F		4		2	5	(0)	3	
G				3	1	4	(0)	

1. Find the shortest paths from A.
2. Find the shortest paths to G.
3–7. Resolve the corresponding exercises in 9.3.
8. Find the shortest paths from A in the foregoing table, altered so that $c_{FA}=-1$, $c_{DG}=2$.
9. In the absence of negative cycles, let n_r denote the number of nodes other than A having shortest paths from A consisting of at most r arcs. Show that $n_r \geq r$ for $r=1,\ldots,p-1$.

9.7 ALL THE SHORTEST PATHS

This algorithm was given by Floyd (1962); see also Warshall (1962) and Murchland (1965). It is independent of 9.6 but uses the same minimal assumptions that all $c_{ii}=0$ and there are no negative cycles ((3) of 9.6). Otherwise the arc lengths c_{ij} are arbitrary real numbers or $+\infty$. To find all the shortest path lengths d_{ij} for a network of p nodes requires p^3 comparisons of two numbers, thus:

$$d_{ij}^{r}=\min\left(d_{ij}^{r-1}, d_{ir}^{r-1}+d_{rj}^{r-1}\right) \tag{1}$$

for $i, j, r=1,\ldots,p$. The calculation begins with the $d_{ij}^{0}=c_{ij}$ and ends with the $d_{ij}^{p}=d_{ij}$. If $c_{ij}=c_{ji}$ then $d_{ij}^{r}=d_{ji}^{r}$, and few more than $p^3/2$ comparisons are required.

To prove that the d_{ij}^{p} from (1) are indeed the shortest path lengths d_{ij}, we use induction on r to show that for every $r=0,\ldots,p$, the d_{ij}^{r} are the shortest path lengths in which only the nodes $1,\ldots,r$ occur as intermediate nodes. Certainly this is true for $r=0$, since $d_{ij}^{0}=c_{ij}$ is the length of a single arc from i to j without any intermediate nodes. Now (1) simply chooses the better of the two alternatives that exist when node r is made available in addition to nodes $1,\ldots,r-1$. If the node r is not used (in effect this is always true if i or $j=r$ since all $d_{ii}^{r}=0$), the minimum length remains d_{ij}^{r-1} as before; if node r is used once, it breaks the path into two parts whose minimum lengths are d_{ir}^{r-1} and d_{rj}^{r-1}.

At the stage r there is no opportunity for node r to be repeated. However, to prevent one of the nodes $1,\ldots,r-1$ from occurring on both sides of r we must appeal to the absence of negative cycles to permit the elimination of the loop thus formed. This insures that $d_{ir}^{r-1}+d_{ri}^{r-1}\geq 0$ and hence all $d_{ii}^{r}=0$. If negative cycles are present, they will be revealed by some $d_{ii}^{r}<0$, but then the d_{ij}^{p} are not shortest path lengths in general.

Table 9.7 gives an example of the algorithm. Another algorithm that is equally efficient but involves more complicated formulas has been developed by G. B. Dantzig; it is described in Dreyfus (1969).

The straightforward application of (1) is for the construction of a mileage table; this also permits the reduction of a transshipment problem to a transportation problem as in 11.5(c).

Arbitrage in foreign currencies is a curious application of this problem. Let γ_{ij} = the number of units of currency i required to purchase one unit of currency j. If

TABLE 9.7 All the Shortest Paths

$i \backslash j$	$A=1$	$B=2$	$C=3$	$D=4$	$E=5$	$F=6$	$G=7=j$
A	0*	7 $2C$	-1	5	∞ $3C$	∞ $15B$ $10CB$ $9D$	∞ $7CE$
B	4 $2D$	0*	2 $1DA$	3	∞ $6C$ $2D$	8 $7D$	∞ $6DE$
C	9 $7B$ $5BD$	3	0*	7 $6B$	4	∞ $11B$ $10BD$	∞ $11BD$ $8E$
D	-1	2 $1AC$	4 $-2A$	0*	-1	4	5 $3E$
E	∞ $14C$ $1D$	∞ $10C$ $3DAC$	7 $0DA$	2	0*	9 $6D$	4
F	∞ $9B$ $5D$ $4ED$	5	∞ $7B$ $4DA$ $3EDA$	6 $5E$	3	0*	7
$G=i$	∞ $3D$	∞ $5DAC$	∞ $2DA$	4	7 $3D$	2	0*

The first entry in each cell is the arc length $c_{ij}=d_{ij}^0$. The last (smallest) entry is the shortest path length $d_{ij}=d_{ij}^7$ or d_{ij}^G; the letters beside it specify the intermediate nodes. Let i, $j \neq r$ in $d_{ij}^r = \min(d_{ij}^{r-1}+0, d_{ir}^{r-1}+d_{rj}^{r-1})$ = the smaller sum of entries on the two diagonals of the rectangle two of whose corners are d_{ij}^{r-1} and $d_{rr}^{r-1}=0$. As each $r=A, B, \ldots, G$ is processed, $c_{rr}=d_{rr}^{r-1}=0$ is starred. Superceded entries may be deleted. For example, $d_{CA}=5$ with path $CBDA$.

the exchange is made via currency r rather than directly, the corresponding factor is $\gamma_{ir}\gamma_{rj}$, which might be smaller than γ_{ij}. For a complete optimization one could use an analogue of (1), namely

$$\delta^r_{ij} = \min\left(\delta^{r-1}_{ij}, \delta^{r-1}_{ir}\delta^{r-1}_{rj}\right) \tag{2}$$

with $\delta^0_{ij} = \gamma_{ij}$ and $\delta^p_{ij} = \delta_{ij}$, or else apply (1) itself to $c_{ij} = \log \gamma_{ij}$ to obtain $d_{ij} = \log \delta_{ij}$. Here $c_{ij} + c_{ji}$ would be positive but close to zero. A longer cycle might possibly have a negative length (also close to zero), which would permit arbitragers to make a profit, but their activities would cause the γ_{ij} to change and eliminate the negative cycle.

EXERCISES 9.7

Use (1) as in Table 9.7 to find all the shortest paths, or a negative cycle, for the given values of c_{ij}.

1.

↗	A	B	C	D	E
A	0	4	9	7	2
B	3	0	2	5	1
C	1	8	0	2	4
D	9	4	2	0	5
E	6	7	1	3	0

2.

↗	A	B	C	D	E
A	0	−1	3	0	1
B	2	0	1	5	4
C	5	1	0	2	−1
D	3	2	5	0	6
E	1	4	2	3	0

3.

A	B	C	D	E
0	3	4	1	2
1	0	2	3	1
2	1	0	4	−1
1	−3	3	0	3
3	1	2	4	0

4.

↗	A	B	C	D	E	F	G
A	0	3	−1	2	∞	∞	∞
B	4	0	1	5	∞	−2	∞
C	2	3	0	1	6	∞	∞
D	1	2	4	0	3	5	−1
E	∞	∞	1	6	0	2	3
F	∞	4	∞	2	5	0	3
G	∞	∞	∞	3	1	4	0

9.8 MAXIMUM-CAPACITY PATHS

In this section we introduce the parameter e_{ij} to represent the extreme (maximum) amount of something (material, electricity, energy, people, etc.) that arc (i, j) can transport from i to j in a specified period of time. In other words the variable x_{ij} representing the flow through the directed arc is subject to the restriction $0 \leq x_{ij} \leq e_{ij}$ while the algebraic value of the flow lies between $-e_{ji}$ and e_{ij}. Then for some i and j it is natural to wish to calculate e^*_{ij}, the largest capacity of any one path from i to j, and f_{ij}, the maximum flow from i to j when all possible paths are utilized simultaneously. Such problems were treated by Ford and Fulkerson (1956, 1962) and Pollack (1960). Only the e^*_{ij} are considered here.

All the algorithms in this chapter that derive shortest path lengths d_{ij} from the arc lengths c_{ij} can be converted into algorithms for calculating the e^*_{ij} from the e_{ij}. The following changes have to be made.

(a) The capacity of a path is the *minimum* of the capacities of its arcs, whereas the lengths are *added*. (See 9.3(b).)

(b) The capacity is to be maximized over alternate paths, whereas the length is minimized in 9.3, 9.6, 9.7, and (1) of 9.4. Hence the $e^*_{A[r]}$ in the analogue of 9.3 are calculated in decreasing order of magnitude.

(c) All $e_{ii}=e^*_{ii}=+\infty$ (whereas $c_{ii}=d_{ii}=0$) and for (other) nonexistent arcs $e_{ij}=0$ (whereas $c_{ij}=+\infty$). Put $e_{ij}=+\infty$ when arc (i, j) exists and has no capacity restriction.

(d) Although the interpretation requires $e_{ij}\geq 0$, this is not mathematically necessary in 9.3 or anywhere else; thus there is no reason to use 9.6. Neither is there any analogue for the absence of negative cycles in 9.6 and 9.7.

(e) The full capacities of paths that share arcs of finite capacity will often not be simultaneously realizable because the *sums* of the flows must not exceed the capacities of the common arcs. No such interference exists between shortest paths of infinite capacity.

(a) usually results in different paths even in the absence of (b). For example, let $e_{ij}=c_{ij}$ and $c_{12}=c_{23}=2$, $c_{13}=3$, the other $c_{ij}=0$. Then the longest path from 1 to 3 has length 2+2 (via node 2) but $e^*_{13}=3$ (via a single arc). This contrasts with the case of the longest-spanning tree in (5) of 9.2.

The formulas corresponding to (1) and (3) of 9.3, (1) of 9.4, and (1) of 9.7 are

$$e^*_{A[r+1]}=\max_{i\in S_r}\min\left[e^*_{Ai},\ \max_{j\in \bar{S}_r} e_{ij}\right]\leq e^*_{A[r]} \tag{1}$$

where $r=1,\ldots, p-1$ and $e^*_{AA}=+\infty$;

$$e^*_{A[r+1]}=\max_{j\in \bar{S}_r} e^r_{Aj} \qquad (r=1,\ldots, p-1) \tag{2}$$

with $e^1_{Aj}=e_{Aj}$ and $e^r_{Aj}=\max[e^{r-1}_{Aj},\min(e^*_{A[r]}, e_{[r]j})]$;

$$e^*_{hj}=\max_{h\leq i<j}\min\left(e^*_{hi}, e_{ij}\right) \text{ (nodes suitably indexed);} \tag{3}$$

$$e^r_{ij}=\max\left[e^{r-1}_{ij},\min\left(e^{r-1}_{ir}, e^{r-1}_{rj}\right)\right] \tag{4}$$

where $i, j, r=1,\ldots, p$; $e^0_{ij}\triangleq e_{ij}$; $e^p_{ij}=e^*_{ij}$.

Note that e_{Aj}^r has different values in (2) and (4) because the intermediate nodes are chosen from $S_r - \{A\}$ and $\{1,\ldots,r\}$, respectively. In solving exercises the same format may be used as for the corresponding shortest path algorithm.

The proof of the validity of (1) and (2) is completely analogous to that of 9.3. Evidently $e_{AA}^* = +\infty$ and $e_{A[2]}^* = \max_{j\neq A} e_{Aj}$ are the two largest possible capacities from node A. For an inductive proof we assume that

$$+\infty = e_{AA}^* \geq e_{A[2]}^* \geq \cdots \geq e_{A[r]}^* \tag{5}$$

are the r largest of the e_{Ai}^*. In a maximum-capacity path from A to any node k in $\bar{S}_r$, let i be the last node of S_r that is encountered and let it be followed by j in $\bar{S}_r$. Then by (a) above

$$e_{Ak}^* = \min(e_{Ai}^*, e_{ij}, e_{jk}^*) \leq \min(e_{Ai}^*, e_{ij}) \leq e_{A[r+1]}^* \tag{6}$$

the second inequality following from (1). Since (1) shows that $e_{A[r+1]}^*$ is the capacity of some path from A to a node of $\bar{S}_r$, (6) shows that it is the largest of all such capacities. Also $e_{A[r]}^* \geq e_{A[r+1]}^*$ by the inductive assumption that (5) are the r largest. This completes the proof of (1); (2) is derived from it as in 9.3. The proofs for (3) and (4) are also analogous to those indicated for shortest paths in 9.4 and 9.7.

By interchanging the operations of maximizing and minimizing (or replacing e_{ij} by $e_0 - e_{ij}$), one can apply the foregoing methods to a problem mentioned by Fulkerson (1966)—to find a path that minimizes the maximum altitude (or any other quantity) encountered along the path.

EXERCISES 9.8

1–3. Set $e_{ii} = +\infty$ and $e_{ij} = c_{ij}$ for $i \neq j$ in the corresponding exercises of 9.7. Determine:

(a) The maximum-capacity paths from A by (1).

(b) All the maximum-capacity paths by (4).

4. In Exercise 7 of 9.4, find the maximum-capacity paths from A, C or K (i.e., from $\{A, C, K\}$).

5. In Exercise 5 of 9.4, find the maximum-capacity paths from A.

REFERENCES (Chapter Nine)

Arrow-Karlin (1958)
Baker (1974)
Bellman (1958)
Bellman-Cooke-Lockett (1970)
Bellman-Dreyfus (1962)
Cahn (1948)
Coffman (1976)
Conway-Maxwell-Miller (1967)
Dantzig (1957, 1960, 1966)
Dijkstra (1959)
Dreyfus (1960, 1969)
Edmonds (1967)
Floyd (1962)
Ford (1956)
Ford-Fulkerson (1962)
Fulkerson (1961, 1966)
Gower-Ross (1969)
Harary (1969)

Jackson (1956)
Johnson (1954–1959)
Karp (1972)
Kelley (1961)
Kershenbaum-Van Slyke (1972)
Klee (1980)
Kruskal (1956)
Lawler (1976)
Lewis-Papadimitriou (1978)
Manne (1963a, b)
Mitten (1959)
Moore (1959)
Müller-Merbach (1970)
Murchland (1965)
Muth-Thompson (1963)
O'Brien (1969)
Pollack (1960)
Pollack-Wiebenson (1960)
Prim (1957)
Rial-Glover-Karney-Klingman (1979)
Warshall (1962)
Welsh (1976)
Zahn (1971)

CHAPTER TEN

Dynamic Programming, Decisions and Games

10.1 EXAMPLES OF DYNAMIC PROGRAMMING

This method of optimization was developed by Richard E. Bellman and others; a list of references follows this chapter. Space permits only a brief introduction in this book.

The algorithms in the preceding chapter include some examples and relatives of dynamic programming; this section and the next are devoted to brief descriptions of eight more examples. The general discussion is in 10.3, which might be studied concurrently. Here we simply note that, as in Chapter 9, dynamic programming problems are optimized in groups or families whose members are distinguished by the values of certain parameters called the *stage* and the *state*. The stage is characterized by a sequential or timelike quality, although in this book it is restricted to nonnegative integer values r. If possible, only one variable, called the *decision variable*, is optimized for each combination of values of the stage and the state.

(a) The *largest of* n *numbers* a_i is usually calculated by finding the maximum $\bar{z}(r)$ of the first $r=2,3,\cdots$ successively as follows, beginning with $\bar{z}(1)=a_1$:

$$\bar{z}(r)=\max_{1\leq i\leq r} a_i=\max\{\bar{z}(r-1),a_r\} \tag{1}$$

where the last member is the larger of the two quantities. Here the only independent variable is the stage r; since the state does not vary, it does not require any symbol or description. However, when (1) occurs as a subroutine in a larger problem, then the i for which $a_i=\bar{z}(r)$ presumably corresponds to a decision variable (which may be a state variable also) in the larger problem. Unless the a_i themselves are calculated recursively, they may be taken in any order.

(b) A simplified version of the first example in Bellman (1954) is the following. Let each node j of a network have a total value $\bar{z}_j(0)$, which is not subject to accumulation or transportation in the course of moving from one node to another. Let $\bar{z}_i(r)$ be the largest such value that can be obtained by starting at node i and traversing exactly r adjacent arcs, which may be directed and are not necessarily distinct. Finally when $r\geq 1$ let j be the node to which one first goes from i and let T_i be the set of possible values of j (the successors of i). T_i could include i itself if the application required this. Then $\bar{z}_i(1)$, $\bar{z}_i(2),\cdots$ can be calculated from

$$\bar{z}_i(r)=\max_{j\in T_i}\bar{z}_j(r-1) \qquad (r\geq 1) \tag{2}$$

Here r is the stage and i and j are consecutive values of the state variable. The nodes i, j, etc. might represent positions of employment, states of competence as the result of training, or military objectives.

The general idea is that to get the optimal result from stage r (and node i) it is sufficient (and usually also necessary) to get the optimal result from the intervening stage $r-1$ and node j (whatever it may be), and then use (1) or (2) to optimize the decision variable ($=j$ in (2)). The essential fact is that $\bar{z}(r)$ in (1) or $\bar{z}_i(r)$ in (2) is a *nondecreasing function* of $\bar{z}(r-1)$ or $\bar{z}_j(r-1)$. This will be discussed in more detail in 10.3.

(c) This simple example is one for which it is easy to determine the amount of calculation required; it can be regarded as a longest-path problem as in 9.4 and Figure 10.4A. Let

$$\bar{z}_h = \max_{1\le i\le m} \max_{1\le j\le n} \left(a_{hi} + b_{ij}\right) \qquad \text{for } 1\le h\le l \tag{3}$$

In the absence of additional information, any method will have to evaluate lm values of a_{hi} and mn values of b_{ij}. However, a naive method of complete enumeration in (3) would require lmn additions and $l(mn-1)$ comparisons (1) whereas the equivalent formulation

$$\bar{z}_h = \max_{1\le i\le m} \left(a_{hi} + \max_{1\le j\le n} b_{ij}\right) \tag{4}$$

requires only lm additions and $m(n-1)+l(m-1)$ comparisons. There are two stages corresponding to the two suboptimizations in (4). In the first stage, i is the state and j the decision variable; in the second stage, h is the state and i the decision variable.

(d) The (linear) *knapsack problem* calculates

$$\bar{z}_r(s) = \max(\pi_1 x_1 + \cdots + \pi_r x_r) \text{ subject to}$$

$$p_1 x_1 + \cdots + p_r x_r \le s, \text{ all } x_j = \text{nonnegative integers.} \tag{5}$$

The algorithm has been described by Dantzig (1957), Gilmore and Gomory (1965, p. 97), and Hu (1969, p. 311). It requires that each p_j be a positive integer. One could assume that s is a positive integer (if not, replace s by $[s]$) and each $\pi_j > 0$ (if not, $x_j = 0$ is clearly optimal). One may interpret $p_j x_j$ as the weight or volume and $\pi_j x_j$ as the value of x_j units of type j placed in a knapsack or other conveyance whose maximum capacity is s.

If $s=0$ one must have $\mathbf{x}=\mathbf{0}$ and hence $\bar{z}_r(0)=0$. If $s<0$ there is no feasible solution and it is convenient to define $\bar{z}_r(s) = -\infty$. If $r=1$ and $s\ge 0$, the largest possible value for the only variable x_1 is $[s/p_1]$, the largest integer not exceeding s/p_1; thus $\bar{z}_1(x) = \pi_1[s/p_1]$. Let n and s^0 denote the largest values of r and s that are of interest. Then these special cases permit the successive evaluation of

$$\begin{array}{ll} \bar{z}_1(1),\ldots,\bar{z}_1(s^0), & \bar{z}_2(1),\ldots,\bar{z}_2(s^0) \\ \cdots\cdots\cdots\cdots\cdots\cdots, & \bar{z}_n(1),\ldots,\bar{z}_n(s^0) \end{array} \tag{6}$$

by setting $r=2,\ldots,n$ in

$$\bar{z}_r(s)=\max\{\bar{z}_{r-1}(s),\bar{z}_r(s-p_r)+\pi_r\} \tag{7}$$

In (7) the larger of two quantities is being taken. The first of these corresponds to $x_r=0$, in which case the problem with variables $x_1,\ldots,x_r$ is no different from that with $x_1,\ldots,x_{r-1}$ and the same s. The second quantity corresponds to $x_r>0$; in this case it is easy to see that $\mathbf{x}$ yields optimal $\bar{z}_r(s)$ if and only if $x_1,\ldots,x_{r-1}$, x_r-1 yields optimal $\bar{z}_r(s-p_r)=\bar{z}_r(s)-\pi_r$, whence $\bar{z}_r(s)=\bar{z}_r(s-p_r)+\pi_r$. That is, x_r is reduced by 1, and s and $\bar{z}_r$ are adjusted to compensate for it.

To determine an optimal $\mathbf{x}^*$ one calculates $\mu_r(s)$, defined as the largest j such that $x_j>0$ in the (selected) optimal solution yielding $\bar{z}_r(s)$. If there is none (because $\mathbf{x}=\mathbf{0}$), set $\mu_r(s)=0$. Evidently $\mu_1(s)=0$ for $s<p_1$ and $\mu_1(s)=1$ for $s\geq p_1$. The other $\mu_r(s)$ are determined in the order of increasing r (perhaps as in (6)) by

$$\begin{aligned} \mu_r(s)&=\mu_{r-1}(s) \qquad &&\text{if } \bar{z}_r(s)=\bar{z}_{r-1}(s)\\ \mu_r(s)&=r &&\text{if } \bar{z}_r(s)>\bar{z}_{r-1}(s)\end{aligned} \tag{8}$$

corresponding to $x_r=0$ and $x_r>0$, respectively.

Finally the component x_j^* $(j=1,\ldots,n)$ in the optimal $\mathbf{x}^*$ yielding $\bar{z}_n(s^0)$ is equal to the number of (consecutive) times that the integer j appears in the nonincreasing sequence

$$\mu_n(s^0)\triangleq r_1,\qquad \mu_n\left(s^0-p_{r_1}\right)\triangleq r_2,\qquad \mu_n\left(s^0-p_{r_1}-p_{r_2}\right)\triangleq r_3,\cdots \tag{9}$$

Here $x_{r_1}^*$ is reduced by 1, then $x_{r_2}^*$ is reduced by 1, etc., and we have $r_1\geq r_2\geq r_3\geq\cdots$, although in general $s_1>s_2$ does not imply $\mu_n(s_1)\geq\mu_n(s_2)$.

In a computer it suffices to store the values of $\bar{z}_r(s)$ and $\mu_r(s)$ for $s=\min p_r$ to s^0 and for only one or two current values of r; then (7) and (8) are used simultaneously. In solving exercises by hand it may be more convenient to calculate all the $\bar{z}_r(s)$ first and then the $\mu_r(s)$. In either case (9) and $\mathbf{x}^*$ are the last to be calculated.

Table 10.1 solves an example in which $\mathbf{p}=(12,11,9,8)$, $\boldsymbol{\pi}=(16,14,11,9)$, and $s^0=28$. The variables x_1,x_2,x_3,x_4 have been indexed in the order of decreasing values of π_r/p_r. As $s\to\infty$, $x_r^*\to\infty$ only if $\pi_r/p_r=\max_j\pi_j/p_j>0$. Evidently the method is particularly suitable for computer programming or for generating solutions for various r and s.

Some or all of the x_r may be limited to the values 0 and 1. For each such r, (7) must be replaced by

$$\bar{z}_r(s)=\max\{\bar{z}_{r-1}(s),\bar{z}_{r-1}(s-p_r)+\pi_r\} \tag{10}$$

and (9) is replaced by

$$\mu_n(s^0)\triangleq r_1,\qquad \mu_{n_1}\left(s^0-p_{r_1}\right)\triangleq r_2,\qquad \mu_{n_2}\left(s^0-p_{r_1}-p_{r_2}\right)\triangleq r_3,\cdots \tag{11}$$

TABLE 10.1 The Knapsack Problem

$\mathbf{p}=$	12	11	9	8	12	11	9	8	12	11	9	8
$\boldsymbol{\pi}=$	16	14	11	9					16	14	11	9
s	$\bar{z}_1(s)\cdots\bar{z}_4(s)$				$\mu_1(s)\cdots\mu_4(s)$				$\mathbf{x}^*$ (for $r=4$)			
7	0	0	0	0	0	0	0	0	0	0	0	0
8	0	0	0	9	0	0	0	4	0	0	0	1
9	0	0	11	11	0	0	3	3	0	0	1	0
10	0	0	11	11	0	0	3	3	0	0	1	0
11	0	14	14	14	0	2	2	2	0	1	0	0
12	16	16	16	16	1	1	1	1	1	0	0	0
13	16	16	16	16	1	1	1	1	1	0	0	0
14	16	16	16	16	1	1	1	1	1	0	0	0
15	16	16	16	16	1	1	1	1	1	0	0	0
16	16	16	16	18	1	1	1	4	0	0	0	2
17	16	16	16	20	1	1	1	4	0	0	1	1
18	16	16	22	22	1	1	3	3	0	0	2	0
19	16	16	22	23	1	1	3	4	0	1	0	1
20	16	16	25	25	1	1	3	3	0	1	1	0
21	16	16	27	27	1	1	3	3	1	0	1	0
22	16	28	28	28	1	2	2	2	0	2	0	0
23	16	30	30	30	1	2	2	2	1	1	0	0
24	32	32	32	32	1	1	1	1	2	0	0	0
25	32	32	32	32	1	1	1	1	2	0	0	0
26	32	32	32	32	1	1	1	1	2	0	0	0
27	32	32	33	33	1	1	3	3	0	0	3	0
28	32	32	33	34	1	1	3	4	0	1	1	1

where $n_k = r_k - 1$ if x_{r_k} is restricted to 0 or 1, and $n_k = r_k$ otherwise. Thus all the values of $\mu_r(s)$ now have to be available, unless the largest indices are reserved for the 0–1 variables; then one could revert to (9) (with a smaller value of n) after the 0–1 variables have been determined and eliminated.

EXERCISES 10.1

Let $r=0$ to 4 in 1 and 2.

1. Apply (b) to Figure 10.3A (but with undirected arcs) and $\bar{z}_{ij}(0)=i+(i-j)^2$.
2. Apply (b) to the network of Exercise 5 of 9.4 (with unlabeled directed arcs) and $\bar{z}_j(0)=3, 2, 5, 6, 4, 8, 7, 9$.
3. How is (b) changed if the given node values can be transported and accumulated as the arcs are traversed? How could this new problem be interpreted as a longest-path problem?
4. Find $\max_i \max_j [(i-h+1)^2 - 3(i-j+2)^2] = \bar{z}(h)$ for $h, i, j = 1, 2, 3, 4$.
5. Repeat 4 with $h, i, j =$ real number in $[1,4]$.
6. Show that (a) with $\bar{z}(n) \geq 0$ is a special case of (d).

In 7 and 8 use (d) to find **x** for $r=3$; $1\le s\le 18$.

7. $\mathbf{p}=(8,5,2)$, $\boldsymbol{\pi}=(7,4,1)$.

8. $\mathbf{p}=(7,3,2)$, $\boldsymbol{\pi}=(19,8,5)$.

10.2 DYNAMIC PROGRAMMING MODELS

This section discusses four more examples of greater generality and then tabulates 13 recursion formulas derived in 9.3 through 10.4. The identifying letter (e) extends the series (a) to (d) of 10.1.

(e) A nonlinear knapsack problem is

$$\bar{z}_r(s_r)=\max[f_1(x_1)+\cdots+f_r(x_r)] \tag{1}$$

subject to $p_1x_1+\cdots+p_rx_r\le s_r$.

The problem is simpler if the x_j, p_j and s_r are nonnegative integers as in (d).

(1) is solved for $r=n$ by way of n stages yielding solutions for $r=1,2,\dots,n$ in succession. For $r=1$,

$$\bar{z}_1(s_1)=\max_{0\le x_1\le[s_1/p_1]} f_1(x_1) \tag{2}$$

which is evaluated by (1) of 10.1. Evidently each of the x_j could be optimized individually in this way if the constraint $\mathbf{px}\le s_r$ were not present. As it is, x_r can still be optimized in stage r if we observe that assigning a value to x_r implies the constraint

$$p_1x_1+\cdots+p_{r-1}x_{r-1}\le s_r-p_rx_r\triangleq s_{r-1} \tag{3}$$

for the preceding variables. However, the problem with variables $x_1,\dots,x_{r-1}$, the constraint (3) and the parameter s_{r-1} has already been solved in stage $r-1$. Thus it suffices to calculate

$$\bar{z}_r(s_r)=\max_{0\le x_r\le[s_r/p_r]}[\bar{z}_{r-1}(s_r-p_rx_r)+f_r(x_r)] \tag{4}$$

for $0\le s_r\le s_n$ and $r=1,2,\dots,n$. We define $\bar{z}_0(s_0)\equiv 0$ so that (4) agrees with (2) when $r=1$.

In solving exercises a double-entry table is constructed for each intermediate stage; this gives values of $\bar{z}_{r-1}(s_r-p_rx_r)+f_r(s_r)$ as a function of s_r and x_r, from which the optimal x_r in (4) is determined as a function of s_r. This permits the final determination of $x_n,\dots,x_1$ as functions of s_n, as members of the sequence

$$s_n, x_n, s_{n-1}\triangleq s_n-p_nx_n, x_{n-1},\dots,x_2, s_1\triangleq s_2-p_2x_2, x_1 \tag{5}$$

(f) The *smoothing problem* is discussed in Bellman and Dreyfus (1962), pp. 104–113, in terms of integer numbers of items in inventory. In order to illustrate the applicability of dynamic programming to problems involving *noninteger variables*, we mention the otherwise analogous application to a problem of seasonal employment discussed by Hillier and Lieberman (1967), pp. 254–259. The object is to find

$$\min z=(x_2-b_1)^2+(x_3-x_2)^2+(x_4-x_3)^2+(b_1-x_4)^2 \\ +c(x_2-b_2)+c(x_3-b_3)+c(x_4-b_4) \tag{6}$$

subject to $x_r \geq b_r$ for $r=2,3,4$.

Here x_r is the employment level in season r, b_r is the required minimum level, and $x_1=x_5=b_1=\max b_r$. The squared terms represent the costs of hiring and firing and the linear terms are the costs of excess employment. The recursion for z can be performed in either of the orders 234 or 432. The first gives

$$z_r(x_{r+1})=\min_{b_r \leq x_r \leq b_1}\left[(x_{r+1}-x_r)^2+c(x_r-b_r)+z_{r-1}(x_r)\right] \tag{7}$$

for $r=2,3,4$ and $z_1(x_2)=(x_2-b_1)^2$. Here $z_r(x_{r+1})$ for $r=2,3,4$ is the minimum of the terms in (6) involving any of $x_2,\ldots,x_r$, so that $z_4(b_1)$ is the minimum actually required in (6).

It is assumed that part-time employment is possible and so the x_r need not be integers. In any case elementary calculus will be useful in evaluating (7). In general the bounds on the x_r have the effect of requiring several different quadratic expressions to represent (7), according to which of several intervals contains x_{r+1}. Thus the method remains somewhat combinatorial and essentially numerical.

(g) *Action-timing* (R. A. Howard, 1971, p. 935 ff.) may be described as follows. The amount of a reward or payoff is determined in a random manner, as by the blindfold drawing of a chip on which the amount is written. The chip is examined and replaced after each drawing, and the probability q_i of drawing the amount a_i is known to the player. The latter is allowed to draw any number of times from 1 to n but only his last drawing determines his reward. How can he maximize the average value of his last drawing?

This is a multistage game against nature (cf. 4.7, 10.4) in which nature's mixed strategy $\mathbf{q}$ is known. The problem is solved by finding $\bar{z}(r)$, the (average) value of the game when a choice of $1,\ldots,r$ drawings is offered. Let $x(r+1)$ denote the amount drawn on the $r+1$st drawing from the last (after which r drawings remain). Evidently an optimal strategy is to choose the larger of $x(r+1)$ and $\bar{z}(r)$ on each occasion, the choice of $x(r+1)$ implying that the last r drawings are not used. This gives

$$\begin{aligned}\bar{z}(r+1) &= \sum_{i=1}^{m} q_i \max\left[a_i, \bar{z}(r)\right] \\ &= \sum_i \{q_i a_i : a_i \geq \bar{z}(r)\} + \bar{z}(r)\sum_i \{q_i : a_i < \bar{z}(r)\}\end{aligned} \tag{8}$$

as the value of $r+1$ drawings. Here $\bar{z}(1)=\Sigma_1^m q_i a_i$. There is no variable state unless (10) below is used.

(h) *A Markov decision process* takes place in n steps between consecutive stages r. As in (g) the average payoffs are calculated from $r=0$ to $r=n$ but the decisions are to be executed in the reverse order n to 1. At each stage r the process is in one of m states indexed by i, and a decision is made (in general depending on i and r) to select an alternative k. Then there is a known probability $q_{ij}^k(r)$ of going to state j at stage $r-1$ and receiving the known payoff $a_{ij}^k(r)$. The (average or expected) value of being in state i at stage r is

$$\bar{z}_i(r)=\max_k \sum_j q_{ij}^k(r)\left[a_{ij}^k(r)+\bar{z}_j(r-1)\right] \tag{9}$$

Equation (9) asserts that k is chosen to maximize the average of the sum of the payoff received and the value of the resulting state. If the values $\bar{z}_j(0)$ of the states j at the final stage 0 are known, then (9) permits the successive calculation of $\bar{z}_i(r)$ for all i and for $r=1,2,\ldots,n$. Howard (1960) discussed this as well as asymptotic formulas, continuous time and discounting.

TABLE 10.2 Recursion Formulas

9.3 $d_{1[r+1]}=\min_{i\in S_r}(d_{1i}+\min_{j\in \bar{S}_r} c_{ij})\geq d_{1[r]}$ with $S_r=\{1,[2],\ldots,[r]\}$. $c_{ij}\geq 0$.

9.4 $d_{1j}=\min_i\{d_{1i}+c_{ij}:1\leq i<j\}$. Acyclic. Arc$(i,j)\Rightarrow i<j$.

9.6 $d_{1j}^{r+1}=\min_{1\leq i\leq m}(d_{1i}^r+c_{ij})$. $1\leq j\leq m$; $1\leq r\leq m-2$. Uses $\leq r+1$ arcs.

9.7 $d_{ij}^r=\min(d_{ij}^{r-1}, d_{ir}^{r-1}+d_{rj}^{r-1})$, $i, j, r=1,\ldots,m$, via nodes in $\{1,\ldots,r\}$.

(a) $\bar{z}(r)=\max\{a_i:1\leq i\leq r\}=\max(\bar{z}(r-1), a_r)$.

(b) $\bar{z}_i(r)=\max\{\bar{z}_j(r-1):(i,j)$ is an arc$\}$
$=$largest $\bar{z}_k(0)$ obtainable via r arcs from i.

(c) $\bar{z}_h=\max_{1\leq i\leq m}\left(a_{hi}+\max_{1\leq j\leq n} b_{ij}\right)$ for $1\leq h\leq l$.

(d) $\bar{z}_r(s)=\max_{\mathbf{x}}\{\boldsymbol{\pi}\mathbf{x}:\mathbf{px}\leq s$; integer $x_j\geq 0\}$
$=\max[\bar{z}_{r-1}(s), \bar{z}_r(s-p_r)+\pi_r]=-\infty$ for $s<0$.

(e) $\bar{z}_r(s)=\max_{\mathbf{x}}\left\{\sum_1^r f_j(x_j):\mathbf{px}\leq s,\text{ integer } x_j\geq 0\right\}$
$=\max_{x_r}\{\bar{z}_{r-1}(s-p_r x_r)+f_r(x_r):0\leq x_r\leq[s/p_r]\}$

(f) $z_r(x_{r+1})=\min\{(x_{r+1}-x_r)^2+c(x_r-b_r)+z_{r-1}(x_r): x_r\geq b_r\}$

(g) $\bar{z}(r)=\sum_1^m q_i\max(a_i, \bar{z}(r-1))$. (Choice of r draws).

(h) $\bar{z}_i(r)=\max_k\sum_j q_{ij}^k(r)[a_{ij}^k(r)+\bar{z}_j(r-1)]$. ($r$ steps left).

10.4 $\bar{z}_i=\max_j(\bar{c}_{ij}+\bar{z}_j)$ if i is a decision node □,
$=\sum_j q_{ij}(\bar{c}_{ij}+\bar{z}_j)$ if i is a chance node ○. Arcs (i,j).

One can get (g) as a special case of (h) in which there is a state i for each a_i and an absorbing state 0. There are two alternatives, either to continue drawing ($k=1$) or to stop ($k=0$). Then $q_{i0}^0(r)=1$, $a_{i0}^0(r)=a_i$, $q_{ij}^1(r)=q_j$, $q_{00}^k(r)=1$, $\bar{z}_j(0)=-\infty$, and the other $q_{ij}^k(r)$ and $a_{ij}^k(r)$ are zero. (9) becomes

$$\bar{z}_i(r)=\max\left[a_i, \sum_j q_j \bar{z}_j(r-1)\right] \tag{10}$$

and this reduces to (8) with $\bar{z}(r)=\Sigma_j q_j \bar{z}_j(r)$.

Table 10.2 collects four recursion formulas from Chapter 9, namely 9.3(1), 9.4(1), 9.6(5), and 9.7(1); the eight examples (a) to (h) of 10.1 and 10.2; and the general decision network of 10.4.

EXERCISES 10.2

1. Reformulate (f) for the (nonperiodic) case of four consecutive months with minimum required employment of b_1, b_2, b_3, b_4 and an estimated voluntary resignation rate of 5 percent per month. The values in the preceding and the succeeding months are known to be b_0 and b_5 exactly.
2. A traveler has time to reach any of three widely separated gasoline stations along the route before running out of gas. Each station limits itself to the prices 51, 54, and 59 cents with independent probabilities .2, .5, .3, respectively. What is the traveler's optimal strategy and minimum average cost?
3. A player has a choice of tossing a die 1,2,... or n times. Let $x=$ the number of spots on the last toss. If $x=1$ or 2, the player wins x dollars, and if $x\geq 3$ he loses x dollars. For what values of n is this a desirable game to play?
4. Use DP to max $\bar{z}=x_1(x_1-5)^2+2x_2(6-x_2)+x_3^2(5-x_3)$ subject to $\mathbf{x}$ integers $\geq \mathbf{0}$, $x_1+x_2+x_3\leq 5$.
5. Use DP to max $\bar{z}=2(x_1-2)^2+(x_2-1)(x_2-2)+2(x_3-2)(4-x_3)$ subject to $\mathbf{x}$ integers $\geq \mathbf{0}$, $x_1+x_2+x_3\leq 5$.
6. Use DP to max $\bar{z}=f_1(x_1)+f_2(x_2)+f_3(x_3)$ subject to $\mathbf{x}$ integers $\geq \mathbf{0}$, $2x_1+3x_2+5x_3\leq s$ and $2\leq s\leq 11$.

$x=$	0	1	2	3	4	5
$f_1(x)=$	0	5	6	12	14	15
$f_2(x)=$	0	5	9	14		
$f_3(x)=$	0	10	16			

7. In an advertising campaign a full-page ad is to appear a total of six times in one or more of five papers that are published in different cities with subscription lists of 3, 4, 5, 7, 9 (times 10^4). It is assumed that each subscriber each day has an independent probability of .5 of noticing the ad (if it is printed). What distribution of the six ads will reach the most subscribers? (i) Formulate this as one of the problems of this section, (ii) Independently, devise an easy shortcut method of solution. (Hint: Place each ad in sequence, in the medium where it will be most effective.)

8. **A**(2×20), **B**(20×10), **C**(10×15), **D**(15×12), **E**(12×3) are matrices of nonzero elements with the indicated dimensions $m\times n$. (i) In how many ways, such as (**A**((**BC**)**D**))**E**, can **ABCDE** be evaluated? (ii) Devise a dynamic programming procedure to determine which grouping requires the fewest (scalar by scalar) multiplications. (Müller-Merbach, 1970, 10.4).

10.3 REMARKS ON DYNAMIC PROGRAMMING (DP)

Although linear programming (LP) is not a prerequisite for this chapter, a comparison of DP with the simplex method of LP may nevertheless form a useful beginning. The following list begins with one or two likenesses and then proceeds to some differences.

(i) Both methods are iterative; usually *each iteration optimizes a single independent scalar variable*. This optimization is *tentative*, being subject to continual revision in LP and to later extension and selection in DP.

(ii) *Enumeration* enters in different ways. The iterations enumerate some of the finite but large set of basic solutions in LP, but in DP they typically enumerate the independent variables, which have fixed identities in DP but not in LP. In addition DP often works with integer-valued variables and enumerates the feasible values of each. The set of all the feasible combinations of integer values is thereby enumerated implicitly, not explicitly.

(iii) LP requires *linearity* of the objective function and the constraints but the latter may be very numerous. DP requires *special structure* rather than linearity, and it can handle very few constraints (e.g., 0 or 1) other than upper and lower bounds on individual variables.

(iv) It is only in DP that the optimality of the subproblem solutions at each iteration has an essential role in guaranteeing the optimality of the later solutions. This is called the *optimality principle*, which is considered more closely at the end of this section. Beside its computational uses, it is a powerful theoretical tool for deriving functional equations, of which (1) of 9.6 is a simple example. LP relies on different ideas, namely, the monotonic variation of z or z^0 and the associated rules of $\min c_j$ or $\min g_j < 0$.

(v) The subproblems are usually of greater practical interest in DP because they form a family whose members are distinguished by particular values of significant *parameters*. In LP the subproblems are not related to any parameter unless special methods such as those of Chapter 8 are employed; otherwise the only subproblems for which optimal solutions are obtained are artificial ones in which (in Phase 2) some of the nonbasic variables are held at zero. (To be sure, this remark may apply equally well to the DP parameter r in problems such as (a), (d), (e), (f) of 10.1 and 10.2.)

Some of the parameters of a family of DP problems, such as the arc lengths c_{ij}, do not change within the family and so might be described as *stationary*. The parameters that do change within the family are called the stage and the state. Within the limited scope of this chapter and this book, the *stage* (denoted by r) is a nonnegative integer that usually starts at 0 or 1 and increases by unit amounts as the calculation of optimal values of the objective function proceeds. (In other applications it could be a real variable). Usually the suboptimizations of the current stage r depend only on the optimal values of the objective function at the immediately preceding stage $r-1$.

The remaining independent parameters that vary within the family are called the *state variables*; in this chapter they are denoted by i, j, h, s or (if s does not appear) x_r. The *decision variables* are those determined by suboptimization; at stage r they are often state variables from stage $r-1$. Although the whole problem may involve many decision variables, it is essential to have very few (e.g., one) of the decision variables determined at each stage as a function of the state variable; the latter also is preferably at most one dimensional (a scalar). This is made possible by the optimality and decomposition principles discussed below. In some applications there may be only one stage or one state (i.e., no stage or no state variable, respectively), or a decision may consist merely in selecting the larger of two numbers.

In network problems the distinction between the stage and the state may not always be clearcut. In 9.4 and 10.4 the node index j could be classified as a state (as it is in the other examples), or it might be regarded as a nonserial stage, which would be at most a trivial example of the nonserial systems of Nemhauser (1966). When the network has a regular structure as in Figure 9.5A, 10.3A and 10.4A, B, it is natural to introduce a stage variable of the standard sort, but still it will usually not label individual nodes. Thus in 9.5, the stage r labels *pairs* of nodes, which are further distinguished by the state of the warehouse (utilized or vacant). In applying 9.4 to the square acyclic network in Figure 10.3A with node labels ij, either i or j or $i+j$ could be used as a stage variable, although $i+j$ permits the optimizations to use only data from the preceding (not the current) stage.

In the recursion formulas for the objective functions in Table 10.2, the stage number r (if it occurs at all) is always increasing (by 1) from one iteration to the next; the notation is chosen to ensure this. However, such consistency becomes impractical if one finds it necessary to iterate sometimes in one direction and sometimes in the reverse. The coexistence of the forward pass (2) (with increasing node indices) and the backward pass (3) (with decreasing indices) in the critical path algorithm of 9.4 may serve as one example even though we choose to regard i and j there as states rather than stages. In 9.4 and 9.5 the nodes are actually indexed more or less in chronological order.

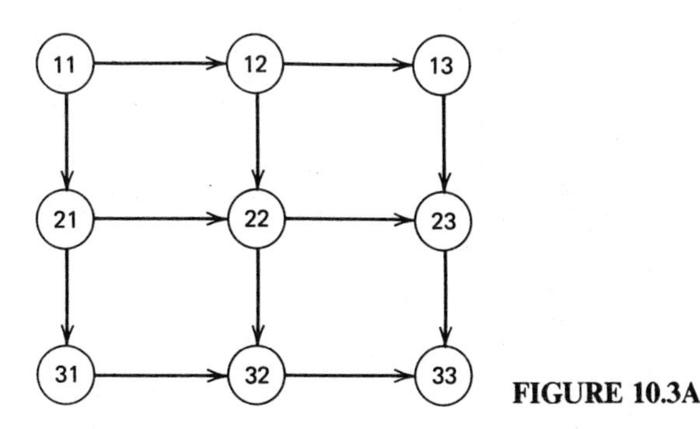

FIGURE 10.3A

Another example may arise when each end (of the chain of stages) in turn is assigned a fixed state while the state at the other end is a variable parameter; e.g., one may want to know all the shortest paths from node 1 and all those to node p. Here the essential fact is that if it is possible, it will be much more efficient to begin the iterations for the objective function values at the end having a fixed state. Otherwise any sequence may be usable as in 9.7 and (a), (d), (e) of 10.1 and 10.2; or two sequences of opposite senses may be available as in 9.5 (a) or 10.2 (f); or only one sequence may be relevant as in the following paragraph.

In probabilistic problems such as 10.2(g), (h) and 10.4 as well as extensive games and some other applications such as 10.1(b), the suboptimizations usually occur in the *reverse of the chronological order* in which the decisions will actually be executed. In these cases the stage number r defined above will count the number of decisions to be executed in the *future* of real time. Many authors regard this as the typical situation and accordingly refer to stage 0 or 1 as the last or final stage. In any case, if the decisions are not recorded at each stage along with the values of the objective function, the latter can be used to reconstruct the decisions in the order of decreasing stage numbers, beginning with any desired state and stage. For example, consider (2) of 10.1, namely

$$\bar{z}_i(r) = \max_{j \in T_i} \bar{z}_j(r-1) \tag{1}$$

If i and r are given, then inspection of the values of $\bar{z}_j(r-1)$ for $j \in T_i$ determines the optimal state j at stage $r-1$; the values of $\bar{z}_k(r-2)$ for $k \in T_j$ determine the state k at stage $r-2$, etc. These ideas are illustrated in Figure 10.3B.

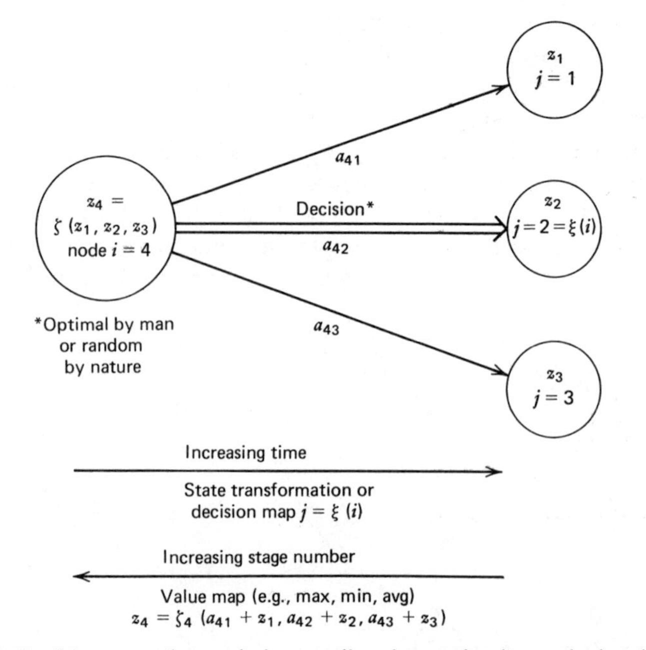

FIGURE 10.3B Decisions may be made in one direction and values calculated in the other.

A sequence of decisions for stages $r,\ldots,1$ is called an r-stage *policy*. The *principle of optimality* is often said to state that the optimality of any r-stage policy with $r>1$ implies the optimality of the $(r-1)$-stage policy obtained from it by omitting stage r and its decision. This is usually true, and the exceptions as in 9.8 are rather harmless. A more precise and more relevant statement as in (4) following is that optimality for stage $r-1,\ldots,1$ plus optimization at stage r implies optimality for stages $r,\ldots,1$, provided the objective function has a suitable form.

Section 9.8 concerns paths of maximum capacity. In Figure 10.3C, ABC is a path of maximum capacity 2, but BC has capacity 2 whereas BDC has capacity 3. However, this phenomenon causes no difficulty in practice unless *all* the optimal policies are wanted. Though the principle overlooks the path ABC, it finds $ABDC$, which also has the capacity 2, and it remains precise so far as the value of the objective function is concerned.

The principle of *decomposition* of Mitten (1964) (see also Nemhauser, 1966) analyzes the principle of optimality as follows. Let the problem be to calculate

$$\bar{z}_n(s_n) = \max_{x_n,\ldots,x_1} g_n(s_n, x_n; \ldots; s_1, x_1) \tag{2}$$

where the s_r are state variables, the x_r are the decision variables (scalars or vectors), and $s_{r-1} = \sigma_r(s_r, x_r)$. If $s_{r-1} = x_r$, this generalizes 10.2 (f).

It is assumed that for $r = 2, 3, \ldots, n$

$$g_r(s_r, x_r; s_{r-1}, x_{r-1}; \ldots; s_1, x_1) \equiv \phi_r[s_r, x_r; g_{r-1}(s_{r-1}, x_{r-1}; \ldots; s_1, x_1)] \tag{3}$$

where for each feasible fixed value of (s_r, x_r), ϕ_r is a nondecreasing function of its last argument g_{r-1}. Then it can be shown that the following decomposition principle is implied:

$$\begin{aligned} &\max_{x_r,\ldots,x_1} g_r(s_r, x_r; \ldots; s_1, x_1) \\ &= \max_{x_r} \phi_r\left[s_r, x_r; \max_{x_{r-1},\ldots,x_1} g_{r-1}(s_{r-1}, x_{r-1}; \ldots; s_1, x_1)\right] \\ &= \max_{x_r} \phi_r[s_r, x_r; \bar{z}_{r-1}(\sigma_r(s_r, x_r))] = \bar{z}_r(s_r) \end{aligned} \tag{4}$$

with $\bar{z}_1(s_1) = \max_{x_1} g_1(s_1, x_1)$ and $r = 2, 3, \ldots, n$.

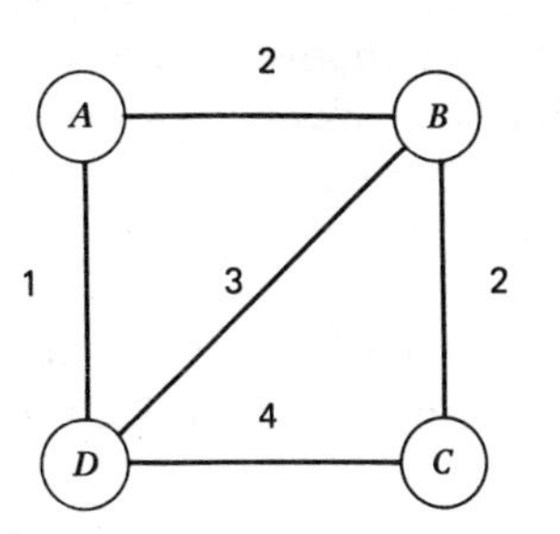

FIGURE 10.3C Maximum-capacity counterexample.

The following are four important forms for the objective functions $g_r(s_r, x_r; \dots ; s_1, x_1)$ such that (4) is satisfied:

$$f_r(s_r, x_r) + \cdots + f_1(s_1, x_1) \tag{5}$$

$$f_r(s_r, x_r) \cdots f_1(s_1, x_1) \qquad \text{with all } f_r \geq 0 \tag{6}$$

$$\min_r f_r(s_r, x_r) \tag{7}$$

$$\max_r f_r(s_r, x_r) \tag{8}$$

It also possible for these operations to be intermingled in certain ways as in

$$f_4(s_4, x_4) + f_3(s_3, x_3) \min[f_2(s_2, x_2), f_1(s_1, x_1)] \tag{9}$$

10.4 DECISION NETWORKS

The figures of this section demonstrate that many of the examples of 10.1 and 10.2 that do not involve networks nevertheless lend themselves to a network representation, which is optimized as in (4) of 9.4 or a generalization thereof. The same is true of multimove games against nature (as in 4.7) or an opposing player (as in 4.1). In the literature the resulting networks are usually assumed to be rooted trees, but *acyclic directed networks* as in 9.4 work just as well for most purposes and may permit use of a more compact diagram.

Each arc (i, j) is directed in the sense of ongoing time (real or fictitious) and represents a transition from state i to state j; it yields an incremental payoff $\bar{c}_{ij}$ ($= -c_{ij}$ if c_{ij} is a cost) to the decisionmaker. It is usually convenient to assume that for each set of values of the parameters defining the problem there is only one initial node (having no incoming arcs) but possibly many *terminal nodes* (having no outgoing arcs). In the figures the latter are represented as heavy dots labeled with known payoffs $\bar{z}_j$ to the decisionmaker.

The nonterminal nodes are of two types. At a *decision node* (represented by a square), the outgoing arcs (i, j) correspond to alternatives among which the decisionmaker chooses to maximize his payoff (when and if he arrives at that node). At a *chance node* (a circle) the outgoing arcs (i, j) represent alternatives chosen at random by nature; the corresponding conditional probabilities q_{ij} are assumed to be known or estimated. For each chance node i one has $q_{ij} \geq 0$ and $\Sigma_j q_{ij} = 1$. These probabilities are used to calculate the average or expected value obtained at the chance node.

Raiffa (1968) used this convention of squares and circles, which recalls the predominance of rounded forms in nature. It is needed and used only in this section and the remainder of the chapter.

The recursion formulas for the values $\bar{z}_i$ of the nodes i are as follows:

$$\bar{z}_i = \max_{j \in S_i} (\bar{c}_{ij} + \bar{z}_j) \text{ if } i \text{ is a decision node } \square \tag{1}$$

$$\bar{z}_i = \sum_{j \in S_i} q_{ij}(\bar{c}_{ij} + \bar{z}_j) \text{ if } i \text{ is a chance node } \bigcirc. \tag{2}$$

Here S_i denotes the set of nodes immediately following node i, so that (i, j) is a directed arc if and only if $j \in S_i$. (1) is essentially the same as the calculation of $\bar{t}_i$ in (4) of 9.4. To be sure, they differ in notation, interpretation and initial values; also we have no use here for the companion formula (2) of 9.4, and many or all of the $\bar{c}_{ij}$ may be zero in practice.

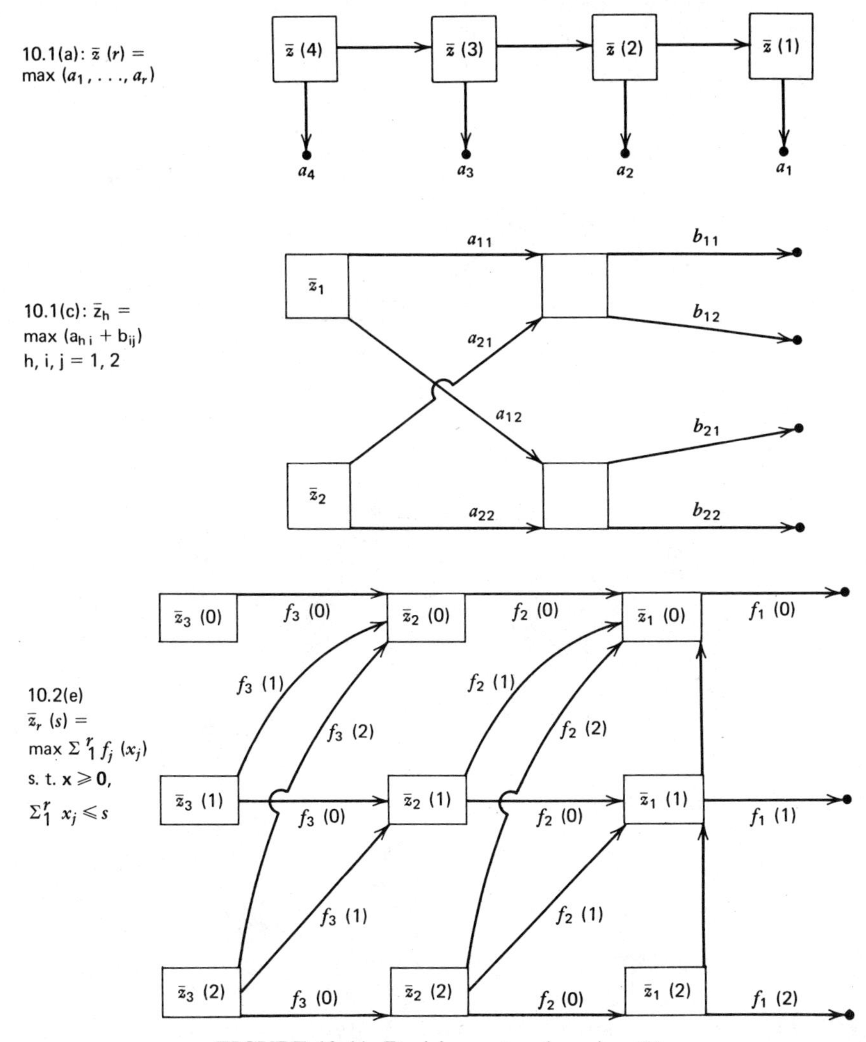

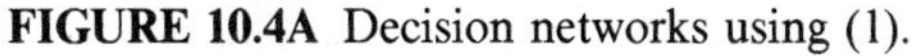

FIGURE 10.4A Decision networks using (1).

The S_i notation accommodates any desired system of node indices: increasing integers or alphabetic letters as in (2) of 9.4, decreasing integers as in (4) of 9.4, or other designations as in Figure 10.4D. The examples (a), (c), (e), (g), (h) from 10.1 and 10.2 all have imbedded in their networks in Figures 10.4A, B and C an arbitrarily variable number of serial stages. Hence their node numbers may be assumed to increase in the same sense as the stage numbers in 10.1 and 10.2, which is opposed to the directions assigned to the arcs; thus $i>j$ in these cases. On the other hand, in the remaining examples of this section there is no freely variable number of serial stages; then if the node indices have any ordering at all, it seems more natural to let them increase in the (chronological) direction assigned to the arcs, so that $i<j$ as in (4) of 9.4.

In Figures 10.4A and B the nodes are labeled with their payoff values calculated by (1) or (2) from right to left. Each arc is labeled with the amount (called $\bar{c}_{ij}$

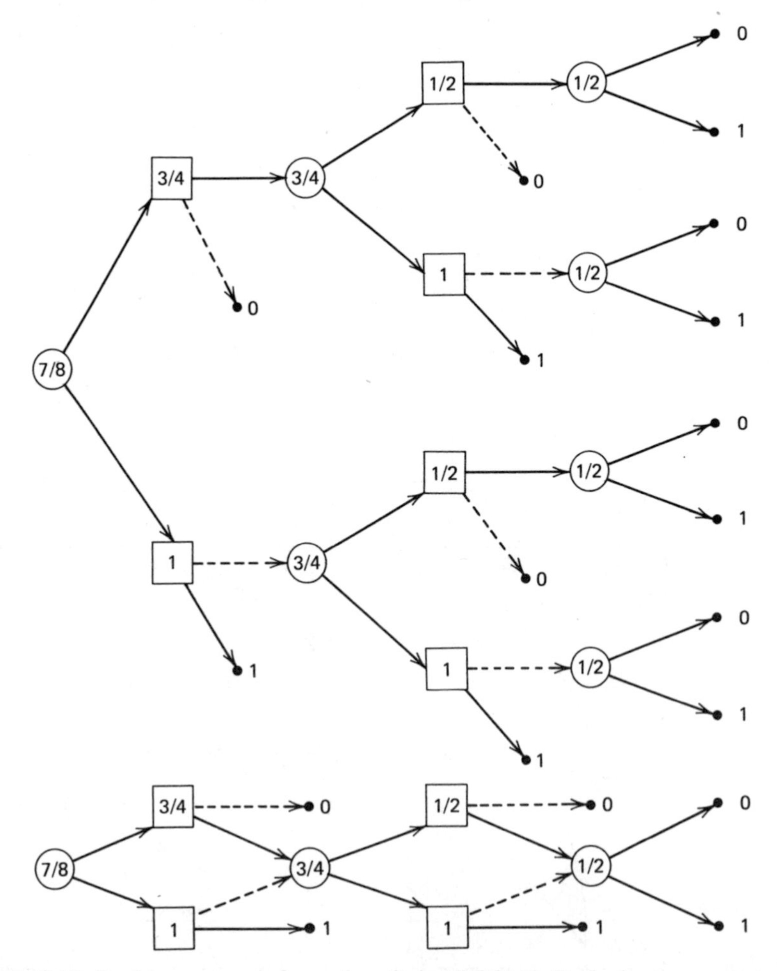

FIGURE 10.4B Decision network for action timing [10.2(g)]. Dashed arcs are never chosen. Calculate values right to left thus: circles get average values; squares get maximum values.

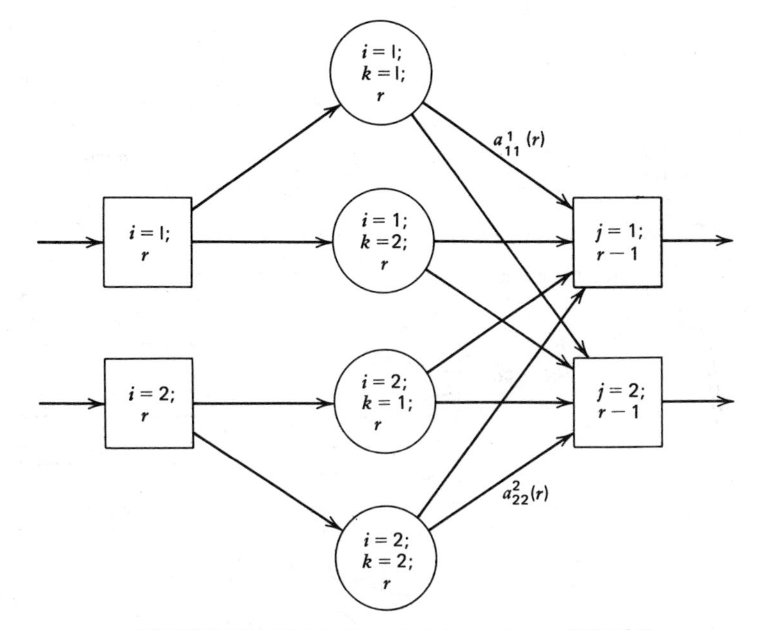

FIGURE 10.4C Markov decision network [10.2(h)].

above) it contributes to the payoff at its initial node, but zero values are omitted. Figure 10.4A refers to (a), (c), (e) of 10.1 and 10.2, which have no chance nodes. Examples (c) and (e) have several initial nodes corresponding to different values of s or h. Figure 10.4B shows a decision tree and the corresponding more compact acyclic network for (g) of 10.2, with $n=3$ and $x(r)=0$ or 1 with equal probabilities. The optimal policy is obvious: never accept 0 so long as another drawing remains. Figure 10.4C shows the part of the Markov decision network corresponding to a unit change in the stage number r for the case of two Markov states $i, j=1,2$ and two decision alternatives $k=1,2$ in (9) of 10.2(h).

A Decision Tree Using Bayes' Theorem

A typical sort of decision tree, as in Raiffa (1968), is one having four arcs in each path; the decisionmaker chooses his experimental procedure (e.g., the sample size or the information source), nature or chance determines the result of the experiment, the decisionmaker makes his final decision, and eventually nature reveals the other relevant circumstances, about which the experiment was intended to supply some information. These "states of nature" include human opinions and actions and the quality of production processes as well as accidents, weather, pestilence, the effects of new discoveries and inventions, etc.

Rather than work with various sets of symbols in different problems, we shall adopt the following standard notation in identifying the nodes and the arcs leading to each, in chronological order:

A is the initial node.
$B, \overline{B}$ mean to buy or not to buy better clues.

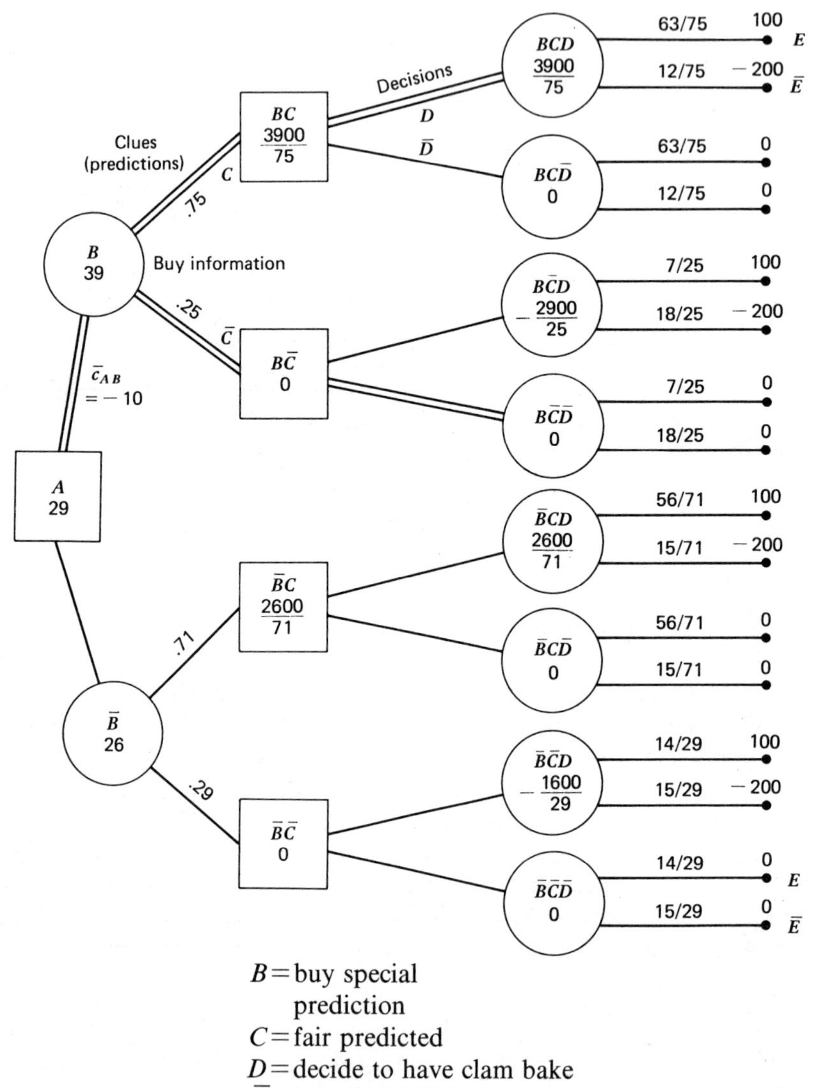

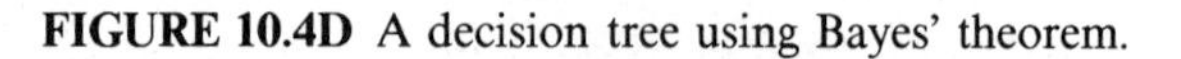
FIGURE 10.4D A decision tree using Bayes' theorem.

$C, \bar{C}$ are clues (observations or predictions) that tend to favor the truth of E and $\bar{E}$, respectively.

$D, \bar{D}$ are decisions to go ahead or to stand pat.

$E, \bar{E}$ are end events (states of nature as finally observed) that tend to favor D and $\bar{D}$, respectively. More generally, the number of alternatives (here two) need not be the same at each stage, but increasing indices j, k, l could have a common significance for C_j, D_k and E_l.

Example

The Rotary Club is considering having a clambake next Saturday. If they go ahead (D) they will make \$100 in case of fair weather E and they will lose \$200 in case of bad weather $\bar{E}$. The probability of the correctness of the routine prediction $\bar{B}$ is .8 or .5 according as the actual weather proves to be fair or not. The corresponding probabilities can be raised to .9 and .6 by paying \$10 for a special prediction B. The records show that fair weather prevails .7 of the time at this place and season. Which prediction should be used and what is the resulting average profit?

Figure 10.4D shows the decision tree, with the arrowheads omitted from the right-hand ends of the directed arcs. First it is necessary to get the probabilities into the appropriate form. By multiplying the given conditional probabilities by the absolute probability (.7 or .3) of the conditioning event E or $\bar{E}$ (fair or not fair) one obtains the following probabilities for the simultaneous occurrence of the events labeling the row and the column, given B and $\bar{B}$, respectively:

	Fair $=E$	$\bar{E}$	E	$\bar{E}$	
$C=$Predict fair	$.9\times.7$	$.4\times.3$	$.8\times.7$	$.5\times.3$	(3)
$\bar{C}=$Predict not fair	$.1\times.7$	$.6\times.3$	$.2\times.7$	$.5\times.3$	
Condition	B prediction		$\bar{B}$ prediction		

Adding the pairs of products in the same row gives the probabilities .75 and .25 for fair and not fair predictions C and $\bar{C}$ given B, and probabilities .71 and .29 given $\bar{B}$. These are the probabilities required for the four arcs in the second stage from the left in Figure 10.4D. Finally dividing these sums into their respective rows in (3) gives the probabilities for the 16 arcs in the right-hand stage, by an application of Bayes' theorem in elementary probability theory.

To illustrate the calculation of the average payoffs or values, consider the path $A\bar{B}CD$ from the initial node A to the chance node $\bar{B}CD$ and thence to two terminal nodes of values 100 and -200.

(i) First one must use the probabilities $.56/(.56+/15)$ and $.15/(.56+.15)$ determined above as weights in calculating the average

$$100\cdot 56/71+(-200)\cdot 15/71=2600/71 \tag{4}$$

which is entered inside the circle as the value of node $\bar{B}CD$. A similar calculation gives zero as the value of node $\bar{B}C\bar{D}$; the payoffs but not the probabilities are affected by the change from D to $\bar{D}$.

(ii) The value of the decision node $\bar{B}C$ is $2600/71$, the maximum of the values $2600/71$ and 0 of its successors $\bar{B}CD$ and $\bar{B}C\bar{D}$.

(iii) The value of the chance node $\bar{B}$ is the average

$$.71 \cdot 2600/71 + .29 \cdot 0 = 26$$

of the values of its successors $\bar{B}C$ and $\bar{B}\bar{C}$.

(iv) The value of the decision node A is 29, the maximum of the values 39–10 and 26 derived from its successors B and $\bar{B}$. This -10, the negative of the cost of the special prediction, is the $\bar{c}_{ij}$ associated with the arc from A to B and the only nonzero $\bar{c}_{ij}$ in the problem.

Thus \$29 is the maximum average profit, obtained from the following optimal policy: buy the special prediction B; go ahead (D) if it predicts fair weather C; cancel ($\bar{D}$) if it predicts bad weather $\bar{C}$. ■

EXERCISES 10.4

1. Verify that the application of (1) to Figure 10.4A yields results agreeing with (a), (c) and (e) of Table 10.2.
2. Use a network like the second in Figure 10.4A to evaluate $\max_{i,j,k} (3h-4)i^2 - 3(i-j)^2 + |2j-k|$ with $h, i, j, k = 1$ or 2.
3. Resolve the example on the assumption that fair weather brings a \$200 profit and bad weather brings a \$100 loss.
4. Resolve the example with a \$100 profit, a \$200 loss, a \$15 fee for B, and fair weather prevailing .6 of the time.

Optimize the following decision networks in which the numbers on the arcs are the q_{ij} (if between 0 and 1) and the $\bar{c}_{ij}$ (usually omitted when 0).

5.

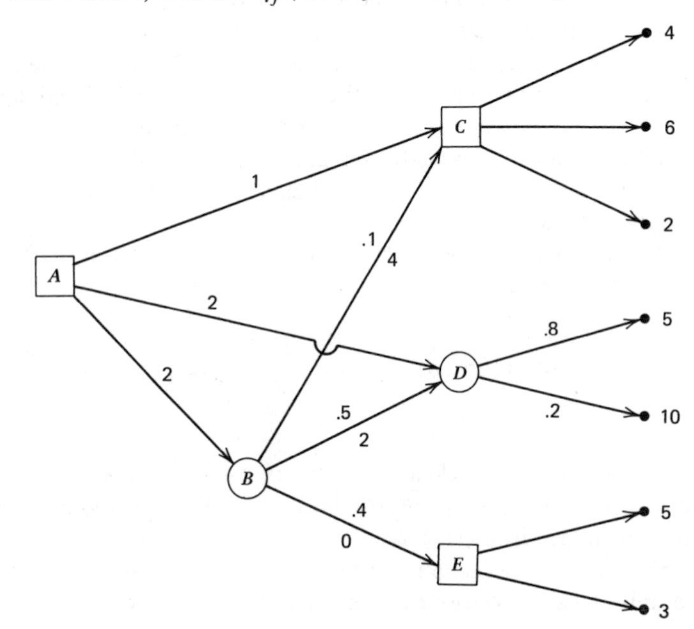

6.

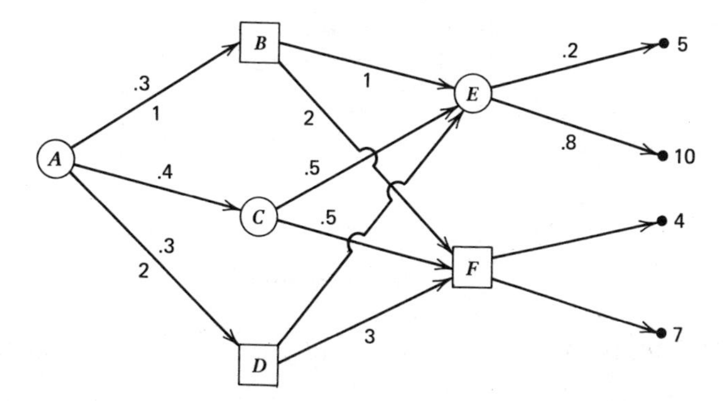

7. A buyer knows that his supplier makes equal use (50% each) of two production lines E, $\bar{E}$, which produce bad items in proportions 10% and 30% respectively. The items can be bought individually or in cartons of six, all six from the same line. The buyer gains 5¢ for each good item he buys and loses 20¢ for each bad item; thus the average gain in buying a carton is 15¢ for E and -15¢ for $\bar{E}$. Should he buy a carton without testing ($\bar{B}$) or buy and test (B) one item in the carton before deciding whether to buy (D) the other five items? Testing adds 1¢ to the cost. The tree is shown with the answer for **8**.

8. Resolve 7 with 60% of the production from E.

9. A firm has to decide whether to market a new product. If it becomes a fad the profit is estimated at $\$20\times10^4$; otherwise there is a loss of $\$2\times10^4$. The firm's success rate on such ventures has been 10%. At a cost of $\$1\times10^4$ it can employ a market survey agency, which has recognized profitable projects with 70% reliability and unprofitable projects with 80% reliability. Find the optimal strategy and $\bar{z}^*$.

10. The following is a typical "tree" problem of elementary probability; solve it as a special case of the problems of this section. Urn i $(i=1,2)$ contains g_i green balls and r_i red balls. An urn is chosen, a ball is drawn from it (both at random), and the ball is put in the other urn. If a ball is now drawn at random from the latter, what is the probability that a green ball is obtained at each drawing?

10.5 ZERO-SUM EXTENSIVE GAMES

Like Chapters 4 and 5, Sections 10.5, 10.6 and 10.7 are concerned with two-person games in which the payoffs to the two players add to zero or are not linearly related, respectively. However, in Chapters 4 and 5 the players I and J were limited to one move each, whereas extensive games may have more moves, and be represented as trees or acyclic networks. The study of extensive games began essentially with Von Neumann and Morgenstern (1944). The various methods of solution (a) to (d) are associated with the concepts of perfect information, decomposition, normalization (reduction to the matrix games of Chapter 4), and behavior strategies, respectively. The ideas of information sets, perfect recall and threats are also relevant. Except perhaps in (a) below, acquaintance with 4.1 and 4.2 is assumed.

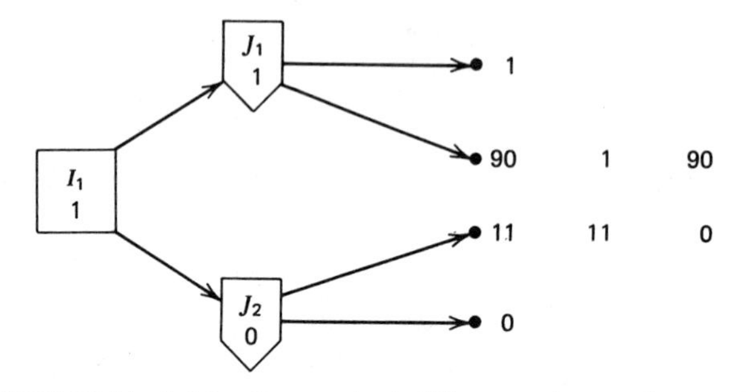

FIGURE 10.5A A 2 by 2 game $i \to j$. (I's pure strategy known to J).

(a) *Perfect information* means simply that like the decision maker in 10.4, each player knows precisely the network node at which he is situated when he has to move (make a decision). As a simple example, Figure 10.5A shows the tree diagram and payoffs for the game (1) of 4.2 in the version $i \to j$ in which I discloses his pure strategy to J before J moves. The payoff 1 at node I_1 is the maximin value in pure strategies. The nodes where J has the choice will be distinguished by cutting off the lower corners of the corresponding rectangles.

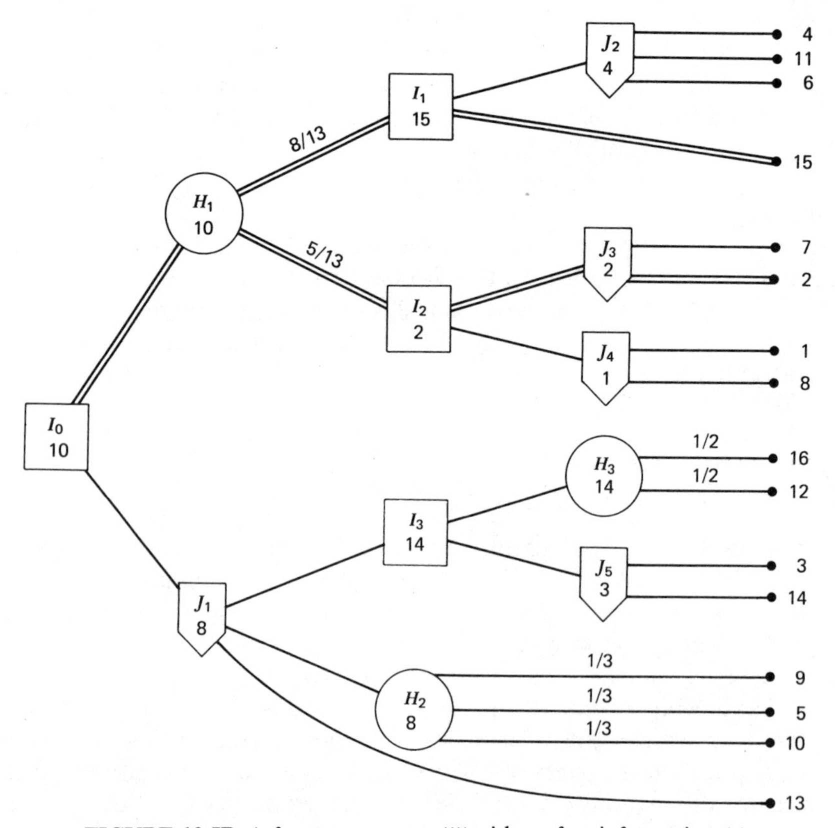

FIGURE 10.5B A four-move game *ijij* with perfect information (a).

A larger game in which there are as many as four unconcealed moves in the sequence *ijij* is shown in Figure 10.5B. As in 10.4 the payoffs at the terminal nodes (the small black discs or dots) at the right are given and the other payoffs are calculated backward from right to left. However, (1) and (2) of 10.4 are augmented by the inclusion of (2) below, and $\bar{z}_i$ is replaced by z_k:

$$z_k = \max_{l \in S_k} z_l \qquad \text{if } I \text{ chooses at node } k \text{ (a square)} \tag{1}$$

$$z_k = \min_{l \in S_k} z_l \qquad \text{if } J \text{ chooses at node } k \text{ (a shield)} \tag{2}$$

$$z_k = \sum_{l \in S_k} q_{kl} z_l \qquad \text{if node } k \text{ is a chance node (a circle)} \tag{3}$$

Here S_k denotes the set of nodes l immediately following node k (on its right), (k, l) being a directed arc, and q_{kl} is the probability that chance at k chooses the arc (k, l).

Arc values c_{kl} could be included as in (1) and (2) of 10.4 but in game theory these are usually eliminated by including them in the terminal payoffs when required; then the arcs can be labeled with the symbols for the pure strategies or with probabilities. Since the arcs are always directed toward the terminal nodes (the heavy dots), the arrowheads indicating the direction can usually be omitted.

The solution procedure based on (1) to (3) can be used to show that a game with perfect information can always be solved in terms of pure strategies. This is considered further in 10.7.

Imperfect Information

An *information set* is a set of nodes at which a player has to move, knowing only that he is located somewhere in the set. Thus he must choose the same strategy for every node in the set, and the number of pure strategies (outgoing arcs) must be the same for each of these nodes. It is normally assumed that a directed path never intersects the same information set in more than one node. (Exercise 9 of 10.6 is an exception.) In the figures, information sets containing more than one node are enclosed by dashed lines. If there are no such, every node being the sole member of its own information set, we have perfect information. The remainder of 10.5 and 10.6 is devoted to cases in which perfect information is lacking, and (1) to (3) are therefore not applicable.

Matrix games in the usual $\{I, J\}$ version with simultaneous moves illustrate a lack of perfect information. The two trees in Figure 10.5C represent the same game in which each player has $m = n = 2$ pure strategies. In the information set $\{J_1, J_2\}$ the nodes J_1 and J_2 are distinguished by I's choice of $i = 1$ or $i = 2$ at I_1, but this choice is assumed not to be known to J when he chooses $j = 1$ or 2.

(b) The *decomposition* of extensive games is described by Kuhn (1953, pp. 203–209). Let the node K be the sole member of its information set, and suppose that the subnetwork containing (only) the directed paths from K contains all of every information set it intersects. Then Kuhn proved the plausible assertion that to calculate the value of the whole game one may first calculate the value z_K of the

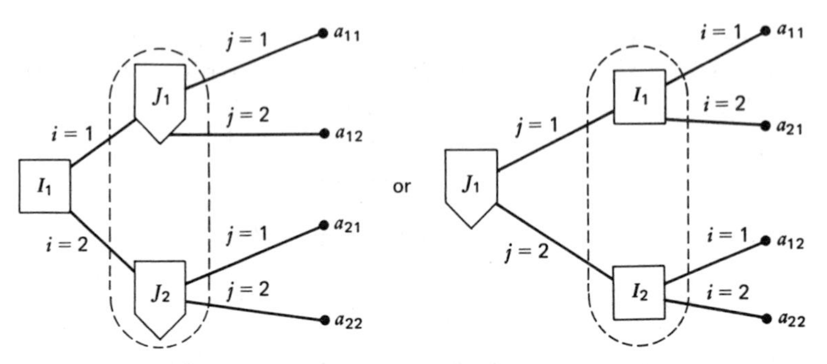

FIGURE 10.5C A 2 by 2 matrix game $\{I, J\}$.

subgame and then replace its subnetwork by the single node K (its initial node or root) with value or payoff z_K.

The following multistage matrix game illustrates this principle. Let I and J choose simultaneously and repeatedly among three objects **2,3,4** whose values are 2,3,4, until the three objects have been claimed or assigned. If the players choose different objects, each receives the object of his choice. If the players choose the same object, it is given to the player who has previously acquired the smaller total value; however, if both choose the same object initially, their choice is disregarded and **4** is arbitrarily assigned to I.

If **4** is assigned to I, the players again choose simultaneously, this time between the remaining objects **2** and **3**. Since the sum of the final payoffs is always $2+3+4=9$, it suffices to consider only I's payoffs as in a zero-sum game. This gives the matrix

	2	**3**
2	4	6
3	7	4

(4)

for which the initial node K is I_1 in Figure 10.5D. Note that if I chooses the same as J in the second round (in (4)), he will be forced to do the same in the third round, with only one object remaining; in neither round will he get any more than the **4** he already has. If I and J choose differently in (4), I will obtain the object of his choice and acquire a total value of 6 or 7. By Exercise 9 of 4.2, I and J should choose 3 with probabilities .4 and .6, respectively in the second round. This gives 5.2 as the value of (4) and for the node I_1.

It is now possible to set up the payoff matrix for the game in the first round beginning at I_0:

	2	**3**	**4**
2	5.2	6	5
3	3	5.2	5
4	4	4	5.2

(5)

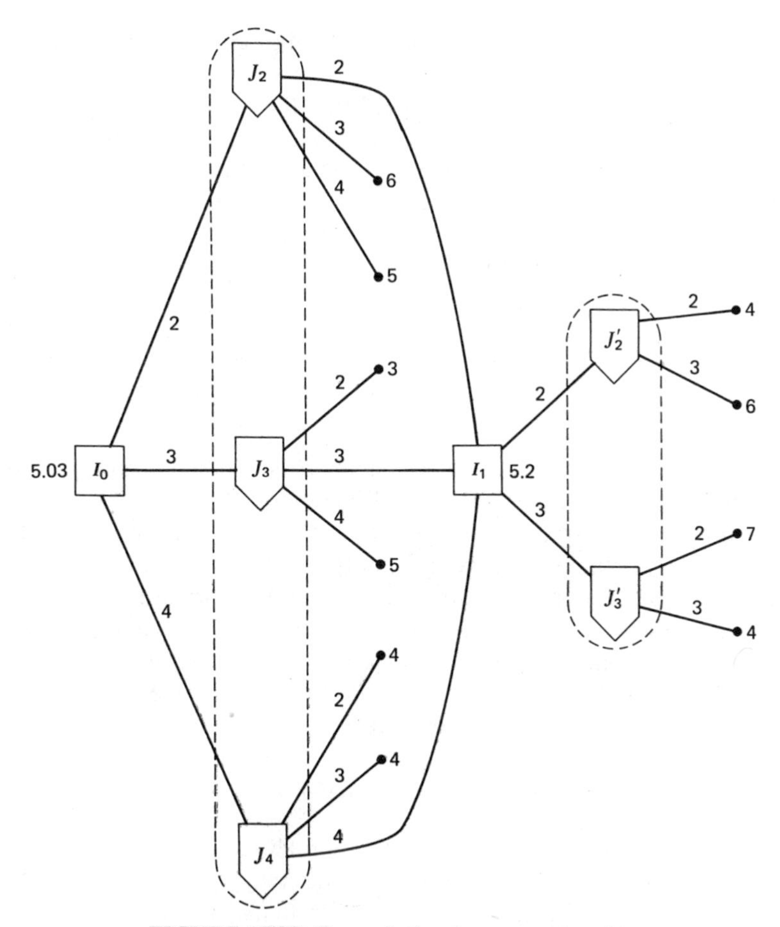

FIGURE 10.5D Example for decomposition (b).

In the diagonal where I and J choose the same object, the payoff is the 5.2 calculated above. Otherwise the one who chooses the more valuable object is at a disadvantage because he is unable to acquire the last remaining object.

In (5) the second row and second column may be deleted because they are inferior to the first; thus **3** is never chosen in the first round. Then Section 4.2 shows that I will choose **2** and J will choose **4**, each with probability $6/7$, yielding a value of $35.2/7 = 5.03$ for I's average payoff in (5) and in the whole game.

(c) The *normalization* of an extensive game is its reduction to a matrix game, which can be solved as in Chapter 4. Each pure strategy for a given player in the matrix game must specify his choice of a pure strategy in each of his information sets in the extensive game. Thus if he has l information sets presenting $n_1, n_2, \ldots, n_l$ alternatives respectively (the numbers of outgoing arcs from each node in the set), then he has $n_1 n_2 \cdots n_l$ pure strategies in the matrix game. It may be possible to eliminate some of these, either in the tree or in the matrix, because they yield payoffs that are the same as or less desirable than those produced by other pure strategies.

Nevertheless, the number of pure strategies will be large for all but the simplest games.

In 4.1 and elsewhere we have emphasized that the sequence in which the players move is important; the versions $i \to j$, $j \to i$ and $\{I, J\}$ may have different values, a_0, a^0 and a^*. The same is true where chance moves are concerned. Suppose that one of the games **A** and **B** is chosen by tossing a coin. If the players know what this choice is before they play, the resulting value is $(\text{val}\,\mathbf{A} + \text{val}\,\mathbf{B})/2$. Otherwise all they can do is play the game $(\mathbf{A}+\mathbf{B})/2$, which may have a different value, thus:

$$\begin{bmatrix} 0 & 0^* \\ 1 & -1 \end{bmatrix} + \begin{bmatrix} 1 & -1 \\ 0 & 0^* \end{bmatrix} = \begin{bmatrix} 1 & -1 \\ 1 & -1^* \end{bmatrix} \tag{6}$$

$$\text{Values:} \qquad 0 \qquad + \qquad 0 \qquad \neq \qquad -1$$

Now the advantages of normalization, cumbersome though it may be, are that it is always applicable and it eliminates the difficulties just mentioned. In the normalized game it is always correct to average the payoffs before playing (as in $(\mathbf{A}+\mathbf{B})/2$) and to play with simultaneous moves (i.e., in the version $\{I, J\}$), because this avoids giving any additional information to either player; the correct amount of information is already embodied in the definitions of the pure strategies. Of course the $\{I, J\}$ version is the only one that can require mixed optimal strategies.

As a simple example consider a 2 by 2 matrix game in the $i \to j$ version, whose network is like Figure 10.5A and whose value by (1) and (2) is known to be

$$a_0 = \max\left[\min(a_{11}, a_{12}), \min(a_{21}, a_{22})\right] \tag{7}$$

Since I has only one information set, in the normalized game he has only the two pure strategies $i=1$ and $i=2$ as before. Since J has two information sets, he has four pure strategies $j'j''=11, 12, 21, 22$, which means that J chooses j' in his first set (following $i=1$) and j'' in his second (following $i=2$). This gives the normalized game

	11	12	21	22
1	a_{11}	a_{11}	a_{12}	a_{12}
2	a_{21}	a_{22}	a_{21}	a_{22}

(8)

For example, consider the derivation of the entry a_{22} in column 12. The row 2 determines that $i=2$ and puts J in his second information set. Therefore he chooses $j=2=$ the second of the column indices 12.

The game (8) is expected to have (7) as its $\{I, J\}$ value; in fact, because of the perfect information, it has (7) as a saddle point. To prove this independently it suffices to note that the $i \to j$ value $\max_i \min_j$ of (8) is immediately expressible as (7), and the $j \to i$ value $\min_j \max_i$ reduces to (7) when one uses the general relation

$$\min\left[\max(a, b), \max(a, c)\right] = \max\left[a, \min(b, c)\right] \tag{9}$$

Alternatively (8) can be reduced to a 1 by 1 matrix having (7) as its only entry by using 4.3(ii); the first two columns can be replaced by the column $a_{11}, \min(a_{21}, a_{22})$, etc.

(d) *Behavior strategies* provide a fourth method for solving zero-sum extensive games, which will be described in the following section. The game of Primitive Poker considered there also supplies a more substantial example of normalization.

(e) Ponssard (1975a) gives a method for solving games with *almost perfect information*—perfect except for two initial chances moves; the result of one of these is known only to I, the other only to J. Ponssard (1975) gives a reduction to a linear program for a three-move game of this sort. These methods can be extended by combining them with (b) (decomposition).

EXERCISES 10.5

1, 2. Draw networks as in Figures 10.5A, C and compare the $i \to j$ and $j \to i$ versions for (4) of 4.1; for (8) of 4.2.

3. Solve Figure 10.5B with the terminal payoffs reversed in sign.

4. Solve Figure 10.5B with J's nodes assigned to chance (with equally likely arcs) and the chance nodes assigned to J.

5. Solve the example of (b) with objects of value 1, 2, 4.

6. Solve (b) modified thus: if I and J choose the same object in the first round, that object is given to I.

7. Two players start with a pile of n matches and take turns removing either 1 or 2 matches. The one who removes the last match wins. Show that the first player to move has a winning strategy if and only if n is not divisible by 3.

8. (Game of Nim) Two players start with two piles having two matches in each pile and take turns removing either 1 or 2 matches from a single pile. Which player can be sure of getting the last match? How?

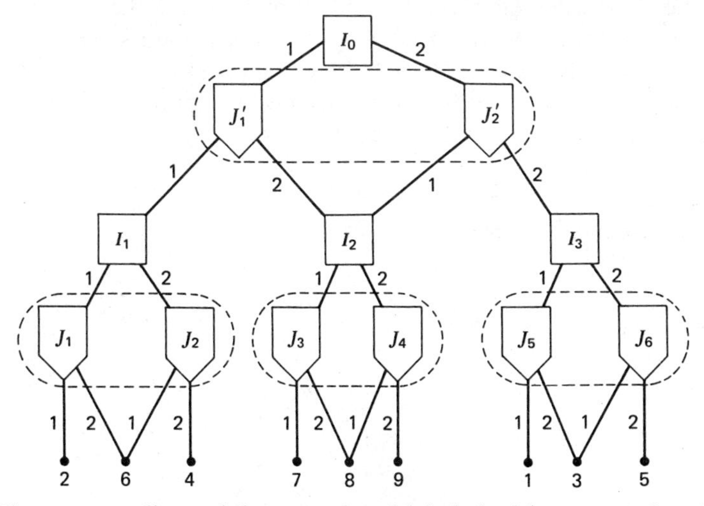

(The arcs are directed downward and labeled with strategy 1 or 2.)

FIGURE 10.5E

9. Solve (g) of 10.2 if the numbers $a_i = 0, 1, 3$ are drawn with equal probabilities for a maximum of five independent drawings; a_i is the amount paid by J to I; I has the choice (to accept the drawing or to continue) on drawings 1 and 3, and J has the choice on drawings 2 and 4.

10. Solve the game in Figure 10.5E by the method of (b).

11. Set up the matrix for Figure 10.5E with information sets $\{I_0\}$, $\{J_1', J_2'\}$, $\{I_1, I_2, I_3\}$, and $\{J_1, J_2, J_3, J_4, J_5, J_6\}$ as in (c).

12. Set up the matrix for Figure 10.5E with information sets $\{I_0\}$, $\{J_1', J_2'\}$, $\{I_1\}$, $\{I_2, I_3\}$, $\{J_1, J_2, J_3, J_4\}\{J_5, J_6\}$. Show that it can be reduced to

$$\begin{matrix} 2 & 6 & 8 \\ 6 & 4 & 8 \\ 8 & 9 & 3. \end{matrix}$$

10.6 POKER AND BEHAVIOR STRATEGIES

As examples it is feasible to use only very rudimentary versions of poker. The symmetric game considered here, which we call *Primitive Poker*, combines some of the simplest assumptions from Kuhn (1950) and Gillies, Mayberry, and Von Neumann (1953).

From a deck of three cards, numbered 1, 2, 3, each of the two players is dealt one card without replacement. Thus there are six equally likely deals $(i, j) = (1,2), \ldots, (3,2)$ with I getting the first card i and J getting j. Card 3 always wins, and card 1 loses unless the other player passes. Each player knows his own card and the nature of the deck but not his opponent's card.

Each player independently of the other must decide whether to bet \$1 or \$2. If both bet the same amount, the cards are compared and the winner receives the amount of the bet from the loser. If they bet different amounts, then the one who bet \$1 must decide whether to "pass" or to "see." To pass is to forfeit the \$1 already committed without comparing the cards; to see is to raise the bet to \$2 and compare cards. Thus three alternatives are available:

a: bet \$1; pass if the opponent bets \$2.
b: bet \$1, choose to see if the opponent bets \$2.
c: bet \$2.

The following matrix gives the payoffs to I.

	a	*b*	*c*
a	± 1	± 1	-1
b	± 1	± 1	± 2
c	1	± 2	± 2

(1)

Here the $\pm$ is interpreted as $+$ or $-$ according as I or J has the higher card.

A pure strategy for either player must specify an alternative a, b, c as a function of the card $1, 2, 3$ that the player possesses. There are $3^3 = 27$ such functions, but this number can be reduced to 4 if one does not require that *all* optimal strategies be discovered. In view of the identical status of I and J it suffices to consider I's situation. If he has card 3, the plus sign must be chosen in (1) and row c is seen to dominate the others; a player who has card 3 cannot lose anything by betting \$2 and he may gain. If I has card 1, the minus sign must be chosen in (1) and row b is dominated by row a. Finally, if I has card 2, row b is dominated by c. There are two cases: if J has card 3, he will choose c and the payoff is -2 for I's b or c; if J has card 1, the plus sign is required in (1) and row b is dominated by c.

Thus each player can be limited to four pure strategies: *aac*, *acc*, *cac*, *ccc*. Here *aac*, for example, means the player chooses a, a, or c according as he has card 1, 2, or 3. The payoffs for the 96 combinations of 4 strategies for I, 4 strategies for J, and 6 deals $(3,2),\ldots,(1,2)$ of one card to each player are tabulated in Table 10.6. Since the deals are equally likely, the sum ((2) in the table) of the six corresponding 4 by 4 matrices is six times the payoff matrix for the game. The division by six is unimportant since the strategies are unaffected and the value of the symmetric game is necessarily zero in any case ((i) of 4.3).

Inspection or the methods of 4.5 show that each player has $(1/3, 1/3, 1/3, 0)$, $(2/3, 0, 0, 1/3)$, or convex combinations thereof as optimal probability vectors; (12) of 4.2 is sufficient to confirm this. However, all these vectors admit a unique interpretation in terms of a so-called behavior strategy because they imply that c is chosen $1/3$ of the time in the first two positions and all the time in the third position

TABLE 10.6 Primitive Poker Payoffs

	aac	*acc*	*cac*	*ccc*	*aac*	*acc*	*cac*	*ccc*	*aac*	*acc*	*cac*	*ccc*
	Deal $(i, j)=(1,2)$				$(1,3)$				$(2,3)$			
aac	−1	−1	−1	−1	−1	−1	−1	−1	−1	−1	−1	−1
acc	−1	−1	−1	−1	−1	−1	−1	−1	−2	−2	−2	−2
cac	1	−2	1	−2	−2	−2	−2	−2	−1	−1	−1	−1
ccc	1	−2	1	−2	−2	−2	−2	−2	−2	−2	−2	−2
	$(2,1)$				$(3,1)$				$(3,2)$			
aac	1	1	−1	−1	1	1	2	2	1	2	1	2
acc	1	1	2	2	1	1	2	2	1	2	1	2
cac	1	1	−1	−1	1	1	2	2	1	2	1	2
ccc	1	1	2	2	1	1	2	2	1	2	1	2

Sum $= 6\mathbf{A}$ (2)

	aac	*acc*	*cac*	*ccc*
aac	0	1	−1	0
acc	−1	0	1	2
cac	1	−1	0	−2
ccc	0	−2	2	0

	a	c	
a	± 1	-1	\$1 pass
c	1	± 2	\$2 bet

$(+ \text{ if } i>j;\ - \text{ if } i<j)$

(i, j) is the deal: I gets card i, J gets card j. *aac* means bet \$1 with card 1 or 2; bet \$2 with card 3.

(with card 3); that is,

$$\begin{aligned}[(a,a,c)+(a,c,c)+(c,a,c)]/3\\ =[2(a,a,c)+(c,c,c]/3\\ =[(2a+c)/3,(2a+c)/3,c]\end{aligned} \qquad (3)$$

In fact an outside observer of repeated plays of this game could only infer the behavior strategy (the last member of (3)); he could not tell which of the optimal probability vectors, if any, the players had in mind.

To summarize, an optimal way to play this game is to bet \$2 when you have card 3; when you have card 1 or 2 you bet \$1 (and pass if your opponent bets \$2) with probability 2/3 and bet \$2 with probability 1/3. (The practice of sometimes betting \$2 on the worthless card 1 is an example of bluffing. See Exercises 1 and 2.) The resulting average payoff is zero as it must be for a symmetric game ((i) of 4.3). This result will now be derived more directly.

A *behavior strategy* (already referred to at (3)) assigns a probability to each alternative at each of the player's information sets; thus the probabilities for each set have the sum 1. Probabilities for different information sets (of the same or different players) are assumed to be independent and hence combined by multiplication. The result is that the probabilities of the $n_1 n_2 \cdots n_l$ pure strategies mentioned in (c) of 10.5 are expressed in terms of $(n_1-1)+(n_2-1)+\cdots+(n_l-1)$ probabilities, a substantial reduction in the number of parameters. The price for this is that the objective function is now a nonlinear polynomial function of the parameters, and the resulting game could have $a_0<a^0$ and hence no unique value, as discussed below.

To apply this to Primitive Poker, let p_1 or p_2 be the probability that I bets \$2 when he has card 1 or 2 respectively, and let q_1 and q_2 be the corresponding probabilities for J. Since these probabilities are independent, the average payoff is one-sixth of the sum obtained by multiplying the (i, j) element of (2) in Table 10.6 by $u_i y_j$ and summing over i and j, with

$$\begin{aligned}\mathbf{u}&=[(1-p_1)(1-p_2),(1-p_1)p_2,p_1(1-p_2,p_1p_2]\\ \mathbf{y}&=[(1-q_1)(1-q_2),(1-q_1)q_2,q_1(1-q_2),q_1q_2]\end{aligned} \qquad (4)$$

This reduces to

$$z(p_1,p_2,q_1,q_2)=[(3p_2-1)(3q_1-1)-(3p_1-1)(3q_2-1)]/18 \qquad (5)$$

from which the previous minimax values

$$p_1^*=p_2^*=q_1^*=q_2^*=1/3 \qquad (6)$$

can be obtained. The necessary and sufficient condition is analogous to (2c) of 4.5, namely for all p_1, p_2, q_1, q_2 in $[0,1]$

$$z(p_1,p_2,q_1^*,q_2^*)\le z(p_1^*,p_2^*,q_1^*,q_2^*)\le z(p_1^*,p_2^*,q_1,q_2), \qquad (7)$$

which is satisfied since all three terms are zero.

A tree diagram for this game is shown in Figure 10.6A. By multiplying the terminal payoffs by the indicated probabilities one obtains another derivation of (5) and (6). Note that J has three information sets corresponding to J's card $j=1,2,3$; that for $j=2$ has to be assembled from two separated parts in the figure.

For the sake of comparison two trivial modifications of this game may be mentioned. If neither player knows the card he has been dealt, each has but one

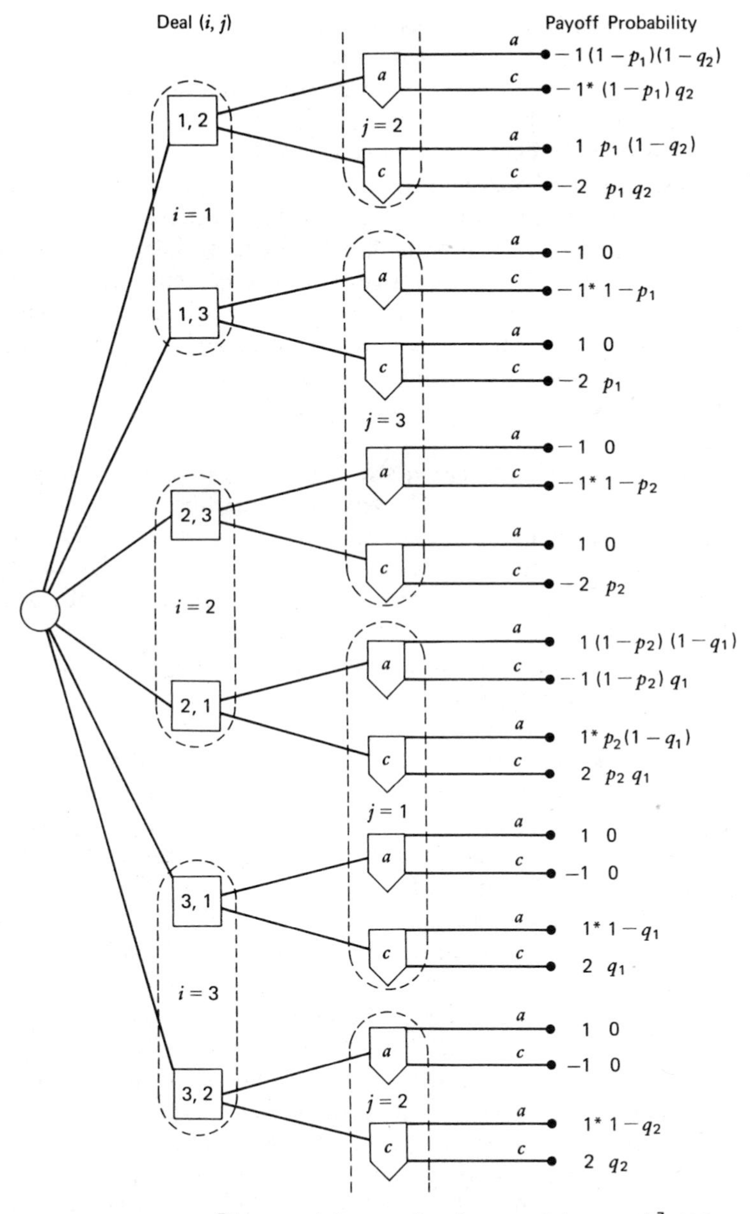

Expected value $=\left[(3p_2-1)(3q_1-1)-(3p_1-1)(3q_2-1)\right]/18$

FIGURE 10.6A Tree for primitive poker.

information set and the sets of four adjacent payoffs in Figure 10.6A have to be averaged over the six deals. This gives

$$\begin{array}{ccc} & a & c \\ a & 0 & -1 \\ c & 1 & 0^* \end{array} \tag{8}$$

which has value 0 and optimal strategies c (always bet \$2). If each player knows both cards in the deal, each has six information sets, one for each deal. Now a 2 by 2 game has to be solved for each deal; the resulting saddle points are marked by asterisks in Figure 10.6A. The values average out to 0 and an optimum strategy is c (bet \$2) when you have the higher card and a (bet \$1) when you have the lower card.

A more realistic game of poker is considered by Zadeh (1977).

A tree diagram for a game is said to have *perfect recall* if every information set M for I has the following property (and similarly for J and his information sets): all the directed paths from the initial node that intersect M must intersect the same previous information sets for I (in the same order) and require identical choices by I within each preceding set. Intersections and choices at nodes or within information sets where the other player or chance decides are irrelevant for this test; this unrestricted variability is what permits M to contain more than one node without losing perfect recall.

Figure 10.6B shows the information sets that must be separated to give perfect recall for that tree. Suppose that $\{C_2, C_3\}$ was an information set for I. Since arrivals at C_2 and C_3 require different choices $i=1$ and $i=2$ by I at his earlier node A_1, the set $\{C_2, C_3\}$ would imply that I forgets his choice at A_1 after he makes it. This is possible if I is actually a team of two or more noncommunicating individuals (as in bridge), each of whom is responsible for a single information set. Of course perfect recall is automatic if each player has but one move (at most one decision node on any directed path) or if there is perfect information.

Theorem 4 of Kuhn (1953) implies that *behavior strategies are always adequate for solving a game with perfect recall.* They are adequate for some other games too (Exercise 8). Kuhn (p. 211) gives an example of a card game in which they are not adequate. Figure 10.6C shows another such example in network and in matrix form, with the row probabilities derived from I's behavior strategies, namely, choose 2 with probability p_1 at I_1 and p_2 at $\{I_2, I_3\}$. The nonconvex set of row vectors $(\overline{w}_1, \overline{w}_2)$ that I can thus produce with p_1 and p_2 in $[0,1]$ is the ruled area in Figure 10.6D. We now show that within this restricted area, the game does not satisfy the condition minimax = maximin.

Recall from 4.2 or 4.4 that a player who knows his opponent's move does not need to use mixed strategies. Thus if J had to reveal his move first he would use his usual optimal strategy $(1/2, 1/2)$ and I would use either of his pure strategies 12 or 21 to yield the minimax $a^0 = 2$. Note that the use of an *optimal pure strategy* does not require any memory, since either member of I's team can determine it at any time by analyzing the game tree. To be sure, in this case we have a tie between rows 12 and 21 and hence need a universal convention that the smallest index among those equally good will be chosen; this selects 12.

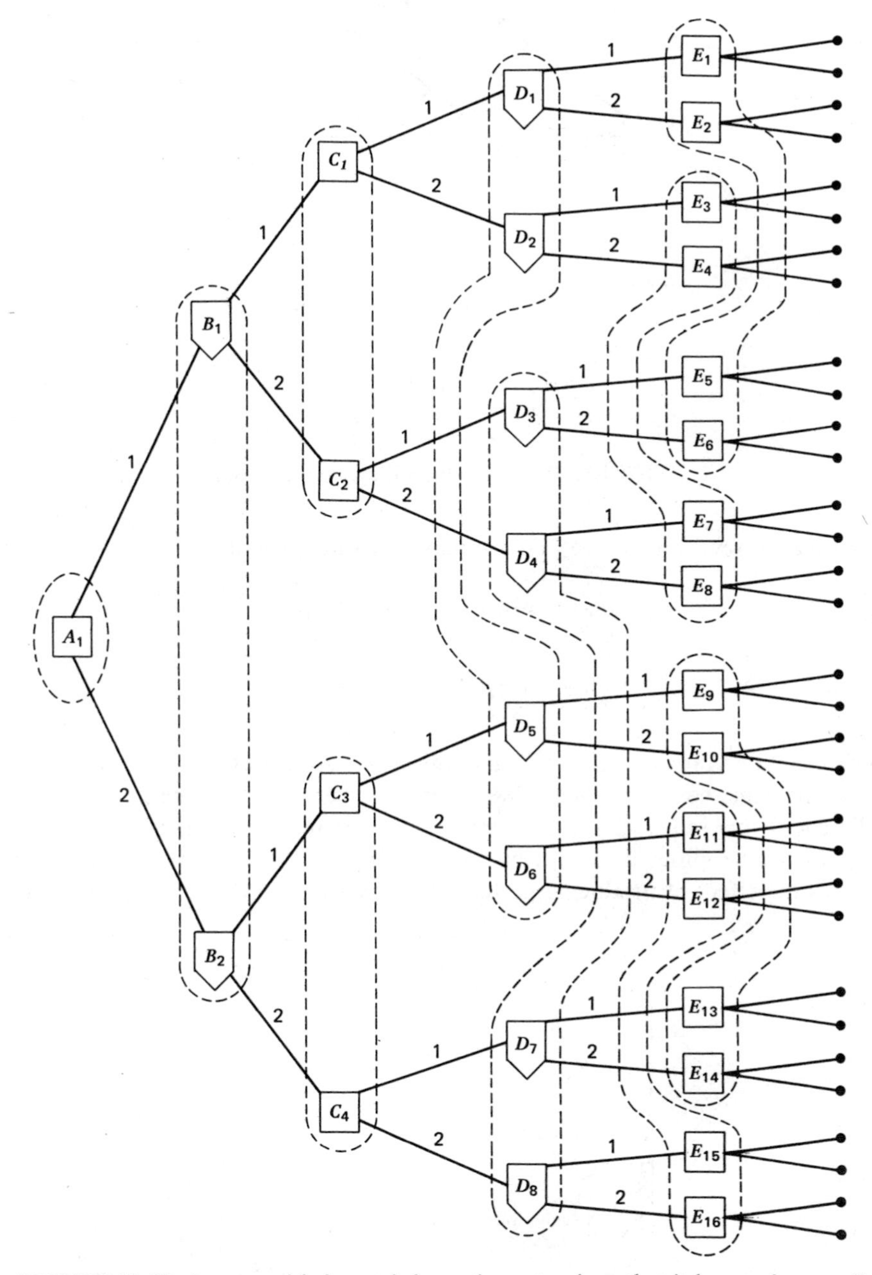

FIGURE 10.6B A game with fewest information sets, given that it has perfect recall.

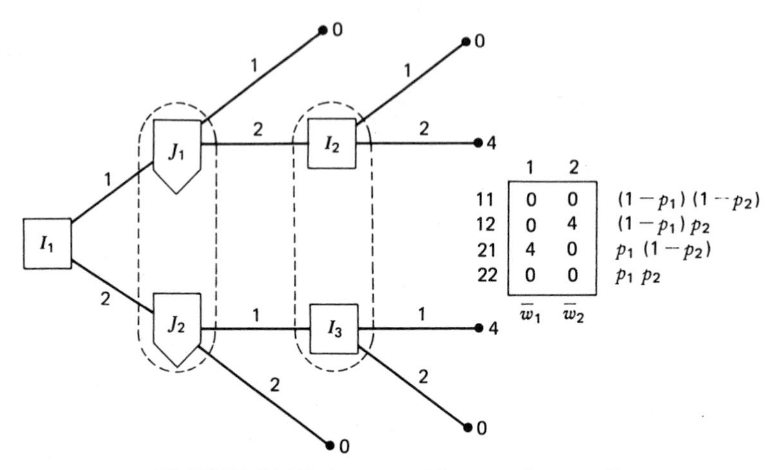

FIGURE 10.6C A game without perfect recall.

If I had to reveal his move first the maximin value would be

$$a_0 = \max_{p_1, p_2} \min[4p_1(1-p_2), 4(1-p_1)p_2]$$

$$= \max_{p_1 \geq p_2} 4(1-p_1)p_2 = \max_p 4(p-p^2) = 1 \tag{9}$$

which is obtained for $p_1 = p_2 = 1/2$ and any of J's strategies. Since this differs from $a^0 = 2$, the game has no unique value in terms of behavior strategies.

To avoid this impasse it is customary to revert to the normalized form and to allow *any* probabilities to be attributed to the rows of the matrix in Figure 10.6C. This has the effect of producing perfect recall by splitting the information sets into

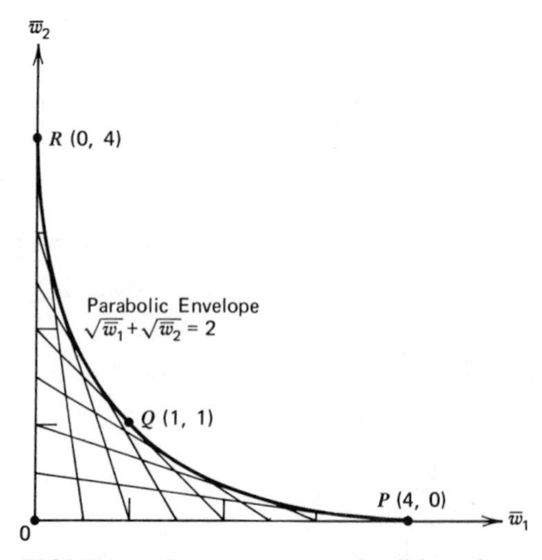

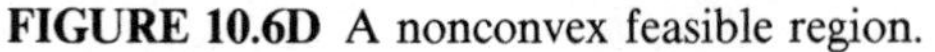
FIGURE 10.6D A nonconvex feasible region.

$\{I_1\}$, $\{J_1, J_2\}$, $\{I_2\}$, $\{I_3\}$; in terms of behavior strategies for I this requires three parameters p_1, p_2, p_3.

This need not really change the game any more than the introduction of mixed strategies did in 4.2; both describe one or more random selections of a pure strategy. An apparent exception occurs when the members of I's team have not made and cannot make (in a manner unknown to J) even one prearranged choice between the rows 12 and 21 so as to exclude 11 and 22. Then the value $\mathbf{u}_0\mathbf{A}\mathbf{y}^0 = 1$ seems inevitable, where $\mathbf{u}_0$ is optimal for I when I moves first, and $\mathbf{y}^0$ is optimal for J when J moves first. If the value $\mathbf{u}_0\mathbf{A}\mathbf{y}^0$ was not unique in the first instance, one might try to make it so by reapplying the solution procedure to the game whose strategies are the possible values of (optimal) $\mathbf{u}_0$ and $\mathbf{y}^0$.

EXERCISES 10.6

1. The value of bluffing: resolve Primitive Poker with I unwilling ever to bet \$2 on card 1.
2. The value of counterbluffing: when I has card 2 he might argue that it is futile for him to bet \$2, since J then has card 1 or 3 and knows for certain whether he will win or lose. Solve the resulting game.
3. Let the bets of \$1 and \$2 in Primitive Poker be replaced by α and γ respectively. Show that $p_1 = q_1 = (\gamma - \alpha)/(\gamma + \alpha)$, $p_2 = q_2 = (3\alpha - \gamma)/(\alpha + \gamma)$ if $0 < \alpha < \gamma < 3\alpha$, and $p_1 = q_1 = p_2 = q_2 = 0$ if $\gamma > 3\alpha$.
4. Keep the bets of \$1 and \$2 but assume that when J has card 2 or 3 he is prevented from knowing which it is; thus the information sets for $j = 2$ and $j = 3$ in Figure 10.6A are merged. (This is only slightly harder.) Show that $\mathbf{p}^* = (0, 1/3)$, $z^* = 1/9$.
5. Normalize Figure 10.6E. (Define the pure strategies and write out the game matrix.)
6. Normalize Figure 10.6E with I deciding at C_0 and chance deciding at I_1 with probabilities $1/2, 1/2$.
7. In Figure 10.6B we want to minimize the number of information sets without losing perfect recall. Then how are the sets changed (a) if A_1 is made a chance node; (b) if $\{B_1\}$, $\{B_2\}$ are information sets?

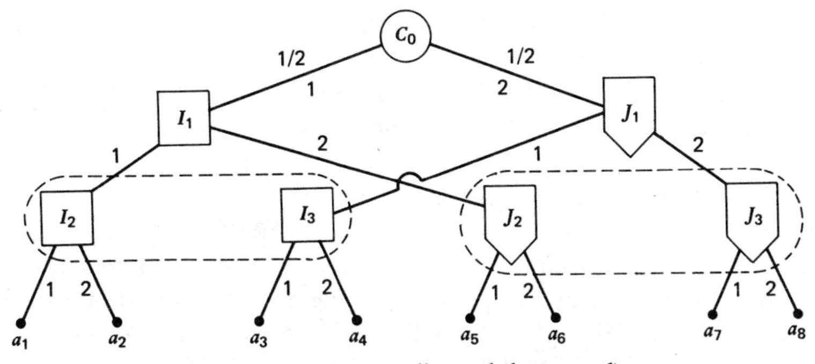

FIGURE 10.6E (arcs directed downward).

8. Let I have two information sets (one following the other), each of which presents two alternatives with behavior probabilities $1-p_1, p_1$ and $1-p_2, p_2$. These generate the probability vector

$$[(1-p_1)(1-p_2), (1-p_1)p_2, p_1(1-p_2), p_1p_2] \tag{10}$$

for the four corresponding pure strategies, which have (row) payoff vectors denoted by $\mathbf{a}^{11}, \mathbf{a}^{12}, \mathbf{a}^{21}, \mathbf{a}^{22}$. Show that (10) produces all the payoff vectors that an unrestricted mixed strategy $\mathbf{u}$ would produce if any one of the following five equations holds:

$$\mathbf{a}^{11} + \mathbf{a}^{22} = \mathbf{a}^{12} + \mathbf{a}^{21} \tag{11}$$

$$\mathbf{a}^{11} = \mathbf{a}^{12}, \qquad \mathbf{a}^{11} = \mathbf{a}^{21}, \qquad \mathbf{a}^{22} = \mathbf{a}^{21}, \qquad \mathbf{a}^{22} = \mathbf{a}^{12} \tag{12}$$

9* (Advanced topic.) The merger of the two information sets in the preceding exercise may be interpreted as requiring $p_1 = p_2$ in (10). If J's situation is similar, behavior strategies lead one to consider a game such as

	$y_1 = (1-q)^2$	$y_2 = 2q(1-q)$	$y_3 = q^2$
$u_1 = (1-p)^2$	1	0	0
$u_2 = 2p(1-p)$	0	0	0
$u_3 = p^2$	0	0	1

with the indicated probability vectors. Discuss its solution and contrast it with the result of normalization.

10.7 VARIABLE-SUM EXTENSIVE GAMES

In this section the payoffs α and β to I and J respectively are specified separately and each tries to maximize his own; minimization is absent from the original problems but may enter later. The formulation of this section is required if and only if the possible payoffs α and β do not all lie on some straight line $c_1\alpha + c_2\beta + c_3 = 0$ with $c_1c_2 > 0$.

Suppose that I has to choose among $(\alpha, \beta) = (1,3)$, $(2,2)$, $(2,1)$. Since his payoff is the first component, he rejects $(1,3)$. In an actual game I's choice between $(2,2)$ and $(2,1)$, each of which gives him 2 units, could lead into a maze of arguments; for simplicity we shall require him to perform a secondary maximization for the benefit of the second player. Thus I chooses $(2,2)$; in the terminology of 6.6, it is the lexicomaximum of $(1,3)$, $(2,2)$, $(2,1)$.

Now consider an extensive game with *perfect information*. With some loss of cogency as discussed below, (1) to (3) of 10.5 may be replaced by

$$(\alpha_k, \beta_k) = \underset{l \in S_k}{\text{lexico-max}} (\alpha_l, \beta_l) \qquad \text{if } I \text{ chooses at } k \text{ (a square)} \qquad (1)$$

$$(\beta_k, \alpha_k) = \underset{l \in S_k}{\text{lexico-max}} (\beta_l, \alpha_l) \qquad \text{if } J \text{ chooses at } k \text{ (a shield)} \qquad (2)$$

$$\alpha_k = \sum_{l \in S_k} q_{kl} \alpha_l, \qquad \beta_k = \sum_{l \in S_k} q_{kl} \beta_l, \qquad (3)$$

if node k is a chance node (a circle) at which there is a probability q_{kl} that the arc to l is chosen. Here S_k denotes the set of nodes l immediately following node k (on its right), (k, l) being a directed arc. The components are not actually reversed in using (2), which merely directs that the β_l (in their usual second position) are to be maximized, and then any ties resolved by maximizing α_l.

Thus a unique noncooperative solution, which may be called the *network solution*, has been obtained without normalizing to obtain a bimatrix version of the game. There is a general theorem, true for any number of players (Kuhn, 1953, p. 209; McKinsey, 1952, p. 130), that the network solution given by (1) to (3) for a game with perfect information is an equilibrium point in pure strategies for the normalized game. To be sure, in the variable-sum case the latter may have other equilibria, pure or mixed, and all these equilibria may be *inefficient* in the sense that one or both players can get more without either getting less.

For example consider the $i \to j$ and $j \to i$ versions of (17) of 5.2 in network and normalized forms in Figure 10.7A. The $i \to j$ version has its network solution (2, 3)

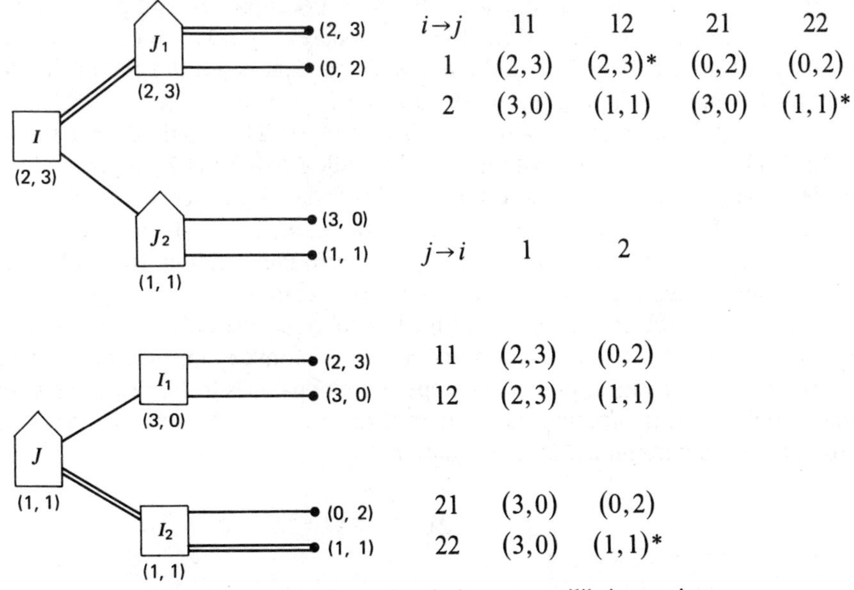

FIGURE 10.7A Network solutions as equilibrium points.

and also (1, 1) as equilibrium points; (2,3) is *not* an equilibrium for the 2 by 2 bimatrix game with simultaneous moves. The $j \rightarrow i$ version has the inefficient network solution (1, 1). Here I prefers (1, 1) to (0,2) because $1>0$ in the first position, and J prefers (1, 1) to (3,0) because $1>0$ in the second position.

Real Estate Ticktacktoe

I and J are heirs or buyers who have to divide four city lots between them. The lots lie side by side in the linear order **9, 8, 6, 7**. Their estimated individual values 9, 8, 6, 7 are equal to their identifying numbers, but two adjacent lots having the same owner are assumed to have a total value 4 units greater than if they were separated. The arrangement is that I and J will each choose two lots, one at a time in the order *ijij*. Of course J has only one choice (the one remaining lot) on the fourth move.

In Figure 10.7B the pairs of numbers between 15 and 21 are the total payoffs to I and J, respectively. In the right-hand column of (effectively terminal) nodes, each of the six possible pairs is listed twice, together with two boldface digits indicating which lots I has choosen. In the column of twelve preceding nodes the two boldface digits indicate the first two moves (the first lot chosen by each player). I_1 is the initial node and nodes **6, 7, 8, 9** correspond to I's first choice.

To illustrate the calculation consider the dashed path in Figure 10.7B. The terminal node **97** yields the payoffs (16, 18) because it assigns to I two separated lots of value $9+7=16$ and assigns to J two adjacent lots **6** and **8** of collective value $6+8+4=18$. At node **96**′ this (16, 18) is rejected in favor of (21, 17), whose first component is larger. At node **9** the pair (16, 18) is restored because its second component is larger. At the initial node I_1 the solution returns to (21, 17) (obtained via different paths) because it has the largest first component.

The optimum path connects nodes I_1, **8**, **86**′ or **87**′, and **89**, all of which have the payoffs (21, 17). The most valuable lot **9**, because it occupies an end position, is not chosen until the third move. The resulting noncooperative solution (21, 17) happens to be efficient (in Exercises 3, 4, and 6 it is not), but that is not the end of the story. J may try to improve his position by threatening I.

J's threat is his option to choose **9** on his first move. This would change the final payoffs to (18, 16), so that J would lose 1 unit while I lost 3 units. In practice many considerations may influence I's reaction to this threat. However, the *Nash cooperative solution* of 5.3 suggests that a side payment of 1 unit from I to J (or a corresponding randomized joint strategy), to shift the solution from (21, 17) to (20, 18), would be an appropriate price for J's cooperation.

In the present problem this is determined as follows. The only efficient solutions are (21, 17) and (17,21) and probability mixtures (convex combinations) thereof. Since the line through these points has slope -1, the payoffs to I and J are regarded as measured in terms of comparable units (i.e., in 5.3, $\theta=1-\theta=1/2$). The Nash solution (α, β) is determined by the equations

$$\alpha+\beta=17+21=21+17=38$$

$$\alpha-\beta=2 \tag{4}$$

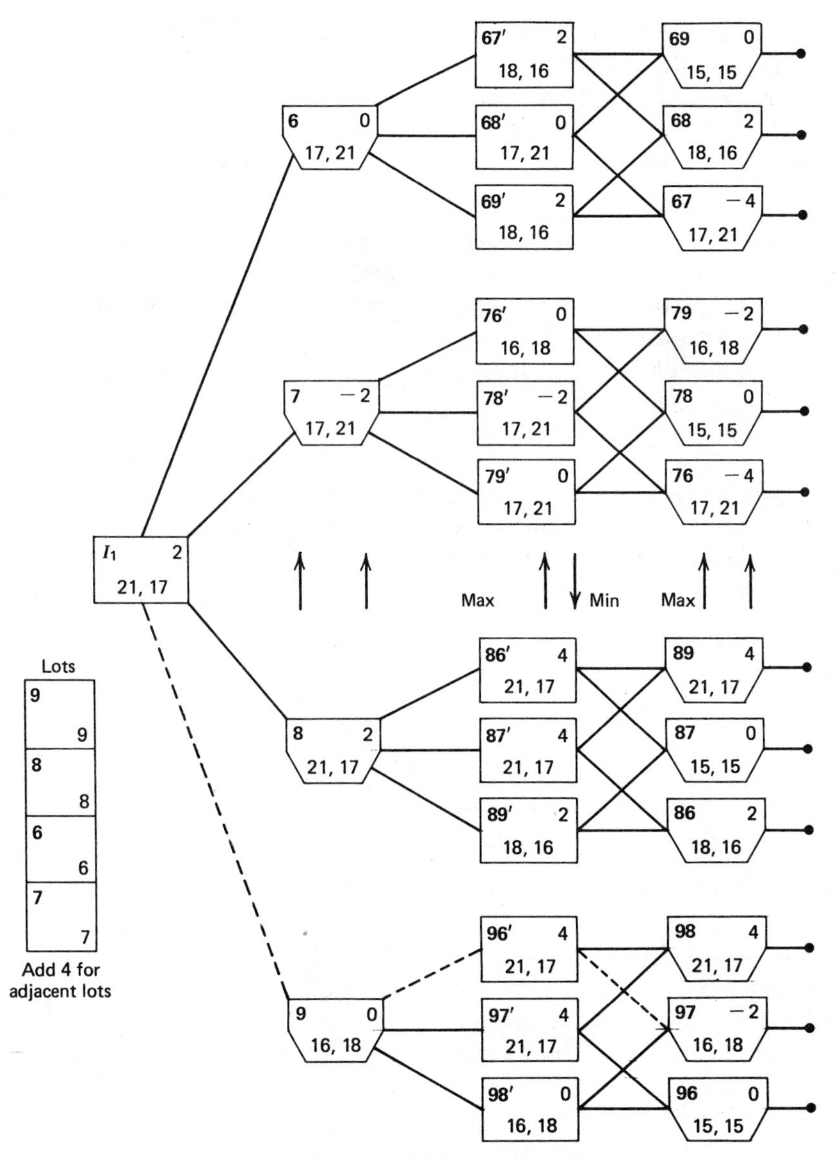

FIGURE 10.7B A variable-sum network game.

whose solution is (20,18). The value $\alpha-\beta=2$ is the value of the zero-sum game solved in the upper right corners of the node boxes in Figure 10.7B, beginning with terminal payoffs that are the differences of the terminal payoffs to I and J. With the new solution the exercise of J's threat would hurt J as much as I (2 units each). Note that the Nash solution makes no use of the (1) to (3) of this section, but only the (1) to (3) of 10.5.

The existence of the Nash solution in 5.3 was proved only for the version $\{I, J\}$ of a bimatrix game, but normalization as in (c) of 10.5 shows that it is available for extensive games in general. Since the $\{I, J\}$, $i \to j$ and $j \to i$ solutions of a matrix game (e.g., using the differences of the payoffs as in (4)) are usually different, the same is true of the corresponding Nash solutions of a bimatrix game.

Imperfect Information

If the game is not already the $\{I, J\}$ version of a bimatrix game, normalization may again be invoked to prove the existence of the Nash cooperative solution, although (as in the example foregoing) it may be possible to calculate the solution without normalization. In fact the Nash solution requires only the solution of *zero-sum* difference games such as (3) of 5.3, and any of the methods (a) to (e) of 10.5 may be used when applicable, although the zero-sum game may have to be solved repeatedly with various trial values of θ.

The usefulness of decomposition (method (b)) is not much impaired by the fact that the same value of θ must be used throughout the game in any one trial solution (and in the final solution). This method of substitution of values is valid also (under the same conditions) for equilibrium points as in 5.4 (proved by Kuhn, 1953) and for the pessimistic maximin solution (16) of 5.3.

The methods for solving the various types of extensive games may be summarized thus:

Information	Perfect	Imperfect
Zero sum	10.5(a)*	10.5(b) to (e)
Variable sum, noncooperative	(1)–(3) of 10.7* (Figure 10.7B)	10.5(b), (c); 5.4 give ≥ 1 equilibrium
Variable sum, Nash cooperative	Use 5.3, 10.5(a) for θ, α, β (Fig. 10.7B)	Use 5.3, 10.5(b) to (e) for θ, α, β (10.7)

Pure strategies suffice in the starred cases. In the zero-sum cases no cooperation is possible and the equilibrium point (α^*, β^*) is unique.

In 5.2 it was suggested that an equilibrium point $(\hat{\alpha}, \hat{\beta})$ that dominates all other equilibria be regarded as the noncooperative $\{I, J\}$ solution of a bimatrix game, and it was noted that some games (of the types C, C', C'') had no such solution. Hence the same can be asserted of the more general category of variable-sum noncooperative games with imperfect information listed in the table above. Furthermore an equilibrium point may be unique and efficient and still not coincide with the Nash solution (Exercises 8 to 12 of 5.3).

EXERCISES 10.7

Find the (noncooperative) network solution and (if you have studied it) the Nash cooperative solution. In 1 and 2 draw networks as in Figure 10.7A and solve the $i \to j$ and $j \to i$ versions.

1. $(2,2)$ $\quad (13,-3)$
$\quad (-3,13)$ $\quad (-12,-12)$

2. $(1,0)$ $\quad (0,1)$
$\quad (0,1)$ $\quad (1,0)$

3.–5. Solve the real estate ticktacktoe for four lots of values 7, 9, 3, 5 and a bonus of 1, 3, or 5, respectively, for any pair of adjacent lots.

6, 7. Four parcels of land whose net values are 0, 3, 5, 6 are arranged in a square as shown, with I controlling territory to the east and west and J controlling territory north and south. As in Figure 10.7B, I and J take turns in selecting a parcel. If one of the players obtains a connecting corridor, he receives a bonus of 5 (Exercise 6) or 7 (Exercise 7).

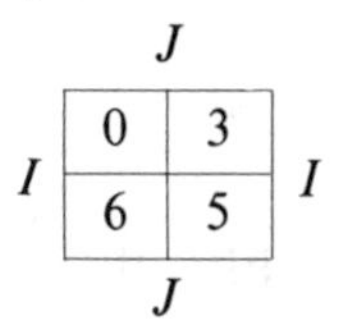

8. I and J attribute different values α and β, respectively, to each of four objects, which they take turns in selecting: $(\alpha, \beta) = (1,3)$, $(2,2)$, $(3,4)$, $(4,1)$. Assume that both values are known to both players.

9. Solve the game in Figure 10.7C. (The network is acyclic.)

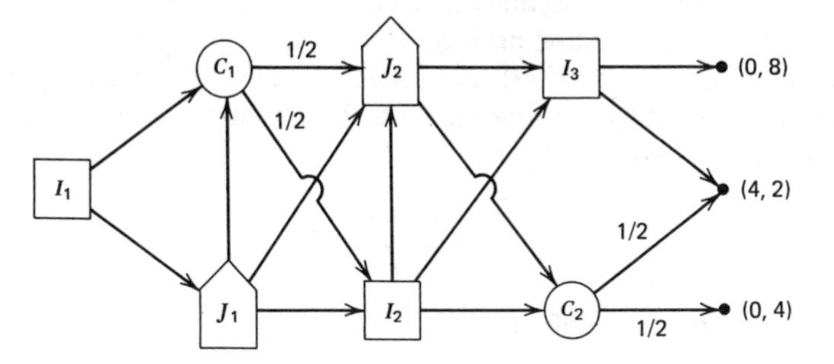

REFERENCES (CHAPTER 10)

Aris-Rutherford-Nemhauser-Wilde (1964)
Beckmann (1968)
Bellman (1954–1961, 1968)
Bellman-Cooke-Lockett (1970)
Bellman-Dreyfus (1962)
Bellman-Kalaba (1965)
Bertelé-Brioschi (1972)
Dantzig (1957)
Derman (1970)
Dresher-Tucker-Wolfe (1957)
Dreyfus (1965)
Gillies-Mayberry-Von Neumann (1953)
Gluss (1972)
Hadley (1964)
Hillier-Lieberman (1967)
Howard, R. A. (1960–1971)
Karlin (1966)
Keeney-Raiffa (1976)
Kuhn (1950–1953)
Mitten (1964)
Müller-Merbach (1970)
Nemhauser (1964–1966)
Ponssard (1975, 1975a)
Raiffa (1968)
Raiffa-Schlaifer (1961)
Schlaifer (1969)
Stanford (1976)
Thrall-Coombs-Davis (1954)
Vajda (1972)
Wagner (1969)
White (1969)
Zermelo (1912)

CHAPTER ELEVEN

Simplex Transportation Algorithms

11.1 THE TRANSPORTATION PROBLEM AND ITS DUAL

Neither the easiest nor the most difficult topic, the simplex transportation algorithms have been left until last partly because they lead into generalizations and alternate methods for which there is no space in this volume, and partly because of the illumination they receive from earlier material, primarily 2.4, 3.1 to 3.3, 9.1 and Chapter 6, and secondarily 7.2 or 7.3, 9.7, 8.2 and 8.3. Nevertheless, despite references to various earlier sections, the only indispensable prerequisite for this chapter is 9.1. The transportation problem (also called the distribution problem) was formulated by Kantorovich (1939) and Hitchcock (1941), applied by Koopmans (1949, 1951), and solved by the simplex method of Dantzig (1951c, 1963).

The problem is to find

$$\min z = \sum_{i=1}^{m} \sum_{j=1}^{n} c_{ij} x_{ij} \tag{1}$$

subject to the constraints $x_{ij} \geq 0$ and

$$\sum_{j=1}^{n} x_{ij} = a_i \qquad (i=1,\ldots,m)$$

$$\sum_{i=1}^{m} x_{ij} = b_j \qquad (j=1,\ldots,n) \tag{2}$$

Other constraints are introduced in 11.5 and 11.6. For consistency one evidently must have

$$\sum_{i=1}^{m} a_i = \sum_{j=1}^{n} b_j \left(= \sum_{i=1}^{m} \sum_{j=1}^{n} x_{ij} \right) \tag{3}$$

and all $a_i \geq 0$ and $b_j \geq 0$. Then the problem always has the feasible solution $x_{ij} = a_i b_j / \Sigma_i a_i$ and hence also an optimal solution, since

$$z \geq \left(\min_{i,j} c_{ij} \right) \sum_{i=1}^{m} \sum_{j=1}^{n} x_{ij} = \left(\min_{i,j} c_{ij} \right) \sum_{i=1}^{m} a_i$$

Sometimes it may be convenient to let $j = m+1, \ldots, m+n$ instead of $1, \ldots, n$.

For some mathematical purposes it is necessary to allow some a_i or b_j to be zero. However, in practice one expects all $a_i>0$ and $b_j>0$; for example $a_m=0$ would imply $x_{mj}=0$ for all j, and these variables and the equation involving a_m may as well be dropped from the problem. The c_{ij} are often nonnegative as in the following paragraph but there is no necessity for this; thus z could be maximized by minimizing its negative.

It has been estimated that possibly half the linear programs solved in practice are of the transportation form, although many of these involve applications other than transportation. However, according to the interpretation from which the problem takes its name, the index i identifies an origin or source at which a supply a_i of some commodity (the same for all sources) is available, and the index j identifies a destination at which a demand b_j is required to be satisfied. The variable x_{ij} is the amount shipped from i to j at a cost of $c_{ij}x_{ij}$. As illustrated in Figure 11.2A, the equations (2) and their indices i or j correspond to the nodes of a network, while each variable x_{ij} and its cost $c_{ij}x_{ij}$ corresponds to a directed arc from source i to destination j.

In applied problems the constraints (2) may be given in a different form such as

$$\sum_{j=1}^{n-1} x_{ij} \le a_i \qquad (i=1,\ldots,m)$$

$$\sum_{i=1}^{m} x_{ij} = b_j \qquad (j=1,\ldots,n-1) \tag{4}$$

with

$$\sum_{i=1}^{m} a_i \ge \sum_{j=1}^{n-1} b_j$$

This takes the form (1) to (3) as soon as one defines

$$x_{in} = a_i - \sum_{j=1}^{n-1} x_{ij} \qquad (i=1,\ldots,m)$$

$$b_n = \sum_{i=1}^{m} a_i - \sum_{j=1}^{n-1} b_j \tag{5}$$

Here the slack variable x_{in} represents the surplus quantity left over at source i; b_n, the total of the m such surpluses, may correspond to a "dummy" destination n. One may take $c_{in}=0$ for $i=1,\ldots,m$ unless there is a preference as to where surpluses occur or actual shipment to a storage site is required.

In Exercise 11 the reader is asked to treat the mathematically equivalent case in which $\sum_{i=1}^{m-1} a_i \le \sum_{j=1}^{n} b_j$ and there are shortages x_{mj} at the n destinations that may or may not be subject to penalties $c_{mj}x_{mj}$ included in z. Other cases will be treated in (3) of 11.5.

Because of the constraints (2) it is impossible to regard all mn of the x_{ij} as independently varying. The obvious procedure is to use (2) to determine some of the

x_{ij} (called the *basic variables*) in terms of the others and so attempt to eliminate the basic x_{ij} from the problem; in so doing one must remember the conditions $x_{ij} \geq 0$ still apply to all the x_{ij}, basic and nonbasic. When the nonbasic (independent) variables are set equal to zero, the result is called a *basic solution*.

The effect that any such elimination has on z can be obtained by multiplying the $m+n$ equations in (2) by parameters $u_1, \ldots, u_m, v_1, \ldots, v_n$, respectively, and adding the results to (1). Each x_{ij} occurs only twice in (2), namely in the two equations containing that particular a_i and b_j, which have multipliers u_i and v_j. Hence the result is

$$z = \sum_{i=1}^{m} \sum_{j=1}^{n} (u_i + v_j + c_{ij}) x_{ij} - \sum_{i=1}^{m} a_i u_i - \sum_{j=1}^{n} b_j v_j \tag{6}$$

The last two sums are constants if the u_i and v_i are regarded as such, and so we have proved the following, which may be of use as a preliminary step to make the c_{ij} more nearly comparable with one another.

THEOREM 1 An optimal solution for the x_{ij} remains optimal if each c_{ij} is replaced by

$$c'_{ij} = u_i + v_j + c_{ij}$$

where the u_i and v_j are any $m+n$ constants.

It turns out to be possible to choose the u_i and v_j in (6) so that

$$u_i + v_j + c_{ij} = 0 \qquad \text{when } x_{ij} \text{ is basic} \tag{7}$$

$$u_i + v_j + c_{ij} \triangleq c'_{ij} \geq 0 \qquad \text{for all } i \text{ and } j \tag{8}$$

(7) eliminates the basic variables from z; then (8) shows that z will be minimized by setting the nonbasic $x_{ij} = 0$ (their smallest permissible value), provided this makes the basic $x_{ij} \geq 0$. Then (6) reduces to

$$\min z \triangleq z^* = \sum_{i=1}^{m} \sum_{j=1}^{n} c_{ij} x^*_{ij} = -\sum_{i=1}^{m} a_i u^*_i - \sum_{j=1}^{n} b_j v^*_j \tag{9}$$

If the u_i and v_j are not restricted by (7) but only by (8) (which are called the dual constraints), one can only conclude from (6) that

$$z \geq -\sum_{i=1}^{m} a_i u_i - \sum_{j=1}^{n} b_j v_j \triangleq \overline{w} \tag{10}$$

Then to obtain equality in (10) as in (9) it would be necessary to *maximize* the dual objective function $\overline{w}$ subject to (8). This is called the *dual* of the transportation problem.

Equations (1), (2), (8) and (10) are all summarized in the Tucker schema that was introduced in 2.4 and (after a somewhat similar discussion) in 3.1. In the present case, with $m=2$ and $n=3$, it takes the form (11).

(11)

	x_{11}	x_{12}	x_{13}	x_{21}	x_{22}	x_{23}	1	
u_1	1	1	1	0	0	0	$-a_1$	$=0$
u_2	0	0	0	1	1	1	$-a_2$	$=0$
v_1	1	0	0	1	0	0	$-b_1$	$=0$
v_2	0	1	0	0	1	0	$-b_2$	$=0$
v_3	0	0	1	0	0	1	$-b_3$	$=0$
1	c_{11}	c_{12}	c_{13}	c_{21}	c_{22}	c_{23}	0	$=z\min$
	$\Vert$	$\Vert$	$\Vert$	$\Vert$	$\Vert$	$\Vert$	$\Vert$	
	c'_{11}	c'_{12}	c'_{13}	c'_{21}	c'_{22}	c'_{23}	$\overline{w}\max$	

($x_{ij}\geq 0$ and $c'_{ij}\geq 0$ at optimality; u_i and v_j not sign restricted.) The *rows* in the rectangle give the coefficients of the x_{ij} in the first row, thus reproducing the primal equations (1) and (2). For example, the first row in the rectangle gives $1\cdot x_{11}+1\cdot x_{12}+1\cdot x_{13}-1\cdot a_1=0$, which is (2) with $i=1$, $n=3$. The *columns* within the rectangle give the coefficients of the u_i and v_j in the first column, thus reproducing the dual equations (8) and (10). For example, the first column in the rectangle gives $1\cdot u_1+1\cdot v_1+c_{11}=c'_{11}\geq 0$, which is (8) with $i=1, j=1$.

Most writers define the u_i and v_j (then called shadow costs or implicit costs as in 3.6) with signs opposed to those used above, and so avoid the minus signs in (10). However, the present usage is consistent with the remainder of this book, and in solving exercises the addition of signed quantities is much more reliable and convenient than subtraction.

THEOREM 2 The initial matrix $\mathbf{A}^0$ of coefficients 1 and 0 (illustrated by (11)) has the rank $m+n-1$.

Proof The matrix $\mathbf{A}^0$ has $m+n$ rows and mn columns. Its rank is the maximum number of linearly independent rows; by Theorem 4 or 6 of 2.7, this is the number of basic x_{ij} that can be solved for. Since the sum of the m equations for sources (with constant terms $a_i,\ldots,a_m$) is equal to the sum of the n equations for destinations at least one row (any one) must be dropped to produce independence, and the rank cannot exceed $m+n-1$. If it is less than $m+n-1$, the $m+n$ equations must satisfy some additional linear relation, say with coefficients u_i^0 and v_j^0. This gives certain terms of (6), namely $\Sigma_i\Sigma_j(u_i^0+v_j^0)x_{ij}\equiv 0$; thus one has $u_i^0=-v_j^0$ for *all* i and j. Evidently this implies that all u_i^0 and all $-v_j^0$ are equal to a single constant, and it follows that the linear relation already mentioned is the only one existing.

The rest of this section is optional.

THEOREM 3 If $\mathbf{\Lambda}$ is any square submatrix of the initial matrix $\mathbf{A}^0$ of coefficients (as in (11) and Theorem 2), then its determinant $|\mathbf{\Lambda}|=0$, 1 or -1. $\mathbf{A}^0$ is then said to have the (total) unimodularity property.

Proof If every column of $\mathbf{\Lambda}$ contains two 1's, then its rows are still linearly dependent as in Theorem 2; the deleted elements in its columns are all zero. Then the determinant is zero by Theorem 10 of 7.7. The same is true if some column of $\mathbf{\Lambda}$ consists entirely of zeros. If there is a column of $\mathbf{\Lambda}$ containing exactly one 1, expanding $|\mathbf{\Lambda}|$ in terms of the elements of this column and their cofactors (Theorem 6 of 7.7) shows that $|\mathbf{\Lambda}|$ is equal to plus or minus the determinant of a smaller submatrix. Since the theorem is true for the one-by-one submatrices (i.e., the elements), it is true in general by induction.

Remark Theorems 8 and 12 of 7.7 and 7.8 show that ratios of determinants suffice for solving equations and for obtaining the elements of any matrix obtained from $\mathbf{A}^0$ by pivoting as in 2.4. Thus Theorem 3 implies that these elements are limited to the values 1, -1 and 0, and the x_{ij}, u_i and v_j can be integers if the a_i, b_j and c_{ij} are integers. Then the pivot rules of 2.4 or 2.5, which are identical in this case, show that in pivoting the changes in the x_{ij} are limited to three values $\pm\Delta x$ and 0, and the changes in the u_i, v_j and c'_{ij} are limited to $\pm\Delta c$ and 0. These facts will be established independently in the next two sections.

EXERCISES 11.1

Write out the Tucker schema and the dual problem for 1 and 2.

1. $\mathbf{a}=(3,8)$, $\mathbf{b}=(5,6)$, $c_{11}=2$, $c_{12}=3$, $c_{21}=7$, $c_{22}=4$.
2. $\mathbf{a}=(4,5,6)$, $\mathbf{b}=(7,8)$, $c_{31}=5$, $c_{32}=6$, others in 1.
3. Find $\mathbf{x}^*, \mathbf{u}^*, \mathbf{v}^*$ in 1 by enumerating four basic solutions.
4. Find z^* in 2, given that $\mathbf{u}^*=(3,2,0)$, $\mathbf{v}^*=(-5,-6)$.
   Use the solution in 3 to optimize 5 and 6.
5. $\mathbf{a}=(5,6)$, $\mathbf{b}=(3,8)$, $c_{11}=2$, $c_{12}=7$, $c_{21}=3$, $c_{22}=4$.
6. $\mathbf{a}=(3,8)$, $\mathbf{b}=(5,6)$, $c_{11}=0$, $c_{12}=1$, $c_{21}=3$, $c_{22}=0$.

Formulate 7 to 15 as transportation problems or duals thereof.

7. Maximize $\bar{z}=5x_1+2x_2+3y_1+4y_2$ subject to
   $\mathbf{x}\geq\mathbf{0}$, $\mathbf{y}\geq\mathbf{0}$, $x_1+x_2=6$, $y_1+y_2=8$, $x_1+y_1=9$.
8. Minimize $z=x_1+3x_2+2x_3+6y_1+3y_2$ subject to
   $\mathbf{x}\geq\mathbf{0}$, $\mathbf{y}\geq\mathbf{0}$, $2x_1+4y_1=5$, $3x_2+y_2=9$, $2x_1+3x_2+x_3=10$, $4y_1+y_2=6$.
9. Minimize $z=x_1+6x_2+3y_1+8y_2$ subject to
   $\mathbf{x}\geq\mathbf{0}$, $\mathbf{y}\geq\mathbf{0}$, $x_1+x_2\leq 7$, $y_1+y_2\leq 9$, $x_1+y_1=5$, $x_2+y_2=4$.
10. Maximize $\bar{z}=x_1+6x_2+3y_1+8y_2$ with no sign restrictions, $x_1+x_2\leq 7$, $y_1+y_2\leq 9$, $x_1+y_1\leq 5$, $x_2+y_2\leq 4$.
11. Minimize $z=\Sigma\Sigma c_{ij}x_{ij}$ subject to all $x_{ij}\geq 0$,
    $\Sigma_{j=1}^{n}x_{ij}=a_i$, $\Sigma_{i=1}^{m-1}x_{ij}\leq b_j$, $\Sigma_{i=1}^{m-1}a_i\leq\Sigma_{j=1}^{n}b_j$.
12. Warehouses at (0,0) and at (24,15) (Cartesian coordinates in a plane) have 35 and 50 items on hand respectively. There are requirements for 28, 15 and 33 items at (0,22), (44,0) and (24,45), respectively. Assume that the transportation cost per item is proportional to (a) the airline distance or (b) the distance along roads that run only north-south and east-west.

13. A tour operator has to get a party of 16 students, 21 working adults and 29 senior citizens from A to B. The four available modes of transportation have space for 25, 20, 15, and 10 persons, respectively. The effective costs c_{ij} may depend on the passengers' probable desire for speed and comfort as well as the posted prices for tickets.

14. A farmer has three kinds of seed, enough for 110, 70 and 150 acres, respectively. The farm consists of three parcels of 100, 120 and 140 acres, respectively, each parcel having its own type of soil. The farmer cannot obtain more seed or land but can estimate the probable return a_{ij} per acre for each combination of seed and soil.

15. A firm wishes to hire two typists, four clerks and a switchboard operator. It has ten applicants, each of whom (i) has been given numerical proficiency ratings t_i, c_i, s_i for the three kinds of work.

16. How is Theorem 3 changed if one considers arbitrary matrices having 1, -1 and 0 as elements with at most two nonzeros in each column?

11.2 BASIC SOLUTIONS, TRIANGULARITY AND TREES

These three concepts are equivalent for transportation problems, as shown in Theorem 1 following. To illustrate the first two, let us undertake to solve (11) of 11.1 for $m+n-1=4$ basic variables $x_{13}, x_{21}, x_{22}, x_{23}$ when the nonbasic variables are zero. After putting $x_{11}=x_{12}=0$, omitting the first row as redundant (this is equivalent to putting $u_1=0$), and rearranging the remaining rows and columns of (11), one obtains (1).

	x_{13}	x_{23}	x_{21}	x_{22}	1	
v_3	1	1	0	0	$-b_3$	$=0$
u_2	0	1	1	1	$-a_2$	$=0$
v_1	0	0	1	0	$-b_1$	$=0$
v_2	0	0	0	1	$-b_2$	$=0$
1	c_{13}	c_{23}	c_{21}	c_{22}	0	$=z$
	$\|$	$\|$	$\|$	$\|$	$\|$	
	0	0	0	0	$\bar{w}$	

(1)

In the bottom line the c'_{ij} are replaced by zeros in accordance with (7) and (8) of 11.1. The system of equations (1) is said to be *triangular* because all the coefficients in the rectangle are zero on one side of the diagonal.

In practice one does not use (1) but rather the more compact arrays

0	0	x_{13}	a_1
x_{21}	x_{22}	x_{23}	a_2
b_1	b_2	b_3	

(2′)

—	—	c_{13}	0
c_{21}	c_{22}	c_{23}	u_2
v_1	v_2	v_3	

(2″)

which are especially convenient for representing the row and the column equations of (1) respectively. For example, the primal equation $x_{13}+x_{23}=b_3$ is the first row in (1) and the third column in (2′). The replacement of u_1 by 0 is now explicit in (2″); this has no effect on the c'_{ij}, x_{ij} or z.

A triangular system of equations such as (1) indicates specific orders in which the equations are used and the variables are determined; the x_{ij} are determined from right to left and the equations are used from bottom to top. One begins with the "shortest" row v_2, in which only one x_{ij} has a nonzero coefficient; it makes $x_{22}=b_2$. Then row v_1 makes $x_{21}=b_1$. Substituting these values in rows u_2 and v_3 gives

$$\begin{aligned} x_{23}&=a_2-x_{21}-x_{22}=a_2-b_1-b_2=b_3-a_1 \\ x_{13}&=b_3-x_{23}=b_1+b_2+b_3-a_2=a_1 \end{aligned} \tag{3}$$

respectively. The nonbasic variables $x_{11}=x_{12}=0$. This primal solution is feasible if $b_3 \geq a_1$. Evidently these operations can be performed equally well in (2′), although the latter allows more options. In particular it includes the row involving a_1, which was omitted in (1); this can determine $x_{13}=a_1$ immediately.

The column (dual) equations for u_i and v_j in (1) are also triangular. The equations are solved from left to right and the variables are determined from top to bottom. The solution can be performed equally well in (2″).

$$\begin{aligned} &\text{Column } x_{13} \text{ (or cell } c_{13} \text{ in (2″)) makes } v_3=-c_{13}. \\ &\text{Column } x_{23} \text{ makes } u_2=-v_3-c_{23}=c_{13}-c_{23}. \\ &\text{Column } x_{21} \text{ makes } v_1=-u_2-c_{21}=c_{23}-c_{13}-c_{21}. \\ &\text{Column } x_{22} \text{ makes } v_2=-u_2-c_{22}=c_{23}-c_{13}-c_{22}. \end{aligned} \tag{4}$$

The omission of row u_1 is equivalent to putting $u_1=0$.

The fact that all the terms appearing in these solutions have coefficients ± 1 illustrates Theorem 3 of 11.1 and Theorem 2 following. Note that the primal and the dual solution procedures traverse the diagonal of (1) in opposite directions, although many other sequences are possible.

Since $c'_{ij}=u_i+v_j+c_{ij}$, these solutions correspond to the tableau

1	x_{11}	x_{12}	1
x_{22}	0	1	$-b_2$
x_{21}	1	0	$-b_1$
x_{23}	-1	-1	a_1-b_3
x_{13}	1	1	$-a_1$
$-z$	$c_{11}-c_{21}$ $+c_{23}-c_{13}$	$c_{12}-c_{22}$ $+c_{23}-c_{13}$	$b_1c_{21}+b_2c_{22}$ $+a_1c_{13}+(b_3-a_1)c_{23}$

(5)

Some sets of $m+n-1$ variables x_{ij} cannot be bases. For example, consider $x_{11}, x_{12}, x_{21}, x_{22}$ in (11) of 11.1. This makes $x_{13}=x_{23}=0$ and the last row equation (involving b_3) contains no variables and cannot be satisfied unless $b_3=0$. Even if $b_3=0$, the remaining equations are still linearly dependent because every column still contains two 1's, as in the original $m+n$ rows. Hence a basis must produce a column having only one 1, like x_{13} in (1). This column determines v_3, after which the row v_3 and the column x_{13} can be deleted. Then there must be another column having only one 1, etc. Thus the triangularity is established.

If a set of basic x_{ij} is not already specified, it can easily be determined one variable at a time so as to be nonnegative as well as basic. This will be described in the first part of 11.4.

In general the **X** and the **C** arrays (2′), (2″) and their corresponding equations are

$$\begin{array}{ccc|c} x_{11} & \cdots & x_{1n} & a_1 \\ \vdots & & \vdots & \vdots \\ x_{m1} & \cdots & x_{mn} & a_m \\ \hline b_1 & \cdots & b_n & \end{array} \qquad (6')$$

$$\sum_{j=1}^{n} x_{ij} = a_i \qquad (7)$$

$$\sum_{i=1}^{m} x_{ij} = b_j \qquad (8)$$

$$\text{All } x_{ij} \geq 0 \qquad (9)$$

$$\min z = \sum_{ij} c_{ij} x_{ij} \qquad (10)$$

$$\begin{array}{ccc|c} c_{11} & \cdots & c_{1n} & u_1 \\ \vdots & & \vdots & \vdots \\ c_{m1} & \cdots & c_{mn} & u_m \\ \hline v_1 & \cdots & v_n & \end{array} \qquad (6'')$$

$$u_i + v_j + c_{ij} \triangleq c'_{ij} = 0 \quad \text{when } x_{ij} \text{ is basic;} \qquad (11)$$

$$u_i + v_j + c_{ij} \triangleq c'_{ij} \geq 0 \qquad (12)$$

Equations (7) to (10) state the primal form of the problem and (11) and (12) the dual. Equations (6′) and (6″) define the problems initially when the x_{ij}, u_i and v_j are symbols and the a_i, b_j and c_{ij} are given data. In the course of the simplex algorithm, tableaus similar to (6′) and (6″) are produced in which the numerical values of the x_{ij}, u_i and v_j are specified. If the c_{ij} are replaced by c'_{ij}, these tableaus can be superimposed within a single rectangle since in each position ij either $x_{ij}=0$ or $c'_{ij}=0$ by (11). The term "column equations" hereafter will mean the primal equations (8), not the dual column equations as in (1).

Transportation Networks and Basic Trees

Figure 11.2A shows a transportation network with three sources ($a_i = 4,2,9$, represented by circles) and three destinations ($b_j = 5,7,3$, represented by squares). The arcs are labeled with their unit costs c_{ij}. The solid lines indicate the *tree* of arcs used by the (basic) optimal solution. The values of its x_{ij} are enclosed in polygons pointing in the direction of shipment or flow; the other x_{ij} are nonbasic and zero. One may verify that these flows reduce all the node values to zero. For example, the requirement 7 at one node is satisfied by the incoming flows $1+2+4$.

Figure 11.2B displays a larger example in the style of Figure 11.2A but with nonbasic arcs omitted. In either figure, *the fact that each arc has exactly two nodes* (*a source and a destination*) *as endpoints mirrors the occurrence of each* x_{ij} *in exactly two*

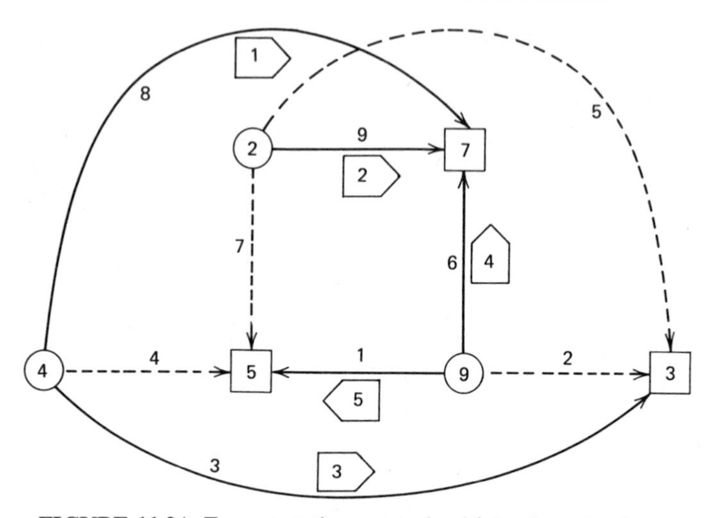

FIGURE 11.2A Transportation network with basic optimal tree.

primal equations, or equivalently, the occurrence of exactly one u_i and one v_j in each dual equation. Note that even if the tree was undirected and unlabeled, its nodes could be divided in a unique fashion into complementary subsets whose sizes would determine the unordered pair $\{m, n\}$, because every (mixed) path encounters origins and destinations in alternation.

For the basic flow pictured in Figure 11.2B, Figure 11.2B′ shows the corresponding pattern of the basic x_{ij} in the tableau (6′). The figure formed by connecting the cells of basic x_{ij} by means of horizontal and vertical lines will be called an *espalier*,

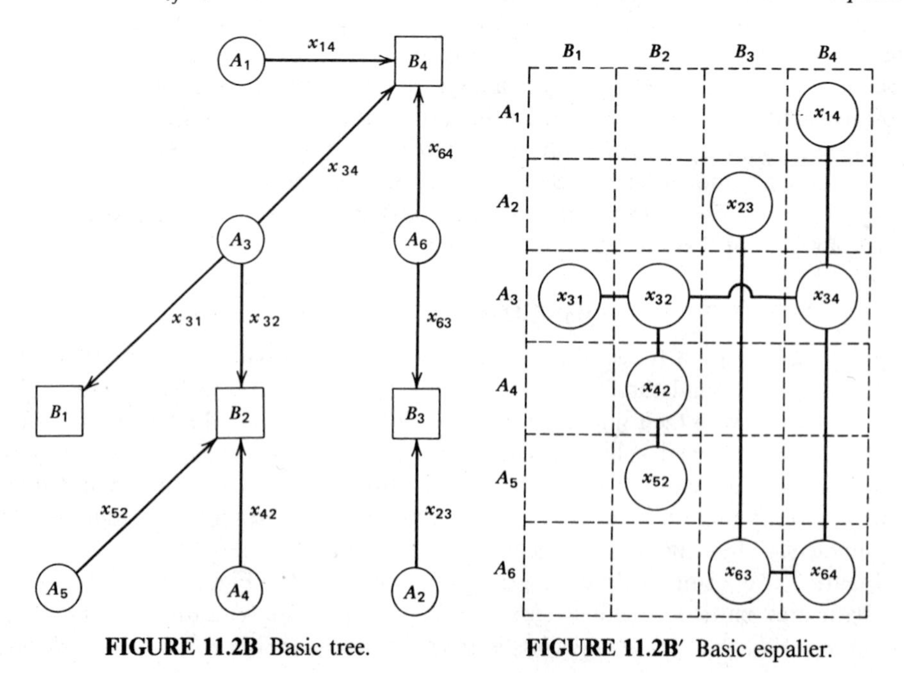

FIGURE 11.2B Basic tree.

FIGURE 11.2B′ Basic espalier.

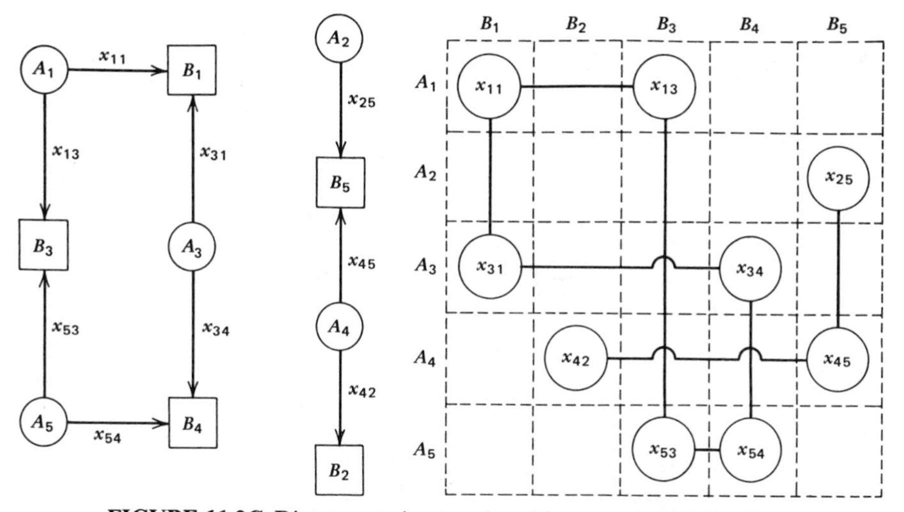

FIGURE 11.2C Disconnected networks with loops. **FIGURE 11.2C′**

which is a tree growing on a trellis. Figures 11.2C and C′ provide a similar comparison for a group of x_{ij} that lacks *both* of the properties (connectedness and the absence of loops) that, by (d_1) of 9.1, are jointly necessary and sufficient to insure that a graph is a tree. The four figures illustrate the easily confirmed fact that the flow network of nodes A_i and B_j and arcs x_{ij} on the left is a tree if and only if the tabular network with "nodes" x_{ij} on the right is a tree or espalier. The latter is a tree only because we agree to suppress certain potential "arcs," such as $\{x_{32}, x_{52}\}$ in Figure 11.2B′. The espalier is useful because it is before one's eyes in the tableau (6′), which easily accommodates nonbasic variables as well as basic.

The following is the main result of this section. $|S|$ denotes the number of elements in the set S.

THEOREM 1 If S is a set of variables x_{ij} in the m by n transportation problem (7), (8), (11), the following conditions are equivalent.

(a) The matrix of coefficients of the x_{ij} in S is triangular and nonsingular as illustrated in (1); $|S|=m+n-1$.

(b) S is a basis, meaning that the values (possibly negative) of its x_{ij} are uniquely determined by (7) and (8) for all assignments of values to the nonbasic x_{ij}.

(c) After any one of them is arbitrarily specified, the values of the remaining u_i and v_j are uniquely determined by

$$u_i + v_j + c_{ij} = 0 \qquad \text{when} \qquad x_{ij} \text{ is in } S \tag{13}$$

for all assignments of values to the c_{ij}.

(d) S is connected and representative (the latter means each row p has some x_{pj} in S and each column q has some x_{iq} in S); $|S| \leq m+n-1$.

(e) S forms an espalier as in Figure 11.2B′; $|S|=m+n-1$.

(f) S is free of loops; $|S| \geq m+n-1$.
(g) The arcs in S and the $m+n$ nodes $A_1, \ldots, A_m, B_1, \ldots, B_n$ form a tree as in Figure 11.2B.

Proof Statements (d), (f) and (g) are equivalent by the theorem of 9.1. (e) is equivalent to these because a tabular network as in Figures 11.2B′, C′ is connected, representative and free of loops if and only if the corresponding flow network as in Figures 11.2B, C has the same properties. The rest of the proof essentially shows that a tree permits a unique solution and other networks do not. More specifically, the following implications suffice.

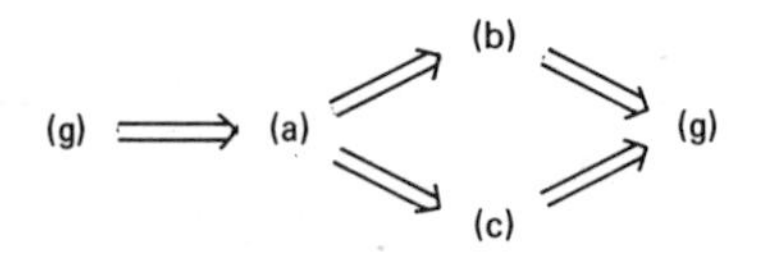

(e) would serve as well as its equivalent (g) in these proofs.

The implication (g)⇒(a) is a by-product of (g)⇒(b) (solving for basic x_{ij}) or (g)⇒(c) (solving for the u_i and v_j) (see above). In either case the procedure works because there is exactly one path joining any two nodes or arcs. At each step one node and one arc are eliminated; i.e., one variable is determined so as to satisfy one equation. Since the number of nodes exceeds the number of arcs by 1, the extra node is eliminated initially when one u_i or v_j is arbitrarily specified. As for the x_{ij} solution, it begins with a node and arc at one end of some path; at the last step one has left one arc and its two nodes, one of which represents a redundant equation. In either case the order in which the nodes and arcs are eliminated determines acceptable orders for the rows and the columns (as in (1)) so that a triangular matrix of coefficients is obtained.

(a)⇒(b) and (c) is obvious; see the discussion of (1).

(b)⇒(g) and (c)⇒(g) will be proved in contrapositive form. If (g) is false, the network is not a tree; it is not connected or it has a loop. If not connected, it may still be possible to determine the basic x_{ij} if $\Sigma a_i = \Sigma b_j$ happens to hold for each subtree; however, this condition is lost if some nonbasic x_{ij} (connecting nodes in different subtrees) is given a nonzero value, and so (b) is false. As for the u_i and v_j, one of them can be assigned an arbitrary value in each subtree, thus violating (c). If the network has a loop, the x_{ij} in its arcs can be alternately increased and decreased by an arbitrary Δx as in Figure 11.3, and (b) is false; while a linear combination of the equations (13), alternately added and subtracted, for the nodes $k, l, p, \ldots, s$ in the loop gives

$$c_{kl} - c_{pl} + c_{pq} - c_{rq} + \cdots - c_{ks} = 0$$

which contradicts the free choice of values of the c_{ij} in (c). To summarize the four cases, if the network is not a tree, the number of possible sets of values for the selected variables is either 0 or ∞ instead of the 1 required for a (uniquely determined) basic solution. Exercise 10 illustrates these ideas.

Condition (d) with $|S|=m+n-1$ is probably easiest to apply. In a small array a visual scan of the espalier as in Figure 11.2B′ will verify that it is connected and representative. In a larger problem one likewise checks that there are $m+n-1$ variables x_{ij} in the purported basis S, and then marks an arbitrary one of the $m+n$ rows and columns (nodes), say row 1. Next mark the columns j such that x_{1j} is in S, then the rows i such that x_{ij} is in S with j having just been marked, then more columns, etc. The $m+n-1$ variables x_{ij} in S form a possible basis if and only if it is possible to mark all m rows and n columns. The procedure is like the solution for the u_i and v_j without the arithmetic.

The following table summarizes the correspondences among the arcs and nodes of the flow network (e.g., Figure 11.2A) and the variables and constraints of the primal and the dual linear programs.

Network Elements	Primal	Dual
m origin nodes A_i	$\Sigma_j x_{ij}=a_i$	Variable u_i
n destination nodes B_j	$\Sigma_i x_{ij}=b_j$	Variable v_j
$m+n-1$ arcs in tree	Basic $x_{ij}\geq 0$	$u_i+v_j+c_{ij}=0$
$(m-1)(n-1)$ other arcs	Nonbasic $x_{ij}=0$	$u_i+v_j+c_{ij}\geq 0$

Since one of the $m+n$ primal equations and one of the dual variables u_i or v_j is redundant, the effective number is $m+n-1$, equal to the number of basic x_{ij}.

THEOREM 2 (i) If the a_i and b_j are all integers, then all the x_{ij} in every basic solution are integers (in fact each basic x_{ij} is equal to the difference between a sum of certain a_i and a sum of certain b_j), and the optimal x_{ij}^* can be chosen to be integers.

(ii) If the c_{ij} are all integers, then all the u_i and v_j in every basic solution can be integers, and the optimal u_i^* and v_j^* can be chosen to be integers.

Proof The basic solutions are or can be integral because no division is required in solving the triangular sets of linear equations. According to Theorem 4(c) of 3.3, if an optimal solution exists at all, as it does here, it can always be obtained as a basic solution; the algorithms of 11.4 and 11.6 will give an independent proof. The relation $\Sigma a_i=\Sigma b_j$ insures that the parenthetical assertion can be realized in two different ways if at all. Each x_{ij} is determined by a row equation (involving a_i) and substituted in a column equation (involving b_j), or vice versa, and these cases alternate as in Figure 11.2B above. Each such substitution changes the signs of the a_i and b_j already appearing in the solution. Since two such sign changes accompany the addition of each new a_i or b_j, all solutions have the form

$$x_{kl}=\pm\left(\Sigma' a_i-\Sigma'' b_j\right)$$

where Σ' and Σ'' are partial sums.

EXERCISES 11.2

1. Draw a network like Figure 11.2A for $m=2$, $n=4$.

In 2 to 4 draw graphs as in Figure 11.2B, B′ or C, C′. Which are bases? (Only ij is given for the x_{ij} in S.)

2. $m=n=4$; $S=\{23,12,43,24,33,21,13\}$
3. $m=n=4$; $S=\{42,14,23,34,43,31,22\}$
4. $m=3$, $n=5$; $S=\{23,33,21,14,34,15\}$

In 5 to 8 find which are bases without drawing graphs.

5. $m=n=4$; $S=\{14,32,21,43,24,33,23\}$
6. $m=6$, $n=4$; $S=\{51,32,14,24,33,62,43,54,63\}$
7. $m=3$, $n=5$; $S=\{12,52,43,51,42,23,13\}$
8. $m=4$, $n=3$; $S=\{43,32,22,13,42,21,23\}$
9. Solve for the basic x_{ij} in Figure 11.2B; for the u_i and v_j with $v_4=0$.
10. To illustrate the proof that (b) or (c) implies (g), show that the values of (i) x_{11}, x_{22} (which are not connected) or of (ii) $x_{11}, x_{12}, x_{21}, x_{22}$ (which form a loop) are not uniquely determined by the remaining x_{ij} (if any) in the transportation problem with $m=n=2$ and all $a_i=b_j=1$. (iii) Comment on the determination of the u_i and v_j in these cases.

11.3 PIVOT STEPS AND PRICING METHODS

As in 2.4, 3.3 and 7.3, a pivot step changes the identity of one of the x_{ij} in the basis, so as to progress toward an optimal solution; the new basic variable $x_{ll'}$ replaces the old $x_{kk'}$. In 3.3 this involved recalculating the entire tableau typified by (5) of 11.2, including the matrix **A** and the vectors **b** and **c**. Now a special feature of the transportation algorithm is that the matrix **A** is never recorded explicitly. Indeed we shall find, as in the last paragraph of 11.1, that only the numbers 0, 1 and -1 occur as elements of **A**, either initially or after a basic solution is obtained; thus **A** may be replaced by suitable logical instructions in the algorithm. To be sure, the general methods of 7.2 and 7.3 record **A** only once, in its initial form; and in fact the present method shares with them the operation of pricing out, the u_i and v_j being equivalent to the π_i or c_i^r. However, the present method is much simpler, in that respect being comparable with 3.3; there is no recording of inverse matrices $\mathbf{B}_r^{-1}$ as in 7.2 or of $\boldsymbol{\eta}^r$ vectors as in 7.3, but rather positions of basic x_{ij}.

The essential operations of pivoting relate to the basic tree illustrated in Figures 11.2B and B′. We use the latter form (the espalier or tabular network) in the figures but speak of arcs x_{ij} and nodes i or j as in the flow network (Figures 11.2A or B).

Updating X and $c'_{ll'}$ in the Δx Loop

When $x_{ll'}$ is nonbasic, its value is zero; when it enters the basis it acquires some value denoted by $\Delta x \geq 0$. It is necessary to express the old basic variables as

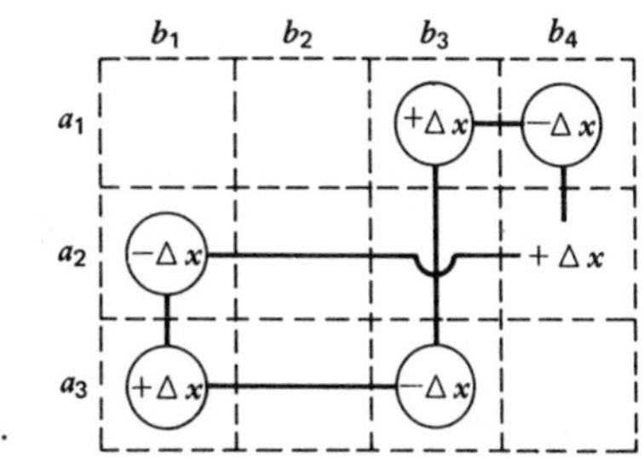

FIGURE 11.3 Δx loop.

functions of the parameter Δx. Since the old basic tree is connected to all $m+n$ nodes, it contains a (mixed) path joining the nodes l and l'; the path together with the arc ll' forms a loop. In each row and column (at each node) this loop has either two x_{ij} (arcs) or none. Since a fixed sum a_i or b_j is specified for the x_{ij} in row i or column j, when $x_{ll'}$ increases by Δx, its two neighbors (say $x_{hl'}$ and $x_{lh'}$) in the loop must decrease by Δx, their neighbors must increase by Δx, etc. The adjustments are $+\Delta x$ and $-\Delta x$ alternately around the loop as in Figure 11.3, where $l=2$, $l'=4$. To obtain a basic solution, some $-\Delta x$ must reduce its $x_{kk'}$ to zero, so that $x_{kk'}$ can leave the basis, and if the smallest such value is selected, no x_{ij} will become negative.

The loop is constructed by labeling $x_{ll'}$ with $+$ or P and the basic x_{ij} with $+$, $-$ or 0 (P, M or Z) as dictated by the triangular solution procedure for the x_{ij} illustrated by (3) of 11.2. Those labeled 0 or Z do not belong to the loop, which may be completed before all the basic x_{ij} are labeled. Computer programs use more elaborate labeling; see the references at the end of 11.4.

The reasoning by which the loop was constructed shows that the column of coefficients of $x_{ll'}$ in a tableau such as (5) of 11.2 has entries ± 1 corresponding to labels $\mp\Delta x$ in the loop, and 0 in the rows of basic variables not in the loop.

The rest of this section reduces the amount of calculation but is not indispensable in 11.4 to 11.6.

The effect of the adjustments in the Δx loop is to increase z by a (negative) amount given by Δx times

$$\begin{aligned} c'_{ll'} &= c'_{ll'} + (-c'_{lp} + c'_{qp} - c'_{qr} + \cdots - c'_{sl'}) \\ &= c_{ll'} - c_{lp} + c_{qp} - c_{qr} + \cdots - c_{sl'} \end{aligned} \tag{1}$$

In Figure 11.3 specifically this is $c_{24} - c_{14} + c_{13} - c_{33} + c_{31} - c_{21}$. The right-hand member has the same value in every tableau because the u_i and v_j always cancel out when one substitutes $c'_{ij} = c_{ij} + u_i + v_j$. The parenthesized terms are zero because they correspond to currently basic x_{ij}. Thus (1) shows how the $\pm\Delta x$ loop can be used to calculate an isolated value of $c'_{ll'}$ without using any u_i or v_j. This *stepping-stone or loop pricing* is analogous to (3) of 8.1 (dual pricing out in row z). It has possible value as an auxiliary procedure. Candidates for the role of $c'_{ll'}$ are the most infeasible c'_{ij} previously obtained by (2) and (3) following. *If (1) proves to be feasible, the loop must be abandoned.* In any case (1) may be applied to the next most infeasible c'_{ij}, or a new major cycle begun with (2) and (3). In Phase 1b in 11.5 or 11.6 the latest $g_{ij}=1$ for currently infeasible x_{ij} must be used in (1). The x_{ij} are updated after each application of (1) that yields an infeasible $c'_{ll'}$. Exercise 7 of 11.6 gives another application.

Pricing Out Via the Δc Partition

In 11.1 it was explained that the elimination of the basic variables from z resulted in the replacement of c_{ij} by

$$c'_{ij} = c_{ij} + u_i + v_j \tag{2}$$

The u_i and v_j were determined by assigning an arbitrary value to one of them and then solving the triangular equations

$$u_i + v_j + c_{ij} = 0 \qquad \text{when } x_{ij} \text{ is basic} \tag{3}$$

This method is indispensable; there is no other equally good way to get the algorithm started, and the method can be used at every iteration. However, it is seldom efficient to solve the same linear system from scratch at every iteration; it is better to use an incremental procedure like the Δx loop, but not so limited as (1).

To get the loop, an additional variable $x_{ll'}$ was introduced. No additional u_i or v_j is available, but an equivalent procedure is to eliminate one of the dual equations; naturally $u_k + v_{k'} + c_{kk'} = 0$ is the one to drop since $x_{kk'}$ is leaving the basis. This allows $c'_{kk'}$ to vary and splits the basic tree T into two subtrees T' and T'' such that k and k' belong to different subtrees, as do l and l'. Since T' and T'' have no node in common, they must lie in complementary sets of rows M' and M'' and complementary sets of columns N' and N''. However, if $x_{kk'}$ is the only basic variable in its row or column, as in the first iteration in Phase 2 in Table 11.4B below, then one of the subtrees consists of a single node, the corresponding espalier is empty, one of the sets M', M'', N', N'' is empty, and another consists of a single element $k = l$ or $k' = l'$.

The subtrees still constrain the u_i and v_j so that the only variation possible is to add $\Delta c', \Delta c'', -\Delta c', -\Delta c''$ when the node index is in M', M'', N', N'' respectively. Subtracting $\Delta c''$ from the u_i in M and adding $\Delta c''$ to the v_j in N gives the equivalent adjustments $\Delta c, 0, -\Delta c, 0$ with $\Delta c = \Delta c' - \Delta c''$. Thus possible new values for the u_i, v_j and c'_{ij} (the latter being unique) are the following:

$$\begin{array}{c|cc|c} & N' & N'' & \\ \hline M' & c_{ij}+u_i+v_j & c_{ij}+u_i+v_j+\Delta c & u_i+\Delta c \\ M'' & c_{ij}+u_i+v_j-\Delta c & c_{ij}+u_i+v_j & u_i \\ \hline & v_j-\Delta c & v_j & \end{array} \tag{4}$$

where we may suppose that $k \in M'$ and $k' \in N''$. In the uncapacitated problem considered here, $c'_{ll'}$ and $c'_{kk'}$ are both increased by $|c'_{ll'}|$, while $x_{ll'}$ and $x_{kk'}$ get adjustments $+\Delta x$ and $-\Delta x$ respectively. Ford and Fulkerson (1957) used (4) as part of a different transportation algorithm.

In a small exercise the smaller of the subespaliers is easily found by inspection. Then $\Delta c = |c'_{ll'}|$ is entered in the corresponding rows and columns, with minus signs on the right or below, whichever serves to change $c'_{ll'}$ from negative to zero.

In general a mere marking of the rows and columns of one of the subtrees will suffice; these are the u_i and v_j to be adjusted. Since they are adjusted in opposite senses, the corresponding c'_{ij} in the subtree, in marked rows *and* columns, are not adjusted. This marking involves only a portion of the scanning of basic cells that occurs in solving (3) for the u_i and v_j (as in (4) of 11.2) or in testing for connectedness as in (d) after Theorem 1 of 11.2. It is simpler because the arithmetic is postponable, there are many zeros, and one scans and labels only a subtree (hopefully the smaller), not the whole tree; the latter is true because $x_{ll'}$ is treated as basic in (3) but nonbasic in (4). One can begin by marking row l or column l' (not both), whichever will result in marking k or k' more quickly, by traversing the shorter portion of the Δx loop between kk' and ll'.

In view of the pivot rules of 2.4 or 2.5, it is clear that in the row kk' of a tableau such as (5) of 11.2, the coefficient of x_{ij} is $+1$ when (4) adjusts c'_{ij} in the same sense as $c'_{kk'}$, -1 when the senses are opposite, and 0 when c'_{ij} is unchanged. Here ll' might be undefined, and labeling would begin with k or k'. Thus the Δx loop determines the elements in the column ll' of $\mathbf{A}$ and the Δc partition determines the elements in the row kk' of A. These correspond to the forward and the backward transformations of 7.3.

The adjustments (4) may be used for updating $\mathbf{u}$ and $\mathbf{v}$ or $\mathbf{C}$, but there is probably no need to do both. Thus we have four methods for pricing out:

(a) The triangular solution (2), (3) is indispensable initially and may be used on other occasions in Phase 1b of 11.6 or in combination with (d). Thus it is the only universal method but as such it is not the most efficient for man or computer.

(b) The Δc partition for one-step updating of $\mathbf{u}$ and $\mathbf{v}$ only in (4) permits a computer program to be choosy about how many of the c'_{ij} it calculates in (2).

(c) The Δc partition for one-step updating of $\mathbf{C}'$ only in (4) is recommended for solving pencil-and-paper exercises. Since one of x_{ij} and c'_{ij} is always zero in uncapacitated problems, both can be recorded in a single m by n matrix, the (basic) x_{ij} being distinguished by circling. However, note that numerical errors will generally go undetected until a special check procedure is performed.

(d) The Δx loop pricing (1) for a few selected c'_{ij} is rendered superfluous by (b) or (c) but might work well in combination with periodic applications of (a). It is not clear how this would compare with (a)+(b) for computer programs.

These procedures do not determine the current value of z, which is not needed. It is found at the end of the calculation by using (9) of 11.1.

There is more in 11.7, in the text and in Exercise 11.

EXERCISES 11.3

For the given $\mathbf{a}, \mathbf{b}, \mathbf{C}$, with $u_2 = 0$ and x_{ij} basic where c_{ij} is parenthesized, solve for $\mathbf{X}, \mathbf{u}, \mathbf{v}$ as in (3) and (4) of 11.2 and find c'_{ij} from (2). Then apply Figure 11.3 and (4) to find how these quantities are changed by the pivot step that exchanges x_{ij} in the starred positions.

1.

	8	5	5	5
9	(10)	5*	8	(6)
7	(6)	6	(7)*	5
7	4	(1)	(2)	3

x_{12} replaces x_{23}.

2.

	1	1	1	1
1	9	(6)	3	(5)
1	7	7	(2)	4*
1	(10)	11	(5)	8
1	15	(12)*	(7)	10

11.4 UNCAPACITATED TRANSPORTATION ALGORITHMS

The broad outlines of the simplex method for these problems are the same as for the general simplex methods early in Chapters 6 and 7. Phase 1 produces a basic solution that is feasible in the primal; i.e., it has $x_{ij} \geq 0$ and row and column sums equal to a_i and b_j, respectively. Phase 2 produces more basic feasible primal solutions with accompanying basic infeasible dual solutions. The value of z decreases (or remains constant) in Phase 2 until its minimization is signaled by the achievement of dual feasibility (all $c'_{ij} = c_{ij} + u_i + v_j \geq 0$).

Phase 1 A feasible basic solution for the x_{ij} is written down by repetitions of steps 1 through 4, beginning with all $a'_i = a_i$ and $b'_j = b_j$.

1. Select some indices p and q that are still available and set $x_{pq} = \min(a'_p, b'_q)$ (the smaller of the two). To minimize z one prefers to have c_{pq} = the smallest available c_{ij}.

2. If $a'_p > b'_q$, then the other x_{iq} $(i \neq p)$ are nonbasic (zero) and the qth column equation in (6′) or (8) of 11.2 is satisfied. This column (equation and variables) is marked so as to be excluded from further consideration in Phase 1. In the remainder of the array of x_{ij}, a'_p is replaced by $a'_p - b'_q$ (perhaps only mentally) to take account of the term $x_{pq} = b'_q$ which has disappeared.

3. Similarly, if $b'_q > a'_p$, the other x_{pj} $(j \neq q)$ are nonbasic (zero) and the pth row of x_{ij} is excluded after b'_q has been replaced by $b'_q - a'_p$.

4. If $a'_p = b'_q = x_{pq}$ (basic), some x_{pj} $(j \neq q)$ or x_{iq} $(i \neq p)$ having a small cost is also made basic, with the value zero; this will be possible except when row p and column q are the only ones not yet excluded. In any case row p and column q are now both excluded. In hand calculation the value 0 for any x_{ij} is written in explicitly only if it is basic; if x_{ij} is nonbasic the space is left blank for the present.

5. The above steps are repeated until all the rows and columns have been excluded. Evidently $m+n-1$ basic x_{ij} will be determined, in agreement with Theorem 1 of 11.2; the -1 arises at the very last step, when the last row and column are both excluded but only one basic variable is determined. At all times the a'_i and b'_j are positive like the a_i and b_j, and the relation $\Sigma a'_i = \Sigma b'_j$ holds, since $\Sigma a_i = \Sigma b_j$ initially and at each step the same quantity is subtracted from each side.

Any feasible basic solution can actually be obtained by the above procedure; Theorem 1 of 11.2 shows that some basic variable x_{pq} (at the end of some path in the tree) in any basis is equal to one of its marginal totals, and feasibility requires that it be the smaller of a_p and b_q.

In solving exercises one can scan all available c_{ij} to find the smallest, but in large problems this is too much even for a computer. Instead (if $m \leq n$) the $\min_i c_{ij}$ in (the remainder of) each successive column is used. This gives n basic x_{ij} and $m-1$ rescans of columns complete the basis.

As a possible refinement, let c^0_{ij} denote the raw values which are adjusted as follows:

$$u^0_i = -\min_j c^0_{ij}, \qquad v^0_j = -\min_i \left(c^0_{ij} + u^0_i\right), \qquad c_{ij} = c^0_{ij} + u^0_i + v^0_j \tag{1}$$

Then all $c_{ij} \geq 0$ and at least one in every row and column is zero. The many ties that thus result for $\min c_{ij}$ may be broken thus. Define p_{kl} and q_{kl} to be the numbers of $c_{kj}=0$ in row k and $c_{il}=0$ in column l of each $c_{kl}=0$, not respectively but so that $p_{kl} \leq q_{kl}$. Then one seeks $\min_{k,l}\{p_{kl}: c_{kl}=0\}$ and among these the $\min_{k,l} q_{kl}$. If ties still remain, choose k and l to maximize the smallest nonzero c_{kj} or c_{il} in the row k or column l of each $c_{kl}=0$ having minimal values of p_{kl} and q_{kl}. This last step is essentially the method of Vogel in Reinfeld and Vogel (1958). It tends to be effective but time-consuming and therefore its use should be restricted. The object is to have high-cost rather than low-cost routes excluded from further consideration. It immediately gives the optimal solution for the "hard" assignment problem with $c_{ij}=(i-1)(j-1)$ proposed by Machol and Wien (1977).

Phase 2 The more substantial phase of the algorithm always begins by calculating the values of the variables u_i, v_j in the basic dual solution that is described as being complementary to the basic primal solution produced in Phase 1. This is accomplished by solving the equations

$$u_i + v_j + c_{ij} = 0 \qquad \text{when } x_{ij} \text{ is basic} \tag{2}$$

thereby eliminating the basic x_{ij} from z in (6) of 11.1. The value of one u_i or v_j is assigned arbitrarily and the remaining $m+n-1$ values are determined one at a time by the $m+n-1$ equations (2) as in (4) of 11.2 or the example below.

Equation (6) of 11.1 also shows that

$$c'_{ij} \triangleq u_i + v_j + c_{ij} \tag{3}$$

are the coefficients of the (nonbasic) x_{ij} remaining in z. The nonbasic x_{ij}, being zero in the basic solution, can only increase, not decrease. Hence if all $c'_{ij} \geq 0$, z cannot be decreased and an optimal solution has been obtained. The corresponding value of $z^* \triangleq \min z$ is obtained by substitution in (9) of 11.1, both expressions being used as a check.

In the absence of optimality one calculates

$$c'_{ll'} \triangleq \min c'_{ij} < 0 \tag{4}$$

In exercises the minimum in (4) is taken over all the $(m-1)(n-1)$ nonbasic positions. In solving large problems on computers it has been found preferable to scan the rows or columns of c'_{ij} one after another and to define $c'_{ll'}$ to be the most negative (if any) in the current row or column. In 6.3 the criterion (4) was criticized and improved upon for use in general linear programs, but it is ideal for transportation problems because all the x_{ij} are measured in the same units (there is only one commodity) and their coefficients a_{ij} are perfectly comparable, being always 1, -1 or 0.

Equations (3) and (4) imply that z will be decreased if $x_{ll'}$ can be increased from its value of zero. In any case its new value is denoted by Δx and the effect on the basic variables is indicated by the Δx loop illustrated in Figure 11.3. In order to decrease z as much as possible while keeping the basic $x_{ij} \geq 0$ we define

$$\Delta x = x_{kk'} = \min\{\text{basic } x_{ij} \text{ in loop with label } - \text{ or } M\} \tag{5}$$

Old status:	Nonbasic $x_{ll'}=0$	Basic $x_{kk'}=\Delta x$
New status:	Basic $x_{ll'}=\Delta x$	Nonbasic $x_{kk'}=0$

One may have $\Delta x>0$ or $\Delta x=0$, but *in either case the variable* $x_{kk'}$ *leaving the basis must have a minus label*. Otherwise when $x_{kk'}$ became nonbasic it would have $c'_{kk'}<0$ and so be eligible to enter the basis again.

Ties frequently occur in determining ll' and kk'. Since it is generally desirable for basic x_{ij} to have small c_{ij}, in solving exercises one chooses the smallest $c_{ll'}$ tied in (4) and the largest $c_{kk'}$ tied in (5).

The new basic solution is obtained by adding and subtracting Δx around the loop as indicated by the labels. Then one could return to (2) and (3), but (4) of 11.3 is faster for exercises and (4) or (1) of 11.3 is faster for computer programs. In any case the conditions $c'_{ll'}<0$ and (5) of this section apply at every iteration. At the end (2) and (3) should be used again to check the optimality.

The proof of the convergence of the simplex method depends on the continual decrease in z, which would prevent any basis from recurring. However, if $\Delta x=0$ there is no decrease, and convergence is delayed or conceivably prevented altogether. The occurrence of $\Delta x=0$ (a form of *degeneracy*) can be avoided by *perturbing* the problem by replacing a_1 by $a_1+n\varepsilon$ and each b_j by $b_j+\varepsilon$, where $\varepsilon>0$ is sufficiently small ($\varepsilon<(2n)^{-1}$ if all a_i and b_j are integers). All these parameters could then be multiplied by a suitable integer to avoid fractions. In the most degenerate case, when all $a_i=b_j=1$, a better device will be described at the end of this section. In any case it is said that no failure to converge has been observed for transportation problems.

Figure 11.4 illustrates the significance of a change in the basis and the dual solution without any change in the numerical values of the x_{ij}. It shows the feasible region and the optimal solution

	c_{ij}		u_i		x_{ij}		a_i
(−2)	(−1)	(0)	0	2	3	0	5
0	0	(0)	0			4	4
2	1	0	$=v_j$	2	3	4	$=b_j$

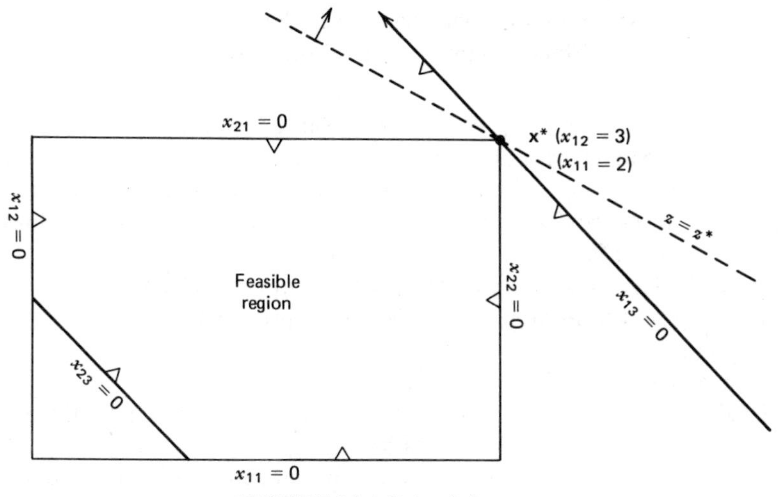

FIGURE 11.4 Primal degeneracy.

(in which x_{21}, x_{22} are nonbasic) for a transportation problem. The solution

(−2)	(−1)	0	2	2	3		5
(0)	0	(0)	0	0		4	4
0	−1	0		2	3	4	

(6)

(in which x_{13}, x_{22} are nonbasic) is not optimal because $u_2+v_2+c_{22}=-1$; nevertheless all the x_{ij} have the same numerical values as before. The figure shows that the constraints $x_{13}\geq 0$, $x_{22}\geq 0$ are not sufficient to define the optimum. Until $x_{21}\geq 0$ is brought in by a change of basis, z could seemingly be decreased by moving northwest along $x_{13}=0$.

The algorithm is summarized in Table 11.4A and applied to an example in Table 11.4B, the c'_{ij} being calculated by (2) and (3) in the first iteration of Phase 2 and by (4) of 11.3 thereafter. After the basic x_{ij} are determined in Phase 1 in the sequence $A, B, \ldots, G$, one sets $u_3=0$. Then

$$
\begin{aligned}
v_1&=-(c_{31}+u_3)=-4 & v_4&=-(c_{34}+u_3)=-3\\
v_5&=-(c_{35}+u_3)=-3 & u_1&=-(c_{11}+v_1)=-5\\
v_2&=-(c_{12}+u_1)=-2 & u_2&=-(c_{22}+v_2)=-2\\
v_3&=-(c_{23}+u_2)=-1 & &
\end{aligned}
$$

The most infeasible $c'_{ij}=c_{ij}+u_i+v_j$ in the first Phase 2 tableau is $c'_{ll'}=c'_{15}=6-5-3=-2$. Since an increase in $x_{15}=0$ will decrease z, position 15 is labeled P. To preserve $b_5=2$, $x_{35}=2$ is labeled M. $x_{34}=3$ gets Z to preserve $b_4=3$. $x_{31}=1$ gets P to preserve $a_3=6$. $x_{11}=5$ gets M to preserve both $b_1=6$ and $a_1=9$. Then $\Delta x=$ the smallest x_{ij} labeled M, which is $x_{kk'}=x_{35}=2$, now labeled $\overline{M}$. Adding $\Delta x=2$ to $x_{15}=0$ and $x_{31}=1$ (labeled P), subtracting it from $x_{35}=2$ and $x_{11}=5$ (labeled M), and leaving the other four basic x_{ij} unchanged gives the basic x_{ij} (and the nonbasic

TABLE 11.4A Summary of the (Uncapacitated) Simplex Transportation Algorithm

---

Phase 1. Let $c_{pq} = \min c_{ij}$ over ij not yet excluded. Let basic $x_{pq} = \min(a'_p, b'_q) =$ maximum feasible value. If $x_{pq} = a'_p$, exclude row p and replace b'_q by $b'_q - a'_p$. If $x_{pq} = b'_q$, exclude column q and replace a'_p by $a'_p - b'_q$. If both are done, set lowest-cost x_{pj} or $x_{iq} = 0$ (basic). This gives $m+n-1$ values of basic x_{ij}; others are 0.

Phase 2. 1. Set any one u_i or $v_j = 0$; calculate the others from $u_i + v_j + c_{ij} = 0$ wherever c_{ij} is marked to show that x_{ij} is basic (optional except in the first and the last iterations).

2. (Pricing Out). Get $c'_{ij} = u_i + v_j + c_{ij}$ or use (4) of 11.3.

3. If all $c'_{ij} \geq 0$, check by (7) to (9) of 11.1; *stop* (optimal). Otherwise $c'_{ll'} = \min c'_{ij} < 0$ gets label P. Break ties by $\min c_{ij}$.

4. Show the effect of an increase in $x_{ll'}$ by labeling the basic variables P (plus, increase), M (minus, decrease), or Z (zero, no change). P and M will alternate around a loop.

5. Find $x_{kk'} = \min\{\text{basic } x_{ij} \text{ labeled } M\} = \Delta x$. It leaves the basis as $x_{ll'}$ enters. Use $\max c_{ij}$ to break ties.

6. Record the new basic x_{ij} in a new tableau: add Δx in P positions and subtract Δx in M positions. Return to step 1 or 2. Steps 1 to 6 correspond to steps 5, 6, 1, 2, 3, 4 in Table 7.3; here one does not begin with an artificial basis.

TABLE 11.4B Transportation Example

	c_{ij} (Phases 1 and 2)								Basic x_{ij} (Phase 1)				
	G	F	A	D	C	u_i		a\b	6	4	7	3	2
G	9G	7F	5	8	6	−5	−5	9	5G	4F			
B	11	4B	3A	5	8	−2	−3	7		0B	7A		
E	4E	8	6	3D	3C	0	−1	6	1E			3D	2C
	−4	−2	−1	−3	−3	= initial v_j							
	−3	−1	0	−2	−1	= final v_j = initial + $\Sigma(\pm \Delta c)$							

Basic x_{ij} determined and equations satisfied in the order $A, B, \ldots, G$. Small c_{ij} preferred. Example (E): $x_{31} = 1 = \min(6-3-2, 6)$ satisfies row 3.

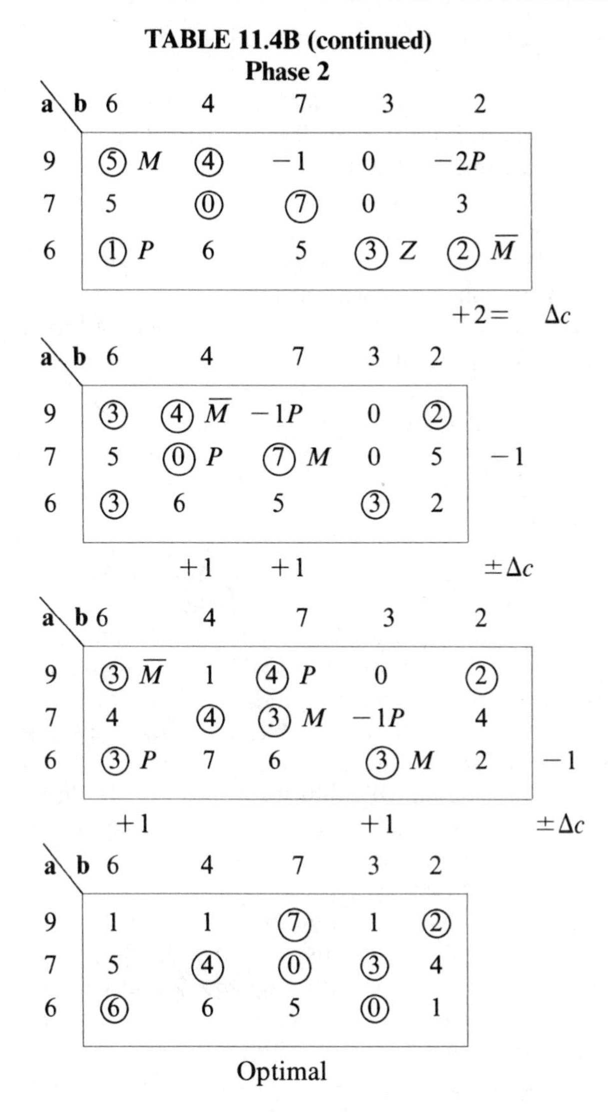

TABLE 11.4B (continued)

Phase 2

a \ b	6	4	7	3	2	
9	⑤ M	④	-1	0	$-2P$	
7	5	⓪	⑦	0	3	
6	① P	6	5	③ Z	② $\overline{M}$	
					$+2=$	Δc

a \ b	6	4	7	3	2	
9	③	④ $\overline{M}$	$-1P$	0	②	
7	5	⓪ P	⑦ M	0	5	-1
6	③	6	5	③	2	
		$+1$	$+1$			$\pm\Delta c$

a \ b	6	4	7	3	2	
9	③ $\overline{M}$	1	④ P	0	②	
7	4	④	③ M	$-1P$	4	
6	③ P	7	6	③ M	2	-1
	$+1$			$+1$		$\pm\Delta c$

a \ b	6	4	7	3	2
9	1	1	⑦	1	②
7	5	④	⓪	③	4
6	⑥	6	5	⓪	1

Optimal

Within the rectangles are basic x_{ij} (circled) and $c'_{ij} = c_{ij} + u_i + v_j$. P, M mark loop of $\pm\Delta x$ ($= x_{kk'}$ at $\overline{M}$) initiated at $c'_{ll'} = \min c'_{ij}$. $\Delta c = |c'_{ll'}|$ added and subtracted as shown to update the c'_{ij}. $\Delta x = 2, 4, 3$ respectively. $z^* = 4\cdot 6 + 4\cdot 4 + 5\cdot 7 + 5\cdot 3 + 6\cdot 2 = 9\cdot 5 + 7\cdot 3 + 6\cdot 1 + 6\cdot 3 + 4\cdot 1 + 3\cdot 2 + 2\cdot 1 = 102$ by (9) of 11.1.

x_{35}) in the next tableau. $\Delta c = |c'_{ll'}| = |c'_{15}| = 2$ must be added either in row 1 or column 5 to make $c''_{15} = 0$. The latter choice requires no further adjustments since x_{35}, the only old basic variable in column 5, is leaving the basis and so can have $c''_{35} > 0$.

In the second Phase 2 tableau in Table 11.4B one has $\Delta c = |c'_{13}| = 1$. When $x_{kk'} = x_{12}$ is dropped from the basis, inspection shows one is left with two disconnected espaliers consisting of x_{15}, x_{11}, x_{31}, x_{34} and x_{22}, x_{23}. Since the latter is smaller, we choose its row 2 to get the cost adjustment -1 and its columns 2 and 3 to get $+1$; these signs are chosen so that $c'_{13} = -1$ is increased to 0.

As a check, z^* is calculated in two ways ((9) of 11.1).

If one prefers to use step 1 in every Phase 2 iteration, the formats of Tables 11.4B and C are still applicable, except that there is no Δc, only the negative c'_{ij} have to be recorded, and the u_i and v_j have to be recorded beside the initial c_{ij} at every iteration. The c_{ij} should be written in ink and the values circled in pencil in the currently basic positions.

Assignment Problem

If one has all $a_i = b_j = 1$, then all $x_{ij} = 1$ or 0 by Theorem 2 of 11.2. This special case is called the assignment problem because of the application in which $-c_{ij}$ or $100 - c_{ij}$ or the like is a measure of the speed or efficiency of the individual or machine i in performing the job j (both of which are regarded as unique), and it is desired to make the best assignment of individuals to jobs. These problems have $m = n$ and $n-1$ of the $2n-1$ basic variables are zero at all times; thus iterations having $\Delta x = 0$ as in Figure 11.4 are particularly frequent.

The solution of the assignment problem in Table 11.4C has been shortened by omitting the details of Phase 1, allowing several nonoverlapping Δx loops in a single tableau, and using the idea of Barr, Glover and Klingman (1977). They allow no basic $x_{ij} = 0$ in (say) the first row and exactly one in each of the other rows, thus eliminating many possible basic solutions from consideration.

Such a solution can be obtained initially by choosing basic $x_{2r} = 1$ and $x_{2r'} = 0$ to have the lowest costs in row 2, and deleting row 2 and column r; choosing basic $x_{3s} = 1$ and $x_{3s'} = 0$ and deleting row 3 and column s, and so on to row n; finally $x_{1h} = 1$ is uniquely determined in row 1. Thus in step 4 of Phase 1, one always chooses some basic $x_{pj} = 0$, not x_{iq}. If $x_{ll'}$ replaces $x_{kk'}$ in the basis, one requires $k = l$ (both variables in the same row). Obviously this preserves the special character of the solution; it is easy to see that $k = l$ is always permissible in this method though not necessarily in other methods; and convergence to an optimal solution is assured without the use of any other device to deal with degeneracy.

To prove the last statement it suffices to show that no basis can be repeated, since the number of bases is finite. No basis can repeat before and after a pivot step having $\Delta x > 0$ because the step produces an irrevocable decrease in z. When $\Delta x = 0$ one has $x_{kk'} = 0$ leaving the basis; this implies $k \neq 1$. Now any path from the node A_1 (row 1) must have $x_{1h} = 1$, $x_{ih} = 0$, $x_{ij} = 1$, $x_{kj} = 0, \ldots$ on it successive arcs. Hence when (4) of 11.3 is used to update the c'_{ij}, the removal of any $x_{kk'} = 0$ will put A_1 and A_k in different subtrees. Thus one can always leave u_1 unchanged while the other u_i are either increased (as u_k is) or left unchanged. This shows that no basis can repeat during a succession of pivot steps having $\Delta x = 0$.

TABLE 11.4C Assignment Problem (Phase 2).

Sequence of tableaus is $\begin{matrix} 1 & 2 \\ 3 & 4 \end{matrix}$ **All** $a_i = b_j = 1$

c_{ij}					u_i	
2	[6]	9	8	8	−6	−6
[1]	(2)	3	6	6	−2	−5
3	9	(2)	4	[1]	2	−5
4	5	(4)	[2]	1	0	−5
2	(6)	[10]	10	2	−6	−6
1	0	−4	−2	−3	= initial v_j	
4	3	3	3	4	= final v_j	

c'_{ij}					
$-3p$	$[0]\bar{m}$	−1	0	−1	3
$[0]m$	$(0)p$	−3	2	1	
6	11	$(0)P$	4	$[0]M$	−7
5	5	(0)	[0]	−2	−7
−3	(0)	$[0]\bar{M}$	2	$-7P$	
		7	7	7	

$\Delta x = 1, 1.$ $\Delta c = 7, 3.$

					u_i
[0]	3	9	10	9	−3
$(0)M$	$[0]P$	4	9	8	−3
−1	4	$[0]p$	4	$(0)m$	
−2	−2	$(0)\bar{m}$	[0]	$-2p$	2
$-3P$	$(0)\bar{M}$	7	9	[0]	
3	3		−2		

$\Delta x = 0, 0.$ $\Delta c = 3, 2.$

[0]	3	6	5	6
(0)	[0]	1	4	5
2	7	[0]	2	(0)
3	3	2	[0]	(0)
(0)	3	7	7	[0]

Optimal

c_{ij} in brackets denotes $x_{ij} = 1$; in parentheses, $x_{ij} = 0$ but basic. P, M and p, m mark loops of $\pm\Delta x$ (two in each of two tableaus) initiated at $c'_{ll'} < 0$. $\Delta x = x_{kk'}$ at $\bar{M}$ or $\bar{m}$. Except $i = 1$, each row always has exactly one parenthesis (basic $x_{ij} = 0$). $\Delta c = |c'_{ll'}|$ added and subtracted as shown to update the c'_{ij}.

$$z^* = 2+2+2+2+2 = 6+5+5+5+6-4-3-3-3-4 = 10$$

In this method the only purpose of the Δx loop is to choose between the two possibilities for k' that exist (except in row 1). The labeling methods of Srinivasan and Thompson (1972) often determine k' without even constructing the loop.

References

Some references on the labeling of basic x_{ij} and the selection of small costs in Phase 1 and step 3 of Phase 2 in computer programs are Glover, Karney and Klingman (1972; 1974 with Napier), Srinivasan and Thompson (1972, 1973), and Murty (1976, p. 309ff.).

EXERCISES 11.4

Optimize 1 to 8. In 1 to 5, $\begin{matrix} \mathbf{C} & \mathbf{a} \\ \mathbf{b} & \end{matrix}$ are given. In the assignment problems 6 to 8 all $a_i = b_j = 1$, and only $\mathbf{C}$ is given.

1.

1	3	2	8	3
6	2	9	7	7
2	4	5	3	7
8	7	4	2	9
7	9	4	6	

2.

11	17	13	9	3
6	11	8	2	7
2	10	5	1	10
4	4	6	6	

3.

				a_i
12	7	10	7	6
11	4	8	3	5
9	2	5	1	2
10	0	2	1	7
6	5	2	7	

4.

					a_i
6	12	11	17	9	10
16	4	7	10	3	10
8	6	0	4	11	10
7	0	8	6	5	10
5	6	6	11	12	

5.

					a_i
13	13	10	12	7	13
12	11	19	8	4	12
29	6	14	15	15	4
8	19	18	9	3	5
3	9	6	7	9	

6.

22	10	23	25	24
17	6	19	18	20
26	16	29	28	30
31	21	32	35	34
13	3	12	14	15

7.

12	6	17	3	5	11	8
16	13	5	10	6	13	8
2	10	16	13	2	7	13
11	7	10	4	12	3	5
7	12	8	9	10	14	6
4	15	2	6	4	3	13
20	9	2	15	15	5	10

8.

1	2	1	0	3	2
10	12	1	0	7	4
7	9	2	0	8	6
11	11	1	0	9	5
0	0	0	0	0	0
4	6	3	0	5	6

$(a_i = b_j = 1)$

9. Interpret Exercise 15 of 11.1 as an assignment problem.

10. A Latin square as used in agricultural experimentation consists of an approximately square piece of ground divided into n^2 smaller squares in which n varieties (A, B, C, D for an illustration) are planted so that each variety occurs exactly once in each row and in each column. Usually one of the many possible such arrangements is chosen at random. However, the experimenter may be interested in the range of variation between the best and the worst possible selection of n squares for each of two varieties. Formulate as a pair of assignment problems.

11.5 ADDITIONAL CONSTRAINTS AND VARIABLES

The simplest form of additional constraint is $x_{ij}=0$ when the arc ij is nonexistent or unusable. This restricted problem, treated in (a) below, provides an introduction to the minimization of an infeasibility form in Phase 1b. More generally, the a_i, b_j and x_{ij} may represent the quantities available, required and transported *per unit time*. Then one may need constraints of the form

$$0 \leq x_{ij} \leq e_{ij} \tag{1}$$

where the extreme value e_{ij} measures the maximum carrying capacity of the highway, railroad, pipeline, electrical transmission line, fleet of trucks or planes, etc., that runs from i to j. Finally, if several arcs make use of a common facility, the sum of the corresponding x_{ij} may have an upper bound. By introducing additional variables, part (b) reduces some of these problems to part (a), and the so-called transshipment problem is reduced to transportation in part (c). The bottleneck transportation problem is considered in the exercises.

The general algorithm for (1) is reserved for 11.6 but it may be remarked here that if $e_{ij} \geq \min(a_i, b_j)$, one may as well let $e_{ij} = \infty$ (no additional restriction); and a necessary but not sufficient condition for (1) to permit a feasible (hence an optimal) solution is obviously

$$\sum_{j=1}^{n} \min(e_{ij}, a_i, b_j) \geq a_i \qquad (i=1,\ldots,m)$$

$$\sum_{i=1}^{m} \min(e_{ij}, a_i, b_j) \geq b_j \qquad (j=1,\ldots,n) \tag{2}$$

(a) Let each $e_{ij} = 0$ or ∞. Thus certain arcs and their x_{ij} (like the x_i^0 in 6.1) are *artificial* in the sense that they do not exist physically or are unusable for legal, personal or mathematical reasons as in (3) below, but otherwise the problem is still uncapacitated. This will be called the *restricted transportation problem*. If it does have feasible solutions, they could be singled out by giving sufficiently large values to the costs c_{ij} of the artificial arcs; thus $e_{ij} = 0$ and $c_{ij} = \infty$ are equivalent conditions for (dispensable) artificial arcs.

Phase 1 of 11.4, now called Phase 1a, is used initially while attempting to avoid artificial arcs. (It may be helpful to make early choices of basic x_{ij} in the rows and columns containing artificial arcs.) If this attempt is successful it is followed by Phase 2, which is essentially the same as in 11.4. The artificial arcs are usually ignored altogether; since their x_{ij} are known to be zero, their c_{ij} have no sign restriction and they are never brought into the basis. However, an artificial x_{ij} with value zero may remain in the basis at the end of Phase 1. Then if it acquires a + or − label in a Δx loop (Figure 11.3), Δx must be zero but the artificial x_{ij} can leave the basis, no matter whether it is labeled $+(P)$ or $-(M)$.

If Phase 1a does not produce a feasible solution, as in 6.1, the objective function z is temporarily replaced by z^0, which is the sum of those artificial x_{ij} that initially have positive (infeasible) values. The minimization of z^0 by the Phase 2 procedure of 11.4 is called Phase 1b; this minimum will be zero if and only if a feasible solution exists. In any case when the minimum is attained at the end of Phase 1b, one will have $g'_{ij} \triangleq u_i + v_j + g_{ij} \geq 0$ wherever x_{ij} is not artificial, but the reduction of the artificial variables to zero can be observed directly. Here $g_{ij} = 1$ if x_{ij} is artificial and initially positive, $g_{ij} = 0$ if x_{ij} is not artificial, and g_{ij} is undefined (this position is ignored) if x_{ij} is artificial and initially zero.

A restricted transportation problem is solved in Table 11.5. Here $z^0 = x_{41}^0$, the only initially positive artificial variable. The values of the u_i and v_j are not necessarily restricted to 1, −1 and 0 in Phase 1b (e.g., let $x_{11}, x_{22}, x_{33}, x_{12}, x_{23}$ be basic with the first three infeasible).

TABLE 11.5 Restricted Transportation Problem

Phase 1b with g_{ij}						u_i^0	Phase 2 with c_{ij}					$=v_j^0$
							−1	0	0	0	1	
(0)	(0)	·	(0)	(0)	0	−8	(9)	7	·	(8)	(7)	−8
0	(0)	(0)	0	·	0	−4	(5)	(4)	(4)	10	·	−4
0	·	·	·	(0)	0	−6	11	·	·	·	(5)	−6
(1)	0	0	·	·	−1	−6	·	8	(6)	·	·	−6
0	0	0	0	0		$v_j^*=$	−1	1	0	0	1	u_i^*

a\b	6	4	7	3	8			6	4	7	3	8
9	② P	③ $\overline{M}$	·	③	①		9	⑤ M	−1 P	·	③	①
8	0	① P	⑦ M	0	·	−1	8	① P	④ $\overline{M}$	③	6	·
7	0	·	·	·	⑦		7	4	·	·	·	⑦
4	④ M	−1	−1 P	·	·		4	·	2	④	·	·
		1	$1=\Delta g$		$\Delta x=3$				$1=\Delta c$		$\Delta x=4$	
9	⑤	1	·	③	①		9	①	④	·	③	①
8	−1 P	④	④ M	−1	·	Δx	8	⑤	1	③	6	·
7	0	·	·	·	⑦	$=1$	7	4	·	·	·	⑦
4	① $\overline{M}$	0	③ P	·	·		4	·	3	④	·	· OPT.

g_{ij} (in Phase 1b) or c_{ij} (in Phase 2) in parentheses where x_{ij} is basic. Basic x_{ij} circled amid g'_{ij} (Phase 1b) or $c'_{ij}=c_{ij}+u_i+v_j$ (Phase 2). Adjustments are $+\Delta x$ at P and $-\Delta x$ at M. Dotted positions are not usable ($x_{ij}=e_{ij}=0$).

$$z^*=9\cdot1+5\cdot5+7\cdot4+4\cdot3+6\cdot4+8\cdot3+7\cdot1+5\cdot7$$
$$=9\cdot8+8\cdot4+7\cdot6+4\cdot6-6(-1)-4\cdot1-8\cdot1=164 \qquad \begin{bmatrix} 2 & 6 \\ 1,3 & 5,7 \\ 4 & 8 \end{bmatrix}$$

The six tableaus are used in the sequence 1 to 8:

It is desirable to use blank spaces or dots in the tables to reduce the clutter of symbols. In the answers for 11.4 the only way to do this was to let them represent nonbasic $x_{ij}=0$. In 11.5 and 11.6 more simplification is achieved by using blanks or dots (or lines or crosses in larger displays) to represent the artificial $x_{ij}=e_{ij}=0$, and using the symbol ϕ for nonbasic $x_{ij}=0$ for which $e_{ij}>0$. Infeasible values of any x_{ij} may be marked #.

The rest of this section is optional.

(b) Dantzig (1963) and A. S. Manne showed how a transportation problem involving a small number of bounds on individual x_{ij} or on certain partial sums of x_{ij} in the same row or column can be reduced to a restricted problem. If two bounded partial sums involve sets of variables in the same row or column, these sets should be disjoint or else one contained in the other, to preserve the triangularity of the basis.

For example, let $m=n=3$ and suppose the upper-bound constraints are

$$x_{11}+x_{12}\leq e_4 \qquad x_{11}\leq e_5 \qquad x_{22}\leq e_6$$

with slack variables $y_4, y_5, y_6 \geq 0$, respectively. Then the constraints can be represented by (3),

		x_{13}	x_4			$=a_1$	(3)
x_{21}		x_{23}			x_{22}	$=a_2$	
x_{31}	x_{32}	x_{33}				$=a_3$	
	x_{12}		y_4	x_{11}		$=e_4$	
x_{11}				y_5		$=e_5$	
	x_{22}				y_6	$=e_6$	
$\Vert$	$\Vert$	$\Vert$	$\Vert$	$\Vert$	$\Vert$		
b_1	b_2	b_3	e_4	e_5	e_6		

in which the blank squares represent artificial variables. Each repeated variable x_{11}, x_{22} automatically has the same value at each occurrence, because of the repetition of e_4, e_5, e_6 as row and column totals. Note that the fourth row and column imply that $x_4 = x_{11}+x_{12}$. This x_4, all the y_j, and all but some one occurrence of each x_{ij} will have zero costs.

(c) A *transshipment problem* results from transportation when the shipment paths are allowed to consist of more than one arc. The sources and destinations are indexed in a single sequence $1,2,\ldots,m+n$ or more. There are two ways in which it can be reduced to a transportation problem again. If there are no capacity restrictions, it suffices to determine the shortest path lengths d_{ij} as in 9.3 or 9.7 and to use them in the transportation algorithm in place of the c_{ij}.

Otherwise an $s>0$ is chosen large enough to avoid the occurrence of negative shipments and each node is split in two. A source supplying a_i becomes a source supplying a_i+s and a destination requiring s; a destination requiring b_j becomes a source of s and a requirement of b_j+s; and every other node becomes a source of s and a requirement of s. This is illustrated in Figure 11.5. The new arcs usually have zero costs. Note that one has $a_h + \Sigma_{i\neq h} x_{ih} = \Sigma_{j\neq h} x_{hj}$ at a source A_h, and

$$\sum_{i\neq h} x_{ih} = \sum_{j\neq h} x_{hj} + b_h \text{ at a destination } B_h. \tag{4}$$

EXERCISES 11.5

In 1 to 8 refer to 1 to 8 in 11.4. Optimize.

1. Require $x_{22}\leq 4$, $x_{43}+x_{44}\leq 3$ (see (3)).
2. Require $x_{32}+x_{33}\leq 4$, $x_{31}+x_{32}+x_{33}\leq 6$.

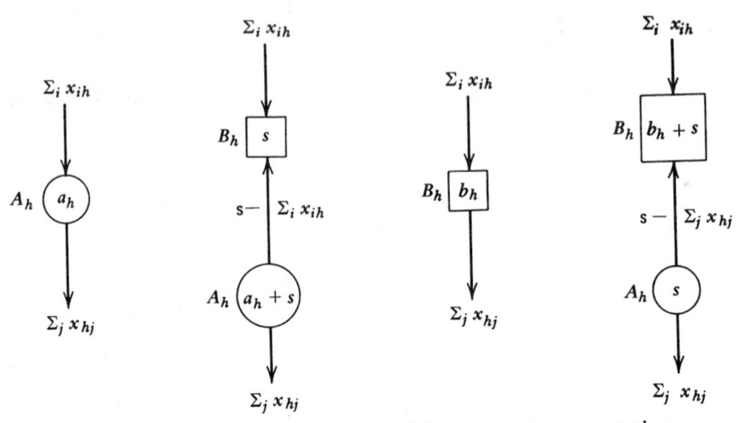

FIGURE 11.5 Reduction of transshipment to transportation.

3. Require $x_{11}, x_{24}, x_{34}, x_{42}, x_{43} = 0$ (artificial).

4. Require $x_{11}, x_{13}, x_{15}, x_{25}, x_{33}, x_{34} = 0$.

5, 6. Begin Phase 1b with the following infeasible basic x_{ij} (· or # = artificial; ϕ is nonbasic $x_{ij} = 0$).

ϕ	·	·	4	9	13
3	9	ϕ	·	·	12
ϕ	·	1#	3	ϕ	4
·	0	5	ϕ	·	5
3	9	6	7	9	34

1	·	0	ϕ	ϕ
·	·	1	·	0
·	ϕ	ϕ	1	·
·	ϕ	ϕ	ϕ	1#
0	1	·	0	·

7. Require $x_{ij} = 0$ (artificial) where $4 \le c_{ij} \le 11$.

8. Require $x_{ij} = 0$ where $c_{ij} = 1$ or 2 or $5 \le c_{ij} \le 8$.

9. Explain how a restricted assignment problem could arise in a personnel (employment) office or in a marriage bureau (introduction service).

10. Explain how the restricted assignment algorithm could be used to determine which of the $n!$ terms in the expansion of an n by n determinant (a_{ij}) has the largest absolute value.

Formulate 11 to 13 as restricted transportation problems.

11. At ports P_1, P_2, P_3 there are 3, 2, 4 tankers available; at ports P'_1, P'_2, P'_3 there are 2, 4, 3 tankers required. Requirements on arrival times preclude sending more than one tanker from P_3 to P'_2. The tankers are identical.

12. A baker's cost for producing bread on the next four days (which include a holiday weekend) is c_k per box ($k = 1, 2, 3, 4$). He has contracted to supply quantities q_k and has the capacity q'_k. Bread can be produced in advance with a penalty of p per box per day for loss of freshness.

13. Two manufacturers ($i = 1, 2$) in time periods $j = 1, 2, 3$ are prepared to supply not more than n_{ij} units at a cost of c_{ij} per unit. The purchaser needs r_j units in period j; he can store unused units at a cost of s per unit per period and expects to be able to sell unused units in period 4 for a price of p per unit.

Exercises 14 to 19 refer to the *bottleneck transportation problem*.

14. Let c_{ij} denote the *time* required to ship any positive quantity x_{ij} from i to j, and assume that $m+n-1$ or more such shipments can be made simultaneously. Write down a symbolic expression for the minimum time $c^{\#}$ required to distribute the supplies a_i to satisfy the demands b_j.
15. How does 14 affect Theorem 1 of 11.1 and the use of basic solutions?
16. Show how to find bounds for $c^{\#}$ $(c_0 \leq c^{\#} \leq c^0)$. [cf. (2)].
17. Modify the ordinary transportation algorithm to solve 14.

Use 14 to 17 to find c_0, $c^{\#}$ and c^0 (from x^*_{ij}).

18. Exercise 2 of 11.4.
19. $m=n=5, a_i=b_j=1, c_{ij}=(i-1)(j-1)$ (Machol and Wien, 1977).

11.6 THE CAPACITATED TRANSPORTATION ALGORITHM

Some of the ways in which the capacity constraints $x_{ij} \leq e_{ij}$ change the algorithms of 11.4 and 11.5(a) are the following:

(a) In each interior position e_{ij} is included as a second entry on the right; thus in Phase 2 one has

$$\begin{array}{llll} \text{Interior cell} & c'_{ij} \text{ or basic } \textcircled{x_{ij}} & e_{ij} & \\ \text{Marginal cell} & [\pm \Delta c] & a_i \quad \text{or} \quad b_j & \end{array} \tag{1}$$

In Phase 1b one has g'_{ij} and $\pm \Delta g$ in place of c'_{ij} and $\pm \Delta c$. When nonbasic $x_{ij} = e_{ij}$ as explained in (c) following, g'_{ij} or c'_{ij} is recorded on the left and e_{ij} is circled (like the basic x_{ij}) to show that it supplies the value of that x_{ij}.

(b) In an effort to avoid having $x_{ij} > e_{ij}$, the basic x_{ij} are determined initially (in the procedure now called Phase 1a) by tentative applications of the formula

$$x_{pq} = \min\left(a'_p, b'_q, e_{pq}\right) \tag{2}$$

If necessary, as when x_{pq} is the last undetermined variable in its row or column; one reverts to the previous formula

$$x_{pq} = \min\left(a'_p, b'_q\right) \tag{3}$$

accepting $x_{pq} > e_{pq}$ if that is necessary to satisfy the row or column equation.

For the sake of variety, not because of $x_{ij} \leq e_{ij}$, Phase 1a in this section scans only one column of c_{ij} for each basic x_{ij} that is determined. This is acceptable for solving exercises and much to be preferred in computer programs. A similar reduction in scanning could be effected in finding min g'_{ij} in Phase 1b and min c'_{ij} in Phase 2.

(c) The concept of nonbasic x_{ij} and their optimality conditions must be generalized. If (2) makes $x_{pq}=e_{pq}<\min(a'_p, b'_q)$, neither a row nor a column equation is thereby satisfied, and so x_{pq} cannot be regarded as basic. Indeed it resembles the nonbasic $x_{ij}=0$ in that it can change in only one direction, namely, away from the bound whose value it has. When the permissible change is a decrease from e_{pq}, the optimality condition also must be reversed, namely $c'_{pq}\leq 0$, so as to prevent any further decrease in z. This situation produces some fairly obvious changes in the determination of the variables $x_{ll'}$ and $x_{kk'}$ to be exchanged, which will be noted later. Bounded variables were considered in 6.5 also.

As usual the nonbasic values 0 and e_{pq} may also be assumed by basic variables in degenerate cases. In fact Dantzig (1963) requires the basic status for x_{pq} in Phase 1a whenever (2) makes $x_{pq}=e_{pq}=\min(a'_p, b'_q)$. This expedites the formation of the basis and generalizes the case $e_{ij}\geq\min(a_i, b_j)$; this does not differ from $e_{ij}=\infty$, in which case $x_{ij}>0$ would always be basic. However, if e_{pq} is very small, the nonbasic status is preferable.

Phase 1a The object is to determine $m+n-1$ basic x_{ij} connecting all $m+n$ of the nodes $i=1,\ldots,m$ and $j=1,\ldots,n$; feasible values are desirable but not necessary. To minimize z it is desirable to look for small costs. Let the costs as first given be denoted by c^0_{ij} and suppose that they are stored one column after another, possibly in the order of increasing values of

$$\sum_i \min\,(a_i, b_j, e_{ij})-b_j \tag{4}$$

In a single pass through these data one can calculate $u^0_i=-\min_j c^0_{ij}$ for $i=1,\ldots,m$. Then the columns $q=1,\ldots,n$ are processed in the order of storage by the following steps.

1. Calculate $c_{iq}=c^0_{iq}+u^0_i$ or $c^0_{iq}+u^0_i+v^0_q$ for $i=1,\ldots,m$. The $v^0_q=-\min_i(c^0_{iq}+u^0_i)$ make the c_{ij} more useful in breaking ties in Phases 1b and 2.

2. Find $c_{pq}=\min_i c_{iq}$. Here some i are permanently excluded in Phase 1a because their row equations have been satisfied, and others are excluded in column q because x_{iq} has been determined to be e_{iq}.

3. Let $x_{pq}=\min(a'_p, b'_q, e_{pq})=$ the maximum amount currently feasible, unless this is found to make the sum a_p or b_q unattainable; then $x_{pq}=\min(a'_p, b'_q)$.

4. If $x_{pq}=e_{pq}<\min(a'_p, b'_q)$, then x_{pq} is nonbasic; no row or column is permanently excluded but $i=p$ is excluded in column q. Otherwise x_{pq} is basic; row p is excluded from Phase 1a if $x_{pq}=a'_p\leq b'_q$ and at least one other x_{iq} is of undetermined status, while column q is excluded in the other cases.

5. a'_p-x_{pq} and b'_q-x_{pq} are new values of a'_p and b'_q.

6. Return to step 2 if the status of some x_{iq} is still undetermined. Otherwise if $q<n$ return to step 1 with q increased by 1. *End Phase* 1*a* if $q=n$ and the status of every x_{in} has been determined; then go to Phase 1b if some x_{ij} is infeasible and go to Phase 2 otherwise.

Phase 1b Pivot steps are performed to convert an infeasible solution obtained in Phase 1a into a feasible solution, if necessary and possible. This is done by minimizing the sum of (the absolute values of) the departures from feasibility. With Phase 1a as described above, the same result is obtained by minimizing the sum of those $x_{ij}>e_{ij}$, provided each such x_{ij} is dropped from the sum as soon as it is reduced to e_{ij}.

Thus Phase 1b differs from Phase 2 below in having the c_{ij} replaced by $g_{ij}=1$ when $x_{ij}>e_{ij}$ and $g_{ij}=0$ elsewhere; the c'_{ij} are replaced by $g'_{ij}=u_i+v_i+g_{ij}$. One must replace $g_{ij}=1$ by $g_{ij}=0$ when x_{ij} becomes feasible, as in the z^I of 6.4. Since this could detract from the usefulness of (4) of 11.3 in computer programs, a version of Phases 1a and 1b analogous to 6.1 will be described at the end of this section.

Phase 1b ends when a feasible solution is obtained or the $g'_{ij}\geq 0$ when $x_{ij}=0$ and $g'_{ij}\leq 0$ when $x_{ij}=e_{ij}$. If a feasible solution is obtained, go to Phase 2; otherwise stop.

Phase 2 Pivot steps are performed to minimize $z=\Sigma_i\Sigma_j c_{ij}x_{ij}$. Ignore positions for which $e_{ij}=0$ and x_{ij} is nonbasic.

1. Set any one u_i or $v_j=0$. Calculate the others from $u_i+v_j+c_{ij}=0$ in the $m+n-1$ basic positions, which have c_{ij} in parentheses. (Optional except in the first and last iterations.)

2. Record $c'_{ij}=u_i+v_j+c_{ij}$ for nonbasic x_{ij} with $e_{ij}>0$, perhaps using (4) of 11.3. $\pm\Delta g$ or $\pm\Delta c$ is bracketed.

3. For the variable x_{ij} the optimality conditions are

$$\text{Basic } x_{ij} \text{ in } [0, e_{ij}] \qquad c'_{ij}=u_i+v_j+c_{ij}=0 \tag{5}$$

$$\text{Nonbasic } x_{ij}=0 \qquad c'_{ij}=u_i+v_j+c_{ij}\geq 0 \tag{6}$$

$$\text{Nonbasic } x_{ij}=e_{ij} \qquad c'_{ij}=u_i+v_j+c_{ij}\leq 0 \tag{7}$$

If all are satisfied, check by step 1, $\Sigma_j x_{ij}=a_i, \Sigma_i x_{ij}=b_j$; *stop*. If (6) or (7) is violated, mark violations with a colon and define

$$|c'_{ll'}|=\max\{|c'_{ij}| \text{ violating (6) or (7)}\} \tag{8}$$

Variable $x_{ll'}$ will become basic except as noted in step 5.

4. Label $x_{ll'}$ with P (plus, increasing) or M (minus, decreasing) according as it equals 0 or $e_{ll'}$. Indicate the effect of the change in $x_{ll'}$ on the basic variables by labeling them P, M or Z (zero, no change). The labels P and M will alternate around a $\pm\Delta x$ loop as in Figure 11.3.

5. In the Δx loop find

$$\Delta x=\min\{x_{ij} \text{ labeled } M \text{ and } e_{ij}-x_{ij} \text{ labeled } P\}. \tag{9}$$

Choose kk' so that $\Delta x=x_{kk'}$ labeled M or $e_{kk'}-x_{kk'}$ labeled P. This *must* be observed even if $\Delta x=0$, so that $c'_{kk'}$ acquires a feasible value. Then $x_{ll'}$ replaces $x_{kk'}$ in the basis except when $k=l, k'=l', \Delta x=e_{kk'}=e_{ll'}>0$. In this exceptional case there is no change in the identities of the basic x_{ij}, only in their values. In particular $x_{ll'}$ changes from 0 to $e_{ll'}$ or vice versa in step 6 so that without changing its value, $c'_{ll'}$ becomes feasible because its optimality condition changes from (6) to (7) or vice versa. The values of the c'_{ij}, u_i, v_j do not change and so steps 1 and 2 are omitted in the next cycle.

6. Calculate the new values of x_{ij} in the loop by adding and subtracting Δx as indicated by the labels P and M. Return to step 1 or 2.

Notes on *breaking ties* in choosing $x_{ll'}$ and $x_{kk'}$ in Phases 1b and 2: To maximize Δx one prefers large e_{ij} for basic x_{ij}, hence $\max e_{ll'}$ and $\min e_{kk'}$. To favor dual feasibility one prefers a small $c_{ll'}$ or $c_{kk'}$ among those whose x_{ij} have or would receive a P label, and a large $c_{ll'}$ or $c_{kk'}$ among those labeled M, even when the value of z is not thereby affected. Also see Exercise 7.

As a partial check at the end of the algorithm, calculate

$$z^* = \sum_{i=1}^{m} \sum_{j=1}^{n} c_{ij}^{0} x_{ij}^{*}$$

$$= \sum_{i=1}^{m} \sum_{j=1}^{n} c'_{ij} x_{ij}^{*} - \sum_{i=1}^{m} a_i\left(u_i^0 + u_i^*\right) - \sum_{j=1}^{n} b_j\left(v_j^0 + v_j^*\right) \tag{10}$$

in both ways. $c'_{ij} x^*_{ij} = 0$ except when nonbasic $x^*_{ij} = e_{ij}$.

The algorithm is applied to an example in Table 11.6A. In the first large tableau (called T_4), the $x_{ij} \neq 0$ are determined in the order $x_{21} = 3, x_{31} = 1, x_{41} = 1, x_{32} = 4, x_{13} = 5, x_{33} = 6, x_{23} = 1, x_{14} = 4, x_{44} = 4, x_{24} = 2$. Of these only x_{44} exceeds its capacity and all but x_{21}, x_{31} and x_{13} are basic. Also see Table 11.6B.

An Alternate Phase 1

To facilitate the use of (4) of 11.3 in Phase 1b, by avoiding changes in the g_{ij} as the x_{ij} become feasible, one may require that all x_{ij} have feasible values unless they are

TABLE 11.6A Capacitated Transportation Problem

c^0_{ij}				u^0_i	$v^1_j =$	2	0	0	−4	u^*_i	g_{ij}				
·	6	5	3	−3	4	·	2	0	(0)	−1	·	0	0	(0)	0
2	7	5	4	−2	−1	0	4	(1)	2	−2	0	0	(0)	(0)	0
1	2	3	5	−1	0	0	(0)	(0)	(4)	−2	0	(0)	(0)	0	0
5	·	9	4	−4	−3	(1)	·	(3)	0	−3	(0)	·	0	(1)	−1
0	−1	−2	0	$=v^0_j$	u^1_i	2	2	1	−2	$=v^*_j$	1	0	0	0	

Phase 1b

—	—	0	1	0	⑤	④	5	[1]	9
1:	③	0	2	① M	6	② $\bar{P}$	2		6 T_4
1:	①	④	4	⑥	6	0	3		11
①	5	—	—	−1: P	4	④ M	3	[1]	5
[−1]	5		4		12	[−1]	10	$\Delta g=1$,	$\Delta x=0$

TABLE 11.6A (Continued)

—	—	1	1	1:	⑤	④	5	9
0	③	0	2	①	6	−1	②	6
0	①	④	4	⑥ M	6	−1: P	3	11
①	5	—	—	⓪ P	4	④ $\overline{M}$	3	5
$b_j =$ 5		4		12		10		$\Delta x = 1$

Phase 2

$\textcircled{x_{i1}}$		c'_{i2}	e_{i2}					Δc	a_i
—	—	6	1	4: M	⑤	④ $\overline{P}$	5	[−4]	9
1:	③	3	2	①	6	−3	②		6
2:	①	④	4	⑤ P	6	① M	3		11
①	5	—	—	①	4	−7	③		5
5		4		12		10			$\Delta x = 1$

—	—	2	1	④	5	−4	⑤		9
1:	③	3	2	①	6	−3	②		6
2: M	①	④	4	⑥ $\overline{P}$	6	⓪	3	[−2]	11
① P	5	—	—	① M	4	−7	③		5
		[2]				[2]			$\Delta x = 0$

—	—	4	1	④	5	−2	⑤	[−1]	9
1: M	③	5	2	① P	6	−1	②	[−1]	6
①	1	④	4	−2	⑥	⓪	3		11
① P	5	—	—	① $\overline{M}$	4	−5	③		5
				[1] = Δc					$\Delta x = 1$

—	—	3	1	④	5	−3	⑤	9
②	3	4	2	②	6	−2	②	6
①	1	④	4	−1	⑥	⓪	3	11
②	5	—	—	1	4	−5	③	5
$b_j =$ 5		4		12		10		

Optimum

$$z^* = \Sigma_i \Sigma_j c^0_{ij} x^*_{ij} = 2\cdot 2 + 1\cdot 1 + 5\cdot 2 + 2\cdot 4 + 5\cdot 4 + 5\cdot 2 + 3\cdot 6 + 3\cdot 5 + 4\cdot 2 + 4\cdot 3$$
$$= \Sigma_i \Sigma_j c'_{ij} x^*_{ij} - \Sigma_i a_i (u^*_i + u^0_i) - \Sigma_j b_j (v^*_j + v^0_j) = -1\cdot 6 - 3\cdot 5$$
$$-2\cdot 2 - 5\cdot 3 + 9(3+1) + 6(2+2) + 11(1+2) + 5(4+3) + 5(0-2) + 4(1-2) + 12(2-1)$$
$$+ 10(0+2) = 106.$$ Also see Table 11.6B.

artificial ($e_{ij}=0$), as in 6.1. To do this one discards (3), uses (2) throughout Phase 1a, and defines the artificial variables

$$x_{i0}=a_i-\sum_{j=1}^{n} x_{ij}, \qquad x_{0j}=b_j-\sum_{i=1}^{m} x_{ij}$$

$$z^0=-x_{00}=x_{10}+\cdots+x_{m0}=x_{01}+\cdots+x_{0n} \tag{11}$$

They occupy an additional row and column (the last) in the tableaus and have $a_0=b_0=0$. Those that are zero are kept at zero and ignored as in 11.5(a). The objective function z^0 to be minimized in Phase 1b has various equivalent forms in (11); it is simplest to take all $g_{ij}=0$ except $g_{00}=-1$.

In Table 11.6A the effect is to change three of the tableaus (T_3, T_4, T_5) as indicated in Table 11.6B. At the last pivot step in Phase 1b the *three* remaining artificial variables leave the basis, two equations (involving a_0 and b_0) are dropped, and one variable x_{34} enters the basis.

EXERCISES 11.6

1.–4. Optimize 1 to 4 of 11.4 with $e_{11}=e_{24}=0$, other $e_{ij}=4$.

5.–6. Optimize beginning with the given x_{ij} (marked # where they violate $x_{ij}\le e_{ij}$). The c_{ij} in the lower right are in parentheses in basic positions.

5.

	1 4	1	2 (1)	1	1 (7)		1 0	2
2	2 5		2 5	3	3 (9)		3 3	5
	3 9	1	1 6	2	2 12	1	1 (6)	4
2	2 (5)	4 #	3 (3)		2 8	2	3 (1)	8
4		6		6		3		

6.

	5 2	3	3 (2)	3	3 (1)	4	4 (3)	10
2 #	0 (–)	3	4 (4)	3	3 2		2 6	8
4	4 (3)	3	3 4		1 5	X	X	7
5 #	3 (1)	X	X		1 3		2 5	5
11		9		6		4		

TABLE 11.6B Notes for Table 11.6A

( ) mark basic positions in tables of c_{ij} or g_{ij}; g_{ij} and u_i^1, v_j^1 initiate Phase 1a and Phase 2 respectively.

○ encloses all values of x_{ij} except nonbasic $x_{ij}=0$ (omitted).

[ ] enclose $\pm\Delta g$ or $\pm\Delta c$ (see (4) of 11.3).

Other numbers: g'_{ij} or c'_{ij} on left; e_{ij} on right; u_i, v_j or a_i, b_j outside.

: marks infeasible values of the reduced costs g'_{ij} or c'_{ij}. P, M, Z denote $+\Delta x, -\Delta x, 0$. $\bar{P}$ or $\bar{M}$ marks $x_{kk'}$ leaving basis. Steps in Phase 2: u_i, v_j; c'_{ij}; infeasible $c'_{ll'}$; P, M loop; $x_{kk'}$; $\pm\Delta x$.

In Phase 1b replace c_{ij} by $g_{ij}=1$ (x_{ij} infeasible) or 0.

	$T_1T_2T_3$	
	T_4	Given c_{ij}^0 in T_1; e_{ij}, a_i, b_j in
Tableaus		T_4 to T_9. Sequence:
in Table	T_5	Phase 1a: u_i^0, v_j^0 in T_1; c_{ij} in T_2;
	T_6	x_{ij} in T_4.
11.6A:	T_7	Phase 1b: T_3, T_4, T_5.
	T_8	Phase 2: u_i^1, v_j^1 in T_2; T_6 to T_9;
	T_9	u_i^*, v_j^* in T_2.

Alternate form of T_3 to T_5 in Phase 1b (artificial x_{i0}, x_{0j} in last column and row.):

·	0	0	(0)	·	0		
0	0	(0)	(0)	·	0		
0	(0)	(0)	0	·	0	g_{ij}	T_3
(0)	·	0	0	(0)	−1		
·	·	·	(0)	(−1)	0		
1	0	0	0	1			

○

—	—	0	1	0	⑤	④	5	—	—		9	
1:	③	0	2	① M	6	② $\bar{P}$	2	—	—	[−1]	6	
1:	①	④	4	⑥	6	0	3	—	—	[−1]	11	T_4
①	5	—	—	−1: P	4	−1	③	① M	0		5	
—	—	—	—	—	—	① M	0	−① P	0		0	
	5	[1]	4	[1]	12		10		0		$\Delta x=0$	
—	—	1	1	1:	⑤	④	5	—	—		9	
0	③	0	2	①	6	−1	②	—	—		6	
0	①	④	4	⑥ M	6	−1: P	3	—	—		11	T_5
①	5	—	—	⓪ P	4	−1	③	① $\bar{M}$	0		5	
—	—	—	—	—	—	① $\bar{M}$	0	−①$\bar{P}$	0		0	

$\Delta x=1$. Drop last row and column in forming T_6 in Table 11.6A.

7. (a) Use (1) of 11.3 to show that of the three infeasible $g'_{ij}=\pm 1$ available for the first Phase 1b pivot in Table 11.6A, g'_{31} has the least potential for increasing z. (b) Resolve this example by bringing x_{31} into the basis initially. (One pivot step may be saved.)

11.7 SENSITIVITY ANALYSIS

For general linear programs this topic was treated in 8.1 and 8.2. If those sections have not been studied, it is suggested that 8.2 be read at least through (iii). We begin with the ranging procedures and number the equations (1) to (6) as in Table 8.2D. No counterparts will be given for the other equations of 8.2 because here the values of the a_{ij}^0 are fixed at 1 or 0; thus there is no reason to range them.

The range of Δz will sometimes be omitted here because it is always obtained by multiplying the range of Δs by $\partial z^*/\partial s$, with attention to the signs. The costs given initially are denoted by c_{ij}^0 while $c_{ij}^*=c_{ij}^0+u_i^*+v_j^*$ and x_{ij}^* are optimal values. Primal variables in the optimal basis will be denoted by $x_{kk'}$ and the others by $x_{ll'}$.

In general $\partial z^*/\partial x_{ll'}\neq c_{ll'}^0$ because $x_{ll'}$ cannot vary independently of the other x_{ij}; first one must eliminate the basic x_{ij} as in (6) to (8) of 11.1. Then with optimal values of u_i and v_j, (6) gives

$$\partial z^*/\partial x_{ll'}=c_{ll'}^*=\quad c_{ll'}^0+u_l^*+v_{l'}^*\geq 0 \tag{1}$$

for $0\leq\Delta x_{ll'}\leq\max\Delta x$. Here $\max\Delta x$ is determined by the Δx loop in Figure 11.3, formed by the x_{ij} in the optimal basis together with the nonbasic $x_{ll'}$.

In the evaluation of $\partial z^*/\partial x_{kk'}$, the basic $x_{kk'}$ can vary only as a result of an increase in some nonbasic $x_{ll'}$, which is chosen to increase the cost at the minimum rate $c_{ll'}^*\geq 0$. Now the discussion of (4) of 11.3 shows how, given a basis and a particular basic $x_{kk'}$, the totality of all positions ij is divided into three categories Z, $\bar{K}$ and K, defined by the following conditions on ij:

Z: $\quad \partial x_{kk'}/\partial x_{ij}=0 \qquad (i \text{ in } M', j \text{ in } N' \text{ or } i \text{ in } M'', j \text{ in } N'')$

$\bar{K}$: $\quad \partial x_{kk'}/\partial x_{ij}=-a_{kk',ij}=1 \qquad (i \text{ in } M'', j \text{ in } N')$

K: $\quad \partial x_{kk'}/\partial x_{ij}=-a_{kk',ij}=-1 \qquad (i \text{ in } M', j \text{ in } N'')$

The third category K is so designated, and is recognized in practice, by virtue of the fact that it contains the basic position kk', the only one outside Z. Now

$$\frac{\partial z^*}{\partial x_{kk'}}=\frac{\partial z^*}{\partial x_{ij}}\bigg/\frac{\partial x_{kk'}}{\partial x_{ij}}=\pm\frac{\partial z^*}{\partial x_{ij}}=\pm c_{ij}^*$$

where the equalities come from the calculus, (4) of 11.3, and (1) above, respectively.

Thus

$$\partial z^*/\partial x_{kk'} = \{\min c_{ij}^* : ij \text{ in } \bar{K}\} \qquad \text{for} \qquad \Delta x_{kk'} \geq 0$$

$$= -\min\{c_{ij}^* : ij \text{ in } K, ij \neq kk'\} \qquad \text{for} \qquad \Delta x_{kk'} \leq 0 \tag{2}$$

Equation (2) determines two values of $ij = ll'$. The corresponding upper bounds $\max \Delta x_{ll'} = \max|\Delta x_{kk'}|$ are the respective values of $\max \Delta x \geq 0$ in the loop of Figure 11.3.

Since the transportation problem requires $\Sigma a_i = \Sigma b_j$, a single a_i or b_j cannot be ranged. Instead we use $\delta_{ll'}$ to imply the replacement of a_l by $a_l + \delta_{ll'}$ and $b_{l'}$ by $b_{l'} + \delta_{ll'}$, and show that

$$\partial z^*/\partial \delta_{kk'} = c_{kk'}^0 \qquad \text{for} \qquad \delta_{kk'} \geq -x_{kk'}^* \tag{3}$$

$$\partial z^*/\partial \delta_{ll'} = -u_l^* - v_{l'}^* = c_{ll'}^0 - c_{ll'}^* \tag{4}$$

Equation (4) is true whether $x_{ll'}$ is nonbasic or not, because the dual solution $\mathbf{u}^*, \mathbf{v}^*$ does not depend on $\delta_{ll'}$ as long as the ij in the optimal basis do not change. Then replacing a_l by $a_l + \delta_{ll'}$ and b_l by $b_{l'} + \delta_{ll'}$ in (9) of 11.1 gives z^* as a linear function of $\delta_{ll'}$; the coefficient of $\delta_{ll'}$ is $-u_l^* - v_{l'}^*$ as (4) asserts.

When $x_{kk'}$ is basic, (3) follows from (4) by putting $ll' = kk'$ and $c_{kk'}^* = 0$. In this case the primal equations are obviously still satisfied if $x_{kk'}^*$ is replaced by $x_{kk'}^* + \delta_{kk'}$, and the solution remains optimal if the latter is nonnegative. Obtaining the range for $\delta_{ll'}$ in (4) when $x_{ll'}$ is nonbasic requires two applications of the loop of Figure 11.3, one conventional with $\Delta x > 0$, and one with $\Delta x \leq 0$. In the latter case $\delta_{ll'} \geq 0$ and $-\delta_{ll'} = \Delta x = x_{ll'} \leq 0$, with $|x_{ll'}|$ as large as possible without making any other x_{ij} negative.

A similar analysis shows that

$$\Delta a_k = -\Delta a_l = \alpha_{kl} \text{ implies } \partial z^*/\partial \alpha_{kl} = u_l^* - u_k^*$$

$$\Delta b_k = -\Delta b_l = \beta_{kl} \text{ implies } \partial z^*/\partial \beta_{kl} = v_l^* - v_k^* \tag{4'}$$

The range for $\alpha_{kl} \geq 0$, for example, is determined by the Δx loop (Figure 11.3) initiated by entries of $+\Delta x$ in row l and $-\Delta x$ in row k, both in the $m+1$st column occupied by the a_i. (Neither of these will be needed in getting a limit on Δx, however.) A loop must be formed because two basic variables Δa_k and Δa_l and only one equation ($\Delta a_k + \Delta a_l = 0$) have been added. As before the signs reflect the fact that the a_i and b_j occur alone in the right-hand sides of the equations, and the range is $0 \leq \alpha_{kl} \leq \max \Delta x$. The same loop with signs reversed determines $\max|\alpha_{kl}|$ when $\alpha_{kl} \leq 0$. If $\Delta x = \beta_{kl}$, Δx replaces b_l and $-\Delta x$ replaces b_k.

For ranging Δc_{ij}^0 we have

$$\Delta z = 0 \qquad \text{for} \qquad \Delta c_{ll'}^0 \geq -c_{ll'}^* \tag{5}$$

$$\partial z^*/\partial c_{kk'}^0 = x_{kk'}^* \tag{6'}$$

Here (6′) is derived from $z^*=\Sigma_i\Sigma_j c_{ij}^0 x_{ij}^*$, since the x_{ij}^* are unaffected by sufficiently small changes in the c_{ij}^0. In the nonbasic position ll' we have $\partial z^*/\partial c_{ll'}^0 = x_{ll'}^* = 0$, and $c_{ll'}^* = c_{ll'}^0 + u_l^* + v_{l'}^*$ is replaced by $c_{ll'}^* + \Delta c_{ll'}^0 \geq 0$, so that (5) results. In (6′) the effect of $\Delta c_{kk'}^0$ on the c_{ij}^* is given by (4) of 11.3, and the range of validity of (6′) is found to be

$$-\min_{\bar{K}} c_{ij}^* \leq \Delta c_{kk'}^0 \leq \min_{K} \{c_{ij}^* : ij \neq kk'\} \tag{6''}$$

These endpoints are the same as the values of $-\partial z^*/\partial x_{kk'}$ in (2) above. This relation preserves the latter's change of sign at $\Delta x_{kk'}=0$ in (2), which is necessary for $\Delta x_{kk'}=0$ to be optimal.

A few remarks may suffice for the procedures analogous to those in 8.1. The need to reconcile the deletion of one row (a source) or one column (a destination) with the requirement $\Sigma a_i = \Sigma b_j$ leads to the use of a dummy node as in (4) and (5) of 11.1, probably with changes in its costs. Changes in the c_{ij}^0 for this or any reason can be handled by recalculating the u_i, v_j and c_{ij}' with the new c_{ij}^0 and the old basis, and then reoptimizing if necessary by Phase 2. Changes in **a** and **b** require a new solution from the beginning, but if the changes are small the solution can be shortened by using or attempting to use the existing set of basic x_{ij} in performing Phase 1 of 11.4, with or without the acceptance of infeasible values of x_{ij} during Phase 1.

Examples

Table 11.7A summarizes the formulas (1) to (6″) and gives the optimal tableau (7) that will be used in the examples. In Table 11.7B each of these formulas is applied to one kk' or ll' in (7). It remains to consider an example in the spirit of 8.1 and an application.

Suppose the values $\Delta c_{11}^0=6$ and $\delta_{21}=-4$ are imposed on (7) simultaneously. If we keep the same basic ij but recalculate the values of x_{ij}, u_i and v_j we obtain

a	**b**=	5	15	5	2	4	−**u**
13		−①	⑫	3	②	2	0
8		−1	③	⑤	6	3	−2
10		⑥	10	13	8	④	−1
	−**v**=	10	6	3	1	8	

(8)

Here $kk'=11$ and (4) of 11.3 has revised the c_{ij}' by subtracting $\Delta c_{11}^0=6$ in columns $N''=\{1,5\}$ (to reduce c_{11}' from 6 back to 0) and adding 6 in row $M'=\{3\}$. The reverse changes have been made in $-\mathbf{u}$ and $-\mathbf{v}$. At first $\delta_{21}=\Delta x_{21} = -4$; to eliminate this we use a Δx loop with $+4$ in positions 21 and 12, and -4 in positions 11 and 22.

There is one primal infeasibility $x_{11}=-1$ and one dual $c_{21}'=-1$. We note that the coefficient of x_{21} accompanying x_{11} is $+1$ (see Exercise 11). According to

8.3(d) this suggests that a dual simplex pivot step is preferable. This step, which uses the Δc partition of Table 11.7B ($kk'=11$), is not quite conventional because $c'_{21}=-1$ is infeasible, but that turns out to be harmless. We need a negative pivot in $\overline{K}$, so that $\Delta c=\min\{10,13,8\}=8=c'_{34}$. This means that $x_{kk'}=x_{11}$ will be replaced by $x_{ll'}=x_{34}$ in the basis. Recalculating the x_{ij}, u_i and v_j for the new basis gives the new optimal solution

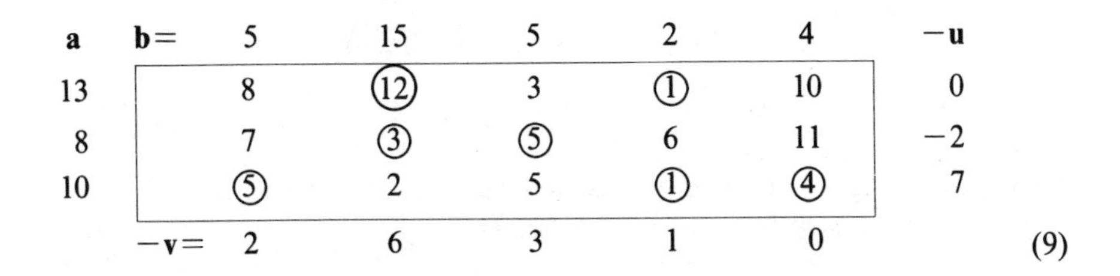

a	**b**=	5	15	5	2	4	−**u**
13		8	(12)	3	(1)	10	0
8		7	(3)	(5)	6	11	−2
10		(5)	2	5	(1)	(4)	7
	−**v**=	2	6	3	1	0	

(9)

The changes in **x** lie on the loop formed by adding x_{34} to the basis in (8). ■

As an application, suppose that (7) represents the shipment of a product from three factories to five markets, and that an increase in the monthly demand $b_1=9$ at the first market is expected. It is necessary to decide which of the factories should increase its production to meet the additional demand. In Table 11.7B we have calculated $\partial z^*/\partial\delta_{11}=4$ for $\delta_{11}\geq-3$ and $\partial z^*/\partial\delta_{21}=2$ for $-3\leq\delta_{21}\leq 8$. Now we also need $\partial z^*/\partial\delta_{31}=c^0_{31}=9$ for $\delta_{31}\geq-6$. Thus $\partial z^*/\partial\delta_{21}=2$ gives the lowest cost per incremental unit transported.

Now the costs of production have to be considered. Suppose these are 7, 10, 6 per unit at the three factories, and that one-time expenditures of 150, 100, 10 per additional unit of monthly demand are also required. Finally a discount rate of 15% per year or 1.25% per month on future expenditures is assumed. Since $a+ar+ar^2+\cdots=a/(1-r)$ with $r=.9875$, the monthly costs of $4+7, 2+10, 9+6$ at the three factories are equivalent to one-time costs that are $1/.0125=80$ times as large, namely 880, 960, 1200. Adding the fixed costs of 150, 100, 10 gives the totals of 1030, 1060, 1210 per additional unit supplied monthly to the first market. Thus the first factory is slated for expansion, for any $b_1>9$. For $\delta_{21}>8$, $\partial z^*/\partial\delta_{21}$ cannot decrease, since the theorem of 8.7 shows that z^* is a convex function of each δ_{ij}.

In conclusion a summary of the rules for using the Δc partition ((4) of 11.3) may be helpful. The partition is determined by the current basis together with one particular $x_{kk'}$ in that basis. By definition $k\in M'$, $k'\in N''$, $K=M'N''=\{ij: i\in M', j\in N''\}$, and $\overline{K}=M''N'=\{ij: i\notin M', j\notin N''\}$. The c'_{ij} having ij in K are increased by Δc and those in $\overline{K}$ are decreased by Δc. In a primal pivot step $ll'\in K$ and $\Delta c=-c'_{ll'}$; in a dual pivot step $ll'\in\overline{K}$ and $\Delta c=c'_{ll'}=\min_{\overline{K}}c'_{ij}$. In both cases $\Delta c\geq 0$ except when $x_{ll'}$ is at its upper bound in 11.6.

In (2) and (6″) of this section one uses both $\min_{\overline{K}}c'_{ij}$ and $\min_{K'}c'_{ij}$, where $K'=K-\{kk'\}$; decreases are contemplated (but not actually effected) within each of $\overline{K}$ and K in turn. In (2) an increase in $x_{kk'}$ in row kk' must be produced by the same increase in an $x_{ll'}$ having a coefficient -1 in $\overline{K}$. In (6″) an increase in $c^0_{kk'}$ must

TABLE 11.7A Summary of Formulas; Example

Let $x_{kk'}$ belong to the optimal basis while $x_{ll'}$ does not.

$$\partial z^*/\partial x_{ll'} = c^*_{ll'} = c^0_{ll'} + u^*_l + v^*_{l'} \geq 0 \tag{1}$$

for $0 \leq \Delta x_{ll'} \leq \max \Delta x$ from Figure 11.3.

$$\begin{aligned} \partial z^*/\partial x_{kk'} &= \min\{c^*_{ij} : ij \text{ in } \bar{K}\} \qquad \text{for} \qquad \Delta x_{kk'} \geq 0 \\ &= -\min\{c^*_{ij} : ij \text{ in } K, ij \neq kk'\} \qquad \text{for} \qquad \Delta x_{kk'} \leq 0 \end{aligned} \tag{2}$$

Each minimizing $ij = ll'$ gives $\max|\Delta x_{kk'}| = \max \Delta x \geq 0$ in Figure 11.3.
Let a_l and $b_{l'}$ each be increased by $\delta_{ll'}$. Then

$$\partial z^*/\partial \delta_{kk'} = c^0_{kk'} \qquad \text{for} \qquad \delta_{kk'} \geq -x^*_{kk'} \tag{3}$$

$$\partial z^*/\partial \delta_{ll'} = -u^*_l - v^*_{l'} = c^0_{ll'} - c^*_{ll'} \qquad \text{(see Exercise 10)} \tag{4}$$

Range for $\delta_{ll'}$ from Figure 11.3, with $\Delta x = -\delta_{ll'}$.

$$\begin{aligned} \Delta a_k = -\Delta a_l = \alpha_{kl} \text{ implies } \partial z^*/\partial \alpha_{kl} = u^*_l - u^*_k \\ \Delta b_k = -\Delta b_l = \beta_{kl} \text{ implies } \partial z^*/\partial \beta_{kl} = v^*_l - v^*_k \end{aligned} \tag{4'}$$

Figure 11.3 has $+\Delta x = \alpha_{kl}$ or β_{kl} in the (exterior) position of a_l or b_l.

$$\Delta z = 0 \qquad \text{for} \qquad \Delta c^0_{ll'} \geq -c^*_{ll'} \tag{5}$$

$$\partial z^*/\partial c^0_{kk'} = x^*_{kk'} \qquad \text{for} \qquad -\min_{\bar{K}} c^*_{ij} \leq \Delta c^0_{kk'} \leq \min_{K} \{c^*_{ij} : ij \neq kk'\} \tag{6''}$$

In (2) and (6″), $\bar{K} = M''N'$ and $K = M'N''$ in (4) of 11.3.
If $\bar{K}$ is empty, $\min_{\bar{K}} = +\infty$.

Example

a	b=9	15	5	2	4	−u*
13	③	⑧	3	②	8	0
12	5	⑦	⑤	6	9	−2
10	⑥	4	7	2	④	5
	−v*=4	6	3	1	2	

(7)

The rectangle encloses values of optimal basic x^*_{ij} (circled), where $c^*_{ij} = 0$; and c^*_{ij}, where nonbasic $x^*_{ij} = 0$. The original $c^0_{ij} = c^*_{ij} - u^*_i - v^*_j$; e.g., $c^0_{21} = 5 - 2 + 4$, $c^0_{31} = 0 + 5 + 4$.

TABLE 11.7B Some Ranges for the Example (7)

(1) shows that $\partial z^*/\partial x_{21} = c^*_{21} = 5$. The loop formed by x_{21} has $x_{21} = \Delta x$, $x_{22} = 7 - \Delta x$, $x_{12} = 8 + \Delta x$, $x_{11} = 3 - \Delta x$. Thus $\max \Delta x = 3$, which makes $x_{11} = 0; 0 \le x_{21} \le 3; 0 \le \Delta z \le 5 \cdot 3 = (\partial z^*/\partial x_{21}) \max \Delta x$.

(2) with $x_{kk'} = x_{11} = 3$ requires that (7) be treated like (4) of 11.3 to obtain the new c'_{ij}, thus:

$$\begin{array}{ccccc} \Delta c & 0 & 3 & 0 & 8+\Delta c \\ 5+\Delta c & 0 & 0 & 6 & 9+\Delta c \\ 0 & 4-\Delta c & 7-\Delta c & 2-\Delta c & 0 \end{array}$$

Here Δc has the coefficient $+1$ in K, -1 in $\bar{K}$ and 0 in Z. Then (2) gives $\partial z^*/\partial x_{11} = \min\{4,7,2\} = 2 = c^*_{34}$ for $\Delta x_{11} \ge 0$, and $-\min\{5,8,9\} = -5 = -c^*_{21}$ for $\Delta x_{11} \le 0$. The loop formed by x_{34} has $\Delta x = 2$; that formed by x_{21} has $\Delta x = 3$ (see (1) above).

$$\text{Thus } \frac{\partial z^*}{\partial x_{11}} = \begin{matrix} 2 \\ -5 \end{matrix} \qquad \text{or} \quad \Delta x_{11} \in \begin{matrix} [0,2] \text{ or} \\ [-3,0] \end{matrix} \qquad \Delta z \in \begin{matrix} [0,4] \text{ or} \\ [0,15] \end{matrix}$$

(3) shows that $\partial z^*/\partial \delta_{11} = c^0_{11} = 0 + 0 + 4 = 4$ for $\delta_{11} \ge -3$.

(4) with $\delta_{ll'} = \delta_{21}$ shows that $\partial z^*/\partial \delta_{21} = -u^*_2 - v^*_1 = -2 + 4 = 2$. The calculation under (1) above is used, and extended to consider negative values of Δx. For the latter, $x_{ll'} = \Delta x$ is ignored, and $x_{12} = 0$ gives $\Delta x = -8$. Thus $\partial z^*/\partial \delta_{21} = 2$ for $-3 \le \delta_{21} \le 8$ and $-6 \le \Delta z \le 16$. $\Delta x = -\delta_{ll'}$ because $x_{ll'}$ and $\delta_{ll'}$ are on opposite sides of their two equations in row l and column l'.

(4′) with $\Delta b_k = -\Delta b_l = \beta_{kl}$ makes $\partial z^*/\partial \beta_{25} = v^*_5 - v^*_2 = -2 + 6 = 4$. If $\beta_{25} = \Delta x$ one has $+\Delta x$ in positions b_5, 31, 12 and $-\Delta x$ in positions 35, 11, b_2. Thus $-\min\{6,8\} = -6 \le \beta_{25} \le 3 = \min\{3,4\}$ is the range.

(5) shows that $\Delta z = 0$ for $\Delta c^0_{21} \ge -c^*_{21} = -5$.

(6) shows that $\partial z^*/\partial c^0_{11} = x^*_{11} = 3$ for $-2 \le \Delta c^0_{11} \le 5$, the negatives of the values that (2) determined for $\partial z^*/\partial x_{11}$.

be offset by the subtraction from z of a multiple of row kk'; then coefficients $+1$ in K are relevant to the preservation of $c'_{ij} \ge 0$.

EXERCISES 11.7

For each indicated parameter s find $\partial z^*/\partial s$ and the interval for Δs within which this $\partial z^*/\partial s$ is valid. Use (7) in 1 to 6.

1. x_{24}, x_{25}, x_{33}

2. x_{14}, x_{22}, x_{23}

3. $\delta_{14}, \delta_{22}, \delta_{23}$

4. $\delta_{24}, \delta_{25}, \delta_{33}$

5. $c^0_{24}, c^0_{25}, c^0_{33}$

6. $c^0_{14}, c^0_{22}, c^0_{23}$

Use the optimal tableau at the right to range

7. $x_{34}, x_{13}, \delta_{13}, \delta_{34}, c^0_{34}, c^0_{13}$

8. $\delta_{11}, x_{41}, c^0_{23}, x_{12}, \delta_{43}$

a	**b**= 7	8	15	5	−**u***
13	③	②	⑧	4	3
12	2	·	⑦	⑤	−4
4	④	5	·	7	−2
6	9	⑥	3	·	1
	−**v*** = 5	0	6	7	

(As in 11.5, a dot means $e_{ij}=0$; ignore these positions.)

9. Use one dual simplex step to reoptimize the tableau above if $a_1=13, b_3=15, x_{13}=8$ are each reduced by 9.

10. *Transportation paradox.* Let the sources A_1, A_2 and the destinations B_1, B_2 be located on a straight highway in the order A_2, B_1, A_1, B_2 with one mile between consecutive points. Let $a_i=i, b_j=j, c_{ij}=$miles from A_i to B_j. Show that $\partial z^*/\partial\delta_{11}=-1$, so that increasing a_1 and b_1 reduces the total transportation cost. How can this be?

11. Verify independently that Figure 11.3 and (4) of 11.3 always agree in their determination of $a_{kk',ll'}=0$, 1 or -1, the coefficient of $x_{ll'}$ in the equation in which basic $x_{kk'}$ occurs.

11.8 ONE PARAMETER IN a, b AND C

In Exercise 13 of 11.5 the data were treated as constants, but the cost of storage (transportation through time) properly includes interest charges on the value of the stored items, whether money is actually borrowed for the purpose or not. Since the interest rate may change during the period of storage, it is useful to have the solution as a function of the interest rate (then called the parameter).

The a_i and b_j also may depend on a parameter, say t. For example, one may have $a_1=t, a_2=\alpha-t$, and the others constant. This means that a total supply of $a_1+a_2=\alpha$ is required from the first two sources, but it remains to be seen how this total will be divided between them. ((4′) of 11.7 gave a partial answer.)

More generally, t may be a measure of time or economic activity, or a parameter interpolating between two production vectors **a** or between two extreme scenarios. Thus the demands b_j and the costs c_{ij} may change with t at various rates; of course some decision has to be made about how the change in Σb_j is distributed among the supplies a_i to keep $\Sigma a_i=\Sigma b_j$. It may well be that decisions about how to route shipments can be made on a month-to-month basis rather than all at once, parametrically, but the firm is still likely to be interested in estimating how the total cost z^* depends on t.

Reading the first part of 8.3 (prior to (a)) is suggested as an introduction to the general ideas of this section. The parameter t is assumed to enter the data linearly so that

$$a_i(t)=a_{i0}+a_{i1}t, \qquad b_j(t)=b_{j0}+b_{j1}t$$

$$c_{ij}(t)=\gamma_{ij}+\delta_{ij}t, \qquad x_{ij}(t)=\xi_{ij}+\eta_{ij}t \tag{1}$$

$$p_{ij}+q_{ij}t=x_{ij} \text{ if basic (circled)}$$

$$=c'_{ij} \text{ otherwise}$$

Variable z^* turns out to be a continuous piecewise linear or quadratic function of t, as it did in Figure 8.3.

To begin we need an optimal solution for some particular value of t; this is possible if and only if $\mathbf{a}(t)\geq\mathbf{0}$ and $\mathbf{b}(t)\geq\mathbf{0}$. If this value of t is 0, the result is that the constant terms a_{i0}, b_{j0}, γ'_{ij} and ξ_{ij} determine the basis changes, but the ensuing arithmetic operations are applied to the binomials in (1). Next we must determine for what values of t the resulting basis is optimal. (Only) for this purpose we need not distinguish between the basic x_{ij} (circled) and the c'_{ij}, since all have the same feasibility condition $p_{ij}+q_{ij}t\geq 0$ (see (1)). Intersecting the resulting semi-infinite intervals gives

$$t' \triangleq -\min_{ij}\{p_{ij}/q_{ij}: q_{ij}>0\} \leq t \leq \min_{ij}\{p_{ij}/(-q_{ij}): q_{ij}<0\} \triangleq t'' \tag{2}$$

as the interval of optimality. As usual the minimum over an empty set is interpreted as $+\infty$. Of course (2) insures that $\mathbf{a}(t)\geq 0$ and $\mathbf{b}(t)\geq 0$. Its left member is the same as $\max_{ij}\{-p_{ij}/q_{ij}: q_{ij}>0\}$.

When t moves outside the interval $[t', t'']$ in (2), one or more x_{ij} and/or c'_{ij} become negative. If it is only one c'_{ij}, say $c'_{ll'}$, that is negative, optimality is restored for some t outside $[t', t'']$ by bringing $x_{ll'}$ into the basis; this is a primal simplex step as in Figure 11.3 and Section 11.4. If it is only one x_{ij}, say $x_{kk'}$, that is negative, optimality is restored by removing $x_{kk'}$ from the basis (a dual simplex pivot step, illustrated by (8) and (9) of 11.7), unless $x_{kk'}$ was the only positive basic variable in row k or the only one in column k'. In the latter case a_k or $b_{k'}$ becomes negative and there is no optimal solution for $t<t'$ or $t>t''$ as the case may be.

This qualification also applies when one $c'_{ll'}$ and one $x_{kk'}$ are the first to become negative and do so simultaneously. Then one does not exchange $x_{kk'}$ and $x_{ll'}$, but rather determines the coefficient $a_{kk',ll'}=1$, -1 or 0 (see Exercise 11 of 11.7). If this is 1, a dual simplex step (with pivot $=-1$) is used to remove $x_{kk'}$ from the basis; if it is -1, a primal simplex step (with pivot $+1$) is used to bring $x_{ll'}$ into the basis. The former case occurs in (8) and (9) of 11.7.

In other more complicated cases of ties in the course of determining either of the variables to be exchanged, it may be necessary to restate the x_{ij} and c'_{ij} by substituting $t=t''+\varepsilon$ or $t'-\varepsilon$, where t'' or t' is a number and ε is a symbol such that $0<\varepsilon\ll 1$, to choose the first of the two variables to be exchanged to produce the greatest change $|\Delta z|$ in z, and to perform more than one pivot step, perhaps using Phase 1b of 11.5 as well as Phase 2 of 11.4. This is explained in more detail in 8.3.

In order to determine the indices of the second of the two variables to be exchanged, the transition value of $t=t'$ or t'' must be substituted in the Δx loop in the primal step or in the c'_{ij} in $\overline{K}$ (defined in the fourth paragraph of 11.7) in the dual step. (However, the value of Δx or Δc must remain a function of t, if such it is.) Like t' or t'' itself, these values are exceptions to the rule (Theorem 2 of 11.2) that all the tableaus have integer entries if the first one does. The lower line in each cell provides space for these values as well as for the labels P and M in the Δx loop and the letter k or l that identifies $x_{kk'}$ (leaving the basis) and $x_{ll'}$ (entering the basis).

The new optimal tableau will determine the optimal solution for another t-interval adjacent to the preceding. In practice one should first increase t as many times as may be relevant, and then if necessary return to the initial tableau and decrease t as many times as may be relevant. The maximum number of optimal bases is usually much smaller than the large but finite number available. According to the theorem of 8.3, the set of values of t that admit an optimal solution is always convex; i.e., it is empty, or the whole real line, or a finite or semi-infinite interval. Thus there is never any need to jump over any interval in t, or even an isolated point, though it may be necessary to jump over one or more bases as in Exercise 4 of 8.3.

Example

The problem in Table 11.8 has already been optimized for $t=0$, but the elements are given as functions of t. Formula (2) or inspection of the 12 functions giving the $x_{ij} \geq 0$ and $c'_{ij} \geq 0$ shows the given basis is optimal for $0 \leq t \leq 1$. For $t<0$, x_{11} becomes negative; since it is the only basic variable in its row, this implies that a_1 becomes negative. Thus no feasible solution exists for $t<0$.

For $t>1$, c'_{13} becomes negative; thus a primal simplex pivot step begins with $ll'=13$. It determines the loop $x_{13}=\Delta x$, $x_{11}=3t-\Delta x$, $x_{21}=5-4t+\Delta x$, $x_{23}=5-3t-\Delta x$. One puts $t=t''=1$ to find that $x_{23}=\min\{x_{11}=3, x_{23}=2\}=x_{kk'}$ leaves the basis, but concludes that $\Delta x=5-3t$, not 2. In (4) of 11.3, Δc is chosen to increase c'_{13} from $1-t$ to 0.

Making the Δx and Δc adjustments gives the second tableau. It is optimal for $1 \leq t \leq 1.2$; x_{33} becomes negative when $t>1.2$. Thus a dual simplex pivot step begins with $kk'=33$ and $\overline{K}=\{14,24\}$. One puts $t=t''=1.2$ to find that $c'_{ll'}=\min\{4-2t=1.6, 2+3t=5.6\}=c'_{14}$, but concludes that $\Delta c=4-2t$, not 1.6. Now x_{14} forms the loop $x_{14}=\Delta x$, $x_{13}=5-3t-\Delta x$, $x_{33}=6-5t+\Delta x$, $x_{34}=4-3t-\Delta x$, in which $\Delta x=5t-6$ so as to increase $x_{kk'}=x_{33}$ from $6-5t$ to 0. (In the dual simplex this loop always includes $x_{kk'}$, labeled P, and $\Delta x=|x_{kk'}|$).

The third tableau is optimal for $1.2 \leq t \leq 1.25$; x_{34} becomes negative for $t>1.25$. Since it is the only basic variable in its row, a_3 becomes negative and no feasible solution can exist for $t>1.25$. The x_{ij} are checked by comparing their row and column sums with a_i and b_j. The c'_{ij} are checked by adding up the Δc adjustments in each row or column, reversing the signs, and applying the resulting adjustments to the c'_{ij} in the last tableau. The result should be the c'_{ij} in the first tableau. Of course the set of basic ij must satisfy Theorem 1(d) of 11.2. Either the c^0_{ij} or $\mathbf{u}^*, \mathbf{v}^*$ (both omitted in Table 11.8) are needed to calculate z^* by (9) of 11.1.

TABLE 11.8 Parametric Transportation Problem

The rectangles enclose values of basic $x_{ij} \geq 0$ (circled), where $c'_{ij} = 0$, and values of $c'_{ij} \geq 0$, where nonbasic $x_{ij} = 0$.

a	b=$5-t$	$2+3t$	$11-8t$	$4-3t$		Δc coefficients (−1 in $\bar{K}$)	
$3t$	($3t$) circled $3\ M$	$2+3t$	$1-t$ $l\ 0\ P$	$5-3t$		0 0	+ +
$12-4t$	$5-4t$ $1\ P$	$2+3t$	$5-3t$ $k\ 2\ \bar{M}$	$3+2t$		0 0	+ +
$10-8t$	$3-t$	$6+2t$	$6-5t$	$4-3t$	$1-t$	− −	0 0
Optimal for $0 \leq t \leq 1$			$-1+t$	$-1+t$	Δc		

Primal step: x_{13} into basis, x_{23} out. (as in (4) of 11.3)

$3t$	$-5+6t$	$2+3t$	$5-3t$ M	$4-2t$ $l\ 1.6\ P$		0 0 0	−	$\bar{K}$
$12-4t$	$10-7t$	$2+3t$	$-1+t$	$2+3t$ 5.6		0 0 0	−	
$10-8t$	$4-2t$	$7+t$	$6-5t$ $k\ \bar{P}$	$4-3t$ M	$4-2t$	+ + +	0	
Optimal for $1 \leq t \leq 1.2$				$-4+2t$	Δc			

Dual step: x_{33} out of basis, x_{14} in.

$3t$	$-5+6t$	$2+3t$	$11-8t$	$-6+5t$	
$12-4t$	$10-7t$	$2+3t$	$-1+t$	$-2+5t$	
$10-8t$	$8-4t$	$11-t$	$4-2t$	$10-8t$	$-5+3t$
Optimal for $1.2 \leq t \leq 1.25$			$1-t$	$5-3t$	Check

No feasible solution for $t<0$ ($a_1<0$) or $t>1.25$ ($a_3<0$). Let $u_i^* = v_i^* = -(i-1)^2 - (3-i)^2 t$ in the first tableau. Then $z^* = -\Sigma a_i u_i^* - \Sigma b_j v_j^* = 134 - 54t + 4t^2, 139 - 62t + 7t^2, 115 - 30t - 3t^2$. For example, $134 - 54t + 4t^2 - (1-t)(10-8t) - (-1+t)(11 - 8t + 4 - 3t) = 139 - 62t + 7t^2$.

EXERCISES 11.8

Optimize for all possible values of t, given x_{ij} (circled), c'_{ij}. To get z^*, let $\mathbf{u}^* = \mathbf{v}^* = -(1, 2+t, 3-t)$ in the given tableau.

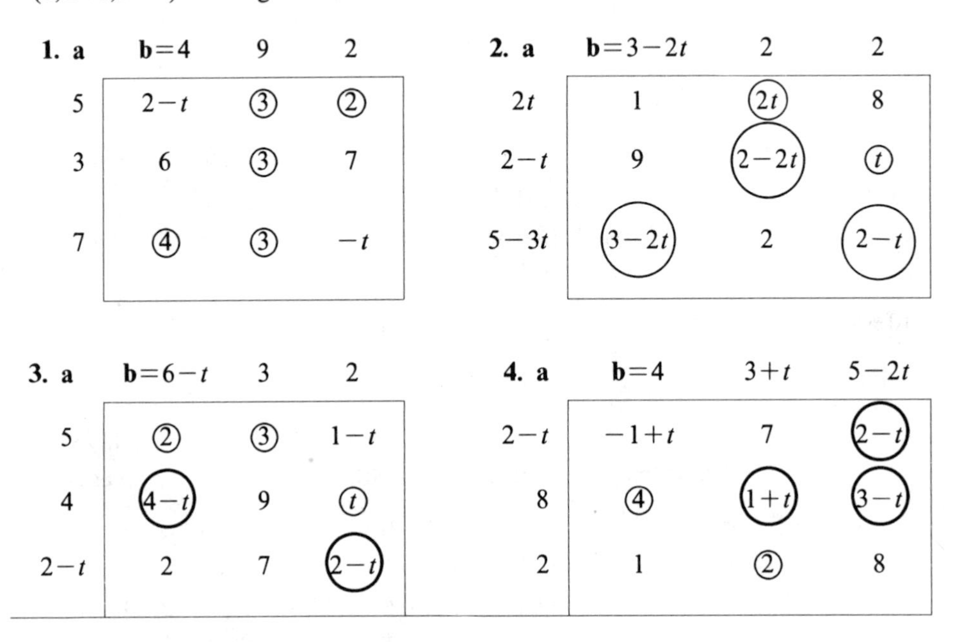

1.

a	b=4	9	2
5	$2-t$	(3)	(2)
3	6	(3)	7
7	(4)	(3)	$-t$

2.

a	b=$3-2t$	2	2
$2t$	1	($2t$)	8
$2-t$	9	($2-2t$)	(t)
$5-3t$	($3-2t$)	2	($2-t$)

3.

a	b=$6-t$	3	2
5	(2)	(3)	$1-t$
4	($4-t$)	9	(t)
$2-t$	2	7	($2-t$)

4.

a	b=4	$3+t$	$5-2t$
$2-t$	$-1+t$	7	($2-t$)
8	(4)	($1+t$)	($3-t$)
2	1	(2)	8

REFERENCES (Chapter 11)

Barr-Glover-Klingman (1977)
Dantzig (1951c, 1957, 1963)
Glover-Karney-Klingman (1972)
Glover-Karney-Klingman-Napier (1974)
Hitchcock (1941)
Kantorovich (1939)
Koopmans (1949–1951)
Machol-Wien (1977)
Murty (1976)
Reinfeld-Vogel (1958)
Schell (1955)
Srinivasan-Thompson (1972–1973)
Szwarc (1971)

Answers to Exercises

Section 1.5

1. $x_1^* = 13/7$, $x_2^* = 2/7$, $z^* = 32/7$.
2. $u_1^* = 5$, $u_2^* = 2$, $\bar{w}^* = 8$.
3. No feasible solution.
4. $\bar{w}$ is unbounded (can be made as large as you like).

5, 6. $x_1^* = 8$, $x_2^* = 6$; $\bar{z}^* = 50$; $z^* = 100$.

7. Unbounded unless $\mathbf{x} \geq \mathbf{0}$; then $\mathbf{x}^* = (0, 50/3)$.

Section 2.1

1. (a) Vector space must contain origin $\mathbf{0}$.
   (b) Its equations must have constant terms $= 0$.
2. (a) Plane $s\mathbf{x} + t\mathbf{y}$ through $\mathbf{0}, \mathbf{x}, \mathbf{y}$. (b) Infinite line $(1-t)\mathbf{x} + t\mathbf{y}$.
3. $(12, -1, -5)$.
4. $s = 2$, $t = -1$.
5. (a) Line $2x_1 = x_2$. (b) Point $(1, 2)$.
6. (a), (b). Line $2x_1 = x_2$.
7. (a) Whole x_1, x_2 plane. (b) Line $x_1 + x_2 = 3$.
8. (a) Plane $x_3 = 0$. (b) Line $x_1 + x_2 = 1$, $x_3 = 0$.
9. (a) All triples. (b) Plane $x_1 + x_2 + x_3 = 1$.
10. $(2/3, 1, 0)$, $(-4/3, 0, 1)$.
11. $(-5/2, 1, 0, 0)$, $(1/2, 0, 1, 0)$, $(-3/2, 0, 0, 1)$. Many other bases exist.
12. Yes, because $(-2, 0, -2) = -(1, 1, 2) - (1, -1, 0)$.
    No, because $(1, 0, 0)$ is no such linear combination and the space is only two dimensional.

13. Yes, because $(1, 3, 4) = 2(1, 1, 2) - (1, -1, 0)$.

Indeterminate; linear variety might include all triples.

14. A linear variety.

Section 2.2

1. (a) Λ (b) $(0, 0)$.
2. (a) A (b) $(0, 1)$, $(1, 0)$.
3. (a) A (b) $(0, 0)$, $(1, 0)$.
4. Nonconvex; $\{(x_1, x_2): x_2 = 0\}$.
5. (a) A (b) None.
6. (a) X (b) $(0, 0)$.
7. Nonconvex; $\{(x_1, x_2): 0 \leq x_1 = x_2 \leq 1\}$.
8. (a) X' (b) None.
9. (a) X (b) None.
10. (a) C (b) $\{(x_1, x_2): x_1^2 + x_2^2 = 1\}$ (the circumference).
11. No, because extreme point $(1, 1, 2) = (1/2)(1, -1, 0) + (1/2)(1, 3, 4)$.
    Yes, because $(1, 0, 1) = (1/2)(1, 1, 2) + (1/2)(1, -1, 0)$.
12. $(0, 0)/3 + (0, 3)/3 + (3, 0)/3 = (1, 1)$.
13. Triangle with vertices $(5, 0)$, $(0, 0)$, $(0, 4)$.

Section 2.3

1. -10; the second.
2. $(1,3,-2)[\mathbf{x}-(1,1,1)]=(x_1-1)+3(x_2-1)-2(x_3-1)=0.$
3. $(-8,9)=-3(2,-1)+2(-1,3)$ or $-8x_1+9x_2=-3(2x_1-x_2)+2(-x_1+3x_2)$
4. $(\mathbf{u}+\mathbf{v})(\mathbf{x}+\mathbf{y})$.
5. 13.
6. 90°.
7. 0°.
8. 120°.
9. 180°.
10. $\{(1,0,-3,2),(0,1,-4,-5)\}$,
 $\{(3,4,1,0),(-2,5,0,1)\}$.
11. $\{(5,-7,1,0,0),(-4,7,0,1,0),(-3,4,0,0,1)\}$,
 $\{(1,0,-5,4,3),(0,1,7,-7,-4)\}$.
12. $\{(12,6,-3,4)\}$,
 $\{(1,-2,0,0),(1,0,4,0),(1,0,0,-3)\}$.

Section 2.4

1.

1	x_1	x_2
x_3	3	4
x_4	-2	5

$u_1=3u_3-2u_4$
$u_2=4u_3+5u_4$

2.

1	u_2	u_4	u_5
u_1	-3	2	1
u_3	2	-5	-4

$x_1-3x_2+2x_4+x_5=0$
$x_3+2x_2-5x_4-4x_5=0$

3.

1	x_3	x_4	x_5
x_1	1*	3	-2
x_2	1	2	1

1	x_1	x_4	x_5
x_3	1	3	-2
x_2	-1	-1*	3

1	x_1	x_2	x_5
x_3	-2	3	7
x_4	1	-1	-3

$\mathbf{x}=(-2,-4,1,1,1)$ satisfies all the row equations.

4. $x_1 \quad -5x_3+4x_4+3x_5=0$
$x_2+7x_3-7x_4-4x_5=0$

1	x_3	x_4	x_5
x_1	-5*	4	3
x_2	7	-7	-4

1	x_1	x_4	x_5
x_3	$-.2$	$-.8$	$-.6$
x_2	1.4	-1.4	.2*

1	x_1	x_4	x_2
x_3	4	-5	3
x_5	7	-7	5

$\mathbf{u}=(1,1,2,-3,-1)$ satisfies all the column equations.

5.

x_1	x_2	x_3	x_4
3	4	1	0
-2	5	0	1

x_1	x_2	x_3	x_4	x_5
1	0	1	3	-2
0	1	1	2	1

6.

	x_3	x_4	x_5	
u_1	1	3	-2	$=-x_1$
u_2	1	2	1	$=-x_2$
	‖	‖	‖	
	u_3	u_4	u_5	

1	x_1	x_2
x_3	-1	-1
x_4	-3	-2
x_5	2	-1

Section 2.5

1.

Check	1	y_1	y_2	y_3
3	x_1	0	3	−1
9	x_2	2	2*	4
−5	x_3	−3	2	−5

	−6	x_1	x_2	y_3
−21	y_1	2	−3	−14
−6	y_2	−2	0	2
−21	x_3	10	−9	−16*

Check	2	y_1	x_2	y_3
−21	x_1	−6*	−3	−14
9	y_2	2	1	4
−28	x_3	−10	−2	−18

	−16	x_1	x_2	x_3
−7	y_1	−18	13	14
−23	y_2	−2	−3	−2
−21	y_3	10	−9	−6

2.

	1	y_1	y_2	y_3
5	x_1	3*	−1	2
2	x_2	0	2	−1
2	x_3	2	−3	2

	2	x_1	y_2	x_3
6	y_1	2	4	−2
0	x_2	−2	−3*	3
−4	y_3	−2	−7	3

	3	x_1	y_2	y_3
5	y_1	1	−1	2
6	x_2	0	6	−3
−4	x_3	−2	−7	2*

	−3	x_1	x_2	x_3
−9	y_1	1	−4	−3
0	y_2	−2	2	3
6	y_3	−4	7	6

If other pivots are used, the rows and columns may appear in other orders in the final tableau.

3. $x_1=(-2y_1-2y_2+3y_3)/11$, $x_2=(5y_1+5y_2-2y_3)/11$, $x_3=(4y_1+15y_2-6y_3)/11$.

4. Note that to replace a label δx_h by x_h again, one divides by δ if x_h is a row label and multiplies by δ if x_h is a column label.

Section 2.6

1.

−2	−4	−3
−3	−6	−4
2	$4\frac{1}{2}$	3

Det=2

2.

22/4	20/4	4/4
−3/4	−2/4	0
−26/4	−24/4	−4/4

Det=4

3. No inverse

Det=0

4.

1	3/4	0	0
0	−1/4	0	0
0	2/4	1	0
0	5/4	0	1

Det =−4

5. $\mathbf{y}=(1,3,-1)$.

6.

Check	1	x_1	x_2	x_3	x_4	1
−5	0	1*	−4	−1	2	−3
−1	0	−2	1	3	−1	−2
−7	0	−3	−2	5	1	−8

Check	1	x_2	x_3	x_4	1
−5	x_1	−4	−1	2	−3
−11	0	−7*	1	3	−8
−22	0	−14	2	7	−17

Check	−7	x_3	x_4	1
−9	x_1	11	−2	−11
−11	x_2	1	3	−8
0	0	0	−7*	7

The only possible value for x_4 if 1; hence it must be basic in the final solution and $\{x_1, x_2, x_3\}$ cannot all be basic as requested. Pivoting on the −7 gives $x_1=(-13+11x_3)/7$, $x_2=(-5+x_3)/7$, $x_4=1$.

7. $y_1=(13-5y_3)/7$, $y_2=(11-y_3)/7$.

8. Inconsistent. **9.** Rank=2. **10.** Rank=3.

Section 2.7

1. Solve four equations: $2=2u_1+u_2+3u_3$, etc. $\mathbf{u}=(2,1,-1)$.

2.

3	-1	4	2	1
0	16	-10	7	-4
0	0	54	-1	-52

3.

2	1	-3	5	-1
0	-7	7	-3	9
0	0	0	0	14

4. Show the rank (=maximum number of possible exchanges)=number of vectors. $m>n\geq$rank.

5. (a) Every row of $\mathbf{A}_2$ is linearly dependent on rows of $\mathbf{A}_1$; the row space of $\mathbf{A}_1$ contains that of $\mathbf{A}_2$.
(b) $\mathbf{A}_1$ and $\mathbf{A}_2$ have identical row spaces.
(c) Pivots in $\mathbf{A}_1$ (part of $\mathbf{A}$) reduce $\mathbf{A}_2$ to $\mathbf{0}$.

6. Rank of (15) is $3=n-\rho=5-2$.

7. The line has rank $1=n-\rho=3-\rho$ so $\rho=2$=the number of linearly independent equations. Solving them gives $\mathbf{x}=(1,2,-1)$ or any multiple thereof.

8. With $K=L=\{1,2\}$, (6) is Image: $\{(7,0,7),(0,7,14)\}$ or $\{(1,0,1),(0,1,2)\}$ Kernel: $\{(3,2,7,0),(-7,-7,0,7)\}$

-7	x_1	x_2	y_3	y_4
y_1	2	-1	3	-7
y_2	-1	-3	2	-7
x_3	7	14	0	0

9.

u_1	u_2	u_3
2*	-1	-2
-1	2	7

2	u_2	u_3
u_1	-1	-2
0	3*	12

3	u_3
u_1	3
u_2	12

gives $\mathbf{u}=(1,4,-1)$ as the orthogonal complement of the first space; i.e., the latter is the solution set of $x_1+4x_2-x_3=0$. Similarly the second basis satisfies $2x_1+3x_2-x_3=0$. The intersection satisfies both equations. The required basis consists of $\mathbf{x}=(1,1,5)$.

x_1	x_2	x_3
1*	4	-1
2	3	-1

1	x_2	x_3
x_1	4	-1
0	-5	1*

1	x_2
x_1	-1
x_3	-5

10. Extend a basis of k vectors for $X\cap Y$ to bases for X and Y by adjoining m and n vectors, respectively. Then rank $(X\cap Y)=k$, rank $X=k+m$, rank $Y=k+n$, rank$(X+Y)\leq k+m+n$. To get equality, note that any dependence among the $k+m+n$ vectors must involve some of m and some of n. Combining these gives a vector in X but not in Y equal to a vector in Y, which is a contradiction.

Section 3.1

1.

1	y_1	y_2	y_3	1
x_1	2	5	-1	-8
x_2	-1	3	4	-10
$-z$	-3	-1	2	0

$\mathbf{x},\mathbf{y},\mathbf{u},\mathbf{v}\geq\mathbf{0}$

	y_1	y_2	y_3	1	
u_1	2	5	-1	-8	$=-x_1$
u_2	-1	3	4	-10	$=-x_2$
1	-3	-1	2	0	$=z$
	‖	‖	‖	‖	
	v_1	v_2	v_3	$\overline{w}$	

Maximize $\overline{w}=-8u_1-10u_2$ subject to
$u_1\geq 0$, $u_2\geq 0$,
$5u_1+3u_2-1=v_2\geq 0$,
$2u_1-u_2-3=v_1\geq 0$,
$-u_1+4u_2+2=v_3\geq 0$.

2.

1	x_1	x_2	x_3	1
x_4	-4	3	-2	8
x_5	2	-1	1	4
$-z$	-1	1	1	0

$(\mathbf{x}\geq 0)$

3.

1	u_3	u_4	u_5	$\overline{w}$
u_1	-3	-1	-1	1
u_2	-1	-1	-4	2
1	5	3	6	0

$(\mathbf{u}\geq 0)$

4. (6) is transposed; the a_{ij}, b_i, c_j, d change signs; $\min(-\overline{w})$.

Section 3.2

1. The solution set X' is the infinite line $x_1+x_2=2$. The feasible region X^+ is the part of X' between $(0,2)$ and $(2,0)$. The optimal solution set $X^*=\{(0,2)\}$.
2. $X'=$ the infinite plane $x_1+x_2+x_3=3$.
   $X^+=$ the triangle with vertices $(3,0,0)$, $(0,3,0)$, $(0,0,3)$. (These are the basic feasible solutions.)
   $X^*=$ the line segment joining $(3,0,0)$ and $(0,3,0)$.
3.

1	x_1	x_2	x_3	1
0	-1^*	4	2	-5
0	2	-1	3	-4

-1	x_2	x_3	1
x_1	4	2	-5
0	-7^*	-7	14

-7	x_3	1
x_1	-14	21
x_2	-7	14

or

1	x_3	1
x_1	2	-3
x_2	1	-2

which has the basic solution $(3,2,0)$.

4. Not basic because the nonzero variables x_2, x_3, x_4 cannot be solved for. In a basic solution the zero variables must uniquely determine the others as in Theorem 8 of 2.7, but here $(0,1,6,5)$ is also a solution.
5. Basic; one equation is redundant.

Section 3.3

1. Nothing applicable.
2. Dual feasible; $k=1$, $l=4$.
3. Theorem 1(a_2); $z^*=+\infty$.
4. Primal feasible; $kl=13$.
5. Optimal
6. Theorem 1(a_1); $z^*=-\infty$.
7. $\mathbf{x}_{3,4,5}$ makes $x_1=-1$; infeasible. $\mathbf{u}_{1,2}=(2,3)$ makes $\mathbf{u}_{3,4,5}=(0,1,1)$ (feasible) and $\bar{w}=24$. So $z^*\geq$ feasible $\bar{w}=24$ by (13).
8. $9\leq z^*\leq 23$ (both are feasible).
9. (a) No; $\mathbf{ux}\neq 0$. (b) No; $z=-11$, $\bar{w}=-7$.
   (c) No; column equations not satisfied.
10. (a) No; column equations not satisfied. (b) No; $x_3<0$. (c) Optimal.
11. Let $x_3=x_1+x_2-3$, $x_4=x_1+3x_2-5$. The basic solutions are $(0,0,-3,-5)$, $(0,3,0,4)$, $(0,5/3,-4/3,0)$, $(3,0,0,-2)$, $(5,0,2,0)$, $(2,1,0,0)$, obtained by setting enough $x_h=0$ to permit the equations to be solved. Only three of these are feasible; $\mathbf{x}^*=(0,3,0,4)$ gives min $z=3$.
12. The feasible basic solutions are $(0,0,0,5,7)$, $(0,0,7/3,5,0)$, $(0,5/2,0,0,9/2)$, $(0,5/2,3/2,0,0)$, $(5,0,0,0,7)$, $(5,0,7/3,0,0)$. The last gives $z^*=-22/3$.
13. $\mathbf{x}_{3,4,5}=(0,1,0)$ and $\mathbf{u}_{1,2}=(1,0)$ are feasible. (They make the basic variables nonnegative.) Hence an optimal solution exists.
14. (a) No; $u_5<0$. (b) No; $z=4$, $\bar{w}=3$. (c) No; $u_3x_3\neq 0$.
    (d) Could be optimal but not basic. (Not enough $x_h=0$).
15. (a) and (c) could be optimal although (a) is not basic.
    (b) Not optimal; $x_1<0$. (d) $z-\bar{w}\neq\mathbf{ux}$; not optimal
16. No; for any $\mathbf{A}$, $\mathbf{x}_{1,2,3}=(0,0,0)$ makes $\mathbf{x}_{4,5,6}=(-2,3,-1)$ not $(0,0,1)$.
17. The validity of the algorithms depends on the primal-dual invariance theorem of 2.4, the finite number of nonrepeated basic solutions, the nonexistence of a solution when a pivot cannot be found (Theorem 1 of 3.3), and the optimality of the basic solution when $\mathbf{b}\leq\mathbf{0}$ and $\mathbf{c}\geq\mathbf{0}$.

Section 3.4

1. First pivot is $a_{14}=2$.

7	x_3	x_1	x_2	1
x_4	10	4	-1	-5
x_5	1	-1	2	-4
$-z$	29	13	2	-32

$\mathbf{x}^*=(0,0,0,5,4)/7$. $z^*=-32/7$.
$\mathbf{u}^*=(13,2,29,0,0)/7$.

2. First pivot is $a_{23}=2$.
   Column x_5 becomes
   $(-5,-1,-1)/2$. $z^*=-\infty$.

3. First pivot is $a_{15}=-3$.

7	x_4	x_1	x_3	1
x_5	−4	−3	−1	−7
x_2	−21	7	7	−7
x_6	2	−2	−3	−7
$-z$	7	7	7	21

4. First pivot is $a_{26}=-1$.

6	x_4	x_1	x_2	1
x_5	−11	−1	−4	−9
x_6	−7	1	−2	−3
x_3	13	5	8	27
$-z$	47	1	10	21

No feasible $\mathbf{x}$ in row x_3.

5. First pivot is $a_{25}=2$.

8	u_1	u_2	u_6	$\overline{w}$
u_4	2	0	−2	−8
u_5	3	4	1	−16
u_3	−5	−4	−15	−16
1	8	8	8	−40

6. In anywhere from 0 (multiply the inequalities by 1, −2, 0 and add) to 3 pivot steps, one finds a negative column. No $\mathbf{u}$ is feasible.

7. First pivot is $a_{15}=-1$.

1	u_2	u_4	u_1	$\overline{w}$
u_5	1	1	1	−2
u_3	−1	0	−2	−2
1	2	1	5	8

8. First pivot is $a_{14}=-1$.

2	u_1	u_2	u_6	1
u_4	−4	−1	−7	−4
u_5	−2	−1	−5	−2
u_3	6	2	10	2
1	8	3	23	8

No $\mathbf{x}$ is feasible in row 3.

9. First pivot is $a_{35}=3$.

5	x_2	x_3	1
x_1	−3	−4	−2
x_4	3	−1	−3
x_5	−1	2	−4
$-z$	1	3	−11

10.

1	y_1	y_2	100
$10x_1$	−3	−5	4
$10x_2$	−7	−5	6
$-.01z$	2	5	0

The dual row labels are $.1u_1$, $.1u_2$, 100.

Section 3.5

4.

5		1	(2)	−3	(3)
	5	x_4	x_2	x_5	1
(−1)	x_1	−2	5	1	−5
(1)	x_3	1	5	2	−5
3	$-z$	3	10	1	15

6.

3		−4	−3	(3)	(2)
	3	x_2	x_1	x_5	1
(1)	x_4	2	1	−1	−11
(0)	x_3	1	−1	−2	−1
2	$-z$	2	1	8	−5

8.

14		4	1	1
	14	$10x_2$	$10x_3$	100
8	$10x_1$	−16	−6	−42
1	y_1	4	−2	−28
1	y_2	−1	4	−14
3	$-.1z$	2	6	−56

10.

8		−3	1		
	8	x_1	x_2	x_6	1
	x_4	2	0	−2	−6
	x_5	3	4	1	−5
−4	x_3	−5	−4	−15	−21
0	$-z$	8	8	8	−16

11.

7		(2)			(0)
	7	x_4	x_1	x_3	1
(1)	x_5	−4	−3	−1	−7
	x_2	−21	7*	7	−7
(−1)	x_6	2	−2	−3	−7
	$-z$	8	−1	2	0

7	x_4	x_2	x_3	1
x_5	−13	3	2	−10
x_1	−21	7	7	−7
x_6	−4	2	−1	−9
$-z$	5	1	3	−1

12. (a) z^* is recorded separately near d^0. (b) Not for checking. Initial $\delta=1$ not copied. Use $-c_M^0$ on left. (c) No nonbasic variable=0. Distinguish primal and dual values in (a) and (b). Use values of b_i in (b), of x_i in (a) and (c).

Section 3.6

(all $x_i \geq 0, y_j \geq 0$)

1.

1	y_1	y_2	y_3	1
x_1	−2.00	−2.50	−2.80	1200
x_2	−.09	−.08	−.06	38
$-z$	1	1	1	0

Model 2; y_j = hours on job j.

2.

1	y_1	y_2	y_3	1
x_1	0	7	0.2	−15
x_2	1.5	2	0	−12
$-z$	−95	−360	−8	0

Model 1; y_j = 1000's of item j.

3.

1	y_1	y_2	y_3	1
x_1	−18	−10	−6	3
x_2	−60	−85	−50	15
x_3	6	5	7	−2
$-z$	8	7	3	0

Modified Model 2;
y_j = units of feed j.

4.

1	y_1	y_2	y_3	1
x_1	.17	.24	.29	−4300
x_2	.27	.25	.19	−3900
x_3	.30	.26	.25	−4700
$-z$	−1.0	−1.6	−2.0	0

Model 1; y_j = yards of cloth j.

5. Let y_{ij} = bags shipped from $i=1,2$ to $j=3,4,5$, and x_i = bags not shipped from $i=1,2$. Transportation model.

1	y_{13}	y_{23}	y_{14}	y_{24}	y_{15}	y_{25}	1
x_1	1	0	1	0	1	0	−1400
x_2	0	1	0	1	0	1	−3500
0	−1	−1	0	0	0	0	1200
0	0	0	−1	−1	0	0	1500
0	0	0	0	0	−1	−1	1800
$-z$	270	300	170	440	330	320	0

6. Let y_j = the number of lots using the indicated packing plan.

	y_1	y_2	y_3	y_4
40	3	0	1	4
50	2	2	0	0
60	0	2	3	1

1	y_1	y_2	y_3	y_4	1
x_1	3	0	1	4	−500
x_2	−2	−2	0	0	400
x_3	0	2	3	1	−300
$-z$	−1	−1	−1	−1	0

Modified Model 1.

7. Let $y_1, \ldots, y_5$ be the proportions purchased (by weight) and y_0 = the proportion of pure water added, if any. y_0 is the only slack variable. Model 2a.

1	y_1	y_2	y_3	y_4	y_5	1
y_0	1	1	1	1	1	−1
0	0	12	20	50	0	−17
0	0	0	4	0	100	−2
0	3	2	0	0	0	−1
$-z$	.30	1.00	1.65	4.00	.25	0

Section 3.7

$(\mathbf{x}\geq 0, \mathbf{y}\geq 0, \bar{\mathbf{x}}\geq 0)$

1. Let y_j = hours of operation of P_j

1	y_1	y_2	1
x_1	14	10	−70
x_2	9	18	−70
x_3	−15	−26	100
x_4	−25	−13	100
$-z$	60	80	0

2. Proportions x_0 of water, y_1, y_2, y_3 of milk.

1	y_1	y_2	y_3	1
x_0	1	1	1	−1
0	3	4	5	−4
x_2	−6	−5	−4	3
x_3	7	5	6	−6
$-z$	6.80	7.50	8.30	0

3. Let y_{i1} = proportion of time that P_i produces WS and y_{i2} = proportion of time that P_i produces MS.

1	y_{11}	y_{12}	y_{21}	y_{22}	1
x_1	1	1	0	0	−1
x_2	0	0	1	1	−1
x_3	30	0	50	0	−35
x_4	0	25	0	40	−35
$-z$	−40	−55	−75	−95	0

This is Model 1.

4. Let y_1, y_2, y_3 = crude oil purchased in 1000 barrels. Let $x_i, \bar{x}_i$ = excess or deficiency of G_i in 1000 gallons.

$-z$	y_1	y_2	y_3	x_1	x_2	$\bar{x}_1$	$\bar{x}_2$	1
0	7	6	12	−1	0	1	0	−58
0	12	7	8	0	−1	0	1	−73
1	370	260	410	−15	−15	23	19	−2721

5. Let y_1, y_2, y_3 = purified water produced in 1000 gallons.

1	y_1	y_2	y_3	1
x_1	−1	−1	−1	40
x_2	1	2	5/2	−50
x_3	−2	−1	1	0
$-z$	5	4	1	0

The third constraint is for the salt:

$$.01y_2+.03y_3 \leq .02(y_1+y_2+y_3).$$

6. Let y_{i1}, y_{i2} = amount of M_i used to make A, AA.

y_{i0} = amount of M_i not used (1000-pound unit)

$$6y_{11}+5y_{21}+3y_{31}\geq 4(y_{11}+y_{21}+y_{31}), \text{ etc.}$$

1	y_{11}	y_{21}	y_{31}	y_{12}	y_{22}	y_{32}	1
y_{10}	1			1			−50
y_{20}		1			1		−40
y_{30}			1			1	−70
x_{11}	−2	−1	1				0
x_{21}	−4	−1	1				0
x_{12}				−1	0	2	0
x_{22}				−2	1	3	0
$-z$	−33	−33	−33	−42	−42	−42	0

7. Let y_{ij} = tons of M_i shipped via T_j.

1	y_{11}	y_{21}	y_{12}	y_{22}	1
0	1		1		−30
0		1		1	−40
x_{11}	1	1			−35
x_{21}			1	1	−50
x_{12}	40	15			−1000
x_{22}			40	15	−1000
$-z$	15	15	12	12	0

Section 4.1

1. Saddle points at $ij=21,23$.
2. $a^0=a_{33}$; $a_0=a_{21}$.
3. $a^0=a_{12}$; $a_0=a_{32}$.
4. Any saddle point a^* or a' is both a minimax and a maximin. Thus (11) implies $a^*\le a'$ and $a'\le a^*$.
5. (a) $a_0=1$ (b) $a^0=7$ (c) $a^*=4$; $u^*=y^*=(.5,.5)$ as in matching pennies.
6. (a), (b), (c). Saddle point. $a^*=a_{12}=2$.
7. (a), (b), (c). Saddle point. $a^*=a_{21}=3$. There are mixed optimal strategies too: $\mathbf{u}^*=(0,1)$; $\mathbf{y}^*=(1,0)$ or $(.2,.8)$ or points between; $a^*=3$.
8. $\begin{matrix} 0 & 1 & -1 \\ -1 & 0 & 1 \\ 1 & -1 & 0 \end{matrix}$
9. $\begin{matrix} 150 & -300 & -400 \\ -200 & 300 & -400 \\ -200 & -300 & 350 \end{matrix}$

Section 4.2

1. $\mathbf{u}^*=(.5,.5)$, $\mathbf{y}^*=(0,.25,0,.75)$, $a^*=1.5$
2. $\mathbf{u}^*=(.4,.6)$, $\mathbf{y}^*=(0,.5,.5,0)$, $a^*=1$
3. $\mathbf{u}^*=(.6,0,.4,0)$, $\mathbf{y}^*=(.5,.5)$, $a^*=2$
4. $\mathbf{u}^*=(2/3,0,0,1/3)$, $\mathbf{y}^*=(1/3,2/3)$, $a^*=5/3$
5. No. This is equivalent to the mixed strategy (.2,.3,.5), the average of the other two.
6. (b) and (c) are optimal.
7. (b) is optimal.
8. Optimal if 115 is changed to 105.

Section 4.3

1.

i	$j=$ 2	3
1	4	1
2	−3	6

$\mathbf{u}^*=(.75,.25,0)$
$\mathbf{y}^*=(0,5,7)/12$
$\mathbf{a}^*=2.25$

2. $\begin{matrix} -3 & 3 \\ 0 & -1 \\ 5 & -4 \end{matrix}$ $(j=2,3)$ Now the average of rows 1 and 2 dominates row 2. $a^*=.2$, $\mathbf{u}^*=(.6,0,.4)$, $\mathbf{y}^*=(0,7,8,0)/15$.

6. No. Column 2 dominates column 3. $a^*=-1/3$, $\mathbf{u}^*=(2,1,0)/3$ or $(0,2,1)/3$, $\mathbf{y}^*=(2,0,1)/3$.
7. No. Row 1 dominates row 3. $a^*=.5$, $\mathbf{u}^*=(.5,.5,0)$, $\mathbf{y}^*=(.75,0,.25)$ or $(.5,.5,0)$.
8. Not if $ab\neq 0$, although $\mathbf{u}^*=\mathbf{y}^*=(0,.5,.5)$, $a^*=0$. If $a,b>0$, $\mathbf{y}^*=(b,0,a)/(a+b)$ also.
9. Yes. Exchange the first two rows; also the last two.
10. Each a_{ij} is paired with an $a_{kl}=-a_{ij}$. Only if these are zero is $ij=kl$.
11. The senses are opposite. In this section only one player is optimizing in each member, so that max≥min. In 4.1, (1) corresponds to (11); the first (right-hand) optimization determines the sense of the inequality.
12. $a^*=-1$ for $t\le -1$; $a^*=t$ for $-1\le t\le 2$; $a^*=2$ for $2\le t$. (Always a saddle point.)
13. $a^*=1+(t+1)\operatorname{val}\begin{bmatrix} 7 & 0 \\ 1 & 4 \end{bmatrix}=2.8t+3.8$ for all t.
14. $a^*=1$ for $t\le -1$; $a^*=2$ for $2\le t$; $a^*=(t^2+2)/3$ for $-1\le t\le 2$.

Section 4.4

1. $\mathbf{u}^*=(0,.4,.6)$, $\mathbf{y}^*=(.6,0,.4)$, $a^*=.2$, $kl=23$ or 33.
2. $\mathbf{u}^*=(1,0,2)/3$, $\mathbf{y}^*=(0,1,2,0)/3$, $a^*=-1/3$, $kl=12$ or 13.
3. $\mathbf{u}^*=(.7,.3,0)$, $\mathbf{y}^*=(.3,.7,0)$, $a^*=2.1$, pivot $kl=12$.
4. $\mathbf{u}^*=(2,0,1)/3$, $\mathbf{y}^*=(.5,0,.5)$, $a^*=3$, $kl=11$ or 13.
5. $\mathbf{u}^*=(0,1,0)$, $\mathbf{y}^*=(0,0,1,0)$, $a^*=2$, pivot $kl=23$.
6. $\mathbf{u}^*=\mathbf{y}^*=(6,5,2)/13$, $a^*=10/13$, $kl=13$, 22 or 31.
7. Replace a^0 by $\max_i a_{iq}$ and a_0 by $\min_j a_{pj}$ in the description of the algorithm. If one of these happens to be $a_{pq}\neq 0$, it is the best pivot.
9. $a_0 \triangleq \max_i \min_j a_{ij} \leq \max_u \min_j \Sigma_i u_i a_{ij} = a^*$ because the mixed strategies $\mathbf{u}$ include the pure strategies i. Similarly $a^* \leq a^0$.

Section 4.5

1. $\mathbf{u}^*=(.5,0,.5)$, $\mathbf{y}^*=(0,.5,.5)$, $a^*=.5$.
2. $\mathbf{u}^*=(0,0,2,1)/3$, $a^*=0$.
3. $\mathbf{u}^*=(0,0,7,2)/9$, $a^*=14/3$. (Prefer large a_{i3}.)
4. $\mathbf{u}^*=(147,116,19)/282$, $\mathbf{y}^*=(0,30,118,134)/282$, $a^*=-110/282$.
5. $u_j^* = y_j^* = a_{jj}^{-1}/\Sigma a_{ii}^{-1}$.
6. $\mathbf{u}^*=(4,5,3)/12$, $\mathbf{y}^*=(1,1,1)/3$, $a^*=2$.
7. $\mathbf{u}^*=(1,3,0)/4$, $\mathbf{y}^*=(13,7,0)/20$, $a^*=-1/4$. For 3 by 3, $\sigma\mathbf{u}=(34,121,-5)$, $\sigma\mathbf{y}=(95,53,2)$.
8. $\mathbf{u}^*=(4,2,3)/9$, $\mathbf{y}^*=(5,1,3)/9$, $a^*=2/9$.
9. (a) $1\leq a^*\leq 4$ (b) $5/3\leq a^*\leq 16/3$.
10. $25=18$ (for tradj $\mathbf{A}$)$+2$ (for $\mathbf{u}$)$+2$ (for $\mathbf{y}$)$+3$ (for a^*).
    $82=12$ (for all the 2 by 2 minors in rows 1 and 2)
    $+12$ (ditto in rows 3 and 4)$+3\cdot 16$(tradj)$+10$. Can save 1 in a^*.
11. Let $\mathbf{u}$, $\mathbf{y}$ and $\mathbf{u}^*$, $\mathbf{y}^*$ come from $\mathbf{A}$ and $\mathbf{A}_{33}$. Show
    $\mathbf{uAy}=\mathbf{uAy}^*=a^*+u_3(a_{31}y_1^*+a_{32}y_2^*-a^*)$,
    $=\mathbf{u}^*\mathbf{Ay}=a^*+y_3(u_1^*a_{13}+u_2^*a_{23}-a^*)$.

If the parenthesized expressions are not zero, they have opposite signs by (2d) and so $u_3y_3\leq 0$.

13.

$t\in(-\infty,-6]$	$[-6,-5]$	$[-5,-1],[0,\infty)$	$[-1,0]$
$a^*=1+t/2$	$t+4$	$(3t+10)/(t+10)$	$(8+t)/(8-t)$
$j=$ 1,2	2	2,3	1,3

14.

$t\in(-\infty,-1]$	$[-1,2]$	$[2,3]$	$[3,5]$	$[5,\infty)$
$a^*=$ 0	$2(1+t)/3$	t	$6-t$	$(t+1)/(2t-4)$
$j=$ 3	1,3	1	2	2,3

15. This is one less than the maximum number of optimal tableaus, since the initial tableau is not counted. Every *new* selection of m basic variables gives a kernel determined by the variables entering and those leaving the basis.
16. $\mathbf{u}^*=(0,1,1)/2$, $\mathbf{y}^*=\mathbf{u}^*$ or $(1,1,1)/3$, $a^*=0$.
17. $\mathbf{A}=\begin{bmatrix} 1 & 0 & 3 \\ 1 & 3 & 0 \end{bmatrix}$ with $1/3\leq u_1^*\leq 2/3$ and $\mathbf{y}^*=(1,0,0)$.

Section 4.6

1.–6. See Section 4.4. For primal feasibility ($m\leq n$), (a) suggests $a=-3,-2,0,0,0,0$.
7. Solve the linear program to get $\mathbf{u}^*=(.4,1.8)$, $\mathbf{y}^*=(1,0,1)$, $z^*=-3$. Then (14) gives $(2,9,5,0,5,5)/26$ as the common optimal strategy for the two players. The value is zero.

1	y_1	y_2	y_3	1
x_1	2	-2	1	-3
x_2	-1	3	2	-1
$-z$	1	-1	-4	0

8. The (skew) symmetry means $z^* = d = 1$, if there is an optimal solution, which there is. Replacing $d = 1$ by 0 gives a symmetric game with optimal strategy $(0,1,2,5)/8$.

9.

−1	−1	−1	4
1	2	2	−6
3	2	1	−8 (sic)

I uses $(5,2,1)/8$
J uses $(2,0,2,1)/5$
by (11). Value=0.

10. Since all b_i and c_j have the same sign, dividing each row by $|b_i|$ and each column by $|c_j|$ will give a tableau like (5). (See Example 4 of 3.4.)

11. Differences: (a) u_i', y_j' in 4.6 are merely proportional, not equal, to the probabilities u_i, y_j in 4.4. (b) The addition of a to each a_{ij} can be avoided in 4.4 but not necessarily in 4.6. (c) The label $-z$ is in an unusual position in 4.4, requiring an extra pivot step. (d) In 4.4, $b_i = -c_j = +1$ (all "infeasible"); in 4.6, $b_j = c_j = \pm 1$ (**b** or **c** infeasible, not both). Similarities: The remainder of the general layout; $d = 0$; same initial pivot position; methods of 3.3 suffice.

Section 4.7

Note: **y** is given for (a) Bayes, (b) minimax loss, (c) regret.

1.

30	60
80	60

(a) (1,0) (go to library)
(b) (0,1) (don't go) (c) (.6,.4)

2.

	Buy	Don't	Prob.	
\$400 prize	−399	0	.001	(a) (0,1) (use this)
\$100 prize	−99	0	.001	(b) (0,1)
No prize	1	0	.998	(c) (399,1)/400 (no; ignores probabilities)
Average	.5	0		

3. Take negative transpose of given matrix.
(a) (1,0) (b) (0,1) (c) (.5,.5). Averages −2.5, −2.4.

4.

	Buy	Rent	Hotel	Prob.	
1 month	21	5	4	.6	(a) (0,1,0)
6 months	26	15	24	.3	(b) (1,0,0) (no)
30 months	50	63	120	.1	(c) (12,17,0)/29=**y*** (13,0,16)/29=**u***
Average	25.4	13.8	21.6		

(b) is invalidated by the failure to consider J's renumeration and other aspects of his life-style in comparison with what they might be elsewhere (obviously this could be difficult); (a) and (c) are unaffected.

5.

−19	0
1	0

(a) (1,0) (undertake the project)
(b) (0,1) (don't) (c) (19,1)/20

6.

−3	−1
0	−1

(a) (1,0) (have operation)
(b) (0,1) (don't) (c) (2,1)/3.

7. (a) (0,0,1,0) (b) (1,0,2,0)/3 (c) (0,0,1,0)

8. Let r_{kl} = saddle point. Since all $r_{ij} \geq 0$ and $\min_j r_{ij} = 0$, (7) of 4.1 gives $0 \leq \max_i r_{il} = r_{kl} = \min_j r_{kj} = 0$. Thus $r_{il} = 0$ for all i and column l in **A** is dominated by all the other columns.

9. (b) and (f): $\min_j[(1-\theta)\min_i a_{ij} + \theta \max_i a_{ij}]$ (Hurwicz, 1951),
(b) and (c): $\min_y \max_i \Sigma_j [a_{ij} - (1-\theta)\min_l a_{il}] y_j$ $(0 \leq \theta \leq 1)$.

10. Are large losses feared or large gains coveted? Are the true losses actually of the form $a_{ij} + \varepsilon_i$ with undetermined ε_i as in Exercise 4? Is nature's strategy $\mathbf{u}^0$ well determined, guessed or unknown? Does J prefer a pure or a mixed strategy?

	α	z^*	$\mathbf{y}^*$
11.	$[5.5,6]$	$30-4\alpha$	$(12-2\alpha,-11+2\alpha)$
12.	$1/3$	5	$(1,0,0)$
	$[1/3,7/12]$	$9-12\alpha$	$(2-3\alpha,0,3\alpha-1)$
	$7/12$	2	$(.25,0,.75)$
	$[7/12,16/15]$	$24(3-\alpha)/29$	$(16-15\alpha,-21+36\alpha,34-21\alpha)/29$
	$16/15$	1.6	$(0,.6,.4)$
13.	$[0,2/9]$	-1	$(0,0,1)$
	$(2/9,2/3]$	$-9\alpha/2$	$(0,3\alpha,2-3\alpha)/2$
	$[2/3,8/9]$	-3	$(0,1,0)$
	$(8/9,4/3]$	$1-9\alpha/2$	$(3\alpha,0,4-3\alpha)/4$
	$[4/3,\infty)$	-5	$(1,0,0)$ (minimin (17))

The minimizing $i=3,2,2,1,1$, respectively. The three objective functions (for $i=1,2,3$) can all be entered in the same tableau if desired; for some values of α, several may have the same optimal basis. When z^* is constant, use the smallest corresponding value of α (0, 2/3 or 4/3); α has to be used as the independent parameter in the algorithm, but thereafter it may be advantageous to regard α as a discontinuous function of z^*. The minimax solution (7) has α in $[0,2/15]$, $z^*=1-9\alpha/2$, $\mathbf{y}^*=(3\alpha,0,4-3\alpha)/4$.

14.

	Insured	Not	Probability	
No accident	50	0	.995	(a) (0,1)
Destroyed	150	4000	.005	(b) (1,0)
Average	50.5	20		(c) (77,1)/78

Section 5.1

1. (3,3) eff., eq.
2. (2,1), (3,−2) eff. $\mathbf{u}^*=(.5,.5)$, $\mathbf{y}^*=(1/3,2/3)$.
3. (2,3), (3,2) eff., eq. $\mathbf{u}^*=\mathbf{y}^*=(.5,.5)$ gives (1.5,1.5).
4. All efficient. $\mathbf{u}^*=(3,5)/8$, $\mathbf{y}^*=(1,3)/4$ gives (15,15)/4.
5. (3,2) eff., eq.; (1,0) eq.; $\mathbf{u}^*=(.5,.5)$, $\mathbf{y}^*=(5/6,1/6)$.
6. (1,0), (0,1) eff.; (1,0) at $ij=11$ is eq.
7. (1,6), (8,1) eff.; (7,0), (0,5) eq.
8. (2,9), (5,8), (6,7), (7,0) eff.; (3,2) eq.
9. The linear relation is $2\alpha+\beta=5$, which has negative slope. The zero-sum game **A** has a saddle point at $ij=12$. This gives the unique and efficient equilibrium (2,1).
10. (−1,−1) (0,2) (1,1)
(2,0) (0,0) (2,0)
(1,1) (0,2) (−1,−1)

    Conventional wisdom seems to favor the central location for both stores; $i=2$ and $j=2$ are dominant strategies but yield the inefficient equilibrium (0,0). $ij=12,32,21,23$ also satisfy the definition of equilibrium. $ij=13$ or 31 yields the efficient compromise (1,1), but then a third firm would probably open in T_2. Similar considerations are observable in politics.
11. (1,1) (2,1) (1,1)
(2,1) (1,1) (−1,2)
(2,−1) (1,1) (1,1)

    $ij=12$ gives the dominant equilibrium (2,1), which is efficient. Each campaigns where he already has the advantage. But 5.3 gives (9,7)/6.
12. Here is one view of the (symmetrical) game.

    Marry; willing to be deferential (6, 6) (4, 9) (−1,0) (3, 9)
Marry on terms of equality (9, 4) (5, 5) (−1,0) (−3,−2)

Decline to marry	(0, −1) (0, −1) (0,0) (0, −1)
Marry with intent to dominate	(9, 3) (−2, −3) (−1,0) (−9, −9)

The symmetrical efficient point is $[(9,4)+(4,9)]/2=(6.5,6.5)$; according to the circumstances, the players take turns in asserting equality, while the other is prepared to be deferential.

Section 5.2

	$i\to j$	$j\to i$	P	S	M	Q	Pure eq.
1.	(2,2)	(2,2)	−	−	+	+	(2,2), (1,1).
2.	(2,3)	(3,2)	−	−	−	+	(1,1).
3.	(3,2)	(2,3)	−	−	−	−	(2,3), (3,2).
4.	(3,5)	(5,3)	−	−	−	−	None.
5.	(3,2)	(3,2)	+	−	+	+	(3,2), (1,0).
6.	(1,0)	(1,0)	−	−	+	+	(1,0) at 11.
7.	(7,0)	(0,5)	−	−	−	−	(7,0), (0,5).
8.	(3,2)	(3,2)	−	+	+	+	(3,2).

9. B'	.3	.5		**10.** D	.31	−.81		**11.** D'	−.1	−.5	
3	−.9	.1		9	.56	−.06		3.5	.9	−.3	
12. C''	.1	.3		**13.** B	.75	−.25		**14.** C	.3	−.1	
6.5	.5	−.9		4	.25	−.75		21	.5	−.7	

15. Let (α',β') and (α'',β'') be the endpoints of Ω' with $\alpha'\le\alpha''$; they are efficient. If $\alpha_0\le\alpha'$ then $(\alpha_0,\beta_0)\le(\alpha',\beta')$ because β' is the largest feasible β. If $\beta_0\le\beta''$ then $(\alpha_0,\beta_0)\le(\alpha'',\beta'')$ because α'' is the largest feasible α. If $\alpha_0>\alpha'$ and $\beta_0>\beta''$ and (α_0,β_0) is feasible, any ray of positive slope through (α_0,β_0) will intersect Ω' in an efficient point that dominates (α_0,β_0). A proof that holds equally well for more than two players is obtained by applying Zorn's lemma or the Hausdorff maximal principle to the partially ordered set of feasible $(\alpha,\beta)\ge(\alpha_0,\beta_0)$. (The feasible points form a closed set Ω, containing its boundary Ω'.)

16. It suffices to consider ij such that $(a_{ij},b_{ij})\neq(a_{kl},b_{kl})$. If kl dominates all these ij, none of them is efficient; neither can any such ij dominate kl, which therefore is efficient. Conversely, a unique efficient (α,β) dominates all the feasible points by Exercise 15.

17. $\max_j b_{ij}=b_{il}$; then $\max_i a_{il}=a_{kl}$: $i\to j$ at kl.
$\max_i a_{ij}=a_{kj}$; then $\max_j b_{kj}=b_{kl}$: $j\to i$ at kl.

18. When the $i\to j$ and $j\to i$ solutions coincide at kl, the first maximization of each (by the second player to move) shows kl is an equilibrium. Also any equilibrium pq will be selected by the first maximizations. Then the selection of kl in the second maximizations shows that $a_{kl}\ge a_{pq}$ (by $i\to j$) and $b_{kl}\ge b_{pq}$ (by $j\to i$). Thus kl is the dominant pure equilibrium.

19. kl is an equilibrium by 17. and 18. Let $(\alpha^*,\beta^*)=(\mathbf{uAy},\mathbf{uBy})$ be another equilibrium so that $\alpha^*\ge\Sigma_j a_{kj}y_j$ while S gives $\Sigma_i\Sigma_j u_i a_{kj}y_j=\Sigma_j a_{kj}y_j=\max_i\Sigma_j a_{ij}y_j\ge\Sigma_i\Sigma_j u_i a_{ij}y_j=\alpha^*$. To avoid getting $\alpha^*>\alpha^*$ one needs $a_{kj}=a_{ij}$ when $u_iy_j>0$. Similarly $b_{il}=b_{ij}$ when $u_iy_j>0$. Thus to avoid ties one needs $u_iy_j=0$ except when $i=k, j=l$, and (α^*,β^*) is at kl.

20. Here the condition is that relabeling the two rows should have the effect of interchanging a_{ij} and b_{ji} (the operation that leaves (7) unchanged). This leads to $a_{11}=b_{21}=a_{22}=b_{12}$ and $b_{11}=a_{21}=b_{22}=a_{12}$ and the form $\begin{matrix}(a,b) & (b,a)\\ (b,a) & (a,b)\end{matrix}$ in place of (7). Subtracting $(a+b)/2$ from each payoff and dividing by $(a-b)/2$ gives the normalized form $\begin{matrix}(1,-1) & (-1,1)\\ (-1,1) & (1,-1)\end{matrix}$.

21. $I\to J$ means that I informs J of I's $\mathbf{u}$, after which J chooses $j=l$. Thus l as a function of $\mathbf{u}$ is determined by $\Sigma_i u_i b_{il}=\max_j\Sigma_i u_i b_{ij}$, and I wants to maximize $\alpha=\Sigma_i u_i a_{il}$ over $\mathbf{u}$ subject to $\mathbf{u}\ge\mathbf{0}$, $\Sigma u_i=1$,

$$\sum_i u_i b_{ij}\le\sum_i u_i b_{il}\qquad\text{for } j=1,\ldots,n.$$

These are n linear programs, one for each l, each having n constraints. In (17) one has $l=1$ and $\alpha=2+u_2$ when $u_2<1/2$, and $l=2$ and $\alpha=u_2$ when $u_2>1/2$. Thus I chooses $u_2=0.5^-$, J chooses

$l=1$, and the $I \to J$ payoffs are $(2.5^-, 1.5^+)$. For $J \to I$, I prefers $i=k=2$ for all y_2; hence J prefers $y_2=1$ to obtain $(1,1)$, the same as $j \to i$ and the unique equilibrium $\{I,J\}$. Thus $I \to J$, $J \to I$ and $\{I,J\}$ do not necessarily yield identical payoffs for bimatrix games as they did for matrix games.

22. The ij values in the axis are (a) 11, 22; (b) 11, 12, 14, 21, 22, 41; (c) 11, 12, 21; (d) 11, 22.

Section 5.3

1. (a) $(3,2)$ at $ij=12$. (b) $(13/6, 13/4)$. (c) $(2,3)$ at $ij=11$.
(d) $(41,49)/18$ between $(2,3)$ and $(3,2)$; $\theta^{\#}=0.5$.
$\mathbf{u}=(8,1)/9$, $\mathbf{y}=(4,0,5)/9$ in (3).

2. (a) $(4,3)$ at $ij=22$. (b) $(98,132)/31$.
(c) $(742,993)/248$ between $(3,4)$ and $(1,5)$.
(d) $(19,23)/6$ between $(3,4)$ and $(4,3)$; $\theta^{\#}=0.5$.
$\mathbf{u}=(5,7)/12$, $\mathbf{y}=(0,2,1)/3$ in (3).

3. $(2,3)$ at $ij=22$; $\theta^{\#}=0.6$; $\mathbf{u}=\mathbf{y}=(0,1,0)$ for (a) to (d).

4. $(1,9)$, $(4,7)$, $(9,1)$, $(10,-1)$ eff. $\theta=2/5, 6/11, 2/3$.

5. Adjacent pairs of efficient (a_{ij}, b_{ij}) determine the slopes in Ω' that are trial values of $-\theta/(1-\theta)$ in (3).

6. Let $\theta=1/2$. The given solution satisfies (2) by construction, since $\theta a_{ij}+(1-\theta)b_{ij}=(a_{ij}+a_{ji})/2$. It satisfies (3) because the payoffs $\theta a_{ij}-(1-\theta)b_{ij}=(a_{ij}-a_{ji})/2$ define a skew-symmetric matrix (a symmetric zero-sum game, whose value is zero). It satisfies (1) because $\alpha^{\#}=\beta^{\#}=(a_{kl}+a_{lk})/2$ is the midpoint of the line segment (with slope $=-1$) joining the extreme points $(a_{kl}, b_{kl})=(a_{kl}, a_{lk})$ and $(a_{lk}, b_{lk})=(a_{lk}, a_{kl})$.

7. $(9,7)$, $\theta^{\#}=2/3$. J has no effective threat; if he rejected $(9,7)$ he would lose 7 while I would lose only 2.

8. Approx. $(8.5, 8.1)$; $\theta^{\#}=.2$. J's attempt at extortion doesn't get far. If J goes from $(9,8)$ to $(1,7)$, I may go on to $(0,0)$.

9. $(121/14, 13/4)$, $\theta^{\#}=7/9$. J seems to have a moral right to $(7,9)$ but his threat is weak. Compare 10.

10. $(53/7, 13/4)$, $\theta^{\#}=7/15$. I seems to have a better claim to $(9,2)$ than he did in 9. However, if $a_{11}=-27$ instead of 0, which would not change $\alpha^{\#}, \beta^{\#}$, this game would be the same as 9 with the a_{ij} rescaled.

11. $(6.75, 7.5)$, $\theta^{\#}=2/3$.

12. $(8,9)$. J's threat is both reasonable and fully effective. The trial value $\theta=8/9$ gives infeasible $(5.06, 32.5)$. $\theta^{\#}=9/16$.

13. $(0, 2.25)$.

14. $(16/7, 3)$.

15. (a) $(a,0)$ if $\theta a \geq (1-\theta)b$, otherwise $(0,b)$.
(b) $(\sigma/2\theta, \sigma/2(1-\theta))$ where $\sigma=\max(\theta a, (1-\theta)b)$.
(c) $(\alpha,\beta)=(1-\theta,\theta)ab/(\theta a+(1-\theta)b)$ by solving $\theta\alpha-(1-\theta)\beta=\delta(\theta)=0$ and $b\alpha+a\beta=ab$ (on Ω'). (d) $(\alpha^{\#},\beta^{\#})=(a,b)/2$ with $\theta=b/(a+b)$.

16. For (2) we must show

$$\max_{ij}\left[\theta' a'_{ij}+(1-\theta')b'_{ij}\right]=\max_{ij}\frac{c_1 d_1\left[\theta a_{ij}+(1-\theta)b_{ij}\right]+c_0 d_1\theta+c_1 d_0(1-\theta)}{d_1\theta+c_1(1-\theta)}$$

$$=\left[c_1 d_1(\theta\alpha+(1-\theta)\beta)+c_0 d_1\theta+c_1 d_0(1-\theta)\right]\left[d_1\theta+c_1(1-\theta)\right]^{-1}$$

$$=\theta'\alpha'+(1-\theta')\beta'$$

The proof is similar for (3). The efficiency property in (1) is clearly unaffected.

Section 5.4

	$\mathbf{y}^*$	$\mathbf{u}^*$	α^*	β^*
1.	(.5,.5,0)	(.75,.25)	2.5	2.25
2.	(0,3,4)/7	(.4,.6)	16/7	3
	(.8,.2,0)	(2,1)/3	2.4	3
	(1,0,0)	(1,0)	3	4
3.	(.75,0,.25)	(.75,0,.25)	1.25	1.5
	(0,1,0)	(0,1,0)	2	3
	(6,5,2)/13	(3,2,1)/6	10/13	1
4.	(21,22,24)/67	(23,17,24)/64	344/67	151/32

5. $(y_2^*, u_2^*)=(0,0),(1,1),(2,1)/3$.

6. $=(0,1)$ to $(0,0)$ to $(1,0)$.

7. $=(.5,0)$ to $(1,0)$.

8. $=(0,1)$ to $(0,.8)$ to $(1,.8)$ to $(1,0)$.

9. $\left.\begin{array}{l}\mathbf{y}^*=(.5,.5,0) \text{ to } (0,2,1)/3 \\ \mathbf{u}^*=(.5,.5,0) \text{ to } (0,2,1)/3\end{array}\right\}$ Form the Cartesian product.

10. $nC(m+n-1,m-1)=(m+n-1)!/(m-1)!(n-1)!$ because the $\mathbf{x},\mathbf{y},a$ basis determines the other, except for a selection among n possible nonbasic pairs.

Section 5.5

In 1 to 5 values of $(\mathbf{I}-\mathbf{A})^{-1}$ and $\mathbf{y}$ are given.

1. $\begin{bmatrix} 1.2 & .267 \\ .4 & 1.2 \end{bmatrix}$ (1.467, 1.6) **2.** $\begin{bmatrix} 1 & 0 \\ 0 & 1.25 \end{bmatrix}$ (1, 1.25) **3.** $-\frac{10}{11}\begin{bmatrix} 8 & 3 \\ 9 & 2 \end{bmatrix}$ Nonproductive

4. $\begin{bmatrix} 47 & 10 & 13 \\ 9 & 70 & 11 \\ 7 & 10 & 53 \end{bmatrix}\Big/40$ (7,9,7)/4 **5.** $\begin{bmatrix} 10 & 30 & 30 \\ 0 & 10 & 0 \\ 0 & 0 & 10 \end{bmatrix}$ (70, 10, 10)

6. Putting all $y_j=1$ proves the first part. Then the second part implies $\mathbf{A}^T$ is productive, but $(\mathbf{I}-\mathbf{A})^{-1}$ and $(\mathbf{I}-\mathbf{A}^T)^{-1}$ are transposes and have the same nonnegative elements. This is evidently related to Theorem 5 of 4.5 with $a=0$.

Section 5.6

1. $\mu=5/3$, $\mathbf{u}=\mathbf{y}=(.5,.5)$.

2. $\mu=4/3$, $\mathbf{u}=(.5,.5)$, $\mathbf{y}=(0,.5,.5)$ or $(5,6,0)/11$.

3. $\mu=0.4$, $\mathbf{u}=(1,0,0)$, $\mathbf{y}$ arbitrary but $y_1>0$; or
$\mu=1$, $u_3=y_1=0$, $u_2y_2>0$; or
$\mu=2$, $y=(0,0,1)$, $\mathbf{u}=$ any convex combination of
(1,0,0), (0,1,0), (0,4,7)/11, (13,40,200)/253 with $u_3>0$.

4. $\mu=2$, $\mathbf{u}=(8,6,5)/19$, $\mathbf{y}=(1,1,1)/3$, $-\lambda^3+.3\lambda^2+.07\lambda+.015=0$

5. $\mu=2$, $\mathbf{u}=(0,0,1)$, $\mathbf{y}=$ any convex combination of (0,0,1), (0,1,5)/6, (1,0,10)/11. Process 3 produces surpluses of goods 1 and 2, which permit some optimal operation of the inefficient processes 1 and 2. No matter what the prices, process 3 is always more profitable than the others.

6. People (+cattle?)+grain (seed)→grain.
Cows+grain+people→cows, steers and milk.
Cows+milk→cows, steers and milk.
Steers+grain+people→steers.
Steers+milk→steers.
People+grain→people.
People+milk→people.
People+steers (beef)→people.
People+cows (beef)→people.
The actual diets of people and animals will be mixtures of the pure diets indicated. Bulls are essential but quantitatively insignificant.

8. Write (1′) of 4.5 as $\min_j \Sigma_i u_i^* a'_{ij} \le a' \le \max_i \Sigma_j a'_{ij} y_j^*$ where $\mathbf{u}^*$ and $\mathbf{y}^*$ are optimal for **A**. (2a) of 4.5 gives
$\min_j \Sigma_i u_i^* a_{ij} = a^* = \max_i \Sigma_j a_{ij} y_j^*$. For all i one has
$\Sigma_j a'_{ij} y_j^* \le \Sigma_j a_{ij} y_j^* + \Sigma_j |a'_{ij} - a_{ij}| y_j^*$ whence
$a' \le \max_i \Sigma_j a'_{ij} y_j^* \le a^* + \max_{ij} |a'_{ij} - a_{ij}|$. Adjoining a similar proof shows that
$|a' - a^*| \le \max_{ij} |a'_{ij} - a_{ij}|$.

9. $\mu = 2/3$, $\mathbf{u} = (0,1,0)$, $y_1 = 0$, $4/11 \le y_3 < 1$.
$\mu = 4/3$, $\mathbf{y} = (0,0,1)$, $u_1 = 0$, $3/8 \le u_2 < 1$.
In both cases, a saddle point at $ij = 23$.
Note that (9) gives $1/2 \le \mu \le 3$.

10. Since (9) gives $3/4 \le \mu \le 5/3$ and $a_{ij} = b_{ij} = 0$ never occurs, there is no saddle point. The solution for $\mu = 1$ suggests $K = L = \{1,3\}$ for the kernel. The determinant gives $6\mu^2 + 8\mu - 19 = 0$ or $\mu = 1.2336$, $\mathbf{u} = (.5982, 0, .4018)$, $\mathbf{y} = (.5670, 0, .4330)$, which satisfies all the requirements.

Section 5.7

1. 2^{n-1} values of $v(S/\bar{S}) = -v(\bar{S}/S)$, the value of the zero-sum game each of whose payoffs is the total for S minus the total for the complementary coalition $\bar{S}$, except that $v(N/\phi) = v(N) =$ the maximum total payoff for N (all the players). Each $v(S/\bar{S})$ occurs in the formula for ϕ''_h; if each S contains h, the coefficient of $v(S/\bar{S})$, namely $(s-1)!(n-s)!/n!$, is the number of permutations of $S - \{h\}$ times the number of permutations of $\bar{S}$, divided by the number of permutations of N.

$$\phi''_1 = v_{12}/2 + v_{1-2}/2 \qquad (n=2)$$

$$= v_{123}/3 + (v_{12-3} + v_{13-2})/6 + v_{1-23}/3 \qquad (n=3)$$

$$= v_{1234}/4 + (v_{123-4} + v_{124-3} + v_{134-2})/12$$

$$+ (v_{12-34} + v_{13-24} + v_{14-23})/12 + v_{1-234}/4 \qquad (n=4)$$

where $v_{12-34} = v(\{1,2\}/\{3,4\})$, etc. The divisor of each group of terms is n times the number of terms in the group.

2. $(5,3,3,1)/12$. This is $[q;\mathbf{w}] = [5;3,2,2,1]$ in 5.8.

3. $(1,1,1,1)/4$. **4.** $(3,5,2)/2$. **5.** $(8,14,20)$.

6. $(1,1,1)$, three times the values for $[q;\mathbf{w}] = [2;1,1,1]$.

7. $(5,4,3)/2$. **8.** $(3,3,2)$. **9.** $(7,10,7)/3$.

10. $(3,13,10)/2$. Four strategies would allow a member to approve both changes; five would indicate a preference between the two approved alternatives. However, one would have to adopt a procedure to resolve indeterminate cases.

11. $(4,4,5)$. **12.** $(8,8,8)/3$.

13. Only that concerning dummies.

14. $\begin{bmatrix} (v_{12} - v_2, v_2) & (v_1, v_2) \\ (v_1, v_2) & (v_1, v_{12} - v_1) \end{bmatrix}$

15. $(v_{12} + v_{1-2}, v_{12} - v_{1-2})/2$ in every cell.

16. New values are $v_{1-2} = 0$ and $v_1 = -1$ respectively.

Section 5.8

A superscript indicates the number of identical components.

Problem	ρ	ϕ	δ	β
1. [9; 8, 5, 2, 1]	$(.743, .086^3)$	$(9,1,1,1)/12$	$(.943, .019^3)$	$(7,1,1,1)/10$
2. [16; 9, 7, 7, 6]	$(.456, .237^2, .069)$	$(5,3,3,1)/12$	$(.568, .2045^2, .023)$	$(5,3,3,1)/12$
3. [5; 4, 2, 2, 1]	$(.571, .143^3)$	$(3,1,1,1)/6$	$(.750, .083^3)$	$(3,1,1,1)/6$
4. [5; 4, 2, 1, 1, 1]	$(.727, .068^4)$	$(6,1,1,1,1)/10$	$(.924, 019^4)$	$(7,1,1,1,1)/11$
5. [16; 9, 8, 4, 3, 2]	$(.428, .317, .111^2, .033)$	$(23,18,8,8,3)/60$	$(.544, .329, .060^2, .007)$	$(9,7,3,3,1)/23$
6. [10; 8, 8, 1, 1, 1]	$(.308^2, .128^3)$	$(9,9,4,4,4)/30$	$(.364^2, .091^3)$	$(2,2,1,1,1)/7$
7. 2 of 3, 1 of 2	$(.224^3, .164^2)$	$(14,14,14,9,9)/60$	$(.257^3, .114^2)$	$(3,3,3,2,2)/13$
8. 2 of 3, 1 of 3	$(.256^3, .077^3)$	$(4,4,4,1,1,1)/15$	$(.308^3, .025^3)$	$(7,7,7,2,2,2)/27$
9. 2 of 3, 3 of 5	$(.1587^3, .1048^5)$	$(5,5,5,3^5)/30$	$(.172^3, .097^5)$	$(4,4,4,3^5)/27$
10. 5; 3, 2, 1, 1	$(.530, .304, .083^2)$	$(7,3,1,1)/12$	$(.694, .250, .028^2)$	$(5,3,1,1)/10$
11. 2 of Mm^2, 2 of $M\mu^2$	$(.352, .162^4)$	$(2,1,1,1,1)/6$	$(.500, .125^4)$	$(2,1,1,1,1)/6$

$$
\begin{array}{ll}
w_1+w_2+w_4 \geq q & w_1+w_2+w_5 \geq q \\
w_1+w_3+w_4 \geq q & w_1+w_3+w_5 \geq q \\
w_1+w_2+w_3 < q & w_1+w_4+w_5 < q
\end{array}
$$

The first two lines minus twice the last gives $0 > 0$.

12. [5; 3, 2, 2, 1]. As in 2, $\phi = \beta = (5,3,3,1)/12$.

13. Exercises 11 and 12 and the case of two houses with identical memberships are relevant. It is no matter that some u_i may have to equal some v_j, but the proof assumes that each common member is willing to vote Yes in one house and No in the other.

14. The middle voter in each house is the only one in that house who can be pivotal, no matter which way the permutation is read; one is pivotal in the direct permutation and the other in the reverse. Since reversal gives the same set of $n!$ permutations with the same probabilities, the two houses get equal numbers of pivots and equal powers.

15. Apart from relabeling of members, there are five cases, two of which make the three members equally powerful:

Minimal Winning Coalitions	$[q; \mathbf{w}]$	β
One of 3 members	[3; 1, 1, 1]	$(1,1,1)/3$
Three of 2 members	[2; 1, 1, 1]	$(1,1,1)/3$
Two of 2 members	[3; 2, 1, 1]	$(3,1,1)/5^\dagger$
One of 2 members	[4; 2, 2, 1]	$(1,1,0)/2$
One of 1 member	[3; 3, 1, 1]	$(1,0,0)$

†Here $\phi = (4,1,1)/6$, $\rho = (.6306, .1847^2)$; elsewhere $\phi = \rho = \beta$.

16. In 3 and 4, $1/6<(1+1)/10$ or $(1+1)/11$.
In 7 and 12, $5/12>(14+9)/60$ or $(3+2)/13$.
To show that member h loses no power when he absorbs the role of h', change the winning coalitions W as follows. If h and $h'\in W$, let $W-\{h'\}$ be winning; if $h'\in W$ but $h\notin W$, let W be winning if and only if $W-\{h'\}$ is winning. This deprives h of no pivot, gives him no more than he will have in his new position, and makes h' a dummy who can be eliminated without affecting ϕ or β.

17. (a) $n_0\doteq\sqrt{n}$ yields max $n_0/(n_0^2+n-n_0)$.
(b) $n_0\doteq n/\sqrt{2}$ yields max $n_0/[n_0^2+(n-n_0)^2]$.

18. Let $b_i=\Sigma_h\{\beta_h\in N_i\}$ so that $\gamma(N_i)=b_i^2/\Sigma_j b_j^2$.
Let $b_0^2=b_3^2+b_4^2+\ldots$. We have to show

$$(b_1^2+b_2^2)/(b_1^2+b_2^2+b_0^2)\leq(b_1+b_2)^2/\left[(b_1+b_2)^2+b_0^2\right]$$

by subtracting from $1=1$ the obvious relation

$$b_0^2/(b_1^2+b_2^2+b_0^2)\geq b_0^2/\left[(b_1+b_2)^2+b_0^2\right].$$

19. $\gamma^{(1)}=\beta$; $\gamma^{(11)}=\beta^2/(\beta^2+\bar\beta^2)$; $\gamma^{(51)}=\beta^2/(\beta^2+5\bar\beta^2/5^2)$;
$\gamma^{(4)}=\beta(\beta+.4\bar\beta)/[(\beta+.4\bar\beta)^2+3\bar\beta^2/5^2]$;
$\gamma^{(1x)}=10(\beta/10)^2/(.1\beta^2+\bar\beta^2)$; $\gamma^{(5x)}=10(\beta/10)^2/(.1\beta^2+5\bar\beta^2/5^2)$.

20. $\phi^{(1)}=\phi^{(5x)}=\phi$. $\phi^{(11)}=1/2$ (two permutations of M^5, m^{10})
$\phi^{(51)}=1/6$ (when m^{10} is in last of six possible positions)
$\phi^{(1x)}=4/11$ (M^5 in first 4 of 11 positions among the m's)
$\phi^{(4)}=(1/4)(10/66)=5/132$ (M^2m^{10} in last of 4 positions among M, M, M; 10 permutations of M, M, m, m, m, preceding m^7; 66 permutations of M^2m^{10}).

21. By (15) each $\partial P/\partial p_h$ has at most $1+2^{n-1}$ values, from 0 to 1 in steps of 2^{1-n}; for $n=10$ (the smallest value for which the proof works) this gives less than 513^{10} or 10^{28} possible values for $\nabla P(1/2,\ldots,1/2)$. There are $\binom{10}{5}/2=126$ partitions of N into $5+5$ members, and each partition supplies a binary choice for a minimal W (winning coalition). Thus there are at least 2^{126} or 10^{37} possible sets of W. A similar proof holds for ϕ alone or in combination with ∇P.

22. $\alpha(m^l)=2^{-5}$ for $l\geq 7$ (for UNSC by (16));

$$=2^{l-15}\left[\binom{10-l}{0}+\ldots+\binom{10-l}{3}\right]\quad\text{for } 4\leq l\leq 6$$

$$=2^{l-15}\left[\binom{10-l}{4-l}+\ldots+\binom{10-l}{3}\right]\quad\text{for } 1\leq l\leq 4$$

$\alpha(m^lM^5)=1$ for $l\geq 4$

$$=1-2^{l-10}\left[\binom{10-l}{0}+\ldots+\binom{10-l}{3-l}\right]\quad\text{for } l\leq 3$$

$\alpha(m^lM^L)=2^{L-5}\alpha(m^lM^5)$ for $1\leq L\leq 5$

l	0	1	2	3	4	5	6	7–10
$\alpha(m^l)$	0	.0051	.0102	.0154	.0205	.0254	.0293	.0312
$\alpha(m^lM^5)$	.8281	.9102	.9648	.9922	1	1	1	1

23. If N is the only winning coalition, (16) gives $\alpha(S)=2^{-n+s}$ for $1\leq s\leq n$ and $\alpha(S)=0$ when S is empty. Putting $v(S)=\alpha(S)$ in (5) of 5.7 gives

$$\phi_h=n^{-1}+\sum_{s=1}^{n-1}\binom{n-1}{s-1}(2^{-n+s}-2^{-s})(s-1)!(n-s)!/n!$$

$$=n^{-1}+n^{-1}\sum_{s=1}^{n-1}(2^{-n+s}-2^{-s})=1/n$$

the same as the Shapley-Shubik value obtained from (12). Then equality also holds when the winning coalitions are a single W_0 and its supersets, because the presence or absence of the dummy members in $N-W_0$ does not affect the satisfaction of the hypotheses of Theorem 1 of 5.7. Equality still holds if W_0 is the *only* winning coalition, because the effect of its supersets can be eliminated by successive subtraction or addition in order of increasing size of the supersets. (This inconsistency with our model does not invalidate the algebra we need.) Equality holds in general because each W enters (16) at most once, in a different term $\pm 2^{-n+s}$, so that the formation of the appropriate linear combination completes the proof. (This procedure is like that on p. 310 of Shapley (1953), but we do not need to work out all the formulas.)

Section 6.1

1. Initial tableau

Check	1	x_1	x_2	x_3	1
2	x_4^0	3	-2	1*	-1
-3	x_5	-1	1	1	-5
4	$-z$	5	-3	1	0
0	$-z^0$	-3	2	-1	1

Pivots 43, 52. $z^*=-1/3$.

Final tableau

Check	3	x_1	x_5	x_4^0	1
-4	x_3	1	2	1	-11
-5	x_2	-4	1	-1	-4
1	$-z$	2	1	-4	-1

$\mathbf{x}^*=(0,4,11,0,0)/3$
$\mathbf{u}^*=(2,0,0,-4,1)/3$

2.

	1	x_3	x_4	x_5	1
-3	x_1^0	3	1	0	-8
2	x_2^0	1	2*	-1	-1
1	$-z$	1	-1	0	0
4	$-z^0$	-4	-3	1	9

Pivots 24, 15, $z^*=-8$.

	1	x_3	x_2^0	x_1^0	1
-8	x_5	5	-1	2	-15
-3	x_4	3	0	1	-8
-2	$-z$	4	0	1	-8

$\mathbf{x}^*=(0,0,0,8,15)$
$\mathbf{u}^*=(1,0,4,0,0)$

3. $\mathbf{x}^*=(5,0,0,1,0,0)$
$\mathbf{u}^*=(0,2,1,0,7,10)/3$
$z^*=-9$
Pivots 52, 61, 24.

4. $\mathbf{x}^*=(0,0,0,5,3)$
$\mathbf{u}^*=(5,-1,11,0,0)$
$z^*=-4$
Pivots 13, 24, 35.

5. $\mathbf{x}^*=(0,0,0,6,4,0,0,1)$
$\mathbf{u}^*=(8,-3,0,0,0,6,3,0)$
$z^*=-26$
3 or 4 pivots.

6. $\mathbf{x}^*=(0,0,0,5,2,0,0,1)$
$\mathbf{u}^*=(-3,11,0,0,0,24,3,0)/2$
$z^*=-15$
2 to 4 pivots.

Section 6.2

The final tableaus are the following.

1.

1	x_4	x_6^0	x_3	x_5^0	1
x_1	-2	1	-3	-1	-3
x_2	-7	2	-7	-3	-4
$-z$	1	-1	0	0	5

Optimal despite negative column initially.

2.

3	x_5^0	x_4^0	x_3	1
x_2	2	1	-9	-5
x_1	1	-1	-6	-1
$-z$	3	3	-9	-9

Unbounded; $x_3 \to \infty$.

3.

1	x_3	x_4^0	x_1^0	x_6^0	1
x_5	2	1	1	2	-3
x_2^0	0	-3	-2	-1	0
$-z$	2	-3	-4	-1	12
$-z^0$	0	3	3	1	0

Optimal; x_4 and x_6 treated as if artificial.

4.

3	x_1^0	x_4	x_5	1
x_3	1	-4	1	-2
x_2^0	2	-2	-1	-7
$-z$	-1	7	2	2
$-z^0$	1	2	1	7

Infeasible;
$\min z^0=7/3>0$.

5. Min $z^0 \geq 0$. Also since g_l is a negative sum of a_{il}, $g_l < 0$ implies that some $a_{il} > 0$.
6. The initial basic dual solution $\mathbf{c}$ is feasible. Use (2) to permit the application of Theorem 4(c_3) of 3.3.
7. $c_j^* > 0$ when x_j is not artificial, so that any feasible change in the nonbasic variables will increase z. Otherwise try to increase some x_j having $c_j^* = 0$. In 1, $\mathbf{x} = (6, 11, 1, 0, 0, 0)$ is also optimal but not basic.
8. The "if" part is obvious; a sum of products of nonnegative numbers is again nonnegative. For the forward implication ("only if"), consider the schema

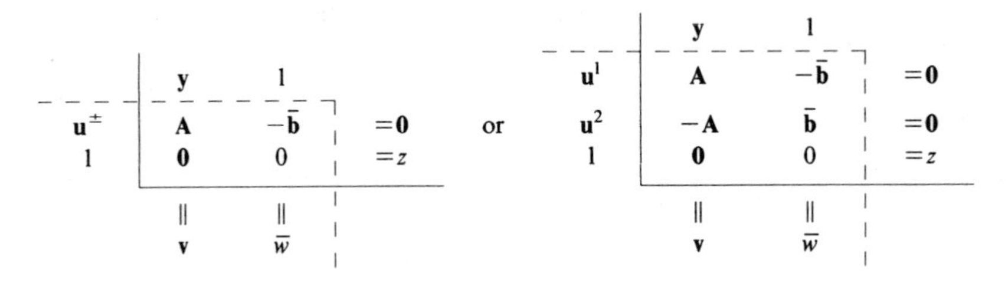

The second form uses (2) and assures us that Theorem 4 of 3.3 is applicable. Now Farkas's hypothesis states that $\bar{w} = -\mathbf{u}^{\pm}\bar{\mathbf{b}} \leq 0$ for every feasible $\mathbf{u}^{\pm}$, and $\mathbf{u}^{\pm} = \mathbf{0}$ is obviously feasible. Thus the dual problem has an optimal solution with max $\bar{w} = 0$, and Theorem 4 of 3.3 asserts that the primal problem has an optimal solution with min $z = 0$. This requires $\mathbf{Ay} - \bar{\mathbf{b}} = \mathbf{0}$ to have a solution $\mathbf{y} \geq \mathbf{0}$, as the lemma concludes. This has applications in nonlinear programming. Applying the lemma to $[\mathbf{A}, -\mathbf{A}]$ and $\mathbf{b}$ proves Theorem 7 of 2.7 for $\mathbf{A}$, $\mathbf{b}$ and $\mathbf{y}^{\pm}$.

Section 6.3

1. $\mathbf{x}^* = (0,0,3,4,0)$, $z^* = -29$, $\mathbf{u}^* = (5,3,0,0,1)$. Pivots 24, 13.
2. $\mathbf{x}^* = (0,1,0,1,0)$, $z^* = -1$, $\mathbf{u}^* = (1,0,2,0,1)$. Pivot 14.
3. Use (3). Pivots 61, 54.
4. Use (5). Pivots 15, 24.
5. $\mathbf{x}^* = (0,0,1.8,.4,0)$, $z^* = -2.2$, $\mathbf{u}^* = (1.0,.6,0,.8,0,.4)$. Pivots 25, 16, 63.
6. $\mathbf{x}^* = (12,0,3,0,0)/11$, $z^* = 36/11$. Use (5). $\mathbf{u}^* = (0,24,0,-25,13)/11$. Pivots 52, 21, 43.
7. $\mathbf{x}^* = (3,0,2,0,0)$, $z^* = -1$, $\mathbf{u}^* = (0,1,0,0,1)$. Pivots 52, 41, 23.
8. Let $c_j/(-c_l) < t < (-g_j)/g_l$ if $g_l > 0$, $-c_l/c_j > t^{-1} > g_l/(-g_j)$ if $c_j > 0$, and $t = 1$ if $c_j = g_l = 0$. Then $tc_l + c_j < 0$, $tg_l + g_j < 0$.

Section 6.4

1. $\mathbf{x}^* = (3,3,1,0,0,2)$, $z^* = 2$, $\mathbf{u}^* = (0,0,0,1,2,0)$. Pivot 46.
2. $\mathbf{x}^* = (0,3,9,0,13,5)/7$, $z^* = 23/7$, $\mathbf{u}^* = (-1,0,0,5,0,0)/7$. Pivots 15, 46.
3. $\mathbf{x}^* = (0,6,7,3,0,5)$, $z^* = -5$, $\mathbf{u}^* = (1,0,0,0,1,0)$. Pivot 16.
4. $\mathbf{x}^* = (0,0,2,1,2,1)$, $z^* = -4$, $\mathbf{u}^* = (-1,5,0,0,0,0)/3$. Pivots 16, 25.
5. $\mathbf{x}^* = (0,0,1,6,4,0)$, $z^* = -26$, $\mathbf{u}^* = (8,3,0,0,0,6)$. Pivots 16, 24, 65.
6. $\mathbf{x}^* = (0,0,1,5,2,0)$, $z^* = -15$, $\mathbf{u}^* = (3,11,0,0,0,24)/2$. Pivots 24, 15.
7. Pivot 64.
8. Pivot 61 by (5) of 6.3.
9. Row 4 indicates $l = 6$ and column 6 indicates $k = 1$.

Section 6.5

1.

1	x_3^1	x_4	x_5^1	1
x_1^4	2	$-1^{\#}$	1	-3
x_2	1	1	1	-6
$-z$	-1	-3	-2	0

1	y_3^1	y_1^4	y_5^1	1
x_2	-3	-1	-2	0
x_4	2	1	1	-4
$-z$	7	3	5	-15

$\mathbf{x}^* = (4,0,1,4,1)$, $\mathbf{u}^* = (-3,0,-7,0,-5)$, $z^* = -15$.

Refl. x_1^4; exch. y_1^4, x_4; refl. x_3^1, x_5^1.

2.

1	x_3^3	$x_4^{2\#}$	x_5^1	1
x_1^0	2	2	3	−8
x_2^0	2	3	1	−10
$-z$	3	1	2	0

4	x_1^0	y_4^2	x_2^0	1
x_3^3	−1	−7	3	−8
y_5^1	−2	−2	2	−4
$-z$	−1	13	−5	32

$\mathbf{x}^*=(0,0,2,2,0)$, $\mathbf{u}^*=(-1,-5,0,-13,0)/4$, $z^*=8$.
Refl. x_4^2, x_5^1; exch. x_1^0, x_3^3; x_2^0, y_5^1 (Phase 1).

3. $\mathbf{x}^*=(0,1,3,3,0)$, $\mathbf{u}^*=(2,0,-1,0,3)$, $z^*=15$.
Refl. x_3^3; exch. x_1, x_4^3.

4. $\mathbf{x}^*=(1,0,4,5,0)/4$, $\mathbf{u}^*=(0,1,-13,0,3)/4$, $z^*=-17/4$.
Exch. x_1^2, x_3^1; refl. x_3^1; exch. y_3^1, x_4; x_2, x_1^2.
Alternatively, since $e_3=1$ is small, exch. x_2, x_4; refl. x_3^1.

5. $\mathbf{x}^*=(0,9,0,0,2,2)$, $\mathbf{u}^*=(2,-1,7,5,0,0)/3$, $z^*=4$.
Refl. x_5^3; exch. x_1, x_6^3; refl. x_2^9; exch. y_2^9, y_5^3.

6. $\mathbf{x}^*=(0,0,13,14,14,9)/7$, $\mathbf{u}^*=(-4,-5,0,-14,-4,0)/7$, $z^*=50/7$.
Refl. x_4^2, x_5^2; exch. x_1^0, x_3^2; x_2^0, x_6^2 (Phase 1). Dual simplex:
refl. x_1^0, x_2^0; exch. y_2^0, x_4^2; refl. x_5^2; exch. y_1^0, x_3^2; refl. x_4^2; exch. y_4^2, x_6^2.

Section 6.6

1, 2. The optimal tableaus are

2	x_6	x_3	x_2	1
x_1	−2	2	−4	0
x_4	5	−2	3	0
x_5	1	0	1	−2
$-z$	1	0	3	−6

Pivots 35, 26, 64.

2	x_3^0	x_6	x_2^0	1
x_1^0	1	−8	−3	0
x_5	−1	6	1	0
x_4	1	2	1	−2
$-z$	3	4	3	−6

Pivots 34, 26, 65.

3. No integer multiple of (0, 1) exceeds (1,0).

Section 7.1

1. $\begin{bmatrix} -3 & -4 & -18 \\ 14 & 7 & 19 \\ 23 & 5 & -5 \end{bmatrix}$ It is not invertible because its rank ≤ 2 because its factors have only two rows or columns.

$\begin{bmatrix} 2 & -17 \\ 20 & -3 \end{bmatrix}$ This also happens to be of rank 2, which makes it invertible.

6. Multiplying left to right gives $[15\quad -9]\begin{bmatrix} 6 & 1 & -3 \\ 5 & 2 & 4 \end{bmatrix}=[45\quad -3\quad -81]$ with 12 multiplications. If the answer to 1. was not available, multiplying right to left would require 27 multiplications.

7. (**AB**)**C** requires $mp(n+q)$ multiplications;
A(**BC**) requires $nq(m+p)$. Five ways for **ABCD**

9.

1.5	−2.5	0
−1	2	0
4.5	−9.5	1

10.

−.3	.1	0
.1	.3	0
1.4	.2	1

	11.			12.			13.		
	1	1	0	1	−1	0	1	3	0
$\mathbf{B}_1=$	0	−3	0	0	2	0	0	4	0
	0	0	1	0	−2	1	0	5	1
	1	1/3	0	1	1/2	0	1	−3/4	0
$\mathbf{B}_1^{-1}=\mathbf{E}^1=$	0	−1/3	0	0	1/2	0	0	1/4	0
	0	0	1	0	1	1	0	−5/4	1
	−1	1	0	1	−1	0	2	3	0
$\mathbf{B}_2=$	−1	−3	0	−1	2	0	−1	4	0
	3	0	1	0	−2	1	1	5	1
	−3/4	−1/4	0	2	1	0	4/11	−3/11	0
$\mathbf{B}_2^{-1}=$	1/4	−1/4	0	1	1	0	1/11	2/11	0
	9/4	3/4	1	2	2	1	−9/11	−7/11	1
	−3/4	0	0	2	0	0	4/11	0	0
$\mathbf{E}^2=$	1/4	1	0	1	1	0	1/11	1	0
	9/4	0	1	2	0	1	−9/11	0	1

Use the last row of $\mathbf{B}_2^{-1}$ for pricing out the initial tableau.

14. See (10) of 2.4.

Section 7.2

1. $\mathbf{c}^r=(11,9,-2,13,-1,0,0)/8$, $z=-29/8$.

2. $\mathbf{g}^r=(14,8,0,0,-19,0,-55,-9)/13$, $z^0=120/13$.

	$\mathbf{x}^*$	$\mathbf{u}^*$	z^*
3.	(0,0,0,2,0,0,0,13)/3	(23,17,1,0,34,1,4,0)/3	−28/3
4.	(0,0,1,0,0,0,0,18)/7	(70,55,0,4,107,4,15,0)/7	−107/7
5.	(0,0,1,0,1,0,0)	(6,5,0,8,0,1,1)/4	−4
6.	(17,0,5,0,1,0,0,0)	(0,22,0,4,0,8,−1,0)	−12
7.	(0,0,0,0,2,0,1,3)	(2,1,−1,1,0,2,0,0)	−12
8.	(0,0,0,0,0,2,3)	(2,2,36,13,14,0,0)	−4
9.	(0,0,0,5,1)/4	(9,3,19,0,0)/4	−5/4

10. Step 2 (updating **b**) is essentially identical with the procedure of 3.5(b). In step 5, $\pi_{\text{II}}^r=-\mathbf{c}_{M^r}^0\mathbf{B}_r^{-1}$ applies 3.5(b) to as much of the current tableau as is available, namely $\mathbf{B}_r^{-1}=\mathbf{A}'_{M^0}$, and the same change of sign occurs in both presentations. Step 6 (pricing out) is not included in 3.5(b); it does not involve any further changes in sign.

Section 7.3

1. $\eta^1,\eta^2=$		(1*,1,−5)/3	(2,3*,−1)/4	
(a) FT	→	(1,2,3)	(1,7,4)/3	(6,7,3)/4
BT	(−17,5,12)/4	(1,5/4,3)	(1,2,3)	←
(b) FT	→	(1,−1,1)	(1,−2,−2)/3	(0,−2,−2)/4
BT	(−3,−1,2)/2	(1,−1/2,1)	(1,−1,1)	←
(c) FT	→	(−1,4,1)	(−1,11,8)/3	(6,11,7)/4
BT	(−5,9,4)/4	(−1,9/4,1)	(−1,4,1)	←

The answers are in the first and last columns.

2.

r	M^r	$\mathbf{b}^r$	η^{r+1}	Forward 1	Forward 2	Forward 3	Backward $\boldsymbol{\pi}^1$	Backward $\boldsymbol{\pi}^2$	Backward $\boldsymbol{\pi}^3$
							0	5/13	14/40
				$\mathbf{a}^0_{.5}$	$\mathbf{a}^0_{.6}$ ↓	$\mathbf{a}^0_{.4}$	0	0	6/40
							2/3	7/13	22/40
	x_1	−3	−1/3	1	4	0	0	5/13	14/40
0	x_2	−1	0	0	1	3	0	0	6/40
	x_3	−4	1/3*	3*	−1	1	2	2	80/40
	x_1	−5/3	3/13*	←	13/3*	−1/3	↑	1	1
1	x_2	−3/3	−3/13		3/3	9/3	$-\mathbf{c}^0_{M^1}$	0	6/40
	x_5	−4/3	1/13		−1/3	1/3		2	2
	x_6	− 5/13	1/40			−1/13		↑	1
2	x_2	−8/13	13/40*		←	40/13*		$-\mathbf{c}^0_{M^2}$	1
	x_5	−19/13	−4/40			4/13			2
	x_6	−16/40							↑
3	x_4	− 8/40							$-\mathbf{c}^0_{M^3}$
	x_5	−56/40							

3. $\boldsymbol{\pi}^3\mathbf{a}^0_{.8} + c^0_8 = (0,1)(2,1) - 2 = -1 = c^3_8 < 0$; hence make x_8 basic.
FT: $\mathbf{a}^0_{.8} = (2,1) \to (5,1)/4 \to (5,4)/11 \to (3,4) = \mathbf{a}^3_{.8}$.
$\boldsymbol{\eta}^4 = (1^*, -4)/3$. $\mathbf{b}^4_{8,4} = (-13, -2)/3$.
BT: $-\mathbf{c}^0_{8,4} = (2,1) \to (-2,3)/3 \to (-2,19)/3 \to (1,19)/3 \to (1,4)/3$.

4. FT: $\mathbf{a}^0_{.8} = (3,2) \to (3,1)/2 \to (6,7)/11 \to (5,7) = \mathbf{a}^3_{.8}$
$\boldsymbol{\eta}^4 = (-5, 1^*)/7$. $\mathbf{b}^4_{3,8} = (-1, -18)/7$.
BT: $(-1,6) \to (-7,11)/7 \to (-7,72)/7 \to (4,72)/7 \to (4,15)/7$.

3.–9. See 3 to 9 of 7.2.

10. Let $\mathbf{e}'$ and $\mathbf{f}'$ be obtained from $\mathbf{e}$ and $\mathbf{f}$ by subtracting 1 from the pivotal components. Then $\mathbf{E}^1\mathbf{E}^2 = \mathbf{E}^2\mathbf{E}^1$ iff $e_q\mathbf{f}' = f_p\mathbf{e}' = \mathbf{0}$ iff $e_q = f_p = 0$ or $\mathbf{e}' = \mathbf{0}$ or $\mathbf{f}' = \mathbf{0}$.

11. $\mathbf{E}^2\mathbf{E}^1$ gives $\boldsymbol{\eta} = \mathbf{e}^{\#} + e_p\mathbf{f}$ = the FT of $\mathbf{e}$ by $\mathbf{f}$ [see (2)].

12. $\boldsymbol{\eta}^1 = (1, 12^*)/5$, $\boldsymbol{\eta}^2 = (5^*, 8)$.

Section 7.4

$\mathbf{y}$ is the column to the right of $\mathbf{T}$ and $\mathbf{x}$ is the row below.

1.

3	−2	0	2	−1
0	1	1	−3	0
−1/3	−2/3	−1/3	2/3	5/3
2/3	7/3	13	−3	−24
3	13	11	8	

2.

5	−4	1	11
−2/5	−3/5	17/5	52/5
3/5	−22/3	76/3	239/3
(149,	37,	239)/76	

3.

1	0	3	2	5
−1	2	4	2	0
3	1/2	−11	−6	−4
2	−1/2	6/11	36/11	−108/11
5	−1	2	−3	

4.

−2	0	−1	1	2
−1/2	3	3/2	−1/2	2
1	−1/3	2	5/6	−7/3
−1/2	1/3	−1	3/2	2
(19,	63,	−62,	48)/36	

5.

b	**B**			**U=L^T**			**y**	Check
−12	6	...	2.449	.8165	26.13	27.76	−4.899	52.26
105	2	64	...	7.958	5.864	130.3	13.70	157.87
−39	64	68	1060	...	18.52	34.68	.4689	53.67
2001	68	1060	2132	21244	...	47.75	7.025	54.77

$$z(\text{calc.}) = -.8302 - .5043v - .2502v^2 + .14713v^3$$

$$= -12.233,\ -5.542,\ -.723,\ -1.663,\ -.623,\ 8.783.$$

The minimized sum of the squares of the residuals is 10.835.

Section 7.5

1. 17 and 36 respectively; when **B** is symmetric, (4) becomes $\Sigma_1^m r(m-r+1) = m(m+1)(m+2)/6$ operations (r for each of $m-r+1$ elements in row r of **U**), of which m are square roots.

2. The numbers are m^3 and $(m^3-m)/3$ respectively. For the latter the numbers are the same as in (4), element by element.

3. To find (**L**, **U**), $\mathbf{U}^{-1}$, $\mathbf{L}^{-1}$ and $\mathbf{U}^{-1}\mathbf{L}^{-1}$ requires

$$m^3 = (m^3-m)/3 + m(m+1)(m+2)/6 + m(m-1)(m-2)/6 + (m^3-m)/3.$$

4. Let λ = the number of pairs of variables exchanged, namely the x_i with i in K becoming nonbasic and those with i in L becoming basic. The triangular decomposition of the submatrix $\mathbf{\Lambda}$ of $\mathbf{A}^0$ lying in rows K and columns L requires $(\lambda^3-\lambda)/3$ operations by (4); $\mathbf{x}_L^*$ and $\mathbf{u}_K^*$ require λ^2 each; $\mathbf{x}_{\bar{K}}^*$ requires $(m-\lambda)\lambda$; $\mathbf{u}_{\bar{L}}^*$ requires $(n-\lambda)\lambda$; and $z^* = \bar{w}^*$ requires λ. The total is $\lambda[m+n+(\lambda^2+2)/3]$ without any checking.

5. This reduces to verifying $(\varepsilon_1+\varepsilon_2)x_1x_2 \le (x_1+x_2)\max(\varepsilon_2 x_1, \varepsilon_1 x_2)$. Extend by mathematical induction.

6.–7. Ignore products of relative errors, or let $y = x_1^{p_1} \dots x_k^{p_k}$,

$$\ln y = p_1 \ln x_1 + \dots + p_k \ln x_k,\quad \frac{dy}{y} = \frac{p_1\,dx_1}{x_1} + \dots + \frac{p_k\,dx_k}{x_k}.$$

$$\text{Relative error} \doteq \left|\frac{dy}{y}\right| \le \left|\frac{p_1\,dx_1}{x_1}\right| + \dots + \left|\frac{p_k\,dx_k}{x_k}\right| = |p_1|\theta_1 + \dots + |p_k|\theta_k.$$

Section 7.6

1.

0	3/5	−1	−18/5*	31/5
3*	−1 U	1	2	1
−1/3	5/3*	−5/3	7/6	−2/3
1/3	1/5	1*	−19/10	19/5

2.

0	1 U	−2*	−1	2
3*	−1/2	−1/2	3/2	2
−1/3	5/6	1	2*	−7/3
1/3	3/2*	−1/2	−1	2

$$(63,\ 48,\ 19,\ -62)/36 = \mathbf{x} = (63,\ 48,\ 19,\ -62)/36$$

3.

$$\begin{bmatrix} 8 & 0 & 0 & 0 \\ 2 & 9 & 8 & 0 \\ 4 & 3 & 2 & 0 \\ 7 & 6 & 4 & 5 \end{bmatrix} = \begin{bmatrix} 8 & 0 & 0 & 0 \\ 2 & 1 & 0 & 0 \\ 4 & 0 & 1 & 0 \\ 7 & 0 & 0 & 1 \end{bmatrix} \begin{bmatrix} 1 & 0 & 0 & 0 \\ 0 & 9 & 8 & 0 \\ 0 & 3 & 2 & 0 \\ 0 & 0 & 0 & 1 \end{bmatrix} \begin{bmatrix} 1 & 0 & 0 & 0 \\ 0 & 1 & 0 & 0 \\ 0 & 0 & 1 & 0 \\ 0 & 6 & 4 & 1 \end{bmatrix} \begin{bmatrix} 1 & 0 & 0 & 0 \\ 0 & 1 & 0 & 0 \\ 0 & 0 & 1 & 0 \\ 0 & 0 & 0 & 5 \end{bmatrix}$$

4.

$$\begin{bmatrix} 6 & 7 & 2^\beta \\ 1^\alpha & 5 & 9 \\ 8 & 3^\gamma & 4 \end{bmatrix} = \begin{bmatrix} 1 & 6 & 0 \\ 0 & 1^\alpha & 0 \\ 0 & 8 & 1 \end{bmatrix} \begin{bmatrix} 1^\beta & 0 & 0 \\ 0 & 1 & 0 \\ 17/13 & 0 & 1 \end{bmatrix} \begin{bmatrix} 1 & 0 & -23 \\ 0 & 1 & 5 \\ 0 & 0 & -90/13^\gamma \end{bmatrix}$$

$$\times \begin{bmatrix} 1 & 0 & 0 \\ 0 & 1 & 0 \\ 0 & 0 & 1^\gamma \end{bmatrix} \begin{bmatrix} -52^\beta & 0 & 0 \\ 9 & 1 & 0 \\ 0 & 0 & 1 \end{bmatrix} \begin{bmatrix} 1 & 0 & 0 \\ 0 & 1^\alpha & 0 \\ 0 & 0 & 1 \end{bmatrix} \begin{bmatrix} 0 & 0 & 1^\beta \\ 1^\alpha & 0 & 0 \\ 0 & 1^\gamma & 0 \end{bmatrix}$$

$$= \begin{bmatrix} 1^\beta & 6 & 0 \\ 0 & 1^\alpha & 0 \\ 17/13 & 8 & 1^\gamma \end{bmatrix} \begin{bmatrix} 0 & -23 & -52^\beta \\ 1^\alpha & 5 & 9 \\ 0 & -90/13^\gamma & 0 \end{bmatrix}$$

$$\mathbf{T} = \begin{bmatrix} 6 & -23 & -52^\beta \\ 1^\alpha & 5 & 9 \\ 8 & -90/13^\gamma & 17/13 \end{bmatrix}$$

Section 7.7

1. As written, $q(\mu)=3$, $q(\nu)=2$; minus sign.
2. To obtain the second determinant change rows to columns in the first and then exchange two pairs of rows. To obtain the third from the second, add column 2 to column 1.
3. The determinants are zero; two columns are proportional, or all columns add to **0**.
4. $cd(af-be)$, -3, $a_{11}a_{22}a_{33}a_{44}$.
5. $$\begin{vmatrix} 2 & 3 & 12 \\ 6 & -12 & 16 \\ 2 & -5 & 6 \end{vmatrix} = -68 \qquad \begin{vmatrix} 2 & 3 & 2 \\ 7 & 5 & 4 \\ 5 & 0 & 1 \end{vmatrix} = -1$$
6. $$\begin{bmatrix} 1/a & -f/ab & (df-be)/abc \\ 0 & 1/b & -d/bc \\ 0 & 0 & 1/c \end{bmatrix} \qquad \begin{bmatrix} -13 & -21 & 10 \\ -2 & -4 & 2 \\ 3 & 5 & -2 \end{bmatrix} \div 2$$
$$\begin{bmatrix} d/(ad-bc) & -b/(ad-bc) & 0 & 0 \\ -c/(ad-bc) & a/(ad-bc) & 0 & 0 \\ 0 & 0 & y/(uy-vx) & -v/(uy-vx) \\ 0 & 0 & -x/(uy-vx) & u/(uy-vx) \end{bmatrix}$$
7. $\mathbf{x}=(149,37,239)/76$.
8. Subtract column 3 from the others to get
$$\begin{vmatrix} 5y & 4y & x_1-3y \\ -y & y & x_2 \\ 2y & -3y & x_3+y \end{vmatrix} = y^2(6y+x_1+23x_2+9x_3)$$
9. $\frac{1}{2}\begin{vmatrix} 5-1 & 3-1 \\ 4-2 & 7-2 \end{vmatrix} = 8$
10. $-\frac{1}{6}\begin{vmatrix} 2 & 1 & 0 \\ 3 & 0 & 1 \\ 0 & 2 & 3 \end{vmatrix} = \frac{13}{6}$
11. Show that $|\mathbf{B}|$ is irreducible because each b_{ij} must occur in exactly one factor, and if b_{11} occurs in a given factor, then all b_{i1} and b_{1j} and hence all b_{ij} occur in that factor. Theorem 9 of 7.7 and Theorem 1 of 7.1 show that $|\mathbf{AB}|=0$ whenever $|\mathbf{B}|=0$ (or $|\mathbf{A}|=0$). The lemma gives $|\mathbf{AB}|=c|\mathbf{A}|\cdot|\mathbf{B}|$ where obviously $c=1$.

Section 7.8

1. $-(-3), \begin{vmatrix}1 & 2\\ -3 & -1\end{vmatrix}, \begin{vmatrix}-3 & -1\\ 0 & 3\end{vmatrix}, \begin{vmatrix}1 & -1 & 2\\ -3 & -1 & -1\\ 0 & 3 & 1\end{vmatrix}$, all over $\begin{vmatrix}1 & -1\\ -3 & -1\end{vmatrix}$

2. $-1, -\begin{vmatrix}1 & 21\\ -1 & 18\end{vmatrix}, \begin{vmatrix}2 & -1\\ -2 & 0\end{vmatrix}, \begin{vmatrix}-1 & 1 & 21\\ 2 & -1 & 18\\ -2 & 0 & -42\end{vmatrix}$, all over $\begin{vmatrix}-1 & 1\\ 2 & -1\end{vmatrix}$

3. Let $\bar{K}$ represent the row $-z$. Then the occurrence of zeros in $\mathbf{B}_r^{-1}$ and $\mathbf{c}_{M'}^0$ permits the transforms of $\mathbf{A}_{11}, \mathbf{A}_{12}, \mathbf{A}_{21}$ and $\mathbf{A}_{22}$ in (4) to be associated with steps 4, 2, 5 and 6 in Table 7.2′.

4. Corollary 1 on page 312 with $\gamma=3,2,1,0$ and $A=3$ gives

$$\begin{vmatrix}3 & -2 & 0\\ 0 & 1 & 1\\ -1 & 0 & -1\end{vmatrix}=-1, \qquad 3\cdot\begin{vmatrix}3 & -2\\ 0 & 1\end{vmatrix}=9, \qquad 3^2\cdot 3=27, \qquad 3^3=27$$

5. $\mathbf{x}$ is an arbitrary multiple of the cofactors of any row of $\mathbf{A}$, the various rows having proportional vectors of cofactors by Theorem 13.

6. For Corollary 1, omit the empty sets $H-L$ and $H'-L'$ outside the block pivots, labeled $\mathbf{w}$ and $\mathbf{z}$. For Corollary 2, omit the empty sets $H'\cap L'$ and $H\cap L$, labeled $\mathbf{v}$ and $\mathbf{y}$. The matrices on the diagonal now must be square.

7. Every 2 by 2 minor is divisible by t, since the matrix of constant terms has rank 1.

8. To solve $\mathbf{Ax}=\bar{\mathbf{b}}$, the rows L and columns L' of $\mathbf{A}$ could be pivoted to give $\mathbf{B}=[\mathbf{A}^{-1},\boldsymbol{\beta}]$. In $\mathbf{B}$ we only want the value of $x_j=\beta_j$, the one-rowed determinant $B(H',H)=B(j,0)$ ($\bar{\mathbf{b}}$ has the index 0). Theorem 14 gives $G=L$, $G'=\{0\}\cup(L'-\{j\})$ and $x_j=B(H',H)=\pm A(G,G')/A(L,L')=\pm|\mathbf{A}_j|/|\mathbf{A}|$ as in Theorem 8.

9.

4	x_3	x_4	x_5
x_1	4*	2	2
x_2	2	3	4

4	x_1	x_4	x_5
x_3	4	2	2
x_2	−2	2*	3

2	x_1	x_2	x_5
x_3	3	−2	−(1/2)
x_4	−2	4	3

A fraction has appeared at $ij=35$ because the initial $\delta=4$ and $\varepsilon=4$ have a common factor.

Section 8.1

1. (b)

3	x_1	x_5	$x_4^{\pm}$	1
x_3	1	2	1	−11
x_2	−4	1	−1	−4
y_4^0	6	−3	−3*	18
$-z$	2	1	−4	−1

3	x_1	x_5	y_4^0	1
x_3	3*	1	1	−5
x_2	−6	2	−1	−10
$x_4^{\pm}$	−6	3	−3	−18
$-z$	−6	5	−4	−25

3	x_3	x_5	y_4^0	1
x_1	3	1	1	−5
x_2	6	4	1	−20
$-z$	6	7	−2	−35

$$\mathbf{C}^{-1}=1/3\begin{bmatrix}-3 & 0 & 0 & 0\\ -4 & 1 & -1 & -4\\ 1 & 2 & 1 & -11\\ 0 & 0 & 0 & -3\end{bmatrix}$$

Row y_4^0 is $(6,-3,-3,18)/3=-(4,-1,2,0)\mathbf{C}^{-1}$ by (5)

	$\mathbf{x}^*$	$\mathbf{u}^*$	z^*	Exchange
1. (a)	(0,4,15,0,0)/3	(2,0,0,−4,1)/3	−1/3	x_2, y_2
(b)	(5,20,0,0,0)/3	(0,0,6,−2,7)/3	−35/3	above
2. (a)	Unbounded as $y_4\to\infty$.			
(b)	(0,0,0,2,3)	(1,1,9,0,0)/4	−2	$y_1^0, x_1^{\pm}$
3. (a)	(1,4,0,0,5/2,0)	(0,0,0,3,0,8)	0	x_5^0, y_5; x_4, x_2
(b)	(0,0,0,40,69,49)/36	(113,81,19,0,0,0)/36	581/36	y_1, x_6

4. (a) (0,0,0,5,16,0) (10,13,0,0,0,4) -53 x_5^0, y_5
(b) (0,0,46,5,14,0)/13 (8,9,0,0,0,22)/13 $-33/13$ $y_1, x_1^{\pm}$

Section 8.2

Extreme values of Δz are equal to $-b_{i^z}c_{j^z}/a_{i^z j^z}$

Parameter s	Range of Δs	Range of Δz	$i^z j^z$	Equations
1. x_1	[0,1/2]	[0,1/6]	51	(1)
x_2	[0,1]	[0,5/3]	52	(1)
x_3	[−2/3,5/6]	[0,2/3], [0,1/6]	36, 51	(2)
x_4	[−1/6,1/6]	[0,1/6], [0,2/3]	51, 36	(2)
x_5	[−1/3,1/6]	[0,1/6], [0,2/3]	51, 36	(2)
x_6	[0,1/6]	[0,2/3]	36	(1)
2. b_1^0	[−2/5,1/2]	[−2/15,1/6]	31, 51	(4)
b_2^0	[−2,1]	[−10/3,5/3]	42, 52	(4)
b_3^0	(−∞,2/3]	0	—	(3)
3. c_4^0	[−4,1]	[−8/3,2/3]	46, 41	(6)
c_5^0	[−4,1/2]	[−4/3,1/6]	56, 51	(6)
c_6^0	[−4,∞)	0	—	(5)
4. a_{36}^0	(−∞,∞)	0	—	(7)
d^0	(−∞,∞)	(−∞,∞)	—	—
5. a_{16}^0	[−12,∞)	0	—	(8′)
a_{35}^0	(−∞,2]	0	—	(9)
6. a_{14}^0	[−1/2,1]	[−2/15,1/6]	31, 51	(14)
a_{25}^0	[−4/3,1/3]	[−4/3,1/6]	56, 51	(14)
a_{15}^0	[−2/3,∞)	[−2/15,1/6]	31, 51	(14)
7. y_1	[−6/17,0]	[0,1.08/17]	13	(2)
y_2	[0,1/3]	[0,14.8/51]	02	(1)
y_3	[−13/51,18/85]	[0,14.8/51], [0,1.08/17]	02, 13	(2)
b_1^0	[−1,10/7]	[−293/119,29.3/17]	01, 31	(4)
b_2^0	(−∞,25/17]	0	—	(3)
c_1^0	(−∞,.18]	(−∞,1.08/17]	13	(6)
c_2^0	(−14.8/17,∞)	0	—	(5)
a_{11}^0	[−9/83,5/3]	[−293/119,1.08/17]	31, 13	(14)
a_{12}^0	(−∞,148/293]	0	—	(8″)
a_{31}^0	[−1,∞)	[−.45/17,1.08/17]	03, 31	(14)
a_{32}^0	[−148/9,∞)	0	—	(8′)
a_{01}^0	(−∞,1/6]	0	—	(9)

8. $\partial\bar{w}^*/\partial u_p = b_p$. $0 \le u_p \le \min_j\{c_j/(-a_{pj})\colon a_{pj}<0\}$.
$-b_p c_{p'}/a_{pq'} \le \Delta\bar{w} = b_p u_p \le 0$. $\partial\bar{w}^*/\partial u_q = b_i/a_{iq}$.
$\partial\bar{w}^*/\partial u_q = \min_i\{b_i/a_{iq}\colon a_{iq}<0\} \ge 0$ (u_q decreasing)
or $-\min_i\{-b_i/a_{iq}\colon a_{iq}>0\} \le 0$ (u_q increasing).

Section 8.3

The interval for t, the optimal basis and $z^*(t)$ are given.

1. (−∞,.5], 12, 0; [.5,2.0], 32, $-1+2t$; [2,∞), 34, $-5+4t$.

2. [1,2], 12, 0; [2,15/7], 32, $(-2+3t-t^2)/4$; [15/7,3], 31, $(6-5t+t^2)/3$.

3. [−1/2,1/8], 15, $1+2t$; [1/8,3/4], 13, $(3-4t)/2$; [3/4,9/7], 23, $3-4t$.
Updated $\mathbf{c}_{1N}=(-8,3,2,2)$.

4. $(-\infty,2]$, 12, 0; ϕ, 15, no opt.; $[2,29/9]$, 45, $(-12+8t-t^2)/5$.
Two pivot steps (one in each phase) are required to get from one interval to the other; ties are broken as in (a).

5. $[-2/3,0]$, 42, $-3t+2t^2$; $[0,3/2]$, 12, 0; $[3/2,2]$, 14, $(6-7t+2t^2)/4$.
When t becomes negative, (d) is used with pivot a_{14}; $pl=13$.

Section 8.4

The interval for t, the optimal basis and $z^*(t)$ are given.

1. $[1/7,1/3]$, 31, $(-3-5t+7t^2)/5$;
$[1/3,2/3]$, 35, $-(2+t)/3$;
$[2/3,\infty)$, 32, $(-4+4t-7t^2)/5$. $\mathbf{b}^0\le\mathbf{0}$ for $t\ge 1/2$.

2. $[2,\infty)$, 43, $-2+2t-t^2$. $t^0=0$.

3. $(-\infty,-2]$, 34, $(-14+26t-12t^2)/3$; $[-2,4/3]$, 35, $-10+10t-2t^2$. $t^0=1.7$.

4. $[1/5,1/2]$, 24, $-1+3t-2t^2$;
$[1/2,5/4]$, 21, $(-6+14t-4t^2)/7$;
$[5/4,5/3]$, 31, $(1+t)/3$; $[5/3,\infty)$, 41, $(-1+t^2)/2$. $t^0=0$.

5. $[1.7,17/6]$, 645, $24.25-19t+4t^2$; $[17/6,3/4]$, 643, $9.8-5.4t+t^2$.
$\theta_2'=4$, $\theta_2''=2$, $t^0=3$.

6. $(-\infty,-1]$, 52, $(-10+5t)/3$; $[-1,0]$, 54, $-4+t$. $\mathbf{b}^0\le\mathbf{0}$ for all t; val $\mathbf{A}=0$.

7. $[-.9,.3]$, 65, $(-54+37t-10t^2)/4$; $[.3,2.4]$, 35, $(-84+26t-5t^2)/7$;
$[2.4,16]$, 34, $(-12+2t-5t^2)/5$; $[16,\infty)$, 64, $(-108+8t-5t^2)/5$.
$\theta_2'=3$, $\theta_2''=2$, $2\le t^0\le 3$.

8. $[1,\infty)$, 34, $12(-1+t)/11$. $t^0=2$.

9. val $\mathbf{A}<0$ implies $\mathbf{Ay}<\mathbf{0}$ for some probability vector $\mathbf{y}$. A sufficiently large multiple of $\mathbf{y}$ will be a feasible $\mathbf{x}$. Conversely, if $\mathbf{Ax}\le-\mathbf{b}<\mathbf{0}$, $\mathbf{x}/\Sigma x_i$ will be a probability vector insuring val $\mathbf{A}<0$.

Section 8.5

1. No b_i or c_j is ever pivotal. Complications really begin when even one pivotal a_{ij} depends on t as in (11) of 8.2.

The interval for t, the optimal basis and $z^*(t)$ are given.

2(a). $(-.25,.70]$, 215, $(8+12t)/(1+4t)$
$(-\infty,-5]\cup[.70,\infty)$, 214, $(19+8t)/(5+t)$

2(b). $[-1,-1/2]$, 254, $(7+4t)/(2+t)$
$[-1/2,-1/3]$, 234, $(4-2t)/(1-t)$
$[-1/3,\infty)$, 214, $(19+8t)/(5+t)$

3. $(-\infty,1/3]$, 253, -1; $[1/3,1]$, 453, $-2/(1+3t)$; $[1,\infty)$, 423, $-1/2$

4. $(-\infty,-1]$, 154, -3;
$[-1,1/4]$, 354, $(-11+4t)/(7+2t)$;
$[1/4,1)$, 352, $(-2+4t)/(1-t)$

5. $(-\infty,-2]$, 54, $(-11+7t)/(3-t)$; $[-2,\infty)$, 34, -5

6. $(-\infty,-4]\cup[2,\infty)$, 32, $(2+3t)/(2+t)$; $(-2,-1]$, 14, $(3+4t)/(2+t)$;
$(-1,2]$, 12, t

7. As in the proof of Theorem 6 of 7.7, any l by l determinant in (3) of 7.8, having $a_{ij}^0+\alpha_{ij}^0t$ as its ij element, equals the sum of 2^l determinants, each having $\mathbf{a}_{.j}^0$ or $\boldsymbol{\alpha}_{.j}^0t$ in column j. Thus the power to which t is raised equals the number of columns of α_{ij}^0 in the determinant, and the latter will be zero unless these columns are linearly independent (Theorem 9 of 7.7). Hence the power of t cannot exceed the rank of the matrix of the α_{ij}^0.

Section 8.6

2. If t_1 and/or t_2 occurs in $\mathbf{b}$, the equations $b_1^0 = b_2^0 = 0$ determine either a point of intersection or a common direction (slope) that is the same in every tableau, because every b_i is some linear combination of b_1^0 and b_2^0; likewise for $\mathbf{c}$.
3. $(-\infty, -2]$, 43, $2-t^2$; $[-2,0]$, 13, $-t^2/2$; $[0,1]$, 14, t.
4. Must have $-1 \leq t_1 \leq 1$ (basis 12). Optimal bases 23, 13, 14, 42 (counterclockwise order) surround 43, which is $-1/3 \leq t_1 \leq 1/2$; $3/5 \leq t_2 \leq 4/5$.
5. Optimal bases 13 $[(1+t_1)/3 \leq t_2 \leq -.2+t_1]$ and $23(-.2+t_1 \leq t_2 \leq -1+2t_1)$ converge at $(t_1, t_2) = (.8, .6)$.
6. Optimal bases 32, 34, 14, 12 cover the whole plane and surround $(1, 1)$.
7. Must have $1-.5t_1 \leq t_2 \leq 2t_1 - 2$. Optimal basis 24 for $t_2 \leq 0$, 14 for $0 \leq t_2$.
8. Must have $t_2 \leq \min(1+t_1, 2.5-.5t_1)$ (basis 12). Optimal bases 13, 23 converge at $(1,2)$; 23, 24, 12 surround $(2, 1)$; 34, 14 touch neither point. Basis 34 gives $z^* = -11 + 14t_1 - t_2 - 5t_1^2 + 2t_1t_2$.

Section 8.7

1. T is convex because $f(\mathbf{t}^0) \leq 0$ and $f(\mathbf{t}^1) \leq 0$ implies that both members of (1) are nonpositive, and so $(1-s)\mathbf{t}^0 + s\mathbf{t}^1$ is in T.

The interval for t, the optimal basis, and $z^*(t)$ are given.

2. $[0, 0.5]$, 12, 0; $[0.5, 0.8]$, 42, $(1-3t+2t^2)/3$;
   $[0.8, 1]$, 43, $(7-13t+5t^2)/5$; $[1, 4/3]$, 43, $(4-6t+t^2)/5$.
   (The linear approximation changes; the basis 43 does not.)
3. $(-\infty, 1/3]$, 43, $(-2+11t-17t^2+12t^3-3t^4)/3$;
   $[2, \infty)$, 13, $-2+7t-9t^2+5t^3-t^4$
4. $[-1, -1/3]$, 41, $-(1+t)(1-t)^2$
   $\left[-1/3, (4-\sqrt{19})/3\right]$, 43, $(-6-6t-16t^2+4t^3)/5$ (see after 6)
   $\left[(4-\sqrt{19})/3, 3\right]$, 13, $(1+t)(-3-4t+t^2)/2$
5. $(-\infty, -1] \cup [2, 3]$, 12, 0; $[-1, -.7844] \cup [3, 3.2877]$, 14, $(3-t)(1-t^2)/2$
   (roots of $3t^3 - 11t^2 + t + 9 = 0$).
6. $\left(0, \sqrt{3}/3\right]$, 14, $(t+t^{-1})/2$; $\left[\sqrt{3}/3, \infty\right)$, 34, $(6t+4t^{-1})/9$.

Sample optimal tableaus for 4 and 6

Check	5	x_2	x_1	1	t
6	x_4	1	2	-3	1
2	x_3	2	-1	-1	-3
-3	1	7	-1	-6	-8
-9	t	6	-8	2	-14
9	t^2	-1	3	-2	4

Check	9	x_1	x_2	t	t^{-1}
4	x_3	-2	-1	-3	1
-2	x_4	1	-4	-3	-5
25	$-z$	1	5	6	4

Section 8.9

1. $(\mathbf{x}, \bar{\mathbf{z}}) = (1,0,0;1,1)$ and $(0,1,0;3,1)/2$ and the edge joining them are efficient. $\boldsymbol{\lambda} = (1,1)$.
2. The ray $(1+x_2, x_2, 0; 3+2x_2, 1-x_2)$ is efficient for all $x_2 \geq 0$. $\boldsymbol{\lambda} = (1,2)$.
3. The final tableau of the efficiency test for the given tableau is shown. Since $v_3 = 0$ (and $v_1 = v_2 = 0$ because nonbasic), $\boldsymbol{\lambda} = (3,1,0) + (2,2,2) = (5,3,2)$ makes $\boldsymbol{\lambda}\mathbf{C}' = \boldsymbol{\lambda}\mathbf{C} = \mathbf{0}$. Every feasible point is efficient.

2	λ_1'	λ_2'	v_3
v_1	3	1	-4
v_2	1	1	-2
λ_3'	5	3	0
$\mathbf{e}\bar{\mathbf{z}}$	3	1	0

4. $\text{Max}(6\bar{z}_1+11\bar{z}_2+2\bar{z}_3)=25$ on the convex hull of $(\underline{0},2,\underline{0},1,\underline{0},\underline{0};2,1,1)$, $(0,0,2,1,0,1;3,2,5)/2$, $(0,0,0,8,2,5;23,7,30)/11$. No other efficient points. Any one of the $\underline{0}$ can be regarded as basic.
5. Except initially, every extreme point is efficient, namely
   $E_0(0,0,0,6,3,2;5,6,16)/17$
   $E_1(6,0,0,0,3,2;-1,12,10)/11$ $\quad E_4(0,1,1,1,0,0;1,-1,1)/2$
   $E_2(3,3,0,0,0,1;1,3,-1)/4$ $\quad E_5(0,0,2,2,1,0;1,0,6)/5$
   $E_3(0,3,0,3,0,1;4,0,2)/7$ $\quad E_6(2,0,2,0,1,0;-1,2,4)/3$
   Faces $E_0E_1E_2E_3$ $\quad E_0E_3E_4E_5$ $\quad E_0E_5E_6E_1$ $\quad$ (max $\boldsymbol{\lambda}\bar{\mathbf{z}}$
   $\boldsymbol{\lambda}$ (36,19,5) $\quad$ (32,1,13) $\quad$ (14,8,16) $\quad$ =22 on each)

Section 9.1

1. $p=5, q=8$. Connected graph like Figure 9.1A with node 34 and its four arcs omitted. Undirected because pivot steps are reversible.
2. $p=12, q=14$. Oriented mixoconnected graph with several loops but no cycles. Node 1 is a root.
3. $p=9, q=8$. The undirected graph consists of a loop connecting the eight outer squares; the center square is not connected to any other. Thus there are two connected components.
4. $p=8, q=19$. Oriented mixoconnected graph with many loops but no cycles.
5. $p=4, q=4$. Oriented mixoconnected graph with one loop and no cycles.

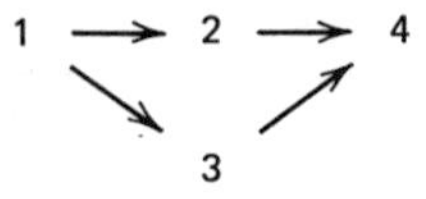

6. Six directed or three undirected arcs; two cycles or one loop; biconnected. Each property is equivalent to $t\geq 0$.

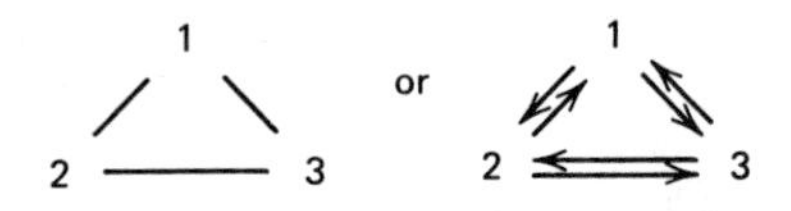

7. Oriented tree with root at node 1.
8. Directed connected network; (1,2) and (2,1) form a cycle.
9. Undirected tree.
10. Oriented tree; root at 3.
11. Mixed path between nodes 1 and 4.
12. Cycle 12341.
13. Directed path from 1 to 4.
14. For any nodes i and j, one has either j accessible from i or vice versa, but not both. This determines the order of the nodes in the path.
15, 16. The incidence relations permit one to trace the cycle or loop.
17. If $G_k \cap G_l \neq \phi$, there is a node i in both G_k and G_l. If j is any other node in G_k (or G_l), there is a path in G_k (or G_l) joining i and j. Thus $j \in G_l$ (or G_k) also, and $G_k = G_l$.

Section 9.2

1. $ED\underset{A}{B}CFGKH$ 37
2. $DA\underset{\underset{G}{E}}{F}BC$ 36
3. $D\overset{A}{\underset{G}{B}}ECF$ 33
4. Six trees. 21
5. $CB\underset{A}{D}FEG$ 68
6. $BA\overset{C}{\underset{K}{D}}FEGH$ 7
7. All arcs incident on node 1.
8. The path connecting nodes $1,2,\ldots,p-1,p$.
9. $ACEDF$ has "length" .09, reliability $(.99)^2(.98)^2(.97)$.
10. $B\overset{B}{D}E, ACFGKH$
11. $BC, DAFEG$

12. $\begin{matrix} A\ B \\ CDGKH \\ F\ E \end{matrix}$ 66

13. $AFEGDCB$ 43

14. 32 cases; 12+4 have $q=0$, 12+1 have $q=1$, 3 have $q=2$.

15. In the proof of the lemma, adding arc $\{l, \bar{l}\}$ and dropping arc $\{k, \bar{k}\}$ does not necessarily preserve the property of being a rooted tree or a path.

16. If either does not belong, add it to T_*. The resulting loop has at least three arcs and hence can be broken by dropping some longer c_{ij}. Or use (b) and note that $\{k, l\}$ and $\{p, q\}$ can be chosen first without forming a loop.

17. Let $(k, l) \in D_*$ and let S include the root 1 and the nodes that can be reached from 1 via $D_* - (k, l)$. Then (k, l) is a shortest arc from a node of S to l. This does not determine D_*, as Figure 9.2E shows.

Section 9.3

One accent indicates the origin; two, the destination.

	A	B	C	D	E	F	G	H	K
1. B HKFCA′DE G	0	8	7	6	18	8	11	20	17
2. BDGK′HEFCA	26	16	11	14	7	8	7	3	0
3. (D) B CA (D)GE′F H K	19	10	4	11	0	1	4	6	10
4. G ADE″HKFC B	18	11	17	12	0	16	14	4	7
5. A (E)FBC′G(E) D	19	5	0	17	22	18	19		
6. AFEG″C B D	12	20	19	9	3	7	0		
7. A BDC′FK EH′G	15	9	0	7	4	1	5	0	10

8. $d_{12} = c_{13} + c_{32} = 0$ but the algorithm gives $c_{12} = 1$.

Section 9.4

1.

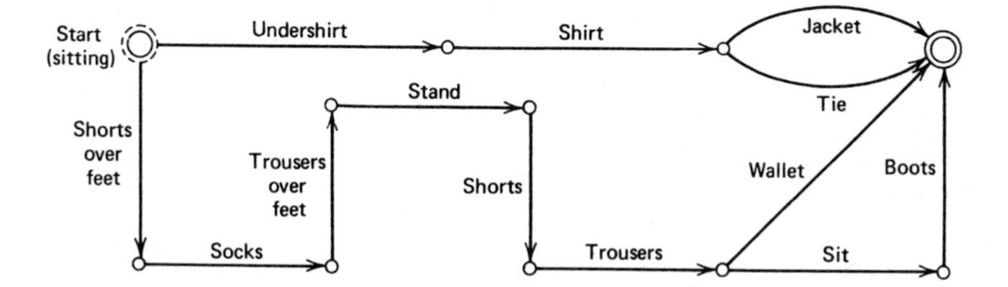

2.

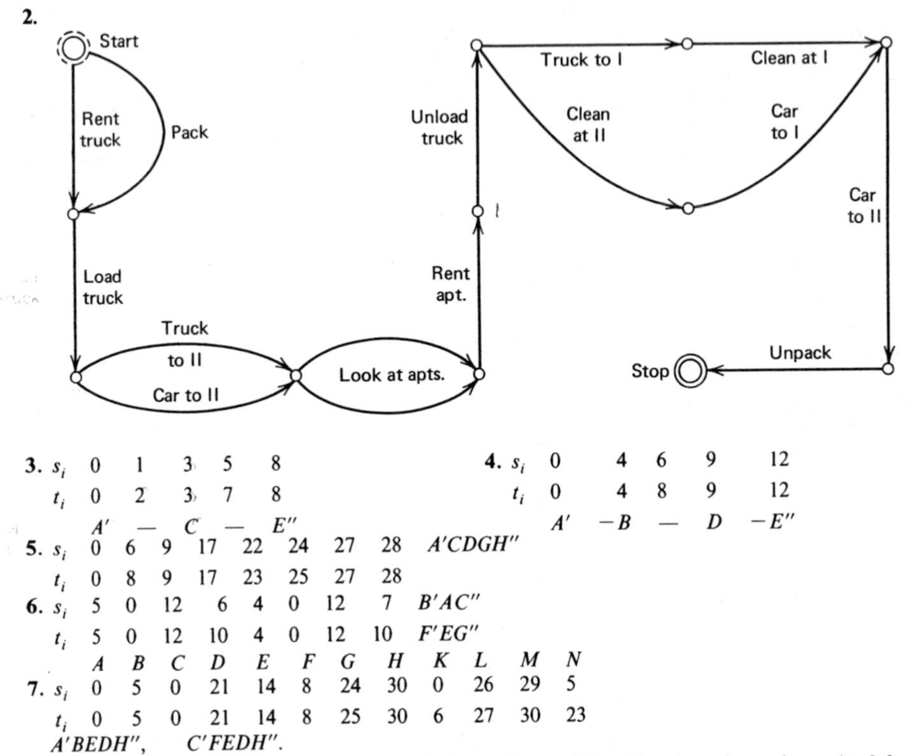

3.

s_i	0	1	3	5	8
t_i	0	2	3	7	8
	A'	—	C	—	E''

4.

s_i	0	4	6	9	12
t_i	0	4	8	9	12
	A'	$-B$	—	D	$-E''$

5.

s_i	0	6	9	17	22	24	27	28	$A'CDGH''$
t_i	0	8	9	17	23	25	27	28	

6.

s_i	5	0	12	6	4	0	12	7	$B'AC''$
t_i	5	0	12	10	4	0	12	10	$F'EG''$

7.

	A	B	C	D	E	F	G	H	K	L	M	N
s_i	0	5	0	21	14	8	24	30	0	26	29	5
t_i	0	5	0	21	14	8	25	30	6	27	30	23

$A'BEDH''$, $C'FEDH''$.

8. Index the nodes as in the second paragraph of 9.4. Set $s_i=0$ for all start nodes and $s_i=t^*\geq 0$ for all end nodes. Define the other s_i arbitrarily so that $i<j$ implies $s_i\leq s_j$, and define $c_{ij}=s_j-s_i\geq 0$ when (i,j) is an arc. Every arc lies on a path of length t^* from some start node to some end node, and all the members of (5) are equal.

Section 9.5

1. (a) $-b_1+d_1-b_2-c_3+d_3-b_4+d_4=11$
(b) $5-1+2+4=10$

2. (a) $-b_1+d_1-b_2+d_2+0-b_4+d_4+0=11$
(b) $2+7+1+3-2=11$

3. (a) $-b_1+d_1-b_2-c_3-c_4+d_4-b_5+d_5+0=15$
(b) $3+2+3+11+2-1=20$

4. $z_m=29, 24.5, 22, 21.2, 21, 20.8^*, 21.1, 21.9$
$m=$ 1 2 3 4 5 6 7 8

5. $z_m=265, 215, 187, 175, 168, 163.5, 162.9^*, 165.6$

6. $v_0=0$ while $u_0=-\infty$ or undefined. Selling involves no increase in the subscript; the arc (u_r, v_r) has length d_r.

7. $\bar{v}_r=\max(\bar{v}_{r+1}, \bar{u}_{r+1}-b_{r+1})$
$\bar{u}_r=\max(\bar{v}_r+d_r, \bar{u}_{r+1}-c_{r+1})$
provided $\bar{v}_n=0$, $\bar{u}_n=d_n(1-\theta_n)$
$\bar{v}_{n-1}=\max(-b_n\theta_n, \bar{u}_n-b_n)$
$\bar{v}_0=\max[\bar{v}_1+d_0\theta_0, \bar{u}_1-b_1(1-\theta_0)-c_1\theta_0]$

Section 9.6

	A	B	C	D	E	F	G	
								A'CDGE
1.	0	2	−1	0	0	0	−1	B
								F
								E
2.	−1	−1	0	−1	1	1	0	ACDFB
								G''

3.–7. See the answers for 3 to 7 of 9.3.

8. *ACBF* is a negative cycle of length −1.

9. Let $(A, i_1, \ldots, i_m)$ be a shortest path containing the maximum number m of arcs. Since $(A, i_1), (A, i_1, i_2), \ldots$ are shortest paths, one has $n_r \geq r$ for $1 \leq r \leq m$. For $m < r \leq p-1$ one has $n_r = p-1 \geq r$.

Section 9.7

The shortest path lengths d_{ij} and the intermediate nodes are given.

1.

	A	B	C	D	E
A	0	4	3E	5E	2
B	3	0	2	4C	1
C	1	5A	0	2	3A
D	3C	4	2	0	5
E	2C	6CA	1	3	0

2.

A	B	C	D	E
0	−1	0B	0	−1BC
1CE	0	1	1CEA	0C
0E	−1EA	0	0EA	−1
3	2	3B	0	2BC
1	0A	1AB	1A	0

3. Negative cycle *ADBA* of length −1.

4.

	A	B	C	D	E	F	G
A	0	2C	−1	0C	0CDG	0CB	−1CD
B	1FD	0	0FDA	0F	0FDG	−2	−1FD
C	2	3	0	1	1DG	1B	0D
D	1	2	0A	0	0G	0B	−1
E	3C	4C	1	2C	0	2	1CD
F	3D	4	2DA	2	2DG	0	1D
G	4D	5D	2E	3	1	3E	0

Section 9.8

The maximum capacities e^*_{ij} and the intermediate nodes are given.

1.

	A	B	C	D	E
A	∞	8C	9	7	5D
B	5D	∞	5DA	5	5D
C	5BD	8	∞	5B	5BD
D	9	8AC	9A	∞	5
E	6	7	6A	6A	∞

2.

A	B	C	D	E
∞	2CDE	3	2C	2CD
5DC	∞	5D	5	5D
5	2DE	∞	2	2D
5C	4E	5	∞	6
4BDC	4	4BD	4B	∞

3.

	A	B	C	D	E
A	∞	3	4	4C	3CD
B	3DE	∞	3D	3	3D
C	3DE	3DEA	∞	4	3D
D	3E	3EA	3	∞	3
E	3	3A	3A	4	∞

4. The complete paths are 5*AB*, 12*AD*, 6*CFE*, 8*CF*, 3*ADG*, 11*KH*, 2*ADGL*, 6*CFEM*, 5*KN*.

5. From *A*: ∞, 6, 9, 8*C*, 7*C*, 3*CD*, 8*CD*, 2*CDF*.

Section 10.1

1.

$r \backslash ij$	11	12	13	21	22	23	31	32	33
0	1	2	5	3	2	3	7	4	3
1	3	5	3	7	4	5	4	7	4
2	7	4	5	4	7	4	7	4	7
3	4	7	4	7	4	7	4	7	4
4	7	4	7	4	7	4	7	4	7

2.

$r\setminus i$	A	B	C	D	E	F	G	H
0	3	2	5	6	4	8	7	9
1	5	6	6	8	8	9	9	—
2	6	8	8	9	9	9	—	—
3	8	9	9	9	9	—	—	—
4	9	9	9	9	—	—	—	—

3. $\bar{z}_i(r) = \max_{j \in T_i} [\bar{z}_i(0) + \bar{z}_j(r-1)]$,

but this allows the node values to be accumulated repeatedly whenever the node is traversed repeatedly. $\bar{z}_k(r) = \bar{z}_k(0) +$ the longest path with $c_{ij} = \bar{z}_j(0)$, repetitions being allowed among the r arcs.

4.

i	$=1$	2	3	4
$j(i)$	$=3$	4	4	4
$\max_j[-3(i-j+2)^2]$	0	0	-3	-12

$i\setminus h$	1	2	3	4
1	1	0	1	4
2	4	1	0	1
3	6	1	-2	-3
4	4	-3	-8	-11
$\max_i$	6	1	1	$4 = \bar{z}(h)$

5. Need $j = i+2$ for $1 \le i \le 2$
$j = 4$ for $2 \le i \le 4$. The larger of
$\max_{1 \le i \le 2} (i-h+1)^2$ and $\max_{2 \le i \le 4} [(i-h+1)^2 - 3(i-2)^2]$ is
$3(3-h)^2/2$ for $1 \le h \le 5-\sqrt{6} = 2.55$,
$(h-2)^2$ for $5-\sqrt{6} \le h \le 4$.

6. Let $s=1$, $\mathbf{p} = (1,1,\ldots,1)$, $\boldsymbol{\pi} = \mathbf{a}$.

	$s=1$	2	3	4	5	6	7	8	9	10	11	12	13	14	15	16	17	18
7.	0	0	0	0	0	0	0	1	1	0	0	0	1	1	0	2	2	1
	0	0	0	0	1	1	1	0	0	2	2	2	1	1	3	0	0	2
	0	1	1	2	0	0	1	0	0	0	0	1	0	0	0	0	0	0
8.	0	0	0	0	0	0	1	0	0	1	0	0	1	2	0	1	2	0
	0	0	1	0	1	2	0	2	3	1	3	4	2	0	5	3	1	6
	0	1	0	2	1	0	0	1	0	0	1	0	0	0	0	0	0	0

Section 10.2

1. $\min z = \sum_{r=0}^{4} [(x_{r+1} - .95x_r)^2 + c(x_r - b_r)]$
subject to $x_r \ge b_r$ for $r=1$ to 4; $x_0 = b_0$; $x_5 = b_5$.
$z_r(x_{r+1}) = \min_{x_r \ge b_r} [(x_{r+1} - .95x_r)^2 + c(x_r - b_r) + z_{r-1}(x_r)]$

2. $z(r) = 54.9, 53.67, 53.136$ for $r = 1,2,3$. Refuse to pay more than 51 cents at the first station ($r=2$ left) or more than 54 cents at the second ($r=1$ left).

3. $\bar{z}(r) = -5/2, -7/6, -5/18, 17/54$ for $r = 1,2,3,4$ and it continues to increase. Do not play unless $n \ge 4$.

4. Stage 1

s	0	1	2	3–5
x_1^*	0	1	2	2
$\bar{z}_1^*$	0	16	18	18

Stage 2

$x_2 \setminus s$	0	1	2	3	4	5
0	0	16	18	18	18	18
1		10	26	28	28	28
2			16	32	34	34
3				18	34	36
4					16	32
5						10
max	0	16	26	32	34	36
x_2						

Stage 3

$\max\{36, 38, 44, 44, 32, 0\}$
$= 44$ at $x_3 = 2$ or 3.
$\mathbf{x}^* = (1,2,2)$ or $(1,1,3)$

5. $\bar{z}_1(s)=8$ for $0 \le s \le 4$; $\bar{z}_1(5)=18$.

$x_2 \backslash s$	0	1	2	3	4	5
0	10	10	10	10	10	20
1		8	8	8	8	8
2			8	8	8	8
3				10	10	10
4					14	14
5						20
$\max_{x_2}$	10	10	10	10	14	20

$$\bar{z}^* = \max_{x_3} \{4,8,10,12,10,4\} = 12. \quad x^* = (0,0,3).$$

6.

s	2	3	4	5	6	7	8	9	10	11
$\bar{z}^*(s)$	5	5	6	10	12	15	15	17	20	22
x_1^*	1	1	2	1	3	1	1	3	1	3
x_2^*	0	0	0	1	0	0	0	1	1	0
x_3^*	0	0	0	0	0	1	1	0	1	1

7. (i) Let $\bar{\mathbf{c}}=(3,4,5,7,9)$. $\text{Max}\,\Sigma_1^5 \bar{c}_r(1-2^{-x_r})$ subject to $\Sigma_1^5 x_r=6$ (nonlinear knapsack (e)).

(ii) The equality of the $p_r \equiv 1$ and the convexity and identity of the functions $1-2^{-x_r}$ justify assigning each successive ad so as to obtain the maximum increment, thus:

$$9/2+7/2+5/2+9/4+4/2+7/4=16.5$$

in units of 10^4 subscribers. $\mathbf{x}=(0,1,1,2,2)$.

8. (i) Let k_r = the number of ways of evaluating a product of r matrices. Since **ABCDE**=**A(BCDE)**=**(AB)(CDE)**=**(ABC)(DE)**=**(ABCD)E**, one has $k_5=k_1k_4+k_2k_3+k_3k_2+k_4k_1=14$, because $k_1=k_2=1$, $k_3=2$, $k_4=k_1k_3+k_2^2+k_3k_1=5$.

(ii)

Matrix	Rows m	Cols. n	Multiplications p
A	2	20	0
B	20	10	0
C	10	15	0
D	15	12	0
E	12	3	0
AB	2	10	$400=2\cdot 20\cdot 10=m(\mathbf{A})\cdot m(\mathbf{B})\cdot n(\mathbf{B})$
BC	20	15	3000
CD	10	12	1800
DE	15	3	540
ABC	2	15	$700=\min(3600,700) \triangleq p[\mathbf{ABC}]$
BCD	20	12	4200=min(4200,6600)
CDE	10	3	990=min(990,2160)
ABCD	2	12	1060=min(9000,2440,1060)
BCDE	20	3	1590=min(1590,4440,4920)
ABCDE	2	3	1132=min(1710,1450,1330,1132)

Thus (((**AB**)**C**)**D**)**E** requires the fewest (1132) multiplications.

$$1132=p[(\mathbf{ABCD})\mathbf{E}]=p[\mathbf{ABCD}]+p[\mathbf{E}]+m(\mathbf{ABCD})\cdot m(\mathbf{E})\cdot n(\mathbf{E})$$
$$=1060+0+2\cdot 12\cdot 3$$

Section 10.4

2. $(h,i,j,k)=(1,1,1,1)$, $(2,2,2,1)$; $\bar{z}_1^*=0$, $\bar{z}_2^*=11$.

3. Go ahead regardless of predictions. $\bar{z}^*=\$110$.

4. Go ahead only if prediction $\bar{B}$ is fair. $\bar{z}^*=\$8$.

	EB	$\bar{E}B$	B	$E\bar{B}$	$\bar{E}\bar{B}$	$\bar{B}$
C	.9×.6	.4×.4	.70	.8×.6	.5×.4	.68
$\bar{C}$	.1×.6	.6×.4	.30	.2×.6	.5×.4	.32

5. Values at A to E are 9, 7, 6, 6, 5.

6. Values at A to F are 10.1, 10, 8, 10, 9, 7.

7. Buy only if tested item is good.
$z^*=0.25$¢/carton

	EB	$\bar{E}B$	B
C	.9×.5	.7×.5	.80
$\bar{C}$	.1×.5	.3×.5	.20

8. Buy without test. (See figure)
$\bar{z}^*=3$¢.

C	.9×.6	.7×.4	.82
$\bar{C}$	.1×.6	.3×.4	.18

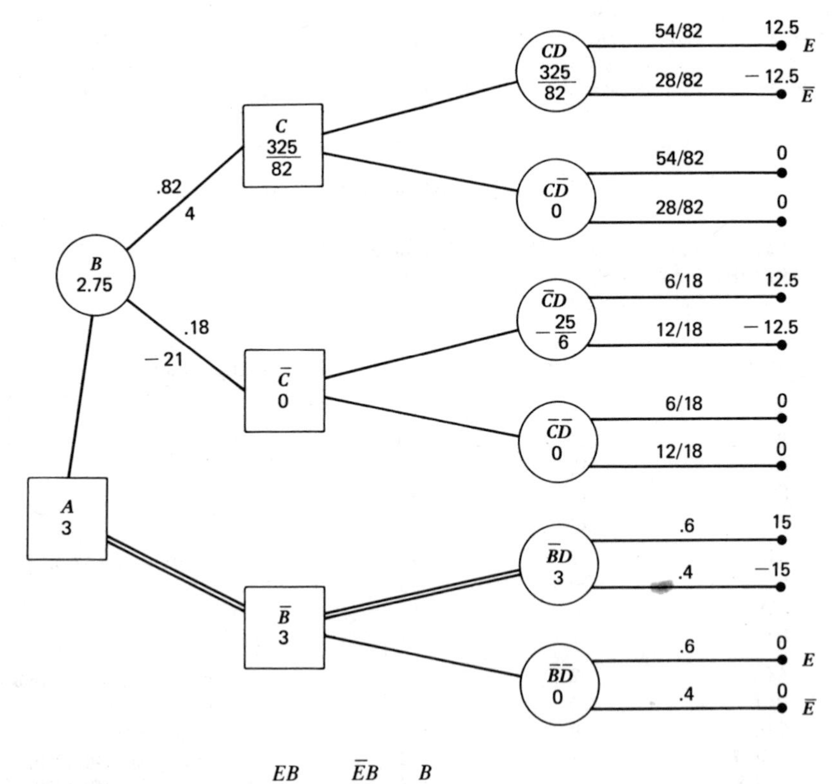

9. Go ahead without survey.
$\bar{z}^*=\$2000$.

	EB	$\bar{E}B$	B	
C	.7×.1	.2×.9	.25	
$\bar{C}$	.3×.1	.8×.9	.75	(all given B)

B is worth \$4, \$14, 1.25¢, 0.75¢, \$8400 in Exercises 3, 4, 7, 8, 9.

10. This tree has $2^3=8$ terminal nodes, but only those two that correspond to drawing two green balls need be considered; they have the value 1 and the other six nodes have the value 0. Each directed path contains three arcs. The probability is the expected value calculated by five applications of (2) with all $\bar{c}_{ij}=0$, namely

$$\tfrac{1}{2}\cdot\frac{g_1}{g_1+r_1}\cdot\frac{g_2+1}{g_2+r_2+1}+\tfrac{1}{2}\cdot\frac{g_2}{g_2+r_2}\cdot\frac{g_1+1}{g_1+r_1+1}$$

Section 10.5

1. $a^0=a_0=-15$.
2. $a^0=40$, $a_0=30$.
3. Value $=-123/13=-11\cdot 8/13-7\cdot 5/13$
4. Value $=10=(12+5+13)/3$
5. $\begin{matrix}4 & 5\\ 6 & 4\end{matrix}$ (4) has value 14/3 $\quad\begin{matrix}14/3 & 5 & 3\\ 2 & 14/3 & 3\\ 4 & 4 & 14/3\end{matrix}$ (5) has value 88/21
6. The value 5 is the saddle point of $\begin{bmatrix}5^* & 6 & 5\\ 3 & 5 & 5\\ 4 & 4 & 5.2\end{bmatrix}$ (5) in which the diagonal elements are the respective values of
   $\begin{matrix}6 & 5\\ 6 & 5\end{matrix}$, $\quad\begin{matrix}3 & 5\\ 7 & 5\end{matrix}$, $\quad\begin{matrix}4 & 6\\ 7 & 4\end{matrix}$ (4)
7. Instead of assigning nodes to I and J, simply label each node with the number of matches remaining and a letter W or L according as optimal play gives a win or a loss to the player who has to move. This gives

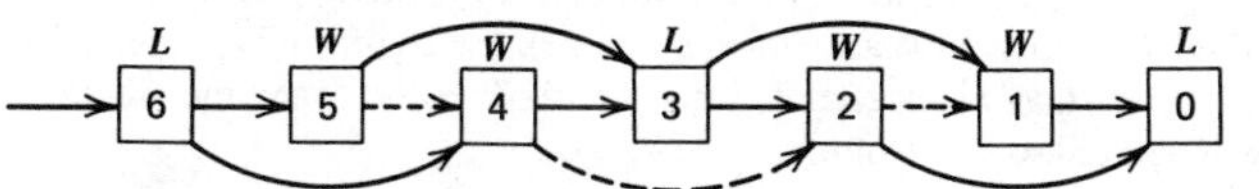

   The dashed arcs are never used in optimal play. Obviously nodes 1 and 2 are winning because all the matches can be taken. Node 3 is losing because it has to be followed by 1 or 2, where the other player wins. Nodes 4 and 5 are winning because they can be followed by 3, where the other player loses, etc.
8. The first player to move loses. (See explanation above.)

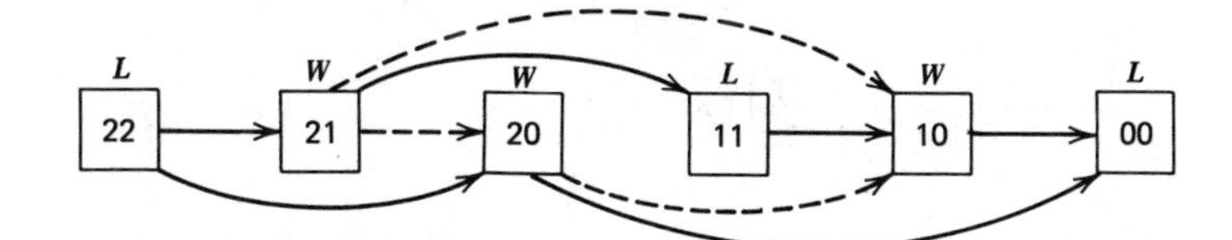

9. Let $z(n)=$ the value when n drawings remain.

   $z(1)=(0+1+3)/3=4/3$ (no choice)
   $z(2)=(0+1+4/3)/3=7/9$ (J minimizes)
   $z(3)=(7/9+1+3)/3=43/27$ (I maximizes)
   $z(4)=(0+1+43/27)/3=70/81$ (J minimizes)
   $z(5)=(70/81+1+3)/3=394/243$ (I maximizes)

10. The subgames $\begin{matrix}2 & 6\\ 6 & 4\end{matrix}$, $\begin{matrix}7 & 8\\ 8^* & 9\end{matrix}$, $\begin{matrix}1 & 3\\ 3^* & 5\end{matrix}$ provide values 14/3, 8, 3 at nodes I_1, I_2, I_3. Then $\begin{bmatrix}14/3 & 8\\ 8 & 3\end{bmatrix}$ gives the final value 6 at I_0.

11.

	11	12	21	22
11	2	6	7	8
12	6	4	8	9
21	7	8	1	3
22	8	9	3	5

12.

	11.	12.	211	212	221	222
111	2	6	7	7	8	8
112	2	6	8	8	9	9
121	6	4	7	7	8	8
122	6	4	8	8	9	9
2.1	7	8	1	3	1	3
2.2	8	9	3	5	3	5

The dots replace irrelevant choices at I_1 and in $\{J_5, J_6\}$.

Section 10.6

In 1 to 4 strategy b remains superfluous, and the given matrix 6A corresponds to (2) in Table 10.6.

1. aac $\quad 0 \quad 1 \quad -1 \quad 0 \quad \mathbf{u}^*=(2,1)/3, 6z^*=-1/3$
acc $\quad -1 \quad 0 \quad 1 \quad 2 \quad \mathbf{y}^*=(2,0,1,0)/3.$
I loses an average of \$1/18 per game by not bluffing.
J bluffs as before ($q_1=y_3+y_4=1/3$), but
$q_2=y_2+y_4=0$.

2. aac $\quad 0 \quad 1 \quad -1 \quad 0 \quad \mathbf{u}^*=(2,1)/3, 6z^*=-2/3$
cac $\quad 1 \quad -1 \quad 0 \quad -2 \quad \mathbf{y}^*=(0,0,2,1)/3$
I loses an average of \$1/9 per game by not counterbluffing (betting \$2 on card 2), which permits J to bluff all the time. J counterbluffs as before ($q_2=1/3$). I's argument is properly false but validated by his belief in it; given that J has card 1, when he bluffs as he sometimes should, he does *not* know whether he will win unless I lets him.

3.

$$\begin{matrix} 0 & \gamma-\alpha & \gamma-3\alpha & 2\gamma-4\alpha \\ \alpha-\gamma & 0 & \gamma-\alpha & 2\gamma-2\alpha \\ 3\alpha-\gamma & \alpha-\gamma & 0 & -2\alpha \\ 4\alpha-2\gamma & 2\alpha-2\gamma & 2\alpha & 0 \end{matrix}$$

This has $\mathbf{u}^*=\mathbf{y}^*=(1,0,0,0)$ if $\gamma>3\alpha$ and $(\gamma-\alpha, 3\alpha-\gamma, \gamma-\alpha, 0)/(\alpha+\gamma)$ if $0<\alpha<\gamma<3\alpha$.

$$p_1=q_1=u_3+u_4 \quad \text{and} \quad p_2=q_2=u_2+u_4.$$

$$z=\frac{\alpha+\gamma}{6}\left[\left(p_2-\frac{3\alpha-\gamma}{\alpha+\gamma}\right)\left(q_1-\frac{\gamma-\alpha}{\gamma+\alpha}\right)-\left(p_1-\frac{\gamma-\alpha}{\gamma+\alpha}\right)\left(q_2-\frac{3\alpha-\gamma}{\alpha+\gamma}\right)\right]$$

4. aac $\quad 0 \quad 1 \quad -1 \quad 0$
acc $\quad 2 \quad 0 \quad 4 \quad 2$
cac $\quad 4 \quad -1 \quad 3 \quad -2$
ccc $\quad 6 \quad -2 \quad 8 \quad 0$
$\mathbf{u}^*=(2,1,0,0)/3$. $\mathbf{y}^*=(1,2,0,0)/3$, $(0,2,0,1)/3$, $(0,5,1,0)/6$ or convex combinations thereof. $\mathbf{p}^*=(0,1/3)$, $\mathbf{q}^*=(0,2/3)$ or $(1/3,1)$, $6z^*=2/3$. In the notation of (3), I uses $[a,(2a+c)/3,c]$; J uses $[a,(a+2c)/3,(a+2c)/3]$ or $[(2a+c)/3, c, c]$. $z=1/9+p_1(2/3-q_2)+(3p_2-1)(q_1-q_2+2/3)/6$.

5.

	11	12	21	22
11	$(a_1+a_3)/2$	$(a_1+a_3)/2$	$(a_1+a_7)/2$	$(a_1+a_8)/2$
12	$(a_2+a_4)/2$	$(a_2+a_4)/2$	$(a_2+a_7)/2$	$(a_2+a_8)/2$
21	$(a_5+a_3)/2$	$(a_6+a_3)/2$	$(a_5+a_7)/2$	$(a_6+a_8)/2$
22	$(a_5+a_4)/2$	$(a_6+a_4)/2$	$(a_5+a_7)/2$	$(a_6+a_8)/2$

6.

	11	12	21	22
11	$(a_1+a_5)/2$	$(a_1+a_6)/2$	$(a_1+a_5)/2$	$(a_1+a_6)/2$
12	$(a_2+a_5)/2$	$(a_2+a_6)/2$	$(a_2+a_5)/2$	$(a_2+a_6)/2$
21	a_3	a_3	a_7	a_8
22	a_4	a_4	a_7	a_8

7. (a) I's sets are $\{E_1, E_2, E_5, E_6, E_9, E_{10}, E_{13}, E_{14}\}$
$\{C_1, C_2, C_3, C_4\}$, $\{E_3, E_4, E_7, E_8, E_{11}, E_{12}, E_{15}, E_{16}\}$.
(b) J's sets are $\{B_1\}$, $\{B_2\}$, $\{D_1, D_2\}$, $\{D_3, D_4\}$, $\{D_5, D_6\}$, $\{D_7, D_8\}$.
8. Let $\boldsymbol{\alpha}=(\alpha_1, \alpha_2, \alpha_3, \alpha_4)$ be any column vector. For (11), let $\mathbf{u}'$ denote the result of replacing p_1 and p_2 in (10) by u_3+u_4 and u_2+u_4, respectively; use $\Sigma u_i=1$ to show that $(\mathbf{u}-\mathbf{u}')\boldsymbol{\alpha}=(u_1u_4-u_2u_3)(\alpha_1-\alpha_2-\alpha_3+\alpha_4)=0$ when (11) holds. If $\mathbf{a}^{11}=\mathbf{a}^{12}$ (i.e., $\alpha_1=\alpha_2$), let $p_1=u_3+u_4$, $p_2=u_4/(u_3+u_4)$.
9. The curved boundary C' in the figure is the locus of the probability vectors of the form $\mathbf{u}=[(1-p)^2, 2p(1-p), p^2]$ or $\mathbf{y}=[(1-q)^2, 2q(1-q), q^2]$. As noted at the end of 10.6, this set can be extended to its convex hull C (the shaded area), although the pure strategy $Q'(0,1,0)$, which would be optimal for J, remains unavailable. On either C' or C one has

$$\min_{\mathbf{y}} \max_{\mathbf{u}} (u_1y_1+u_3y_3) = \min_{\mathbf{y}} \max(y_1, y_3) = \tfrac{1}{4}\left(q=\tfrac{1}{2}\right)$$

$$\max_{\mathbf{u}} \min_{\mathbf{y}} (u_1y_1+u_3y_3) = \max_{\mathbf{u}} \min_{q} \left[u_1(1-q)^2+u_3q^2\right]$$

$$= \max_{\mathbf{u}} \frac{u_1u_3}{u_1+u_3} = \tfrac{1}{8} \text{ on } C' \quad \text{ or } \tfrac{1}{4} \text{ on } \quad C$$

In all cases J may as well limit himself to C', and his minimax strategy is the point $Q(1/4, 1/2, 1/4)$. This is also the maximin strategy for I on C', but it yields the value 1/8, not 1/4. The maximin strategy for I on C is $P(1/2, 0, 1/2)$, which yields the value 1/4, the same as the minimax. This equality of the two values on C could be inferred from either of two general theories, represented by the theorems on pp. 186 and 227 of McKinsey (1952) or Theorem 3 of Kakutani (1941), for any payoff matrix $\mathbf{A}$. Note that the information sets alone (one for each player) do not distinguish between C' and C.

This problem illustrates an effect of the violation of the usual requirement that no directed path intersect the same information set in more than one node: the only pure strategies produced by normalization are (1,0,0) and (0,0,1). The first node encountered is assumed to determine the *pure* strategy for its successors in the set, which thus become superfluous. The corresponding mixed strategies are then limited to the straight line segment APB.

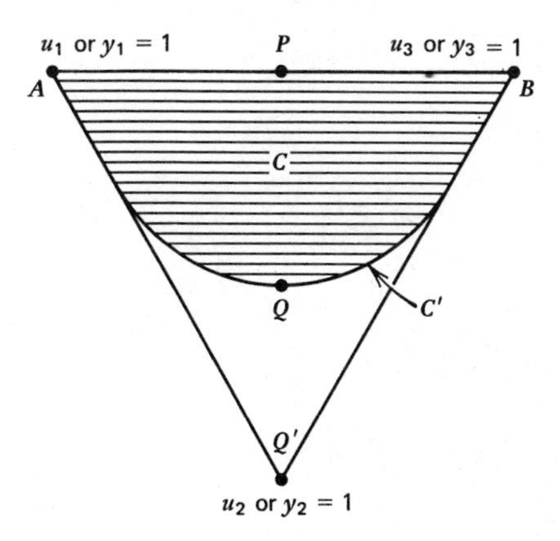

Section 10.7

1. $i\to j$ and $j\to i(2,2)$; Nash $(5,5)=(-3,13)/2+(13,-3)/2$.
2. $i\to j(0,1)$. $j\to i(1,0)$. Equilibrium and Nash $(1,1)/2$.
3. Selections 9,7,5,3 give $(14,10)=(9+5, 7+3)$ in network solution. Nash $(15,11)=(3/4)(17,9)+(1/4)(9,17)$.
4. Selections 9,7,3,5 give $(15,12)=(9+3+3, 7+5)$. Nash $(17,13)=(3/4)(19,11)+(1/4)(11,19)$.
5. Selections 9, 5, 7, 3 give $(21,13)=(9+7+5, 5+3+5)$. Nash $(19.5, 14.5)=(13/16)(21,13)+(3/16)(13,21)$.

6. Selections 6,5,3,0 give $(9,5)=(6+3,5+0)$. Nash $(11.5,7.5)=(7/16)(16,3)+(9/16)(8,11)$.
7. Selections 3,5,0,6 give $(10,11)=(3+0+7,5+6)$. Nash $(12.5,8.5)=(5/16)(18,3)+(11/16)(10,11)$.
8. Selections (3,4), (2,2), (4,1), (1,3) give $(7,5)=(3+4,2+3)$. Nash $(25,26)/4=(1/4)(7,5)+(3/4)(6,7)=(\alpha,\beta)$ from $2\alpha+\beta=19, 2\alpha-\beta=6$.
9. Network solution (3,2.5) is average of (4,2) and $(4,2)/2+(0,4)/2$. Nash $(10/3,3)=(1/6)(0,8)+(5/6)(4,2)$ from $3\alpha+2\beta=16, 3\alpha-2\beta=4$.

Section 11.1

1.

	x_{11}	x_{12}	x_{21}	x_{22}	1	
u_1	1	1	0	0	-3	$=0$
u_2	0	0	1	1	-8	$=0$
v_1	1	0	1	0	-5	$=0$
v_2	0	1	0	1	-6	$=0$
1	2	3	7	4	0	$=z$
	$\geq$	$\geq$	$\geq$	$\geq$	$\parallel$	
	0	0	0	0	$\bar{w}$	$x_{ij}\geq 0$

Dual maximizes $\bar{w}=-(3u_1+8u_2+5v_1+6v_2)$ subject to $u_1+v_1+2\geq 0$, $u_1+v_2+3\geq 0$, $u_2+v_1+7\geq 0$, $u_2+v_2+4\geq 0$. u_i and v_j not sign restricted.

3. The only *feasible* basic solutions are

$$\mathbf{x}=\begin{bmatrix}0 & 3\\ 5 & 3\end{bmatrix} \quad \text{or} \quad \begin{bmatrix}3 & 0\\ 2 & 6\end{bmatrix} \quad \text{with } z=56 \text{ or } 44. \text{ Thus the latter is optimal.}$$

Then $u_1^*+v_1^*+2=u_2^*+v_1^*+7=u_2^*+v_2^*+4=0$. Setting $v_1^*=0$ gives $\mathbf{u}^*=(-2,-7)$, $v^*=(0,3)$, $u_1^*+v_2^*+3=4\geq 0$.

4. $z^*=\bar{w}^*=-(3,2,0)(4,5,6)-(-5,-6)(7,8)=61$.
5. Interchange rows and columns.

$$\mathbf{x}^*=\begin{bmatrix}3 & 2\\ 0 & 6\end{bmatrix}, \quad z^*=44, \quad \mathbf{u}^*=(0,3), \quad \mathbf{v}^*=(-2,-7)$$

6. Just like 3 except z^* is reduced to $44-2a_1-4a_2=6$.

7.

x_1	x_2	6
y_1	y_2	8
9	5	14

$\min(-\bar{z})$

8.

$2x_1$	$3x_2$	8
$4y_1$	y_2	6
5	9	14

$x_3=2$

9.

x_1	x_2	x_3	7
y_1	y_2	y_3	9
5	4	7	16

10. Use the dual. Set $x_1=-u_1$, $y_2=-u_2$, $x_2=-v_1$, $y_1=-v_2$, $\bar{z}=\bar{w}$, $\mathbf{a}=(1,8)$, $\mathbf{b}=(6,3)$, $c_{11}=7$, $c_{22}=9$, $c_{12}=5$, $c_{21}=4$

11. $x_{mj}=b_j-\sum_{i=1}^{m-1} x_{ij} \qquad a_m=\sum_{j=1}^{n} b_j-\sum_{i=1}^{m-1} a_i$

12. **Ca**=(a) **b**

22	44	51	0	35
25	25	30	0	50
28	15	33	9	

(b)

22	44	69	0	35
31	35	30	0	50
28	15	33	9	

16. The proof is unchanged except when every row and column contains exactly two nonzeros. Then the positions of these nonzeros may be regarded as nodes connected by alternately horizontal and vertical arcs. The resulting graph consists of one or more loops, each of which contributes a factor of ± 2 or 0 to the determinant. This is true because the minor determinant for one such loop includes all the nonzero elements in its rows and columns and has only two nonzero terms (1 or -1) in its expansion. For example

$$\begin{vmatrix} 0 & 1 & 1 & 0 & 0 \\ 1 & 0 & 1 & 0 & 0 \\ 1 & 1 & 0 & 0 & 0 \\ 0 & 0 & 0 & 1 & -1 \\ 0 & 0 & 0 & 1 & 1 \end{vmatrix} = \begin{vmatrix} 0 & 1 & 1 \\ 1 & 0 & 1 \\ 1 & 1 & 0 \end{vmatrix} \cdot \begin{vmatrix} 1 & -1 \\ 1 & 1 \end{vmatrix} = 2^2$$

Thus $|\mathbf{A}|=0$ or $\pm 2^q$ in this more general case.

Section 11.2

1. There are eight arcs, with five in the basic tree.
2, 5. Basic tree.
3, 6. Disconnected graph with loop; not basic.
4. Tree but needs some x_{i2}; not basic.
7. A_3 not included; has a loop; not basic.
8. One x_{ij} too many; has a loop; not basic.
9. $x_{14}=a_1, x_{31}=b_1, x_{52}=a_5, x_{42}=a_4, x_{23}=a_2$, $x_{32}=b_2-a_4-a_5, x_{63}=b_3-a_2, x_{64}=a_2+a_6-b_3$, $x_{34}=b_3+b_4-a_1-a_2-a_6$. $v_4=0, u_3=-c_{34}$, $u_6=-c_{64}, v_3=c_{64}-c_{63}, v_2=c_{34}-c_{32}, u_2=c_{63}-c_{64}-c_{23}, u_4=c_{32}-c_{34}-c_{42}, u_5=c_{32}-c_{34}-c_{52}, v_1=c_{34}-c_{31}, u_1=-c_{14}$.
10. (i) Although $x_{11}=x_{22}=1$ is the unique solution when nonbasic $x_{12}=x_{21}=0$, no solution of (7)–(8) has $x_{12}=1, x_{21}=0$. (ii) Nonunique solutions $x_{11}=x_{22}=1$ and $x_{12}=x_{21}=1$ (other $x_{ij}=0$). (iii) One of u_1, u_2, v_1, v_2 is arbitrary and three have to be determined. For this, two equations are not enough and four are too many unless it happens that $c_{11}-c_{12}-c_{21}+c_{22}=0$.

Section 11.3

Before and after pivoting, the tableaus show with $c'_{ij}=0$ replaced by circled basic x_{ij}. u_i and v_j to be adjusted have accents.

	b	
a	**C′**	**u**
	v	

1.

	8	5	5	5	$\Delta x=3$
9	④ M	$-5P$	-3	⑤	-4
7	④ P	0	③ $\overline{M}$	3	0
7	3	⑤ M	② P	6	5′
	-6	$-6'$	$-7'$	-2	$\Delta c=5$

	8	5	5	5	
9	①	③	2	⑤	-4
7	⑦	5	5	3	0
7	-2	②	⑤	1	0
	-6	-1	-2	-2	

2.

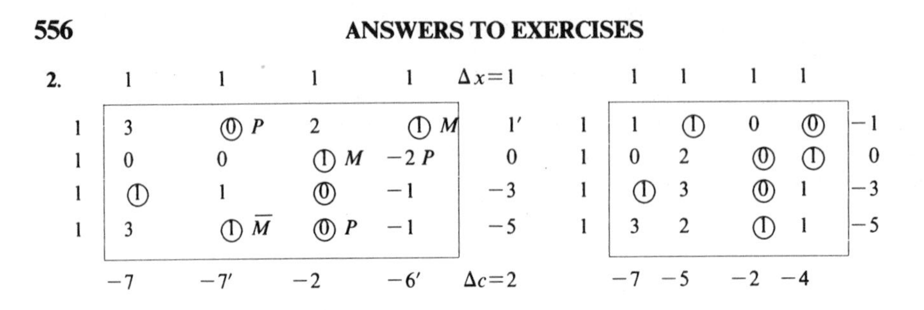

Section 11.4

One may replace u_i, v_j by $u_i + c_0, v_j - c_0$, but still other optimal solutions may exist. Data: $\begin{matrix} \mathbf{X}^* & \mathbf{u}^* \\ \mathbf{v}^* & z^* \end{matrix}$ Dots represent nonbasic $x_{ij} = 0$.

1.

2	·	1	·	−3
·	7	·	·	−2
5	2	·	·	−4
·	·	3	6	−5
2	0	1	3	60

2.

·	3	·	·	−17
·	1	·	6	−11
4	0	6	·	−10
8	0	5	9	112

3.

6	0	·	0	−12
·	·	·	5	− 8
·	·	·	2	− 6
·	5	2	·	− 5
0	5	3	5	93

4.

5	·	3	·	2	−15
·	·	·	·	10	− 9
·	·	3	7	·	− 4
·	6	·	4	·	− 6
9	6	4	0	6	163

5.

·	5	6	·	2	−13
·	·	·	7	5	−10
·	4	·	·	·	− 6
3	·	·	·	2	− 9
1	0	3	2	6	269

6.

·	1	·	·	·	−24
0	0	·	1	·	−20
1	·	·	·	·	−29
0	·	·	·	1	−34
·	·	1	·	0	−15
3	14	3	2	0	100

7.

·	·	·	1	·	·	·	−6
·	·	·	·	1	·	·	−9
1	·	·	·	0	·	·	−5
·	1	·	0	·	0	0	−7
·	·	·	·	·	·	1	−8
·	·	0	·	0	1	·	−7
·	·	1	·	·	·	·	−7
3	0	5	3	3	4	2	29

8.

1	·	·	·	·	·	−2
·	·	·	·	·	1	−4
·	·	·	1	·	·	−5
·	·	1	0	·	·	−5
·	1	·	·	0	0	0
0	·	·	0	1	·	−5
1	0	4	5	0	0	11

9. List 10 separate positions, including 3 unemployed.

10. Find the assignments for which the total fertility is maximized and minimized.

Section 11.5

Nonbasic $x_{ij} = 0$ are dots if $e_{ij} = 0$ and ϕ if $e_{ij} > 0$.

Data: $\begin{matrix} \mathbf{X}^* & \mathbf{u}^* \\ \mathbf{v}^* & z^* \end{matrix}$

1.

x_{11}	x_{12}	x_{13}	x_{14}	·	·	3
x_{21}	·	x_{23}	x_{24}	x_{22}	·	7
x_{31}	x_{32}	x_{33}	x_{34}	·	·	7
x_{41}	x_{42}	·	·	·	x_6	9
·	x_{22}	·	·	y_5	·	4
·	·	x_{43}	x_{44}		y_6	3
7	9	4	6	4	3	33

ϕ	ϕ	3	ϕ	·	·	0
2	·	1	ϕ	4	·	−7
4	ϕ	ϕ	3	·	·	−3
1	5	·	·	·	3	−9
·	4	·	·	ϕ	·	−2
·	·	ϕ	3	·	ϕ	−2
1	2	−2	0	5	9	101

2.

x_{11}	x_{12}	x_{13}	x_{14}	·	·	3
x_{21}	x_{22}	x_{23}	x_{24}	·	·	7
·	·	·	x_{34}	·	x_6	10
·	x_{32}	x_{33}	·	y_5	·	4
x_{31}	·	·	·	x_5	y_6	6
4	4	6	6	4	6	30

ϕ	ϕ	3	ϕ	·	·	−16
ϕ	4	1	2	·	·	−11
·	·	·	4	·	6	−10
·	ϕ	2	·	2	·	− 8
4	·	·	·	2	ϕ	− 8
6	0	3	9	8	10	117

3.

·	4	2	0	−14
4	1	ϕ	·	−11
2	ϕ	ϕ	·	− 9
ϕ	·	·	7	−8
0	7	4	7	121

4.

·	ϕ	·	10	·	−17
ϕ	3	6	1	·	−10
5	3	·	·	2	−12
ϕ	ϕ	ϕ	ϕ	10	− 6
4	6	3	0	1	364

5.

3	·	·	1	9	−21
ϕ	9	3	·	·	−19
ϕ	·	·	4	ϕ	−24
·	ϕ	3	2	·	−18
8	8	0	9	14	402

6.

1	·	0	ϕ	ϕ	−24
·	·	0	·	1	−20
·	1	ϕ	ϕ	·	−28
·	0	1	ϕ	·	−33
0	ϕ	·	1	·	−15
2	12	1	1	0	104

7.

0	·	ϕ	1	·	·	·	−13
1	0	·	·	·	ϕ	·	−17
0	·	ϕ	ϕ	1	·	ϕ	− 3
·	·	·	·	0	1	·	−13
·	1	·	·	·	ϕ	·	−16
·	ϕ	0	·	·	0	1	−13
ϕ	·	1	ϕ	ϕ	·	·	−13
1	4	11	10	1	10	0	51

8.

·	·	·	0	1	·	−9
ϕ	ϕ	·	0	·	1	−9
·	1	·	0	·	·	−9
ϕ	ϕ	·	1	ϕ	·	−9
1	0	0	ϕ	ϕ	ϕ	0
ϕ	·	1	ϕ	·	·	−3
0	0	0	9	6	5	19

10. Let $c_{ij} = -\log|a_{ij}|$. x_{ij} is artificial where $a_{ij} = 0$.

11. Reduce a_3 from 4 to 1, define $a_4 = 3$, require $x_{42} = 0$.

12. **C** **a** = / **b**

c_1	c_1+p	c_1+2p	c_1+3p	0	q_1'
·	c_2	c_2+p	c_2+2p	0	q_2'
·	·	c_3	c_3+p	0	q_3'
·	·	·	c_4	0	q_4'
q_1	q_2	q_3	q_4	$\Sigma(q_i'-q_i)$	

13.

c_{11}	$c_{11}+s$	$c_{11}+2s$	$\min(c_{11}+3s-p,0)$	n_{11}
·	c_{12}	$c_{12}+s$	$\min(c_{12}+2s-p,0)$	n_{12}
·	·	c_{13}	$\min(c_{13}+s-p,0)$	n_{13}
c_{21}	$c_{21}+s$	$c_{21}+2s$	$\min(c_{21}+3s-p,0)$	n_{21}
·	c_{22}	$c_{22}+s$	$\min(c_{22}+2s-p,0)$	n_{22}
·	·	c_{23}	$\min(c_{23}+s-p,0)$	n_{23}
r_1	r_2	r_3	$\Sigma\Sigma n_{ij} - \Sigma r_j$	

14. $c^{\#} = \min_{\mathbf{X}\text{ feasible}} \max_{i,j} \{c_{ij} : x_{ij} > 0\}$

15. Theorem 1 of 11.1 is not valid here; any adjustment of the c_{ij} must leave their order relations ($<$) unchanged. This includes addition of a constant, multiplication by a positive constant, or replacement by integers from 1 to mn representing the relative magnitudes of the c_{ij}. By using loops of $\pm\Delta x$, any feasible **X** can be made basic without creating new $x_{ij} > 0$; thus $c^{\#}$ is not increased by considering only basic solutions. However, basic solutions are less important in this problem.

16. $\max_{ij}\{c_{ij} : x_{ij} > 0\}$ for any feasible set of x_{ij} is an upper bound c^0 for $c^{\#}$. Either of $\max_i \min_j c_{ij}$ and $\max_j \min_i c_{ij}$ is a crude lower bound. A better lower bound (often equal to $c^{\#}$ in small problems) is found as follows. For each $i = 1, \ldots, m$, calculate the partial sums of the b_j (replace by e_{ij} if it is smaller) in the order of increasing value of c_{ij}. Let α_i denote the value of c_{ij} at which the partial sum first equals or exceeds a_i. Define β_j similarly. Clearly $c^{\#} \geq c_0 \stackrel{\Delta}{=} \max\{\alpha_i, \beta_j\}$.

17. Use Phase 1a (and 1b if necessary) to get a low-cost feasible solution as usual. (The x_{ij}^* for the ordinary transportation problem can be used but may be inferior.) If this makes $c^0 = c_0$ in 16, the problem is solved. Otherwise choose a trial value c for $c^{\#}$ satisfying $c_0 \leq c < c^0$. Apply Phase 1b with $x_{ij} = 0$ when $c_{ij} > c^0$ and with $g_{ij} = 1$ when $c_{ij} > c$ and $g_{ij} = 0$ elsewhere. (Other possibilities are $g_{ij} = \max(c_{ij} - c, 0)$ or $g_{ij} = 0, 1, 2, 4, 8, \ldots$ for consecutive c_{ij} increasing from c.) If c is close to c^0, always bring into the basis the $x_{II'}$ having $c_{II'} = \min\{c_{ij}: g'_{ij} < 0\}$. Raise or lower c as required until a feasible solution is obtained such that the x_{pq} satisfying $c_{pq} = \max_{ij}\{c_{ij}: x_{ij} > 0\}$ cannot be reduced to zero. Then $c^{\#} = c_{pq}$. Since the optimal **X** is far from unique, one might be interested in minimizing the total cost of transportation (perhaps with different c_{ij}) subject to $x_{ij} = 0$ when $c_{ij} > c^{\#}$.

18. $\boldsymbol{\alpha} = (9, 6, 2), \boldsymbol{\beta} = (2, 10, 5, 1), c_0 = c^{\#} = 10, c^0 = 17$.

19. $c_0 = 0, c^{\#} = c^0 = 4$.
$x_{ij}^* = x_{ij}^{\#} = 1$ when $i + j = 6$.

Section 11.6

Nonbasic $x_{ij} = 0$ are dots if $e_{ij} = 0$ and ϕ if $e_{ij} > 0$. Nonbasic $x_{ij} = e_{ij} > 0$ are barred.
u_i^*, v_j^* relate to the original c_{ij}^0.

Data:

x_{ij}^*	u_i^*
v_j^*	z^*

1.

·	3	ϕ	ϕ	−3
3	$\bar{4}$	ϕ	·	−8
4	1	ϕ	2	−4
ϕ	1	4	$\bar{4}$	−7
2	0	3	1	84

2.

·	ϕ	1	2	−16
ϕ	4	3	·	−11
4	ϕ	2	$\bar{4}$	− 8
6	0	3	7	121

3.

·	2	ϕ	$\bar{4}$	−16
$\bar{4}$	1	ϕ	·	−13
2	ϕ	ϕ	0	− 9
ϕ	2	2	3	− 9
0	9	7	8	115

4.

·	ϕ	2	$\bar{4}$	$\bar{4}$	−19
ϕ	$\bar{4}$	2	·	$\bar{4}$	−15
4	ϕ	2	$\bar{4}$	ϕ	− 8
1	2	ϕ	3	4	− 7
0	7	8	1	2	261

5.

ϕ	2	ϕ	ϕ	− 6
$\bar{2}$	ϕ	3	ϕ	− 9
1	$\bar{1}$	$\bar{2}$	ϕ	−12
1	3	1	3	− 8
3	5	0	7	103

6.

4	2	3	1	−3
·	$\bar{4}$	$\bar{3}$	1	−6
$\bar{4}$	3	ϕ	·	−5
$\bar{3}$	·	ϕ	2	−5
1	1	2	0	83

7. (a) In terms of the c_{ij}^0 in the Δx loop counterclockwise from $c_{II'}^0$, the reduced cost coefficients of Δx are $9-4+4-5=4$ for $x_{II'} = x_{43}$, $-2+5-4+4=3$ for x_{21}, $-1+5-4+4-5+3=2$ for x_{31}. The latter is preferred as in (1) of 6.3.

Section 11.7

Equations (1) to (6) are used consecutively in 1 to 6 and in 7. The value of $\partial z^*/\partial s$ precedes its interval of validity (for Δs).

1. 6: $0 \leq x_{24} \leq 2 = x_{14}^*$. 9: $0 \leq x_{25} \leq 3 = x_{11}^*$.
7: $0 \leq x_{33} \leq 5 = x_{23}^*$.

2. $-c_{34}^* = -2$: $-x_{14}^* = -2 \leq \Delta x_{14} \leq 0$.
$-c_{21}^* = -5$: $-x_{11}^* = -3 \leq \Delta x_{22} \leq 0$.
$c_{13}^* = 3$: $0 \leq \Delta x_{22} \leq 5 = x_{23}^*$.
$-c_{13}^* = -3$: $-x_{23}^* = -5 \leq \Delta x_{23} \leq 0$.

3. 1: $\delta_{14} \geq -2$. 4: $\delta_{22} \geq -7$. 1: $\delta_{23} \geq -5$.

4. -1: $-x^*_{14} = -2 \leq \delta_{24} \leq 8 = x^*_{12}$. 0: $-x^*_{11} = -3 \leq \delta_{25} \leq 6 = x^*_{31}$.
8: $-x^*_{23} = -5 \leq \delta_{33} \leq 3 = x^*_{11}$.

5. $\Delta z = 0$ for $\Delta c^0_{24} \geq -6$, $\Delta c^0_{25} \geq -9$, $\Delta c^0_{33} \geq -7$.

6. 2: $-\infty < \Delta c^0_{14} \leq 2$. 7: $-3 \leq \Delta c^0_{22} \leq 5$, 5: $-\infty < \Delta c^0_{23} \leq 3$.

7. 7: $0 \leq x_{34} \leq 4 = x^*_{31}$. $-c^*_{43} = -3$: $-x^*_{42} = -6 \leq \Delta x_{13} \leq 0$.
$c^*_{21} = 2$: $0 \leq \Delta x_{13} \leq 3 = x^*_{11}$. 9: $\delta_{13} \geq -8$.
5: $-x^*_{31} = -4 \leq \delta_{34} \leq 3 = x^*_{11}$. 0: $\Delta c^0_{34} \geq -7$.
8: $-2 \leq \Delta c^0_{13} \leq 3$.

8. 8: $\delta_{11} \geq -3$. 9: $0 \leq x_{41} \leq 3 = x^*_{11}$.
7: $-c^*_{14} = -4 \leq \Delta c^0_{23} \leq 2 = c^*_{21}$.
$-c^*_{32} = -5$: $-x^*_{12} = -2 \leq \Delta x_{12} \leq 0$. $c^*_{43} = 3$: $0 \leq \Delta x_{12} \leq 6 = x^*_{42}$.
7: $-x^*_{42} = -6 \leq \delta_{43} \leq 2 = x^*_{12}$.

9. Exchange x_{13} and x_{21} to get optimal solution

a	**b**=7	8	6	5	−**u**
4	②	②	2	6	3
12	①	·	⑥	⑤	−2
4	④	5	·	9	−2
6	9	⑥	5	·	1
	−**v**=5	0	4	5	

10.

	1	2	
1	2	①	0
2	①	①	2
	−1	1	

$z^* = 5$

	2	2	
2	2	②	0
2	②	⓪	2
	−1	1	

$z^* = 4$

$\partial z / \partial \delta_{11} = -u^*_1 - v^*_1 = -1$. Costly x_{22} (with $c_{22} = 3$) is reduced. The absence of slack variables produces this effect.

11. $a_{kk',ll'} = 0$ iff kk' does not occur in the loop formed by ll' in Figure 11.3 iff ll' (and hence its loop) is contained in $M'N'$ or $M''N''$ in (4) of 11.3.
$a_{kk',ll'} = 1$ iff there is an even number of arcs between the arcs kk' and ll' in the loop formed by ll' iff kk' and ll' fall in the same set (called $M'N''$) in (4) of 11.3. This last requires an even number of right-angle turns in going from kk' to ll' in the espalier, and this is the number of intervening arcs. The case $a_{kk',ll'} = -1$ is like the preceding but with "odd" replacing "even"; kk' is in $M'N''$ and ll' in $M''N'$.

Section 11.8

1. Basis 12, 13, 22, 31, 32 for $-\infty < t \leq 0$. $z^* = 60 + 3t$.
Basis 12, 22, 31, 32, 33 for $0 \leq t \leq 2$. $z^* = 60 + t$.
Basis 11, 12, 22, 32, 33 for $2 \leq t < \infty$. $z^* = 68 - 3t$.

2. Basis 12, 22, 23, 31, 33 for $0 \leq t \leq 1$. $z^* = 32 - 14t + 2t^2$.
Basis 11, 12, 23, 31, 33 for $1 \leq t \leq 1.25$. $z^* = 30 - 12t + 2t^2$.
Basis 11, 12, 13, 23, 33 for $1.25 \leq t \leq 1.5$. $z^* = -5 + 16t + 2t^2$.

3. Basis 11, 12, 21, 23, 33 for $0 \leq t \leq 1$. $z^* = 37 - t + t^2$.
Basis 11, 12, 13, 21, 33 for $1 \leq t \leq 2$. $z^* = 37$.
Basis 11, 12, 21, 31, 33 for $-\infty < t \leq 0$. $z^* = 37 - 3t + t^2$.

4. Basis 13, 21, 22, 23, 32 for $1 \leq t \leq 2$. $z^* = 49 - t + 3t^2$.
Basis 11, 21, 22, 23, 32 for $-1 \leq t \leq 1$. $z^* = 47 + 2t + 2t^2$.
Basis 11, 21, 23, 31, 32 for $-1.5 \leq t \leq -1$. $z^* = 46 + t + 2t^2$.
Basis 11, 13, 23, 31, 32 for $-3 \leq t \leq -1.5$. $z^* = 43 + 2t + 4t^2$.

Bibliography

After most items the section or chapter of this book to which it is most relevant is indicated. Broad coverage of mathematical programming (usually linear) is denoted by MP. References relating to each chapter are listed at the end of the chapter (general references after Chapter 1).

Abadie, Jean, ed. (1970). *Integer and Nonlinear Programming*. North-Holland, American Elsevier. MP

Antosiewicz, H. A., ed. (1955). *Proceedings of the Second Symposium in Linear Programming*, vols. 1 and 2. National Bureau of Standards and U.S. Air Force. MP

Aris, Rutherford, G. L. Nemhauser and D. J. Wilde (1964). Optimization of multistage cyclic and branching systems by serial procedures. American Institute of Chemical Engineers Journal 10:913–919. 10.3

Aronofsky, J. S., ed. (1969). *Progress in Operations Research*, vol. 3. John Wiley & Sons. MP

Arrow, Kenneth J. (1951). *Social Choice and Individual Values*. John Wiley & Sons. 5, 8.8

Arrow, K. J., L. Hurwicz and H. Uzawa (1958). *Studies in Linear and Nonlinear Programming*. Stanford University Press. MP

Arrow, K. J. and Samuel Karlin (1958). Price speculation under certainty. Pp. 189–197 of Arrow et al. (1958). 9.5

Aubin, J.-P. (1981). Cooperative fuzzy games. Mathematics of Operations Research 6:1–13. 5.7

Aumann, Robert J. (1961). The core of a cooperative game without side payments. Transactions of the American Mathematical Society 98:539–552. 5

Avis, D. and V. Chvatal (1978). Notes on Bland's pivoting rule. Mathematical Programming Study 8:24–34. (M. L. Balinski and A. J. H. Hoffman, eds.) 6.6

Baker, Kenneth R. (1974). *Introduction to Sequencing and Scheduling*. John Wiley & Sons. 9.4, 10.2

Balinski, Michel L., ed. (1974). *Pivoting and Extensions*. Mathematical Programming Study 1. North-Holland. MP

Balinski, M. L. and R. W. Cottle, eds. (1978). *Complementarity and Fixed Point Problems*. Mathematical Programming Study 7. North-Holland. 5.4

Balinski, M. L. and R. E. Gomory (1963). A mutual primal-dual simplex method. Pp. 17–26 of Graves and Wolfe (1963). 6.4

Balinski, M. L. and Eli Hellerman, eds. (1975). *Computational Practice in Mathematical Programming*. Mathematical Programming Study 4. North-Holland. MP

Balinski, M. L. and C. Lemarechal, eds. (1978). *Mathematical Programming in Use*. Mathematical Programming Study 9. North-Holland. MP

Balinski, M. L. and A. W. Tucker (1969). Duality theory of linear programs. SIAM Review 11:347–377. 3.3, 4.4, 5.6

Banzhaf, John F. III (1965). Weighted voting doesn't work: A mathematical analysis. Rutgers Law Review 19:317–343. 5.8

_____ (1966). Multimember electoral districts—Do they violate the "one man, one vote" principle? Yale Law Journal 75:1309–1338. 5.8

_____ (1968). A mathematical analysis of the electoral college. Villanova Law Review 13:304–332 with comments 13:303, 333–346 and 14:86–96. 5.8

Barr, R. S., F. Glover and D. Klingman (1977). The alternating basis algorithm for assignment problems. Mathematical Programming 13:1–13. 11.4

Bartels, Richard H. (1971). A stabilization of the simplex method. Numerische Mathematik 16:414–434. 7.6

Bartels, R. H. and G. H. Golub (1969). The simplex method of linear programming using LU decomposition. Communications of the Association for Computing Machinery 12:266–268 and 275–278. 7.6

Bartels, R. H., G. H. Golub and M. A. Saunders (1970). Numerical techniques in mathematical programming. Pp. 123–176 of Rosen et al. (1970). 7.6

Bazaraa, M. S. and J. J. Jarvis (1977). *Linear Programming and Network Flows*. John Wiley & Sons. MP

Beale, E. M. L. (1968). *Mathematical Programming in Practice*. Pitman. MP

_____ (1969). Mathematical programming: algorithms. Pp. 135–173 of Aronofsky (1969). MP

_____ (1970). Computational methods for least squares. Pp. 213–227 of Abadie (1970). 2.6, 7.6

_____ (1975). The current algorithmic scope of mathematical programming systems. Mathematical Programming Study 4 (Balinski and Hellerman, 1975), pp. 1–11. 7.6

Beckmann, Martin J. (1968). *Dynamic Programming of Economic Decisions*. Ökonometrie und Unternehmensforschung, vol. 9. Springer-Verlag. 10.2

Bell, D. E., R. L. Keeney and H. Raiffa, ed. (1977). *Conflicting Objectives in Decisions*. John Wiley and Sons. 8.9

Bellman, Richard E. (1954). The theory of dynamic programming. Bulletin of the American Mathematical Society 60:503–516. 10.1

_____ (1957). *Dynamic Programming*. Princeton University Press. 10

_____ (1958). On a routing problem. Quarterly of Applied Mathematics 16:87–90. 9.6

_____ (1961). *Adaptive Control Processes: A Guided Tour*. Princeton University Press. 10

_____, ed. (1963). *Mathematical Optimization Techniques*. University of California Press.

_____ (1968). Stratification and control of large systems with applications to chess and checkers. Information Sciences 1:7–21. 10.5

Bellman, R. E., K. L. Cooke and J. A. Lockett (1970). *Algorithms, Graphs and Computers*. Academic Press. 9.6, 10.4

Bellman, R. E. and S. E. Dreyfus (1962). *Applied Dynamic Programming*. Princeton University Press. 9.5, 10

Bellman, R. E. and Marshall Hall, Jr., eds. (1960). *Combinatorial Analysis*. Proceedings of the 10th Symposium in Applied Mathematics. American Mathematical Society.

Bellman, R. E. and Robert Kalaba (1965). *Dynamic Programming and Modern Control Theory*. Academic Press. 10

Benayoun, R., J. de Montgolfier, J. Tergny and O. Laritchev (1971). Linear programming with multiple objective functions: Step method (STEM). Mathematical Programming 1:366–375. 4.7, 8.8

Ben-Israel, A., A. Charnes, A. P. Hurter and P. D. Robers (1970). On the explicit solution of a special class of linear economic models. Operations Research 18:462–470. 5.5

Bennett, P. G. and M. R. Dando (1979). Complex strategic analysis: A hypergame study of the fall of France. Journal of the Operational Research Society 30:23–32. 5.2

Bennett, P. G., M. R. Dando and R. G. Sharp (1980). Using hypergames to model difficult social issues: An approach to the case of soccer hooliganism. Journal of the Operational Research Society 31:621–635. 5.2

Bertelé, U. and F. Brioschi (1972). *Nonserial Dynamic Programming*. Academic Press. 10.3

Blackwell, David and M. A. Girshick (1954). *Theory of Games and Statistical Decisions*. John Wiley & Sons. 4

Bland, Robert G. (1977). New finite pivoting rules for the simplex method. Mathematics of Operations Research 2:103–107. 6.6

Blin, Jean-Marie (1973). The general concept of multi-dimensional consistency: Some algebraic aspects of the aggregation problem. Pp. 164–178 of Cochrane and Zeleny (1973). 8.9

______ (1977). Fuzzy sets in multiple criteria decision making. Pp. 129–146 of Starr and Zeleny (1977). 8.9

Bodewig, E. (1956, 1959). *Matrix Calculus*. North-Holland. 7

Bohnenblust, H. F., S. Karlin and L. S. Shapley (1950). Solutions of discrete two-person games. Pp. 51–72 in Kuhn and Tucker (1950). 4.5

Bonnardeaux, J., J. P. Dolait and J. S. Dyer (1976). The use of the Nash bargaining model in trajectory selection. Management Science 22:766–777. 5.3

Borel, Emile (1921-27). Three papers on game theory translated by L. J. Savage; comments by Maurice Fréchet. Econometrica 21 (1953) 95–127. 4

Brams, Steven J. (1975). *Game Theory and Politics*. Free Press (Macmillan). 5

______ (1976). *Paradoxes in Politics*. Free Press (Macmillan). 5

______ (1978). *The Presidential Election Game*. Yale University Press. 5

Brown, George W. (1951). Iterative solution of games by fictitious play. Pp. 374–376 of Koopmans (1951). 4.3

Brown, T. A. (1961). Simple paths on convex polyhedra. Pacific Journal of Mathematics 11:1211–1214. 8.8

Buchet, Jacques de (1971). How to take into account the low density of matrices to design a mathematical programming package. Pp. 211–217 of Reid (1971). 7.6

Cahn, A. S., Jr. (1948). The warehouse problem (abstract). Bulletin of the American Mathematical Society 54:1073. 9.5

Charnes, A. and W. W. Cooper (1961). *Management Models and Industrial Applications of Linear Programming*. John Wiley & Sons. MP

Charnes, A., W. W. Cooper and B. Mellon (1952). Blending aviation gasolines. Econometrica 20:135–159. 3.7

Cochrane, J. L. and M. Zeleny, eds. (1973). *Multiple Criteria Decision Making*. University of South Carolina Press. 8.8

Coffman, E. G., Jr., ed. (1976). *Computer and Job-Shop Scheduling Theory*. John Wiley & Sons. 9.4

Conway, R. W., W. L. Maxwell and L. W. Miller (1967). *Theory of Scheduling*. Addison-Wesley. 9.4

Cook, W. D. and L. M. Seiford (1978). Priority ranking and consensus formation. Management Science 24:1721–1732. 5, 8.9

Cooper, Leon and David Steinberg (1974). *Methods and Applications of Linear Programming*. W. B. Saunders. MP

Crout, Prescott D. (1941). A short method for evaluating determinants and solving systems of linear equations with real or complex coefficients. Transactions of the American Institute of Electrical Engineers 60:1235–1240. 7.4

Crowder, H. and J. M. Hattingh (1975). Partially normalized pivot selection in linear programming. Mathematical Programming Study 4 (Balinski and Hellerman, 1975), pp. 12–25. 6.3

Danskin, John M. (1967). *The Theory of Max-Min*. Ökonometrie und Unternehmensforschung, vol. 5. Springer-Verlag. 4

Dantzig, George B. (1949). Programming of interdependent activities II. Econometrica 17:200–211. Reprinted in Koopmans (1951), pp. 19–32. 3

______ (1951a). A proof of the equivalence of the programming problem and the game problem. Pp. 330–335 of Koopmans (1951). 4.4

______ (1951b). Maximation of a linear function of variables subject to linear inequalities. Pp. 339–347 of Koopmans (1951). 3.3, 6.1

______ (1951c). Application of the simplex method to a transportation problem. Pp. 359–373 of Koopmans (1951). 11.4

______ (1955a). Upper bounds, secondary constraints, and block triangularity in linear programming. Econometrica 23:174–183. 6.5

______ (1955b). Optimal solution of a dynamic Leontief model with substitution. Econometrica 23:295–302. 5.5

______ (1956). Constructive proof of the min-max theorem. Pacific Journal of Mathematics 6:25–33. 4.4, 4.5

______ (1957). Discrete variable extremum problems. Operations Research 5:266–277. 9.3, 10.1, 11.4

______ (1960). On the shortest route through a network. Management Science 6:187–190. 9.3

______ (1963). *Linear Programming and Extensions*. Princeton University Press. MP, 4

______ (1963a). Compact basis triangularization for the simplex method. Pp. 125–132 of Graves and Wolfe (1963). 7.6

______ (1966). All the shortest routes in a graph. Technical Report 66-3, Operations Research House, Stanford University. 9.7

______ (1968). Large-scale linear programming. Pp. 77–92 of Dantzig and Veinott (1968), Part 1. Reprinted in pp. 51–72 of Kuhn (1970). 7.3

______ (1969). Sparse matrix techniques in two mathematical programming codes. Technical Report 69-1, Operations Research House, Stanford University. 7.6

Dantzig, G. B., L. R. Ford and D. R. Fulkerson (1956). A primal-dual algorithm for linear programs. Pp. 171–181 of Kuhn and Tucker (1956). 6.3

Dantzig, G. B., S. M. Johnson and W. B. White (1958). A linear programming approach to the chemical equilibrium problem. Management Science 5:38–43. 3

Dantzig, G. B., Alex Orden and Philip Wolfe (1955). The generalized simplex method for minimizing a linear form under linear inequality restraints. Pacific Journal of Mathematics 5:183–195. 6.6

Dantzig, G. B. and A. F. Veinott, Jr., eds. (1968). *Mathematics of the Decision Sciences*, Parts 1 and 2. Lectures in Applied Mathematics, vols. 11 and 12. American Mathematical Society. MP

Debreu, Gerard (1952). A social equilibrium existence theorem. Proceedings of the National Academy of Sciences, U.S.A. 38:886–893. 5.4

Derman, Cyrus (1970). *Finite-State Markovian Decision Processes*. Academic Press. 10.2

Dijkstra, E. W. (1959). A note on two problems in connexion with graphs. Numerische Mathematik 1:269–271. 9.2, 9.3

Dinkelbach, W. (1969). Sensitivitätsanalysen und Parametrische Programmierung. Springer-Verlag. 8

Dixon, Robert G., Jr. (1968). Representation values and reapportionment practice: The eschatology of "one man, one vote." Nomos 10:167–195. 5.8

Doolittle, M. H. (1878). Method employed in the solution of normal equations and the adjustment of a triangulation. U. S. Coast and Geodetic Survey Report, pp. 115–120. 7.4

Dorfman, Robert (1951). Application of the simplex method to a game theory problem. Pp. 348–358 of Koopmans (1951). 4

Dorfman, R., P. A. Samuelson and R. M. Solow (1958). *Linear Programming and Economic Analysis*. McGraw-Hill. 5.6

Dresher, Melvin (1961). *Games of Strategy: Theory and Applications*. Prentice-Hall. 4

Dresher, M., L. S. Shapley and A. W. Tucker, eds. (1964). *Advances in Game Theory*. Annals of Mathematical Studies No. 52. Princeton University Press. 5

Dresher, M., A. W. Tucker and P. Wolfe, eds. (1957). *Contributions to the Theory of Games*, vol. 3. Annals of Mathematics Studies No. 39. Princeton University Press. 10

Dreyfus, Stuart E. (1960). A generalized equipment replacement study. Journal of SIAM 8:425–435. 9.5

______ (1965). *Dynamic Programming and the Calculus of Variations*. Academic Press. 10

______ (1969). An appraisal of some shortest-path algorithms. Operations Research 17:395–412. Reprinted in Geoffrion (1972), pp. 221–238. 9

Dubey, Pradeep (1975). On the uniqueness of the Shapley value. International Journal of Game Theory 4:131–139. 5.7

______ A. Neyman and R. J. Weber (1981). Value theory without efficiency. Mathematics of Operations Research 6:122–128. 5.7

______ and L. S. Shapley (1979). Mathematical properties of the Banzhaf power index. Mathematics of Operations Research 4:99–131. 5.8

Dyer, M. E. and L. G. Proll (1977). An algorithm for determining all extreme points of a convex polytope. Mathematical Programming 12:81–96. 8.8

Eaves, B. Curtis (1971). The linear complementarity problem. Management Science 17:612–634. 5.4

Ecker, J. G. and Nancy S. Hegner (1978). On computing an initial efficient extreme point. Journal of the Operational Research Society 29:1005–1007. 8.8

Ecker, J. G., N. S. Hegner and I. A. Kouada (1980). Generating all maximal efficient faces for multiple objective linear programs. Journal of Optimization Theory and Applications 30:353–381. 8.8

Ecker, J. G. and I. A. Kouada (1975). Finding efficient points for linear multiple objective programs. Mathematical Programming 8:375–377. 8.8

______ (1978). Finding all efficient extreme points for multiple objective linear programs. Ibid. 14:249–261. 8.8

Edmonds, Jack (1967). Optimum branchings. Journal of Research of the National Bureau of Standards, Section B. 71:233–240. Reprinted in Dantzig and Veinott (1968), Part 1, pp. 346–361. 9.2

Evans, J. P. and R. E. Steurer (1973a), A revised simplex method for linear multiple objective programs. Mathematical Programming 5:54–72. 8.8

______ (1973b), Generating extreme points in linear multiple objective programming: Two algorithms and computing experience. Pp. 349–365 of Cochrane and Zeleny (1973). 8.8

Fabian, Tibor (1958). A linear programming model of integrated iron and steel production. Management Science 4:415–449. 3.7

Farkas, J. (1902). Über die Theorie der einfachen Ungleichungen. Journal der Reine und Angewandte Mathematik 124:1–24. 6.2 (ex. 8)

Farquhar, Peter H. (1977). A survey of multiattribute utility theory and applications. Pp. 59–89 of Starr and Zeleny (1977). 8.9

Fieldhouse, Martin (1980). Khachian's algorithm: Fact and fantasy. SIGMAP Bulletin No. 29 of the Association for Computing Machinery, pp. 92–96. 7

Floyd, Robert W. (1962). Algorithm 97: Shortest path. Communications of the Association for Computing Machinery 5:345. 9.7

Ford, L. R., Jr. (1956). Network flow theory. RAND Corp. P-923. 9.6

Ford, L. R., Jr. and D. R. Fulkerson (1962). *Flows in Networks*. Princeton University Press. 9.5, 9.8

Forrest, J. J. H. and J. A. Tomlin (1972). Updated triangular factors of the basis to maintain sparsity in the product form simplex method. Mathematical Programming 2:263–278, 7.6

Forsythe, G. E. and C. B. Moler (1967). *Computer Solution of Linear Algebraic Systems*. Prentice-Hall. 7.5

Frank, A. Q. and L. S. Shapley (1981). The distribution of power in the U.S. Supreme Court. Rand Corporation report N-1735-NSF. 5.8

Fulkerson, D. R. (1961). A network flow computation for project cost curves. Management Science 7:167–178. 9.4

______ (1966). Flow networks and combinatorial operations research. American Mathematical Monthly 73:115–138. Reprinted in Geoffrion (1972). 9

______ and P. Wolfe (1962). An algorithm for scaling matrices. SIAM Review 4:142–146. 7.5

Gács, P. and L. Lovász (1979). Khachian's algorithm for linear programming. STAN-CS-79-750, Department of Computer Science, Stanford University. Also in Mathematical Programming Study 14 (H. König et al., 1981). 7

Gale, David (1956). The closed linear model of production. Pp. 285–303 of Kuhn and Tucker (1956). 5.6

______ (1960). *The Theory of Linear Economic Models*. McGraw-Hill. MP, 4, 5.6

Gale, D., H. W. Kuhn and A. W. Tucker (1950a). On symmetric games. Pp. 81–87 of Kuhn and Tucker (1950). 4.6

______ (1950b). Reductions of game matrices. Ibid. pp. 89–96. 4

______ (1951). Linear programming and the theory of games. Pp. 317–329 of Koopmans (1951). 4

Gale, D. and S. Sherman (1950). Solutions of finite two-person games. Pp. 37–49 of Kuhn and Tucker (1950). 4

Garvin, Walter W. (1960). *Introduction to Linear Programming*. McGraw-Hill. MP

Garvin, W. W., H. W. Crandall, J. B. John and R. A. Spellman (1957). Applications of linear programming in the oil industry. Management Science 3:407–430. 3.7

Gass, Saul I. (1958, 1969). *Linear Programming Methods and Applications*. McGraw-Hill. MP

Gentleman, W. M. and S. C. Johnson (1974). The evaluation of determinants by expansion by minors. Mathematics of Computation 28:543–548. 7.5

Geoffrion, Arthur M. (1968). Proper efficiency and the theory of vector maximization. Journal of Mathematical Analysis and Applications 22:618–630. 8.8

______, ed. (1972). *Perspectives on Optimization*. Addison-Wesley. MP

Gewirtz, A., H. Sitomer and A. W. Tucker (1974). *Constructive Linear Algebra*. Prentice-Hall.

Gill, P. E., G. H. Golub, W. Murray and M. A. Saunders (1974). Methods for modifying matrix factorizations. Mathematics of Computation 28:505–535. 7.6

Gill, P. E. and W. Murray (1970). A numerically stable form of the simplex algorithm. National Physical Laboratory (U.K.) Maths 87. 7.6

______, eds. (1974). *Numerical Methods for Constrained Optimization*. Academic Press. 7.6

Gillies, D. B., J. P. Mayberry and J. von Neumann (1953). Two variants of poker. Pp. 13–50 of Kuhn and Tucker (1953). 10.6

Glover, F., D. Karney and D. Klingman (1972). The augmented predecessor index method for locating stepping-stone paths and assigning dual prices in distribution problems. Transportation Science 6:171–181. 11

Glover, F., D. Karney, D. Klingman and A. Napier (1974). A computation study on start procedures, basis change criteria, and solution algorithms for transportation problems. Management Science 20:793–813. 11

Gluss, Brian (1972). *An Elementary Introduction to Dynamic Programming: A State Equation Approach*. Allyn and Bacon. 10.3

Goldfarb, D. and J. K. Reid (1977). A practicable steepest-edge simplex algorithm. Mathematical Programming 12:361–371. 6.3, 7.3.

Goldman, A. J. and A. W. Tucker (1956). Theory of linear programming. Pp. 53–97 of Kuhn and Tucker (1956). 3

Golub, G. H. and M. A. Saunders (1970). Linear least squares and quadratic programming. Pp. 229–256 of Abadie (1970). 7.6

Gower, J. C. and G. J. S. Ross (1969). Minimum spanning trees and single linkage cluster analysis. Applied Statistics 18:54–64. 9.2

Graves, Robert L. (1963). Parametric linear programming. Pp. 201–210 of Graves and Wolfe (1963). 8.7

Graves, R. L. and Philip Wolfe, eds. (1963). *Recent Advances in Mathematical Programming*. McGraw-Hill. MP

Gray, D. F. and W. R. S. Sutherland (1980). Inverse programming and the linear vector maximization problem. Journal of Optimization Theory and Applications 30:523–534. 8.8

Gröbner, Wolfgang (1956). *Matrizenrechnung*. R. Oldenbourg, Munich. 7

Hadley, G. (1962). *Linear Programming*. Addison-Wesley. MP

______ (1964). *Nonlinear and Dynamic Programming*. Addison-Wesley. 10.2

Harary, Frank (1969). *Graph Theory*. Addison-Wesley. 9.1

Harris, Paula M. J. (1973). Pivot selection methods of the Devex linear programming code. Mathematical Programming 5:1–28 or Mathematical Programming Study 4:30–57 (Balinski and Hellerman). 6.3

Harsanyi, John C. (1963). A simplified bargaining model for the n-person cooperative game. International Economic Review 4:194–220. 5.3, 5.7

______ (1964). A general solution for finite noncooperative games, based on risk-dominance. Pp. 651–679 of Dresher, Shapley and Tucker (1964). 5

______ (1977). *Rational Behavior and Bargaining Equilibrium in Games and Social Situations*. Cambridge University Press. 5, 10.5

Hellerman, Eli and Dennis Rarick (1971). Reinversion with the preassigned pivot procedure. Mathematical Programming 1:195–216. 7.6

______ (1972). The partitioned preassigned pivot procedure (P4). Pp. 67–76 of Rose and Willoughby (1972). 7.6

Hillier, F. S. and G. J. Lieberman (1967). *Introduction to Operations Research*. Holden-Day. MP, 10

Hitchcock, Frank L. (1941). The distribution of a product from several sources to numerous localities. Journal of Mathematics and Physics 20:224–230. 11.1

Householder, Alton S. (1958). Unitary triangularization of a non-symmetric matrix. Journal of the Association for Computing Machinery 5: 339–342. 7.6

______ (1964). *The Theory of Matrices in Numerical Analysis*. Blaisdell Division of Ginn. 7

Howard, Nigel (1966). The mathematics of metagames. General Systems 11:187–200. 5.2

______ (1971). *Paradoxes of Rationality: Theory of Metagames and Political Behavior*. M.I.T. Press. 5.2

_____ (1974). "General" metagames: An extension of the metagame concept. Pp. 261–283 in Rapoport (1974). 5.2

Howard, Ronald A. (1960). *Dynamic Programming and Markov Processes*. M.I.T. Press. 10.2

_____ (1971). *Dynamic Probabilistic Systems*. Vol. 1: Markov Models. Vol. 2: Semi-Markov and Decision Processes. John Wiley & Sons. 10.2

Hurwicz, Leonid (1951). Optimality criteria for decision making under ignorance. Cowles Commission Discussion Paper, Statistics, No. 370. (Also see Luce and Raiffa, 1957, p. 282.) 4.7

Isermann, Heinz (1977). The relevance of duality in multiple objective linear programming. Pp. 241–262 of Starr and Zeleny (1977). 8.8

_____ (1978). Duality in multiple objective linear programming. Pp. 274–285 of Zionts (1978). 8.8

Jackson, J. R. (1956). An extension of Johnson's results on job lot scheduling. Naval Research Logistics Quarterly 3:201–203. 9.4

Johnson, S. M. (1954). Optimal two and three-stage production schedules with set-up times included. Ibid. 1:61–68. 9.4

_____ (1959). Discussion: Sequencing *n* jobs on two machines with arbitrary time lags. Management Science 5:299–303. 9.4

Kakutani, Shizuo (1941). A generalization of Brouwer's fixed point theorem. Duke Mathematical Journal 8:457–459. 4.5, 5, 10.6 (ex. 9).

Kantorovich, L. V. (1939). Mathematical methods in the organization and planning of production. Leningrad State University. Translated in Management Science 6 (1960) 366–422. 11

Karlin, Samuel (1959). *Mathematical Methods and Theory in Games, Programming and Economics*. 2 vols. Addison-Wesley. MP, Games.

_____ (1966). *A First Course in Stochastic Processes*. Academic Press. 5.6, 10.2

Karp, R. M. (1972). A simple derivation of Edmond's algorithm for optimum branchings. Networks 1:265–272. 9.2

Keeney, R. L. and Howard Raiffa (1976). *Decisions with Multiple Objectives: Preferences and Value Tradeoffs*. John Wiley & Sons. 4.7, 8.9, 10.4

Kelley, James E., Jr. (1961). Critical path planning and scheduling: mathematical basis. Operations Research 9:296–320. 9.4

Kemeny, J. G., O. Morgenstern and G. L. Thompson (1956). A generalization of the Von Neumann model of an expanding economy. Econometrica 24:115–135. 5.6

Kemeny, J. G., J. L. Snell and G. L. Thompson (1957). *Introduction to Finite Mathematics*. Prentice-Hall. 5.6

Kershenbaum, A. and R. Van Slyke (1972). Computing minimum spanning trees efficiently. Proceedings of the 1972 Conference at Boston, Association for Computing Machinery. 9.2

Khachian, L. G. (1979). (Algorithm for linear programming). Akademii Nauk SSSR Doklady 224 (no. 5) 1093–1096. See Gács and Lovász (1979), Wolfe (1980), Fieldhouse (1980). 7

Klee, Victor (1968). A class of linear programming problems requiring a large number of iterations. Pp. 65–76 of Dantzig and Veninott (1968), Part 1. 3.4, 6

_____ (1980). Combinatorial optimization: What is the state of the art? Mathematics of Operations Research 5:1–26. 9

Klee, V. and G. J. Minty (1972). How good is the simplex algorithm? Pp. 159–176 of Shisha (1972). 3.4, 6

Klyuyev, V. V. and N. I. Kokovkin-Shcherbak (1965). On the minimization of the number of arithmetic operations for the solution of linear algebraic systems of equations. Technical Report CS24, Computer Science Department, Stanford University. 7.5

Knuth, Donald E. (1968-73). *The Art of Computer Programming*. 3 vols. Addison-Wesley.

Koehler, G. J., A. B. Whinston and G. P. Wright (1975). The solution of Leontief substitution systems using matrix iterative techniques. Management Science 21:1295–1302. 5.5

Koopmans, Tjalling C. (1949). Optimum utilization of the transportation system. Econometrica 17:136–145. 11.1

______, ed. (1951). *Activity Analysis of Production and Allocation*. Cowles Commission Monograph No. 13. John Wiley and Sons. MP, 4

Kristol, Irving (1980). The battle for Reagan's soul. Wall Street Journal, May 16, 1980, editorial page. 5.1(d)

Kruskal, Joseph B., Jr. (1956). On the shortest spanning subtree of a graph and the traveling salesman problem. Proceedings of the American Mathematical Society 7:48–50. 9.2

Kuhn, Harold W. (1950). A simplified two-person poker. Pp. 97–103 of Kuhn and Tucker (1950). 10.6

______ (1953). Extensive games and the problem of information. Pp. 193–216 of Kuhn and Tucker (1953). 10.5

______, ed. (1970). Proceedings of the Princeton Symposium on Mathematical Programming, 1967. Princeton University Press. MP

Kuhn, H. W. and R. E. Quandt (1963). An experimental study of the simplex method. Pp. 107–124 of the Proceedings of the 15th Symposium in Applied Mathematics of the American Mathematical Society: Experimental Arithmetic, High-Speed Computing and Mathematics. 6.3

Kuhn, H. W. and A. W. Tucker, eds. (1950, 1953). *Contributions to the Theory of Games*, vols. 1 and 2. Annals of Mathematics Studies Nos. 24 and 28. Princeton University Press. (Three other volumes in this series are Tucker and Luce, 1959; Dresher et al., 1957, 1964). 4, 5, 10.

______, eds. (1956). *Linear Inequalities and Related Systems*. Annals of Mathematics Study No. 38. Princeton University Press. MP, 5.6

Künzi, H. P., H. G. Tzschach and C. A. Zehnder (1968). *Numerical Methods of Mathematical Optimization*. Academic Press. 6.3

Lawler, Eugene L. (1976). *Combinatorial Optimization: Networks and Matroids*. Holt, Rinehart and Winston. 9

Lemaire, J. (1973). A new value for games without transferrable utilities. International Journal of Game Theory 2:205–213. 5.3

Lemke, Carlton E. (1954). The dual method of solving the linear programming problem. Naval Research Logistics Quarterly 1:36–47. 3.3

______ (1965). Bimatrix equilibrium points and mathematical programming. Management Science 11:681–689. 5.4

______ (1968). On complementary pivot theory. Pp. 95–114 of Dantzig and Veinott (1968), Part 1. 5.4

Lemke, C. E. and J. T. Howson (1964). Equilibrium points of bimatrix games. Journal of SIAM 12:413–423. 5.4

Leontief, Wassily W. (1936). Quantitative input and output relations in the economic system of the United States. Review of Economic Statistics 18:105–125. 5.5

______ (1966). *Input-Output Economics*. Oxford University Press, New York. 5.5

Lewis, H. R. and C. H. Papadimitriou (1978). The efficiency of algorithms. Scientific American, January 1978, pp. 96–109. 9.2

Lin, T. D. and R. S. H. Mah (1977). Hierarchial partition—A new optimal pivoting algorithm. Mathematical Programming 12:260–278. 7.6

Lucas, William F. (1974). Measuring power in weighted voting systems. Technical Report No. 227, Department of Operations Research, Cornell University. 5.8

Luce, R. Duncan and Howard Raiffa (1957). *Games and Decisions*. John Wiley and Sons. 4, 5, 10.5

Luenberger, David G. (1969). *Optimization by Vector Space Methods*. John Wiley and Sons. MP

Machol, R. E. and M. Wien (1977). A hard assignment problem. Operations Research 25:364. 11.4

Manas, M. and J. Nedoma (1968). Finding all vertices of a convex polyhedron. Numerische Mathematik 12:226–229. 8.8

Manne, Alan S. (1963a). *Scheduling of Petroleum Refining Operations*. Harvard Economic Studies, vol. 98. Harvard University Press. 9.4

_____ (1963b), On the job-shop scheduling problem. Pp. 187–192 of Muth and Thompson (1963). 9.4

Markowitz, Harry M. (1957). The elimination form of the inverse and its application to linear programming. Management Science 3:255–269. 7.4

_____ (1959). *Portfolio Selection: Efficient Diversification of Investments*. John Wiley and Sons. 5.1

Mathematical Programming Studies. See M. L. Balinsky, with or without coeditor.

Mattheiss, T. H. (1973). An algorithm for determining irrelevant constraints and all vertices in systems of linear inequalities. Operations Research 21:247–260. 8.8

_____ and David S. Rubin (1980). A survey and comparison of methods for finding all vertices of convex polyhedral sets. Mathematics of Operations Research 5:167–185. 8.8

_____ and B. K. Schmidt (1980). Computational results on an algorithm for finding all vertices of a polytope. Mathematical Programming 18:308–329. 8.8

McDonald, John (1975). *The Game of Business*. Doubleday. 5

McKinsey, J. C. C. (1952). *Introduction to the Theory of Games*. McGraw-Hill. 4, 5, 10

Mensch, A., ed. (1964). *Theory of Games*. NATO Conference at Toulon. American Elsevier. 5

Mills, Harlan D. (1956). Marginal values of matrix games and linear programs. Pp. 183–193 of Kuhn and Tucker (1956). 5.3, 5.6

Mitten, L. G. (1959). Sequencing *n* jobs on two machines with arbitrary time lags. Management Science 5:293–298. 9.4

_____ (1964). Composition principles for synthesis of optimal multistage processes. Operations Research 12:610–619. 10.3

Moore, E. F. (1959). The shortest path through a maze. Annals of the Computation Laboratory of Harvard University 30:285–292. Harvard University Press. 9.6

Motzkin, T. S. (1955). The probability of solvability of linear inequalities. Pp. 607–611 of vol. 2 of Antosiewicz (1955). 6

Motzkin, T. S., H. Raiffa, G. L. Thompson and R. M. Thrall (1953). The double description method. Pp. 51–73 of Kuhn and Tucker (1953). 4, 6, 8.8

Müller-Merbach, Heiner (1970). *Optimale Reihenfolgen*. Ökonometrie und Unternehmensforschung, vol. 15. Springer-Verlag. 9, 10

Murchland, J. D. (1965). A new method for finding all elementary paths in a complete directed graph. London School of Economics Report LSE-TNT-22. 9.7

Murty, Katta G. (1976). *Linear and Combinatorial Programming*. John Wiley & Sons. MP

Muth, J. F. and G. L. Thompson, eds. (1963). *Industrial Scheduling*. Prentice-Hall. 9.4

Myerson, Roger B. (1977a). Graphs and cooperation in games. Mathematics of Operations Research 2:225–229. 5.7

_____ (1977b). Values of games in partition function form. International Journal of Game Theory 6:23–31. 5.7

_____ (1978). Threat equilibria and fair settlements in cooperative games. Mathematics of Operations Research 3:265–274. 5.7

Nash, John F. (1951). Noncooperative games. Annals of Mathematics 54:286–295. 5.1–5.4

______ (1953). Two-person cooperative games. Econometrica 21:128–140. 5

Nemhauser, George L. (1964). Decomposition of linear programs by dynamic programming. Naval Research Logistics Quarterly 11:191–196. 10.3

______ (1966). *Introduction to Dynamic Programming*. John Wiley & Sons. 10.3

Nering, Evar D. (1963). *Linear Algebra and Matrix Theory*. John Wiley & Sons. 2

O'Brien, James J. (1969). *Scheduling Handbook*. McGraw-Hill. 9.4

Orchard-Hays, William (1968). *Advanced Linear Programming Computing Techniques*. McGraw-Hill. MP, 8

Owen, Guillermo (1968). *Game Theory*. W. B. Saunders. 4, 5, 10

______ (1971). Optimal threat strategies of bimatrix games. International Journal of Game Theory 1:3–9. 5.3

______ (1971a). Political games. Naval Research Logistics Quarterly 18:345–355. 5.7, 5.8

______ (1972). Values of games without side payments. International Journal of Game Theory 1:95–109. 5.3

Parthasarathy, T. and T. E. S. Raghavan (1971). *Some Topics in Two-Person Games*. American Elsevier. 4, 5

Peters, R. J. and C. L. J. Van der Meer (1973). Parametric linear programming. Institute for Economic Research, State University of Groningen. 8.6

Pollack, Maurice (1960). The maximum capacity through a network. Operations Research 8:733–736. 9.8

Pollack, M. and W. Wiebenson (1960). Solutions of the shortest route problem—A review. Operations Research 8:224–230. 9

Ponssard, J. P. (1975). A note on the linear programming formulation of zero-sum sequential games with incomplete information. International Journal of Game Theory 4:1–5. 10.5

______ (1975a). Zero-sum games with "almost" perfect information. Management Science 21:794–805. 10.5

Prim, R. C. (1957). Shortest connection networks and some generalizations. Bell System Technical Journal 36:1389–1401. 9.2, 9.3

Raiffa, Howard (1968). *Decision Analysis*. Addison-Wesley. 10.4

Raiffa, H. and Robert Schlaifer (1961). *Applied Statistical Decision Theory*. Studies in Managerial Economics, Graduate School of Business Administration, Harvard University. 10.4

Ralston, Anthony (1965). *A First Course in Numerical Analysis*. McGraw-Hill. 7.4

Rapoport, Anatol (1966). *Two-Person Game Theory: The Essential Ideas*. University of Michigan Press. 4, 5

______ (1967). Escape from paradox. Scientific American, July, 1967, Pp. 50–56. 5.2

______ (1970). *N-Person Game Theory: Concepts and Applications*. University of Michigan Press. 5

______ ed. (1974). *Game Theory as a Theory of Conflict Resolution*. D. Reidel (Dordrecht and Boston). 5

Rapoport, A. and A. M. Chammah (1966). The game of chicken. American Behavioral Scientist 10 (Nov. 1966) 10–28. 5.1

______ (1965). *Prisoner's Dilemma*. University of Michigan Press. 5

Rapoport, A. and M. Guyer (1966). A taxonomy of 2×2 games. General Systems 11:203–214. 5.2

Rapoport, A., M. Guyer and D. G. Gordon (1976). *The 2×2 Game*. University of Michigan Press. 5.1, 5.2

Reid, J. K., ed. (1971). *Large Sparse Sets of Linear Equations*. (Oxford Conference of 1970). Academic Press. 7.6

Reinfeld, N. V. and W. R. Vogel (1958). *Mathematical Programming*. Prentice-Hall. 11.4

Rial, R., F. Glover, D. Karney and D. Klingman (1979). A computational analysis of alternative algorithms and labeling techniques for finding shortest path trees. Networks 9:215–248. 9.3

Rice, John R. (1966). Experiments on Gram-Schmidt orthogonalization. Mathematics of Computation 20:325–328. 7.6

Riker, W. H. and L. S. Shapley (1968). Weighted voting: A mathematical analysis for instrumental judgements. Nomos 10:199–216. 5.8

Robinson, Julia (1951). An iterative method of solving a game. Annals of Mathematics 54:296–301. 4.3

Rose, D. J. and R. A. Willoughby, eds. (1972). *Sparse Matrices and Their Applications*. IBM Research Symposia Series. Plenum Press. 7.6

Rosenthal, R. W. (1976). An arbitration model for normal-form games. Mathematics of Operations Research 1:82–88. 5.7(b)

Roy, Bernard (1971). Problems and methods with multiple objective functions. Mathematical Programming 1:239–266. 4.7, 8.9

Sasieni, M., A. Yaspan and L. Friedman (1959). *Operations Research—Methods and Problems*. John Wiley & Sons. MP

Saunders, Michael A. (1975). A fast stable implementation of the simplex method using Bartels-Golub updating. Technical Report SOL 75-25, Department of Operations Research, Stanford University. 7.6

Savage, L. J. (1951). The theory of statistical decision. Journal of the American Statistical Association 46:55–67. 4.7

Schell, Emil D. (1955). Distribution of a product by several properties. Pp. 615–642 of Antosiewicz (1955), vol. 2. 11

Schlaifer, Robert (1969). *Analysis of Decisions Under Uncertainty*. McGraw-Hill. 4.7, 10.4

Schumann, Jochen (1968). *Input-Output Analyse*. Ökonometrie und Unternehmensforschung, vol. 10. Springer-Verlag. 5.5

Seneta, E. (1973). *Nonnegative Matrices: Introduction to Theory and Applications*. John Wiley & Sons. 5.5, 5.6

Sengupta, S. S., M. L. Podrebarac and T. D. H. Fernando (1973). Probabilities of optima in multiobjective linear programs. Pp. 217–235 of Cochrane and Zeleny (1973). 8.8

Shapley, Lloyd S. (1953). A value for *n*-person games. Pp. 307–317 of Kuhn and Tucker (1953). 5.7

______ (1964). Some topics in two-person games. Pp. 1–28 of Dresher, Shapley and Tucker (1964). 4, 5

______ (1967). Utility comparison and the theory of games. RAND Corp. P-3582 or Colloques Internationaux du Centre National de la Recherche Scientifique, No. 171, Paris, 1969. 5.3

______ (1974). A note on the Lemke-Howson algorithm. Mathematical Programming Study 1:175–189. (Balinski, 1974). 5.4

Shapley, L. S. and M. Shubik (1954). A method for evaluating the distribution of power in a committee system. American Political Science Review 48:787–792. 5.8

Shapley, L. S. and R. N. Snow (1950). Basic solutions of discrete games. Pp. 27–35 of Kuhn and Tucker (1950). 4.5

Shisha, Oved, ed. (1967, 1970, 1972). *Inequalities*. 3 vols. Academic Press.

Shubik, Martin (1959). *Strategy and Market Structure: Competition, Oligopoly and the Theory of Games*. John Wiley & Sons. 5

______, ed. (1964). *Game Theory and Related Approaches to Social Behavior: Selections*. John Wiley & Sons. 5

______ (1975a). *Games for Society, Business and War*. American Elsevier. 5

______ (1975b). *Methods for Gaming*. American Elsevier. 5

______ (1980). *Market Structure and Behavior* (with Richard Levitan). Harvard University Press. 5

Simonnard, Michel (1966). *Linear Programming*. Translated by W. S. Jewell. Prentice-Hall. MP

Singleton, R. R. and W. F. Tyndall (1974). *Games and Programs*. W. H. Freeman. MP

Smith, D. M. and W. Orchard-Hays (1963). Computational efficiency in product form linear programming codes. Pp. 211–218 of Graves and Wolfe (1963). 7.3

Spivey, W. A. and R. M. Thrall (1970). *Linear Optimization*. Holt, Rinehart and Winston. MP

Srinivasan, V. and G. L. Thompson (1972). Accelerated algorithms for labeling and relabeling of trees, with applications to distribution problems. Journal of the Association for Computing Machinery 19:712–726. 11

______ (1973). Benefit-cost analysis of coding techniques for the primal transportation algorithm. Ibid. 20:194–213. 11

Stanford, Robert E. (1976). Analytical solution of a dynamic transaction flow problem. Mathematical Programming 10:214–229. 5.5, 10.2

Starr, M. K. and M. Zeleny, eds. (1977). *Multiple Criteria Decision Making*. Vol. 6, TIMS Studies in the Management Sciences. North-Holland. 4.7, 8.8

Steuer, Ralph E. (1977). An interactive multiple objective linear programming procedure. Pp. 225–239 of Starr and Zeleny (1977). 8.8

Stewart, T. J. (1981). A descriptive approach to multiple-criteria decision making. Journal of the Operational Research Society 32:45–53. 8.9

Stiefel, Eduard L. (1963). *An Introduction to Numerical Mathematics*. Translated by W. C. and C. J. Rheinboldt. Academic Press. 2, 3

Stoer, J. and C. Witzgall (1970). *Convexity and Optimization in Finite Dimensions*. Springer-Verlag. 8.8

Sylvester, J. J. (1851). On the relation between the minor determinants of linearly equivalent quadratic functions. Philosophical Magazine, Series 4, 1:295–305; 415. 7.8

Szwarc, W. (1971). The transportation paradox. Naval Research Logistics Quarterly 18:185–202. 11.7

Thompson, G. L. (1956). On the solution of a game-theoretic problem. Pp. 275–284 of Kuhn and Tucker (1956). 5.6

Thrall, R. M., C. H. Coombs and R. L. Davis, eds. (1954). *Decision Processes*. John Wiley & Sons. 4.7, 10.4

Todd, Michael J. (1974). A generalized complementary pivoting algorithm. Mathematical Programming 6:243–263. 5.4

______ (1976). Comments on a note by Aggarwal. Ibid. 10:130–133. 5.4

______ (1978). Bimatrix games—an addendum. Ibid. 14:112–115. 5.4

Tomlin, J. A. (1972). Modifying triangular factors of the basis in the simplex method. Pp. 77–85 of Rose and Willoughby (1972). 7.6

______ (1975a). An accuracy test for updating triangular factors. Mathematical Programming Study 4:142–145 (Balinski and Hellerman, 1975). 7.6

______ (1975b). On scaling linear programming problems. Ibid. 4:146–166. 7.6

Tucker, Albert W. (1956). Dual systems of homogeneous linear relations. Pp. 3–18 of Kuhn and Tucker (1956). 3, 5.6

______ (1960a). A combinatorial equivalence of matrices. Pp. 129–140 of Bellman and Hall (1960). 7.8

______ (1960b). Solving a matrix game by linear programming. IBM Journal of Research and Development 4:507–517. 4.4

_____ (1963a). Combinatorial theory underlying linear programs. Pp. 1–6 of Graves and Wolfe (1963). 2, 3.3

_____ (1963b). Simplex method and theory. Pp. 213–231 of Bellman (1963). 5.6

_____ (1968). Complementary slackness in dual linear subspaces. Pp. 137–143 of Dantzig and Veinott (1968), Part 1. 5.6

_____ See also Kuhn and Tucker (1950–56); Dresher et al. (1957, 1964); Balinski and Tucker (1969); Gewirtz et al. (1974).

Tucker, A. W. and R. D. Luce, eds. (1959). *Contributions to the Theory of Games*, vol. 4. Annals of Mathematics Study No. 40, Princeton University Press. (For other volumes see Kuhn and Tucker, 1950, 1953; Dresher et al. 1957, 1964.) 5

Vajda, Stephen (1958). *Readings in Linear Programming*. John Wiley & Sons. MP

_____ (1961). *Mathematical Programming*. Addison-Wesley, MP

_____ (1972). *Probabilistic Programming*. Academic Press. MP

Van de Panne, Cornelius (1975). *Methods for Linear and Quadratic Programming*. North-Holland, American Elsevier. MP

_____ (1975a). A node method for multiparametric linear programming. Management Science 21:1014–1020. 8.6

Von Neumann, John (1928). Zur Theorie der Gesellschaftsspiele. Mathematische Annalen 100:295–320. 4.5

_____ (1937), translated as "A model of general economic equilibrium." Review of Economic Studies 13 (1945) 1–9. 5.6

_____ (1954). A numerical method to determine optimum strategy. Naval Research Logistics Quarterly 1:109–115. 4.3

Von Neumann, John and Oskar Morgenstern (1944, 1947, 1953). *Theory of Games and Economic Behavior*. Princeton University Press. 4, 5, 10

Wagner, Harvey M. (1969). *Principles of Operations Research with Applications to Managerial Decisions*. Prentice-Hall. 10

Warshall, Stephen (1962). A theorem on Boolean matrices. Journal of the Association for Computing Machinery 9:11–12. 9.7

Welsh, D. J. A. (1976). *Matroid Theory*. Academic Press. 9

White, Douglas J. (1969). *Dynamic Programming*. Oliver & Boyd; Holden-Day. 10.3

Whittle, Peter (1971). *Optimization Under Constraints*. John Wiley & Sons. MP

Wilkinson, J. H. (1961). Error analysis of direct methods of matrix inversion. Journal of the Association for Computing Machinery 8:281–330. 7.5

_____ (1963). *Rounding Errors in Algebraic Processes*. Prentice-Hall. 7.5

_____ (1965). *The Algebraic Eigenvalue Problem*. Clarendon Press, Oxford.

Williams, John D. (1954). *The Compleat Strategyst*. McGraw-Hill. 4

Willoughby, R. A., ed. (1969). *Sparse Matrix Proceedings*. RA-1, IBM Watson Research Center, Yorktown Heights, N.Y. (1968 Conference). 7.6

Wilson, Robert (1971). Computing equilibria in N-person games. SIAM Journal of Applied Mathematics 21:80–87. 5.4

Wolfe, Philip (1961). An extended composite algorithm for linear programming. RAND Corp. P-2373. 6.4

_____ (1963). A technique for resolving degeneracy in linear programming. SIAM Journal 11:205–211. 6.6

_____ (1965). The composite simplex algorithm. SIAM Review 7:42–54. 6.4

_____ (1980). The ellipsoid algorithm [of Khachian]. Optima (Mathematical Programming Society Newsletter), No. 1. 7

Wolfe, Philip and Leola Cutler (1963). Experiments in linear programming. Pp. 177–200 in Graves and Wolfe (1963). 6.3

Young, H. P. (1978). Power, prices and incomes in voting systems. Mathematical Programming 14:129–148. 5.8

Yu, P. L. and M. Zeleny (1975). The set of all undominated solutions in linear cases and a multicriteria simplex method. Journal of Mathematical Analysis and Applications 49:430–468. 8.8

Zadeh, Lofti A. (1968). Fuzzy algorithms. Information and Control 12:94–102. 8.9

______ (1973). Outline of a new approach to the analysis of complex systems and decision processes. Pp. 686–725 of Cochrane and Zeleny (1973). 8.9

Zahn, C. T. (1971). Graph-theoretical methods for detecting and describing gestalt clusters. Transactions on Computers, C-20, January, 1971, Institute of Electrical and Electronic Engineers. 9.2

Zeleny, Milan (1973). Compromise programming. Pp. 262–301 of Cochrane and Zeleny (1973). 8.8

Zeleny, M. and J. L. Cochrane (1973). A priori and a posteriori goals in macroeconomic policy making. Pp. 373–391 of Cochrane and Zeleny (1973). 4.7, 8.8

Zelinsky, Daniel (1968, 1973). *A First Course in Linear Algebra*. Academic Press. 2

Zermelo, E. (1912). Über eine Anwendung der Mengenlehre auf die Theorie des Schachspiels. Proceedings of the Fifth International Congress of Mathematicians, Vol. 2, pp. 501–504. Cambridge University Press. 10.5

Zionts, Stanley, ed. (1978). *Multiple Criteria Problem Solving*. Vol. 155 of Lecture Notes in Economics and Mathematical Systems. Springer-Verlag. 8.8

Zoutendijk, G. (1970). A product-form algorithm using contracted transformation vectors. Pp. 511–523 of Abadie (1970). 7.3

______ (1976). *Mathematical Programming Methods*. North-Holland. MP

Addenda

Gill, P. E., W. Murray and M. H. Wright (1981). *Practical Optimization*. Academic Press. MP

Gilmore, P.C. and Gomory, R.E. (1965). Multistage cutting stock problems of two and more dimensions. Operations Research 13:94–120. 10.1

Hu, T.C. (1969). *Integer Programming and Network Flows,* Addison Wesley

Scarf, Herbert (1973) (with Terje Hansen). *The Computation of Economic Equilibria*. Yale University Press. This book generalizes 5.4 in many ways, including nonlinear functions, by using matrix equivalents of simplicial subdivisions to approximate a fixed point of a mapping. The change of basis in 5.4 becomes a replacement operation on a "primitive set."

Name Index

Also see the alphabetized (first listed) authors in the bibliography.

Arrow, K. J., 189
Aubin, J. P., 212
Aumann, R. J., 151
Avis, D., 257

Baker, K. R., 413
Balinski, M. L., 58, 102, 124, 196, 244
Banzhaf, J. F., xii, 198-227
Barr, R. S., 490
Bartels, R. H., 300
Beale, E. M. L., 257, 296, 299
Bellman, R. E., 412, 417, 418, 429, 433
Benayoun, R., 382
Ben-Israel, A., 189
Bennett, P. G., viii, 167, 168
Bland, R. G., 257
Blottner, W. O., 322, 421
Bohnenblust, H. F., 129, 130
Bonnardeaux, J., 179
Brams, S. J., 216
Brioschi, F., 562
Brown, G. W., 120
Brown, T. A., 374

Cahn, A. S., 414
Chammah, A. M., 165
Charnes, A., 96, 562
Chvatal, V., 257
Cochrane, J. L., 574
Coffman, E. G., 413
Conway, R. W., 413
Cook, W. D., 382
Cooke, K. L., 418
Coombs, C. H., 572
Cooper, L., 99
Cooper, W. W., 96, 562
Cottle, R. W., 560
Crandall, H. W., 565
Crout, P. D., 283
Crowder, H., 243
Cutler, L., 241, 242, 249

Dando, M. R., viii, 168
Dantzig, G. B., xii, 61, 65, 105, 188, 240, 241, 250, 257, 266, 272, 290, 297, 322, 336, 421, 422, 429, 468, 494
Davis, R. L., 572
Debreu, G., 154, 185
de Montgolfier, J., 561
Dijkstra, E. W., 391, 400, 403
Dolait, J. P., 179
Doolittle, M. H., 283
Dorfman, R., 192
Dreyfus, S. E., 400, 406, 417, 421, 422, 433, 561
Dubey, P., 212, 220
Dyer, J. S., 179
Dyer, M. E., 374-375

Eaves, B. C., 184
Ecker, J. G., 367, 370, 371
Edmonds, J., 398
Evans, J. P., 367

Farkas, J., 239
Farquhar, P. H., 382
Fernando, T. D. H., 571
Flood, M. K., 155
Floyd, R. W., 422
Ford, L. R., 241, 414, 418, 425, 482
Forrest, J. J. H., 300
Forsythe, G. E., 294
Frank, A. Q., 227
Friedman, L., 571
Fulkerson, D. R., 241, 407, 412, 414, 425, 426, 482

Gale, D., 105, 129, 138, 189
Garvin, W. W., 96

Gentleman, W. M., 292
Gewirtz, A., 51, 52
Gill, P. E., 300, 574
Gillies, D. B., 454
Gilmore, P. C., 429, 574
Girshick, M. A., 562
Glover, F., 490, 491, 571
Goldfarb, D., 243
Golub, G. H., 300
Gomory, R. E., 102, 244, 429, 574
Gordon, D. G., 167
Gower, J. C., 396
Graves, R. L., 336, 359
Gröbner, W., 311
Guyer, M., 167

Hadley, G., 266
Hall, M., 561
Hansen, T., 574
Harris, P. M. J., 26, 243, 244, 293
Harsanyi, J. C., 179, 200, 202
Hattingh, J. M., 243
Hegner, N. S., 370, 371
Hellerman, E., 297, 299, 560
Hentges, G., 251
Hillier, F. S., 433
Hitchcock, F. L., 468
Householder, A. S., 300
Howard, N., 167, 168
Howard, R. A., 433, 434
Howson, J. T., 154, 180-187
Hu, T. C., 429, 574
Hurter, A. P., 562
Hurwicz, L., 525, 560

Isermann, H., 371

Jackson, J. R., 413
Jarvis, J. J., 561
John, J. B., 565
Johnson, S. C., 292
Johnson, S. M., 413, 563

Kakutani, S., 553
Kalaba, R., 561
Kantorovich, L. V., 468
Karlin, S., 129, 130, 194, 560
Karney, D., 491, 571
Karp, R. M., 398
Keeney, R. L., 382, 561
Kelley, J. E., 412
Kemeny, J. G., 190, 198
Kershenbaum, A., 396
Klee, V., 82
Klingman, D., 490, 491, 571
Klyuyer, V. V., 292
Koehler, G. J., 189
Kokovkin-Shcherbak, N. I., 292
Koopmans, T. C., 468
Kouada, I. A., 367, 370, 371
Krisotl, I., viii
Kruskal, J. B., 391
Kuhn, H. W., 5, 105, 138, 242, 449, 454, 458, 463, 466
Künzi, H. P., 243

Laritchev, O., 561
Lawler, E. L., 391, 398
Lemaire, J., 179
Lemarechal, C., 560
Lemke, C. E., 66, 154, 180-187
Leontief, W. W., 188
Lieberman, G. J., 433
Lin, T. D., 297
Lockett, J. A., 418
Lonseth, A. T., xiii
Lovász, L., 565
Lucas, W. F., 216
Luce, R. D., 105, 573

Machol, R. E., 485, 497
McKinsey, J. C. C., 131, 175, 463, 553
Mah, R. S. H., 297
Manas, M., 374
Manne, A. S., 494
Markowitz, H. M., 153, 240
Mattheiss, T. H., 374, 375
Maxwell, W. L., 413
Mayberry, J. P., 454
Mellon, B., 96
Miller, L. W., 413
Mills, H. D., 102
Minty, G. J., 82
Mitten, L. G., 413, 439
Moler, C. B., 294
Moore, E. F., 418
Morgenstern, O., 105, 151, 171, 190, 200, 208, 217, 447
Muller-Merbach, H., 413, 421, 436
Murchland, J. D., 422
Murray, W., 300, 574
Murty, K. G., 491
Myerson, R. B., 211, 212

Napier, A., 491
Nash, J. F., 154, 170-180
Nedoma, J., 374
Nemhauser, G. L., 437, 439, 560

Neyman, A., 212

O'Brien, J. J., 413
Orchard-Hays, W., 255, 322, 336, 350, 572
Orden, A., 257, 272
Owen, G., 179, 211, 223, 227

Papadimitriou, C. H., 568
Parsons, T. D., 309
Podrebarac, M. L., 571
Pollack, M., 425
Ponssard, J. P., 453
Prim, R. C., 391
Proll, L. G., 374, 375

Quandt, R. E., 242

Raghavan, T. E. S., 570
Raiffa, H., 105, 382, 440, 443, 561, 569
Rao, M. R., 421
Rapoport, A., 165, 167
Rarick, D., 297, 299
Reid, J. K., 243
Reinfeld, N. V., 485
Rice, J. R., 299
Robers, P. D., 562
Robinson, J., 121
Rosenthal, R. W., 202
Ross, G. J. S., 396
Roy, B., 382
Rubin, D. S., 375

Samuelson, P. A., 192
Saunders, M. A., 300
Savage, L. J., 141
Scarf, H., 574
Schlaifer, R., 570
Schmidt, B. K., 375
Schumann, J., 189
Seiford, L. M., 382
Seneta, E., 189
Shapley, L. S., xii, 119, 121, 129, 130, 131, 171, 179, 198-227, 564, 571
Sharp, R. G., 168
Sherman, S., 129
Shubik, M., 213-227
Sitomer, H., 51
Snell, J. L., 190, 198
Snow, R. N., 131
Solow, R. M., 192
Spellman, R. A., 565
Srinivasan, V., 491
Stanford, R. E., 189
Steinberg, D., 99
Steinitz, E., 48
Steuer, R. E., 367, 382
Stewart, T. J., 382
Stiefel, E. L., 41, 48
Stoer, J., 365
Sutherland, W. R. S., 566
Sylvseter, J. J., 310

Tergny, J., 561
Thompson, G. L., 190, 198, 491, 569
Thrall, R. M., 569, 572
Todd, M. J., 181, 182, 187
Tomlin, J. A., 251, 293, 297, 300
Tucker, A. W., xii, 5, 28, 48, 51, 58, 102, 105, 124, 138, 155, 196, 244, 307, 309, 311, 564, 566, 568
Tyndall, W. F., 572
Tzschach, H. G., 568

Uzawa, H., 560

Van der Meer, C. L. J., 570
Van Slyke, R. M., 396
Veinott, A. F., 563
Vogel, W. R., 485
Von Nuemann, J., 115, 121, 127, 151, 171, 190, 200, 208, 217, 447, 454

Warshall, S., 422
Weber, R. J., 212
Welsh, D. J. A., 391
Whinston, A. B., 568
White, D. J., 417
White, W. B., 563
Wiebenson, W., 570
Wien, M., 485, 497
Wilde, D. J., 560
Wilkinson, J. H., 283, 294
Williams, J. D., 106
Willoughby, R. A., 571
Wilson, R., 154, 185
Witzgall, C., 365
Wolfe, P., 241, 242, 244, 249, 257, 259, 564, 566
Wright, G. P., 568
Wright, M. H., 574

Yaspan, A., 571
Yen, J. Y., 421
Yu, P. L., 371

Zadeh, L, A., 458
Zahn, C. T., 397
Zehnder, C. A., 568
Zeleny, M., 371, 562, 572
Zoutendijk, G., 278

Subject Index

Action timing, 433, 442
Acyclic networks, 388-389, 407-417, 440-467
Addenda (bibliographic), 574
Adjugate determinant, 311
Affine subspace (linear variety), 13, 19, 61
All-integer pivot procedures, 34, 43, 45, 312-313, 352
Almost-complementary solutions, 182
Altruistic strategies, 155, 166, 169
Applications of linear programming, 1-5, 88-99
  assignment, 490-491
  blending, 92, 96-98
  bottleneck transportation, 497, 557-558
  diet, 94
  economic growth, 190-198
  electric power, 319-322, 329-330, 333, 344-345
  fertilizer mixtures, 2, 27, 78, 88-92
  games, 123-127, 134-137
  investment, 142-144
  model formulation, 97
  Models, 1, 2, 2a, 88-93
  multiple objectives, 364-383
  packaging, 94
  sand and gravel, 95
  sign conventions, 95-96
  spacecraft, 99, 179
  transportation, 468ff., 507
  transshipment, 495-496
  see Decision making; Dynamic programming; Games; Networks
Arbitrage in currencies, 422-423
Arcs of graph, 386-387, 409
Artificial variables (=0), 230, 246
  bimatrix equilibrium, 184
  exclusion rule, 234
  transportation, 493, 502
Assignment problem, 490-491
  $c_{ij}=(i\text{-}1), (j\text{-}1)$, 485, 497
Assortment, 417
Asymptotic voting powers, 223-225
Axis of game, 168

b $\leqslant$ **0** feasible, 100-103
Back substitution, 39, 284-285
Backward transformation, 275, 278, 288-300, 316, 482-483
Banzhaf value, 213-227
  probability gradient, 221
  Shapely-Shubik comparison, 220
Basic solution, 61, 470
  extreme point, 72
  tree, 475-479
Basic variables, 25, 28, 53, 60, 61, 232, 470-479
Basis of vestor space, 10, 11, 48-52
  canonical, 24, 41, 51
  dual, 26
  of linear program, 28, 61, 266
  transportation, 473-479
Bayes solution, 140-146, 376-380, 443-446
Behavior strategies, 455-458, 461
Bicameral legislatures, 217-218, 220, 227
Bimatrix games, 149-187. *See also* Games
Bimatrix model of economic growth, 190-198
Blending problems, 92, 96-98
Blocking variables, 63-66, 251-252
Block pivot, 306-313
Bluffing, 456, 461, 552
Bottleneck transportation, 497, 557-558
Bounded variables, 249-256
  blocking variables, 251-252
  dual-bounded, 255-256
  reflection, 251-254
  transportation, 497-503

Calculus and LP, 5-6
  uses, 177, 221-223, 323ff., 504ff.
Canonical form of LP, 57
Capacities of arcs, 396, 424-426, 497-503
Central solutions, 195
Change column or row, 86-87, 315-322

Characteristic function of game, 199-212, 218
  v(s) *vs.* v(s/s̄), 171, 199-209
  corresponding normal form, 207-209
Checking: Gaussian elimination, 285
  optimal LP solutions, 71, 84-85
  optimal strategies, 116, 128
  parametric programming, 341, 353
  pivot steps, 35
  revised simplex, 274
  tableau equivalence, 87
  voting power, 215
  Z^0 (infeasibility form), 233
Checklist of topics, xix
Chicken game, 156, 168
Cholesky factorization, 299
Circling (nonconvergence), 257
Clambake decisions, 140, 443-446
Classification of games, 105-106, 160-164, 466
Cluster analysis, 396
Coalitions in games, 199-212, 218-223
Complementary, k-almost, 182
Complementary slackness: games, 129, 181
  linear programs, 61, 68-71
  multiple objectives, 371
  strong form, 195-196
  transportation, 475
  types of variables, 237, 255
Composite LP methods, 244
Concave function, 73, 145, 358
Connected graphs, 387-389
Constraints: = or ⩽, 55, 229
Continuous games, 175
Convex: combination, 17, 150, 366-367, 371
  cone, 16, 19
  function, 19, 73, 145, 358
  hull, 17-18, 150, 371
  set, 15-19, 72, 129, 150, 338, 358, 371
Convexity, lack of, 186, 344, 350, 359, 367, 458-461
Cooperative Nash solution, 159, 170-179, 464-466
Cramer's rule, 304, 314 (ex. 8)
Crashing in LP, 279
Critical events, paths, tasks, 407-413
Cycle in graph, 388
  negative, 419, 422, 425

Δa,..., Δx (ranging), 322-335, 504-510
Δc partition, 482-483, 504-510
Δx loop, 481, 510 (ex. 11)
$|\Delta z|$ max, 79, 339
Decision making, 140-147, 376-383, 433-446
  acyclic networks, 440-467
  Bayes solutions, 140, 376-380, 443-446
  clambake, 140, 444-446
  decision networks, 440-467
  dynamic programming, 428-446
  games against nature, 140-147
  insurance, 141, 147
  investment, 107-108, 142-144, 319-345, 376-380
  Markov process, 434-435, 443
  minimax loss or regret, 141-142, 382
    Bayes constrained, 143-144
  minimin loss, 145
  multiple objectives, 376-383
  sensitivity analysis, 319-345, 507
  shortest trees and paths, 390-426
  variables, 437-438
  voting, 213-228, 382, 387
  see Applications of linear programming; Games
Decomposition: in dynamic programming, 439-440
  of games, 449-451
Degeneracy, 257-259, 367, 486, 490
Dependence: among chapters, xiii
  linear, 48, 304, 365
Determinantal equations: economic growth rate, 194-197
  utility comparison (Nash value), 177
Determinants, 300-313
  adjugate, 311
  Cramer's rule, 304
  evaluation, 38, 292, 303, 306
  exact division, 306, 312-313
  expansions, 303, 308
  matrix inversion, 38, 263, 304
  minors and rank, 304, 308
  multiplication, 303
  pivoting tableaus, 306-313
  proportional minors, 308, 309-312
  two by two, 34, 118-119
  unimodular, 471-472
Diet problem, 94
Dimension, *see* Rank of matrix
Directed graphs, 387-389
  acyclic, 407-417, 440-467
Discounting future payments, 417, 507
Distribution problem, 468-514. *See also* Transportation
Domination; bimatrix equilibria, 159, 186-187
  points (collective), 151, 159
  strategies (individual), 120, 159, 368
Double precision, 293-294
Dual bases, spaces, systems of equations, 22-28
Duality, 5

Duality *(Cont'd)*
clues to, 23-25, 30-31, 55-58
games, 123-124, 129, 181
importance of, 71
linear programs, 55-58, 61, 67-71, 91-92
multiple objectives, 371
optimality, used to prove, 68-71
transportation, 470, 475
unsymmetrical, 52
variables types of, 237, 255
Dual simplex algorithm, 66
bounded variables, 254
transportation, 507
Dummies (in games), 204, 209
Dynamic programming (DP), 428-443
action timing, 433, 442
decision networks, 440-446
decomposition principle, 439-440
extensive games, 447-467
knapsack problems, 429-432
LP comparison, 436
Markov process, 434-435, 443
opposed sequences, 438
optimality principle, 429, 439-440
shortest trees and paths, 390-426
smoothing, 433
summary (table), 434
variables, types of, 428, 436-440

Economic models: arbitrage in currencies, 422-423
bimatrix growth, 190-198
input-output, 90
investment, 142-144, 319-345, 376-380
Leontief, 188-189
see also Applications of linear programming
Efficiency of algorithm, 82, 242-243, 249, 290-292, 375, 568
Efficient points, 151, 364ff.
extreme points, 366, 369-373, 381
in games, 151-179, 463
in multiple objectives, 364-383
tests for, 174-175, 367-371
Eigenvalues, 194
Electric power, 319-322, 329-330, 333, 344-345
Elementary matrices, 265, 272-279, 295-299, 325-328
Enumeration: kernels, 129-133, 185-187, 195
vertices, 74 (ex. 11, 12), 373-376
see Dynamic programming
Equilibria: bimatrix games, 154, 156, 159, 167-168, 180-187
extensive games, 463
generalized (axis), 167-168
matrix games, 117, 128
nonconvex, 186
n-person games, 154
odd number of, 163, 184
pure *vs.* mixed, 154, 187
Equivalence: of games and LP, 109, 123-127, 134-139
of tableaus, 31, 35, 87
Errors, control of, 293-300. *See also* Checking
Espalier (tree), 476-477
Essential games, 209
Eta vectors, 265, 272-278, 291
dispensable, 278, 291
Euclidean distance and angle, 22
Exact division, 34, 45, 306, 312-313, 352
Excess profits $\leqslant 0$, 191
Exchange of variables, 27-38, 67
full exchange, 50, 36-53
Expected value, 111-118. *See also* Probability
Explicit-inverse method, 268-272
Extensive games, 447-466. *See also* Games
Extreme points, 17
basic feasible solutions, 72-73
efficient, 366, 369-373, 381
inadequate, 73, 383
of sets of equilibrium strategies: bimatrix games, 185
matrix games, 132

Factor analysis, 227, 382
Farkas's Lemma, 239, 534
analogous theorem, 53
Feasibility: for games, 150
for LP, 61, 68, 255
with parameter, 337, 346-348, 511
Fertilizer problems, 2, 27, 78, 88-92
Fictious play, 120-122
Float (scheduling), 410-412
Flowcharts: multiple objectives, 369
two-phase simplex, 232
Forest (loopless graph), 396
Forward transformation, 273, 278, 288-300, 316, 481
Fully exchanged tableaus, 50, 36-53
Fuzzy games, algorithms, 212, 562, 574

Games: bimatrix, 149-187, 462-466
axis, 168
Chicken, 156, 168
classification, 160-164, 466
domination, 151, 159
economic growth, 190-198
efficient points, 151
equilibria, 154, 156, 159, 180-187, 463

Games (Cont'd)
generalized, 167-168
extensive, 462-466
hypergames, 167
implication diagram, 159
marriage game, 158, 526-527
maximin security value, 164, 179
metagames, 167
Nash cooperative solution, 170-179, 464-466
noncooperative solution, 159, 463-464
ordered moves, 166, 167, 170 (ex. 21)
Prisoner's Dilemma, 155, 166, 167
relevance, viii-x
spacecraft trajectories, 179
threats, 167, 168, 171, 176, 464
continuous, 175, 553
extensive, 447-466
behavior strategies, 455-458, 461
bluffing, 456, 461, 552
decomposition, 449-451
information sets, 449-462
minimax not unique, 456-462 (ex. 9), 553
nim, 453
nonconvex region, 460
normalization, 451-453
perfect information, 448-449, 453
perfect recall, 458-462
primitive poker, 454-458, 461
variable-sum, 462-466
matrix, 105-146
checking solution, 116, 128
dominant strategies, 120
economic growth, 190-198
equilibria, 117, 128
extensive, 447-461
fictious play, 120-122
graphical solution, 114-116
graph of **uAy**, 117
investment, 107-108, 142-144
kernels, 129-133, 195
Lloyd's of London, 117
LP equivalence, 109, 123-127, 134-139
matching pennies, 4, 110
maximin solution, 106-110
minimax theorem, 127-129
mixed startegies, **u, y,** 110-113
mutual confidence, 118
against nature, 140-146
ordered moves, 106-110
parametric, 123, 129, 134, 143-147, 176-177, 190-198
pure strategies, *i, j,* 105-110
relevance, 117, 118
saddle point, 106-109
scaling problem, 117
solution by LP, 123-127, 134-137
strictly determined, 108
symmetric, 119, 138
two pure strategies, 114-115, 118-119
value a*, 106, 116, 127, 346, 349 (ex.9), 368
n-person, 198-227
Banzhaf value, 213-227
characteristic function, 199-209
normal form exhibited, 207-209
coalitions, 199-212, 218-223
dummies, 204, 209
efficient points, 151-179. 204-206
equilibria, 154, 185
fuzzy, 212
inessential, 209
maximin security level, 206
multiple objectives, 364
Shapley (Shubik) value, 198-227
voting games, 213-227
Games against nature, 140-146
Gaussian elimination, 39, 282-289
Gauss-Seidel emthod, 294
g coefficients of Z^0, 231
Givens matrices, 300
Gram-Schmidt orthogonalization, 299
Graphical solutions: critical paths, 412, 415-416
decision tree, 444
dynamic programming, 441-443
efficient points, 151, 152, 172, 173, 381
extensive games, 448-467
game classification, 161
linear programs, 2, 80, 83, 89, 240, 487
matrix games, 114-117
parametric programming, 344, 356, 357, 381
shortest paths, 401
shortest spanning tree, 391
Graphs, 385-389. *See also* Networks

Homogeneous linear equations, 9-12, 20-26, 48-52
Homogeneous sign convention, 95-96, 100-103
Householder matrices, 300
Hypergames, 167

Imputation (game payoffs), 205-206
Individual rationality, 179, 205
Inessential games, 209
Infeasibility forms: Z^0, 230-234, 240
Z^1, 245-246

Information, perfect, 448-449, 453, 463-466
Information sets, 349-462
Input-output models, 90
economic growth, 190-198
Leontief, 188-189
Insurance decisions, 141, 147
Integer pivot procedures, 34, 43, 45, 312-313, 479
Interest rates, 191, 417, 507
Interior points, 6, 365-366
Invariance of rank, 264
of solution sets, 31, 50, 84-87
Inverse matrix, 36-38, 263-264, 290-292
explicit-inverse method, 268-272
product form of inverse, 272-282, 290-291
reinversion, 279, 288-289, 294
Invertible mapping, 51
Investment decisions, 107-108, 142-144, 319-345, 376-380
Involutory matrix, 26

Jordan elimination, 38-39, 229, 272, 290-291

Kernel: of equilibrium, 186-187
of linear map, 20, 41, 51
of matrix game, 129-133, 195
Knapsack problems, 429-432

Latin square, 492
Lattice, polyhedral, 373-374
Least squares, 41, 289
Leontief models, 188-189
Lexicographic ordering, 248, 257-259, 339, 360, 382, 486, 490
Linear, piecewise, 144, 344, 360-363, 511
Linear algebra, 8-53, 261-268, 300-313
basis, 11, 23-25, 28, 53, 266
convex sets, 15-19
determinants, 300-313
duality, 23-25, 30-31, 52, 55-58, 71
extreme points, 17, 72-73
linear equations, 38-46, 282-294, 304
linear (in)dependence, 48, 304
linear maps, 20, 51
linear variety, 12-14
matrix factorization, 265-266, 282-292, 299-300
matrix inversion, 36-38, 263-264, 290-292, 304
matrix multiplication, 261-262
orthogonality, 22-26, 68-71
pivot steps, 27-38, 47, 67, 263-266, 306-313
rank, 48-52, 262-264, 304, 308
scalar product, 20-22
vector space, 9-12, 48-49
Linear (in)dependence, 48, 304, 365
Linear equations, 8-12
dual systems, 25, 28
homogeneous, 9-12, 20-26, 48-52
in kernels, 129-133, 185-187
in linear programs, 55, 229-235
solution of, 38-46, 282-294
determinants, 131, 304, 314 (ex. 8)
elementary methods, 41-46
Gaussian elimination, 39, 282-289
Gauss-Seidel (relaxation), 294
Jordan elimination, 38-39, 229, 272, 290-291
kernels, 131, 185
number of operations, 290-292
symmetric, 285, 289 (ex. 5)
triangular, 282-292, 473-474
Linear form or functional, 20
Linear (in)dependence, 48, 304, 365
Linear map, 20, 51
Linear programming (LP), 1-6, 55-78, 229-238
applications, 1-5, 88-99, 142-144, 319, 468, 495-496
duality, 55-58, 67-71, 91-92
errors, control of, 35, 66-67, 71, 84-85, 293-300
games (matrix), 109, 123-127, 134-139
bimatrix equilibria, 180-187
multiple objectives, 364-383
optimality conditions, 68, 71, 255
parameters, 315-363, 504-513
pivot steps, 27-38, 64-67
sensitivity, 86-87, 91-92, 315-335, 504-509
simplex: two-phase, 60, 229-237
revised, 268-282, 288-292
simplex algorithm, 62-79
solutions (feasible, basic), 61
tableau, 27-32, 57-60, 95-96, 100-103
transportation, 468-513
variables, types of, 230, 236-237, 251, 255
Linear variety, 13, 19, 61
Lloyd's of London, 117
Logical variables (i in M^0), 268
Loops in graph, 387, 481

Markov decisions, 434-435, 443
Marriage game, 158, 526-527
Matching pennies, 4, 110, 119, 170 (ex. 20)
Matrix, 23-53, 261-268
echelon, 49
elementary, 265, 272-279, 295-299, 325-328

Matrix *(Cont'd)*
factorization, 265-266, 282-292, 299-300
games, 105-146
Givens and Householder, 300
input-output, 90, 188-198
inverse, 36-38, 263-264, 290-292, 304
involutory, 26
multiplication, 261-262
nonnegative, 189-214

involutory, 26
multiplication, 261-262
nonnegative, 189, 194
operations, 24-26, 32, 261-268
orthogonal, 299-300
productive, 189
rank, 32, 40, 51, 262, 264, 304, 308
singular, 264
skew-symmetric, 119, 138-139, 196
transpose, 21, 24-28
triangular, 282-292, 473-474
unit, 24, 263
Matroids, 391
Maximin (security value), 106-110, 127, 164, 179, 206
Metagames, 167
Minimax loss or regret, 141-145, 382
Minimax theorem, 127-129
inapplicable, 456-462 (ex.9), 553
Minimin loss, 145
Mixed strategies **u, y**, 110-113
relevance, 117, 118, 461
Mixoconnected, 388, 409, 476
Model formulation, 95-97
Models 1, 2, 2a, 88-93
Multiple column selection, 279
Multiple objectives, 364-383
algorithm, phases 1 to 3b, 369-373
efficient faces (test), 371
efficient points, 364-373
connected, 373
not convex, 367
enumeration of vertices, 373-376
examples and variations, 376-383
games: matrix, 368
n-person, 364
games against nature, 140-146, 364
nonlinear utility, 145, 376-380
Multiplication:
Multipliers π, 267-268, 270, 273-275

Nash cooperative solution, 159, 170-179, 464-466
Nature: games against, 140-146
states of, 140, 443-446
Negative cycles, 419, 422, 425
Negative transpose, x, 24-28
Networks, 385-427, 440-467, 475-481
acrylic, 388-389, 407-417, 440-467
basic trees, 475-479
capacities, 424-426, 497-503
connected, 387-389
critical events, paths, tasks, 407-413
decisions, 440-446
extensive games, 447-467
graphs, 385-389
noncooperative game solution, 463-464
scheduling, 407-413
shortest paths (all), 422-424
shortest paths from one node, 400-421
shortest spanning tree, 390-398
transportation, 475-481
transshipment, 495-496
trees, 388-389, 475-479
warehouse problem, 414-416
Node of graph, 386
decision mode, 440, 444, 449, 463
iterative formulas at, 425, 434, 437, 440-441, 449, 463
scheduled event, 409
transshipment, 496
Noncooperative solution, 159, 180-187. *See also* Equilibria
Nonlinear problems, 5, 177, 194-197, 357-363, 432-433, 456, 497
nonconvexity, 186, 344, 350, 359, 367, 458-461
Nonlinear utility, 117, 145, 168, 378-383
Nonnegative matrices, 189, 194
Nonserial stages, 437, 562
Normal-form game from given characteristic function, 207-209
extensive game, 451-453
Notations, x-xi, 59-60, 100-103. *See also* Tableau; Variables
Numbers of operations: determinants, 292
game kernel, 132
linear equations, 290-300

One-to-one mapping, 51
Onto (mapping), 51
Optimality, regions of, 61, 356-357, 381
Optimality conditions: games, 127-129
linear programs, 68, 71, 255
parameters, 337, 511
Optimality principle in DP, 429, 439-440
Oriented graphs, 387-389
Orthogonal: complements, 22-26

Orthogonal *(Cont'd)*
LP solution sets, 68-71
matrices, 299-300
vectors, 22

π multipliers, 267-268, 270, 273-275
π pivot, 33-34
Packaging problem, 94
Paradoxes: minimax not unique, 456-462 (ex. 9)
Prisoner's Dilemma, 155, 167
residual errors, 294
transportation, 510, 572
voting, 387
see Convexity, lack of
Parametric games, 123, 129, 134, 143-147, 176-177, 190-198
Paremetric programming, 315-363, 504-513
electric power, 319-345
investment decision, 142-146
multiple objectives, 365-366, 371-373, 376-382
nonlinear parameter, 357-363
parameter in **b** and **c**, 336-348
parameter in one column, 346-353
sensitivity, 86-87, 91-92, 315-335
several parameters, 354-357
transportation, 504-513
Partition of players N, 211-212, 223, 226-227
Pass (through data), 279, 288, 409, 497-498
Path in graph, 386-388
shortest, 400-424
Payoff, 106, 149
diagram, 149-152, 172-173
expected (average), 112, 117
nonlinear utility, 117, 145, 168, 378-383
Perfect information, 448-449, 453, 463-466
Perfect recall, 458-462
Permutations: determinants, 301-302
metagames, 167
pivot steps, 38, 306
players (Shapley value), 119, 209-214, 219, 227
Phases 1 and 2, 229-235. *See also* Multiple objectives; Simplex methods
Piecewise linear, 144, 344, 360-363, 511
Pivotal voter, 213-216
Pivot steps, 27-38, 47, 67
applications, numerical (list), 47
arithmetic (all-integer), 34, 43, 45, 312-313, 352
arithmetic (computer), 29-30
block pivot, 306-313
checking, 35, 66-67
determinants, 38, 306-313
duality, 30-31
fully exchanges tableaus, 50, 36-53
linear equations, 38-46, 282-288
matrix inversion, 36-38
matrix multiplication, 264-268
parametric programming, 339-340
positioning for sparseness or size, 295-299
proving theorems, 48-53, 68
revised simplex, 272-275
sequence, 38, 124, 282, 295-299, 306
simplex algorithm, 64-67
solution sets unchanged, 31, 50, 84-87
transportation, 480-483, 507, 510 (ex. 11)
updating, 261, 266-268, 272-278, 288-300, 316
Pivotal voter, 213-216
Poker, primitive, 454-458, 461
Polyhedral lattice, 373-374
set, 16-19
Positioning of pivots, 295-299
Preferences, 167, 168, 382
Pricing out, 267-268, 270, 273-275, 316, 481-483
Primal-dual: identification, 59
invariance, 31, 50, 84-87
method, 241
Prisoner's Dilemma, 155, 166, 167
one-sided, 166-167
trilemma, 166
Probability, 111-118
a priori (Bayes), 140-146, 376-380
Bayes Theorem, 443-446
expected value, 112, 149-154, 199
at nodes, 433-435, 438, 441-446, 449, 456-457
Lloyd's of London, 117
mixed strategies, 111-118
Permutations (Shapley), 199, 209-211, 219, 227
reliability, 396, 406
simplex of vectors, 113
voting approval gradient (=Banzhaf value), 220-223
Product form of inverse, 272-282, 290-291
Productive matrix, 189
Product of determinants, 303
of matrices, 261-262, 436
Product of vectors: matrix or tensor **xxu**, 21, 32, 327-328
scalar ux, 21
Pure strategies i, j, 105-110, 167
two alternatives, 114-115, 118-119, 160-164

Quasifeasibility in Phase 1, 230
  abandoned, 244-249

Ranging, 322-335, 504-510
Rank of matrix, 32, 40, 51, 262, 264, 304, 308
  of Shapley formulas, 210-211
  of solution set, 52
  of transportation matrix, 471
  of vector space, 11, 23, 48-49
Rationality: collective, *see* Efficiency of algorithm
  individual, 179, 205
Ratio pricing, 239-244
Recall, perfect, 458-462
Reflection of variables, 251-254
Regression analysis, 41, 223, 227, 289
Regret, minimax, 141-142, 382
Reinversion, 279, 288-289, 294
Reliability of network, 396, 406
Replacement of equipment, 416-417
Requirements space, 266
Residuals, 41, 289, 294
Revised simplex method, 268-282, 288
Revising optimal solutions, 86-87, 91-92, 177, 315-384, 504-514
Rooted graphs, 389

Saddle point, 106-109
Sand and gravel problem, 95
Scalar product, 21
Scaling (linear): matrices, 78, 297
  payoffs, 171
Scheduling networks, 407-413
  n jobs on two machines, 411
Schema, Tucker, 28, 124
Sensitivity analysis, 86-87, 91-92, 315-335
  change row or column, 315-322
  electric power, 319-322, 329-330, 333, 344-345
  games, 177
  parametric programming, 336-363, 510-514
  ranging, 322-335, 504-509
  transportation, 504-513
Set operations, 15
Shapley-Shubik value, 213-227
  comparison with Banzhaf, 220
Shapley value, 198-227
  from axioms, 204-205
  general games, 198-212
  imputation, 205-206
  original *vs.* modified, 171, 200-202
  partition of players N, 211-212, 223, 226-227
  from permutations, 199-200, 211
  unsymmetrical, 209-212, 227
    degrees of freedom, 210
  voting games, 213-227
Shortest trees and paths, 390-424
Side payments, 150-151, 170, 173, 176, 198-212, 220
Sign restrictions, 229-230, 236-237, 255
Signs, of b^i, c_j, a_{ij}, 67, 95-96, 100-103, 109
Simplex algorithms (Phase 2), 61-67
  assignment problem, 490-491
  checking, 35, 71, 84-85
  degeneracy, 257-259
  dual, 66, 254, 507
  examples, 75-82
  finite convergence, 67, 248, 257-259
  max $|\Delta z|$, 79, 339
  revising optimal solutions, 86-87, 91-92, 177
  transportation, 488-491
  see Pivot steps; Tableau
Simplex methods, 60, 229-239
  artificial variables, 230-239, 246
  bounded variables, 249-256
  composite methods, 244-249
  crashing, 279
  explicit inverse, 268-271
  freer row selection, 244-249
  infeasibility forms, Z^I, 245-246
    Z_0, 230-234, 240
  multiple objectives, 364-383
  parametric programming, 336-363
  ratios for column selection, 239-244
  reinversion, 279, 288-289, 294
  revised, 268-282, 288
  sensitivity analysis, 315-335
  two-phase, 60, 229-235
  see Transportation, 468-483. 484-513
Simplex of probability vectors, 113
Slack variables (basic), 55-57, 469
Smoothing problem, 433
Solution set, 8, 19, 61
  invariant, 31, 50, 84-87
  orthogonal, 22-26, 68-71
  rank, 52, 210, 471
Spacecraft applications, 99, 179
Spanning set, 11, 48
Sparseness of matrices, 295-299
Spread (asked-bid), 256
Stage in DP, 436-438
Standard form of equations, 52, 232
State in DP, 437-438
State of nature, 140, 443-446
Steiner network problem, 398
Strategies for games, 105-113
Strictly determined game, 108
Structual variables (j in N^o), 268

Symmetry: bimatrix games, 160
  linear equations, 285, 289 (ex. 5)
  matrix games, 119, 138
  Shapley value, 199, 204, 209-212, 227

Tableau (LP), 27-32, 264
  condensed, 27-28, 81
  equivalence class, 31, 35, 87
  extended, 32, 81
  fully exchanged, 50, 36-53
  negative transpose, x, 24-28
  parametric, 341, 350-351
  sign conventions, 95-96, 100-103
  transportation, 475
  Tucker schema, 28, 124
  see Pivot steps; Variables
Threats, 167, 168, 171, 176, 464, 528
Traffic problems, 406, 424-426
Transferable utility, 150-151, 170, 173, 176, 198-212
Translation of vector space, 13
Transportation, 468-513
  assignment, 490-491
  basic trees, triangularity, 473-479
  bottleneck, 497, 557-558
  capacitated, 497-503
  Δx loop, 481, 510 (ex. 11)
  Δc partition, 482-483, 504, 510
  degeneracy, 486-487, 490-491
  dual, 470, 479
  espalier, 476-477
  integer solutions, 479
  paradox, 510, 572
  parameters, 510-514
  pivot steps: dual simplex, 507
    pricing, 480-483
  rank m+n-1, 471
  restricted, 493-495
  sensitivity, 504-509
  simplex (two-phase) algorithms
    capacitated, 497-503
    restricted, 493-494
    unrestricted, 484-490
  tableaus, 475, 479, 497
  transshipment, 495-496
  trees, triangular bases, 473-479
  Tucker schema, 471
  unimodular determinants, 417-472
  Vogel's method, 485
Transpose $\mathbf{U}^t$ or $\mathbf{A}^t$, 21-22
  negative, x, 24, 25
Transshipment, 495-496
Trees, 388-389
  decisions and games, 440-467
  shortest spanning, 390-398
  transportation bases, 475-479
  see Networks
Triangular equations: B=LU, 282-292
  transportation, 210, 473-474
Tucker schema, 28, 124

Uncertainty, decision-making under, 140-147, 376-380
Unimodular determinants, 471-472
United Nations Security Council, 216-217, 226
Unrestricted variables, 236-239, 471
Updating LP data, 261, 266-268, 272-278, 288-300, 316
  transportation, 480-483
  see Pivot steps
Utility, intrinsic common scale (Nash), 170, 174
  nonlinear, 117, 145, 168, 378-383
  scale changes, 171
  transferable, 150-151, 170, 173, 176
  types of, 168-169

Value, expected, 111-118, 433-457. *See also* Probability
Value of game, 106, 116, 127, 198, 346, 349 (ex. 9)
  Banzhaf, 213-227
  linear transformations, 119, 171
  Nash, 170-179, 464-466
  parameter making it zero, 176-177, 190-198
  Shapley (Shubik), 198-227
  two by two matrix, 118-119
  zero for symmetric game, 119
  see Games
Variables, 55-61, 255, 436-437, 468-471
  artificial, 230, 234, 246, 493,184
  basic, 25, 60-61, 232, 473-479
  bounded, 249-256, 497-503
  complementary, 61, 68-71, 237, 255, 475
  dynamic programming, 436-440
  five types (primal and dual), 255
  games, 123-124, 134-135
  LP tableau, 27-32, 55-61, 264
  primal, 59
  sign restrictions, 229-230, 236-237, 255
  slack (basic), 55-57, 469
  transportation, 468-471
  unrestricted, 236-239, 471
  Z^0, Z^I (infeasibility forms), 231, 245
Vector (space), 9-12, 20-22, 48-49, 261-262
  axioms, 11
Vertices: as basic solutions, 72

Vertices *(Cont'd)*
  enumeration of, 74 (ex. 11, 12), 373-376
  of graph, 386
Vogel's method, 485
Voting, preferential, 382
Voting paradox, 387
Voting power, 213-228
  asymptotic results, 223-225
  bicameral legislatures, 217-218, 220, 227
  enumeration of cases, 215, 216
  partition of players N, 223, 226-227
  pivot (deciding voter), 213-216
  probability gradient, 221-223
  Shapley *vs.* Banzhaf value, 220
  UN Security Council, 216-217, 226
  winner takes all, 225

Warehouse problem, 414-416
Winner takes all, 225

Z^0, Z^I (infeasibility), 231, 245
Zero-sum two-person, games, 105-139
  extensive, 447-461
  see Games, matrix